ARTIFICIAL TRANSMISSION LINES FOR RF AND MICROWAVE APPLICATIONS

WILEY SERIES IN MICROWAVE AND OPTICAL ENGINEERING

KAI CHANG, Series Editor
Texas A&M University

A complete list of the titles in this series appears at the end of this volume.

ARTIFICIAL TRANSMISSION LINES FOR RF AND MICROWAVE APPLICATIONS

FERRAN MARTÍN

WILEY

Published by John Wiley & Sons, Inc., Hoboken, New Jersey
Published simultaneously in Canada

Library of Congress Cataloging-in-Publication Data:

Martín, Ferran, 1965–
Artificial transmission lines for RF and microwave applications / Ferran Martín.
pages cm.
Includes bibliographical references and index.
ISBN 978-1-118-48760-0 (hardback)
1. Radio lines. 2. Microwave transmission lines. I. Title.
TK6565.T73M37 2015
621.3841′3–dc23

2015007897

Set in 10/12pt Times by SPi Global, Pondicherry, India

Printed in the United States of America

10 9 8 7 6 5 4 3 2 1

1 2015

To Anna, Alba and Arnau

CONTENTS

PREFACE

Transmission lines and waveguides are essential components in radiofrequency (RF) and microwave engineering for the guided transmission of electromagnetic (EM) energy (power and information signals) between two points. Moreover, transmission lines and waveguides are key building blocks for the implementation of passive and active RF/microwave devices of interest in wireless communications (filters, diplexers, splitters, couplers, amplifiers, oscillators, mixers, etc.). In planar technology, low-cost devices can be fabricated by etching patterns (a set of transmission lines and stubs providing certain functionality) in a printed circuit board (PCB), avoiding the use of lumped components, such as capacitors, inductors, or resonators. Transmission line-based circuits are usually designated as distributed circuits, since transmission lines can be described by a network of distributed parameters. In certain designs (e.g., amplifiers and mixers), it is necessary to combine distributed and lumped active elements, such as diodes or transistors. Nevertheless, the main relevant aspect of distributed components is their capability to mimic lumped-reactive elements or a combination of them (e.g., resonators or even more complex reactive circuits). It is thus possible to design fully planar functional devices on the basis of the distributed approach, or to minimize the number of lumped elements (unavoidable in certain components, e.g., active circuits) in the designs.

Distributed circuits have two main drawbacks: (1) their dimensions scale with frequency, and (2) transmission lines exhibit very limited design flexibility. Typically, the required transmission lines in the designs have a length of the order of the wavelength, which means that dimensions may be too extreme if the operating frequencies are moderate or low. Concerning the second aspect, distributed circuits are designed by means of transmission lines with certain phase (at the operating frequency) and characteristic impedance. The phase varies linearly with the length of the line and frequency (to a first-order approximation transmission lines are dispersionless). Therefore, the functionality of distributed circuits is limited to a certain band; namely, the required nominal phase is lost if frequency deviates from the operating value, which means that distributed circuits are bandwidth limited by nature.

The limitations of ordinary transmission lines as building blocks for device design are in part originated from the fact that these lines exhibit a limited number of free parameters for design purposes (by excluding losses, ordinary lines are described by a distributed network with two reactive elements in the unit cell). However, by truncating the uniformity (in the longitudinal direction), by etching patterns in the ground plane, by loading the line with reactive elements, or by using a combination of these (or other) strategies to increase the degrees of freedom of the lines, many possibilities to reduce device size, to improve performance, or to achieve novel functionalities, are open. This book has been mainly conceived to introduce and study alternatives to ordinary lines for the design and implementation of RF/microwave components with superior characteristics in the above cited aspects. We refer to these lines as artificial transmission lines, and the term is as wide as the number of strategies that one can envision to improve the size, performance, or functionality of ordinary lines. The book is devoted to the analysis, study, and applications of artificial lines mostly implemented by means of a planar transmission line (host line) conveniently modified (e.g., with modulation of transverse dimensions, with etched patterns in the metallic layers, and with reactive loading), in order to achieve certain functionality, superior performance, or reduced size. Nevertheless, it will be shown that in certain artificial waveguiding structures, such as electroinductive and magnetoinductive delay lines, the host line is not present. Waveguide-based components are not included in this book, entirely focused on artificial transmission lines in planar technology. Obviously, it is not possible to cover all the material available in the literature, related to the topic of artificial transmission lines, in a single book. Necessarily, the contents of this book are influenced by the personal experience and background of the author. However, many RF/microwave devices and applications of artificial transmission lines reported by other researchers are included in this book, or properly referenced.

The book is devoted to readers that are already familiar with RF/microwave engineering. The aim of writing this book has been to provide an up-to-date state of the art in artificial transmission lines, and an in-depth analysis and study of those aspects, structures, devices, and circuits that are more relevant (according to the criterion of the author) for RF/microwave engineering, including design guidelines that can be useful to researchers, engineers, or students involved in the topics covered by this book. Nevertheless, Chapter 1 is dedicated to the fundamentals of planar transmission lines for coherence and completeness, since most of the concepts of this chapter are used in the subsequent chapters, and are fundamental to understand the principles and ideas behind the design and applications of artificial transmission lines.

Chapter 2 is focused on artificial transmission lines based on periodic structures, where periodicity plays a fundamental role and is responsible for the presence of band gaps in the transmission spectrum of these lines. The Floquet analysis (leading to the concept of space harmonics), complemented by the coupled mode theory (from which useful expressions for the design of periodic artificial lines are derived), and the transfer matrix method (useful to obtain the dispersion relation of the fundamental space harmonic), are included in the chapter. The last part is devoted to the applications, which have been divided into those of periodic nonuniform transmission lines (e.g., harmonic and spurious suppression), and those of reactively loaded lines, where not only the reflection properties of periodic structures but also the inherent slow-wave effect associated to reactive loading, are exploited.

Chapters 3 and 4 are dedicated to artificial transmission lines inspired by metamaterials, or based on metamaterial concepts. The importance of these artificial lines in this book has forced the author to separate the fundamentals/theory and applications into different chapters in order to avoid an excessive chapter length. Thus, Chapter 3 is focused on the theory, circuit models, and main implementations of metamaterial transmission lines, whereas Chapter 4 deals with the applications. Many applications of metamaterial transmission lines are based on the superior controllability of the characteristic impedance and dispersion of these lines, as compared to ordinary lines, related to the presence of reactive elements loading the line. Indeed, metamaterial transmission lines have opened a new way of "thinking" in the design of microwave components, where tailoring the dispersion diagram, and not only the characteristic impedance, is the key aspect (we may accept that metamaterial transmission lines have given rise to microwave circuit design on the basis of impedance and dispersion engineering). The further controllability of the relevant line parameters (phase constant and characteristic impedance) in metamaterial transmission lines, as compared to ordinary lines, has a clear parallelism with the further controllability of the constitutive parameters (permittivity and permeability) in effective media metamaterials (periodic artificial structures exhibiting controllable EM properties, different from those of the materials which they are made). Indeed, we can define an effective permittivity and permeability in metamaterial transmission lines despite that these lines are one-dimensional structures, and we can design the lines in order to support backward (or left-handed) wave propagation (as occurs in metamaterials with simultaneous negative effective permittivity and permeability). However, whereas in effective media metamaterials periodicity and homogeneity (satisfied if the period is much smaller than the wavelength) are necessary conditions to properly define an effective permeability and permittivity, periodicity, and homogeneity are not requirements for impedance and dispersion engineering with metamaterial transmission lines.

The former metamaterial transmission lines were implemented by loading a host line with series capacitors and shunt inductors (CL-loaded approach), or by loading the host lines with electrically small resonators, formerly used for the implementation of bulk effective media metamaterials (metamaterial resonators). This latter approach has been called resonant-type approach. Both approaches are included in this book (and many other latter developments), but special emphasis is put on the resonant-type approach. Moreover, in Chapter 4 there are several applications where, rather than the controllability of the impedance and dispersion of the artificial lines, the working principle is the resonance of a transmission line (host line) loaded with metamaterial resonators (these lines are designated as transmission lines with metamaterial loading in Chapter 4). Since metamaterial transmission lines are inspired by metamaterials, an introduction to these artificial media and the former implementation are included in Chapter 3. Chapter 3 includes also a section devoted to study the main electrically small resonators useful for the synthesis of metamaterials and microwave circuits based on them (resonant-type approach). In Chapter 4, the applications include enhanced bandwidth components, multiband components, filters and diplexers, active devices with novel functionalities (e.g., distributed amplifiers), novel antennas (e.g., leaky wave antennas and antennas for RFID tags), microwave sensors, and so on.

In Chapter 5, the focus is on reconfigurable components based on tunable artificial lines and nonlinear transmission lines. Several materials, components, and technologies (including varactors, RF-MEMS, ferroelectrics, and liquid crystals) for the implementation of tunable components are introduced. Then the chapter focuses on the design of tunable artificial transmission lines and their applications, mostly, although not exclusively, devoted to filters. The last part of the chapter deals with the topic on nonlinear transmission lines, structures that support the propagation of solitons and are of interest for harmonic multiplication.

Finally, other advanced transmission lines or, more generally, waveguiding structures are presented and studied in Chapter 6, including applications. The covered topics are electroinductive and magnetoinductive wave delay lines, common-mode suppressed differential lines, lattice network-based transmission lines, transmission lines loaded with non-Foster components, and metamaterial-based substrate-integrated waveguides. Grouping these topics in a single chapter does not obey to a thematic reason, but to the fact that most of them have been recently proposed and/or are still under development, or even to the fact that they are very specific to be included in the previous chapters (e.g., the electroinductive and magnetoinductive wave delay lines and the substrate-integrated waveguides).

It is the author's hope that the present manuscript constitutes a reference book in the topic on artificial transmission lines and their RF and microwave applications, and that the book can be of practical use to researchers, students and engineers involved in RF and microwave engineering, especially to those active in planar circuit and antenna design.

Ferran Martín
Santa Maria d'Oló (Barcelona)
September 2014

ACKNOWLEDGMENTS

This book is the result of an intensive research activity on the topic of artificial transmission lines carried out by the author and his research group (Centre d'Investigació en Metamaterials per a la Innovació en Tecnologies Electrònica i de les Comunicacions - CIMITEC) at the Universitat Autònoma de Barcelona, and also by many other researchers worldwide, with whom the author has had the privilege to collaborate or interact. It is impossible to include a complete list of all the people that have made possible to write this book. Nevertheless, I must express my most sincere gratitude to several colleagues, friends, and co-workers that have made invaluable contributions to it. I apologize if I omit somebody that deserves to be acknowledged and is not included in the following list.

First of all, I would like to give special thanks to the current and past members of my Group (Jordi Bonache, Joan García, Nacho Gil, Marta Gil, Francisco Aznar, Adolfo Vélez, Benito Sans, Gerard Sisó, Ferran Paredes, Gerard Zamora, Miguel Durán-Sindreu, Jordi Selga, Jordi Naqui, Paris Vélez, Simone Zuffanelli, Pau Aguilà, Marco Orellana, Lijuan Su, Marc Sans, Ignasi Cairó, Javier Herraiz, David Bouyge, and Anna Cedenilla), and to the visiting professors (Javier Mata) and students (Kambiz Afrooz and Ali Karami-Horestani). It is an honor to be the head of such productive and fruitful research group (a consequence of the continuous and endless effort of the involved people). Many of the ideas and results presented in the book have their origin in the researchers of CIMITEC, and therefore this book also belongs to them. I would like to highlight Jordi Bonache, who has had many brilliant ideas since more than one decade ago, providing very interesting research results and innovative applications on the basis of artificial transmission lines and related concepts. I must also acknowledge the contribution of Jordi Naqui to this book, who edited many figures of the manuscript; Gerard Zamora, for reviewing part of Chapter 4;

and Anna Cedenilla, who was in charge of the permissions for the use of many copyrighted figures. I would not like to forget the support of the administrative staff (headed by Mari Carmen Mesas during many years) and technicians (Javier Hellín) of my department, who are not very visible but are essential for the success of the research activities.

During the recent years, we have had fruitful collaborations with many groups that have contributed to the progress of the topic covered by the book. Among them, I would like to cite the groups of Prof. Francisco Medina (Universidad de Sevilla); Prof. Mario Sorolla (Universidad Pública de Navarra), who passed away in November 2012; Prof. Vicente Boria (Universitat Politècnica de Valencia); Prof. Rolf Jakoby (TU Darmstadt); Prof. Tatsuo Itoh (University of California Los Angeles); Prof. Christophe Fumeaux (University of Adelaide); Dr. Walter de Raedt (IMEC); Prof. Pierre Blondy (XLIM-Université de Limoges); Prof. Didier Lippens (IEMN-Université de Lille); and the groups involved in the Network of Excellence within the VI Framework Program of the European Union, METAMORPHOSE (2004–2008). We have recently been a partner in the collaborative project *Engineering Metamaterials* (2008–2014) of the CONSOLIDER INGENIO 2010 Program (MICIIN-Spain), which has represented an ideal platform for cooperation between the partners, and a continuous source of ideas. Special thanks go to Francisco Medina (Universidad de Sevilla), Christian Damm (TU Darmstadt), Silvio Hrabar (University of Zagreb), and Txema Lopetegi (Universidad Pública de Navarra) for reviewing some parts of the manuscript (Txema Lopetegi has also co-authored some sections of Chapter 2 and two appendixes). I will never forget the extraordinary contribution of my past PhD student (shared with Prof. Mario Sorolla), Francisco Falcone (now associate professor at the Universidad Pública de Navarra), to the topic of metamaterial transmission lines (many of the ideas presented in this book are in part due to him). To end the acknowledgments relative to external collaborations, I would like to mention Prof. Ricardo Marqués (Universidad de Sevilla), who is a well-known authority in the field of metamaterials, and with whom I have had the privilege to cooperate, learn, and co-write a previous book. We have had many stimulating discussions on many topics, including the modeling of artificial lines based on metamaterial concepts. There is no doubt that Prof. Marqués has been a key researcher for the progress and applications of metamaterial transmission lines based on the resonant-type approach.

The research activity that has conducted to the results presented in this book has been funded by several agencies or institutions. Particularly, I would like to express my gratitude to the past Spanish Ministry of Science and Innovation (MICIIN), for supporting our work through the collaborative project *Engineering Metamaterials*, cited earlier (ref. CSD2008-00066), and to the current Spanish Ministry of Economy and Competitiveness (MINECO), for funding our research activities through other national projects. Thanks are also given to the past Spanish Ministry of Industry, Commerce and Tourism (MICyT) for giving us support through several collaborative projects with companies for the development of precompetitive products on the basis of our research activities on topics related to this book. The Government of Catalonia is also acknowledged for giving us support as members of TECNIO (a network of Research and Technology Transfer Centers), and for funding several research projects of CIMITEC. I would also like to express my most sincere gratitude to *Institució Catalana de Recerca i Estudis Avançats* (ICREA) for supporting my work through an ICREA Academia Award (calls 2008 and 2013), and to my university for the continuous support, which includes the *Parc de Recerca UAB-Santander* Technology Transfer Chair. At the European level, I would like to express my gratitude to the European Commission and to the Eureka Program for funding several international projects. I would also like to mention the *Virtual Institute for Artificial Electromagnetic Materials and Metamaterials*, "METAMORPHOSE VI AISBL," for giving us the opportunity to disseminate and promote our research activities, and the Institute of Electrical and Electronics Engineers (IEEE) for elevating me to the grade of IEEE fellow (in acknowledgment to my contributions to the topic of metamaterial-based transmission lines).

Last, but not least, I would like to acknowledge the support of my wife, Anna, who has created the necessary atmosphere to write a long manuscript like this, who has accepted my long absences (this extends also to my children, Alba and Arnau), and who has done her best in favor of the family, myself, and my vocation. I would also like to include in the list many other people who have supported or influenced me, such as my parents (Juan and Rosario), my "second parents" (Josep Maria and Josefina), my grandparents, Carlos, Rut, my parents-in-law (Lina and Josep), and many others that are not in my mind at this moment, but are always present in my heart. Thank you very much!

FERRAN MARTÍN

1

FUNDAMENTALS OF PLANAR TRANSMISSION LINES

1.1 PLANAR TRANSMISSION LINES, DISTRIBUTED CIRCUITS, AND ARTIFICIAL TRANSMISSION LINES

In radiofrequency (RF) and microwave engineering, transmission lines are two-port networks used to transmit signals, or power, between two distant points (the source and the load) in a guided (in contrast to radiated) way. There are many types of transmission lines. Probably, the most well-known transmission line (at least for nonspecialists in RF and microwave engineering) is the coaxial line (Fig. 1.1), which consists of a pair of concentric conductors separated by a dielectric, and is typically used to feed RF/microwave components and to connect them to characterization and test equipment. Other planar transmission lines are depicted in Figure 1.2. There are many textbooks partially or entirely focused on transmission lines and their RF and microwave applications [1–8]. The author recommends these books to those readers interested in the topic of the present book (artificial transmission lines), which are not familiar with conventional (or ordinary) transmission lines. Nevertheless, the fundamentals of planar transmission lines are considered in this chapter for completeness and for better comprehension of the following chapters. As long as waveguides (and even optical fibers) do also carry electromagnetic (EM) waves and EM energy between two points, they can also be considered transmission lines. However, this book is entirely devoted to planar structures; and for this reason, waveguides are out of the scope of this chapter.

Artificial Transmission Lines for RF and Microwave Applications, First Edition. Ferran Martín.

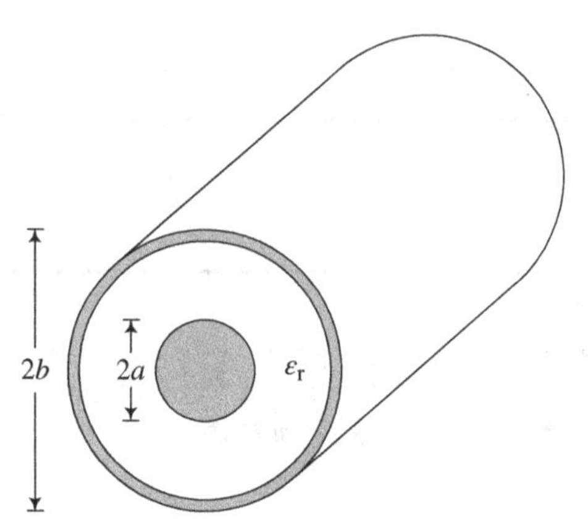

FIGURE 1.1 Perspective three-dimensional view of a coaxial transmission line. The relevant geometry parameters of the line are indicated, and ε_r is the relative permittivity (or dielectric constant) of the dielectric material.

Obviously, there are not transmission lines in natural form.[1] Transmission lines must be fabricated in order to satisfy certain requirements or specifications; in this sense, they are actually artificial (i.e., man-made) structures. However, the term *artificial transmission line* is restricted to a specific type of transmission lines, to distinguish them from the conventional ones.[2] Before discussing the definition and scope of the term *artificial transmission line*, let us now point out the different approaches for the study of planar (conventional) transmission lines. If the physical length of the transmission line is much smaller than the wavelength of the transmitted signals, the voltages and currents in the line are uniform, that is, they do not depend on the position in the line.[3] Under these conditions, the voltages and currents are dictated by the Kirchhoff's current and voltage laws and by the terminal equations of the lumped elements present at the input and output ports of the line, or at any position in the line. This is the so-called lumped element approach, which is generally valid up to about 100 MHz, or even further for planar structures (or circuits) including transmission lines not exceeding the typical sizes of printed circuit boards or PCBs (i.e., various centimeters). At higher frequencies, typically above 1 GHz, the finite propagation velocity of the transmitted signals (of the order of the speed of light) gives rise to variations of voltage and current along the lines, and the lumped circuit approach is no longer valid. At this regime, transmission lines can be analyzed by means of field theory, from Maxwell's equations. However, most planar transmission lines can alternatively be studied and described by means of an intermediate approach between lumped circuits and field equations: the distributed circuit approach. Indeed, for

[1] Exceptions to this are, for instance, the axons, which transmit nerve signals in brain neurons.
[2] Conventional (or ordinary) transmission lines are uniform along the propagation direction (see Fig. 1.2).
[3] Strictly speaking, this is true if losses are negligible. The effects of losses in transmission lines will be discussed later in detail.

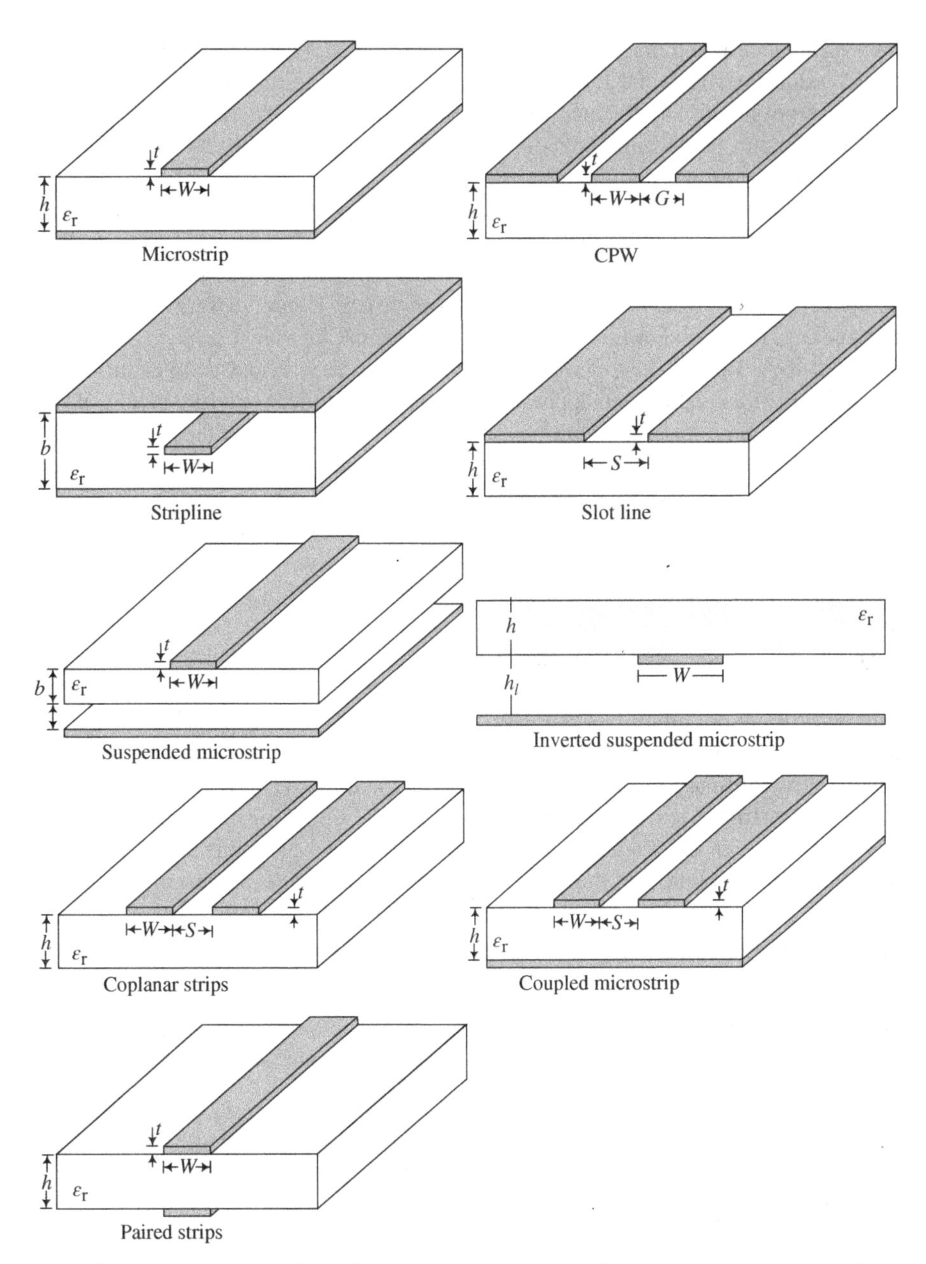

FIGURE 1.2 Perspective three-dimensional view of the indicated planar transmission lines, and relevant geometry parameters. These transmission lines are used for the implementation of distributed circuits, where the shape and transverse dimensions (W, S, G) of the line (or set of lines and stubs) are determined in order to obtain the required line functionality.

transverse electric and magnetic (TEM),[4] or quasi-TEM, wave propagation in planar transmission lines (i.e., the fundamental modes), there is a link between the results inferred from the distributed analysis and field theory. Nevertheless, this connection is discussed and treated in Appendix A, since it is not necessary to understand the contents of the present and the next chapters.

The most intriguing aspect of transmission lines operating at microwave frequencies and beyond is the fact that such lines can replace lumped elements, such as capacitances and inductances, in planar circuits, thus avoiding the use of lumped components which increase cost and circuit complexity. Hence, in RF and microwave engineering, transmission lines are not only of interest for signal or power transmission, but they are also key elements for microwave device and component design on the basis of the distributed approach. Thus, the constituent building blocks of distributed circuits are transmission lines and stubs,[5] which are implemented by simply etching metallic patterns on a microwave substrate (such patterns define a set of transmission lines and stubs providing certain functionality).

Distributed circuits are typically low cost since they are implemented in planar technology. However, the design flexibility, performance, or functionality of planar microwave circuits can be enhanced (and/or their dimensions can be reduced) by loading the lines with reactive elements (not necessarily planar),[6] or by breaking the uniformity of the lines in the direction of propagation, or by considering specific arrangements able to provide certain advantages as compared to ordinary lines. In the context of this book, the term *artificial transmission line* is used to designate these lines with superior characteristics, and to distinguish them from their conventional counterparts (ordinary lines). Hence, notice that the term *artificial transmission line* is not only restricted to designate artificial structures mimicking the behavior of ordinary lines (e.g., an LC ladder network or a capacitively loaded line acting as a slow wave transmission line).[7] In this book, the definition of *artificial transmission line* is

[4] Transmission lines supporting TEM modes require at least two conductors separated by a uniform (homogeneous) dielectric, and the electric and magnetic field lines must be entirely contained in such dielectric. In such modes, the electric and magnetic field components in the direction of propagation are null. A coaxial line is an example of transmission line that supports TEM modes. Microstrip and CPW transmission lines (see Fig. 1.2) are nonhomogeneous open lines, and hence do not support pure TEM modes, but quasi-TEM modes.

[5] Stubs are short- or open-circuit transmission line sections, shunt or series connected to another transmission line, intended to produce a pure reactance at the attachment point, for the frequency of interest.

[6] Notice that this loading refers to line loading along its length, not at the output port (as considered in Section 1.3 in reference to ordinary lines). A line with a load at its output port is usually referred to as terminated line.

[7] Artificial lines that mimic the behavior of ordinary lines are sometimes referred to as synthetic lines. Synthetic lines can be implemented by means of lumped, semilumped, and/or distributed components (combination of transmission lines and stubs). Synthetic lines purely based on the distributed approach (e.g., stub-loaded lines) are out of the scope of this book since they are indeed implemented by combining ordinary lines. Other artificial lines that can be considered to belong to the category of synthetic lines (e.g., capacitively loaded lines) are included in this book; but obviously, it is not possible to include all the realizations of synthetic lines reported in the literature. Artificial lines able to provide further functionalities than ordinary lines (e.g., metamaterial transmission lines with multiband functionality) are not considered to be synthetic transmission lines.

very broad and roughly covers all those lines that cannot be considered ordinary lines. Nevertheless, in many applications of artificial transmission lines, these lines simply replace ordinary lines, and the design approach of microwave circuits based on such artificial lines is similar to the one for ordinary lines, based on the control of the main line parameters. Therefore, in the next subsections, we will focus the attention on the study and analysis of ordinary lines, including the main transmission line parameters, reflections at the source and load (mismatching), losses in transmission lines, a comparative analysis of the most used planar transmission lines, and examples of applications. Most of these contents will be useful in the following chapters. Other useful contents for this chapter and chapters that follow (and in general for RF/microwave engineering), such as the Smith Chart and the scattering S-matrix, are included for completeness in Appendix B and C, respectively.

1.2 DISTRIBUTED CIRCUIT ANALYSIS AND MAIN TRANSMISSION LINE PARAMETERS

Planar transmission lines can be described by cascading the lumped element two-port network unit cell depicted in Figure 1.3, corresponding to an infinitesimal piece of the transmission line of length Δz, and C', L', R', and G' are the line capacitance, line inductance, line resistance, and line conductance per unit length, respectively. R' is related to conductor losses, whereas G' accounts for dielectric losses. From Kirchhoff's circuit laws applied to the network of Figure 1.3, the following equations are obtained:

$$v(z,t)-R'\Delta z\cdot i(z,t)-L'\Delta z\frac{\partial i(z,t)}{\partial t}-v(z+\Delta z,t)=0 \tag{1.1a}$$

$$i(z,t)-G'\Delta z\cdot v(z+\Delta z,t)-C'\Delta z\frac{\partial v(z+\Delta z,t)}{\partial t}-i(z+\Delta z,t)=0 \tag{1.1b}$$

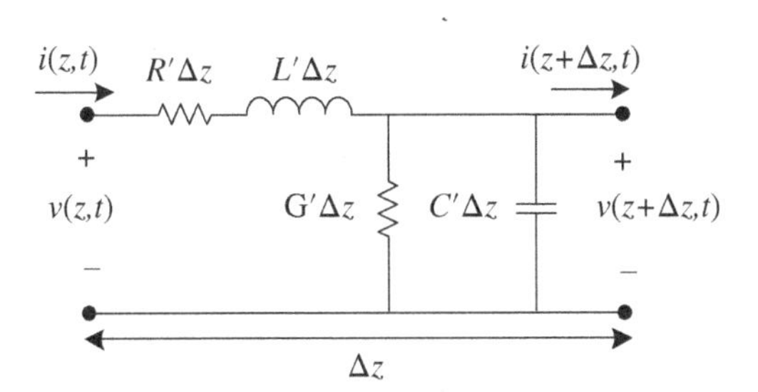

FIGURE 1.3 Lumped element equivalent circuit model (unit cell) of an ordinary transmission line.

By dividing these equations by Δz, and taking the limit as $\Delta z \rightarrow 0$, it follows:

$$\frac{\partial v(z,t)}{\partial z} = -R' i(z,t) - L' \frac{\partial i(z,t)}{\partial t} \tag{1.2a}$$

$$\frac{\partial i(z,t)}{\partial z} = -G' v(z,t) - C' \frac{\partial v(z,t)}{\partial t} \tag{1.2b}$$

Equations 1.2 are known as the telegrapher equations. If we now consider sinusoidal steady-state conditions (i.e., $v(z, t) = V(z)\cdot e^{j\omega t}$ and $i(z, t) = I(z)\cdot e^{j\omega t}$), the time variable in the previous equations can be ignored:

$$\frac{dV(z)}{dz} = -(R' + j\omega L')I(z) \tag{1.3a}$$

$$\frac{dI(z)}{dz} = -(G' + j\omega C')V(z) \tag{1.3b}$$

and the well-known wave equations result

$$\frac{d^2V(z)}{dz^2} - \gamma^2 V(z) = 0 \tag{1.4a}$$

$$\frac{d^2I(z)}{dz^2} - \gamma^2 I(z) = 0 \tag{1.4b}$$

where $\gamma = \alpha + j\beta$ is the complex propagation constant, given by

$$\gamma = \sqrt{(R' + j\omega L')(G' + j\omega C')} \tag{1.5}$$

and α and β are the attenuation constant and the phase constant, respectively. Notice that if conductor and dielectric losses can be neglected ($R' = G' = 0$), $\alpha = 0$, and the phase constant is proportional to the angular frequency and given by

$$\beta = \omega\sqrt{L'C'} \tag{1.6}$$

The general solutions of the wave equations are traveling waves of the form:

$$V(z) = V_o^+ e^{-\gamma z} + V_o^- e^{\gamma z} \tag{1.7a}$$

$$I(z) = I_o^+ e^{-\gamma z} + I_o^- e^{\gamma z} \tag{1.7b}$$

where the first and second terms correspond to wave propagation in $+z$ and $-z$ directions, respectively. By combining (1.3) and (1.7), it follows that the relation between voltage and current for the traveling waves, also known as the characteristic impedance, is given by

$$Z_o = \frac{V_o^+}{I_o^+} = \frac{-V_o^-}{I_o^-} = \sqrt{\frac{R' + j\omega L'}{G' + j\omega C'}} \tag{1.8}$$

For lossless lines, the voltage and current in the line are in phase, and the characteristic impedance is a real number:

$$Z_o = \sqrt{\frac{L'}{C'}} \tag{1.9}$$

Although losses may limit the performance of distributed microwave circuits, losses are usually neglected for design purposes, and the propagation constant and characteristic impedance are approximated by (1.6) and (1.9), respectively. According to (1.6), the dispersion relation β–ω is linear. The phase velocity, v_p, and the group velocity, v_g, are thus identical and given by

$$v_p = \frac{\omega}{\beta} = \frac{1}{\sqrt{L'C'}} \tag{1.10}$$

$$v_g = \left(\frac{d\beta}{d\omega}\right)^{-1} = \frac{1}{\sqrt{L'C'}} \tag{1.11}$$

and the wavelength in the line is given by:

$$\lambda = \frac{2\pi v_p}{\omega} = \frac{2\pi}{\beta} = \frac{2\pi}{\omega\sqrt{L'C'}} \tag{1.12}$$

That is, it is inversely proportional to frequency.[8] Sometimes, the length of a transmission line (for a certain frequency) is given in terms of the wavelength, or expressed as electrical length, $\phi = \beta l$, where l is the physical length of the line, and ϕ is an angle indicating whether distributed effects should be taken into account or not (as a first-order approximation, distributed effects are typically neglected if $\phi < \pi/4$). In many distributed circuits, transmission lines and stubs are $\lambda/4$ or $\lambda/2$ long at the operating frequency, corresponding to electrical lengths of $\phi = \pi/2$ and $\phi = \pi$, respectively.

For plane waves in source-free, linear, isotropic, homogeneous, and lossless dielectrics, the wave impedance, defined as the ratio between the electric and magnetic fields, and the phase velocity, are given by [1, 2] (see Appendix A):

$$\eta = \sqrt{\frac{\mu}{\varepsilon}} \tag{1.13}$$

[8] As will be shown, for artificial transmission lines expressions (1.10–1.12) are not necessarily valid. Indeed, for certain artificial lines, the wavelength either increases or decreases with frequency depending on the frequency regions.

$$v_p = \frac{1}{\sqrt{\mu\varepsilon}} \tag{1.14}$$

where ε and μ are the dielectric permittivity and magnetic permeability, respectively. These expressions, derived from Maxwell's equations, do also apply to TEM wave propagation in planar transmission lines, and therefore the main line parameters can be expressed in terms of the material parameters.[9] Notice that for nonmagnetic materials $\mu = \mu_o$, the permeability of vacuum, and hence the phase velocity can be rewritten in the usual form:

$$v_p = \frac{1}{\sqrt{\mu_o\varepsilon}} = \frac{1}{\sqrt{\mu_o\varepsilon_o\varepsilon_r}} = \frac{c}{\sqrt{\varepsilon_r}} \tag{1.15}$$

where c is the speed of light in vacuum, and $\varepsilon = \varepsilon_o\varepsilon_r$ (ε_o and ε_r being the permittivity of vacuum and the dielectric constant, respectively). However, for open nonhomogeneous lines, such as microstrip or coplanar waveguide (CPW) transmission lines, where pure TEM wave propagation is not possible, the previous expression does not hold. Nevertheless, the phase velocity in open lines can be expressed as (1.15) by simply replacing the dielectric constant of the substrate material, ε_r, with an effective dielectric constant, ε_{re}, which takes into account the presence of the electric field lines in both the substrate material and air[10]:

$$v_p = \frac{c}{\sqrt{\varepsilon_{re}}} \tag{1.16}$$

1.3 LOADED (TERMINATED) TRANSMISSION LINES

A uniform (in the direction of propagation) transmission line is characterized by the phase constant β (or by the electrical length βl), and by the characteristic impedance, Z_o. In a semi-infinitely long transmission line with a traveling wave generated by a source, the characteristic impedance expresses the relation between voltage and current at any transverse plane of the line. If losses are neglected, it follows that the power carried by the traveling wave along the line is given by

$$P^+ = \frac{1}{2}\frac{|V_o^+|^2}{Z_o} \tag{1.17}$$

[9] However, the wave impedance should not be confused with the characteristic impedance, Z_o, of transmission lines supporting TEM waves, which relates the voltage and current in the line and depends not only on the material parameters but also on the geometry of the line (see Appendix A).

[10] See at the end of Appendix A for more details.

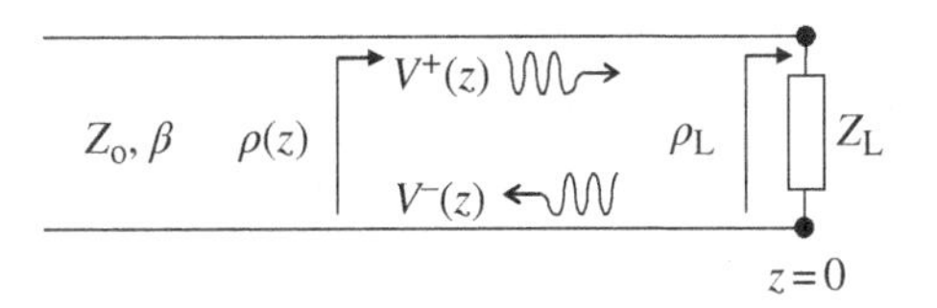

FIGURE 1.4 Transmission line terminated with an arbitrary load, located at $z=0$.

However, if the line is terminated by a load, three different situations may arise: (1) the incident power is completely absorbed by the load, (2) the incident power is completely reflected by the load, and (3) the incident power is partially absorbed and reflected by the load. Let us consider that the impedance of the load is Z_L, that this load is situated in the plane $z=0$ of the line (as Fig. 1.4 illustrates), and that a traveling wave of the form $V^+(z)=V_o^+ \cdot e^{-j\beta z}$ is present in the line. The ratio of voltage to current for such travelling wave is $V^+(z)/I^+(z)=Z_o$. At $z=0$, the relation between the voltage, V_L, and the current, I_L, in the load must satisfy the Ohm law, that is, $V_L/I_L=Z_L$. Since, in general, $Z_L \neq Z_o$, a reflected wave must be generated at $z=0$, so that the Ohm law is preserved. Therefore, the voltage and current in the line can be expressed as follows:

$$V(z)=V_o^+e^{-j\beta z}+V_o^-e^{j\beta z} \tag{1.18a}$$

$$I(z)=\frac{V_o^+}{Z_o}e^{-j\beta z}-\frac{V_o^-}{Z_o}e^{j\beta z} \tag{1.18b}$$

By forcing the Ohm law at $z=0$, it follows that

$$Z_L=\frac{V(0)}{I(0)}=\frac{V_o^+ + V_o^-}{V_o^+ - V_o^-}Z_o \tag{1.19}$$

and the relation between the amplitude of the reflected and the incident wave, also known as reflection coefficient, is

$$\rho_L=\frac{V_o^-}{V_o^+}=\frac{Z_L-Z_o}{Z_L+Z_o} \tag{1.20}$$

From (1.20), it follows that if $Z_L=Z_o$ (matched load), $\rho_L=0$ and the incident power is absorbed by the load (i.e., there are not reflections in the load). Conversely, if the load is an open or a short circuit, the reflection coefficient is $\rho_L(Z_L=\infty)=1$ and $\rho_L(Z_L=0)=-1$, respectively, and the incident power is reflected back to the source. Notice that the incident power is also reflected back to the source for reactive loads, where $|\rho_L(Z_L=j\chi)|=1$, χ being the reactance. Partially reflected and absorbed power occurs for resistive loads not matched to the line, or for complex loads. Notice also that for passive loads ($Z_L = R + j\chi$, with $R > 0$), the modulus of the reflection coefficient is $|\rho_L| \le 1$. This is expected since the reflected power, given by

$$P^- = \frac{1}{2}\frac{|V_o^+|^2}{Z_o}|\rho_L|^2 \tag{1.21}$$

cannot be higher than the incident power (given by 1.17) for passive loads. In microwave engineering, the reflection coefficient is typically expressed in dB and identified as the return loss:

$$RL = -20\log|\rho_L| \tag{1.22}$$

For infinitely long transmission lines or for transmission lines terminated with a matched load, constant amplitude travelling waves are present in the line. However, if a reflected wave is generated in the load plane, a standing wave is generated in the line, where the amplitude is modulated by the modulus of the reflection coefficient. From (1.18a) and (1.20), the voltage in the line can be written as follows:

$$V(z) = V_o^+\left[e^{-j\beta z} + \rho_L e^{j\beta z}\right] \tag{1.23}$$

If we now express the reflection coefficient in polar form ($\rho_L = |\rho_L| \cdot e^{j\theta}$), the voltage in the line can be rewritten as follows:

$$V(z) = V_o^+ e^{-j\beta z}\left[1 + |\rho_L| e^{2j\beta z + j\theta}\right] \tag{1.24}$$

from which it follows:

$$|V(z)|^2 = |V_o^+|^2\left[1 + |\rho_L|^2 + 2|\rho_L|\cos(2\beta z + \theta)\right] \tag{1.25}$$

Equation 1.25 indicates that the amplitude is a maximum ($V_{max} = |V_o^+|[1+|\rho_L|]$) and a minimum ($V_{min} = |V_o^+|[1-|\rho_L|]$) at planes separated by $\lambda/4$, and the ratio between the maximum and minimum voltage in the line, known as voltage standing wave ratio, is given by

$$\text{SWR} = \frac{1 + |\rho_L|}{1 - |\rho_L|} \tag{1.26}$$

As anticipated, the SWR is determined by the reflection coefficient. However, it only depends on the modulus of the reflection coefficient, not on its phase, θ. This means that from the information of the SWR, it is not possible to completely characterize the load. For instance, it is not possible to distinguish between a short circuit, an open circuit, or a reactive load, since the reflection coefficient of these loads has the same modulus ($|\rho_L| = 1$). Nonetheless, in many applications the relevant information is the matching between the line (or the source) and the load in terms of the power transmitted to the load, the phase information being irrelevant.

Although wave reflection in a transmission line is caused by a mismatch between the line and the load, and hence it is ultimately generated at the plane of the load ($z = 0$), the reflection coefficient can be generalized to any plane of the line, as the ratio between the voltage of the incident and reflected wave, that is

$$\rho(z) = \frac{V^-}{V^+} = \frac{V_o^- e^{j\beta z}}{V_o^+ e^{-j\beta z}} = \rho_L e^{2j\beta z} \tag{1.27}$$

where, as expected, $|\rho(z)| = |\rho_L|$.

One important point of terminated lines is the amount of power delivered by a given source to the line. If the source has complex impedance, Z_s, such power is directly characterized by the power wave reflection coefficient, s, given by [9]:

$$s = \frac{Z_{in} - Z_s^*}{Z_{in} + Z_s} \tag{1.28}$$

where the asterisk denotes complex conjugate, and Z_{in} is the impedance seen from the input port of the line, that is, looking into the load (Fig. 1.5). Actually, the power transmission coefficient, which is the relevant parameter for computing the power transmitted to the line for the general situation of a source with complex impedance, is given by

$$\tau = 1 - |s|^2 = 1 - \left|\frac{Z_{in} - Z_s^*}{Z_{in} + Z_s}\right|^2 \tag{1.29}$$

The input impedance, Z_{in}, depends on the distance between the load and the input port (i.e., the plane of the source). This impedance can be simply computed as follows:

$$Z_{in} = \frac{V(-l)}{I(-l)} = \frac{V_o^+ \left(e^{j\beta l} + \rho_L e^{-j\beta l}\right)}{V_o^+ \left(e^{j\beta l} - \rho_L e^{-j\beta l}\right)} Z_o \tag{1.30}$$

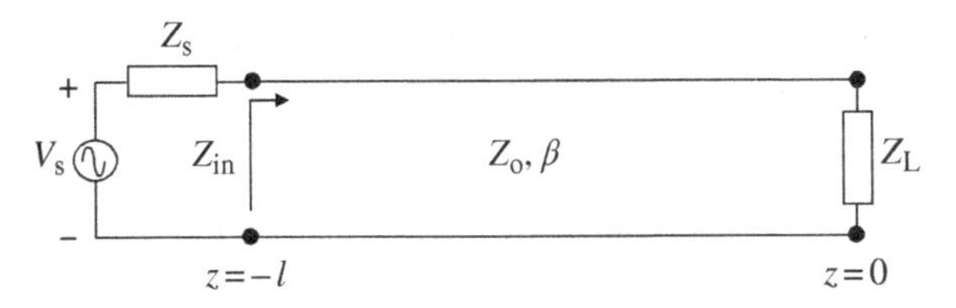

FIGURE 1.5 Transmission line of length l fed by a voltage source and terminated with an arbitrary load, located at $z = 0$.

And after some minor manipulation,

$$Z_{in} = Z_o \frac{Z_L + jZ_o \tan(\beta l)}{Z_o + jZ_L \tan(\beta l)} \tag{1.31}$$

The analysis of (1.31) reveals that in the limit $\beta l \rightarrow 0$, $Z_{in} = Z_L$, as expected, since this regime corresponds to the lumped element approximation discussed in Section 1.1. Expression (1.31) also indicates that the need to model the line as a distributed circuit does not solely depend on frequency (through β), but also on the line length l. Indeed, the key parameter is the electrical length, βl, as anticipated in the previous subsection. From (1.31), it follows that the input impedance is a periodic function with βl, and hence a periodic function with both the line length and the frequency. If the frequency is set to a certain value, the input impedance is a periodic function of period $\lambda/2$ with the line length. This means that the input impedance looking into the load seen from planes separated by a multiple of $\lambda/2$ is identical. From this result, it follows that for a $\lambda/2$ line ($\beta l = \pi$), the input impedance is the one of the load, $Z_{in} = Z_L$.

Let us now consider several cases of particular interest. If the line length is $l = \lambda/4$ ($\beta l = \pi/2$), the input impedance is

$$Z_{in} = \frac{Z_o^2}{Z_L} \tag{1.32}$$

which means that the input impedance is inversely proportional to the load impedance, and hence a $\lambda/4$ transmission line acts as an impedance inverter. This means that a reactive load with inductive/capacitive reactance is seen as capacitive/inductive reactance from the input port; in other words, the sign of the reactance is reversed in $\lambda/4$ lines. From (1.32), it also follows that an open-circuit load is transformed to short-circuit at the input port of a $\lambda/4$ line, and vice versa.

For the general case of open-ended ($Z_L = \infty$) and short-circuited ($Z_L = 0$) lines, the input impedance given by (1.31) takes the following form:

$$Z_{in}(Z_L = \infty) = -jZ_o \cot(\beta l) \tag{1.33a}$$

$$Z_{in}(Z_L = 0) = jZ_o \tan(\beta l) \tag{1.33b}$$

Thus, the input impedances are purely reactive, just as those of lumped reactive elements (inductors and capacitors). For lines satisfying $\beta l < \pi/2$, a short-circuited line resembles an inductor, whereas an open-ended line mimics a capacitor. However, the reactances of lumped inductors and capacitors have different mathematical forms than those of shorted and opened transmission lines. This means that we cannot replace lumped reactive elements with open or shorted lines exhibiting identical behavior. However, by forcing the impedance of a capacitor and inductor to be equal to those of (1.33) we obtain

$$-\frac{j}{\omega C} = -jZ_o \cot(\beta l) \tag{1.34a}$$

$$j\omega L = jZ_o \tan(\beta l) \tag{1.34b}$$

and the previous expressions have solutions at many different frequencies. Let us consider the smallest of these angular frequencies and call it ω_c. If we set the length of the line to be $\lambda/8$ at this frequency (i.e., $\beta l = \pi/4$ at ω_c), the previous expressions take the following form:

$$\frac{1}{\omega_c C} = Z_o \tag{1.35a}$$

$$\omega_c L = Z_o \tag{1.35b}$$

Therefore, if we wish to implement a short-circuited transmission line with the same impedance as an inductor L at frequency ω_c, we set the characteristic impedance of the transmission line to $Z_o = \omega_c L$. Likewise, if we wish to obtain the reactance of a capacitor C at frequency ω_c, we set the characteristic impedance of the transmission line to $Z_o = 1/\omega_c C$. In both cases, the length of the line must be set to $\lambda/8$ at ω_c. Expressions (1.35) are called Richard's transformations [10], and are useful to avoid the use of lumped reactive components in certain microwave circuits such as low-pass filters. However, since Richard's transformations guarantee identical reactances between the lumped and the distributed reactive elements at a single frequency, we cannot expect that the response of a lumped circuit is identical to that of the distributed counterpart.

Obviously, the load of a transmission line can be another transmission line with different characteristic impedance. Let us consider that two transmission lines of characteristic impedances Z_o and Z_1, respectively, are cascaded as shown in Figure 1.6, and that the transmission line to the right of the contact plane ($z = 0$), that is, the line acting as load, is either infinitely long or terminated with a matched load (so that there are not reflected waves in this line). Under these conditions, the load impedance seen by the transmission line to the left of $z = 0$ is simply Z_1. Hence, the reflection coefficient at $z = 0$ is given by

$$\rho = \frac{Z_1 - Z_o}{Z_1 + Z_o} \tag{1.36}$$

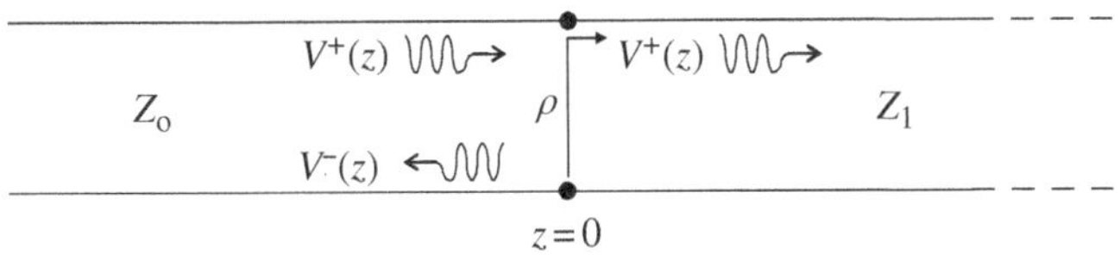

FIGURE 1.6 Cascade connection of two transmission lines with different characteristic impedance.

and the incident wave is partially transmitted to the second line. For $z<0$, the voltage in the line can be expressed as (1.23), whereas to the right of the contact plane, the voltage can be expressed in terms of a transmission coefficient, T, as follows:

$$V(z)=V_o^+Te^{-j\beta z} \tag{1.37}$$

By forcing expressions (1.23) and (1.37) to be identical at $z=0$, the transmission coefficient is found to be

$$T=1+\rho=\frac{2Z_l}{Z_l+Z_o} \tag{1.38}$$

and the transmission coefficient expressed in dB is identified as the insertion loss:

$$IL=-20\log|T| \tag{1.39}$$

Notice that the transmission coefficient, defined by the fraction of the amplitude of the voltage of the incident wave transmitted to the second transmission line (expression 1.37), can be higher than one (this occurs if $Z_l>Z_o$). This result does not contradict any fundamental principle (i.e., the conservation of energy), since the transmitted power is always equal or less than the incident power (however, the amplitude of the voltage of the transmitted wave can be higher than V_o^+).

To end this subsection, let us briefly consider the reflections generated by a source with mismatched impedance that feeds a transmission line with characteristic impedance Z_o and length l (Fig. 1.5). Let us assume that the load of the transmission line is also mismatched, so that a reflected wave is generated by the load once the incident wave reaches the load plane. Once the switch is closed at $t=0$, the following expression must be satisfied at any time $t>0$:

$$V_s=Z_sI(-l)+V(-l) \tag{1.40}$$

Before the reflected wave at the load reaches the plane of the source ($z=-l$), that is, for $t<2l/v_p$, expression (1.40) is written as follows:

$$V_s=Z_s\frac{V_1^+}{Z_o}+V_1^+ \tag{1.41}$$

where V_1^+ is the amplitude of the incident wave generated by the source after the switch is closed, that is,[11]

[11] This expression is valid if the source impedance is purely resistive. However, expression (1.45) is valid for any source impedance (purely resistive, purely reactive, or complex). The reason is that the time domain analysis giving (1.45) can be initiated once the transient associated to reactive or complex load and/or source impedances has expired.

$$V_1^+ = V_s \frac{Z_o}{Z_o + Z_s} \tag{1.42}$$

Once the reflected wave V_1^- reaches the source plane, a reflected wave V_2^+ must be generated by the source in order to satisfy the Ohm law. Hence, (1.40) is expressed as follows:

$$V_s = Z_s \frac{V_1^+ - V_1^- + V_2^+}{Z_o} + V_1^+ + V_1^- + V_2^+ \tag{1.43}$$

By combining (1.41) and (1.43), we obtain the following:

$$\frac{Z_s}{Z_o}\left(V_1^- - V_2^+\right) = V_1^- + V_2^+ \tag{1.44}$$

and the reflection coefficient at the source is found to be

$$\rho_s = \frac{V_2^+}{V_1^-} = \frac{Z_s - Z_o}{Z_s + Z_o} \tag{1.45}$$

which is formally identical to the reflection coefficient at the load.

Obviously, when the reflected wave at the source (V_2^+) reaches the load plane, a new wave (V_2^-) is generated by reflection at the load plane, and the process continues indefinitely until the steady state is achieved. This endless bouncing process converges to a steady state since the amplitude of the reflected waves progressively decreases.[12] Let us calculate, as an illustrative example, the steady-state voltage at $z = -l$ for the structure of Figure 1.5. If the initial wave is designated as V_1^+, the first reflected one at $z = -l$, taking into account the phase shift experienced by the wave along the transmission line, is $V_1^- = V_1^+ \rho_L \cdot e^{-2j\beta l}$ (see Fig. 1.7). The next two are $V_2^+ = V_1^+ \rho_L \rho_s \cdot e^{-2j\beta l}$ and $V_2^- = V_1^+ \rho_L^2 \rho_s \cdot e^{-4j\beta l}$. The steady-state voltage at $z = -l$ is given by the superposition of the left to right (+) and right to left (−) waves, that is[13],

$$V(z=-l) = \sum_i V_i^+ + \sum_i V_i^- = V_1^+ \frac{1}{1 - \rho_i \rho_s} + V_1^+ \frac{\rho_i}{1 - \rho_i \rho_s} \tag{1.46}$$

where V_1^+ is given by (1.42) and $\rho_i = \rho_L \cdot e^{-2j\beta l}$. Introducing (1.42) and (1.45) in (1.46) gives

$$V(z=-l) = V_s \frac{Z_o\left(1 + \rho_L e^{-2j\beta l}\right)}{Z_o + Z_s - (Z_s - Z_o)\rho_L e^{-2j\beta l}} \tag{1.47}$$

[12] This is consequence of the modulus of the reflection coefficients at the source and the load, which is smaller than one for passive loads. However, if the line is loaded with an active load, instability is potentially possible.

[13] To derive (1.46), the identity $1 + x + x^2 + \ldots = 1/(1 - x)$, where $|x| < 1$, has been used.

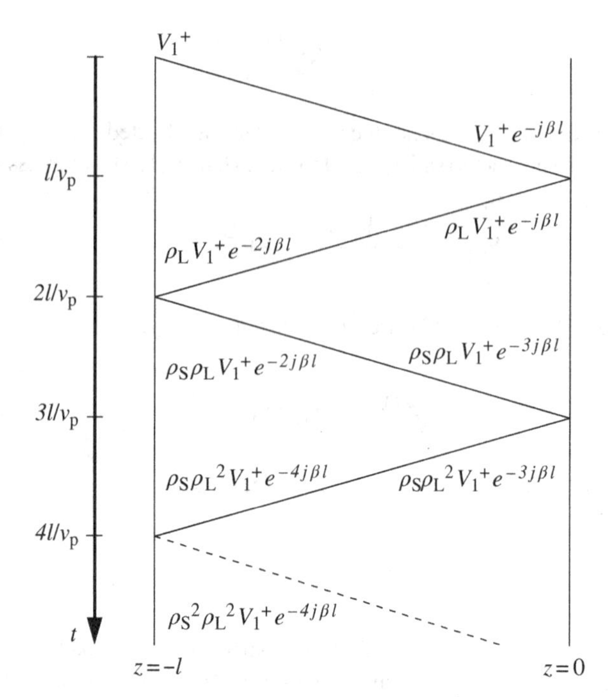

FIGURE 1.7 Bounce diagram corresponding to the example discussed in the text. The vertical axis is the time axis.

which can be expressed as follows:

$$V(z=-l) = V_s \frac{Z_{in}}{Z_{in} + Z_s} \tag{1.48}$$

where Z_{in}, given by (1.30), is the impedance seen from the source plane. As expected, the steady-state voltage at $z=-l$ is simply given by the voltage divider, considering the series connection of the source impedance and the input impedance of the loaded line.

1.4 LOSSY TRANSMISSION LINES

In planar transmission lines of practical interest for RF and microwave circuit design losses are small and they are usually neglected for design purposes (as it was mentioned in Section 1.2).[14] However, although small, losses produce

[14] To guarantee small losses, distributed circuits must be preferably implemented in commercially available low-loss microwave substrates.

attenuation and distortion in the transmitted signals, and the analysis of their effects on wave propagation is of interest. There are three main causes of losses: (1) the finite conductivity of the metals (conductor losses), (2) the dissipation in the dielectric (either caused by the presence of free electrons or by dipole relaxation phenomena), and (3) radiation losses. Although radiation losses may be dominant under some circumstances, transmission lines operating as guided-wave structures must be designed in order to exhibit small radiation. Hence, this loss mechanism is not considered by the moment.[15]

Ohmic (or conductor) and dielectric losses are accounted for by the lumped element circuit model of the transmission line (Fig. 1.3) through the series resistance R' and shunt conductance G', respectively. Let us now calculate the complex propagation constant (expression 1.5) under the low-loss approximation (justified by the reasons explained earlier), namely, $R' \ll \omega L'$ and $G' \ll \omega C'$. The complex propagation constant can be rearranged and written as follows:

$$\gamma = j\omega\sqrt{L'C'}\sqrt{1 - j\left(\frac{R'}{\omega L'} + \frac{G'}{\omega C'}\right) - \frac{R'G'}{\omega^2 L'C'}} \tag{1.49}$$

Neglecting the last term in the square root of (1.49) and applying the Taylor series expansion up to the first order, the complex propagation constant can be approximated by

$$\gamma = j\omega\sqrt{L'C'}\left[1 - \frac{j}{2}\left(\frac{R'}{\omega L'} + \frac{G'}{\omega C'}\right)\right] \tag{1.50}$$

From (1.50), the attenuation constant and the phase constant for low-loss transmission lines can be easily identified:

$$\alpha = \frac{1}{2}\left(\frac{R'}{Z_o} + G'Z_o\right) \tag{1.51}$$

$$\beta = \omega\sqrt{L'C'} \tag{1.52}$$

where Z_o is the lossless characteristic impedance of the line given by (1.9).

[15] Nevertheless, leaky-wave transmission lines are specifically designed to enhance radiation. These lines are the building blocks of leaky-wave antennas (LWAs), as will be shown in Chapter 4.

With regard to the characteristic impedance (expression 1.8), it can be rearranged and written as follows:

$$Z_o = \sqrt{\frac{L'}{C'}} \sqrt{1 - j\frac{\frac{C'\omega}{G'} - \frac{L'\omega}{R'}}{\frac{L'C'\omega^2}{R'G'} - j\frac{L'\omega}{R'}}} \tag{1.53}$$

By virtue of the low-loss approximation, the second term of the denominator in the square root can be neglected, and, by using the first order Taylor series approximation, the characteristic impedance is found to be

$$Z_o = \sqrt{\frac{L'}{C'}}\left[1 - \frac{j}{2}\left(\frac{R'}{\omega L'} - \frac{G'}{\omega C'}\right)\right] \tag{1.54}$$

and this expression can be further simplified to (1.9).[16]

According to these results, it follows that the phase constant and the characteristic impedance of low-loss transmission lines can be closely approximated by considering the line as lossless. The attenuation constant (1.51) has two contributions: one associated to conductor losses (proportional to R'), and one associated to dielectric losses (proportional to G').

Despite for low-loss lines the phase constant can be approximated by a linear function, the effects of dispersion may be appreciable in very long transmission lines, and may give rise to signal distortion. However, there is a special case where the phase constant of lossy transmission lines varies linearly with frequency, and dispersion is not present. Such lines are called distortionless lines and must satisfy the following identity (known as distortionless, or Heaviside, condition)

$$R'C' = G'L' \tag{1.55}$$

In view of the general expression of the complex propagation constant (1.5), the unique way to achieve a linear dependence of β with frequency for a lossy line is to achieve a complex propagation constant of the form $A + j\omega B$ (where A and B are constants). The only way this can be satisfied is if $R' + j\omega L'$ and $G' + j\omega C'$ differ by no more than a constant factor. This means that both the real and imaginary parts must be independently related by the same factor, which leads to (1.55). Under the condition specified by (1.55), the complex propagation constant is[17]

$$\gamma = \frac{R'}{Z_o} + j\omega\sqrt{L'C'} \tag{1.56}$$

[16] Notice that this means to neglect $R'/\omega L'$ and $G'/\omega C'$ in (1.54). However, if these two terms are neglected in (1.50), we find the trivial solution corresponding to the lossless line, with β given by (1.6) and $\alpha = 0$.

[17] Notice that, using (1.55), the attenuation constant can also be expressed as $\alpha = (R'G')^{1/2}$ or $\alpha = G'Z_o$.

and β is given by (1.6). It is interesting to mention that the attenuation constant ($\alpha = R'/Z_o$) does not depend on frequency, which means that all frequency components are attenuated the same factor. This means that distortionless lines are able to transmit pulse signals or modulated signals without distortion (although these signals are attenuated along the line due to losses). It is also simply to demonstrate that the characteristic impedance of distortionless lines is a real constant given by (1.9).

In practical transmission lines, the distortionless condition is not easy to satisfy since G' is usually very small. To compensate this, the line can be loaded with series connected inductances periodically spaced along the line. This strategy leads to an unconventional transmission line that can indeed be considered an artificial transmission line.[18] Nevertheless, the elements of the distributed circuit of a transmission line are not exactly constant (in particular, R' varies weakly with frequency), and the distortionless condition (expression 1.55) is difficult to meet in practice. For planar transmission lines used as building blocks in distributed circuits, where the lines are low-loss and short, dispersion is not usually an issue, except at high frequencies (dozens of GHz) or for very wideband signals.

1.4.1 Dielectric Losses: The Loss Tangent

Although for the design of most planar distributed circuits losses are neglected in the first steps, it is important to simulate their effects before fabrication. In EM solvers, losses are introduced by providing the conductivity of the metallic layers, and the loss tangent ($\tan\delta$) of the substrate material. The loss tangent takes into account dielectric losses, including both conduction losses (due to nonzero conductivity of the material) and losses due to damping of the dipole moments (that represents an energy transfer between the external electric field and the material at microscopic level).

For a dielectric material, the application of an electric field gives rise to the polarization of the atoms or molecules of the medium in the form of electric dipole moments, which contribute to the total displacement flux according to

$$\vec{D} = \varepsilon_o \vec{E} + \vec{P}_e \tag{1.57}$$

where $\vec{P}_e$ is the electric polarization, which is related to the applied electric field through[19]

$$\vec{P}_e = \varepsilon_o \chi_e \vec{E} \tag{1.58}$$

[18] According to the definition of artificial transmission line adopted in this book (see Section 1.1), L-loaded distortionless lines belong to this group, but such lines are out of the scope of this manuscript. As will be seen in Chapter 2, periodic loaded lines exhibit a cut-off frequency. Beyond this frequency, attenuation dramatically increases, and hence these lines may not support the transmission of high-frequency or broadband signals.

[19] It is assumed that the material is linear, isotropic, and homogeneous, that is, the electric susceptibility and permittivity are scalars that do not depend on the position and magnitude of the external field. For anisotropic materials, the relation between $\vec{E}$ and $\vec{P}_e$, or between $\vec{E}$ and $\vec{D}$, is a tensor.

χ_e being the electric susceptibility. Combining the previous expressions, the electric displacement can be written as follows:

$$\vec{D} = \varepsilon_o(1+\chi_e)\vec{E} = \varepsilon\vec{E} \tag{1.59}$$

In vacuum, the electric susceptibility is null, whereas in dielectric materials it is a complex number, where the imaginary part accounts for losses. In low-loss materials, the imaginary part of the susceptibility can be neglected to a first-order approximation. However, the effects of material losses on circuit performance may play a role in distributed circuits.[20] For this reason, losses cannot be neglected in the evaluation of circuit performance (typically inferred from commercially available EM simulators). In lossy materials, the dielectric permittivity is thus a complex number that can be expressed as follows:

$$\varepsilon = \varepsilon_o(1+\chi_e) = \varepsilon' - j\varepsilon'' \tag{1.60}$$

where ε'' is a positive number due to energy conservation [1, 11].

Conduction losses, associated to the presence of free electrons in the dielectric material (it is assumed that the material exhibits a nonzero conductivity), do also contribute to the imaginary part of the complex permittivity. The conduction current density is related to the electric field through the Ohm law:

$$\vec{J} = \sigma\vec{E} \tag{1.61}$$

where σ is the conductivity of the material. By introducing the previous expression in the Ampere–Maxwell law, we obtain

$$\nabla \times \vec{H} = j\omega\varepsilon\vec{E} + \sigma\vec{E} \tag{1.62a}$$

$$\nabla \times \vec{H} = j\omega\left(\varepsilon' - j\frac{\sigma+\omega\varepsilon''}{\omega}\right)\vec{E} \tag{1.62b}$$

From (1.62), it is clear that the effects of conduction losses can be accounted for by including a conductivity dependent term in the imaginary part of the complex permittivity. The loss tangent is defined as the ratio between the imaginary and real parts of this generalized[21] permittivity:

[20] In low-loss microwave substrates, losses are dominated by the finite conductivity of the metal layers. However, in general-purpose substrates, such as FR4, or in high-resistivity silicon (HR-Si) substrates, among others, material losses may significantly degrade the circuit performance.

[21] The complex permittivity including the contribution of σ is usually designated as effective complex permittivity in most textbooks. However, in this book, the term *effective permittivity* is either referred to the permittivity of effective media, or metamaterials, as will be seen in Chapter 3, or it is used to describe wave propagation in quasi-TEM lines by introducing an "averaged" permittivity (and permeability) in the equations governing purely TEM wave propagation (see Appendix A).

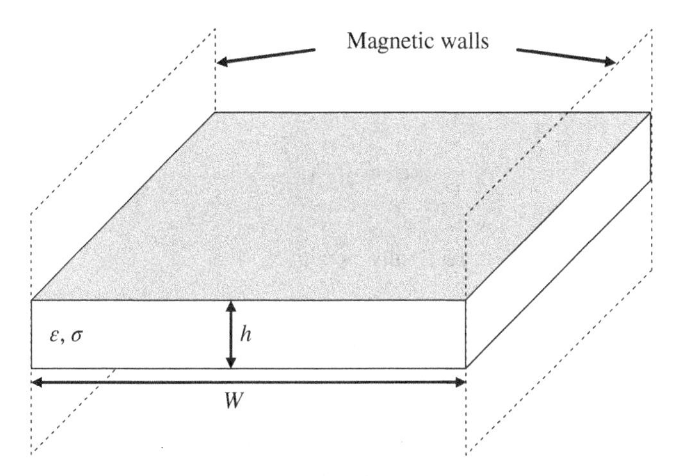

FIGURE 1.8 Parallel plate transmission line with magnetic walls at the edges.

$$\tan\delta = \frac{\sigma + \omega\varepsilon''}{\omega\varepsilon'} \tag{1.63}$$

Low-loss dielectrics are characterized by a small tanδ (typically $10^{-4} - 10^{-2}$).

Let us now try to link the loss tangent to the dielectric contribution of the attenuation constant for low-loss transmission lines. Let us consider a hypothetical transmission line with purely TEM wave propagation, for example, a stripline, or a parallel plate transmission line with magnetic walls at the lateral sides (or with very wide plates to neglect the fringing fields). This simplifies the analysis, and provides compact formulas, which is enough for our purposes. For the case of the parallel plate transmission line (Fig. 1.8), let W and h be the width of the metal plates and the height of the substrate, respectively. The per-unit-length line conductance is related to the dielectric conductivity as[22]

$$G' = \frac{W}{h}\sigma \tag{1.64}$$

which in turn can be expressed in terms of the per unit length capacitance as

$$G' = \frac{C'}{\varepsilon'}\sigma \tag{1.65}$$

[22] We assume that the conductivity in (1.64) is actually the effective conductivity, given by the numerator of (1.63).

The dielectric contribution of the attenuation constant for low-loss lines is given by the second term of the right-hand side of (1.51), and can be written as follows:

$$\alpha_{\mathrm{d}} = \frac{1}{2} G' \sqrt{\frac{L'}{C'}} \tag{1.66}$$

Introducing (1.65) in (1.66), we finally obtain

$$\alpha_{\mathrm{d}} = \frac{1}{2} \frac{C'}{\varepsilon'} \sigma \sqrt{\frac{L'}{C'}} = \frac{1}{2} \frac{\sigma}{\varepsilon'} \frac{\beta}{\omega} = \frac{\beta}{2} \tan\delta = \frac{1}{2} \omega \sqrt{\mu\varepsilon'} \tan\delta \tag{1.67}$$

In nonhomogeneous open lines, such as microstrip lines or CPWs, expression (1.67) is not strictly valid, but it provides a rough approximation of α_{d} by merely introducing the effective permittivity (defined as the effective dielectric constant times the permittivity of vacuum) in the last term. Indeed, for microstrip transmission lines (1.67) rewrites as [12, 13] follows:

$$\alpha_{\mathrm{d}} = \frac{\pi f}{c} \frac{\varepsilon_{\mathrm{r}}(\varepsilon_{\mathrm{re}} - 1)}{\sqrt{\varepsilon_{\mathrm{re}}}(\varepsilon_{\mathrm{r}} - 1)} \tan\delta \tag{1.68}$$

where it has been assumed that $\mu = \mu_{\mathrm{o}}$ and $(\mu_{\mathrm{o}}\varepsilon_{\mathrm{o}})^{-1}$ is the speed of light in vacuum, c. Notice that if $\varepsilon_{\mathrm{r}} = \varepsilon_{\mathrm{re}}$, (1.67) and (1.68) are identical. The analysis of (1.68) also reveals that for high values of ε_{r} (and hence $\varepsilon_{\mathrm{re}}$), expression (1.67) provides a good estimation of α_{d} by introducing in the root the effective permittivity.

The dielectric material (substrate) used in planar transmission lines is characterized by the loss tangent (which accounts for dielectric losses) and by the dielectric constant (which determines the phase velocity). Although these parameters are supplied by the manufacturer with the corresponding tolerances, there are sometimes substantial variations that make necessary the characterization of the material (i.e., the measurement of the dielectric constant and the loss tangent) for an accurate design. The dielectric constant and the loss tangent of a substrate material can be experimentally inferred by means of a microstrip ring resonator configuration (Fig. 1.9) [13–15]. The transmission coefficient exhibits transmission peaks at frequencies that depend on the dielectric constant of the substrate, and the loss tangent is extracted from the quality factor of the resonance peaks along with the theoretical calculations of the conductor losses (see next subsection).

For a ring resonator, resonances occur at those frequencies where the ring circumference is a multiple of the wavelength, λ. The resonance condition can thus be expressed as follows:

$$\lambda = \frac{2\pi r_{\mathrm{m}}}{n} = \frac{v_{\mathrm{p}}}{f_{\mathrm{n}}} = \frac{c}{\sqrt{\varepsilon_{\mathrm{re}}}} \frac{1}{f_{\mathrm{n}}} \tag{1.69}$$

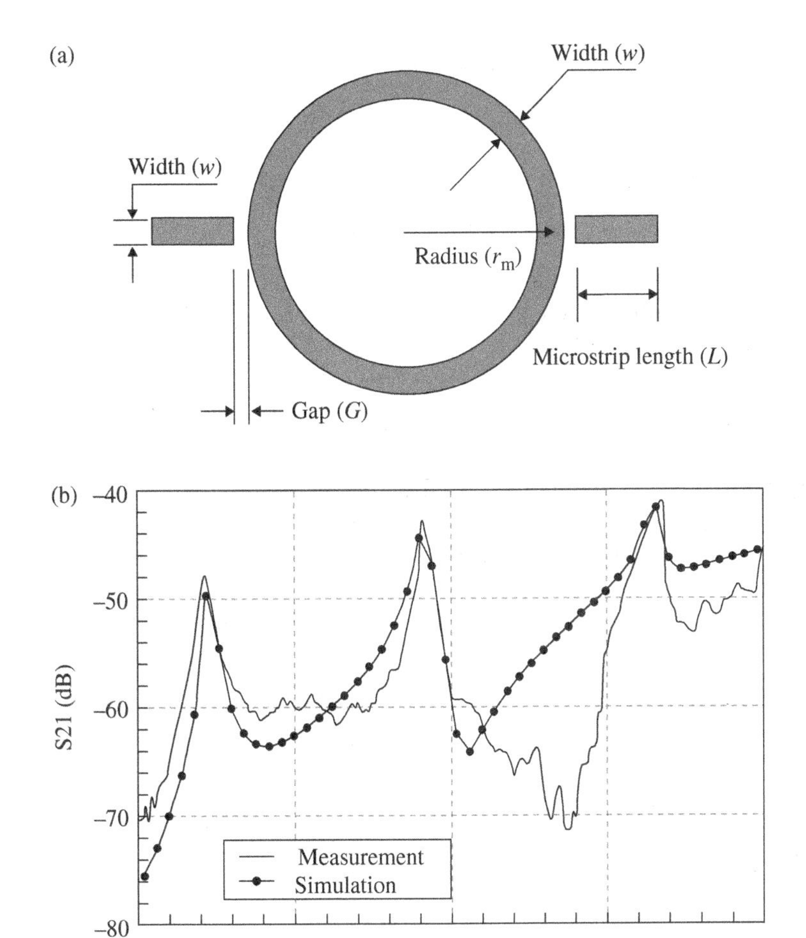

FIGURE 1.9 Microstrip ring resonator configuration used to extract the dielectric constant and loss tangent of the substrate (a), and typical frequency response with transmission peaks (b). Reprinted with permission from Ref. [15]; copyright 2007 IEEE.

where n refers to the nth-order resonance, and r_m is the mean ring radius. The effective dielectric constant is thus [13–15]

$$\varepsilon_{re} = \left(\frac{nc}{2\pi r_m f_n}\right)^2 \tag{1.70}$$

Once the effective dielectric constant is known (the resonance frequencies can be easily inferred from the measured transmission coefficient), the dielectric constant is given by [6, 13]:

$$\varepsilon_r = \frac{2\varepsilon_{re} + M - 1}{M + 1} \tag{1.71}$$

where

$$M = \left(1 + 12\frac{h}{W'}\right)^{-1/2} \tag{1.72}$$

and W' is the effective strip width, given by

$$W' = W + \frac{1.25t}{\pi}\left(1 + \ln\frac{2h}{t}\right) \tag{1.73}$$

In (1.72) and (1.73), h and t are the thickness of the substrate and metal layer, respectively, and W is the strip width. Expressions (1.71–1.73) are valid under the assumption that $W \geq h \gg t$, which is usually satisfied (general expressions are given in [6]).

For the determination of $\tan\delta$, it is first necessary to measure the unloaded quality factor, given by [13, 14]:

$$Q_o = \frac{Q_L}{1 - 10^{-IL/20}} \tag{1.74}$$

where IL is the measured insertion loss at resonance, and the loaded quality factor is given by

$$Q_L = \frac{f_o}{BW_{-3\ dB}} \tag{1.75}$$

$BW_{-3\ dB}$ being the −3 dB bandwidth, which can be easily measured from the transmission coefficient. The total attenuation in the resonator is related to the unloaded quality factor by

$$\alpha = \frac{\pi}{Q_o \lambda} \tag{1.76}$$

where λ is given by (1.69). By subtracting to (1.76) the conduction attenuation constant, given by expression (1.90) (see the next subsection), the dielectric attenuation constant, α_d, can be inferred,[23] and by using (1.68), the loss tangent can be finally obtained.

[23] It is assumed that radiation from the ring is negligible (valid at moderate frequencies) [13]; hence, the contribution of radiation loss to the attenuation constant is null.

Alternatively, the dielectric constant and the loss tangent of thin film un-clad substrates and low-loss sheet materials can be measured by means of specific instrumentation, namely, a split cylinder resonator. It is a cylindrical resonant cavity separated into two halves, one of them being movable in order to accommodate varying sample thicknesses (i.e., the sample is loaded in the gap between the two cylinder halves). Each cylinder half accommodates a small coupling loop, introduced through a small hole, in order to measure the transmission coefficient of the fundamental TE_{011} mode. The principle for the determination of the dielectric constant and the loss tangent is the variation of the resonance frequency and quality factor with loaded and un-loaded cylinder (obviously, the thickness of the sample must be accurately known for a correct measurement). More details on this method are given in Refs. [16, 17].

1.4.2 Conductor Losses: The Skin Depth

Let us consider a conductor material with finite, but high, conductivity (i.e., a low-loss conductor), where the conduction current dominates over the displacement current, or $\sigma \gg |\omega\varepsilon'|, |\omega\varepsilon''|$. The complex propagation constant in such medium, with general expression given by (see Appendix A)

$$\gamma = \alpha + j\beta = j\omega\sqrt{\mu\left(\varepsilon' - j\frac{\sigma + \omega\varepsilon''}{\omega}\right)} \tag{1.77}$$

can be written as follows:

$$\gamma = j\omega\sqrt{\mu\left(-j\frac{\sigma}{\omega}\right)} = \frac{1+j}{\delta_{\mathrm{p}}} \tag{1.78}$$

with

$$\delta_{\mathrm{p}} = \sqrt{\frac{2}{\omega\mu\sigma}} \tag{1.79}$$

From (1.78), it follows that both the phase constant and the attenuation constant are given by

$$\alpha = \beta = \frac{1}{\delta_{\mathrm{p}}} \tag{1.80}$$

Since the attenuation of the fields is given by $e^{-\alpha z} = e^{-z/\delta p}$, where z is the direction of propagation, δ_{p} indicates the distance over which a plane wave is attenuated by a factor of e^{-1}. If a plane wave in vacuum impinges on a conductor material with its surface perpendicular to the wave vector, the wave transmitted to the conductor decays

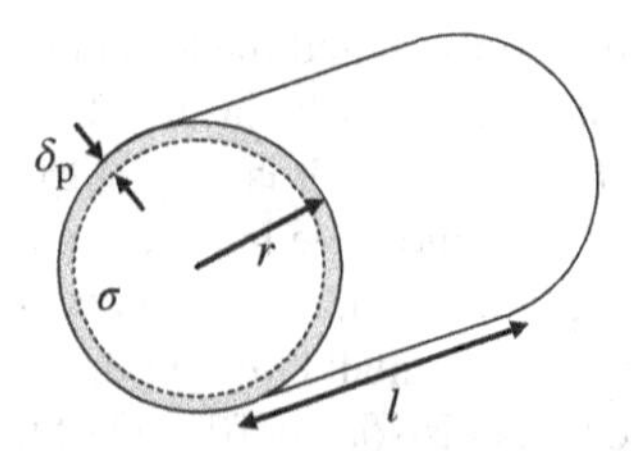

FIGURE 1.10 Cylindrical conductor with conductivity σ. The effective cross section for the calculation of the AC resistance is given by the annular gray region corresponding to one skin depth.

exponentially from its surface. Therefore, δ_p is referred to as penetration depth.[24] It is worth mentioning that (i) δ_p decreases with the square root of the conductivity and frequency, and (ii) the wavelength is given by $\lambda = 2\pi\delta_p$ and is typically much smaller than the wavelength in free space.

Similarly, if a conductor such as a wire or a metallic strip is carrying an AC current, the current tends to concentrate in the conductor surface as frequency increases. This phenomenon is known as *skin effect* and increases the high frequency AC resistance of the conductor since the effective conductor cross section is reduced. To a first-order approximation, the effective cross section of the conductor is limited by the external surface contour and by the curve resulting by reducing such contour by the penetration depth, also known as skin depth. Let us consider as a simple illustrative example a cylindrical conductor with length l and radius r (Fig. 1.10). The AC resistance is given by

$$R_{\mathrm{AC}} = \frac{l}{\sigma A_{\mathrm{AC}}} = \frac{l}{\sigma 2\pi r \delta_p} \tag{1.81}$$

where A_{AC} is the effective conductor cross section.

Expression (1.81) can also be derived by considering the intrinsic impedance of the low-loss conductor, given by[25]

$$\eta = \sqrt{\frac{\mu}{\varepsilon}} = \sqrt{j\frac{\mu}{\sigma}\omega} = (1+j)R_s \tag{1.82}$$

where R_s is the surface resistance

$$R_s = \frac{1}{\sigma\delta_p} = \sqrt{\frac{\mu\omega}{2\sigma}} \tag{1.83}$$

[24] In copper, the most used clad metal for PCB technology, the penetration depth (or skin depth) at 1 and 10 GHz is 2.06 and 0.66 μm, respectively.

[25] Notice that in a good conductor, the phase of the magnetic component of an EM wave propagating through it lags that of the electric component by $\pi/4$.

Considering again a plane wave impinging on the surface of a low-loss conductor, the conduction current density within the conductor is given by[26]

$$\vec{J} = \sigma \vec{E} = \sigma E_o e^{-\frac{z}{\delta_p}} \cdot e^{-j\frac{z}{\delta_p}} \vec{x} \tag{1.84}$$

where E_o is the electric field in the surface, and it has been assumed that the electric field is polarized in the x-direction. The current across a surface normal to $\vec{x}$, of width a and infinitely long in the direction normal to the surface (z-direction) is given by:

$$I = \int_0^\infty dz \int_{y_o}^{y_o+a} dy \left(\sigma E_o e^{-\frac{z}{\delta_p}} \cdot e^{-j\frac{z}{\delta_p}} \right) \tag{1.85}$$

which gives

$$I = a\sigma E_o \frac{\delta_p}{1+j} = \frac{aE_o}{R_s(1+j)} \tag{1.86}$$

The voltage drop between two points separated a distance l in the x-direction is simply $V = E_o l$. Thus, the surface impedance is given by

$$Z_s = R_s(1+j)\frac{l}{a} \tag{1.87}$$

which reduces $R_s(1+j)$ for a square geometry. The term R_s is usually designated as square resistance. For the cylindrical conductor considered earlier, the AC resistance can thus be inferred from the real part of (1.87) by considering $a = 2\pi r$, which gives (1.81).

Deriving the relation between the conduction loss attenuation constant, α_c, and the surface (or square) resistance in most planar transmission lines is not straightforward. Nevertheless, for a parallel plate transmission line, the per-unit-length resistance is roughly given by [12]

$$R' = \frac{2R_s}{W} \tag{1.88}$$

where W is the width of the strips. The previous expression is valid under the assumption that the current density is uniform in the transverse plane and concentrated one

[26] This result is also obtained from Maxwell's equations by calculating the current density distribution in the direction normal to the surface (z-direction) of a semi-infinite conductor, with current density parallel to such surface, and assuming that the current density at the surface is J_x $(z = 0) = \sigma E_o$ (see Appendix D).

skin depth from the interface between the conductors and the substrate. Introducing (1.88) in the first term of (1.51), gives

$$\alpha_c = \frac{R_s}{Z_o W} \tag{1.89}$$

For microstrip lines, closed-form expressions for α_c have been derived in [12]. Specifically, if $W \geq 2h$[27]

$$\alpha_c = \frac{R_s}{Z_o h} \frac{1}{\left\{ \frac{W'}{h} + \frac{2}{\pi} \ln \left[2\pi e \left(\frac{W'}{2h} + 0.94 \right) \right] \right\}^2} \cdot \left[\frac{W'}{h} + \frac{W'/\pi h}{\frac{W'}{2h} + 0.94} \right] \times \left\{ 1 + \frac{h}{W'} + \frac{h}{\pi W'} \left[\ln \left(\frac{2h}{t} + 1 \right) - \frac{1 + t/h}{1 + t/2h} \right] \right\} \tag{1.90}$$

The conduction attenuation constants for microstrip lines satisfying $W/h \leq 1/2\pi$ and $1/2\pi < W/h \leq 2$ are reported in Ref. [12]. It is worth mentioning that (1.90) simplifies to (1.89) in the limit of wide strips ($W/h \gg 1$).

1.5 COMPARATIVE ANALYSIS OF PLANAR TRANSMISSION LINES

The objective of this subsection is to briefly highlight some advantages and limitations of planar transmission lines from a comparative viewpoint. The most used transmission lines for the implementation of planar distributed circuits are microstrip lines and CPWs because no more than two metal levels are needed for their implementation. Striplines are closed and shielded structures, but they require three metal levels and their use is very limited. However, striplines support TEM wave propagation, and the phase velocity in these lines does not depend on their lateral geometry (i.e., the strip width).[28] Conversely, microstrip lines and CPWs are nonhomogeneous open lines that do not support purely TEM waves, but quasi-TEM waves. The dielectric substrate slows the waves down, as compared to the field lines in air, and the field lines tend to bend forward, thus preventing the presence of TEM modes.

[27] If surface roughness is not negligible, its effects can be accounted for in (1.90) by merely including an additional term in R_s, as reported in [13]. In [12], α_c appears multiplied by the factor 8.68 since the units are given in dB/unit length. This factor arises from the conversion from Nepers (Np) to dB. The Neper is a unit in logarithmic scale that uses the natural logarithm. Thus, if an arbitrary variable $F(z)$ is attenuated with position as $F(z) = F_o \cdot \exp(-\alpha z)$, the loss factor expressed in Np is $-\ln[F(z)/F_o] = \alpha z$, and α is said to have units of Np/unit length. The loss factor expressed in dB is $-20\log[F(z)/F_o] = -20\log[\exp(-\alpha z)] = \alpha z 20\log(e) = 8.68\, \alpha z$. Hence, the conversion from Np/unit length to dB/unit length introduces the above number 8.68 in expression (1.90).

[28] For TEM wave propagation in a stripline, the substrate material must be homogeneous and isotropic. Strictly speaking, purely TEM waves do also require perfect conductors. This latter condition cannot be satisfied in practice, but as long as the resistivity of the conductors is low, purely TEM wave propagation can be assumed.

From the viewpoint of shielding, the advantage of microstrip lines over CPWs is the presence of the ground plane in the back substrate side, which effectively isolates the structure from the backside region, and prevents it from potential interference effects caused by other circuits or materials (including metallic holders). CPWs with backside metallization (also known as conductor-backed CPWs) can also be implemented, but this backside metallization may induce leaky wave propagation as a result of the parasitic parallel plate waveguides present at both sides of the CPW axis. As compared to microstrip lines, CPWs without backside ground plane only need a metallic layer for their implementation. This eases fabrication and the shunt connection of lumped elements, since the ground plane is coplanar to the central (conductor) strip, and vias are not necessary. Nevertheless, in asymmetric CPW structures, such as bended lines, or asymmetrically loaded (along its length) lines, or in CPW lines with discontinuities, the parasitic slot mode[29] may be generated and obscure the fundamental mode (and hence degrade device performance). To prevent the presence of the slot mode in asymmetric CPW transmission lines, the ground plane regions must be electrically connected through air bridges, or by means of backside strips and vias (this technique has been applied to many CPW-based artificial transmission lines and microwave devices based on them, as will be seen along this book).

In CPWs, the transverse line geometry (see Fig. 1.2) is determined by the strip, W, and slot, G, widths (and of course by the substrate thickness, h, which is not usually a design parameter). Therefore, the characteristic impedance does not univocally determine the transverse line dimensions (W and G). This flexibility in the lateral geometry can be of interest in some applications. Moreover, for a given substrate thickness, it is possible to achieve higher characteristic impedances in CPW technology as compared to microstrip lines, where the strip width is univocally determined by the impedance value.

Suspended microstrip lines can be implemented by using sustaining posts in order to create an air gap between the ground plane and the substrate, or by means of advanced micromaching technologies [18], where the (lossy) substrate is partly removed by etching. As compared to conventional microstrip lines, suspended microstrip lines are thus low-loss lines. Moreover, because most of the field is in the air gap, higher characteristic impedances can be realized. Additionally, the presence of the air gap reduces the effects of dispersion, and such lines are of special interest in the upper microwave and lower millimeter wave bands. However, despite these beneficial properties, suspended microstrip lines are difficult to implement, and their use is restricted to applications where the required performance justifies the higher fabrication costs, or to monolithic microwave integrated circuits (MMICs). A modification of the suspended microstrip line is the so-called inverted microstrip line, where the conductor strip is placed below the substrate, in contact with the air gap. The advantages and drawbacks are similar to those of the suspended microstrip line.

[29] The slot mode of a CPW is, in general, an undesired mode of odd nature, that is, the symmetry plane of the CPW transmission line is an electric wall (or a virtual ground) for this mode. Conversely, the symmetry plane is a magnetic wall for the fundamental (even) CPW mode.

Slot lines are transmission lines that can be used either alone or in combination with microstrip lines on the opposite side of the substrate [19, 20]. Resonant slots coupled to microstrip lines have been used for the implementation of stop band filters, and slot antennas, consisting of a resonant slot fed by a microstrp line, are very well known [21]. For guided wave applications, radiation must be minimized. This is achieved through the use of high permittivity substrates, which causes the slot-mode wavelength to be small compared to free-space wavelength, and thereby results in the fields being closely confined to the slot with negligible radiation loss. The slot line shares with the CPW the coplanar configuration; hence, slot lines are especially convenient for shunt connecting lumped elements. Like microstrip lines or CPWs, slot lines do not support purely TEM modes. Indeed, the slot mode is markedly a non-TEM mode, and hence the characteristic impedance and the phase velocity in the slot line vary with frequency [19] (in contrast to microstrip lines or CPWs, where the line parameters are constant to a first-order approximation).

Except the slot line, the planar transmission lines considered earlier are single-ended (or unbalanced) lines. However, in applications where high immunity to noise, low crosstalk, and low electromagnetic interference (EMI) are key issues (i.e., in high-speed digital circuits), balanced (or differential) lines, and circuits are of primary interest. In two-conductor unbalanced transmission lines, the conductors have different impedance to ground, as sketched in Figure 1.11a. Such lines are fed by single-ended ports in which there is an active terminal and a ground terminal (i.e., one of the conductors is fed whereas the other is tied to ground potential). One of the conductors transports the signal current and the other acts as the return current path. By contrast, in two-conductor balanced lines (Fig. 1.11b), the conductors have equal potential with respect to ground and are in contra phase, and the currents flowing in the conductors have equal magnitude but opposite direction (each conductor provides the signal return path for the other). Such balanced lines are fed by a differential port, consisting of two terminals, neither of which being explicitly tied to ground. In balanced lines, the conductors have the same impedance to ground, this being the main relevant difference compared to unbalanced lines. Microstrip lines, CPWs, and striplines are examples of unbalanced lines. By contrast, slot lines, or coplanar strips (CPS), are balanced structures by nature. However, these balanced structures can be regarded as either balanced or unbalanced, depending on whether the excitation is balanced or unbalanced, respectively. Two-conductor balanced transmission lines can also be implemented by etching parallel strips at both sides of a dielectric slab

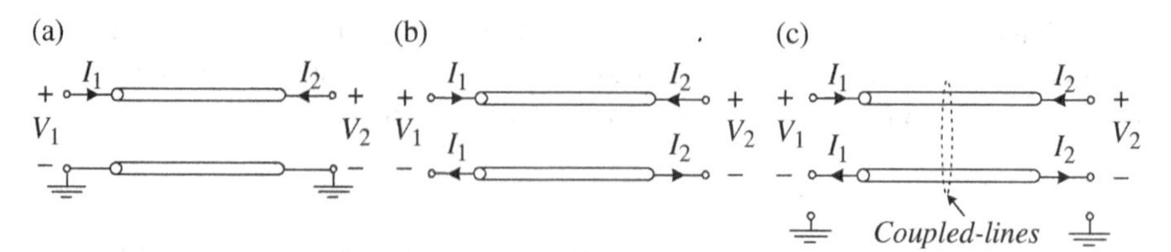

FIGURE 1.11 Schematic of two-port transmission lines. (a) Two-conductor unbalanced line, (b) two-conductor balanced line, and (c) three-conductor balanced line.

(see Fig. 1.2). This paired strips transmission line is useful, for instance, to feed antipodal printed dipole antennas [22]. Notice that the symmetry plane of the paired strips transmission line is an electric wall, and hence a virtual ground. Therefore, this structure can be analyzed by applying symmetry properties, that is, by removing the lower half of the structure, and adding a conducting plate, acting as ground plane, in the backside of the "sliced" substrate. Obviously, the resulting structure is a single-ended microstrip transmission line, and hence the main line parameters are calculated by applying the same formulas. However, the voltage drop across the paired strips is twice the voltage drop in the microstrip transmission line, whilst the current is the same. This means that the characteristic impedance of the balanced paired strips line is twice the characteristic impedance of the microstrip line.

Most practical implementations of balanced lines incorporate a ground plane, or some other global reference conductor. Such differential structures cannot be considered as pure two-conductor systems, since the ground plane becomes the third conductor of a three-conductor line (Fig. 1.11c). Such three-conductor line can be implemented by means of a pair of coupled lines over a ground plane. If the three-conductor line is not balanced due to the ground plane, currents flowing on it can unbalance the currents in the lines. On the contrary, if the three-conductor line is balanced, the active lines carry equal and opposite currents because the impedances of either line to ground are equal (see Fig. 1.11c). For instance, although a microstrip line is unbalanced, a two-port differential microstrip line can be designed by means of symmetric coupled lines (illustrated in Fig. 1.2—coupled microstrip) differentially driven. This balanced line consists of edge coupled lines that can be seen as a CPS line with a ground plane.

1.6 SOME ILLUSTRATIVE APPLICATIONS OF PLANAR TRANSMISSION LINES

There are several textbooks focused on the analysis and design of planar distributed circuits and antennas. Our aim in this Section is not to review all these transmission line applications, but to simply discuss some examples of distributed circuits where the involved transmission lines and stubs operate at different regimes, i.e., have different electrical lengths at the frequencies of interest. This includes planar circuits based on semi-lumped transmission lines (i.e., transmission lines with length $l < \lambda/10$), and based on $\lambda/8$, $\lambda/4$ and $\lambda/2$ lines. Some of the implementations reported in this Section will be later designed by means of artificial transmission lines in order to reduce circuit size, improve circuit performance, or achieve novel functionalities (or a combination of the previous beneficial aspects).

1.6.1 Semilumped Transmission Lines and Stubs and Their Application to Low-Pass and Notch Filters

Let us start by considering the design of circuits on the basis of electrically small planar components, usually referred to as semilumped components. The characteristic dimension[30] of these components is typically smaller than $\lambda/10$ at the frequency of interest. The main relevant characteristic of semilumped components is the fact that they can be described by simple reactive elements, such as inductors, capacitors, or LC resonant tanks, up to frequencies satisfying the semilumped element approximation ($l<\lambda/10$). Let us consider an electrically small ($l<\lambda/10$) section of a transmission line, with electrical length βl and characteristic impedance Z_o. From (1.6), we can write

$$\beta l = \omega l \sqrt{L'C'} \tag{1.91}$$

The previous equation can be expressed in terms of the characteristic impedance and the per-unit-length inductance of the line as follows:

$$\beta l = \omega l \sqrt{L'C'} = \omega l \frac{L'}{Z_o} \tag{1.92}$$

or as a function of the characteristic impedance and the per-unit-length line capacitance according to

$$\beta l = \omega l \sqrt{L'C'} = \omega l Z_o C' \tag{1.93}$$

Thus, the line inductance and capacitance of the considered transmission line section can be expressed as follows:

$$L\omega = Z_o \beta l \tag{1.94a}$$

$$C\omega = \frac{\beta l}{Z_o} \tag{1.94b}$$

Equations 1.94 reveal that if the line is electrically short and Z_o is high, the capacitance of the considered transmission line section can be neglected and hence the line is essentially a series inductance; conversely, for an electrically short low impedance line, the line can be described by a shunt capacitance (the line inductance can be neglected). At first sight, one may erroneously deduce from (1.94) that the electrically short line condition (semilumped approximation) is not a requirement to describe the line by means of a series inductance or a shunt capacitance (very high/low value of Z_o leads to a negligible line capacitance/inductance). However, the line is not free from distributed effects (despite the fact it has an extreme—high or low—characteristic impedance) if it is not electrically short. If the line is electrically short, expressions (1.33) can be approximated by

[30] By characteristic dimension we mean the length (for a transmission line section or stub), the diameter (for a circular semilumped component, such as a circularly shaped split ring resonator or SRR), or the longest side length (for a rectangular planar component, such as a folded stepped impedance resonator or SIR). The analysis of these electrically small resonators and their applications will be considered later.

$$Z_{\text{in}}(Z_{\text{L}}=\infty)=-j\frac{Z_{\text{o}}}{\beta l}=-j\frac{1}{C\omega} \tag{1.95a}$$

$$Z_{\text{in}}(Z_{\text{L}}=0)=jZ_{\text{o}}\beta l=jL\omega \tag{1.95b}$$

where (1.94) has been used. Hence, it is demonstrated that an electrically small transmission line section with extreme characteristic impedance can be described either by a series inductor or by a shunt capacitor, and the element values can be inferred from (1.94).[31]

From the previous words, it follows that by cascading electrically small transmission line sections with high and low characteristic impedance, we can implement a ladder network with series inductances and shunt capacitances, that is, a low-pass filter. The design procedure simply consists of setting the high and low characteristic impedance to implementable values, and determining the line length l of each transmission line section by means of (1.94) [1]. Notice that the higher/lower the characteristic impedance, the shorter the resulting inductive/capacitive transmission line section. After calculation, it is necessary to verify that each transmission line section satisfies the semilumped element approximation. An illustrative example of a stepped impedance low-pass filter (in CPW technology) and its frequency response are presented in Figure 1.12. The device is a ninth-order Butterworth low-pass filter with a cut-off frequency of $f_c = 2$ GHz (the element values of the filter are inferred from impedance and frequency transformation from the low-pass filter prototype [1]). The widths of the central strips and slots for the different sections are obtained by means of a transmission line calculator, once the characteristic impedance of the high- and low-impedance transmission line sections is set. Such calculators incorporate the formulas that link the lateral line geometry to the characteristic impedance, present in most textbooks focused on transmission lines [6].

In microwave engineering, shunt- and series-connected resonators (either distributed or semilumped) are key elements. In particular, shunt-connected series resonators introduce transmission zeros (notches) in the frequency response, which are of interest for harmonic suppression, or to improve the selectivity in microwave filters (elliptic low-pass filters are implemented by means of shunt resonators in order to generate transmission zeros above the pass band of interest). An open-ended shunt stub behaves as a parallel connected series resonator in the vicinity of the frequency that makes the line to be $\lambda/4$ long. However, the stub length can be significantly reduced by considering a stepped impedance topology, as depicted in Figure 1.13 [23]. Such element is known as stepped impedance shunt stub (SISS) and is described by a grounded series resonator. The narrow (high impedance) and wide (low impedance) transmission line sections correspond to the inductance and capacitance, respectively. The admittance of the SISS (seen from the host line) is given by

[31] Alternatively, expressions (1.94) and the semilumped approximation requirement have been derived by considering the equivalent T-circuit model of a transmission line section, where the series and shunt impedance are calculated from the elements of the impedance or admittance matrix [1].

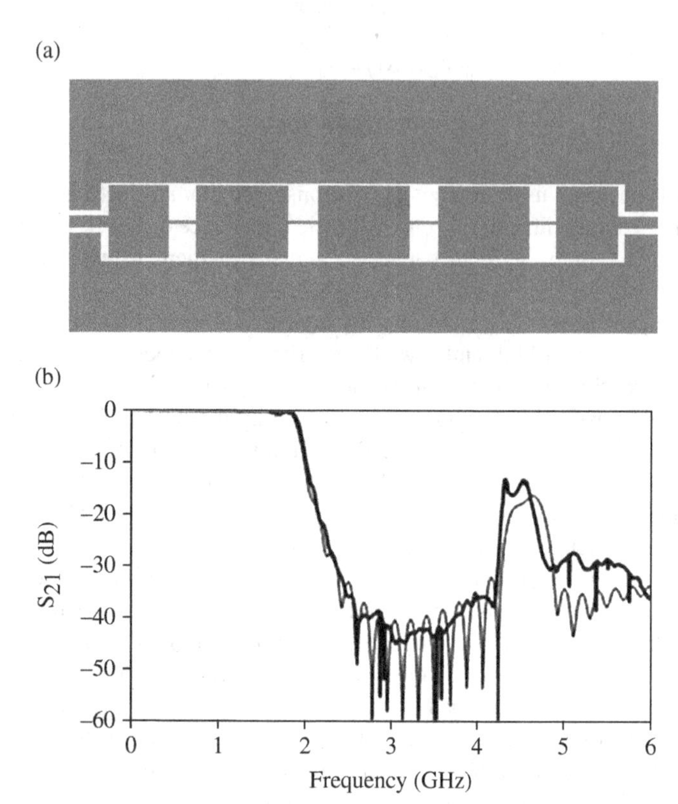

FIGURE 1.12 Order-9 Butterworth stepped impedance low-pass filter (a) and measured (solid line) and EM simulated (thin line) frequency response (b). The filter was fabricated on the *Rogers RO3010* substrate with dielectric constant $\varepsilon_r = 10.2$, and thickness $h = 1.27$ mm. Filter length is 9.4 cm.

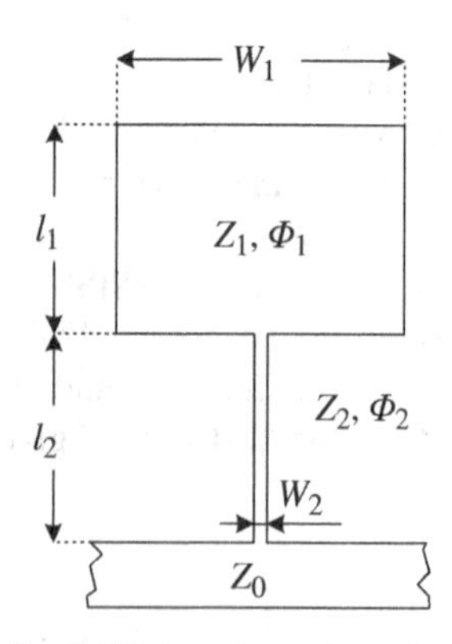

FIGURE 1.13 Topology of the SISS in microstrip technology and relevant dimensions ($Z_2 \gg Z_1$).

$$Y_{\rm SISS} = -j\frac{\tan\phi_1 + K\tan\phi_2}{Z_2\tan\phi_1\cdot\tan\phi_2 - Z_1} \tag{1.96}$$

where ϕ_1 and ϕ_2 are the electrical lengths of the low- and high-impedance line sections, respectively, and

$$K = \frac{Z_1}{Z_2} \tag{1.97}$$

is the impedance ratio of the SISS [24]. At resonance, the denominator in (1.96) vanishes, and the following condition results[32]:

$$K = \tan\phi_1\cdot\tan\phi_2 \tag{1.98}$$

It is obvious from (1.98) that to minimize the total electrical length ($\phi_{\rm T} = \phi_1 + \phi_2$) of the resonator, K must be as small as possible ($K \ll 1$) [24]. The reduction of $\phi_{\rm T}$ is important for two reasons: (1) to reduce the length of the SISS and (2) to be able to describe the SISS by means of a lumped element model (grounded series LC resonator) over a wide frequency band. Under the assumption that the two transmission line sections of the SISS are electrically small, the tangents in (1.96) can be linearized, and the admittance of the SISS is found to be

$$Y_{\rm SISS} = -j\frac{\phi_1 + K\phi_2}{Z_2\phi_1\phi_2 - Z_1} \tag{1.99}$$

This admittance is identical to that of an LC series resonant tank, given by

$$Y_{\rm LC} = -j\frac{\omega C}{LC\omega^2 - 1} \tag{1.100}$$

provided the following mapping is satisfied:

$$C = \frac{l_1}{v_{\rm p1}Z_1} + \frac{l_2}{v_{\rm p2}Z_2} = C_1 + C_2 \tag{1.101}$$

$$L = \frac{Z_2 l_2}{v_{\rm p2}}\frac{\frac{l_1}{v_{\rm p1}Z_1}}{\frac{l_1}{v_{\rm p1}Z_1} + \frac{l_2}{v_{\rm p2}Z_2}} = L_2\frac{C_1}{C_1 + C_2} \tag{1.102}$$

[32] Notice that if $K = 1$, the resonance condition (1.98) rewrites as $\tan(\phi_1 + \phi_2) = \infty$, giving $\phi_{\rm T} = \phi_1 + \phi_2 = \pi/2$, that is, a $\lambda/4$ open stub at resonance. This result is easily inferred by applying the following trigonometric identity to (1.96): $\tan(\phi_1 + \phi_2) = (\tan\phi_1 + \tan\phi_2)/(1 - \tan\phi_1\cdot\tan\phi_2)$.

where v_{p1} and v_{p2} are the phase velocities of the low- and high-impedance transmission line sections, respectively, C_1 and C_2 are the line capacitances, namely,

$$C_1 = \frac{l_1}{v_{p1}Z_1} = C_1' \cdot l_1 \quad (1.103)$$

$$C_2 = \frac{l_2}{v_{p2}Z_2} = C_2' \cdot l_2 \quad (1.104)$$

C'_i $(i = 1, 2)$ being the per unit length capacitances of the lines, and L_2 is the inductance of the high impedance transmission line section, that is,

$$L_2 = L_2' \cdot l_2 \quad (1.105)$$

Notice that although C is dominated by the capacitance of the low impedance transmission line section, C_1, the contribution of C_2 on C may be nonnegligible if either K or l_2 are not very small. It is also interesting to note that L is somehow affected by the capacitive line section, such inductance being smaller than the inductance of the high-impedance transmission line section, L_2. Obviously, to a first-order approximation, the resonator elements are given by $C \approx C_1$ and $L \approx L_2$ (as expected on account of 1.95), although this approximation sacrifices accuracy.

Figure 1.14 shows the photograph of a SISS resonator loading a 50 Ω microstrip transmission line, where $L = 3.7$ nH and $C = 6.9$ pF (the resonance frequency is $f_o = 1$ GHz [23]). The impedance of the inductive line is that corresponding to a line width of 150 μm, namely, $Z_2 = 101.8$ Ω (the structure is implemented on the *Rogers RO3010* substrate with dielectric constant $\varepsilon_r = 10.2$ and thickness $h = 1.27$ mm). For the low-impedance transmission line section, the width (23 mm) guarantees that the first transverse resonance occurs beyond $2f_o$ (see details in Ref. [25]). This gives a characteristic impedance of $Z_1 = 5.8$ Ω. The lengths of the lines, $l_1 = 2.8$ mm and $l_2 = 4$ mm, were derived from (1.101) and (1.102) (actually some optimization was required due to the effects of line discontinuities, not accounted for by the model). These line lengths correspond to electrical lengths of $\phi_1 = 20.7°$ and $\phi_2 = 23.8°$ at $2f_o$. The EM simulation and the measured frequency response of the SISS-loaded line is also depicted in Figure 1.14. The circuit simulation of the same line loaded with a shunt connected LC resonator with the reactive values given above is also included in the figure, for comparison purposes. The agreement is excellent up to $2f_o$, indicating that the semilumped approximation is valid in the considered frequency range. Although, typically, notch filters (i.e., stop band filters with a peaked response, or transmission zero) exhibit narrow stop bands, the structure of Figure 1.14 can be considered to belong to this category (the bandwidth can be controlled by the inductance/capacitance ratio).

To end this subsection devoted to semilumped transmission lines and components, let us briefly consider the stepped impedance resonator (SIR) [26], which consists of a pair of wide transmission line sections sandwiching a narrow strip (Fig. 1.15a). This resonator is electrically smaller than the conventional $\lambda/2$ resonator (in the same form

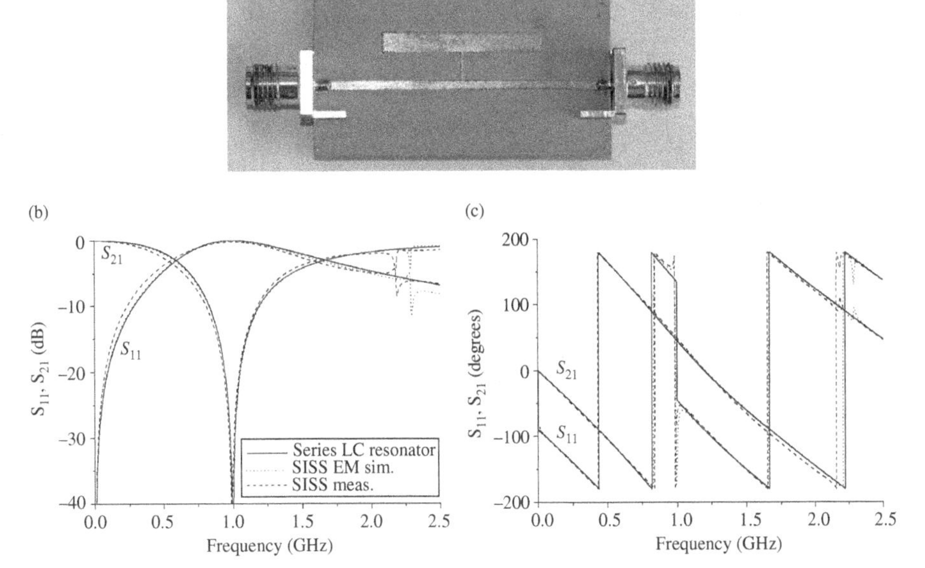

FIGURE 1.14 SISS-loaded microstrip line (a), insertion and return loss (b) and phase response (c). Reprinted with permission from Ref. [23]; copyright 2011 IET.

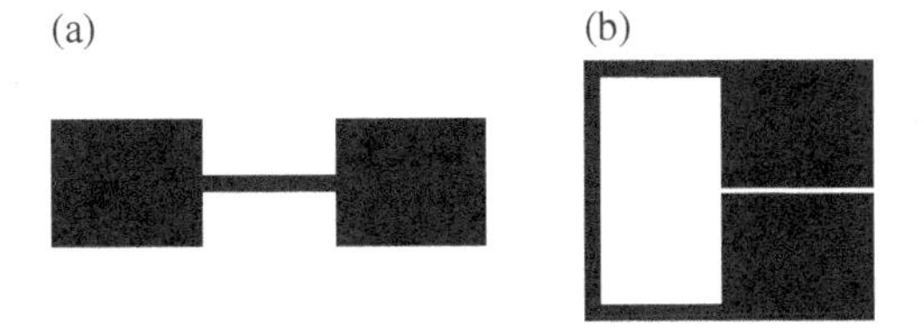

FIGURE 1.15 Topology of a SIR (a), and folded SIR (b).

than the SISS is electrically smaller than the $\lambda/4$ open stub). SIRs are typically driven through electric coupling, though they can also be externally driven by means of a time varying magnetic field by simply folding the SIR topology, as shown in Figure 1.15b. The size of the folded SIR can be further reduced by decreasing the gap distance between the wide transmission line sections, since this introduces an extra capacitance to the structure. SIRs have been used in a wide variety of microwave applications, including applications involving artificial transmission lines (as will be later shown). To illustrate the potentiality of these resonators, an SIR-based order-3 elliptic-function low-pass filter is reported (see further details in Ref. [27]). The filter is based on a CPW transmission line loaded with an SIR etched in the back substrate side. The wide strip sections of the SIR are placed face-to-face with the central strip and ground plane of the CPW host line, resulting in a shunt connected series resonator which introduces a transmission zero. The circuit model of an elliptic low-pass filter

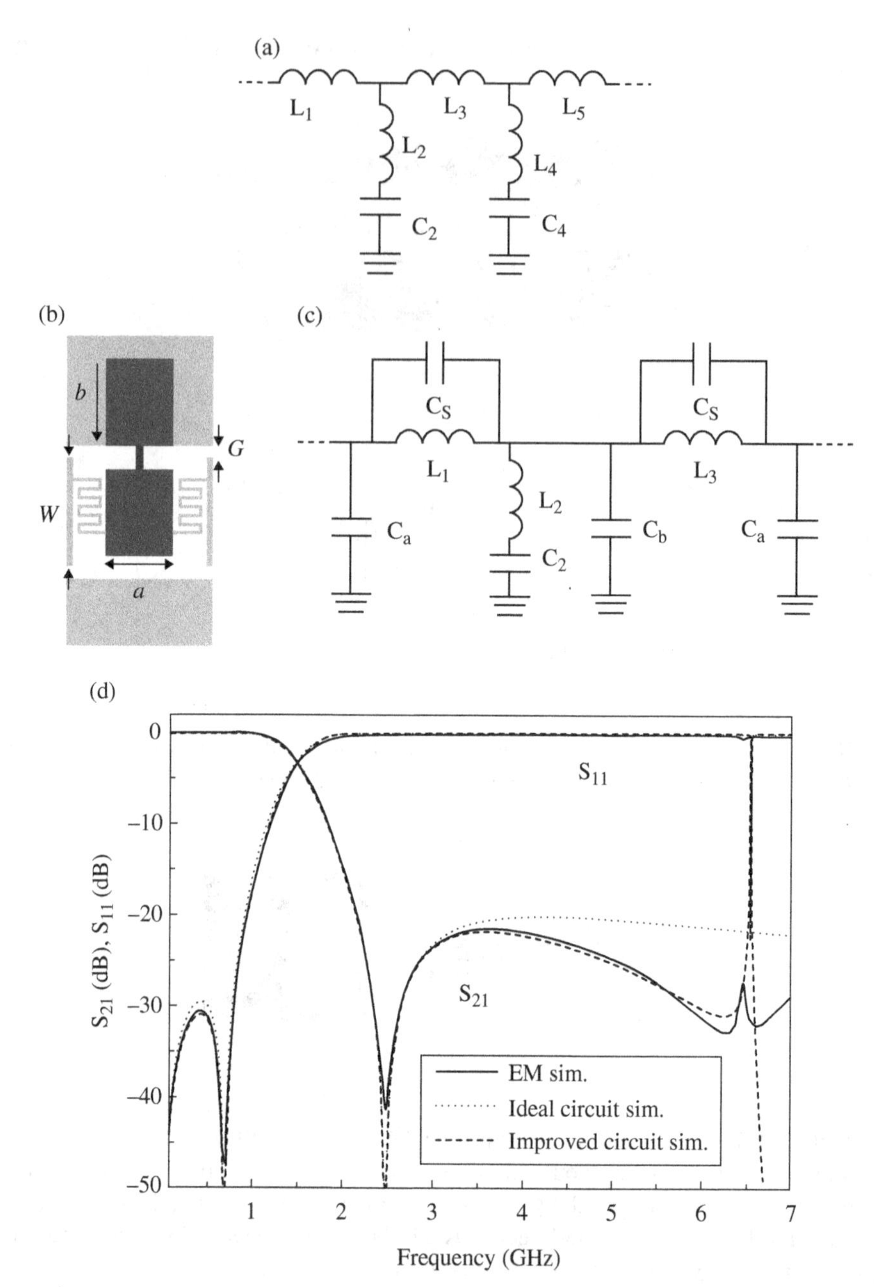

FIGURE 1.16 (a) Low-pass elliptic-function prototype filter with shunt connected series resonators (the circuit correspond to a fifth-order prototype), (b) topology of the SIR-based low-pass filter (order-3), (c) equivalent circuit model including parasitics, and (d) EM response, ideal filter prototype response and circuit response including parasitics. The considered substrate thickness and dielectric constant are $h = 254\,\mu m$ and $\varepsilon_r = 11.2$, respectively. Dimensions are $W = 5$ mm, $G = 0.55$ mm, $a = 3.24$ mm, $b = 3.99$ mm. Back side metal is indicated in black colour. The element values of the ideal prototype filter shown in (a) are $L_1 = L_3 = 4.7$ nH, $L_2 = 1.38$ nH, $C_2 = 2.98$ pF. The element values of the complete circuit model in reference to the circuit shown in (c) are $L_1 = L_3 = 4.7$ nH, $L_2 = 1.65$ nH, $C_2 = 2.5$ pF, $C_a = 0.08$ pF, $C_b = 0.44$ pF, $C_s = 0.115$ pF. With regard to parasitics, C_s models the capacitance associated to the meander, and C_a, C_b are the capacitances from the central strip to the ground plane. Reprinted with permission from Ref. [27]; copyright 2010 IEEE.

based on shunt connected series resonators,[33] the proposed order-3 SIR-based filter, its circuit model including parasitics, and the frequency response, are depicted in Figure 1.16. The remarkable aspects of these SIR-based filters are the small size, and the excellent agreement between the ideal filter response (ideal circuit simulation) and the EM simulation up to frequencies above the transmission zero frequency. At higher frequencies, the parasitics must be included for an accurate description of the filter response (indicated as improved circuit simulation in the figure). Elliptic function low-pass filters using SISS in microstrip technology have also been reported [28].

1.6.2 Low-Pass Filters Based on Richard's Transformations

Let us now consider the potential of $\lambda/8$ open and short-circuit stubs, which are electrically larger than the semilumped components considered in the previous section. According to Richard's transformations, such stubs can be used to replace shunt inductors and capacitors. Let us illustrate their application to the implementation of low-pass filters. The key idea is to force the length of the stubs to be $\lambda/8$ at the filter cut-off frequency, ω_c. Using (1.35), the reactance of the short-circuit and open-circuit stub at ω_c is forced to be identical to that of the inductor and capacitor, respectively. Below that frequency, we also expect a similar reactance because the stubs have roughly a linear dependence with frequency. However, discrepancies between the reactances of the stubs and lumped elements are expected above ω_c.[34] Let us consider the implementation of an order-3 Chebyshev low-pass filter with a cut-off frequency of $f_c = 2$ GHz, and 0.5 dB ripple. From impedance and frequency transformation from the low-pass filter prototype, the series inductances and the shunt capacitance of the filter are $L_1 = L_3 = 6.35$ nH, and $C_2 = 1.74$ pF, respectively. Application of (1.35) gives $Z_{o1,3} = 79.8\ \Omega$ and $Z_{o2} = 45.6\ \Omega$, for the inductive and capacitive stubs, respectively. Since the implementation of series stubs is complex, at least in microstrip technology, let us replace the inductive stubs with shunt stubs. To this end, the Kuroda identity shown in Figure 1.17 is used [1].[35] To apply the Kuroda identity of Figure 1.17, we first cascade a pair of $\lambda/8$ long $50\ \Omega$ transmission line sections at the input and output port of the filter. Notice that this has effects on the phase response of the filter, but not on the magnitude of the insertion and return loss. Application of the Kuroda identity leads to the circuit of Figure 1.18. The layout of this filter, for microstrip technology considering the *Rogers R04003C* substrate with dielectric constant $\varepsilon_r = 3.38$ and thickness $h = 0.81$ mm, is also depicted in the Figure. The EM simulation of the filter response is compared with the ideal Chebyshev filter response, where it can be appreciated that the agreement is excellent up to the

[33] Alternatively, elliptic-function lowpass filters can be implemented by cascading series connected parallel resonators and shunt capacitors.

[34] Since the reactance of the stubs is a periodic function with frequency, the frequency response of the filter is also a periodic function, and spurious (or harmonic) bands are generated.

[35] Kuroda's identities are equivalences between two-port networks containing series- or shunt-reactive elements and transmission line sections (called unit elements), that are used to physically separate transmission line stubs, to transform series stubs into shunt stubs, and to change unrealizable characteristic impedances into implementable ones.

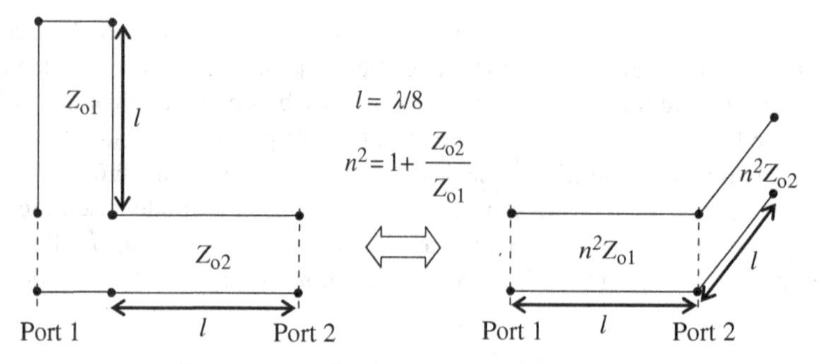

FIGURE 1.17 Kuroda identity used for the design of the filter of Figure 1.18.

cut-off frequency, and the frequency selectivity of the distributed implementation is significantly better, although with the presence of spurious bands (the filter response repeats every 8 GHz), as predicted before.

1.6.3 Power Splitters Based on $\lambda/4$ Lines

Power splitters, combiners, and couplers are fundamental building blocks in RF/microwave engineering. Most of their simplest implementations as distributed circuits are based on $\lambda/4$ lines.[36] Let us, hence, illustrate the application of $\lambda/4$ lines to the implementation of power splitters (it will be later shown that these lines can be replaced with artificial lines in order to reduce splitter size and obtain dual-band functionality). Power splitters are reciprocal[37] three-port networks (or multiport networks if the number of output ports is higher than 2) with a matched input port, that is, there is not power return to this port if the output ports are terminated with matched loads. If the splitter is lossless, the scattering matrix (for the case of a 1:2 device) can be written in the general form [1]:

$$S = \begin{pmatrix} 0 & \alpha & \alpha \\ \alpha & \gamma & -\gamma \\ \alpha & -\gamma & \gamma \end{pmatrix} \tag{1.106}$$

[36] Power splitters can also be implemented by means of lumped resistive elements [1]. Such splitters ideally exhibit an infinite operational bandwidth, but they are lossy. By contrast, distributed splitters can be considered (to a first approximation) lossless, but their functionality is restricted to a certain band in the vicinity of the operational frequency.

[37] Reciprocal networks are defined as those networks verifying that the effects of a source, located at one port, over a load, located at another port, are the same if the source and load interchange the ports where they are connected [1]. In reciprocal networks, the scattering matrix is symmetric.

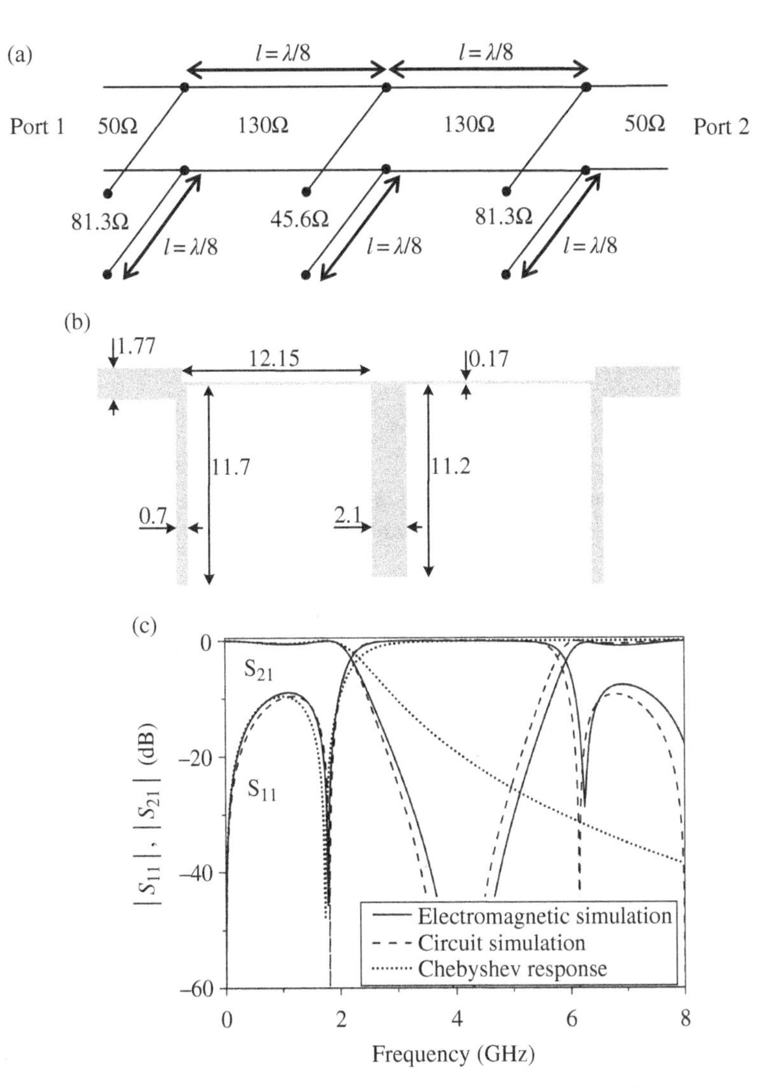

FIGURE 1.18 Schematic (a), layout (b), and frequency response (c) of the low-pass filter based on Richard's transformations. The relevant dimensions (in mm) are indicated. The circuit simulation in (c) was obtained by using a commercial circuit and schematic solver, where the transmission lines and stubs are modeled by the corresponding distributed models.

with $|\alpha| = 1/\sqrt{2}$ and $|\gamma| = 1/2$. The two canonical forms of lossless symmetric power splitters are depicted in Figure 1.19, where the impedances of the inverters ($\lambda/4$ lines), indicated in the figure, are derived by forcing the matching condition for the input port. In both implementations of Figure 1.19, $\alpha = -j/\sqrt{2}$, whereas $\gamma = 1/2$ for Figure 1.19a and $\gamma = -1/2$ for Figure 1.19b. The number of output ports of the splitter

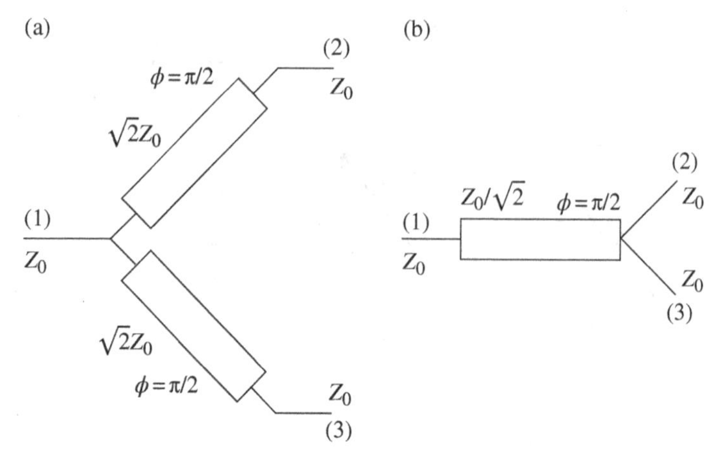

FIGURE 1.19 Canonical forms of the two-output distributed symmetric power splitter. (a) With two inverters and (b) with one inverter.

can be arbitrarily large. In order to preserve matching, the characteristic impedance of the inverters must be $Z_o = 50\sqrt{n}\Omega$ for the structure of Figure 1.19a and $50/\sqrt{n}\Omega$ for the structure of Figure 1.19b, where n is the number of output ports. A 1:3 power divider, corresponding to the configuration of Figure 1.19a, and its EM response are shown in Figure 1.20. The device was designed to be operative at 1.5 GHz, as revealed by the good matching at that frequency.

1.6.4 Capacitively Coupled $\lambda/2$ Resonator Bandpass Filters

The last illustrative example is a bandpass filter based on $\lambda/2$ transmission lines acting as distributed resonators. If a gap is etched on a transmission line, the frequency response is a high-pass type response, and this can be described by series connecting a capacitor to the line. However, if two capacitive gaps are etched in the line, rather than a high-pass response with enhanced rejection in the stop band, the structure exhibits a bandpass response that can be attributed to a resonance phenomenon. At the frequency where the distance between gaps is roughly $\lambda/2$, the forward and backward travelling waves (caused by gap reflections) in the resonator sum up in phase, and small coupling (and hence power transfer) between the feeding line and the resonator is enough to achieve total power transmission (assuming lossless lines) between the input and the output ports. Of course, this resonance phenomenon occurs at frequencies satisfying $l = n\lambda/2$ (with $n = 1, 2, 3, \ldots$).[38] Based on this phenomenon, bandpass filters with controllable response can be implemented (see Ref. [1] for further details on the design of this type of filters). As an example, an order-3 bandpass filter in

[38] Actually, this is the resonance condition for an unloaded resonator. The gaps introduce some phase shift in the reflected waves, and the resonance condition is slightly modified.

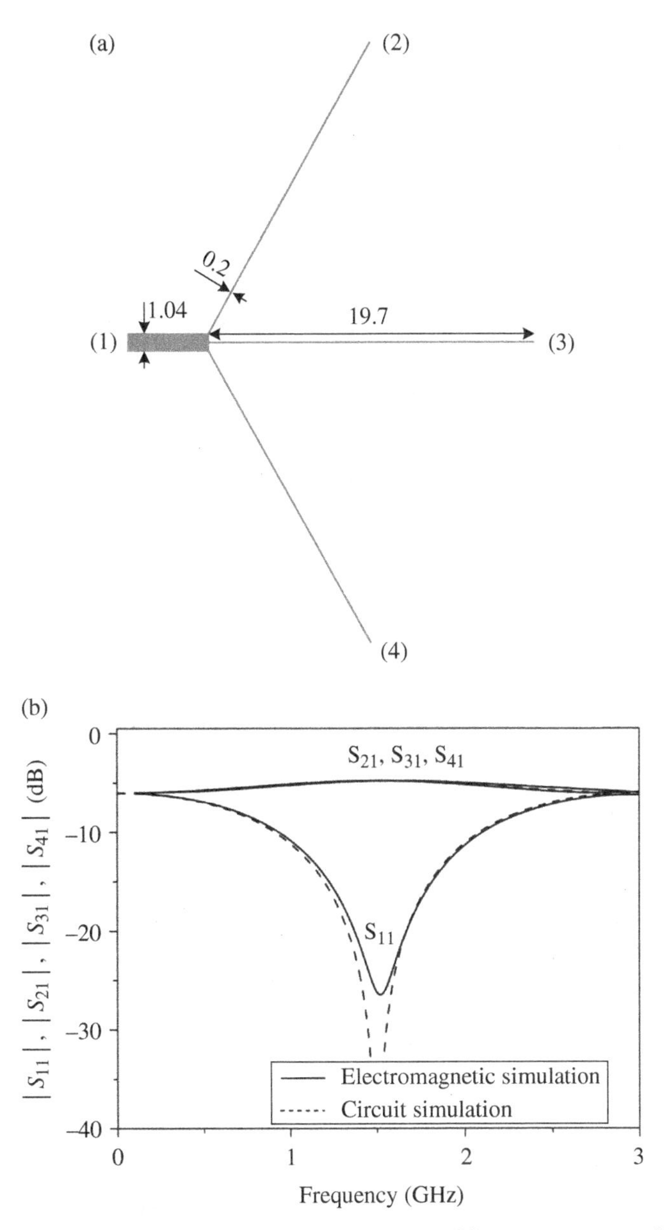

FIGURE 1.20 Example (layout) of a power splitter (a), and frequency response (b). Relevant dimensions (in mm) and device ports are indicated. The width of the three $\lambda/4$ lines gives a characteristic impedance of $Z_o = 50\sqrt{3} = 86.6\ \Omega$. The considered substrate is the *Rogers RO3010* with dielectric constant $\varepsilon_r = 10.2$ and thickness $h = 1.27$ mm.

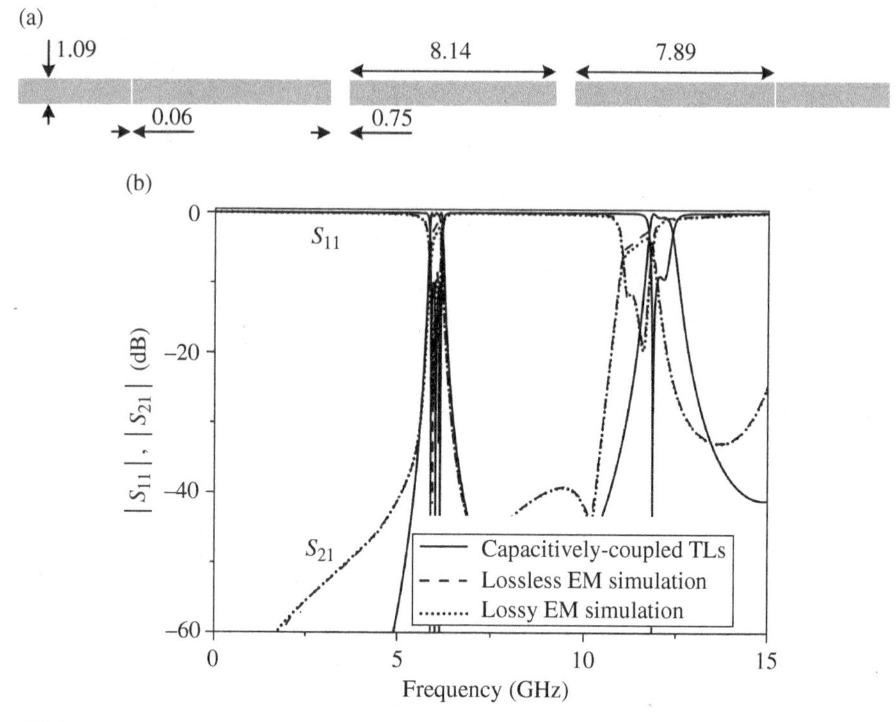

FIGURE 1.21 Example of a capacitively coupled $\lambda/2$ resonator bandpass filter (a) and frequency response (b). Relevant dimensions (in mm) are indicated. The considered substrate is the *Rogers RO3010* with dielectric constant $\varepsilon_r = 10.2$ and thickness $h = 1.27$ mm.

microstrip technology and its frequency response are shown in Figure 1.21. Resonator lengths and inter-resonators distance (i.e., gap space) have been calculated in order to obtain a Chebyshev response with 0.5 dB ripple, central frequency $f_o = 6$ GHz, and 5% fractional bandwidth (it will be shown in Chapter 2 that resonator's length can be reduced by means of slow wave artificial transmission lines).

REFERENCES

1. D. M. Pozar, *Microwave Engineering,* Addison Wesley, Reading, MA, 1990.
2. B. C. Wadell, *Transmission Line Design Handbook,* Artech House, Norwood, MA, 1991.
3. J. C. Freeman, *Fundamentals of Microwave Transmission Lines,* John Wiley, New York, 1996.
4. F. Di Paolo, *Networks and Devices Using Planar Transmission Lines,* CRC Press, Boca Raton, FL, 2000.
5. P. C. Magnusson, A. Weisshaar, V. K. Tripathi, G. C. Alexander, *Transmission Lines and Wave Propagation,* CRC Press, Boca Raton, FL, 2001.
6. I. Bahl and P. Barthia, *Microwave Solid State Circuit Design,* 2nd Edition, John Wiley, New York, 2003.

7. R. K. Mongia, I. J. Bahl, P. Barthia, and J. Hong, *RF and Microwave Coupled-Line Circuits,* 2nd Edition, Artech House, Norwood, MA, 2007.
8. L. Ganesan, and S. S. Sreeja Mole, *Transmission Lines and Waveguides,* 2nd Edition, McGraw Hill, New Delhi, 2010.
9. K. Kurokawa, "Power waves and the scattering matrix," *IEEE Trans. Microw. Theory Technol.*, vol. **MTT-13**, pp. 194–202, 1965.
10. P. I. Richards, "Resistor-transmission-line circuits," *Proc. IRE*, vol **36**. pp. 217–220, 1948.
11. S. Ramo, J. R. Whinnery, and T. Van Duzer, *Field and Waves in Communication Electronics,* 3rd Edition, John Wiley, New York, 1994.
12. R. A. Pucel, D. J. Massé, and C. P. Hartwig, "Losses in microstrip," *IEEE Trans. Microw. Theory Technol.*, vol. **MTT-16**, pp. 342–350, 1968.
13. G. Zou, H. Gronqvist, P. Starski, and J. Liu, "Characterization of liquid crystal polymer for high frequency system-in-a-package applications," *IEEE Trans. Adv. Packag.*, vol. **25**, pp. 503–508, 2002.
14. K. Chang, *Microwave Ring Circuits and Antennas,* John Wiley, New York, 1994.
15. L. Yang, A. Rida, and R. Vyas, M. M. Tentzeris "RFID tag and RF structures on a paper substrate using inkjet-printing technology," *IEEE Trans. Microw. Theory Technol.*, vol. **55**, pp. 2894–2901, 2007.
16. Association Connecting Electronics Industries. IPC TM-650 2.5.5.13. *Relative Permittivity and Loss Tangent Using a Split-Cylinder Resonator.* http://www.ipc.org/TM/2-5-5-13.pdf. Accessed February 5, 2015.
17. M. D. Janezic and J. Baker-Jarvis, "Full-wave analysis of a split-cylinder resonator for nondestructive permittivity measurements," *IEEE Trans. Microw. Theory Technol.*, vol. **47**, pp. 2014–2020, 1999.
18. C. T.-C. Nguyen, L. P. B. Katehi, and G. M. Rebeiz, "Micromachined devices for wireless communications," *Proc. IEEE*, vol. **86**, pp. 1756–1768, 1998.
19. S. B. Cohn, "Slot line on a dielectric substrate," *IEEE Trans. Microw. Theory Technol.*, vol. **MTT-17**, pp. 768–778, 1969.
20. K. C. Gupta, R. Carg, I. Bahl, and P. Barthia, *Microstrip Lines and Slotlines, 2nd Edition,* Artech House, Norwood, MA, 1996.
21. R. Garg, P. Bhartia, I. Bahl, and A. Ittipiboon, *Microstrip Antenna Design Handbook,* Artech House Inc., Norwood, 2001.
22. F. J. Herraiz-Martínez, F. Paredes, G. Zamora, F. Martín, and J. Bonache, "Dual-band printed dipole antenna loaded with open complementary split-ring resonators (OCSRRs) for wireless applications," *Microw. Opt. Technol. Lett.*, vol. **54**, pp. 1014–1017, 2012.
23. J. Naqui, M. Durán-Sindreu, J. Bonache, and F. Martín, "Implementation of shunt connected series resonators through stepped-impedance shunt stubs: analysis and limitations," *IET Microw. Antennas Propag.*, vol. **5**, pp. 1336–1342, 2011.
24. M. Makimoto and S. Yamashita, "Compact bandpass filters using stepped impedance resonators," *Proc. IEEE*, vol. **67**, pp. 16–19, 1979.
25. T. C. Edwards and M. B. Steer, *Foundations of Interconnect and Microstrip Design,* 3rd Edition, John Wiley, New York, 2000.
26. M. Makimoto and S. Yamashita, "Bandpass filters using parallel-coupled stripline stepped impedance resonators," *IEEE Trans. Microw. Theory Technol.*, vol. **MTT-28**, pp. 1413–1417, 1980.

27. M. Durán-Sindreu, J. Bonache, and F. Martín, "Compact elliptic-function coplanar waveguide low-pass filters using backside metallic patterns," *IEEE Microw. Wireless Compon. Lett.*, vol. **20**, pp. 601–603, 2010.

28. G. Matthaei, L. Young, and E.M.T. Jones, *Microwave Filters, Impedance Matching Networks, and Coupling Structures,* Artech House, Norwood, MA, 1980.

2

ARTIFICIAL TRANSMISSION LINES BASED ON PERIODIC STRUCTURES

2.1 INTRODUCTION AND SCOPE

In the framework of RF/microwave engineering, one-dimensional periodic structures are transmission lines and waveguides periodically loaded with identical elements (lumped, semilumped, inclusions, defects, etc.), or with a periodic perturbation in their cross-sectional geometry (nonuniform transmission lines and waveguides).[1] The main relevant properties of one-dimensional periodic structures, which will be reviewed in this chapter, can be useful for the implementation of artificial transmission lines with various functionalities and applications based on them. Specifically, periodic transmission lines exhibit stop/pass bands, and support the propagation of waves with phase velocities lower (slow waves) or higher (fast waves) than the speed of light. Thus, transmission lines based on periodic structures can be applied to the implementation of filters, reflectors, electromagnetic bandgaps (EBGs),[2] slow wave structures

[1] Two-dimensional and three-dimensional periodic structures can also be of interest at RF/microwave (and even at optical) frequencies, in applications such as frequency-selective surfaces, antenna substrates and superstrates (to improve antenna performance), isolators, and so on, but these structures are out of the scope of this book, which is focused on artificial transmission lines.

[2] In analogy with semiconductor crystals (which exhibit forbidden energy bands, or gaps), periodic structures in the optical domain are usually identified as photonic crystals (PCs), or photonic bandgaps (PBGs). This explains the term "electromagnetic bandgap" (EBG) used to designate such structures at RF/microwave frequencies.

Artificial Transmission Lines for RF and Microwave Applications, First Edition. Ferran Martín.

(of interest for device miniaturization), and leaky wave antennas (LWAs), among others.

The purpose of this chapter is to briefly present the Floquet (or Bloch mode) analysis of one-dimensional periodic structures (which will bring us to the concept of space harmonics), and the transfer matrix method, applied to the unit cell, for obtaining the modal solutions (or dispersion curves) of the fundamental space harmonic. This will be important to predict the frequency response of the considered lines based on periodic structures. Moreover, an alternative and complementary analysis for periodic transmission lines with nonuniform cross section, (e.g., microstrip lines with defected ground planes or strip width modulation), based on the coupled mode theory, is also included in this chapter. By means of justified approximations, it will be shown that this analysis provides valuable information (through analytical expressions) relative to the main relevant parameters of these lines (S-parameters, maximum reflectivity and bandwidth of the stop bands, etc.). The periodic transmission lines that will be considered and studied in this chapter through the earlier-cited approaches (the transfer matrix method and the coupled mode analysis) include EBG-based transmission lines (either with defected ground planes or strip width modulation), and transmission lines loaded with reactive elements. The main applications of these lines will also be reviewed in the chapter.

2.2 FLOQUET ANALYSIS OF PERIODIC STRUCTURES[3]

Let us consider a one-dimensional infinite periodic structure (transmission line or waveguide) with period l, and the propagation axis denoted as z. If the time dependence is chosen as $e^{j\omega t}$, and if the cross-sectional dependence is suppressed, the Floquet's theorem states that the fields propagating along the line can be expressed as Bloch waves according to [1–5]:

$$\Psi(z) = e^{-\gamma z} \cdot P(z) \tag{2.1}$$

where γ is the propagation constant, and $P(z)$ is a periodic function with period l:

$$P(z+l) = P(z) \tag{2.2}$$

Thus, the field behavior can be expressed in terms of a fundamental traveling wave, with propagation constant γ, and a standing wave $P(z)$, which repeats in each unit cell

[3] This section is mainly based on a short course given by Arthur A. Oliner, who recently passed away, and who was a pioneer in the topic of periodic structures and leaky waves. His contributions on this topic are so well-written and comprehensible that can be effortlessly understood. Let us consider this section, extracted from this short course (published in Ref. [2]), as a tribute to him.

and represents the local variations due to the periodicity. From (2.1), it follows that the fields at positions separated by one period are related by:

$$\Psi(z+l) = e^{-\gamma(z+l)} \cdot P(z+l) = e^{-\gamma l} \cdot \Psi(z) \tag{2.3}$$

or, in other words, the fields of a Bloch wave repeat at each unit cell terminal, having a propagation factor $e^{-\gamma l}$.

Since $P(z)$ is a periodic function, it can be expanded in a Fourier series as

$$P(z) = \sum_{n=-\infty}^{n=+\infty} P_n e^{-j\frac{2\pi n}{l} z} \tag{2.4}$$

By inserting (2.4) into (2.1), the fields can be expressed as a superposition of traveling waves of the form

$$\Psi(z) = \sum_{n=-\infty}^{n=+\infty} P_n e^{-j\left(\beta + \frac{2\pi n}{l}\right) z} \cdot e^{-\alpha z} \tag{2.5}$$

where the propagation constant has been decomposed into the phase and attenuation constants, $\gamma = \alpha + j\beta$. The components of (2.5) are called space harmonics, in analogy to the harmonic decomposition of a periodic signal in time domain [2, 4]. The phase constants of the space harmonics are thus given by:

$$\beta_n = \beta + \frac{2\pi n}{l}, \quad n = 0, \pm 1, \pm 2, \ldots \tag{2.6}$$

and the phase of the fundamental harmonic ($n = 0$) is simply β. It is important to mention that the space harmonics do not exist independently. Rather than being modal solutions by themselves, they represent individual contributions to the whole field.

Since the phase constants of the space harmonics differ by a constant, it follows that the group velocity is the same for all harmonics, that is,

$$v_{\mathrm{gn}} = \frac{d\omega}{d\beta} \tag{2.7}$$

whereas the phase velocities are given by

$$v_{\mathrm{pn}} = \frac{\omega}{\beta + 2\pi n / l} \tag{2.8}$$

From the space harmonics representation of the periodic structure, it is possible to infer the pass/stop band characteristics inherent to periodicity. The reason is mode interference (or coupling) between modes with similar phase velocities but opposite group velocities (this aspect will be studied in Section 2.4 in detail). This results in stop bands (or gaps) in the frequency response. To gain insight on this effect, let us consider that a transmission line is slightly (periodically) perturbed so that the phase constant β of the fundamental harmonic can be approximated by a straight line (see expression 1.6). This means that the $\beta - \omega$ diagrams (or dispersion curves) of the set of space harmonics, obtained by displacing the dispersion diagram of the fundamental harmonic a quantity $2\pi n/l$, are also straight lines, as depicted in Figure 2.1. This diagram points out that there are points where the straight lines cross, giving rise to mode coupling and hence stop bands, as Figure 2.2 illustrates. Thus, Figure 2.2 depicts the actual dispersion curves of the slightly perturbed transmission line. The gap bandwidths and the specific frequency dependence of the phase constants

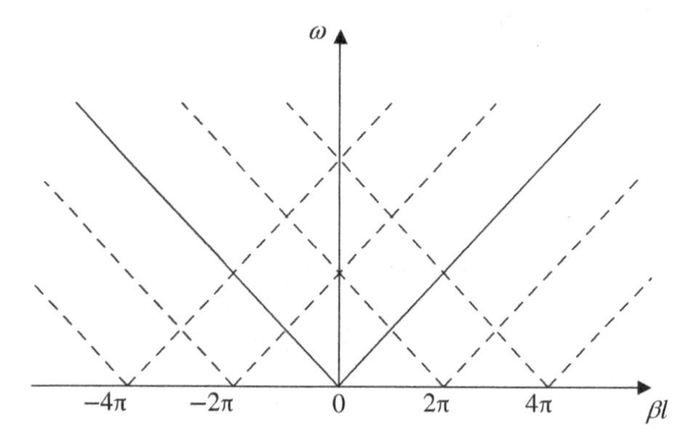

FIGURE 2.1 β-ω diagram of the fundamental harmonic (solid line) and high-order space harmonics (dashed lines) for a transmission line with an infinitesimal periodic perturbation.

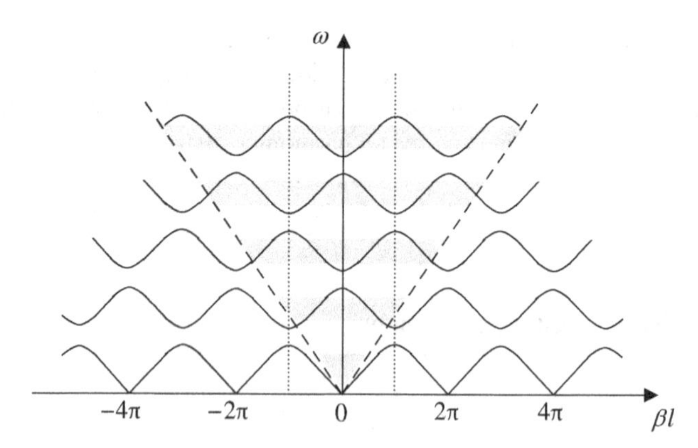

FIGURE 2.2 Dispersion diagram of a one-dimensional periodic structure with space harmonics. The limits of the first Brillouin zone are indicated by vertical dotted lines. The region where leaky wave radiation is possible is delimited by the dashed lines with slopes $-c/l$ and $+c/l$ (and indicated by the gray-white alternating strips).

(identical for all the space harmonics) are related to the nature and magnitude of the perturbation or loading element. Nevertheless, in view of Figure 2.2, the band gaps appear at frequencies satisfying $\beta l = \pi$ (or $l = \lambda/2$), that is,

$$f = \frac{v_{\mathrm{p}}}{2l} \tag{2.9}$$

and its harmonics, v_{p} being the phase velocity of the unperturbed line. Expression (2.9) is known as the Bragg condition, and states that the first stop band caused by the effects of periodicity[4] appears at the frequency satisfying that the period is half the wavelength of the unperturbed line, λ. The presence of additional band gaps depends on the harmonic content of the Fourier expansion (2.4), which in turn depends on the specificities of the loading elements or perturbation (this aspect will also be considered in Section 2.4).

It is remarkable that the part of the $\beta - \omega$ diagram of Figure 2.2 comprised between $-\pi < \beta l < \pi$ is repetitive every $\beta l = 2\pi$ along the abscissa axis. Usually, this region is referred to as the first Brillouin zone, and the curves within this region give full information of the dispersion characteristics of the periodic structure [1, 5]. It is also important to mention that if we consider either an infinitely long or terminated (with a matched load) periodic transmission line (so that reflections are avoided) fed by a source at one end, the waves excited by the source must carry their energy (and hence the group velocity) in the direction against the source (i.e., toward the load if it is present). Thus, in such situation, only the portions of the curves of Figure 2.2 with positive slope (or group velocity) are of interest.[5] Regardless of the propagation direction, there are regions of the dispersion curves where $\beta(\omega) \cdot d\beta(\omega)/d\omega < 0$. In these regions, the phase and group velocities are of opposite sign, and the corresponding waves are called backward waves.[6]

Another relevant aspect of one-dimensional periodic structures is the fact that these structures may radiate if they are open. The reason is that there are portions of the dispersion curves where the phase velocity is higher than the speed of light in vacuum, c, (fast wave regions, see Fig. 2.2). These radiating harmonics are leaky waves, and exhibit properties similar to leaky waves in uniform structures. A detailed analysis of leaky wave radiation in periodic structures is out of the scope of this book (see [6–9] for further details).[7] Therefore, we will simply present a brief and straightforward

[4] Stop bands not related to periodicity may appear at lower frequencies, for instance, by periodically loading a transmission line with resonators coupled to it. In this case, the first stop band is centered at the resonance frequency of the resonant element.

[5] It is assumed that the energy travels toward the positive z-direction.

[6] The concept of backward wave will be considered in detail in Chapter 3. Nevertheless, it is important to distinguish them from backward travelling waves. In backward waves, the energy propagates in the opposite direction to the phase of the waves, and it is therefore an unconventional type of propagation. Thus, for energy transmission in the positive z-direction, the phase of the waves propagates in the negative/positive z-direction for backward/forward waves. Backward/forward travelling waves are simply forward waves propagating in the negative/positive z-direction.

[7] Nevertheless, we dedicate a subsection in Chapter 4 devoted to LWAs based on metamaterial transmission lines.

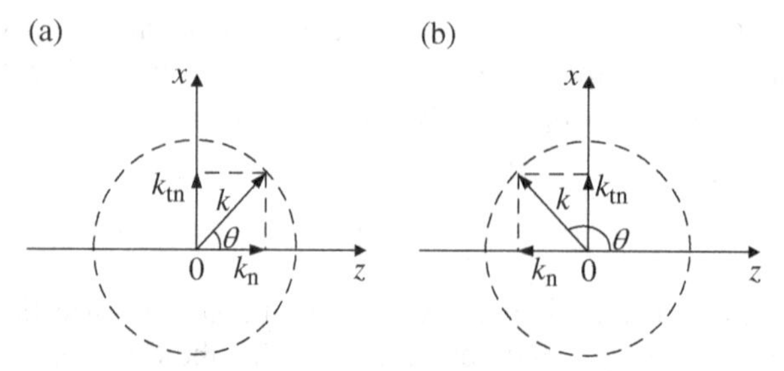

FIGURE 2.3 Diagrams illustrating the generation of a forward (a) and backward (b) leaky wave in a periodic structure.

analysis to understand the phenomenon. Let us consider that the fields in the open periodic structure vary in both the longitudinal and transverse directions. The free-space wavenumber k is related to the parameters of each space harmonic as [2][8]:

$$k^2 = k_n^2 + k_{tn}^2 = \left(\beta + \frac{2\pi n}{l} - j\alpha\right)^2 + k_{tn}^2 \tag{2.10}$$

where it has been assumed that $k_n = \beta_n - j\alpha$ (component in the longitudinal direction) and $k_{tn} = \beta_{tn} - j\alpha_{tn}$ (component in the transverse direction). To graphically illustrate the generation of leaky waves by the fast wave space harmonics, let us assume that the longitudinal attenuation constant α is negligible (i.e., $k_n = \beta_n$). In that case, if $k_n^2 > k^2$ (slow wave regions), k_{tn} is imaginary, and the mode does not leak. Conversely, if $k_n^2 < k^2$ (fast waves), k_{tn} is real and the mode radiates. This situation is illustrated in Figure 2.3, where it can be seen that for $0 < k_n < |k|$ and $-|k| < k_n < 0$, the radiation is forward and backward, respectively. For the specific frequencies where $k_n = 0$, $k_n = |k|$ and $k_n = -|k|$, leaky wave radiation is designated as broadside, endfire and backfire, respectively. Indeed, the direction of the radiated beam can be easily inferred from:

$$\cos\theta = \frac{k_n}{|k|} \tag{2.11}$$

which results from direct inspection to Figure 2.3.

Let us now take into account the attenuation constants in (2.10). Through separation of the real and imaginary parts, we obtain:

$$k^2 = k_n^2 + k_{tn}^2 = (\beta_n^2 - \alpha^2 + \beta_{tn}^2 - \alpha_{tn}^2) - 2j(\alpha\beta_n + \alpha_{tn}\beta_{tn}) \tag{2.12}$$

[8] The free-space wavenumber is denoted as k, rather than k_0 (as usual), to avoid confusion with the fundamental space harmonic. However, in Section 2.4, we recover the usual designation (see expression 2.45) since in that section k is used as a parameter related to the weighting factors of the coupling coefficient (to be defined later), according to expression (2.66).

Since the imaginary part of the free-space wavenumber k is zero, it follows that

$$\alpha_{tn} = -\alpha \frac{\beta_n}{\beta_{tn}} \tag{2.13}$$

Expression (2.13) is interesting because it points out a significant difference between the forward and backward leaky waves. Namely, since the longitudinal attenuation constant α and transverse phase constant β_{tn} are both positive, the transverse attenuation constant α_{tn} is of opposite sign to the longitudinal phase constant of the space harmonic β_n. Thus, if $\beta_n < 0$, α_{tn} is positive, meaning that the radiated fields decay in the transverse direction. Conversely, if $\beta_n > 0$, α_{tn} is negative, and the fields increase in the transverse direction. In this latter case (forward wave radiation), since this wave type would diverge at infinity in the transverse plane if it were defined everywhere, it cannot be spectral, that is, a proper mode. Therefore forward leaky waves are said to be non-spectral, contrary to backward leaky waves, which are called spectral [2]. By introducing (2.13) into (2.12), the relation between k, β_n and β_{tn} is found to be:

$$\beta_n^2 = \frac{k^2 - \beta_{tn}^2 + \alpha^2}{1 - \dfrac{\alpha^2}{\beta_{tn}^2}} \tag{2.14}$$

Notice that although forward (nonspectral) leaky waves are mathematically improper,[9] they have physical meaning and are useful for the implementation of radiating elements (LWAs). In practice, the leaky wave is defined only within a wedge-shaped region determined by the position of the source, where the leaky wave decays at all angles from the source. This solves the above inconsistency between the mathematical solution and the "physical" leaky waves, never exhibiting progressively increasing fields in the transverse direction.

2.3 THE TRANSFER MATRIX METHOD

Although the space harmonics are fundamental to explain several properties of periodic structures (the weight of each harmonic depends on the nature and magnitude of the perturbation), usually the fundamental space harmonic is dominant, and it suffices for the description of many structures (especially in the pass band regions, far enough from the band gap edges). In order to obtain the modal solutions (or dispersion characteristics) of the fundamental space harmonic (or equivalently of the Bloch wave), the transfer matrix method is very useful in situations where the fields at two positions separated by a period (related by Eq. 2.3) can be expressed as mutually dependent through a certain transfer function (or matrix), characteristic of the unit cell structure.

[9] The increasing field in the transverse direction violates the boundary condition at infinity since the mathematical description of leaky waves holds throughout all space.

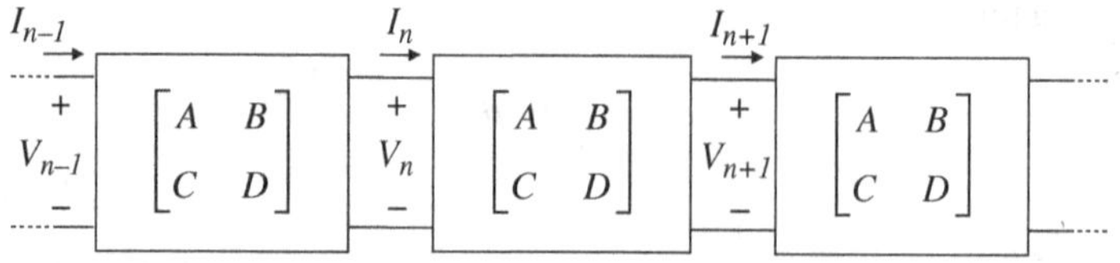

FIGURE 2.4 Periodic structure with unit cell described by the transfer *ABCD* matrix.

2.3.1 Dispersion Relation

Let us consider the periodic structure depicted in Figure 2.4, where the unit cells are represented by boxes, and let us consider that the voltages and currents at the reference planes between adjacent unit cells are well (uniquely) defined[10] and measurable quantities. The voltages and currents on either side of the nth unit cell are related by the *ABCD* matrix (see Appendix C) according to

$$\begin{pmatrix} V_n \\ I_n \end{pmatrix} = \begin{pmatrix} A & B \\ C & D \end{pmatrix} \begin{pmatrix} V_{n+1} \\ I_{n+1} \end{pmatrix} \tag{2.15}$$

On the other hand, according to the Floquet's theorem, the voltages and currents at the n and $n+1$ planes only differ by the propagation factor, that is

$$V_{n+1} = e^{-\gamma l} \cdot V_n \tag{2.16a}$$

$$I_{n+1} = e^{-\gamma l} \cdot I_n \tag{2.16b}$$

From (2.15) and (2.16), it follows that

$$\begin{pmatrix} V_n \\ I_n \end{pmatrix} = \begin{pmatrix} A & B \\ C & D \end{pmatrix} \begin{pmatrix} V_{n+1} \\ I_{n+1} \end{pmatrix} = \begin{pmatrix} e^{\gamma l} \cdot V_{n+1} \\ e^{\gamma l} \cdot I_{n+1} \end{pmatrix} \tag{2.17}$$

[10] Strictly speaking, this uniqueness of voltages and currents is only possible for TEM modes, but it can also be made extensive to quasi-TEM modes. For non-TEM modes, the voltage and the current are not properly (uniquely) defined. However, it is possible to define an equivalent voltage and current that make these variables (and even their ratio, the impedance) useful quantities. To this end, the following considerations are applied [10]: (i) voltage and current are only defined for a particular mode, and are defined so that the voltage and current are proportional to the transverse electric and magnetic fields, respectively; (ii) the equivalent voltages and currents should be defined so that their product gives the power flow of the mode; and (iii) the voltage to current ratio for a single travelling wave should be equal to the characteristic impedance of the mode. Nevertheless, unless otherwise specified, the periodic structures considered throughout this book can be described, to a first-order approximation, by considering only the TEM or quasi-TEM modes. Therefore, expression (2.15) involves well-defined variables in our case.

or

$$\begin{pmatrix} A-e^{\gamma l} & B \\ C & D-e^{\gamma l} \end{pmatrix} \begin{pmatrix} V_{n+1} \\ I_{n+1} \end{pmatrix} = 0 \tag{2.18}$$

Notice that, according to (2.17) and (2.18), the voltages and currents propagating in the line are the eigenvectors, whereas the propagation factor is given by the eigenvalues, or eigenmodes, of the system. For a nontrivial solution, the determinant of the matrix in (2.18) must be zero, namely

$$AD + e^{2\gamma l} - (A+D)e^{\gamma l} - BC = 0 \tag{2.19}$$

Since for a reciprocal system $AD - BC = 1$ (see Appendix C), (2.19) can be expressed as follows[11]:

$$e^{\gamma l} + e^{-\gamma l} = A + D \tag{2.20}$$

and the dispersion relation can be finally written as

$$\cosh(\gamma l) = \frac{A+D}{2} \tag{2.21}$$

In a lossless and reciprocal periodic structure, the right-hand side of (2.21) is purely real. This means that the propagation constant is either purely real ($\gamma = \alpha$, $\beta = 0$) or purely imaginary ($\gamma = j\beta$, $\alpha = 0$).[12] In the first case, the Bloch wave is attenuated along the line, and the corresponding regions define the stop bands of the structure. If $\gamma = j\beta$ and $\alpha = 0$, $\cosh(\gamma l) = \cos(\beta l)$, and (2.21) rewrites as follows:

$$\cos(\beta l) = \frac{A+D}{2} \tag{2.22}$$

Expression (2.22) is thus valid in the propagation regions, where the modulus of the right-hand side is smaller than 1. If the unit cell of the periodic structure is symmetric with respect to the plane equidistant from the input and output ports, $A = D$ and (2.22) can be simplified to

$$\cos(\beta l) = A \tag{2.23}$$

[11] Reciprocity is assumed throughout this chapter.

[12] As it will be shown in Chapter 3, under some circumstances, lossless periodic structures may support modes that appear as conjugate pairs, that is, modes of the form $\gamma = \alpha \pm j\beta$. Such modes are called complex modes.

2.3.2 Bloch Impedance

Another important parameter is the relationship between the voltage and current at any position (plane) of the periodic structure. Such parameter can be inferred from (2.18), namely,

$$\left(A - e^{\gamma l}\right) V_{n+1} + B I_{n+1} = 0 \tag{2.24}$$

and it follows that

$$\frac{V_{n+1}}{I_{n+1}} = -\frac{B}{A - e^{\gamma l}}, \ \forall n \tag{2.25}$$

Expression (2.25) does not depend on the plane where such voltage to current relation is calculated. It resembles the characteristic impedance of a transmission line, defined as the relation between voltage and current for a single propagating wave at any position in the line. However, since the propagating waves in the periodic structure are Bloch waves, it is more convenient to identify the impedance given by (2.25) as the Bloch impedance, Z_{B}. Isolating $e^{\gamma l}$ from (2.20) and introducing it into (2.25), it follows that the Bloch impedance has two solutions:

$$Z_{\mathrm{B}}^{\pm} = -\frac{2B}{A - D \mp \sqrt{(A+D)^2 - 4}} \tag{2.26}$$

one corresponding to forward traveling waves and the other to backward traveling waves. In general, the two solutions of (2.26) are complex. In the propagation regions, $(A + D)^2 < 4$, and the resulting solutions have the same magnitude, identical imaginary part and real parts of opposite sign. In the forbidden (band gap) regions, the two solutions are purely imaginary[13] and exhibit different magnitude, unless the unit cell is symmetric. In this case, $A = D$ and (2.26) is simplified to

$$Z_{\mathrm{B}}^{\pm} = \pm \frac{B}{\sqrt{A^2 - 1}} \tag{2.27}$$

For a lossless and symmetric structure, the two solutions of the Bloch impedance in the allowed regions are real and have opposite signs. Such different signs are indicative of propagation in the forward or backward direction (the negative sign for backward traveling waves is related to the definition of the currents in Fig. 2.4).

Let us discuss the meaning of the complex Bloch impedance that results in the allowed bands of lossless periodic structures with asymmetric unit cells. According

[13] In the forbidden regions, $(A + D)^2 > 4$, hence the denominator in (2.26) is a real number. Since for a lossless structure B is purely imaginary, it follows that the Bloch impedance is purely imaginary in those regions.

to expression (1.8), in a lossless transmission line, the characteristic impedance is purely real. However, the Bloch impedance in lossless periodic structures is complex, if the unit cell is asymmetric. This complex impedance cannot be related to attenuation losses if the periodic line is lossless. The origin of the complex Bloch impedance in lossless periodic structures with asymmetric unit cells is intimately related to the termination of the line. If the unit cell is asymmetric, a finite periodic structure consisting of a certain number of cascaded unit cells is also asymmetric. Therefore, the input impedance of such an asymmetric structure terminated by certain load impedance depends on the locations (left or right) of the source and the load. In particular, if the structure is terminated with the Bloch impedance, the impedance seen from the input port is also the Bloch impedance. However, since the input impedance depends on the position of the source (input port), it follows that the Bloch impedance must be different for forward and backward traveling waves in asymmetric periodic transmission lines. Nevertheless, for an infinite periodic structure, where the effects of terminations vanish, propagation in the forward or backward directions must be undistinguishable. Indeed, if the structure is infinite, it can be described by a cascade of symmetric unit cells, by simply shifting the reference planes, as described in Figure 2.5.[14] Since the Bloch impedance for this symmetric unit-cell-based structure in the propagation regions is real, we can conclude that the complex Bloch impedance in the asymmetric structure is caused by a phase shift in only one of the variables (voltage or current), as compared to the symmetric case (where voltage and current are in phase), as consequence of a displacement of the reference planes (this situation is also illustrated in Fig. 2.5). From this analysis, we can also conclude that for both forward and backward traveling waves propagating in the periodic structure with asymmetric unit cells, the real part of the Bloch impedance must be identical, whereas the imaginary parts must have identical magnitude and different sign (for the validity of this statement, we have assumed that the backward traveling waves exhibit positive current in the backward direction).

To gain insight on the effects of asymmetry, let us consider a periodic structure that can be described by the network of Figure 2.5, where the dashed lines are the reference planes for the symmetric unit cell described by a T-circuit model, whereas the dotted lines are the reference planes of the asymmetric unit cell described by an L-circuit.[15] The elements of the *ABCD* matrix for the unit cells considered in Figure 2.5 can be easily inferred [10]. In particular, for the symmetric unit cell:

$$A = D = 1 + \frac{Z_s}{2Z_p} \tag{2.28a}$$

[14] This statement is based on the fact that any two-port network can be described by means of an equivalent π- or T-circuit. Despite the asymmetry of the equivalent π- or T-circuit of any asymmetric unit cell, by cascading such circuits, it is possible to describe any infinite structure by means of symmetric unit cells by simply shifting the reference planes. However, in a physical periodic structure, it is not necessarily possible to identify a symmetric unit cell.

[15] An L-circuit can be considered a particular case of a T-circuit where one of the series impedances is null.

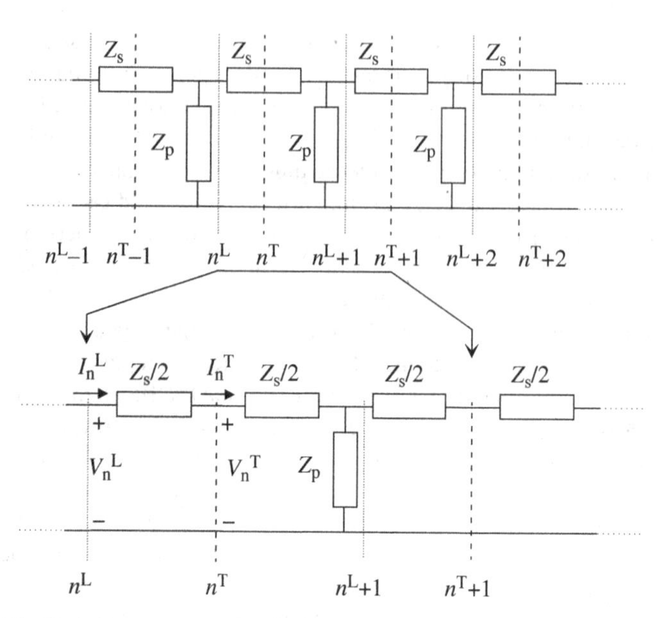

FIGURE 2.5 Periodic structure where the reference planes of the symmetric T-circuit unit cell (vertical dashed lines) and asymmetric L-circuit unit cell (vertical dotted lines) are indicated. The superscripts L or T in the port variables and reference planes are used to differentiate between the asymmetric and symmetric unit cells. Notice that the port currents do not vary by shifting the reference planes $\left(I_n^L = I_n^T\right)$, contrary to the port voltages $\left(V_n^L \neq V_n^T\right)$. To make the reference planes of the symmetric unit cells accessible, it suffices to split the series impedance into two identical impedances equal to $Z_s/2$.

$$B = Z_s + \frac{Z_s^2}{4Z_p} \tag{2.28b}$$

$$C = \frac{1}{Z_p} \tag{2.28c}$$

whereas the following parameters apply to the asymmetric unit cell:

$$A = 1 + \frac{Z_s}{Z_p} \tag{2.29a}$$

$$B = Z_s \tag{2.29b}$$

$$C = \frac{1}{Z_p} \tag{2.29c}$$

$$D = 1 \tag{2.29d}$$

Introducing the elements of (2.28) in (2.27), the Bloch impedance corresponding to the structure described by the symmetric unit cell is found to be

$$Z_B^{\pm} = \pm \frac{Z_s\left(1+\dfrac{Z_s}{4Z_p}\right)}{\sqrt{\left(1+\dfrac{Z_s}{2Z_p}\right)^2-1}} = \pm\sqrt{\frac{Z_s}{2}\left(\frac{Z_s}{2}+2Z_p\right)} \tag{2.30}$$

and it is real in the propagation regions if Z_s and Z_p are purely reactive impedances. For the structure composed by a cascade of the asymmetric L-circuit unit cells, the Bloch impedance is derived from (2.26) and (2.29)

$$Z_B^{\pm} = -\frac{2Z_s}{\dfrac{Z_s}{Z_p}\mp\sqrt{\left(2+\dfrac{Z_s}{Z_p}\right)^2-4}} = \pm\sqrt{\frac{Z_s}{2}\left(\frac{Z_s}{2}+2Z_p\right)}+\frac{Z_s}{2} \tag{2.31}$$

and it is a complex number in the region where wave propagation is allowed. As mentioned before, according to the usual definition of the positive current for forward and backward traveling waves, the two solutions of the Bloch impedance must be actually expressed as follows:

$$Z_B^{\pm} = \sqrt{\frac{Z_s}{2}\left(\frac{Z_s}{2}+2Z_p\right)}\pm\frac{Z_s}{2} \tag{2.32}$$

Expression (2.32) indicates that the real part of the Bloch impedance is identical to that of the symmetric structure (2.30), as expected. Notice that the imaginary part is the impedance that must be cascaded to the truncated periodic line ($+Z_s/2$ and $-Z_s/2$ in the load and source planes, respectively) in order to transform the finite asymmetric periodic structure to a symmetric network. Therefore, the complex Bloch impedance in the propagation regions of lossless asymmetric structures indicates that to match a load/source to the line it is necessary to series connect a reactive impedance, able to compensate the effects of asymmetry, plus the required resistive part, given by the absolute value of (2.30).

In the circuit of Figure 2.5, it is remarkable that the dispersion relation, given by (2.22), is insensitive to the position of the reference planes of the unit cell. Either by using the *A* and *D* parameters given by (2.28) or (2.29), expression (2.22) is found to be

$$\cos(\beta l) = 1+\frac{Z_s}{2Z_p} \tag{2.33}$$

2.3.3 Effects of Asymmetry in the Unit Cell through an Illustrative Example

At this point, it is interesting to provide an example to illustrate the effects of asymmetry of the unit cell in the behavior of the structure. Let us consider the ladder network corresponding to the circuit model of a lossless conventional transmission line, where each unit cell describes a section of the transmission line that must be electrically short (Figure 2.6). For the symmetric unit cell, the Bloch impedance, inferred from (2.30), is

$$Z_B = \sqrt{\frac{L}{C}\left(1 - \frac{\omega^2}{\omega_o^2}\right)} \tag{2.34}$$

where ω is the angular frequency, and $\omega_o = 2/\sqrt{LC}$ is a cutoff frequency. Above this cutoff frequency, the Bloch impedance is purely imaginary, and propagation is not allowed. The network of Figure 2.6 is indeed a low-pass filter, whereas a lossless transmission line is an all-pass structure. The discrepancy is explained because the network of Figure 2.6, where L and C model the per-section inductance and capacitance of the transmission line, is valid at low frequencies (the wavelength must be much larger than the considered line section length). Nevertheless, for frequencies satisfying $\omega \ll \omega_o$, the Bloch impedance coincides with the well-known characteristic impedance of the line (expression 1.9). For the asymmetric unit cell (L-network), the Bloch impedance, given by expression (2.32), is

$$Z_B^{\pm} = \sqrt{\frac{L}{C}\left(1 - \frac{\omega^2}{\omega_o^2}\right)} \pm j\frac{1}{2}L\omega \tag{2.35}$$

As frequency decreases, the imaginary part of the Bloch impedance also decreases. Therefore, at sufficiently low frequencies, where $L\omega/2 \ll (L/C)^{1/2}$, the Bloch impedance that results by considering the asymmetric unit cell is essentially real. This means that in this frequency regime, the behavior of both terminated periodic structures (with cascaded symmetric and asymmetric unit cells) is expected to be similar. This has been verified by comparing the reflection and transmission coefficients (considering a source impedance of 50 Ω) of the 20-cell structures depicted in Figure 2.7, loaded with a resistance of value $R = (L/C)^{1/2} = 50\ \Omega$, which is a matched load at low

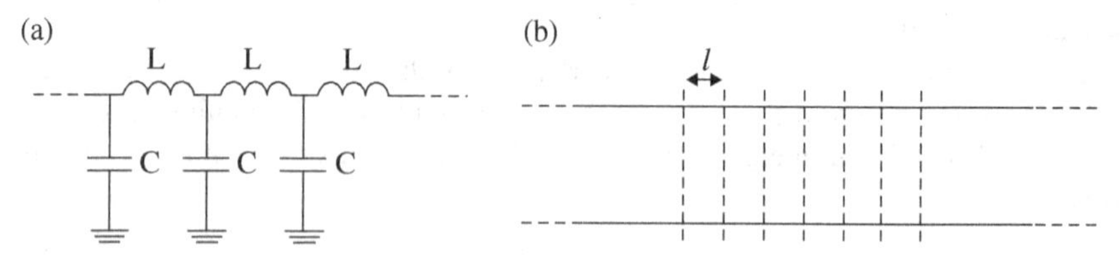

FIGURE 2.6 Model of a losses transmission line (a), where each unit cell describes the finite transmission line sections of length l (b).

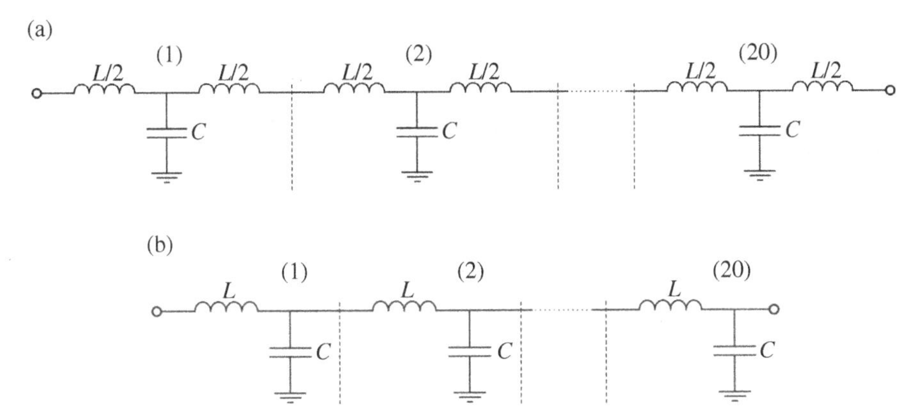

FIGURE 2.7 (a) Symmetric (T-circuit) and (b) asymmetric (L-circuit) networks modeling a 20-section lossless transmission line.

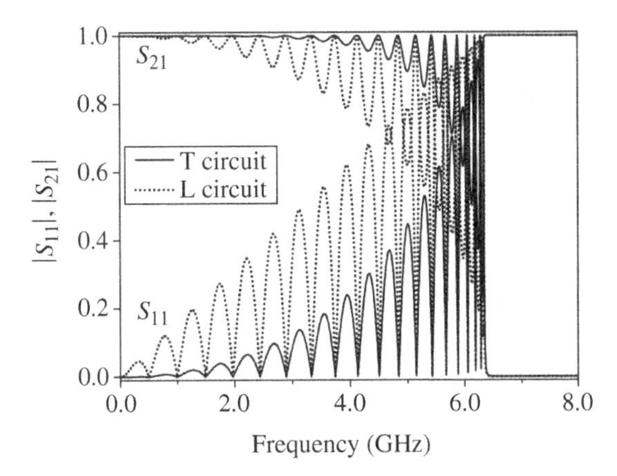

FIGURE 2.8 Reflection (S_{11}) and transmission (S_{21}) coefficients for the structures depicted in Figure 2.7 with $C = 1$ pF and $L = 2.5$ nH.

frequencies (see Fig. 2.8). As expected, at low frequencies all the injected power is delivered to the load ($|S_{11}| \ll 1$ and $|S_{21}| \cong 1$), and the reflection and transmission coefficients are similar for both structures. However, as frequency increases, the Bloch impedance varies and the structures are no longer matched to the source. The injected power is thus partially reflected back to the source (except at certain frequencies[16]), and above ω_o the lines are in the evanescent region, where $|S_{11}| \cong 1$ and

[16] Below the cut-off frequency, there are transmission peaks (reflection zeros) corresponding to phase matching. Namely, at these frequencies, the injected power is delivered to the load (despite that the line impedance is not matched to the ports) since the phase of the line is a multiple of 180°.

$|S_{21}| << 1$. However, the relevant aspect is that as frequency increases, both reflection coefficients and transmission coefficients progressively diverge as consequence of the increasing effects of the imaginary part of the Bloch impedance in the asymmetric periodic structure. If we force the imaginary part of (2.35) to be much smaller than the real part, that is,

$$\frac{1}{2}L\omega << \sqrt{\frac{L}{C}\left(1-\frac{\omega^2}{\omega_o^2}\right)} \tag{2.36}$$

the following condition is obtained:

$$\omega << \frac{1}{\sqrt{2}}\omega_o \tag{2.37}$$

which is indeed more restrictive (but comparable) than the necessary condition for the validity of the networks of Figure 2.7 as circuit models of a lossless transmission line ($\omega << \omega_o$). In summary, this analysis points out that in the long wavelength regime ($\lambda >> l$), both terminated periodic structures of Figure 2.7 can be indistinctly used. This means that both networks are appropriate to describe or model a finite lossless transmission line. However, in general, this identification between the symmetric and asymmetric networks (which simply differ in the location of the reference planes – see Fig. 2.5) cannot be made. In other words, for a correct description of an actual periodic structure by means of a cascade of T- or π-sections, it is fundamental that such sections properly account for the possible asymmetries of the unit cell of the periodic structure (this aspect will be discussed in Chapter 3 in reference to asymmetric planar structures).

2.3.4 Comparison between Periodic Transmission Lines and Conventional Lines

It is important to mention that one-dimensional periodic structures, such as periodic loaded, or perturbed, transmission lines, can be characterized as if they were conventional transmission lines, that is, by means of a characteristic, or Bloch, impedance, and by means of a propagation constant.[17] However, as compared to conventional lines, where dispersion is absent (under ideal conditions), periodic transmission lines exhibit dispersion as well as pass bands and stop bands. Moreover, the Bloch impedance is frequency dependent, whereas the characteristic impedance of a conventional transmission line is solely dependent on the substrate characteristics and cross-sectional geometry of the line. Nevertheless, given a certain frequency in the region of propagation, periodically loaded or perturbed lines behave as conventional lines with identical propagation constant and characteristic impedance at the considered

[17] If the repeating period of the structure is asymmetric, caution must be taken due to the effects of termination.

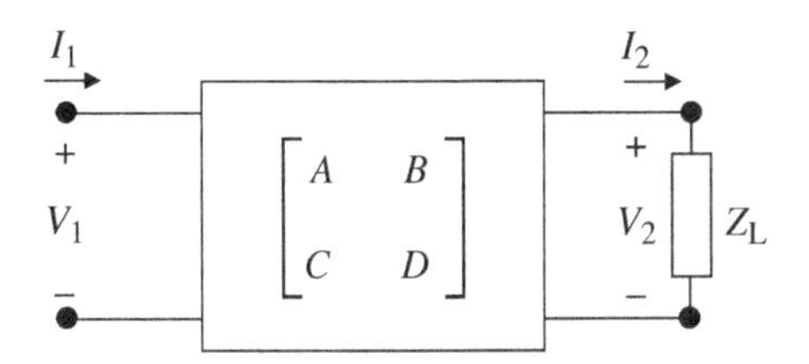

FIGURE 2.9 Arbitrary two-port network terminated with a load impedance Z_L.

frequency. Thus, for instance, if a periodic transmission line is terminated by a matched load at the considered frequency, that is, a load impedance identical to the Bloch impedance, all the injected power will be delivered to the load (as has been demonstrated in the example reported earlier and illustrated in Fig. 2.8—low-frequency region). Similarly, a finite periodic transmission line excited by a harmonic source and terminated by a load, exhibits identical behavior to that of a conventional line with identical electrical length and Bloch impedance. Specifically, if the electrical length of the periodic line is a multiple of π, the impedance seen from the source is the load impedance. These artificial lines based on periodic structures can also be used for the implementation of narrow band impedance or admittance inverters, by simply setting the Bloch impedance to the required impedance of the inverter at the design frequency, and the electrical length to π/2 or an odd multiple of π/2. Indeed, as will be demonstrated in Chapter 3, it is possible to implement impedance inverters by means of a single unit cell (of different type than those considered in this chapter), which is very interesting in terms of device miniaturization.

2.3.5 The Concept of Iterative Impedance

The characteristic impedance of a transmission line can also be defined as the load impedance necessary to "see" the same impedance from the source plane (or input port). In two-port networks, the impedance that satisfies the above requirement is called iterative impedance, Z_I.[18] For a given unit cell, the iterative impedance and the Bloch impedance are identical. To demonstrate this, let us consider that an arbitrary two-port network, described by the *ABCD* matrix, is loaded with an arbitrary impedance Z_L (Fig. 2.9). The voltage and current at the input port can be expressed as follows:

$$V_1 = AV_2 + B\frac{V_2}{Z_L} \tag{2.38a}$$

[18] Notice that the iterative impedance of an asymmetric network is different for ports 1 and 2 (similarly to the Bloch impedance for forward and backward travelling waves in such networks). The iterative impedance should not be confused with the image impedances of a two-port network. If we designate these image impedances as Z_{i1} and Z_{i2}, they satisfy (i) Z_{i1} is the input impedance at port 1 when port 2 is terminated with Z_{i2} and (ii) Z_{i2} is the input impedance at port 2 when port 1 is terminated with Z_{i1} [10]. However, for symmetric networks, the image impedances Z_{i1} and Z_{i2} are identical and coincide with the iterative impedance and Bloch impedance.

$$I_1 = CV_2 + D\frac{V_2}{Z_L} \tag{2.38b}$$

Hence, the input impedance is

$$Z_{in} = \frac{V_1}{I_1} = \frac{A + B/Z_L}{C + D/Z_L} \tag{2.39}$$

If $Z_L = Z_I$, then $Z_{in} = Z_I$, and Z_I can be isolated from (2.39), giving

$$Z_I = \frac{A - D \pm \sqrt{(A+D)^2 - 4}}{2C} \tag{2.40}$$

Although apparently expressions (2.40) and (2.26) are different, simple algebra demonstrates that $Z_B = Z_I$, as anticipated before. Namely, (2.26) can be expressed as follows:

$$Z_B^{\pm} = -\frac{2B}{A - D \mp \sqrt{(A+D)^2 - 4}} \cdot \frac{A - D \pm \sqrt{(A+D)^2 - 4}}{A - D \pm \sqrt{(A+D)^2 - 4}} \tag{2.41}$$

which gives

$$Z_B^{\pm} = \frac{2B\left(A - D \pm \sqrt{(A+D)^2 - 4}\right)}{4(AD-1)} = \frac{A - D \pm \sqrt{(A+D)^2 - 4}}{2C} = Z_I \tag{2.42}$$

where the reciprocity condition $AD - BC = 1$ has been used. Indeed, the term "iterative impedance" is usually restricted to two-port networks with few (or even a single) unit cells, where the propagation of Bloch waves as traveling waves with several wavelengths is not possible. However, it is important to clarify that expressions (2.21) and (2.26) are valid regardless of the number of cells (they are valid even for a single-cell structure). Throughout this book, the terms Bloch impedance and characteristic impedance, in reference to artificial transmission lines, are considered to be synonymous.

2.4 COUPLED MODE THEORY[19]

As anticipated in the introduction, the specific periodic structures that will be analyzed in this chapter are transmission lines loaded with reactive elements, and nonuniform planar transmission lines, namely, transmission lines with periodic perturbation in the conductor strip width or in the ground plane. For the latter case, the coupled mode

[19] This section has been co-authored with Txema Lopetegi (Public University of Navarre, Spain).

theory, and specifically the cross-section method, is a very useful analysis tool [11–15]. It is a complementary approach to the transfer matrix method, able to provide detailed information (under several reasonable approximations) on the main parameters of the periodic nonuniform transmission line, such as the S-parameters, the central frequencies of the stop bands as well as their rejection level and bandwidth, the harmonic content, and so on.

2.4.1 The Cross-Section Method and the Coupled Mode Equations

The basic idea of the cross-section method is that the electromagnetic (EM) fields at any cross section of a nonuniform transmission line can be expressed as a superposition of the forward and backward traveling waves associated to the different modes of an auxiliary (or reference) uniform transmission line with the same cross section as the considered cross section of the nonuniform transmission line.[20] Namely,

$$\Psi = \sum_i a_i \Psi^i + \sum_i \int_0^\infty a_i(k_t)\Psi^i(k_t)dk_t \tag{2.43}$$

where Ψ is either the total electric or magnetic field of the nonuniform transmission line, Ψ^i is the field pattern of the ith discrete mode of the auxiliary uniform transmission line, and a_i are the corresponding weights, or coefficients, which can be interpreted as the complex amplitudes of the discrete modes along the non-uniform transmission line, and are only function of z (the propagation direction) and frequency. Notice that expression (2.43) gives the field expansion of an open transmission line (or waveguide) where, besides the discrete modes, a system of modes with a continuous spectrum is also included (second term of the right-hand side of 2.43). This additional system is in general necessary since the radiated field cannot be represented by the discrete modes. Conversely, in a closed waveguide, the discrete modes (including the propagating and evanescent modes) form a complete and orthogonal system.[21] The coefficients of the above field expansion satisfy a system of integrodifferential equations, known as coupled mode equations. The derivation of the coupled mode equations for the general case of open transmission lines (or waveguides) is not straightforward and is out of the scope of this book (the authors recommend Refs. [12–14] to the interested readers). The interest in this book is to infer simplified coupled mode equations of practical use for the analysis of the considered structures by applying reasonable approximations to the general equations. Thus, the starting

[20] This includes the same permeability and permittivity distributions for nonhomogeneous substrates. Nevertheless, only artificial transmission lines implemented on homogeneous substrates are considered in this book.
[21] The modes of a closed waveguide can be enumerated by a subscript, and the set is composed by an infinite number of discrete modes. However, in an open waveguide, the system of discrete modes is formed by a finite number of modes, all of them in propagation.

point is the expression of the general integrodifferential coupled mode equations that, for an open transmission line or waveguide nonuniform in the propagation direction, take the form[22]:

$$\frac{da_m}{dz}+j\beta_m a_m=\sum_i a_i C_{mi}+\sum_i\int_0^\infty a_i(k_t)C_{mi}(k_t)dk_t \tag{2.44a}$$

$$\frac{da_n\left(\tilde{k}_t\right)}{dz}+j\beta_n\left(\tilde{k}_t\right)a_n\left(\tilde{k}_t\right)=\sum_i a_i C_{ni}^c\left(\tilde{k}_t\right)+\sum_i\int_0^\infty a_i(k_t)C_{ni}^c\left(\tilde{k}_t,k_t\right)dk_t \tag{2.44b}$$

where β_m is the phase constant of the mth discrete mode, C_{mi} is the coupling coefficient between the mth and ith discrete modes, and $\beta_n(k_t)$, $C_{mi}(k_t)$, $C_{ni}^c\left(\tilde{k}_t\right)$, and $C_{ni}^c\left(\tilde{k}_t,k_t\right)$ are the corresponding parameters for the continuous spectrum modes (their expressions can be found in Ref. [14]), and

$$k_t^2=k_o^2-\beta_i^2(k_t) \tag{2.45}$$

with $k_0=\omega\sqrt{\mu_o\varepsilon_o}$, and $\beta_i(k_t)=+\sqrt{k_0^2-k_t^2}$ for $i>0$, $\beta_i(k_t)=-\sqrt{k_0^2-k_t^2}$ for $i<0$. There is one of these coupled mode equations for each mode, including their forward (+) and backward (–) traveling waves.

Expression (2.44a and b) can be substantially simplified if several assumptions are made:

- The first one is to neglect the coupling of energy to the modes of the continuous spectrum. This is reasonable if the considered structures have little radiation losses in the considered frequency bands (however, the continuous spectrum cannot be neglected in applications where devices are designed to radiate, i.e., antennas). This approximation notably simplifies the coupled mode equations since expression (2.44b) can be ignored, and the second term of the right-hand side of (2.44a) has a negligible effect as well.
- The second approximation consists of assuming single-mode operation, that is, the fundamental mode with two associated waves: the forward and backward traveling waves. In structures with periodic defects or patterns in the ground plane, further modes may appear. For instance, in microstrip lines with slots

[22] For the derivation of these equations, it is also assumed that the nonuniform transmission lines are lossless.

in the ground plane, a quasimicrostrip mode[23] and a quasislot mode are possible (at moderate frequencies and for conventional microwave substrate thicknesses, higher-order modes are not present). However, if the periodic microstrip structure is excited by a pure microstrip line, the corresponding mode (quasi-TEM mode) has very small correlation with the quasislot mode, and strong correlation with the quasimicrostrip mode. Therefore, the quasislot mode can be neglected, and single-mode operation is justified.

With the previous approximations, the system of coupled mode equations simplifies to [14, 16]

$$\frac{da^+}{dz} + j\beta^+ a^+ = a^+ C^{++} + a^- C^{+-} \tag{2.46a}$$

$$\frac{da^-}{dz} + j\beta^- a^- = a^- C^{--} + a^+ C^{-+} \tag{2.46b}$$

where the subscripts (+) and (−) designate the forward and backward traveling waves, respectively, associated to the fundamental TEM or quasi-TEM mode. Expression (2.46a and b) can be further simplified to

$$\frac{da^+}{dz} + j\beta a^+ = a^- K \tag{2.47a}$$

$$\frac{da^-}{dz} - j\beta a^- = a^+ K \tag{2.47b}$$

where the fundamental property that the phase constants of the forward and backward traveling wave of a discrete mode are identical except the sign ($\beta^+ = -\beta^- = \beta$) has been used. With regard to the coupling coefficients, notice that due to symmetry considerations, $C^{+-} = C^{-+} \equiv K$. Moreover, $C^{--} = C^{++} = 0$ since the coupled mode Equations 2.46a and 2.46b must give trivial solutions of the form $a^+ \propto e^{-j\beta z}$ and $a^- \propto e^{+j\beta z}$, for the case of a lossless uniform transmission line (where $K = 0$). Expression (2.47a and 2.47b) can thus be expressed in the following compact form:

$$\frac{d}{dz}\begin{bmatrix} a^+ \\ a^- \end{bmatrix} = \begin{bmatrix} -j\beta & K \\ K & j\beta \end{bmatrix} \cdot \begin{bmatrix} a^+ \\ a^- \end{bmatrix} \tag{2.48}$$

[23] By quasimicrostrip mode, we mean the fundamental mode of the line when a defect in the form of a slot in the ground plane is present.

where a single coupling coefficient, K, between the forward and backward traveling waves associated to the fundamental TEM or quasi-TEM mode, plus the phase constant of the auxiliary uniform waveguide, characterize the structure. Notice that both the phase constant and the coupling coefficient are functions of z and frequency, that is, $\beta = \beta(z,\omega)$ and $K = K(z,\omega)$. The phase constant of the operation mode can be calculated as

$$\beta(z,\omega) = \frac{\omega}{c}\sqrt{\varepsilon_{re}(z,\omega)} \tag{2.49}$$

where the explicit dependence of the effective dielectric constant, defined in Chapter 1, on z and ω has been introduced in the square root in order to highlight that, rather than being an averaged dielectric constant, the effective dielectric constant varies with the cross section, which in turn depends on the position along the propagation axis in a nonuniform transmission line. The effective dielectric constant at each cross section (i.e., that of the auxiliary uniform transmission line), and hence the phase constant, can be numerically calculated [10, 17], or inferred by means of most available EM solvers.

The exact expression of the coupling coefficient involves the calculation of the fields of the modes in the auxiliary uniform transmission line associated to the cross section of interest (which varies with z) [14]. Nevertheless, the coupling coefficient can be approximated by a simple expression that depends only on the characteristic impedance, Z_o, of the transmission line according to (see Appendix E):

$$K(z,\omega) = -\frac{1}{2Z_o}\frac{dZ_o}{dz} \tag{2.50}$$

In microstrip or coplanar waveguide transmission lines with periodic perturbation in the conductor strip, Z_o can be point-to-point obtained along the propagation axis by means of available transmission line calculators. If the ground plane is etched, for instance, by periodically drilling holes underneath the conductor strip of a microstrip line (a usual technique for the implementation of EBG-based reflectors [18, 19]), the calculation of Z_o is not so straightforward. Nonetheless, the impedance of expression (2.50) must be inferred from the power and current carried by the mode according to

$$Z_o(z,\omega) = 2\frac{P^+}{|I^+|^2} \tag{2.51}$$

Indeed, the dependence of Z_o with frequency can be neglected to a first-order approximation as it is customarily done in planar transmission lines.

The matrix equation (2.48) is a system of first-order linear differential equations. In order to solve these equations, it is necessary to define the z interval corresponding to the structure under study, delimited by $z = 0$ and $z = L$ (the device length), and appropriate boundary conditions. Such boundary conditions depend on how the ports of the structure are loaded. Typically, the output port is loaded with a matched load, whereas a unitary excitation is applied at the input of the device. Under these conditions, the following boundary values apply:

$$a^+(z=0) = a_{\text{in}}^+ = 1 \tag{2.52a}$$

$$a^-(z=L) = a_{\text{out}}^- = 0 \tag{2.52b}$$

and the coupled mode equations can be easily solved through numerical methods.

2.4.2 Relation between the Complex Mode Amplitudes and S-Parameters

Once the complex amplitudes $a^+(z,\omega)$ and $a^-(z,\omega)$ are inferred, the S-parameters of the device can be straightforwardly calculated. According to the definitions of these parameters, given in Appendix C,

$$S_{11} = \left.\frac{b_1}{a_1}\right|_{a_2=0} = \left.\frac{V_1^- \sqrt{Z_o}}{V_1^+ \sqrt{Z_o}}\right|_{a_2=0} \tag{2.53a}$$

$$S_{21} = \left.\frac{b_2}{a_1}\right|_{a_2=0} = \left.\frac{V_2^- \sqrt{Z_o}}{V_1^+ \sqrt{Z_o}}\right|_{a_2=0} \tag{2.53b}$$

$$S_{22} = \left.\frac{b_2}{a_2}\right|_{a_1=0} = \left.\frac{V_2^- \sqrt{Z_o}}{V_2^+ \sqrt{Z_o}}\right|_{a_1=0} \tag{2.53c}$$

$$S_{12} = \left.\frac{b_1}{a_2}\right|_{a_1=0} = \left.\frac{V_1^- \sqrt{Z_o}}{V_2^+ \sqrt{Z_o}}\right|_{a_1=0} \tag{2.53d}$$

where Z_o is the reference impedance of the ports (it is assumed that both ports have the same impedance), and $z = 0$ and $z = L$ are the reference planes of the input (1) and output (2) ports, respectively. The relations between the voltages associated to the incident (V_1^+, V_2^+) and reflected (V_1^-, V_2^-) waves from the ports and those associated to the forward V^+ and backward V^- traveling waves are as follows:

$$V_1^+ = V^+(z=0) \tag{2.54a}$$

$$V_1^- = V^-(z=0) \tag{2.54b}$$

$$V_2^+ = V^-(z=L) \tag{2.54c}$$

$$V_2^- = V^+(z=L) \tag{2.54d}$$

On the other hand, the voltages associated to the forward V^+ and backward V^- traveling waves must be proportional to the complex amplitudes according to[24]

$$V^+ = Ca^+ \tag{2.55a}$$

$$V^- = Ca^- \tag{2.55b}$$

Therefore, using (2.54) and (2.55), the S-parameters can be expressed as follows:

$$S_{11} = \left.\frac{Ca^-(z=0)}{Ca^+(z=0)}\right|_{a^-(z=L)=0} = \left.\frac{a^-(z=0)}{a^+(z=0)}\right|_{a^-(z=L)=0} \tag{2.56a}$$

$$S_{21} = \left.\frac{Ca^+(z=L)}{Ca^+(z=0)}\right|_{a^-(z=L)=0} = \left.\frac{a^+(z=L)}{a^+(z=0)}\right|_{a^-(z=L)=0} \tag{2.56b}$$

$$S_{22} = \left.\frac{Ca^+(z=L)}{Ca^-(z=L)}\right|_{a^+(z=0)=0} = \left.\frac{a^+(z=L)}{a^-(z=L)}\right|_{a^+(z=0)=0} \tag{2.56c}$$

$$S_{12} = \left.\frac{Ca^-(z=0)}{Ca^-(z=L)}\right|_{a^+(z=0)=0} = \left.\frac{a^-(z=0)}{a^-(z=L)}\right|_{a^+(z=0)=0} \tag{2.56d}$$

Hence, by solving the coupled mode equations with the boundary conditions given by (2.52), the S-parameters can be inferred using (2.56).

[24] The proportionality between the voltages and complex amplitudes of the forward and backward travelling waves can be alternatively expressed as $V^+ = Ca^+$ and $V^- = -Ca^-$. The (–) or (+) sign in the second expression simply depends on the relations between the (*x*,*y*) dependent part of the fields of the forward and backward travelling waves. The (+) sign (i.e., 2.55) results when such relations are taken as $E_x^- = E_x^+$; $E_y^- = E_y^+$; $E_z^- = -E_z^+$ for the electric fields, and as $H_x^- = -H_x^+$; $H_y^- = -H_y^+$; $H_z^- = H_z^+$ for the magnetic fields, following [11, 15]. The (–) sign is obtained if the fields are chosen to satisfy $E_x^- = -E_x^+$; $E_y^- = -E_y^+$; $E_z^- = E_z^+$ and $H_x^- = H_x^+$; $H_y^- = H_y^+$; $H_z^- = -H_z^+$, following [13, 14]. Notice that if the (–) sign is chosen in (2.55b), a (–) sign would appear in (2.56a) and (2.56c), with the result of a more unfamiliar expression for S_{11} and S_{22}. However, this change in the relation between the fields of the forward and backward travelling waves implies also a change in the sign of the coupling coefficient *K*. Notice that the change of sign in *K* and a^- (necessary to leave the S-parameters unaltered) keeps invariant the coupled mode equations (2.48).

2.4.3 Approximate Analytical Solutions of the Coupled Mode Equations

Solving the system of coupled mode equations (2.48) requires in general a numerical method. However, by considering a pair of additional approximations (valid in most cases of practical interest), it is possible to further simplify the equations in such a way that analytical solutions can be obtained, and, from them, relevant parameters of EBG-based planar transmission lines[25] can be inferred. Such needed approximations are [14]:

- The z dependence of the phase constant β (or ε_{re}) is neglected. This means that an averaged value of β (and ε_{re}) must be chosen. As justified in Appendix F, the most adequate value is:

$$\varepsilon_{re} = \left(\frac{1}{l} \int_0^l \sqrt{\varepsilon_{re}(z)} \cdot dz \right)^2 \tag{2.57}$$

  where l is the period of the EBG-based structure.
- The analysis is restricted to a range of frequencies in the vicinity of the central frequency of a certain (nth) rejected band.[26] This approximation is necessary in order to express the coupled mode equations in terms of the nth coefficient of the Fourier expansion of the coupling coefficient, $K(z)$, which is a periodic function of z if the nonuniformity of the transmission line is also periodic.[27]

Let us now see how the previous approximations notably simplify the coupled mode equations. Due to periodicity, the coupling coefficient can be expanded in a Fourier series as follows:

$$K(z) = \sum_{n=-\infty}^{n=+\infty} K_n e^{j\frac{2\pi n}{l} z} \tag{2.58}$$

which in turn can be expressed as follows:

$$K(z) = \sum_{n=1}^{n=+\infty} |2K_n| \cos\left(\frac{2\pi n}{l} z + \arg(K_n) \right) \tag{2.59}$$

[25] As already mentioned in the introduction of this chapter, by EBG-based planar transmission lines, we mean nonuniform periodic lines with either modulation of the conductor strip or ground plane etching.

[26] In general, periodic EBG-based transmission lines exhibit multiple rejection bands, as results from the Floquet analysis of Section 2.2.

[27] Notice that the period of $K(z)$ is also l.

since $K(z)$ is a real function (i.e., $K_{-n} = K_n^* = |K_n| \cdot e^{-j\arg(Kn)}$), and $K_0 = 0$ (the mean value of the coupling coefficient of a periodic structure is null). Let us now express the complex amplitudes as a function of the averaged β as follows [14]:

$$a^+(z) = A^+(z) \cdot e^{-j\beta z} \tag{2.60a}$$

$$a^-(z) = A^-(z) \cdot e^{+j\beta z} \tag{2.60b}$$

Using these variables, the coupled mode equations can be written as

$$\frac{dA^+}{dz} = K(z) \cdot A^- \cdot e^{+j2\beta z} \tag{2.61a}$$

$$\frac{dA^-}{dz} = K(z) \cdot A^+ \cdot e^{-j2\beta z} \tag{2.61b}$$

and introducing (2.58) in (2.61), the following equations result:

$$\frac{dA^+}{dz} = \sum_{n=1}^{\infty} \left(K_n \cdot e^{+j2\left(\beta + \frac{n\pi}{l}\right)z} + K_{-n} \cdot e^{+j2\left(\beta - \frac{n\pi}{l}\right)z} \right) \cdot A^- \tag{2.62a}$$

$$\frac{dA^-}{dz} = \sum_{n=1}^{\infty} \left(K_n \cdot e^{-j2\left(\beta - \frac{n\pi}{l}\right)z} + K_{-n} \cdot e^{-j2\left(\beta + \frac{n\pi}{l}\right)z} \right) \cdot A^+ \tag{2.62b}$$

For the solution of Equations 2.62 over the whole frequency spectrum, it is necessary to keep all the coefficients, K_n, of the series expansion of $K(z)$, unless some of them are negligible or null.[28] However, for frequencies in the vicinity of those frequencies satisfying

$$\beta = \frac{n\pi}{l} \tag{2.63}$$

only one term in the summations of (2.62) is significant; namely, the term K_{-n} in (2.62a), and the term K_n in (2.62b). The reason is that the corresponding exponential has an argument close to zero for these terms, giving a significant contribution to the sum. For the other terms, the exponentials are rapidly varying functions of z, and when these terms, multiplied by A^-, or A^+ (which are slowly varying functions of z), are integrated over several periods, their contribution is negligible. From a physical point

[28] Notice, for instance, that for a sinusoidal variation of $K(z)$, all the coefficients of the series are null, except $K_{\pm 1}$.

of view, this means that at those frequencies satisfying (2.63), a strong coupling (and hence an important transfer of energy) between the forward and backward traveling waves is expected. This energy transfer gives rise to significant reflection of the injected power, with the result of stop bands in the frequency response, caused by the perturbation. Notice that condition (2.63), usually known as phase matching, or resonant coupling, is equivalent to expression (2.9). Therefore, the coupling equations (2.62) provide the mathematical explanation for the presence of band gaps in artificial transmission lines based on periodic structures. Notice that at frequencies far enough from those frequencies satisfying phase matching, all the terms in the right-hand side of (2.62) experiment quick variations with z. When such terms are multiplied by A^- (in 2.62a) or A^+ (in 2.62b), and are integrated over a distance much larger than the period l, their contributions tend to vanish, and A^- and A^+ tend to be constant. Indeed, if the boundary conditions (2.52) are applied, and the considered frequencies are within the pass band regions, we expect that the solutions of (2.62) can be approximated by $A^+(z) = 1$ and $A^-(z) = 0$ (or $a^+(z) = e^{-j\beta z}$ and $a^-(z) = 0$), corresponding to roughly total transmission between the input and output ports. Notice that if $A^-(z) = 0$, it directly results from (2.62a) that $A^+(z)$ is constant with z, and (2.62b) gives a negligible value of $A^-(z)$ if the number of periods of the structure is large enough, this being consistent with our previous prediction.

According to the previous analysis, for the frequency range in the vicinity of the nth rejected frequency band; and for a periodic structure with total length satisfying $L >> l$, the coupled mode equations can be simplified to

$$\frac{dA^+}{dz} = K_n^* \cdot A^- \cdot e^{+j2\Delta\beta z} \tag{2.64a}$$

$$\frac{dA^-}{dz} = K_n \cdot A^+ \cdot e^{-j2\Delta\beta z} \tag{2.64b}$$

where

$$\Delta\beta = \beta - \frac{n\pi}{l} \tag{2.65}$$

$\Delta\beta$ being null at the frequency satisfying the phase matching condition. To solve (2.64), let us express the weighting factors of the coupling coefficient as

$$K_n = jk \tag{2.66}$$

The coupling equations can thus be written as

$$\frac{dA^+}{dz} = -jk^* \cdot A^- \cdot e^{+j2\Delta\beta z} \tag{2.67a}$$

$$\frac{dA^-}{dz} = jk \cdot A^+ \cdot e^{-j2\Delta\beta z} \tag{2.67b}$$

In order to solve the previous equations analytically, the following new variables are introduced [14, 20]:

$$R(z) = A^+(z) \cdot e^{-j\Delta\beta z} = a^+(z) e^{j\frac{n\pi}{T}z} \tag{2.68a}$$

$$S(z) = A^-(z) \cdot e^{j\Delta\beta z} = a^-(z) e^{-j\frac{n\pi}{T}z} \tag{2.68b}$$

and the coupled mode equations can be written as follows:

$$\frac{dR}{dz} = -j\Delta\beta \cdot R - jk^* \cdot S \tag{2.69a}$$

$$\frac{dS}{dz} = j\Delta\beta \cdot S + jk \cdot R \tag{2.69b}$$

The general solutions of (2.69) are of the form[29]:

$$R(z) = R^+ \cdot e^{-\gamma z} + R^- \cdot e^{\gamma z} \tag{2.70a}$$

$$S(z) = S^+ \cdot e^{-\gamma z} + S^- \cdot e^{\gamma z} \tag{2.70b}$$

where

$$\gamma = +\sqrt{|k|^2 - (\Delta\beta)^2} \tag{2.71}$$

Let us now introduce boundary conditions compatible with (2.52), namely, a matched output port. This means that $a^-(z = L) = S(z = L) = 0$, and, using (2.70b), it follows that $S^+ = -S^- \cdot e^{2\gamma L}$. Therefore, $S(z)$ can be expressed as

$$S(z) = S^+ \cdot e^{-\gamma z} \left(1 - e^{-2\gamma(L-z)}\right) \tag{2.72}$$

[29] The calculation of the general solutions of (2.69) is not straightforward. To demonstrate that (2.70), with γ given by (2.71), are the general solutions of (2.69), we first introduce (2.70) in (2.69). This gives two equations where all the terms are multiplied either by $e^{\gamma z}$ or $e^{-\gamma z}$. By equating the corresponding weights of these exponentials, four linear equations in the variables S^+, S^-, R^+ and R^-, which can be expressed in matrix form, are obtained. For a non-trivial solution, it is necessary that the associated determinant is null, and this occurs if (2.71) is satisfied.

and introducing (2.72) in (2.69b), $R(z)$ is found to be

$$R(z) = -\frac{S^+ e^{-\gamma z}}{jk} \cdot \left\{ \gamma \left(1 + e^{-2\gamma(L-z)} \right) + j\Delta\beta \left(1 - e^{-2\gamma(L-z)} \right) \right\} \tag{2.73}$$

Assuming an incident wave at the input port $R(z=0) = R_o$, the integration constant S^+ can be obtained, and expressions (2.72) and (2.73), after some tedious calculations, are found to be

$$R(z) = -\frac{R_o \{ \Delta\beta \sinh(\gamma(z-L)) + j\gamma \cosh(\gamma(z-L)) \}}{\Delta\beta \sinh(\gamma L) - j\gamma \cosh(\gamma L)} \tag{2.74a}$$

$$S(z) = \frac{R_o k \sinh(\gamma(z-L))}{\Delta\beta \sinh(\gamma L) - j\gamma \cosh(\gamma L)} \tag{2.74b}$$

Finally, from (2.68), the solutions of the coupled mode equations (2.47a and 2.47b) under the considered approximations are

$$a^+(z) = -\frac{\{ \Delta\beta \sinh(\gamma(z-L)) + j\gamma \cosh(\gamma(z-L)) \}}{\Delta\beta \sinh(\gamma L) - j\gamma \cosh(\gamma L)} a^+(0) \cdot e^{-j\frac{n\pi}{T} z} \tag{2.75a}$$

$$a^-(z) = \frac{k \sinh(\gamma(z-L))}{\Delta\beta \sinh(\gamma L) - j\gamma \cosh(\gamma L)} a^+(0) \cdot e^{j\frac{n\pi}{T} z} \tag{2.75b}$$

Notice that the above expressions are the approximate solutions of the coupled equations for a periodic EBG structure with length L, with input and output ports placed at $z=0$ and $z=L$, respectively, and excited by an incident wave $a^+(z=0) = a^+(0)$, and with the output port matched, that is, $a^-(z=L) = 0$. Once the approximate solutions of the complex amplitudes $a^+(z)$ and $a^-(z)$ are known, the S-parameters can be inferred using (2.56). This gives

$$S_{11} = \left. \frac{a^-(z=0)}{a^+(z=0)} \right|_{a^-(z=L)=0} = \frac{k \sinh(\gamma L)}{-\Delta\beta \sinh(\gamma L) + j\gamma \cosh(\gamma L)} \tag{2.76a}$$

$$S_{21} = \left. \frac{a^+(z=L)}{a^+(z=0)} \right|_{a^-(z=L)=0} = \frac{j\gamma \cdot e^{-j\frac{n\pi}{T} L}}{-\Delta\beta \sinh(\gamma L) + j\gamma \cosh(\gamma L)} \tag{2.76b}$$

In order to obtain S_{12} and S_{22}, it is necessary to modify the boundary conditions. In this case, the input port is matched, and hence $a^+(z=0) = R(z=0) = 0$. Therefore, using (2.70a), it follows that $R^+ = -R^-$, and $R(z)$ can be expressed as

$$R(z) = R^+(e^{-\gamma z} - e^{\gamma z}) = -2R^+ \sinh(\gamma z) \tag{2.77}$$

Introducing (2.77) in (2.69a), $S(z)$ is found to be

$$S(z) = \frac{2R^+}{jk^*} \cdot \{\gamma\cosh(\gamma z) + j\Delta\beta\sinh(\gamma z)\} \tag{2.78}$$

As proceeded before, in order to determine the integration constant R^+, it is necessary to set the incident wave at the output port to $S(z = L) = S_o$. Substituting the resulting value in (2.77) and (2.78), we obtain

$$R(z) = -\frac{S_o k^* \sinh(\gamma z)}{\Delta\beta\sinh(\gamma L) - j\gamma\cosh(\gamma L)} \tag{2.79a}$$

$$S(z) = \frac{S_o\{\Delta\beta\sinh(\gamma z) - j\gamma\cosh(\gamma z)\}}{\Delta\beta\sinh(\gamma L) - j\gamma\cosh(\gamma L)} \tag{2.79b}$$

and using (2.68),

$$a^+(z) = \frac{k^*\sinh(\gamma z)}{\Delta\beta\sinh(\gamma L) - j\gamma\cosh(\gamma L)} a^-(L) \cdot e^{-j\frac{n\pi}{l}(z+L)} \tag{2.80a}$$

$$a^-(z) = -\frac{\{\Delta\beta\sinh(\gamma z) - j\gamma\cosh(\gamma z)\}}{\Delta\beta\sinh(\gamma L) - j\gamma\cosh(\gamma L)} a^-(L) \cdot e^{+j\frac{n\pi}{l}(z-L)} \tag{2.80b}$$

where $a^-(L) = S_o e^{+jn\pi L/l}$. Finally, using (2.56), S_{22} and S_{12} are found to be

$$S_{22} = \left.\frac{a^+(z=L)}{a^-(z=L)}\right|_{a^+(z=0)=0} = \frac{k^*\sinh(\gamma L) \cdot e^{-j\frac{2n\pi}{l}L}}{-\Delta\beta\sinh(\gamma L) + j\gamma\cosh(\gamma L)} \tag{2.81a}$$

$$S_{12} = \left.\frac{a^-(z=0)}{a^-(z=L)}\right|_{a^+(z=0)=0} = \frac{j\gamma \cdot e^{-j\frac{n\pi}{l}L}}{-\Delta\beta\sinh(\gamma L) + j\gamma\cosh(\gamma L)} \tag{2.81b}$$

Notice that $S_{12} = S_{21}$, and the S-parameters satisfy the unitarity conditions, as expected for reciprocal and lossless networks.

Inspection of (2.75) and (2.80) indicates that the z-dependent part of the wave solutions is an exponential with phase constants given by [3, 14, 21]

$$\beta' = \frac{n\pi}{l} \pm j\gamma = \frac{n\pi}{l} \pm j\sqrt{|k|^2 - (\Delta\beta)^2} \tag{2.82a}$$

for the forward traveling wave, $a^+(z)$, and

$$\beta' = -\frac{n\pi}{l} \pm j\gamma = -\frac{n\pi}{l} \pm j\sqrt{|k|^2 - (\Delta\beta)^2} \tag{2.82b}$$

for the backward traveling wave, $a^-(z)$. Therefore β' is the approximate phase constant of the Bloch waves of the periodic structure. Notice that for $|\Delta\beta| < |k|$, the phase constant has an imaginary part. This region is the forbidden band (bandgap) of the periodic structure. On the contrary, far enough from the phase matching condition ($|\Delta\beta| > |k|$), β' is real and the structure is transparent. In the limit of small perturbations ($k \cong 0$), and for sufficiently large values of $|\Delta\beta|$ (i.e., $|\Delta\beta| >> |k|$), the phase constant can be approximated by

$$\beta' = \frac{n\pi}{l} \pm \Delta\beta\sqrt{1 - \frac{|k|^2}{(\Delta\beta)^2}} \approx \frac{n\pi}{l} \pm \Delta\beta = \frac{n\pi}{l} \pm \left(\beta - \frac{n\pi}{l}\right) = \beta, \text{ or } -\beta + \frac{2n\pi}{l} \tag{2.83a}$$

or

$$\beta' \approx -\frac{n\pi}{l} \pm \Delta\beta = -\frac{n\pi}{l} \pm \left(\beta - \frac{n\pi}{l}\right) = -\beta, \text{ or } \beta - \frac{2n\pi}{l} \tag{2.83b}$$

which is coherent with (2.6). Notice, however, that in (2.83) there are only four terms: the fundamental (forward or backward traveling), and those space harmonics with index $\pm n$. This is expected since this analysis is valid for frequencies in the vicinity of the nth forbidden band.

In order to obtain a more accurate solution for the phase constant of the fundamental space harmonic (notice that 2.82 gives an approximate solution), it is necessary to calculate the complex amplitudes $a^+(z)$ and $a^-(z)$ numerically, and, from them, the S-parameters. Using the transformations (C.22), the *ABCD* matrix can be inferred, and the dispersion relation can be finally derived from (2.22).

2.4.4 Analytical Expressions for Relevant Parameters of EBG Periodic Structures

Apart from the S-parameters, there are other relevant parameters of interest in EBG one-dimensional periodic structures such as the frequency of maximum reflectivity, the value of this maximum reflectivity, and the bandwidth of the frequency band where significant reflection is achieved. The purpose of this subsection is to provide analytical expressions for these parameters [14, 16, 22].

The frequency of maximum reflectivity, $f_{\max}$, is that frequency satisfying $\Delta\beta = 0$, or $\beta_{\max} = n\pi/l$. Therefore,

$$f_{\max} = n\frac{c}{2l\sqrt{\varepsilon_{\rm re}}} \tag{2.84}$$

where c is the speed of light in vacuum. Notice that if $\Delta\beta = 0$, then $\gamma = |k|$, and the value of maximum reflectivity (from 2.76a and 2.81a) is

$$|S_{11}|_{\max} = |S_{22}|_{\max} = \tanh(|k|L) \tag{2.85}$$

Notice that $|S_{11}|_{\max} = |S_{22}|_{\max} \rightarrow 1$ (i.e., 0 dB) as $|k|L \rightarrow \infty$. From 2.76b and 2.81b, the maximum attenuation (minimum transmission) is found to be

$$|S_{21}|_{\min} = |S_{12}|_{\min} = \text{sech}(|k|L) \tag{2.86}$$

It is important at this point to analyze the dependence of the maximum reflectivity, or minimum transmission, with the parameters of the EBG periodic structure. Notice that, as expected, the value of maximum reflectivity depends on the magnitude of the periodic perturbation through $|k|$, which is the modulus of the considered K_n term of the Fourier series expansion of the coupling coefficient (see expressions 2.58 and 2.59). The larger the periodic perturbation, the higher the derivative of the characteristic impedance and hence the amplitude of the coupling coefficient given by (2.50). Thus, increasing the magnitude of the perturbation has the effect of increasing the terms of the series expansion of $K(z)$. On the other hand, an increase in device length, L, means that the structure includes an increasing number of periods, with the effect of more power being transferred to the backward traveling waves and hence more reflection.

Concerning the reflected bandwidth, it is reasonable to consider that it is given by the frequency range between the first reflection zeros around the maximum reflectivity. At first sight, one may erroneously deduce from (2.76a) and (2.81a) that such pair of reflection zeros are obtained by forcing $\gamma = 0$. However, notice that by applying the l'Hôpital's rule, S_{11} and S_{22} tend to a finite value in the limit when $\gamma \rightarrow 0$. Indeed, the frequency corresponding to the first minimum of reflection satisfies

$$\gamma L = \text{j}\pi \tag{2.87}$$

or

$$|k|^2 + \frac{\pi^2}{L^2} = (\Delta\beta)^2 \tag{2.88}$$

and assuming that $\varepsilon_{\rm re}$ does not depend on frequency in the rejected bandwidth (i.e., neglecting dispersion), the bandwidth is found to be

$$\mathrm{BW} \doteq \frac{c|k|}{\pi\sqrt{\varepsilon_{\mathrm{re}}}}\sqrt{1+\left(\frac{\pi}{|k|L}\right)^2} \tag{2.89}$$

2.4.5 Relation between the Coupling Coefficient and the S-Parameters

As discussed in Section 2.4.3, as long as the coefficients of the Fourier series expansion of the coupling coefficient, K_n, are different than zero, stop bands in the vicinity of those frequencies satisfying the phase matching condition (2.63) are expected in EBG-based periodic transmission lines. In other words, the number of reflection bands is intimately related to the harmonic content of the coupling coefficient $K(z)$. For instance, it is expected that for a sinusoidal coupling coefficient, a single reflection band, centered at the frequency given by (2.84) with $n = 1$, appears. Indeed, under conditions of low reflectivity, the relation between S_{11} and $K(z)$ can be easily approximated by the Fourier transform. To demonstrate this, let us express the generalized reflection coefficient in an EBG structure as

$$\rho(z) = \frac{a^-(z)}{a^+(z)} \tag{2.90}$$

and let us consider the coupled mode equations (2.47a and 2.47b), where the only approximation applied is single-mode operation. Using (2.47), the derivative of ρ with z can be expressed as [14]:

$$\frac{d\rho(z)}{dz} = 2j\cdot\beta\cdot\rho(z) + K(z)\left(1-\rho(z)^2\right) \tag{2.91}$$

which is a nonlinear equation that does not have a known general solution. However, if it is assumed that β is not a function of z, and $|\rho(z)|^2 \ll 1$ (low reflectivity conditions), (2.91) can be expressed as [23]

$$\frac{d\rho(z)}{dz}\cdot e^{-2j\beta z} = 2j\cdot\beta\cdot\rho(z)\cdot e^{-2j\beta z} + K(z)\cdot e^{-2j\beta z} \tag{2.92}$$

where the equation has been multiplied by $e^{-2j\beta z}$. The previous equation can be simplified to

$$\frac{d}{dz}\left(\rho(z)\cdot e^{-2j\beta z}\right) = K(z)\cdot e^{-2j\beta z} \tag{2.93}$$

and integrating both sides between $z = 0$ and $z = L$, the following result is obtained:

$$\rho(L)\cdot e^{-2j\beta L} - \rho(0) = \int_0^L K(z)\cdot e^{-2j\beta z}\cdot dz \tag{2.94}$$

If we now assume that the output port is matched, then $\rho(L) = 0$, and $\rho(0) \equiv S_{11}$. Therefore,

$$S_{11} = -\int_0^L K(z)\cdot e^{-2j\beta z}\cdot dz \tag{2.95}$$

and (2.95) resembles the Fourier transform of $K(z)$. If we assume that $K(z) = 0$ for $z < 0$ and for $z > L$ (device with length L, between $z = 0$ and $z = L$), (2.95) can be rewritten as follows:

$$S_{11} = -\int_{-\infty}^{\infty} K(z)\cdot e^{-j2\pi\left(\frac{\beta}{\pi}\right)z}\cdot dz \tag{2.96}$$

and (2.96) is the Fourier transform of $K(z)$ in the variable β/π, that is,

$$S_{11}\left(\frac{\beta}{\pi}\right) = -\mathbf{F}(K(z)) \tag{2.97}$$

Although $K(z)$ is zero outside the interval $[0,L]$, in a first-order approximation we can introduce the series expansion (2.58) of $K(z)$ in (2.96). According to this, the reflection coefficient S_{11} will be significant in the vicinity of those frequencies satisfying

$$\frac{n}{l} = \frac{\beta}{\pi} \tag{2.98}$$

which is identical to the phase matching condition (2.63). Thus, in summary, the harmonic content of the coupling coefficient determines the frequency behavior of the reflection coefficient, and gives the number of rejected bands. This is a fundamental property that can be helpful in certain applications. However, the exact determination of S_{11} from (2.97) in situations where high reflectivity is required (as occurs in most practical applications) is not possible.

2.4.6 Using the Approximate Solutions of the Coupled Mode Equations

The aim of this subsection is to provide two examples to illustrate the use of the approximate analytical solutions of the coupled mode equations and to compare the results with full-wave EM simulations and measurements of the considered structures (further results are given in Ref. [22]). Let us first consider a periodic microstrip transmission line with a coupling coefficient $K(z)$ exhibiting a sinusoidal dependence on z. A rejection band in the vicinity of that frequency satisfying the phase matching condition (i.e., 2.63 with $n = 1$) is expected. However, according to the preceding theory, since $K(z)$ does not have harmonic content, further rejection bands will not appear. In order to determine the variation of the characteristic impedance with the axial position, let us integrate (2.50) with z (the dependence on ω is considered to be negligible). This gives

$$-2\int_0^z K(z')\cdot dz' = \int_{Z_o(0)}^{Z_o(z)} \frac{1}{Z_o'}\cdot dZ_o' \tag{2.99}$$

where integration is performed between $z = 0$ and an arbitrary point, z, within the periodic microstrip line. Solution of (2.99) gives

$$Z_o(z) = Z_o(0)\cdot e^{-2\int_0^z K(z')\cdot dz'} \tag{2.100}$$

Let us now assume that the periodic structure exhibits a continuous variation of the characteristic impedance at the extremes of the device ($z = 0$ and $z = L$)—the characteristic impedance at these extremes being $Z_o(0) = Z_o(L) = Z_{o,ref}$, that is, the reference impedance of the ports and access lines. Moreover, let us consider that the impedance in the periodic region (interval $[0,L]$) experiences variations around $Z_{o,ref}$ and starts increasing from the input port. This forces the coupling coefficient to be expressed as

$$K(z) = -2|K_1|\cdot\cos\left(\frac{2\pi}{l}z\right) = 2|K_1|\cdot\cos\left(\frac{2\pi}{l}z + \pi\right) \tag{2.101}$$

where K_1 is the first term of the Fourier series expansion of $K(z)$ and $\arg(K_1) = \pi$ (the other terms are null). Introducing (2.101) in (2.100), the characteristic impedance is found to be

$$Z_o(z) = Z_{o,ref}\cdot e^{\frac{2|K_1|l}{\pi}\sin\left(\frac{2\pi}{l}z\right)} \tag{2.102}$$

For the present example, the period of the EBG structure and the reference impedance of the ports are set to $l = 1$ cm and $Z_{o,ref} = 50\ \Omega$, respectively, and 10 unit cells are considered, that is, $L = 10l = 10$ cm. In order to avoid extreme characteristic impedances along the line, the coupling coefficient amplitude is set to $|K_1| = 0.64\ \text{cm}^{-1}$. This gives maximum and minimum characteristic impedances of $Z_{o,max} = 1.50Z_{o,ref} = 75\ \Omega$ and $Z_{o,min} = 0.66Z_{o,ref} = 33.3\ \Omega$, respectively. The considered substrate parameters are those of the *Rogers RO3010* substrate, with thickness $h = 1.27$ mm and dielectric constant $\varepsilon_r = 10.2$. In the present example, the periodic perturbation of the characteristic impedance of the line is carried out by varying the strip width,

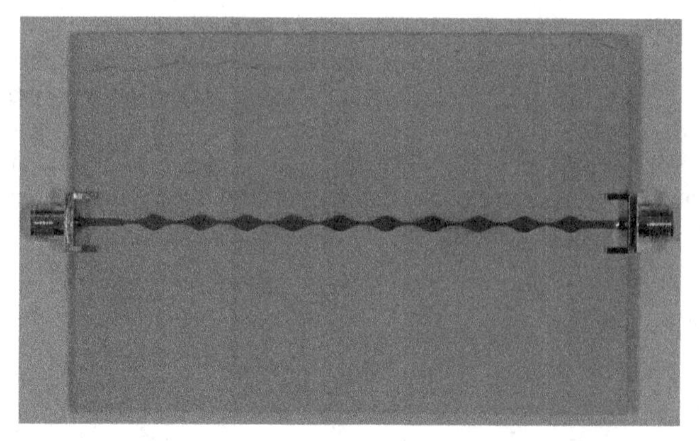

FIGURE 2.10 Photograph of the EBG microstrip line with sinusoidal coupling coefficient.

W, along the line.[30] In practice, the width of the line as a function of the characteristic impedance can be obtained by means of a transmission line calculator. A simple procedure is to calculate the line width at discrete points along the period, and then approximate $W(z)$ by a piecewise function. The photograph of the resulting microstrip EBG structure, including access lines, is depicted in Figure 2.10. The reflection (S_{11}) and transmission (S_{21}) coefficients of the structure inferred from full-wave EM simulation and measurement are compared with those given by expressions (2.76) in Figure 2.11. The analytical solutions (2.76) have been obtained by considering $n = 1$, $k = -jK_1 = j0.64\ \text{cm}^{-1}$, $l = 1$ cm, and β given by

$$\beta = \frac{2\pi f}{c}\sqrt{\varepsilon_{re}} \tag{2.103}$$

with ε_{re} inferred from (2.57). The agreement between the predictions of analytical solutions of the coupled mode theory and the full-wave simulation and experimental results is remarkable. The frequency of maximum attenuation, the value of maximum reflectivity, the value of minimum transmission, and the bandwidth, obtained from expressions (2.84–2.86), and (2.89), respectively, are found to be $f_{max} = 5.66$ GHz, $|S_{11}|_{max} \sim 1$ (i.e., ~0 dB), $|S_{21}|_{min} = 3.32 \times 10^{-3}$ (i.e., −49.6 dB) and BW = 2.55 GHz. As expected, these values are in good agreement with those inferred from the frequency response depicted in Figure 2.11.

It is clear in view of Figure 2.11 that the frequency response of the designed microstrip EBG structure exhibits a single reflection band around f_{max}. No further forbidden bands are present, as one expects from the lack of harmonic content of the considered coupling coefficient, $K(z)$. Conversely, by considering a coupling coefficient given by a square-shaped periodic function of the form indicated in Figure 2.12, forbidden bands centered at the fundamental frequency and its odd-order harmonics are expected, as derived from the harmonic content of such coupling coefficient. From

[30] The characteristic impedance in microstrip transmission lines can also be modulated through ground plane etching, as will be later shown.

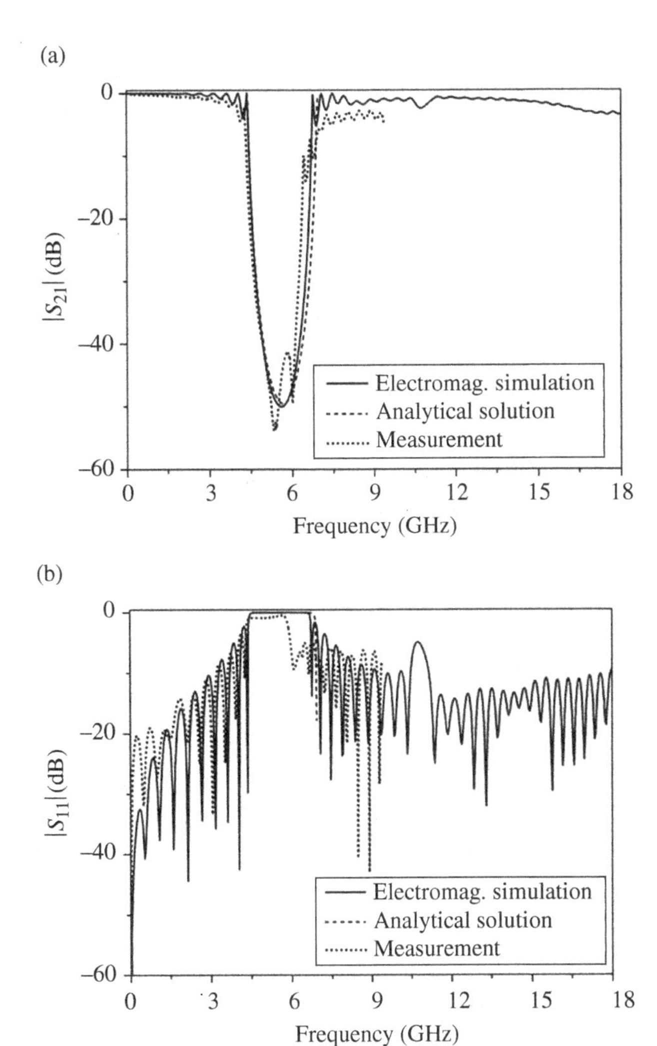

FIGURE 2.11 Frequency response of the EBG structure depicted in Figure 2.10. (a) Insertion loss and (b) return loss. The measurement is depicted up to 10 GHz. The EM simulations have been obtained by excluding ohmic and dielectric losses.

(2.100), the characteristic impedance corresponding to the coupling coefficient of Figure 2.12 can be expressed as a piecewise function in one period, l, as follows:

$$Z_o(z) = Z_{o,\mathrm{ref}} \cdot e^{2Kz}, \qquad z \in \left[0, \frac{l}{4}\right] \tag{2.104a}$$

$$Z_o(z) = Z_{o,\mathrm{ref}} \cdot e^{-2K\left(z-\frac{l}{2}\right)}, \quad z \in \left[\frac{l}{4}, \frac{3l}{4}\right] \tag{2.104b}$$

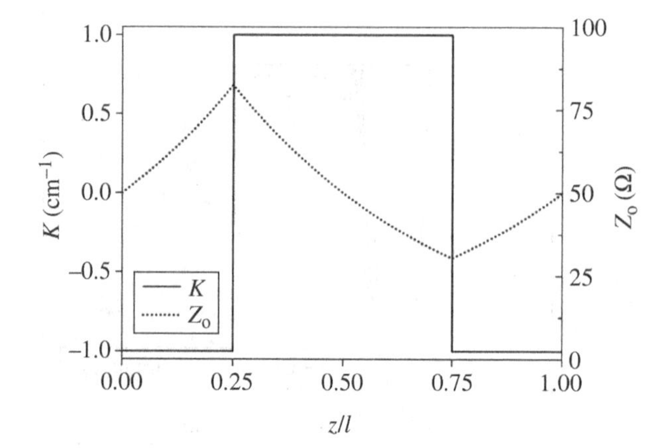

FIGURE 2.12 Dependence of the coupling coefficient and characteristic impedance with the axial position z, corresponding to an EBG microstrip line with a square shaped coupling coefficient.

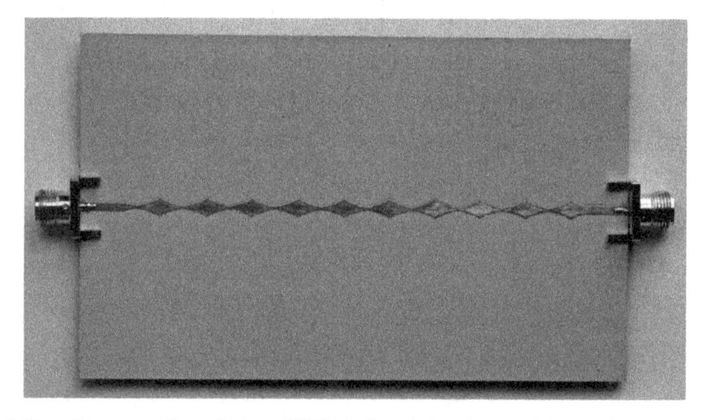

FIGURE 2.13 Photograph of the EBG microstrip line with square-shaped coupling coefficient.

$$Z_o(z) = Z_{o,\text{ref}} \cdot e^{2K(z-l)}, \quad z \in \left[\frac{3l}{4}, l\right] \tag{2.104c}$$

where K is the amplitude of the coupling coefficient (the characteristic impedance is also depicted in Fig. 2.12). The 10-cell microstrip EBG structure shown in Figure 2.13 exhibits a dependence of the characteristic impedance with the axial position given by (2.104), with $l = 1$ cm and $K = 1\ \text{cm}^{-1}$. With these parameter values, the characteristic impedance varies around the reference impedance with extreme values given by $Z_o(l/4) = Z_{o,\max} = 1.65Z_{o,\text{ref}} = 82.4\ \Omega$, and by $Z_o(3l/4) = Z_{o,\min} = 0.61Z_{o,\text{ref}} = 30.4\ \Omega$

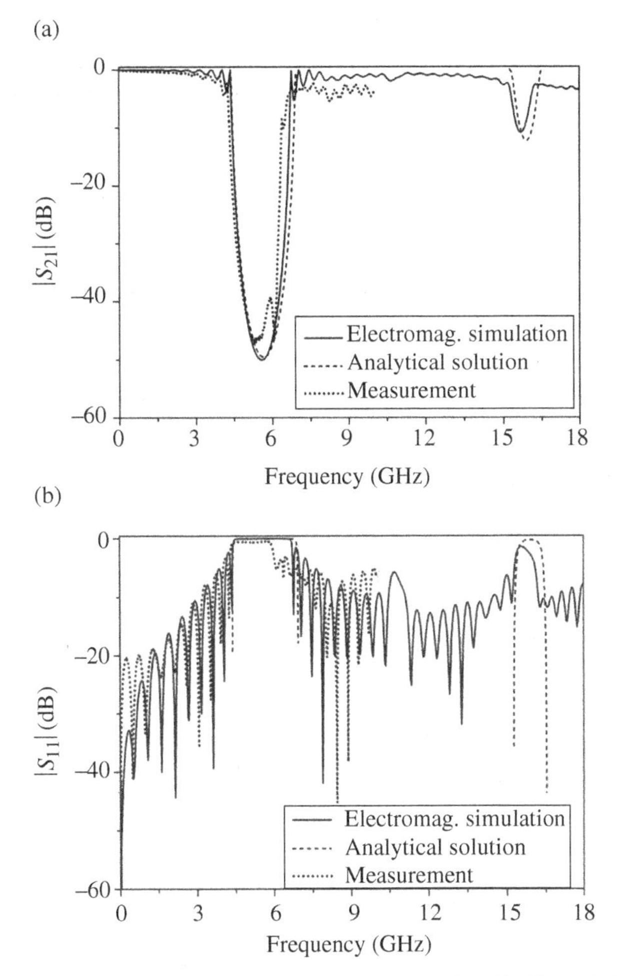

FIGURE 2.14 Frequency response of the EBG structure depicted in Figure 2.13. (a) Insertion loss and (b) return loss. The measurement is depicted up to 10 GHz. The EM simulations have been obtained by excluding ohmic and dielectric losses.

(the considered substrate is also the *Rogers RO3010* with thickness $h = 1.27$ mm and dielectric constant $\varepsilon_r = 10.2$). The broadband frequency response of the designed microstrip EBG structure, depicted in Figure 2.14, confirms that only the fundamental and odd harmonics are reflected, and the frequencies of maximum attenuation are given by (2.84) with n odd. The value of minimum transmission in each forbidden band can be obtained from (2.86), with k given by (2.66), where the terms of the coupling coefficient are given by

$$K_n = -\frac{2K}{n\pi}, \quad n = 1, 3, 5, \ldots \tag{2.105}$$

Thus, the minimum transmission at the first three forbidden bands, inferred from analytical expressions, is −49.3 dB, −12.5 dB, and −5.7 dB; these values are in good agreement with the frequency response of the device, confirming the validity of the theory.

2.5 APPLICATIONS

Several applications of periodic transmission lines, including nonuniform transmission lines and reactively loaded lines, are reviewed in this section.

2.5.1 Applications of Periodic Nonuniform Transmission Lines[31]

As it has been demonstrated along this chapter, periodic transmission lines with conductor strip or ground plane modulation exhibit stop bands and pass bands that can be useful in various applications. These applications include reflectors, high-Q resonators, spurious suppression in planar microwave filters, and harmonic suppression in active circuits, among others. In the next subsections, such applications are discussed and some examples are reported. Moreover, it will be shown that quasi-periodic nonuniform transmission lines are useful to implement chirped delay lines (CDLs).

2.5.1.1 Reflectors Band gap structures were first applied to the optical domain as photonic bandgaps (PBGs) [24], and later used for the implementation of reflectors at microwave frequencies in microstrip technology [18, 25]. The first EBG structures in microstrip technology were indeed implemented by drilling a two-dimensional periodic pattern of holes through the substrate [25, 26]. A simpler strategy to inhibit signal propagation at controllable frequencies consists of partially etching the ground plane as depicted in Figure 2.15 [18]. As reported in Ref. [18], this structure exhibits wider and deeper stopbands than previous designs using the dielectric hole approach [25]. However, the effects of the perturbation are negligible beyond the defect row below the conductor strip [19]. This suggests that EBG reflectors consisting of only the central row of the pattern (one-dimensional EBG) will have similar behavior as a two-dimensional reflector, with the advantage of a considerable reduction in the transverse dimension. Figure 2.16a depicts a one-dimensional and a two-dimensional reflector with a period of $l = 18.9$ mm, giving a Bragg frequency of 3 GHz in the considered substrate (the *Rogers RO3010*, with dielectric constant $\varepsilon_r = 10.2$ and thickness $h = 1.27$ mm). The ratio between the radius of the circles and the period is 0.25. Figure 2.16b indicates that the frequency responses of both structures are almost undistinguishable. Hence, the one-dimensional EBG transmission line suffices to achieve strong reflection at the operating frequency.

[31] This subsection has been co-authored with Txema Lopetegi (Public University of Navarre, Spain).

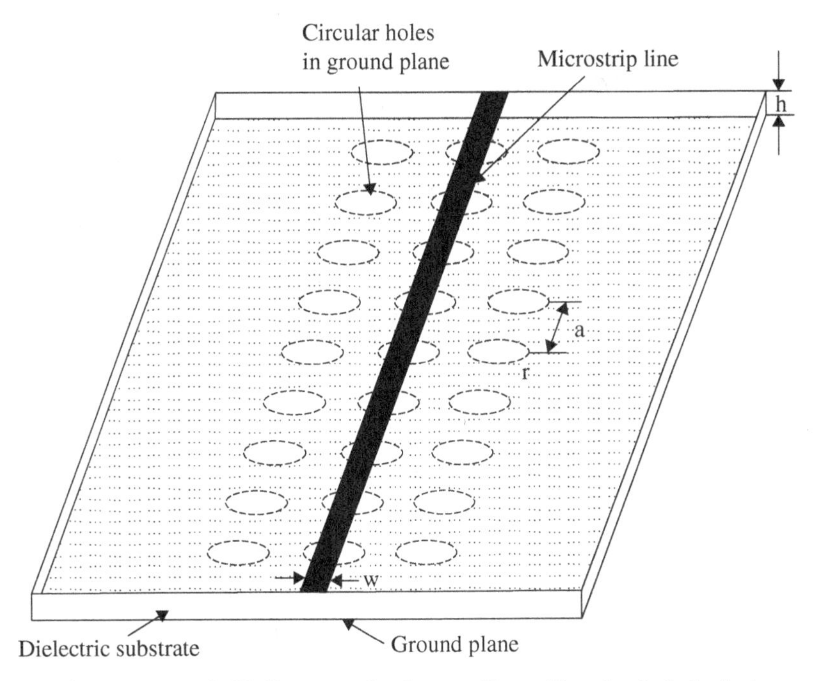

FIGURE 2.15 Microstrip EBG structure implemented by etching circular holes in the ground plane. Reprinted with permission from Ref. [18]; copyright 1998 IEEE.

In order to reduce the ripple in the pass bands, inherent to periodic structures, a tapering technique, consisting of a modification of the radius of each circle according to the position in the structure, can be applied [27]. The improvement of the frequency response achieved by means of the tapering function is related to the progressive matching of the characteristic impedance of the Bloch wave toward the input and output characteristic impedance that the tapering function produces. It is also possible to enhance the bandwidth of EBG microstrip lines by using chirping techniques, formerly used in fiber Bragg gratings [28]. The idea is to linearly distribute the position of the etched holes (center-to-center distance of adjacent holes) along the line [29]. This modulation of the period has the effect of varying the Bragg frequency along the line, with the result of an enhanced bandwidth. The effects of tapering and chirping can be appreciated in Figure 2.17, where the frequency responses of a nonchirped, chirped, and tapered and chirped EBG microstrip lines are compared. Chirping the lines clearly enhances the rejected bandwidth, whereas the pass band ripple level is significantly reduced by tapering the structure.

If a strong rejection level and sharp cutoff is required, the number of periods of the structure has to be necessarily large. To avoid an excessive aspect ratio (length versus width) of the one-dimensional EBG microstrip line, it is possible to bend the line as depicted in Figure 2.18 [30]. The tapering and chirping techniques can be applied to these bended EBG structures as well, in order to enhance the reflection bandwidth and reduce the ripple level in the pass bands [14].

(a)

(b)

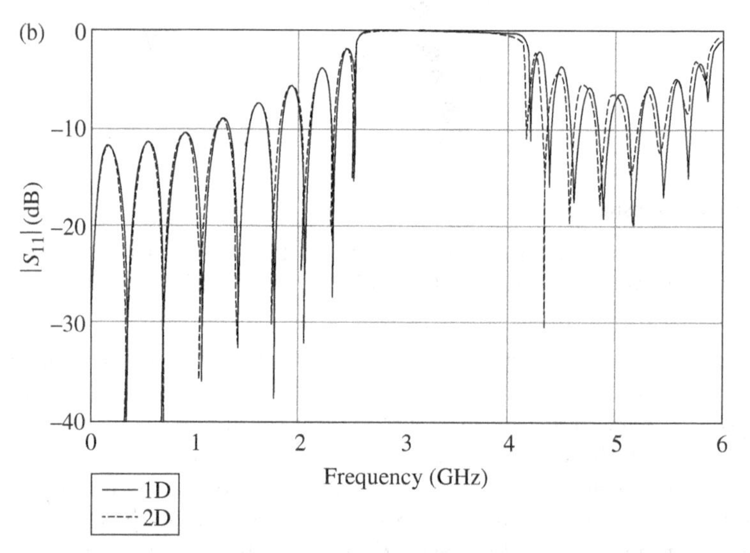

FIGURE 2.16 Two-dimensional and one-dimensional EBG microstrip line (a) and measured reflection coefficient (b). The line width is $W = 1.2$ mm corresponding to a 50 Ω line in the considered substrate. Reprinted with permission from [19]; copyright 1999 John Wiley.

It was shown in Section 2.4.5 that the harmonic content of the coupling coefficient $K(z)$ determines the number of rejected bands in the frequency response of EBG-based transmission lines, and this was verified in Section 2.4.6 by considering two different periodic structures. Specifically, it was demonstrated that a square-shaped coupling coefficient produces stop bands not only at the fundamental frequency but also at its odd harmonics. However, the rejection level decreases with the frequency index; and for certain applications, it may be of interest to achieve two, or even further, rejected bands centered at frequencies not necessarily being harmonics. To implement multiple-frequency-tuned EBG transmission lines, a solution is to add various sinusoidal functions tuned at the design frequencies, as reported in Ref. [31]. Specifically,

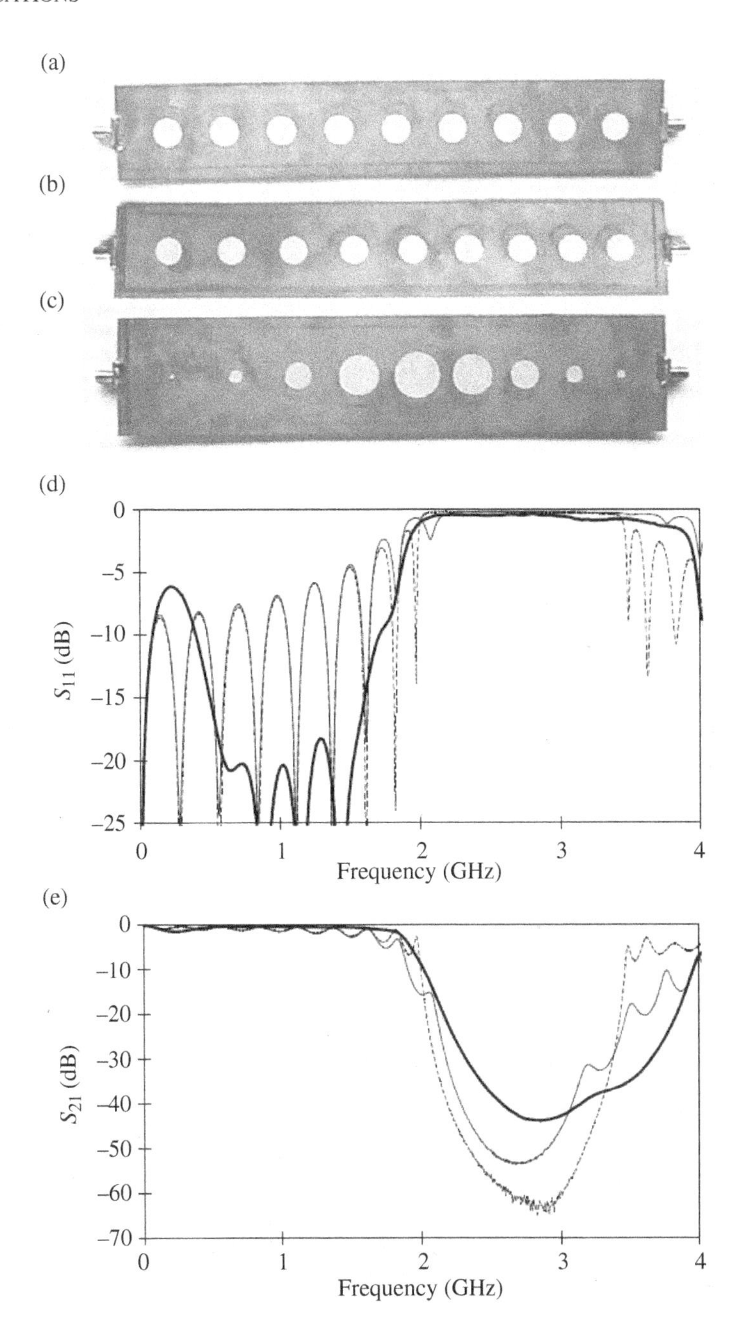

FIGURE 2.17 Photographs of a nonchirped (a), chirped (b), and tapered and chirped (c) EBG microstrip lines, measured return loss (d) and measured insertion loss (e). The traces are dashed line (nonchirped), thin solid line (chirped), and thick solid line (tapered and chirped). Reprinted with permission from Ref. [29]; copyright 2000 John Wiley.

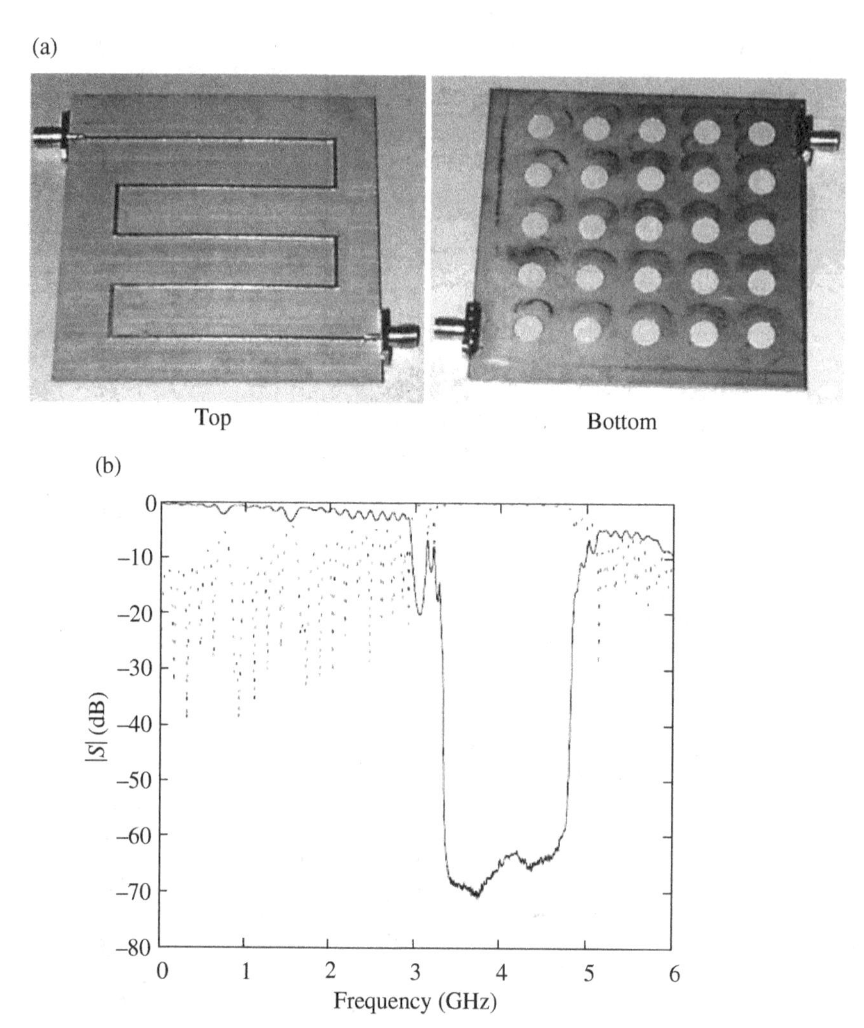

FIGURE 2.18 Photograph of a bended EBG microstrip line (a) and measured insertion (solid line) and return (dashed line) loss (b). The substrate is the *Rogers RO3010* with dielectric constant $\varepsilon_r = 10.2$ and thickness $h = 1.27$ mm. The period of the structure is $l = 15.5$ mm, the ratio between the radius and the period is 0.25 and the strip width is 1.2 mm (corresponding to a 50 Ω line). Reprinted with permission from Ref. [30]; copyright 1999 John Wiley.

the double- and triple-frequency-tuned EBG structures of Ref. [31] are implemented by adding two and three raised-sine functions, respectively. The double-tuned EBG has 9 periods of $l_1 = 23.9$mm and 14 periods of $l_2 = 14.8$ mm, in order to have the rejected frequencies around 3 GHz and 4.5 GHz, respectively. Both sine functions have an amplitude-to-period ratio of $t/l = 0.2$. The triple-frequency-tuned EBG microstrip structure is designed with three added raised-sine functions. The first sine has 9 periods of $l_1 = 23.9$ mm, the second one has 11 periods of $l_2 = 19.35$ mm, and the third one has 14 periods of $l_3 = 14.8$ mm, to obtain their stop bands centered at 3 GHz, 4 GHz,

(a)

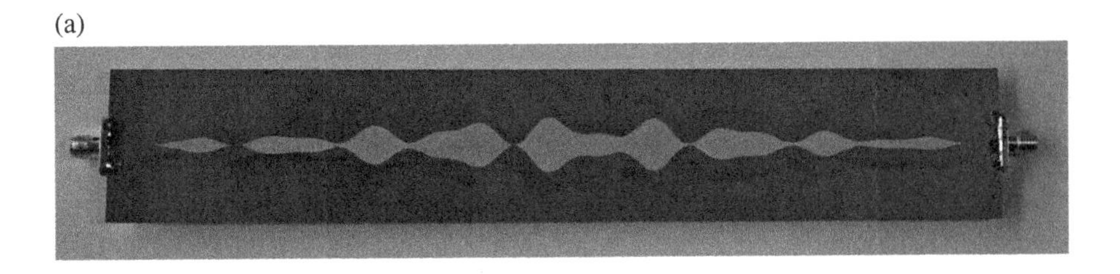

(b)

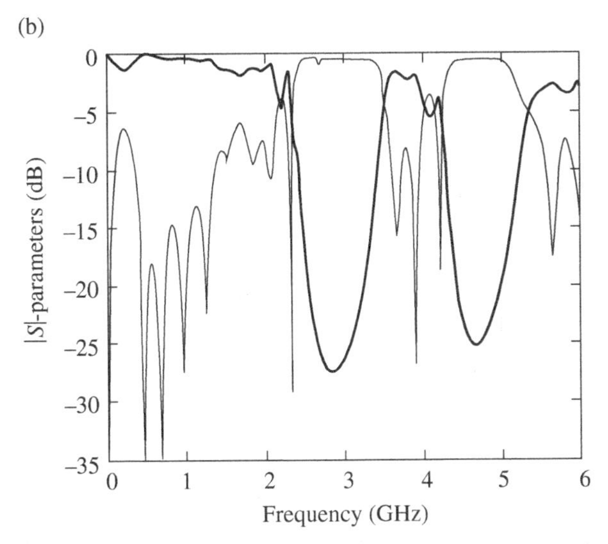

FIGURE 2.19 Photograph (ground plane) of a double-frequency-tuned EBG microstrip line (a) and measured insertion (thick line) and return (thin line) loss (b). The substrate is the *Rogers RO3010* with dielectric constant $\varepsilon_r = 10.2$ and thickness $h = 1.27$ mm. The conductor strip width is $W = 1.2$ mm. Reprinted with permission from Ref. [31]; copyright 2000 IEEE.

and 5 GHz, respectively. In all these cases, the t/l ratio is fixed to 0.15. The amplitude of each particular sine is limited in order to maintain the total amplitude of the ground plane perturbation in such values that radiation is kept in low levels. In both designs, the patterns resulting from the addition of various sine functions are Hamming-windowed to achieve a low-rippled response at the pass band [31]. These double- and triple-tuned EBG microstrip lines, including their frequency responses, are depicted in Figures 2.19 and 2.20, respectively.[32] The results of these figures clearly demonstrate that band gaps at closely spaced frequencies can be achieved. Notice that the patterned functions in the ground plane do not give a combination of pure sinusoidal functions for the coupling coefficient, and, hence, further stops bands at the harmonics of the design frequencies may appear.

[32] The raised sine functions determine the geometry of the slotted ground plane, rather than the coupling coefficient. Nevertheless, for small perturbations, the harmonic content of the rejected bands can be approximated by the Fourier transform of the perturbation geometry.

(a)

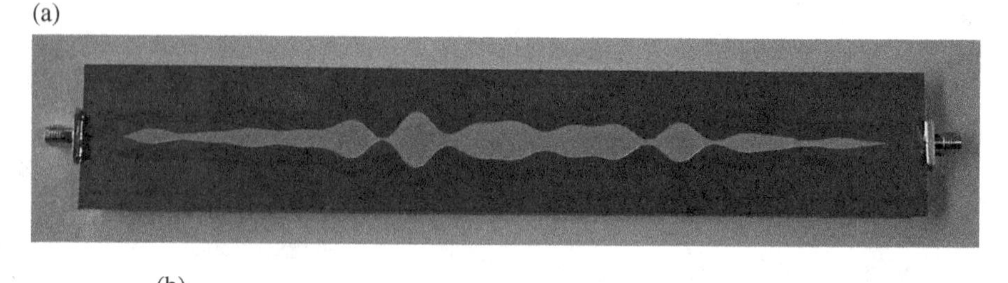

(b)

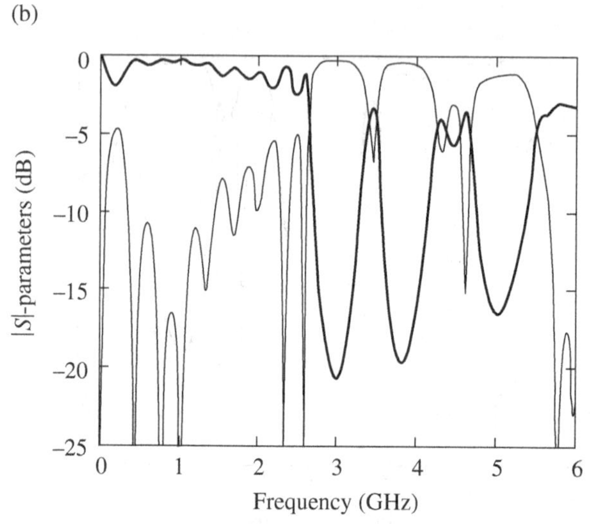

FIGURE 2.20 Photograph (ground plane) of a triple-frequency-tuned EBG microstrip line (a) and measured insertion (thick line) and return (thin line) loss (b). The substrate is the *Rogers RO3010* with dielectric constant $\varepsilon_r = 10.2$ and thickness $h = 1.27$ mm. The conductor strip width is $W = 1.2$ mm. Reprinted with permission from Ref. [31]; copyright 2000 IEEE.

2.5.1.2 High-Q Resonators It is possible to implement Bragg resonators by combining two EBG-based reflectors separated by a microstrip line acting as a resonant cavity [32]. In principle, such resonant cavity can be obtained by removing one cell of the EBG structure, which corresponds to a resonator structure with two uniform EBG reflectors separated by a 180° line at the Bragg frequency of the reflector. However, since the reflectors contribute to the phase, the resonance condition is written as [33] follows:

$$\phi_{\text{ref1}}(f_o) + \phi_{\text{ref2}}(f_o) - 2\cdot\beta(f_o)\cdot L_m = 2\pi n \tag{2.106}$$

where n is an integer that must be chosen so that the resonance frequency lies within the reflected frequency band of the EBG reflectors. The other variables of (2.106) are as follows:

- ϕ_{ref1}: phase of the reflection coefficient of the first reflector
- ϕ_{ref2}: phase of the reflection coefficient of the second reflector
- β: phase constant of the mode of interest in the resonant cavity
- L_{m}: length of the microstrip transmission line that separates the two reflectors

Thus, the length of the resonant cavity must be optimized in order to satisfy (2.106) at the required frequency. To this end, it is necessary to study the reflectors independently and obtain the phase of the reflection coefficient at the design frequency, that is, ϕ_{ref1} and ϕ_{ref2}. Once these phases are known (they are identical if the same reflectors are considered at both sides of the resonant cavity), the electrical length of the line can be inferred, and from the value of the phase constant, the length L_{m} can be determined. Figure 2.21a shows the photograph of the fabricated Bragg reflector reported in Ref. [32], where $\phi_{\text{ref1}} = \phi_{\text{ref2}} = -95.84°$ at the design frequency ($f_{\text{o}} = 4.2$ GHz), which gives $\beta L_{\text{m}} = 84.15°$ for $n = -1$ and $L_{\text{m}} = 6.25$ mm. The considered substrate is the *Rogers RO3010* with dielectric constant $\varepsilon_{\text{r}} = 10.2$, thickness $h = 1.27$ mm, and $\tan\delta = 0.0026$. Hole dimensions and separations are those of Figure 2.16. The measured frequency response of this resonator is depicted in Figure 2.21b, where the resonance is located at the desired frequency, and the measured quality factor is $Q = 129.2$, as reported in Ref. [32], which is a high value for a planar structure. Other high-Q resonators based on a similar concept but implemented in CPW, CPS and slot lines are reported in Ref. [34].

2.5.1.3 Spurious Suppression in Planar Filters One of the most interesting applications of EBG-based transmission lines acting as Bragg reflectors is the suppression of undesired harmonics in planar microwave filters. Indeed, in the EBG-based reflectors discussed before, the periodic perturbation was achieved through ground plane etching. However, it is also possible to modulate the strip width leaving the ground plane unaltered, as was pointed out in Section 2.4.6. This eases fabrication, and it is an interesting solution in those applications where the structure must rest on top of a metallic holder. This latter approach was applied to the design of coupled line microstrip bandpass filters with spurious suppression [35]. In such filters, the width of the coupled lines was perturbed following a sinusoidal law. This modulates the characteristic impedance so that the harmonic pass band of the filter is rejected (first spurious band), while the desired pass band is maintained virtually unaltered. In this "wiggly-line" filter, the period of the perturbation, l, is obtained following:

$$l = \frac{\lambda_{\text{g}}}{4} \tag{2.107}$$

where λ_{g} is the guided wavelength at the design frequency (filter central frequency). According to expression (2.107), it is clear that in each coupled line section exactly a complete period of the perturbation can be accommodated. The perturbation has been implemented in Ref. [35] through the modulation of the outer edge of the coupled

(a)

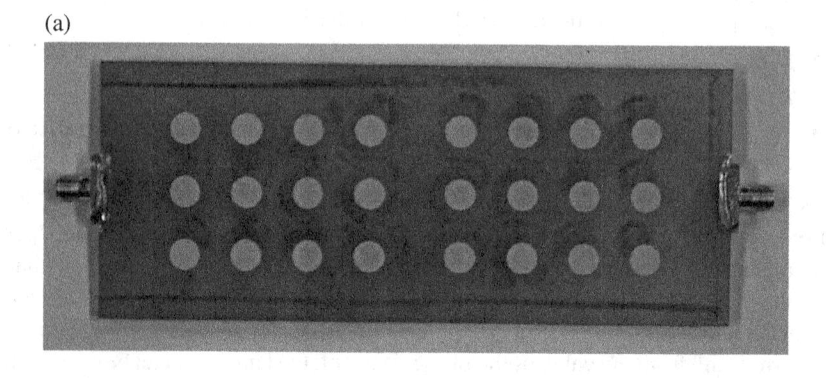

(b)

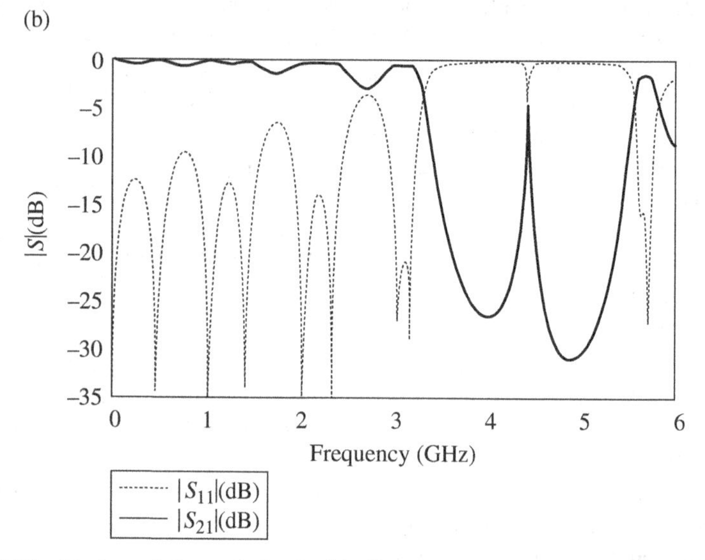

FIGURE 2.21 Photograph (ground plane) of the Bragg resonator (a) and measured frequency response (b). The conductor strip width is $W = 1.2$ mm. Reprinted with permission from Ref. [32]; copyright 1999 Springer.

lines (i.e., leaving the inner edge unaltered). The conductor strip-width variation in each coupled line section (denoted by the index i) is given by

$$W_i(z) = W_i\left(1 + \frac{M(\%)}{100}\cos\left(\frac{2\pi z}{l_i} + \phi\right)\right) \tag{2.108}$$

where W_i is the width of the coupled lines in the conventional implementation, ϕ is the initial phase (0° or 180°), l_i is the period of the sinusoidal perturbation, and M is the strip width modulation parameter.

A prototype device of a "wiggly line" coupled line bandpass filter is depicted in Figure 2.22 [35]. It is an order-3 Butterworth bandpass filter centered at 2.5 GHz with a 10% fractional bandwidth and a strip width perturbation factor $M = 47.5\%$. This filter was implemented on the *Rogers RO3010* substrate with $\varepsilon_r = 10.2$ and $h = 1.27$ mm. The measured frequency response of this device is also depicted in Figure 2.22 jointly with that frequency response which was obtained on an identical filter, but with different modulation factor (i.e., $M = 37.5\%$). The results indicate that in both cases, the pass band of interest is not affected by the strip width modulation, whereas significant rejection of the first spurious band is achieved. This is over 40 dB in the "wiggly line" filter with $M = 47.5\%$.

It has also been implemented a microstrip coupled line bandpass filter with multi-spurious suppression [36]. To this end, the modulation of the widths of the coupled line sections was done by means of different periods, so that very good out-of-band performance was demonstrated, with spurious suppression above 30 dB up to the fourth harmonic. The device and the measured frequency response are depicted in Figure 2.23 (further details are given in Ref. [36]).

2.5.1.4 Harmonic Suppression in Active Circuits Harmonic suppression in active circuits, such as power amplifiers, mixers, oscillators, or active antennas, using EBGs is very interesting in order to improve device performance. By reducing the harmonic content, the power level at the fundamental frequency increases, and the efficiency of the device can be enhanced [37–40]. It has also been demonstrated that EBG structures are useful to reduce the phase noise in microwave oscillators [41]. In this case, the key idea is to increase the quality factor of the resonator by effectively increasing the phase slope of the input matching network through the stopband effect of the periodic structure.

In Refs. [37–39], the EBG structures are cascaded to the output of the devices, whereas in Refs. [40, 41] the periodic structure is integrated with the active device. This latter strategy was also used for the design of an active antenna with improved efficiency, where a multiple-tuned EBG structure was used in order to achieve the suppression of the even and the odd harmonics [14]. The design of EBG-based active circuits is a two-step process: first, the conventional (i.e., without EBG) device is designed according to the procedures described in well-known textbooks [10]; second, the EBG, which is independently designed, is introduced in the active device in order to suppress the harmonic content. The active circuit design (first step) is out of the scope of this book. Therefore, let us review the design of the active antenna reported in Ref. [14], with emphasis on the design of the EBG, as an illustrative example of the potentiality of periodic structures to improve the performance of active circuits.

The active antenna consists of a microwave oscillator directly feeding a patch antenna. The introduction of the EBG within the oscillator circuit has the effect of suppressing the harmonics present in the radiated power spectrum, increasing at the same time the power level at the fundamental frequency. One important aspect of the active antenna reported in Ref. [14] is the use of a double-tuned EBG, which effectively reduces the power level of the first ($2f_o$) and second ($3f_o$) harmonic

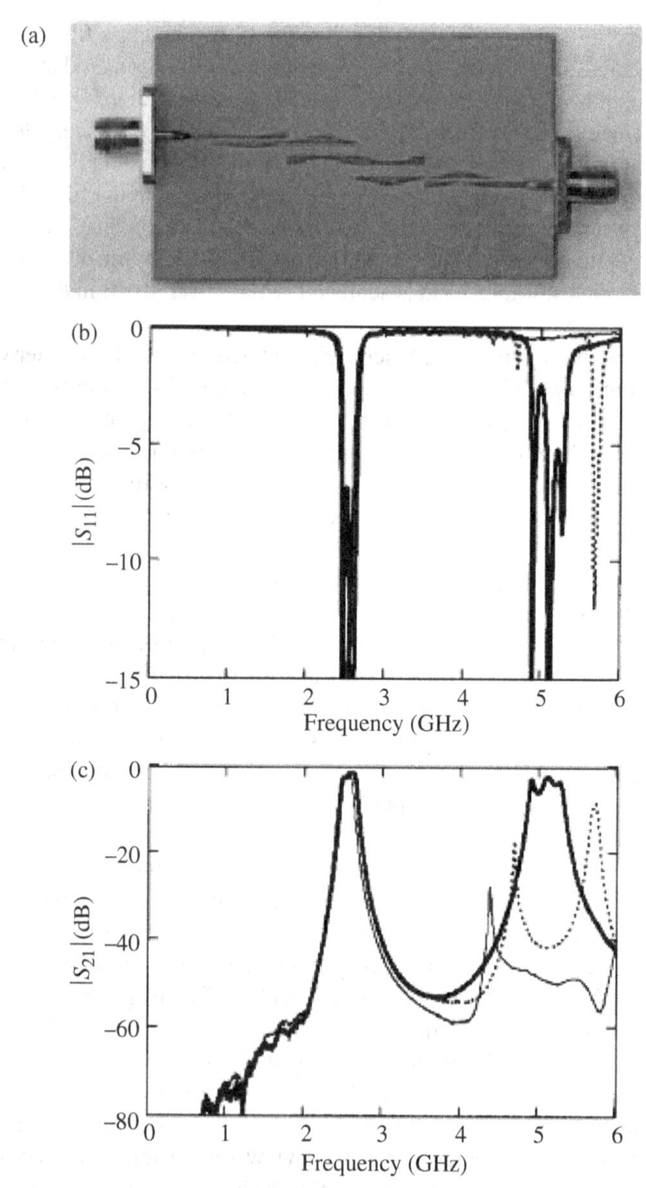

FIGURE 2.22 Coupled line bandpass filter with strip width modulation (a) and measured return (b) and insertion (c) loss. Conventional: thick solid line; "wiggly line" filter with M = 37.5%: dashed line; "wiggly line" filter with M = 47.5%: thin solid line. Reprinted with permission from Ref. [35], copyright 2001 IEEE.

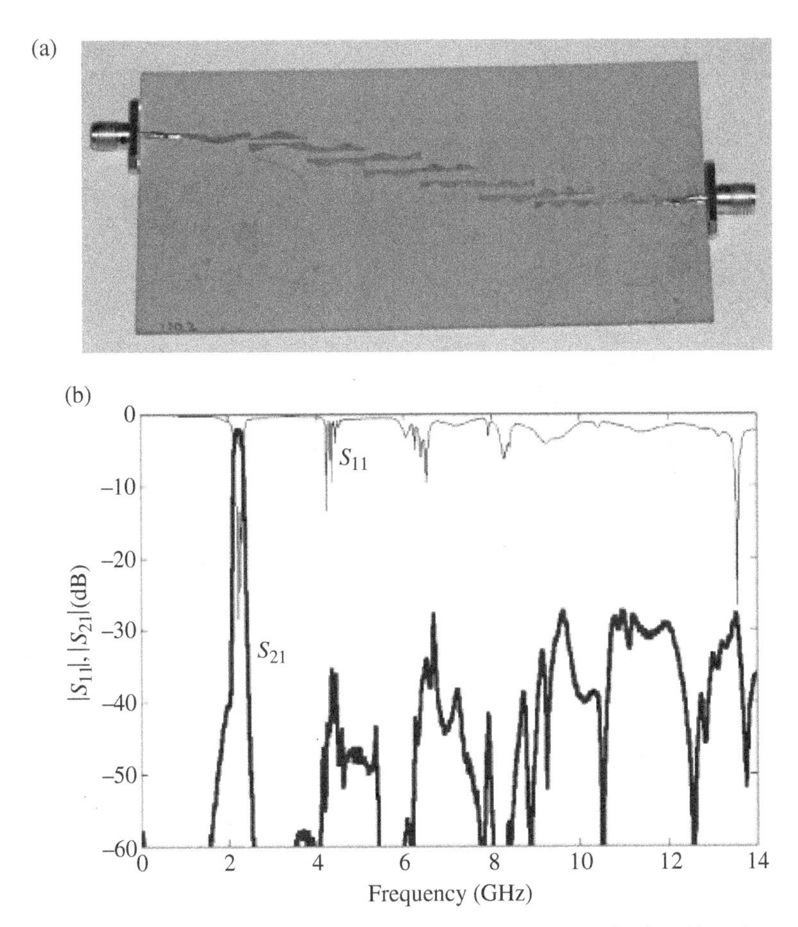

FIGURE 2.23 Coupled line bandpass filter with multispurious rejection (a) and measured frequency response (b). Reprinted with permission from Ref. [36], copyright 2004 IEEE.

(notice that by using a single tuned EBG structure at $2f_o$, the odd-order harmonics, i.e., $3f_o$, $5f_o$, etc., cannot be suppressed). Specifically, the considered EBG is formed by the addition of two raised sinusoids etched in the ground plane, and tuned at $2f_o$ and $3f_o$. The photograph of the classical designed active antenna of Ref. [14] is depicted in Figure 2.24a, where the patch antenna is the terminating load of the oscillator. The considered transistor is the Si bipolar junction transistor (BJT) *AT-42035*, operating in the common-base configuration at $V_{CE} = 8$ V and $I_C = 35$ mA. The oscillation frequency was set to $f_o = 4.5$ GHz, which means that the EBG structure must be able to efficiently suppress those frequencies in the vicinity of 9 GHz and 13.5 GHz. The periods of the raised sinusoids are 6.2 mm (to reject the first harmonic) and 4.1 mm (to reject the second harmonic), whereas the amplitude of the perturbation to period ratio is 0.3 for both raised sinusoids (the *Rogers RO3010* substrate with thickness $h = 1.27$ mm and dielectric constant $\varepsilon_r = 10.2$ was used). Figure 2.24b shows the

(a)

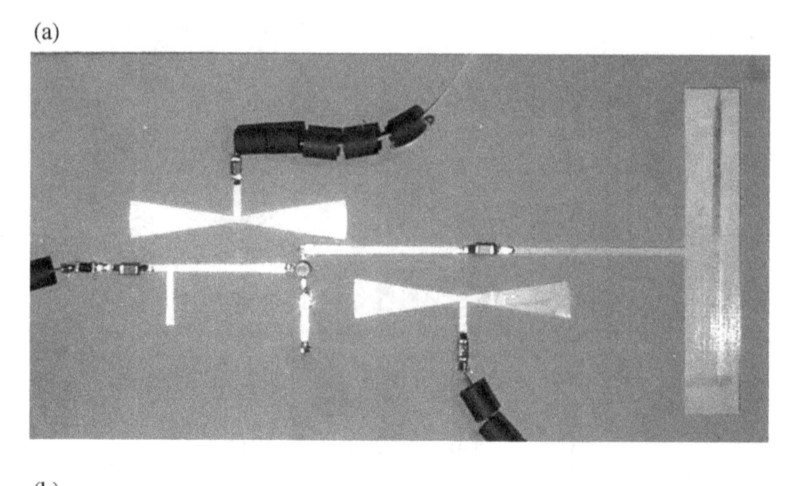

(b)

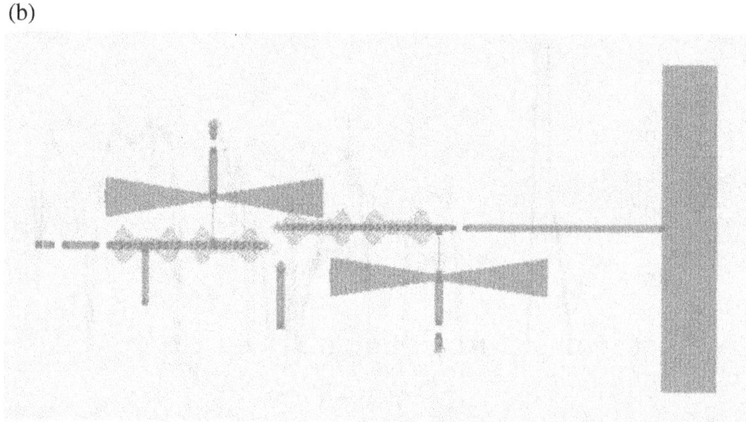

FIGURE 2.24 Photograph of the classical active antenna designed in Ref [14] (a) and layout of the same antenna including the EBG structure to reject the first and second harmonics (b). From Ref. [14]; reprinted with permission from the author.

layout of the active antenna, including the double-tuned EBG, etched in the ground plane, and integrated within the oscillator matching networks. The measured power spectrum of both antennas (with and without EBG), inferred by means of a spectrum analyzer and a rectangular horn antenna optimized for the fundamental oscillation, is depicted in Figure 2.25. For the prototype with EBG, the first and second harmonic are reduced around 20 dB and 10 dB, respectively, whilst an increase of the output power of roughly 2.6 dB at the fundamental frequency is achieved. It is important to highlight that the introduction of the EBG structure within the oscillator circuit did not require a redesign process. Moreover, this approach does not increase the layout area.

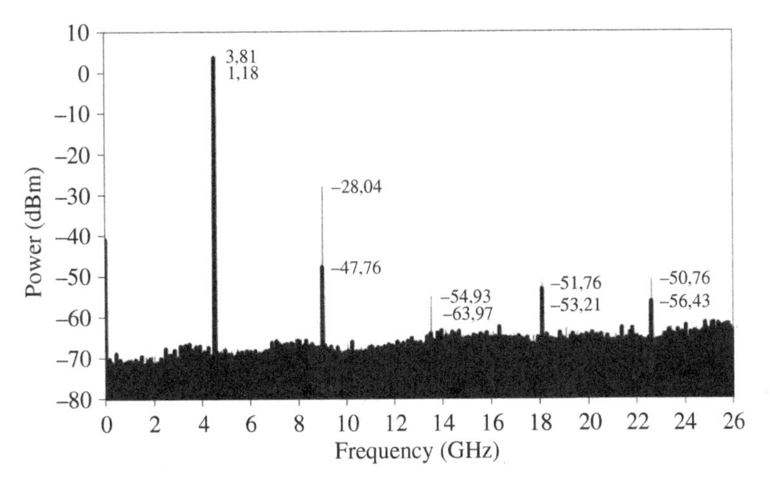

FIGURE 2.25 Measured power spectrum for the classical active antenna (thin line) and for the active antenna with EBG (thick line). From Ref. [14]; reprinted with permission from the author.

2.5.1.5 Chirped Delay Lines A CDL is a quadratic-phase filter whose frequency response $H(\omega)=A_0(\omega)\cdot\exp(-j\cdot\phi_0(\omega))$ is characterized by constant insertion losses, $A_0(\omega)$, and a group-delay, $\dot{\phi}_0=d\phi_0/d\omega$, that varies linearly with frequency in the operation bandwidth. CDLs are extensively used in the emerging field of microwave analog signal processing [42], and several interesting applications have been reported in the last years, for example, signal time compression and magnification systems [43], tunable time-delay systems [44], real-time spectrum analyzers [45] and frequency discrimination systems [42], among others. All these systems work satisfactorily at high frequencies and with wide bandwidths.

CDLs with very high time-bandwidth products (defined as the total delay excursion times the operation bandwidth), over ranges of several GHz, can be implemented by using quasiperiodic EBG structures in microstrip technology. To obtain them, the starting point is a single-frequency-tuned EBG structure, with sinusoidal coupling coefficient, of the type proposed in Section 2.4.6. In order to achieve the required group-delay that varies linearly with frequency, the period of the characteristic impedance perturbation is also varied linearly along the propagation axis, z. The characteristic impedance implemented to obtain the CDL in the reported example is [45, 46]:

$$Z_0(z)=Z_{0,\mathrm{ref}}\cdot e^{A\cdot W(z)\cdot\sin\left(\int\zeta(z)\cdot dz\right)}=Z_{0,\mathrm{ref}}\cdot e^{A\cdot W(z)\cdot\sin\left(\frac{2\pi}{l_0}\cdot z+C\cdot z^2-C\cdot L^2/4\right)} \quad (2.109)$$

where $\zeta(z)=2\pi/l_0+2Cz$ is the spatial angular frequency (that is linearly modulated along z), C is the chirp parameter, and the microstrip line extends from $z=-L/2$ to

$z = L/2$, L being the total device length. The perturbation period of the resulting quasi-periodic EBG structure is

$$l(z) = \frac{2\pi}{\zeta(z)} = \frac{2\pi}{2\pi/l_0 + 2Cz} \tag{2.110}$$

and therefore the frequency locally reflected at position z (calculated through 2.84) is

$$f(z) = \frac{c}{2 \cdot l(z) \cdot \sqrt{\varepsilon_{re}}} = \frac{c}{2 \cdot \sqrt{\varepsilon_{re}}} \cdot \left(\frac{1}{l_0} + \frac{C}{\pi} \cdot z \right) \tag{2.111}$$

where ε_{re} is the effective dielectric constant, c is the speed of light in vacuum, and l_0 is the perturbation period at the center of the device.

As can be seen, the phase-matching condition for resonant Bragg coupling between the forward and backward traveling waves is ideally satisfied at only one position for each frequency. At this position, the propagating wave will be back-reflected. Moreover, since the perturbation is linearly chirped then the mode-coupling location varies linearly with frequency. As a result, the reflection time is also a linear function of frequency. This is equivalent to say that the group delay (in reflection) will vary linearly with frequency, as intended in the CDL. According to (2.111), the CDL implemented following (2.109) will work in reflection with an operation bandwidth:

$$\Delta f = |f(z = L/2) - f(z = -L/2)| = \frac{1}{2\pi} \cdot \frac{c}{\sqrt{\varepsilon_{re}}} \cdot |C| \cdot L \tag{2.112}$$

around a central frequency:

$$f_0 = f(z = 0) = \frac{c}{2 \cdot l_0 \cdot \sqrt{\varepsilon_{re}}} \tag{2.113}$$

and with a group-delay slope, ψ (s/Hz):

$$\psi = \frac{2L\sqrt{\varepsilon_{re}}}{c \cdot \Delta f} = \frac{4\pi \cdot \varepsilon_{re}}{C \cdot c^2} \tag{2.114}$$

The group-delay slope is derived by taking into account that $\Delta f \cdot \psi$ is the time-delay difference between the arrivals at the input of the extreme frequencies of the bandwidth, $f(z = -L/2)$ and $f(z = L/2)$, the former reflected at the input and the latter at the output of the device, respectively (path difference $= 2 \cdot L$).

The three last equations can be inverted to obtain expressions for the main design parameters in (2.109) as a function of the operation bandwidth, Δf (Hz), central frequency, f_0 (Hz), and group-delay slope, ψ (s/Hz):

$$l_0 = \frac{c}{2 \cdot f_0 \cdot \sqrt{\varepsilon_{re}}}$$

$$C = \frac{4\pi \cdot \varepsilon_{re}}{\psi \cdot c^2} \tag{2.115}$$

$$L = \frac{c \cdot |\psi| \cdot \Delta f}{2 \cdot \sqrt{\varepsilon_{re}}}$$

The additional parameters in (2.109) to finish the CDL design are $Z_{0,\text{ref}}$, that corresponds to the characteristic impedance of the ports (the integration constant is fixed to $-C \cdot L^2/4$ to ensure input and output ports with $Z_{0,\text{ref}}$, assuming that L is a multiple of l_0); A, a nondimensional amplitude factor for the perturbation, and $W(z)$, a windowing function for smoother input and output impedance transitions to avoid partial reflections from the extremes of the structure that give rise to different long-path Fabry–Perot like resonances, which cause undesirable rapid ripple to appear around the mean values in the magnitude and group-delay versus frequency patterns degrading the CDL performance. Furthermore, both negative values of C, which imply upper (higher loss) frequencies to be reflected at the beginning, and asymmetric $W(z)$, which can compensate for the longer lossy round trips, can be used to lead to better equalized reflection losses across the operation band.

To demonstrate the proposed design method, a CDL in microstrip technology is reported. The characteristic impedance of the ports is $Z_{0,\text{ref}} = 50\ \Omega$, and the *Rogers RO3010* substrate with dielectric constant $\varepsilon_r = 10.2$, and thickness $h = 1.27$ mm is employed. The central operation frequency is $f_0 = 9$ GHz, and the bandwidth is $\Delta f = 12$ GHz, with a group-delay slope of $\psi = -0.5$ ns/GHz. The design parameters have been calculated using (2.115) to fulfill the response: $l_0 = 6.4$ mm, $L = 50 \cdot l_0$, and $C = -2080\ \text{m}^{-2}$ [45]. A Gaussian asymmetric tapering function is used $W(z) = \exp\left(-4 \cdot ((z - L/4)/L)^2\right)$, and $A = 0.4$. Once all the parameters in (2.109) have been fixed, the impedance variation is implemented as a strip width modulation; see Figure 2.26a. In Figure 2.26b and c, the S_{11} parameter obtained by solving the simplified system of coupled mode equations (2.47a and 2.47b), that is, lossless and single-mode approximation (dotted line), is compared with the simulation employing the commercial software *Agilent Momentum* (thin solid line) and the measurements of the fabricated prototype (thick solid line), showing that the CDL provides the required features of flat magnitude and linear group-delay variation with frequency. The reflection losses are maintained around 3 dB over the entire bandwidth. Finally, the design requirements of central operation frequency, bandwidth and group-delay slope are also properly satisfied.

The main limitation of the CDL structure proposed here is that it operates in reflection, and therefore it requires an additional component to recover the reflected signal. However, this limitation can be avoided by implementing the CDL structure in coupled line technology as explained in Ref. [47].

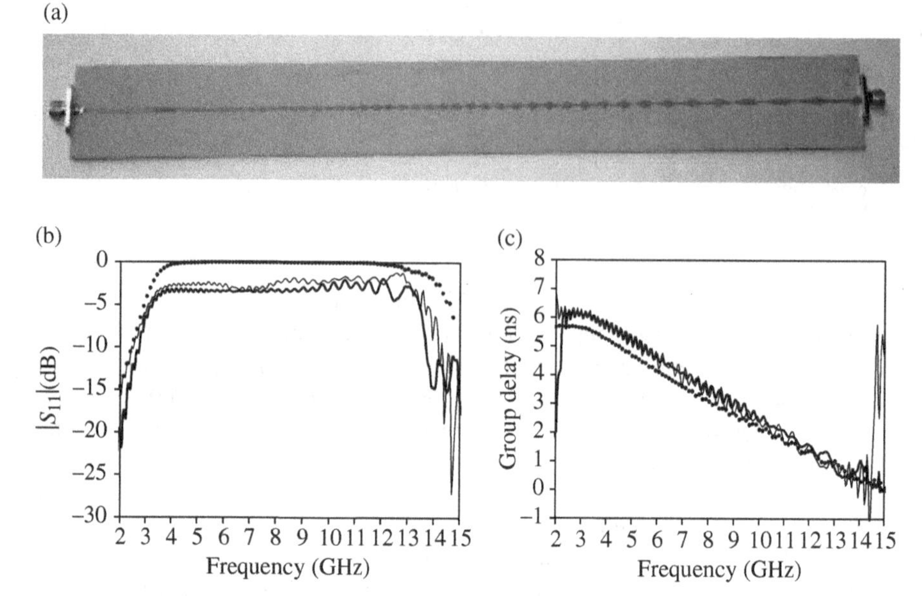

FIGURE 2.26 Photograph of the microstrip CDL (a) and S_{11} parameter of the device: magnitude (b) and group-delay (c) obtained by coupled mode theory (lossless) approximation (dotted line), *Agilent Momentum* (thin solid line) and measurement (thick solid line). Reprinted with permission from Ref. [45], copyright 2003 IEEE.

2.5.2 Applications of Reactively Loaded Lines: The Slow Wave Effect

In the previous applications of EBG-based transmission lines, a periodic perturbation was introduced either in the ground plane or in the conductor strip. However, band gaps in the transmission coefficient of periodic transmission lines can also be achieved by reactively loading the line. Specifically, by periodically loading the line with shunt-connected capacitances, either implemented by lumped elements or with planar semilumped components, significant dispersion arises, pass bands and stop bands appear, and the phase velocity of the lines is reduced. This slow-wave effect is interesting to reduce the size of microwave components. On the other hand, the transmission characteristics of these periodic loaded structures are of interest for the design of planar filters with small size and/or improved performance. All these aspects are discussed in this subsection.

Notice that the coupled mode theory presented in Section 2.4 does not apply to transmission lines periodically loaded with reactive elements. To analyze such artificial lines, the transfer matrix method is a convenient approach. Let us consider a transmission line periodically loaded with shunt-connected capacitances (Fig. 2.27a). The dispersion relation, inferred from (2.22) is:

$$\cos\beta l = \cos kl - \frac{\omega C_{\mathrm{ls}} Z_{\mathrm{o}}}{2} \sin kl \tag{2.116}$$

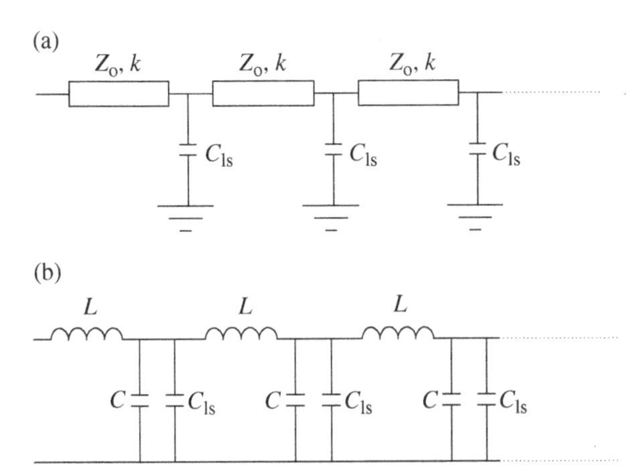

FIGURE 2.27 Transmission line periodically loaded with shunt connected capacitances (a), and lumped-element equivalent circuit model (b).

where k and β are the phase constants of the unloaded and loaded line,[33] respectively, C_{ls} are the loading capacitances, l is the distance between adjacent capacitances (period), and Z_o is the characteristic impedance of the line sections between adjacent capacitances. The dispersion relation is depicted in a Brillouin diagram in Figure 2.28, where $\theta = kl$ is proportional to frequency and $\varphi = \beta l$. This is a representation in a reduced zone, that is, $0 \leq \varphi \leq \pi$. The structure exhibits a low-pass filter-type response with multiple spurious bands (only the first three bands are visible in the diagram of Fig. 2.28). For design purposes, however, a lumped-element circuit model of the capacitively loaded transmission line is convenient. This circuit model is depicted in Figure 2.27b, where C and L are the per-section capacitance and inductance of the line, respectively. This model is valid under the assumption that $C_{ls} >> C$. According to the lumped-element equivalent circuit of the periodically loaded line, the first pass band of the structure is delimited by the following cutoff frequency

$$f_C = \frac{1}{\pi\sqrt{L(C + C_{ls})}} \tag{2.117}$$

the characteristic (or Bloch) impedance of the loaded line at low frequencies is given by

[33] Notice that in this subsection k is used to designate the phase constant of the unloaded line. Hence, do not confuse with the coupling coefficient parameter (introduced in Section 2.4—see expression 2.66), or with the free-space wavenumber (Section 2.2—see expression 2.10). The use of the same symbol to designate different variables is avoided as much as possible throughout this book. However, exceptions are made either to use the usual symbols for certain variables or to be faithful to the sources/references where the symbols and related expressions are introduced.

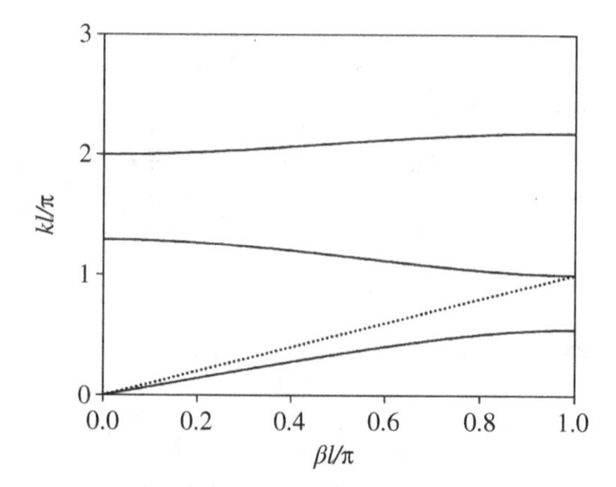

FIGURE 2.28 Typical dispersion diagram of a transmission line periodically loaded with shunt connected capacitances, depicted in a reduced Brillouin zone. The dispersion relation for the unloaded line is also depicted (dotted line). Notice that the smaller slope of the capacitively loaded line at low frequencies (as compared to that of the unloaded line) is indicative of the slow wave effect of capacitively loaded lines.

$$Z_B = \sqrt{\frac{L}{C + C_{ls}}} \tag{2.118}$$

and the lower frequency of the first spurious band is

$$f_S = \frac{1}{2\sqrt{LC}} \tag{2.119}$$

From these equations, the element values can be inferred and the capacitively loaded line can thus be implemented.

The lumped-element equivalent circuit model is only valid in the first band of the structure. Indeed, the circuit model of Figure 2.27b does not exhibit spurious bands. However, the lower frequency of the first spurious band (2.119) can be easily inferred, since $\theta = kl = \pi$ at this frequency. Hence,

$$kl = 2\pi f_S \sqrt{LC} = \pi \tag{2.120}$$

from which (2.119) results.

The cutoff frequency (expression 2.117) can be obtained from (2.116) by considering that $\theta = kl$ is small in the first pass band (which is equivalent to consider that $C_{ls} >> C$). At the cutoff frequency, $\varphi = \beta l = \pi$, and expression (2.116) can be written as

$$-1 = 1 - \frac{(k_c l)^2}{2} - \frac{2\pi f_C C_{ls} Z_o}{2} k_c l \tag{2.121}$$

where the sin and cos functions have been expanded in Taylor series up to the first order, and k_c, given by

$$k_c l = 2\pi f_C \sqrt{LC} \tag{2.122}$$

is the phase constant of the unloaded line at the cutoff frequency. Introducing (2.122) in (2.121), and taking into account that the characteristic impedance of the unloaded line is

$$Z_o = \sqrt{\frac{L}{C}} \tag{2.123}$$

the cutoff frequency can be isolated, and (2.117) is found.

To gain more insight on the validity of the model of Figure 2.27b as C_{ls} increases (as compared to C), the dispersion relations (first band) of three transmission lines loaded with different capacitances have been obtained by means of (2.116), and by applying (2.33) to the unit cell of the circuit of Figure 2.27b. The results, compared in Figure 2.29a, confirm that the dispersion curves inferred from both models are identical in the low-frequency limit, regardless of the value of C_{ls}.[34] This result is expected since at low frequencies the wavelength of the unloaded line is large as compared to the distance, l, between adjacent capacitors. As frequency increases both curves progressively diverge, the difference being a maximum at the cutoff frequency. However, such maximum difference decreases as the difference between C_{ls} and C increases, and the results of Figure 2.29 indicate that for $C_{ls} = 9C$ the relative difference between the cutoff frequencies given by both models is as small as 3.5%. Figure 2.29b compares the Bloch impedance corresponding to the exact (distributed) and approximate (circuit) models for the three considered cases, where it is found that both models tend to give the same values as C_{ls} increases, as expected.

In the next subsections, it is shown that capacitively loaded lines are useful for size reduction (associated to the slow-wave effect) and spurious suppression. Such lines can thus be applied to the design of compact filters with spurious suppression, including lowpass and bandpass filters. The two reported examples are bandpass filters (the application of capacitively loaded lines to lowpass filtering structures can be found in [48–51]).

2.5.2.1 *Compact CPW Bandpass Filters with Spurious Suppression*

By periodically loading a transmission line with shunt capacitances, the phase velocity can be substantially reduced. In the low-frequency limit, the phase velocity of the periodically loaded line, v_{pL}, is given by

[34] Notice that the second and third bands given by the distributed model are also depicted in Figure 2.29a. The presence of these bands is not accounted for by the lumped element model, since it is not valid above the first pass band.

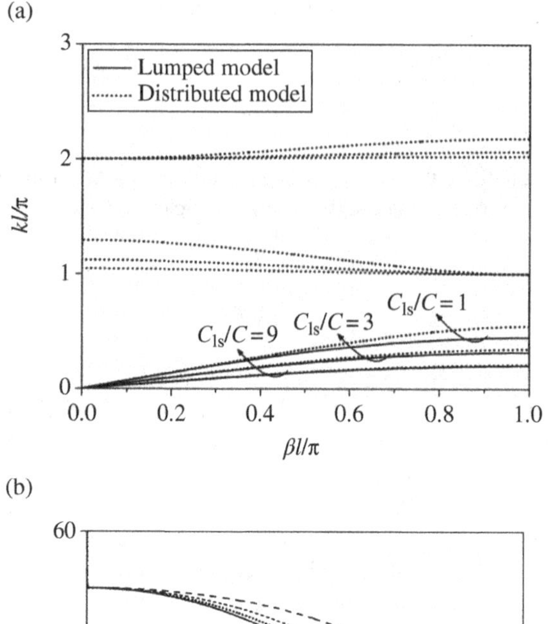

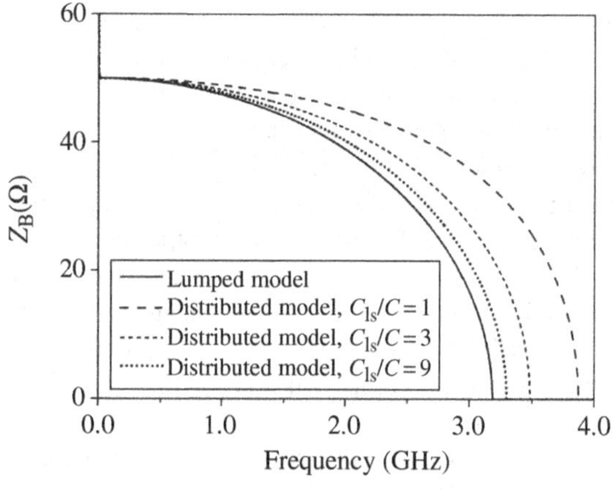

FIGURE 2.29 Dispersion diagram (a) and Bloch impedance (b) of a transmission line periodically loaded with a capacitance C_{ls} of the indicated value. The curves corresponding to the distributed and lumped-element circuit models are indicated in dashed and solid lines, respectively. For the three cases, $L = 5$ nH and $C + C_{ls} = 2$ pF, so that the Bloch impedance at low frequencies and the cutoff frequency (lumped-element model) are identical for the three considered cases. Notice that the values of kl at the cutoff frequency are different since they depend on C (see expression 2.122), which is different for the three considered cases.

$$v_{pL} = \frac{l}{\sqrt{L(C + C_{ls})}} \tag{2.124}$$

The effect of such phase velocity reduction is a decrease of the guided wavelength; hence, significant size reduction of distributed circuits implemented with such periodically loaded lines can be achieved. Such artificial lines, also referred to as

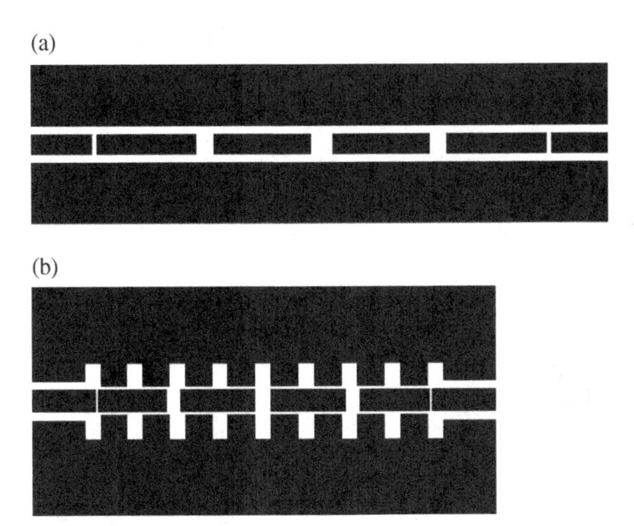

FIGURE 2.30 Layouts of the conventional (a) and periodic-loaded (b) CPW bandpass filters. The layouts are drawn to scale, the length of the conventional filter being 5.6 cm. Reprinted with permission from Ref. [54], copyright 2004 John Wiley & Sons.

slow-wave transmission lines, or simply slow-wave structures, are an alternative to high dielectric constant substrates for device miniaturization. Moreover, since these capacitively loaded lines exhibit stop bands, they can be used to simultaneously reduce the size of distributed circuits and suppress undesired bands (as pointed out before).

In CPW technology, the loading capacitances can be implemented by means of T-shaped planar structures, as reported in Refs [50, 51]. The slow-wave effect can also be achieved by periodically approaching the conductor strip and ground planes. This idea was applied by Görur *et al.* [52, 53] to the design of compact resonators in CPW technology and by Martín *et al.* [54] to the design of compact capacitively (gap) coupled resonator bandpass filters (see the example provided in Section 1.6.4, corresponding to a conventional implementation). The details of the design of these filters can be found in Ref. [54]. Essentially, the periodic perturbation (providing capacitive loading) has been determined to obtain a cutoff frequency beyond the pass band of interest, but below the first harmonic band. In this way, spurious elimination is achieved. Moreover, the line perturbation effectively increases the capacitance of the line, decreases the phase velocity (slow-wave effect), and hence reduces the length of the different resonators forming the filter.[35] The filter described in Ref. [54] is a fourth-order bandpass filter with central frequency $f_o = 6$ GHz and 10% fractional bandwidth. It is depicted in Figure 2.30 and compared with the conventional

[35] A detailed design process of the capacitively loaded slow wave transmission lines using expressions (2.117)–(2.119) is given in reference to the bandpass filter of the next subsection, also based on slow-wave transmission-line sections, but implemented in microstrip technology.

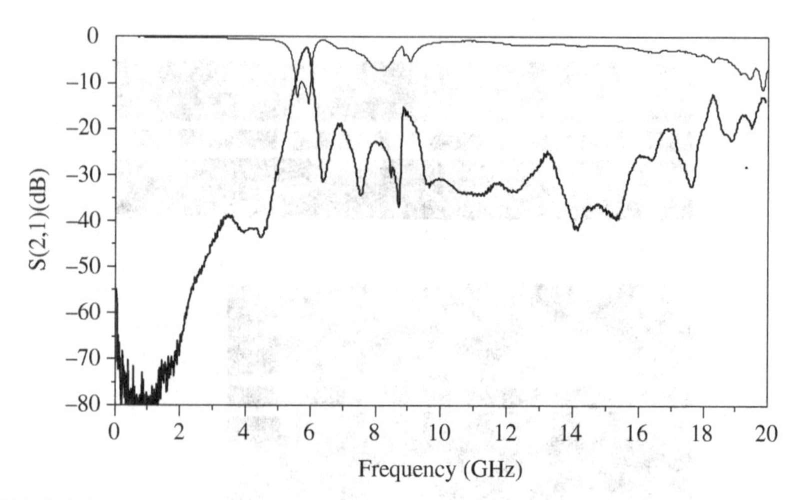

FIGURE 2.31 Measured insertion (bold line) and return (thin line) loss for the periodic-loaded filter of Figure 2.30b. Reprinted with permission from Ref. [54], copyright 2004 John Wiley & Sons.

implementation (designed on the same substrate—*Rogers RO3010* with $\varepsilon_r = 10.2$, $h =$ 1.27 mm—with identical central frequency). Roughly 30% size reduction is achieved by means of the slow-wave periodic structure. The measured frequency response of the device is depicted in Figure 2.31. The device is spurious free in a wide band due to the effects of periodicity. This concept has been applied to the miniaturization of other components, such as hybrids and couplers [55].

Other periodic slow wave structures based on inductive loading [56], or combined inductive and capacitive loading [57–62] have also been reported. The topic of slow-wave transmission lines has attracted much attention in recent years for their potential in size reduction of distributed circuits, and other different approaches (out of the scope of this book) have been reported [63], including slow-wave transmission lines for passive and active CMOS devices operating at millimetre wavelengths [64–69].

2.5.2.2 Compact Microstrip Wideband Bandpass Filters with Ultrawideband Spurious Suppression Capacitively loaded lines can also be implemented in microstrip technology by means of patch capacitances. These structures can also be applied to the design of compact bandpass filters (due to the inherent slow-wave effect). However, since high capacitance values can be achieved by means of capacitive patches, the most relevant aspect of these periodic loaded microstrip lines is the huge achievable stop bands, which are very interesting to suppress spurious bands over a very wide frequency range. As an illustrative example of capacitively loaded microstrip lines, a wideband bandpass filter is considered, where some of the transmission line sections are substituted by patch loaded lines [70]. The filter consists of a cascade of shunt stubs of equal length, alternating with

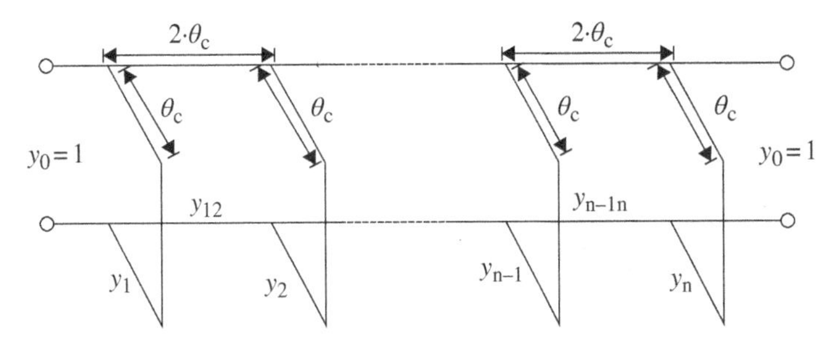

FIGURE 2.32 Schematic of the wideband bandpass filter where y_i and y_{jk} are the normalized characteristic admittances of the stubs and transmission lines, respectively.

transmission lines with twice the stub electrical length. The analysis of this type of filters was done by Levy [71]. A very interesting characteristic of these filters is that by using n stubs, an insertion function of degree $2n-1$ is implemented. These filters are useful to generate wide transmission bands, although wideband spurious are also present in their frequency response. The schematic of the filter is depicted in Figure 2.32. The network shown in Figure 2.32 implements the transfer function described in expression (2.125a) as a function of the normalized frequency variable $\theta = \theta_c f/f_c$ [72]:

$$|S_{21}(\theta)|^2 = \frac{1}{1+\kappa^2 F_n^2(\theta)} \tag{2.125a}$$

with

$$F_n(\theta) = \frac{\left(1+\sqrt{1-x_c^2}\right)\cdot T_{2n-1}\left(\dfrac{x}{x_c}\right) - \left(1-\sqrt{1-x_c^2}\right)\cdot T_{2n-3}\left(\dfrac{x}{x_c}\right)}{2\cdot\cos\left(\dfrac{\pi}{2}-\theta\right)} \tag{2.125b}$$

$$x = \sin\left(\frac{\pi}{2}-\theta\right) \tag{2.125c}$$

$$x_c = \sin\left(\frac{\pi}{2}-\theta_c\right) \tag{2.125d}$$

where $T_n = \cos(n\cdot\cos^{-1}(x))$ and κ is the pass band ripple constant. The bandwidth of the filter is delimited by the frequencies f_c and $(\pi/\theta_c - 1)f_c$; therefore, the bandwidth can be controlled by the value of the angle θ_c.

As an illustrative example, a $n=3$ prototype with $f_c = 1.4$ GHz and $\theta_c = 35°$ (that implies a 4.8 GHz bandwidth) is implemented. To determine the impedance values of the short-circuit stubs and line elements, the tabulated element values supplied by Hong and Lancaster in their text book [72] for optimum distributed highpass

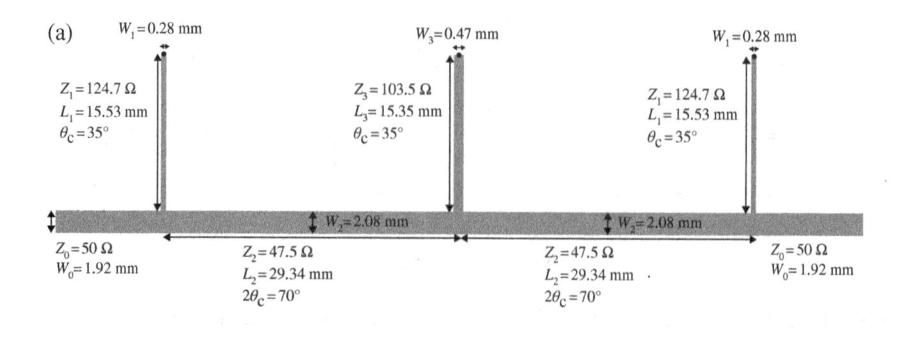

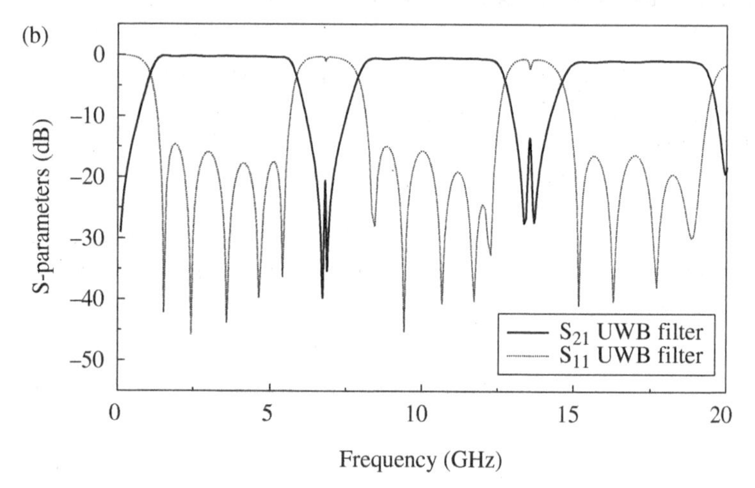

FIGURE 2.33 (a) Layout of the wideband bandpass filter including electrical parameters and geometry of transmission lines and stubs. (b) EM simulation of the frequency response of the filter. Reprinted with permission from Ref. [70], copyright 2006 IEEE.

filters are used. The resulting impedance values are shown in Figure 2.33. The simulation results depicted in Figure 2.33b correspond to the structure described in Figure 2.33a, implemented in an *Arlon* substrate with dielectric constant $\varepsilon_r = 2.4$ and thickness $h = 0.675$ mm. Five reflection zeros can be observed in the transmission bands, such as one expects on account of the $2n - 1$ degree of the filter function ($n = 3$ in our case). The structure of Figure 2.33 is the conventional implementation.

The capacitively loaded structure is obtained by replacing the interstub transmission line sections with patch-loaded lines (with two unit cells). The electrical length of these lines is 70°, whereas the characteristic impedance is 47.5 Ω. The phase shift per cell is given by

$$\phi = 2\pi \cdot f\sqrt{L(C + C_{ls})} \tag{2.126}$$

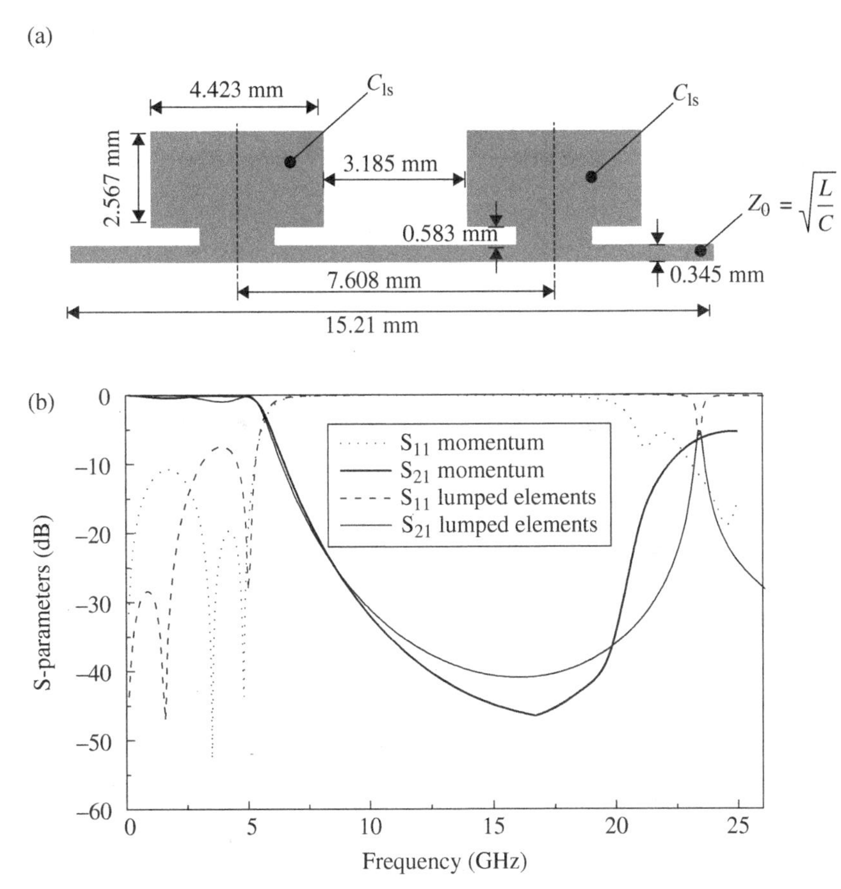

FIGURE 2.34 (a) Layout of the periodic-loaded structure in microstrip technology and (b) electrical and EM simulation. Reprinted with permission from Ref. [70], copyright 2006 IEEE.

By forcing (2.118) and (2.126) to obtain the required electrical characteristics (i.e., $\phi = 35°$ at 1.4 GHz and $Z_B = 47.5\,\Omega$) and by setting f_s (expression 2.119) to $f_s =$ 22 GHz in order to obtain a huge rejection band, the electrical parameters of the patch (capacitively) loaded line can be inferred. The results are as follows: $C = 0.17$ pF, $L =$ 3.1 nH and $C_{ls} = 1.2$ pF. With these values, the impedance of the unloaded line is $Z_o = 123.5\,\Omega$, and the cutoff frequency of the periodic-loaded line is $f_C =$ 5.6 GHz, that is, it lies between the pass band of interest and the first spurious frequency of the conventional filter.

The layout of the patch-loaded transmission line sections was obtained by means of an optimization procedure by using *Agilent Momentum*. It is depicted in Figure 2.34 together with the simulated frequency response. The band gap extends up to more than 20 GHz (as required). Hence, it is expected that by introducing such lines in the filter layout, the frequency response is free of spurious up to very high frequencies.

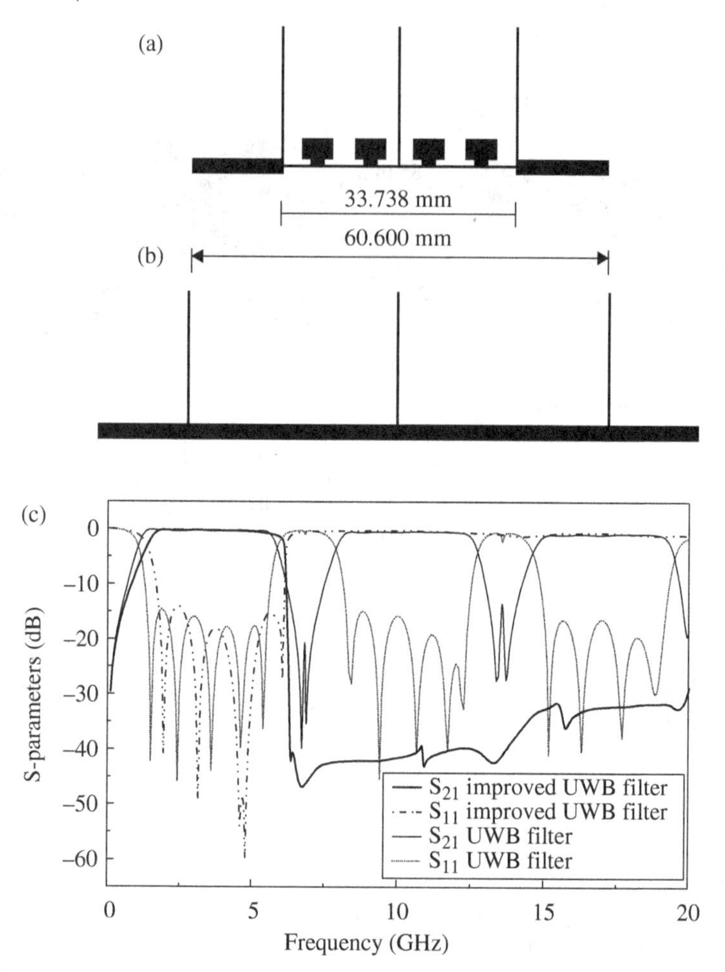

FIGURE 2.35 Layout of the patch-loaded filter (a) standard filter (b) and EM simulation of both filters (c). The filters are designed and implemented using an *Arlon* substrate with $\varepsilon_r = 2.4$ and thickness $h = 0.675$ mm. Reprinted with permission from Ref. [70], copyright 2006 IEEE.

A comparative layout of the conventional and capacitively loaded (patch-loaded) filters and the corresponding simulated frequency responses are depicted in Figure 2.35. The photograph of the fabricated patch-loaded filter is shown in Figure 2.36, jointly with the measured frequency response. The out-of-band performance is good, with a rejection level better than −30 dB up to 20 GHz. In-band losses are lower than 0.9 dB and return losses are better than 10 dB. Finally, approximately 50% length reduction is obtained by using the capacitively loaded filter.

A balanced (differential) version of this type of filters, covering the frequency spectrum assigned for ultrawideband (UWB) communications (3.1–10.6 GHz), is reported in Ref. [73].

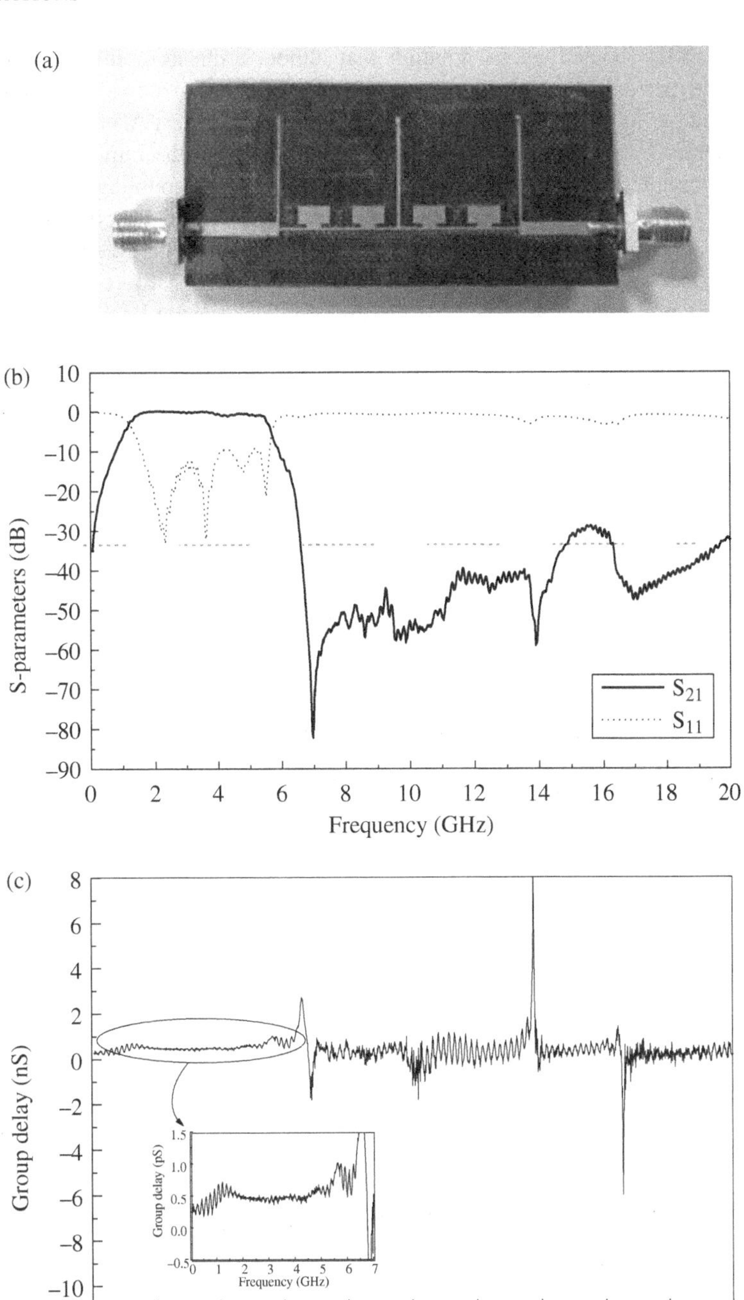

FIGURE 2.36 (a) Picture of the fabricated patch-loaded filter, (b) measured S-parameters, and (c) measured group delay. Reprinted with permission from Ref. [70], copyright 2006 IEEE.

To end this chapter, let us mention that although the most intensive research activity on periodic transmission lines (EBG-based and reactively loaded lines) was carried out in the late nineties and first years of this century, there is still activity on this topic. In particular, EBGs have been combined with coupled lines for the design of coupled line directional couplers with enhanced coupling factor [22, 74, 75]. By properly modulating the common-mode and differential-mode characteristic impedances, it is possible to achieve contra-phase reflection coefficients for the even and odd modes and, consequently, redirect the reflected signal to the coupled port. Another recent application concerns the design of differential lines with common-mode suppression [76]. In this case, the common-mode impedance is periodically modulated, whereas the differential-mode impedance is kept uniform along the line. The result is a band gap for the common-mode, whereas the line is transparent for the differential signals (other techniques for common-mode suppression in differential lines are pointed out in Chapter 6).

REFERENCES

1. L. Brillouin, *Wave Propagation in Periodic Structures: Electric Filters and Crystal Lattices*, 2nd Edition, Dover Publications, New York, 1953.
2. A. A. Oliner, "Radiating periodic structures: analysis in terms of k vs. β diagrams," in *Short Course on Microwave Field and Network Techniques*, Polytechnic Institute of Brooklyn, New York, 1963.
3. A. Yariv and P. Yeh, *Optical Waves in Crystals*, Wiley, New York, 1984.
4. R. E. Collin, *Foundations for Microwave Engineering*, 2nd Edition, McGraw Hill, Singapore, 1992.
5. J. D. Joannopoulos, R. D. Meade, and J. N. Winn, *Photonic Crystals: Molding the Flow of Light*, Princeton University Press, Princeton, NJ, 1995.
6. A. A. Oliner, "Leaky-wave antennas," in *Antenna Engineering Handbook*, R. C. Johnson, Ed. McGraw-Hill, New York, 1993.
7. F. B. Gross, Ed. *Frontiers in Antennas: Next Generation Design & Engineering*, Mc Graw Hill, New York, 2011.
8. J. L. Volakis, Ed. *Antenna Engineering Handbook*, 4th Edition, Mc Graw Hill, New York, 2007.
9. C. A. Balanis, Ed. *Modern Antenna Handbook*, Wiley, Hoboken, NJ, 2008.
10. D. M. Pozar, *Microwave Engineering*, 2nd Ed., Addison-Wesley, Reading, MA, 1998.
11. F. Sporleder and H. G. Unger, *Waveguide Tapers, Transitions and Couplers,* Peter Peregrinus, London, UK, 1979.
12. V. V. Shevchenko, *Continuos Transitions in Open Waveguides,* Golem, Boulder, CO, 1971.
13. B. Z. Katsenelenbaum, L. Mercader, M. Pereyaslavets, M. Sorolla, and M. Thumm, *Theory of Nonuniform Waveguides—the Cross-Section Method*, IEE Electromagnetic Waves Series, 44, The Institution of Engineering and Technology, London, UK, 1998.

14. T. Lopetegi, *Photonic band gap structures in microstrip technology: study using the coupled mode formalism and applications*, PhD Thesis Dissertation, Universidad Pública de Navarra, Pamplona, Spain, 2002.
15. D. Marcuse, *Theory of Dielectric Optical Waveguides*, 2nd Ed., Academic Press, San Diego, CA, 1991.
16. T. Lopetegi, M. A. G. Laso, M. J. Erro, M. Sorolla, and M. Thumm, "Analysis and design of periodic structures for microstrip lines by using the coupled mode theory," *IEEE Microw. Wireless Compon. Lett.*, vol. **12**, pp. 441–443, 2002.
17. I. Bahl and P. Barthia, *Microwave Solid State Circuit Design*, 2nd Edition., Wiley, New York, 2003.
18. V. Radisic, Y. Qian, R. Coccioli, and T. Itoh, "Novel 2-D photonic bandgap structure for microstrip lines," *IEEE Microw. Guided Wave Lett.*, vol. **8**, pp. 69–71, 1998.
19. F. Falcone, T. Lopetegi, and M. Sorolla, "1-D and 2-D photonic bandgap microstrip structures," *Microw. Opt. Technol. Lett.*, vol. **22**, pp. 411–412, 1999.
20. H. Kogelnik and C. V. Shank, "Coupled-wave theory of distributed feedback lasers," *J. Appl. Phys.*, vol. **43**, pp. 2327–2335, 1972.
21. A. Yariv and M. Nakamura, "Periodic structures for integrated optics," *IEEE J. Quantum Electron.*, vol. **13**, pp. 233–253, 1977.
22. I. Arnedo, M. Chudzik, J. Schwartz, I. Arregui, A. Lujambio, F. Teberio, D. Benito, M. A. G. Laso, D. Plant, J. Azaña, and T. Lopetegi, "Analytical solution for the design of planar EBG structures with spurious-free frequency response," *Microw. Opt. Technol. Lett.*, vol. **54**, pp. 956–960, 2012.
23. E. F. Bolinder, "Fourier transforms in the theory of inhomogeneous transmission lines," *Proc. IRE*, vol. **38**, pp. 1354, 1950.
24. E. Yablonovitch, "Photonic band gap structures," *J. Opt. Soc. Am. B* **10**, pp. 283–295, 1993.
25. Y. Qian, V. Radistic, and T. Itoh, "Simulation and experiment of photonic band-gap structures for microstrip circuits," *Proc. Asia-Pacific Microw. Conf.*, Hong Kong, December 1997, pp. 585–588.
26. T. J. Ellis and G. M. Rebeiz, "MM-w ave tapered slot antennas on micromachined photonic bandgap dielectrics," *IEEE MTT-S Int. Microw. Symp. Dig.*, San Francisco, CA, June 1996, pp. 1157–1160.
27. M. J. Erro, M. A. G. Laso, T. Lopetegi, D. Benito, M. J. Garde, and M. Sorolla, "Optimization of tapered bragg reflectors in microstrip technology," *Int. J. Infrared Milli. Waves*, vol. **21**, pp. 231–245, 2000.
28. K. O. Hill and G. Meltz, "Fiber Bragg grating technology fundamentals and overview," *J. Lightwave Technol.*, vol. **15**, pp. 1263–1276, 1997.
29. M. A. G. Laso, T. Lopetegi, M. J. Erro, D. Benito, M. J. Garde, and M. Sorolla, "Novel wideband photonic bandgap microstrip structures," *Microw. Opt. Technol. Lett.*, vol. **24**, pp. 357–360, 2000.
30. F. Falcone, T. Lopetegi, M. Irisarri, M. A. G. Laso, M. J. Erro, and M. Sorolla, "Compact photonic bandgap microstrip structures," *Microw. Opt. Technol. Lett.*, vol. **23**, pp. 233–236, 1999.
31. M. A. G. Laso, T. Lopetegi, M. J. Erro, D. Benito, M. J. Garde, and M. Sorolla, "Multiple-frequency-tuned photonic bandgap microstrip structures," *IEEE Microw. Guided Wave Lett.*, vol. **10**, pp. 220–222, 2000.

32. T. Lopetegi, F. Falcone, and M. Sorolla, "Bragg reflectors and resonators in microstrip technology based on electromagnetic crystal structures," *Int. J. Infrared Milli. Waves*, vol. **20**, pp. 1091–1102, 1999.
33. V. L. Bratman, G. G. Denisov, N. S. Ginzburg, and M. I. Petelin, "FEL's with Bragg reflection resonators: cyclotron autoresonance masers versus ubitrons," *IEEE J. Quantum Electron.*, vol. **19**, pp. 282–296, 1983.
34. T.-Y. Yun and K. Chang, "Uniplanar one-dimensional photonic-bandgap structures and resonators," *IEEE Trans. Microw. Theory Techn.*, vol. **49**, pp. 549–553, 2001.
35. T. Lopetegi, M. A. G. Laso, J. Hernández, M. Bacaicoa, D. Benito, M. J. Garde, M. Sorolla, and M. Guglielmi, "New microstrip wiggly-line filters with spurious passband suppression," *IEEE Trans. Microw. Theory Techn.*, vol. **49**, pp 1593–1598, 2001.
36. T. Lopetegi, M. A. G. Laso, F. Falcone, F. Martín, J. Bonache, L. Pérez-Cuevas, and M. Sorolla, "Microstrip wiggly line band pass filters with multispurious rejection," *IEEE Microw. Wireless Compon. Lett.*, vol. **14**, pp.531–533, 2004.
37. V. Radisic, Y. Qian, and T. Itoh, "Broad-band power amplifier using dielectric photonic bandgap structure," *IEEE Microw. Guided Wave Lett.*, vol. **8**, pp. 13–14, 1998.
38. C. Y. Hang, V. Radisic, Y. Qian, and T. Itoh, "High efficiency power amplifier with novel PBG ground plane for harmonic tuning," *IEEE MTT-S Int. Microw. Symp. Dig.*, Anaheim, CA, June 1999, pp. 807–810.
39. F. R. Yang, Y. Qian, and T. Itoh, "A novel uniplanar compact PBG structure for filter and mixer applications," *IEEE MTT-S Int. Microw. Symp. Dig.*, Anaheim, CA, June 1999, pp. 919–922.
40. Q. Xue, K. M. Shum, and C. H. Chan, "Novel oscillator incorporating a compact microstrip resonant cell," *IEEE Microw. Wireless Compon. Lett.*, vol. **11**, pp. 202–204, 2001.
41. Y. T. Lee, J. S. Lim, J. S. Park, D. Ahn, and S. Nam, "A novel phase noise reduction technique in oscillators using defected ground structure," *IEEE Microw. Wireless Compon. Lett.*, vol. **12**, pp. 39–41, 2002.
42. C. Caloz, S. Gupta, Q. Zhang, and B. Nikfal, "Analog signal processing: a possible alternative or complement to dominantly digital radio schemes," *Microw. Mag.*, vol. **14**, pp. 87–103, 2013.
43. J. D. Schwartz, J. Azaña, and D. V. Plant, "A fully electronic system for the time magnification of ultra-wideband signals," *IEEE Trans. Microw. Theory Techn.*, vol. **55**, pp. 327–334, 2007.
44. J. D. Schwartz, I. Arnedo, M. A. G. Laso, T. Lopetegi, J. Azaña, and D. V. Plant, "An electronic UWB continuously tunable time-delay system with nanosecond delays," *IEEE Microw. Wireless Compon. Lett.*, vol. **18**, pp. 103–105, 2008.
45. M. A. G. Laso, T. Lopetegi, M. J. Erro, D. Benito, M. J. Garde, M. A. Muriel, M. Sorolla, and M. Guglielmi, "Real-time spectrum analysis in microstrip technology," *IEEE Trans. Microw. Theory Techn.*, vol. **51**, pp. 705–717, 2003.
46. M. A. G. Laso, T. Lopetegi, M. J. Erro, D. Benito, M. J. Garde, M. A. Muriel, M. Sorolla, and M. Guglielmi, "Chirped delay lines in microstrip technology," *IEEE Microw. Wireless Compon. Lett.*, vol. **11**, pp. 486–488, 2001.
47. A. Lujambio, I. Arnedo, M. Chudzik, I. Arregui, T. Lopetegi, and M. A. G. Laso, "Dispersive delay line with effective transmission-type operation in coupled-line technology," *IEEE Microw. Wireless Compon. Lett.*, vol. **21**, pp. 459–461, 2011.

48. F. Martín, F. Falcone, J. Bonache, T. Lopetegi, M. A. G. Laso, and M. Sorolla, "New periodic-loaded electromagnetic bandgap coplanar waveguide with complete spurious passband suppression," *IEEE Microw. Wireless Compon. Lett.*, vol. **12**, pp. 435–437, 2002.
49. F. Martín, F. Falcone, J. Bonache, T. Lopetegi, M. A. G. Laso, and M. Sorolla, "Analysis of the reflection properties in electromagnetic bandgap coplanar waveguides loaded with reactive elements," *Prog. Electromag. Res.*, vol. **42**, pp. 27–48, 2003.
50. F. Martín, F. Falcone, J. Bonache, M. A. G. Laso, T. Lopetegi, and M. Sorolla, "New CPW low pass filter based on a slow wave structure," *Microw. Opt. Technol. Lett.*, vol. **38**, pp. 190–193, 2003.
51. F. Martín, F. Falcone, J. Bonache, M. A. G. Laso, T. Lopetegi, and M. Sorolla, "Dual electromagnetic bandgap CPW structures for filter applications," *IEEE Microw. Wireless Compon. Lett.*, vol. **13**, pp. 393–395, 2003.
52. A. Görür, "A novel coplanar slow-wave structure," *IEEE Microw. Guided Wave Lett.*, vol. **4**, pp. 86–88, 1994.
53. A. Görür, C. Karpuz, and M. Alkan, "Characteristics of periodically loaded CPW structures," *IEEE Microw. Guided Wave Lett.*, vol. **8**, pp. 278–280, 1998.
54. F. Martín, J. Bonache, I. Gil, F. Falcone, T. Lopetegi, M. A. G. Laso, and M. Sorolla, "Compact spurious free CPW band pass filters based on electromagnetic bandgap structures," *Microw. Opt. Technol. Lett*, vol. **40**, pp. 146–148, 2004.
55. K. W. Eccleston and S. H. M. Ong, "Compact planar microstripline branch-line and rat-race couplers," *IEEE Trans. Microw. Theory Techn.*, vol. **51**, pp 2119–2125, 2003.
56. L. Zhu, "Guided-wave characteristics of periodic microstrip lines with inductive loading: slow-wave and bandstop behaviors," *Microw. Opt. Technol. Lett.*, vol. **41**, pp. 77–79, 2004.
57. F.-R. Yang, K.-P. Ma, Y. Qian, and T. Itoh, "A uniplanar compact photonic-bandgap (UC-PBG) structure and its applications for microwave circuits," *IEEE Trans. Microw. Theory Techn.*, vol. **47**, pp. 1509–1514, 1999.
58. J. Sor, Y. Qian, and T. Itoh, "Miniature low-loss CPW periodic structures for filter applications," *IEEE Trans. Microw. Theory Techn.*, vol. **49**, pp. 2336–2341, 2001.
59. D. Nesic, "A new type of slow-wave 1-D PBG microstrip structure without etching in the ground plane for filter and other applications," *Microw. Opt. Technol. Lett.*, vol. **33**, pp. 440–443, 2002.
60. S.-G. Mao, M.-Y. Chen, "A novel periodic electromagnetic bandgap structure for finite-width conductor-backed coplanar waveguides," *IEEE Microw. Wireless Compon. Lett.*, vol. **11**, pp. 261–263, 2001.
61. S.-G. Mao, C.-M. Chen, and D.-C. Chang, "Modeling of slow-wave EBG structure for printed-bowtie antenna array," *IEEE Antennas Wireless Propag. Lett.*, vol. **1**, pp. 124–127, 2002.
62. C. Zhou and H. Y. David Yang, "Design considerations of miniaturized least dispersive periodic slow-wave structures," *IEEE Trans. Microw. Theory Techn.*, vol. **56**, pp. 467–474, 2008.
63. J. Wang, B.-Z. Wang, Y.-X. Guo, L. C. Ong, and S. Xiao "A compact slow-wave microstrip branch-line coupler with high performance," *IEEE Microw. Wireless Compon. Lett.*, vol. **17**, pp. 501–503, 2007.

64. N. Yang, C. Caloz, and K. Wu, "Slow-wave rail coplanar strip (RCPS) line with low impedance capability," *Proc. 29th URSI General Assembly*, Chicago, IL, August 2008.
65. D. Kaddour, H. Issa, A.-L. Franc, N. Corrao, E. Pistono, F. Podevin, J.-M. Fournier, J.-M. Duchamp, and P. Ferrari, "High-Q slow-wave coplanar transmission lines on 0.35-μm CMOS process," *IEEE Microw. Wireless Compon. Lett.*, vol. **19**, pp. 542–544, 2009.
66. T. LaRocca, J. Y.-C. Liu, and M.-C. F. Chang, "60 GHz CMOS amplifiers using transformer-coupling and artificial dielectric differential transmission lines for compact design," *IEEE J. Solid State Circ.*, vol. **44**, pp. 1425–1435, 2009.
67. J. J. Lee and C. S. Park, "A slow-wave microstrip line with a high-Q and a high dielectric constant for millimeter-wave CMOS applications," *IEEE Microw. Wireless Compon. Lett.*, vol. **20**, pp. 381–383, 2010.
68. A.-L. Franc, D. Kaddour, H. Issa, E. Pistono, N. Corrao, J.-M. Fournier, and P. Ferrari, "Impact of technology dispersion on slow-wave high performance shielded CPW transmission lines characteristics," *Microw. Opt. Technol. Lett.*, vol. **52**, pp. 2786, 2789, 2010.
69. M. Abdel Aziz, H. Issa, D. Kaddour, F. Podevin, A. M. E. Safwat, E. Pistono, J.-M. Duchamp, A. Vilcot, J.-M. Fournier, and P. Ferrari, "Slow-wave high-Q coplanar striplines in CMOS technology and their RLCG model," *Microw. Opt. Technol. Lett.*, vol. **54**, pp. 650–654, 2012.
70. J. García-García J. Bonache, and F. Martín, "Application of electromagnetic bandgaps (EBGs) to the design of ultra wide band pass filters (UWBPFs) with good out-of-band performance," *IEEE Trans. Microw. Theory Techn.*, vol. **54**, pp. 4136–4140, 2006.
71. R. Levy, "A new class of distributed prototype filters with applications to mixed lumped/distributed component design," *IEEE Trans. Microw. Theory Techn.*, vol. **18**, pp. 1064–1071, 1970.
72. J.-S. Hong and M. L. Lancaster, *Microstrip Filters for RF/Microwave Applications*, Wiley, Hoboken, NJ, 2001.
73. P. Vélez, J. Bonache, J. Mata-Contreras, and F. Martín, "Ultra-wideband (UWB) balanced bandpass filters with wide stop band and intrinsic common-mode rejection based on embedded capacitive electromagnetic bandgaps (EBG)," *IEEE MTT-S Int. Microwave Symp. Dig.*, Tampa, FL, June 2014.
74. M. Chudzik, I. Arnedo, A. Lujambio, I. Arregui, F. Teberio, M. A. G. Laso, and T. Lopetegi, "Microstrip coupled-line directional coupler with enhanced coupling based on EBG concept," *Electron. Lett.*, vol. **47**, pp. 1284–1286, 2011.
75. M. Chudzik, I. Arnedo, A. Lujambio, I. Arregui, F. Teberio, D. Benito, T. Lopetegi, and M. A. G. Laso, "Design of EBG microstrip directional coupler with high directivity and coupling," *Proc. 42th Eur. Microw. Conf.*, Amsterdam, the Netherlands, October 2012.
76. P. Vélez, J. Bonache, and F. Martín, "Differential microstrip lines with common-mode suppression based on electromagnetic bandgaps (EBGs)," *IEEE Antennas Wireless Propag. Lett.*, vol. **14**, pp. 40–43, 2015.

3

METAMATERIAL TRANSMISSION LINES: FUNDAMENTALS, THEORY, CIRCUIT MODELS, AND MAIN IMPLEMENTATIONS

3.1 INTRODUCTION, TERMINOLOGY, AND SCOPE

There is not a universal definition of the term *metamaterial*. However, there is a general agreement that metamaterials are artificial structures exhibiting unusual and controllable electromagnetic (EM), optical, or acoustic properties.[1] Metamaterials are periodic (or quasi-periodic) structures with unit cells (or "atoms") consisting of combinations of metals and/or dielectrics.[2] Rather than from the composition of their constituent elements, the unusual (and sometimes exotic) properties of metamaterials come from their structure. Therefore, by properly engineering or designing these artificial materials, it is possible not only to achieve properties beyond those properties that can be found among natural media,[3] but also to control or tune them, in order to obtain certain requirements, specifications, or performance. Probably, the most popular metamaterial structures (or at least those that have been able to attract the interest of broadcasting media) are those devoted to hide objects (also known

[1] Sometimes, metamaterials are classified according to the nature of the waves to which they interact as acoustic, electromagnetic (including RF, microwave, millimeter-wave, and terahertz), or optical (photonic) metamaterials.

[2] Exceptions to this are random media, although for many researchers such artificial structures do not belong to the domain of metamaterials. On the other hand, by quasiperiodic structures we refer to artificial structures where the period, or any other structure parameter, is modulated in order to obtain variable properties with position, or improved functionalities.

[3] From the Greek language, the prefix "meta" means "beyond," "of a higher order."

Artificial Transmission Lines for RF and Microwave Applications, First Edition. Ferran Martín.

as invisibility cloaks.[4]) These structures are clear examples of metamaterials engineered to achieve unprecedented properties, such as controllable wave guiding, in such a way that scattering is inhibited and the cloak (and the "objects" inside it) cannot be detected.

For many researchers in the field, the term *metamaterial* is restricted only to those artificial periodic structures with unit cell dimensions much smaller than the guided wavelength. In these artificial materials, the incident radiation "sees" the structure as a continuous (or effective) medium, so that it can be described or modeled in terms of effective parameters, such as the effective permittivity, permeability, or refractive index (these effective medium metamaterials operate in the refraction regime). For other researchers, metamaterials do also include those periodic structures exhibiting their properties at wavelengths comparable to the period of the structure (i.e., working in the diffraction regime), such as the photonic crystals (PCs) or electromagnetic bandgaps (EBGs), studied in the previous chapter (particularly EBG-based artificial lines). It is obvious that the properties of effective medium metamaterials and PCs (or EBGs) come from different principles, but a discussion on the scope of the term *metamaterial* is not worthy. Nevertheless, effective medium metamaterials seem to offer a higher degree of flexibility and controllability (as compared to PCs or EBGs) for the design of novel functional devices, or to improve the performance of existing ones. Indeed, the research activity in the field has been progressively dominated by effective medium metamaterials in the recent years, and probably this explains that most researchers consider the operation in the refraction regime as a requirement for designating an artificial periodic structure as a metamaterial.

In Chapter 2, periodic transmission lines based on EBGs or capacitively loaded transmission lines were studied, but they were not designated as metamaterial transmission lines. Essentially, EBG-based lines are periodic structures able to inhibit signal propagation at certain frequency bands, and capacitively loaded periodic lines are not only able to filter certain frequencies, but they also exhibit a slow wave effect useful for device miniaturization. However, the controllability of the dispersion diagram and Bloch impedance of EBG-based and capacitively loaded lines is very limited, as compared to that of artificial transmission lines that mimic effective medium metamaterials. These latter lines, consisting of host lines loaded with reactive elements (inductors, capacitors, resonators, or a combination of them) are usually referred to as metamaterial transmission lines, and become the aim of the present chapter. Despite the fact that metamaterial transmission lines are one-dimensional structures, effective parameters, such as the effective permeability and permittivity, can be defined (and tailored) in such lines, as will be shown later. However, rather than the permeability or permittivity, the interest in these lines is the controllability of the main line parameters, that is, the characteristic impedance and the phase constant (or dispersion). Thanks to this controllability (much superior than in conventional lines), RF/microwave devices with improved performance, or devices with novel functionalities can be implemented, as will be demonstrated in Chapter 4 (several RF/microwave applications of

[4] There are many papers focused on the implementation of invisibility cloaks. The author recommends the seminal paper by Smith and co-workers [1], where a realization at microwave frequencies is reported.

metamaterial transmission lines can also be found in general textbooks covering the topic of electromagnetic metamaterials [2–4]).

Although the first proposed metamaterial transmission lines were implemented by means of periodic structures with many periods [5–8], few cells (and even a single unit cell) sometimes suffice to achieve the required line specifications (this is interesting for device miniaturization) [9]. Therefore, in this book, the periodicity is not a requirement for the consideration of an artificial transmission line with controllable dispersion and characteristic impedance as a metamaterial transmission line. The key aspect is the superior controllability, as compared to conventional lines, of line parameters. If the metamaterial transmission line comprises many periods, the guided wavelength at the frequencies of interest is typically much longer than the unit cell dimensions, as required in effective medium metamaterials. However, if the number of cells is small, or if the artificial transmission line contains a single-unit cell, the long wavelength condition may not be satisfied at the frequencies of interest. In the context of this book, metamaterial transmission lines are artificial lines, consisting of a host line loaded with reactive elements, which allow for further control on phase constant and characteristic impedance, as compared to conventional lines. These lines are useful in the design of circuits based on dispersion and impedance engineering, as will be seen, and homogeneity and periodicity are not relevant [9]. For simplicity purposes, metamaterial transmission lines can be designated as *metalines* and the microwave circuits based on them as *metacircuits*.[5]

For completeness, we will also include in this book (Chapter 4) transmission lines loaded with metamaterial resonators, that is, electrically small resonators coupled to the lines, and formerly used for the implementation of effective medium metamaterials. Such lines are usually identified as transmission lines with metamaterial loading, and their functionality is based on particle resonance, rather than on dispersion and impedance engineering.[6] For this main reason, these lines are included in the next chapter, where several applications will be pointed out.

The chapter begins with a very short overview of effective medium metamaterials, with special emphasis on negative index and left-handed (LH) media, including their synthesis and EM properties. This introduction to negative index and LH media is convenient to properly understand the link between effective medium metamaterials and metamaterial transmission lines, and is of special interest to understand the origin of the so-called resonant-type metamaterial transmission lines [4, 8, 10] (exhaustively

[5] It is important to clarify that there are two- and three-dimensional effective media metamaterials implemented by extending the metamaterial transmission line concept to two and three dimensions. Such artificial structures are called transmission line metamaterials. Indeed, metamaterial transmission lines satisfying the periodicity and homogeneity condition are one-dimensional transmission line metamaterials, but not all the transmission line metamaterials are metamaterial transmission lines. In other words, the terms "metamaterial transmission line" (or metaline) and "transmission line metamaterial" are not pure synonymous.

[6] Notice that metamaterial transmission lines, that is, lines with controllable dispersion and characteristic impedance (above that achievable in conventional lines), can be implemented by means of metamaterial resonators loading the line. However, as long as the functionality of these lines is based on dispersion and impedance engineering, they are classified as metamaterial transmission lines (metalines), rather than as transmission lines with metamaterial loading.

studied in this chapter),[7] and other properties of metamaterial transmission lines as well. The chapter will continue with the introduction of the dual transmission line concept, as a ladder network consisting of a cascade of series capacitances and shunt connected inductances, that is, the dual counterpart of the lossless model of a conventional line. Although dual lines mimic the behavior of LH media (i.e., such lines support backward waves), in practice purely dual transmission lines are difficult (not to say impossible) to implement. As will be shown, these lines are implemented by loading a host line with series capacitors and shunt inductors, which means that at sufficiently high frequencies the line parameters are dominant and the structure behaves as a conventional line at these frequencies. Thus, the synthesis of artificial lines exhibiting backward wave propagation (i.e., LH lines) leads to metamaterial transmission lines actually exhibiting a composite right-/left-handed (CRLH) behavior [11], that is, backward wave propagation at low frequencies and forward (or right handed—RH) wave propagation at high frequencies. CRLH lines are the subject of an exhaustive analysis and study in this chapter. This includes the main CRLH line types, their properties, and their physical implementations. Since the applications of CRLH lines are very extensive and constitute an important part of this book, they are covered in a separate chapter (Chapter 4), together with the applications of transmission lines with metamaterial loading. Thus, Chapter 4, entirely focused on applications, has been mainly conceived as an independent chapter in order to avoid an excessively long Chapter 3. Other advanced artificial transmission lines are left for Chapter 6.

3.2 EFFECTIVE MEDIUM METAMATERIALS

The research activity in the field of metamaterials and their applications has experienced a significant growth since 2000, when the first effective medium metamaterial structure exhibiting negative permeability and permittivity simultaneously was fabricated and characterized [12]. This work and other subsequent works [13–15] confirmed some of the predictions of Veselago [16], made more than three decades before, relative to the properties of hypothetical media exhibiting a simultaneous negative value of the dielectric permittivity and magnetic permeability (such properties will be discussed later). Nowadays, it is accepted that *Metamaterials* constitute a transversal topic within science and technology, where disciplines as diverse as physics, chemistry, electromagnetism, electronics engineering, microwaves, photonics, optoelectronics, acoustics, micro/nanotechnology, and materials science, among others, and applications in fields such as sensors, medical diagnosis, security, antennas and radiofrequency identification (RFID), and so on, are involved. The increasing number of available books [2–4, 17–22], conferences, focused sessions, journal issues, and funded projects worldwide related to metamaterials is indicative of the high interest of these artificial structures in the past years, and many researchers, including the author of this book, refer to the

[7] Since resonant-type metamaterial transmission lines are exhaustively studied in this chapter, the analysis of several electrically small resonators useful for the synthesis of such lines and for the design of microwave circuits based on them, is also included.

exponential growth of the topic starting in 2000 as the "big bang" of *Metamaterials Science and Technology*. Key to this exponential growth was the physical implementation of artificial structures with unprecedented properties, derived from the simultaneous negative value of the effective permittivity and permeability. Such structures have been designated as double-negative (DNG) media [23], negative refractive index (NRI) media [24], backward media [25], Veselago media [26], or left-handed (LH) media [12, 16]. The latter term is the one which will be mainly used along this chapter, although some of the alternative designations will be occasionally used to highlight some medium properties linked to the term.

Although the LH structures (mainly due to their unique and controllable properties) are those artificial media that motivated the intensive research activity in the field of metamaterials from the beginning of the century, single negative (SNG) artificial structures were already pointed out before 2000. Indeed, structures consisting of parallel plates, waveguides, and metallic wires were presented as artificial plasmas scaled at microwave frequencies at the beginning of sixties by Rotman [27]. Since lossless plasmas exhibit a negative permittivity below the so-called plasma frequency, the work by Rotman and subsequent works [28, 29] opened the way to the design of artificial media with negative permittivity (such SNG media exhibiting negative permittivity have been called epsilon negative—ENG—structures). Artificial structures with negative permeability (also called mu negative—MNG—media), consisting of an array of split-ring resonators (SRRs), were also proposed in the late nineties by Pendry *et al.* [30]. However, SNG structures (either ENG or MNG) do also exist in natural form. Thus, low-loss plasmas and metals and semiconductors at optical and infrared frequencies exhibit a negative permittivity below the plasma frequency. Media with negative permeability are less common in nature due to the weak magnetic interaction in most solid state materials. Nevertheless, ferrites magnetized to saturation exhibit a tensor magnetic permeability with negative elements near the ferromagnetic resonance.[8] By contrast, DNG materials do not exist in natural form. The lack of naturally available DNG materials essentially means that the electrodynamics of media having simultaneously negative permittivity and permeability has been obviated for years. Nevertheless, notice that since the propagation constant of a plane wave is given by $k = \omega(\varepsilon\mu)^{1/2}$, a simultaneous change of sign in the permittivity and permeability does not make the medium opaque. In other words, LH media are transparent (provided the wave impedance of the DNG medium is not very different from that of the surrounding medium). However, with the exception of the seminal and meritorious work of Veselago [16], the electrodynamics, main properties, and limitations of DNG media were not an object of interest until 2000, when the physical implementation of LH media became a reality.

3.2.1 Wave Propagation in LH Media

Wave propagation in media with simultaneously negative permittivity and permeability was first analyzed and discussed by Veselago [16] at the end of 1960s, although it

[8] Ferrimagnetic materials are widely used in microwave engineering for the implementation of nonreciprocal devices, such as circulators.

was necessary to wait for more than 30 years to see the first practical realization of an LH medium [12]. In order to study wave propagation in LH media, let us first consider the wave equations (see Appendix A):

$$\nabla^2 \vec{E} + \omega^2 \mu \varepsilon \vec{E} = 0 \tag{3.1a}$$

$$\nabla^2 \vec{H} + \omega^2 \mu \varepsilon \vec{H} = 0 \tag{3.1b}$$

Since expressions (3.1) are not affected by a simultaneous change of sign in ε and μ, it is clear that low-loss LH media must be transparent. In fact, in view of equations (3.1), it can be erroneously concluded that solutions of (3.1) are invariant after a simultaneous change of the signs of ε and μ. However, when Maxwell's first-order differential equations are explicitly considered,

$$\nabla \times \vec{E} = -j\omega\mu \vec{H} \tag{3.2a}$$

$$\nabla \times \vec{H} = j\omega\varepsilon \vec{E} \tag{3.2b}$$

it follows that these solutions are quite different. In fact, for plane-wave fields of the form $\vec{E} = \vec{E}_0 \exp\left(-j\vec{k}\cdot\vec{r} + j\omega t\right)$ and $\vec{H} = \vec{H}_0 \exp\left(-j\vec{k}\cdot\vec{r} + j\omega t\right)$, the above equations become:

$$\vec{k} \times \vec{E} = \omega\mu \vec{H} \tag{3.3a}$$

$$\vec{k} \times \vec{H} = -\omega\varepsilon \vec{E} \tag{3.3b}$$

According to (3.3), for double-positive (DPS) index media (i.e., $\varepsilon > 0$ and $\mu > 0$), $\vec{E}$, $\vec{H}$, and $\vec{k}$ form a RH orthogonal system of vectors. However, for DNG media, expressions (3.3) rewrite as

$$\vec{k} \times \vec{E} = -\omega|\mu| \vec{H} \tag{3.4a}$$

$$\vec{k} \times \vec{H} = \omega|\varepsilon| \vec{E} \tag{3.4b}$$

and $\vec{E}$, $\vec{H}$ and $\vec{k}$ now form an LH triplet,[9] as illustrated in Figure 3.1. However, the direction of the time-averaged flux of energy, which is determined by the real part of the Poynting vector,

[9] For this reason, media with ε and μ simultaneously negative are called left-handed (LH) media [16]. In analogy, artificial media with ε and μ simultaneously positive (or even conventional dielectrics) can be designated as right-handed (RH) media, besides double-positive (DPS) media.

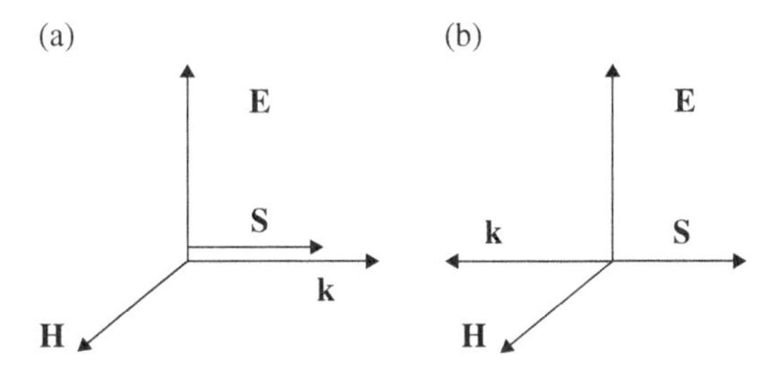

FIGURE 3.1 System of vectors $\vec{E}$, $\vec{H}$, $\vec{k}$, and $\vec{S}$ (depicted as bold symbols) for an ordinary (a) and a LH (b) medium. Reprinted with permission from Ref. [4]; copyright 2008 John Wiley.

$$\vec{S} = \frac{1}{2}\vec{E} \times \vec{H}^* \tag{3.5}$$

is unaffected by a simultaneous change of sign in ε and μ. This means that $\vec{E}$, $\vec{H}$, and $\vec{S}$ form a RH triplet in both types of media (DNG and DPS). Hence, in LH media, $\vec{S}$ and $\vec{k}$ are antiparallel, or, in other words, the direction of energy flow is opposite to the direction of the wave fronts in LH media. Thus, LH media support backward waves; and for this reason, the term *backward media* has also been proposed to designate such media. Nevertheless, the concept of backward waves to describe propagating waves with antiparallel phase and group velocities is not new. Such waves may appear in nonuniform waveguides and periodic structures, and also in the so-called backward-wave (or dual) transmission lines (as will be shown later) [31–33]. However, backward wave propagation in homogenous isotropic media seems to be a unique property of LH media, and the exotic EM properties of such media are intimately related to backward wave propagation, as will be seen in Section 3.2.3.

3.2.2 Losses and Dispersion in LH Media

So far, losses have been neglected in this chapter. In ordinary (RH) media, waves decay as they propagate since the energy is lost due to the effect of losses. However, in LH media, waves grow in the direction of propagation of the wavefronts (i.e., they decay in the direction of propagation of the energy). This result can be obtained, by considering that the real part of the permeability and permittivity are negative, as corresponds to LH media, whilst the imaginary parts are both negative, as results from the well-known complex Poynting theorem and energy conservation considerations [31, 34]. By considering a plane wave with square wave number

$$k^2 = \omega^2 \varepsilon\mu \tag{3.6}$$

it can be deduced that $\mathrm{Im}(k^2) > 0$ in LH media [4]. To demonstrate this, let us express k, ε and μ as

$$k = k' + jk'' \tag{3.7a}$$

$$\varepsilon = \varepsilon' - j\varepsilon'' \tag{3.7b}$$

$$\mu = \mu' - j\mu'' \tag{3.7c}$$

where the explicit negative sign in the imaginary part of ε and μ has been considered, namely, $\text{Im}(\varepsilon) = -\varepsilon''$ and $\text{Im}(\mu) = -\mu''$, ε'' and μ'' being positive quantities. Introducing (3.7) into (3.6), we obtain the following:

$$k'^2 - k''^2 + 2jk'k'' = \omega^2[\mu'\varepsilon' - \mu''\varepsilon'' - j(\mu'\varepsilon'' + \varepsilon'\mu'')] \tag{3.8}$$

From which it follows that $\text{Im}(k^2) = 2k'k'' > 0$ if ε' and μ' are negative. Finally, if $\text{Im}(k^2) > 0$, either

$$\text{Re}(k) > 0 \text{ and } \text{Im}(k) > 0 \tag{3.9a}$$

or

$$\text{Re}(k) < 0 \text{ and } \text{Im}(k) < 0 \tag{3.9b}$$

Notice that (3.9a) corresponds to a backward wave with the energy propagating in the negative direction, whereas (3.9b) corresponds to a backward wave with the energy propagating in the positive direction (according to the usual sign conventions).

An important consequence of the negative values of permittivity and permeability in LH media is that such media are inherently dispersive. Indeed, if negative values of ε and μ are introduced in the usual expression for the time-averaged density of energy in transparent nondispersive media, U_{nd}, given by

$$U_{\text{nd}} = \frac{1}{4}\left(\varepsilon\left|\vec{E}\right|^2 + \mu\left|\vec{H}\right|^2\right) \tag{3.10}$$

a negative density of energy is obtained, which is a nonphysical result. Inspite that any physical media other than vacuum must be dispersive, expression (3.10) can be used as long as the considered medium is weakly dispersive (as occurs in many different cases). The correct expression for the density of energy in the general case (dispersive and nondispersive media) is [35]:

$$U = \frac{1}{4}\left(\frac{\partial(\omega\varepsilon)}{\partial\omega}\left|\vec{E}\right|^2 + \frac{\partial(\omega\mu)}{\partial\omega}\left|\vec{H}\right|^2\right) \tag{3.11}$$

From (3.11), the requirement of positive energy density gives

$$\frac{\partial(\omega\varepsilon)}{\partial\omega} > 0 \quad \text{and} \quad \frac{\partial(\omega\mu)}{\partial\omega} > 0 \tag{3.12}$$

which is compatible with $\varepsilon < 0$ and $\mu < 0$ provided $\partial\varepsilon/\partial\omega > |\varepsilon|/\omega$ and $\partial\mu/\partial\omega > |\mu|/\omega$. This means that any physical implementation of an LH medium must be highly dispersive, in agreement with the low-loss Drude–Lorenz model for ε and μ, which provides negative values for ε and/or μ in the highly dispersive regions, just above the resonances [36].

From the inequalities (3.12), it can also be deduced that the phase and group velocities are of opposite sign in LH media. Namely,

$$\frac{\partial k^2}{\partial\omega} = 2k\frac{\partial k}{\partial\omega} = 2\frac{\omega}{v_p v_g} \tag{3.13}$$

where $v_p = \omega/k$ and $v_g = \partial\omega/\partial k$ are the phase and group velocities, respectively. From (3.6) and (3.12), the following inequality results:

$$\frac{\partial k^2}{\partial\omega} = \omega\mu\frac{\partial(\omega\varepsilon)}{\partial\omega} + \omega\varepsilon\frac{\partial(\omega\mu)}{\partial\omega} < 0 \tag{3.14}$$

provided $\varepsilon < 0$ and $\mu < 0$. Therefore, from (3.13) it is apparent that v_p and v_g have opposite signs, that is, wavepackets and wavefronts travel in opposite directions in LH media.

3.2.3 Main Electromagnetic Properties of LH Metamaterials

Besides backward wave propagation, LH media exhibit further unusual properties, such as a negative refractive index [13, 37–42], subwavelength resolution and focusing [43, 44],[10] inverse Doppler effect, backward Cerenkov radiation [45], and negative Goos–Hänchen shift [46–50], among others. Except backward wave propagation and backward Cerenkov radiation (intimately related to backward leaky wave radiation in LH and CRLH transmission lines), the other exotic properties of LH media are of interest in bulk two-dimensional or three-dimensional metamaterials, rather than in metamaterial transmission lines. Therefore, only the implications of backward wave propagation in LH media relative to negative refraction[11] and backward Cerenkov

[10] The topic of subwavelength resolution and focusing has been an object of an intensive research in the recent years. A complete list of references would be too long for inclusion in a book focused on artificial transmission lines. Therefore, only the seminal work by J.B. Pendry [43] and a recent paper that includes many references on the topic [44] are cited for those readers willing to be introduced in the field.

[11] Although negative refraction is not of interest for metamaterial transmission lines, this topic is succinctly considered in this book since the negative refractive index is probably the most genuine characteristic of an LH medium, and it has been the subject of an intensive research activity, both theoretical and experimental. Moreover, negative refractive index materials realized by extending the metamaterial transmission line concept to two- and three dimensions (2D and 3D transmission line metamaterials) have been proposed.

radiation are considered in this textbook. An exhaustive analysis of the other physical effects derived from backward wave propagation in LH media can be found in the references given earlier, in the seminal paper of Veselago [16], or in several books related to metamaterials (see, e.g., Refs [4, 20]).

3.2.3.1 Negative Refraction In the refraction of an incident beam at the interface between an ordinary material and an LH medium, boundary conditions impose continuity of the tangential components of the wavevector along the interface. Hence, due to backward wave propagation in the LH region, it follows that, unlike in ordinary refraction, the angles of incidence and refraction have opposite signs; or in other words, the incident and refracted beams lie at the same side of the normal, as Figure 3.2 illustrates. From the aforementioned continuity of the tangential components of the wavevectors of the incident and refracted beams, it follows that (Fig. 3.2)

$$\frac{\sin\theta_1}{\sin\theta_2} = \frac{-|k_2|}{k_1} = \frac{n_2}{n_1} < 0 \tag{3.15}$$

which is the well-known Snell's law. In this expression, n_1 and n_2 are the refractive indices of the ordinary and LH media, respectively, given by $n^2 = c^2\varepsilon\mu$. Assuming $n_1 > 0$, from Equation 3.15 it follows that $n_2 < 0$. That is, the sign of the square root in the refractive index definition must be chosen to be negative [16]:

$$n = -c\sqrt{\varepsilon\mu} < 0 \tag{3.16}$$

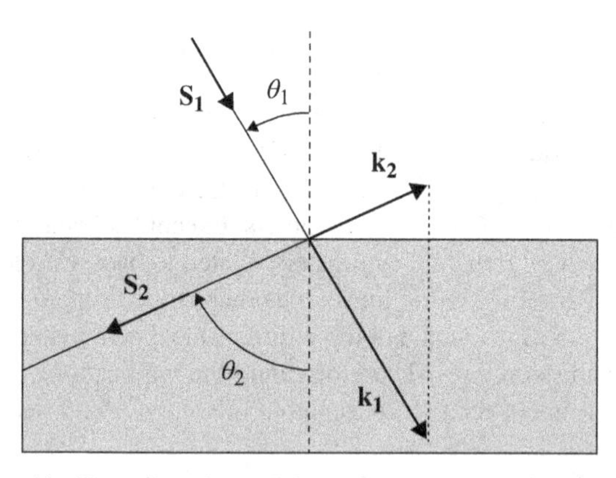

FIGURE 3.2 Graphic illustrating the negative refraction between an ordinary (subscript 1, top) and LH (subscript 2, gray region) medium. Poynting and wavevectors for each medium are labeled as $\vec{S}_1$, $\vec{S}_2$, $\vec{k}_1$, and $\vec{k}_2$, respectively, and depicted as bold symbols in the figure. Negative refraction arises from the continuity of the components of the wavevectors, $\vec{k}_1$ and $\vec{k}_2$, parallel to the interface. The beams propagate along the direction of energy flow, that is, they are parallel to the Poynting vectors. Reprinted with permission from Ref. [4]; copyright 2008 John Wiley.

For this reason, LH media are also referred to as *negative refractive index* (NRI) or *negative refractive* media.

It is important to mention, however, that negative refraction is not exclusive of NRI materials. For instance, it can be achieved at the interface between an ordinary isotropic medium and a hypothetical uniaxial medium with $\varepsilon_\perp > \varepsilon_\parallel$ for very specific angles of incidence and polarization [4], and negative refraction has been demonstrated in PCs [51–53] as well. Thus, the interface between an NRI metamaterial and an ordinary medium exhibits negative refraction, but such unusual effect can be achieved by other means. Light or (more generally) EM radiation bending can be achieved and controlled in anisotropic artificial structures. On the basis of transformation optics, such structures can be engineered in order to exhibit cloaking effects (see, e.g., Ref. [54] and references therein).

3.2.3.2 Backward Cerenkov Radiation Cerenkov radiation is emitted when a charged particle (i.e., an electron) passes through a dielectric medium at a speed, v, greater than the phase velocity of light in that medium (c/n). As the charged particle travels, electrons in the atoms of the medium are displaced, the atoms become polarized, and, to restore equilibrium, photons are emitted. If the particle motion is slow, as compared to the velocity of light in the medium, these photons destructively interfere with each other, and no radiation is detected. Conversely, if the charged particle travels faster than light in the medium the emitted radiation forms a shock wave, similar to the sonic shock waves generated in a supersonic aircraft. This situation is illustrated in Figure 3.3a, where it can be appreciated that the spherical wavefronts radiated by the particles are delayed with regard to the particle motion, with the result of forward Cerenkov radiation. From Figure 3.3a, the radiation angle is given by

$$\cos\theta = \frac{c}{nv} \tag{3.17}$$

In the illustration of Figure 3.3a, it is assumed that the medium surrounding the charged particle is an ordinary medium. Let us now consider that the surrounding medium is LH. Since wave propagation is backward, the spherical wavefronts move inward to the source with velocity $c/|n(\omega)|$, where $n(\omega) < 0$ and the frequency dependence has been explicitly introduced in n to highlight the dispersive nature of the LH medium.[12] Therefore, each wavefront collapses at advanced positions of the particle, as Figure 3.3b illustrates, and the resulting shock wave travels backward forming an obtuse angle with the trajectory of the particle. This angle is given by (3.17) with n replaced with $-|n(\omega)|$.

At this point, it is important to mention that there are similarities between Cerenkov radiation and leaky wave radiation in periodic structures. It was argued in Section 2.2 that open periodic structures may radiate at those frequency regions where the phase velocity of the guided waves; v_p, is higher than the speed of light in vacuum, c,

[12] Since the particle radiates in a broad spectral range, the Cerenkov radiation spectra must exhibit forward and backward wave radiation [45].

(a)

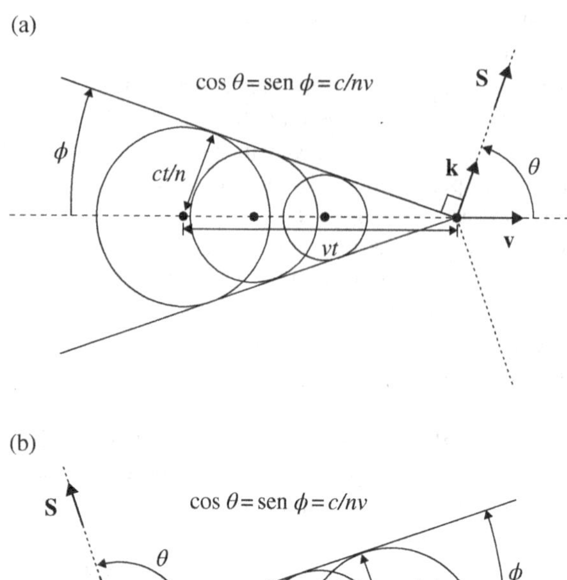

(b)

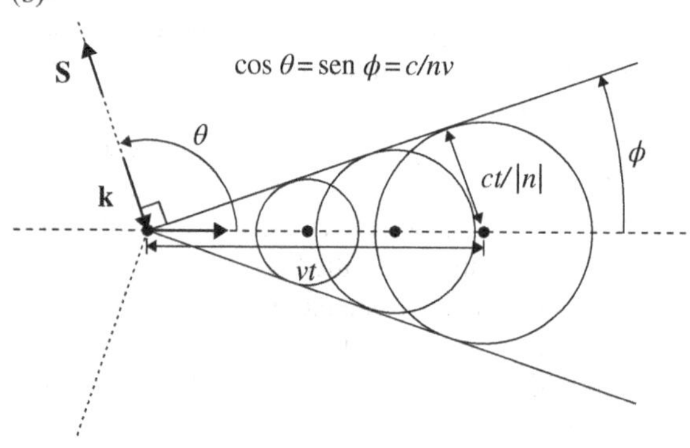

FIGURE 3.3 Illustration of the formation of Cerenkov shock waves in an ordinary medium (a), and in an LH medium (b). In (a), the spherical wavefronts move outward from the source at a velocity c/n; whereas in (b), the spherical wavefronts move inward to the source at a velocity $c/|n(\omega)|$. Reprinted with permission from Ref. [4]; copyright 2008 John Wiley.

(fast-wave regions).[13] This phenomenon can be interpreted in terms of coupling, or energy transfer, between the guided waves and the radiated (leaky) waves. If $v_p > c$ (or c/n in a medium different than vacuum or air), the guided mode can be coupled to a free space (leaky) mode since the wavevector $\vec{k}$ of the space harmonic has a real transversal component, k_{tn}, giving rise to a leaky wave with wavevector $\vec{k}$, as Figure 2.3 illustrates. The Cerenkov radiation can be interpreted in a similar fashion in terms of wave coupling. For superluminal charged particles ($v > c/n$), the fields generated in the transverse direction are responsible for particle radiation since they do not decay, and this can be explained by the real transverse component of the charge wavevector under these conditions. Such coupling can be seen as a conversion

[13] It was assumed in Section 2.2 that the open periodic structure is surrounded by vacuum (or air). However, the analysis of leaky wave radiation in such structures can be generalized to any surrounding medium by simply replacing c with c/n.

between kinetic energy and radiated power. Notice also that (3.17) is equivalent to (2.11), and that these expressions are valid for both forward and backward wave radiation. However, backward Cerenkov radiation occurs because the surrounding medium is LH, whereas backward leaky wave radiation of the space harmonics in periodic structures in certain frequency regions is caused by the backward wave nature of the propagating waves in those regions (the surrounding medium being typically air). For backward Cerenkov radiation, the diagram of Figure 2.3a applies, corresponding to a positive wavevector for the charged particle in the direction of motion. However, notice that since the Poynting vector is in opposite direction to the wavevector, backward wave radiation arises. Forward and backward leaky wave radiation in CRLH transmission lines will be considered in Chapter 4. The phenomenology is essentially the same than leaky wave radiation in periodic structures, although leaky CRLH transmission lines radiate in the fundamental mode.

3.2.4 Synthesis of LH Metamaterials

In this section, the synthesis of the first bulk LH medium, consisting of a combination of metallic posts and split-ring resonators (SRRs) [12], is briefly reviewed, and the SRR is analyzed in certain detail. This will help to understand the origin of the first LH (actually CRLH) transmission line implementation based on SRRs [8], that mimics the one-dimensional medium reported in Ref. [12][14] and constitutes the basis of a set of artificial transmission lines based on the resonant-type approach (such lines and other metamaterial transmission lines will be analyzed in Section 3.5). A complete review of the different reported approaches for the implementation of LH and other artificial media, including two- and three-dimensional structures, would be too long for inclusion in this book. Nevertheless, it is important to mention that implementations of two- and three-dimensional metamaterials based on SRRs [13, 55, 56], as well as structures fabricated by extending the metamaterial transmission line concept to two and three dimensions [57–65], have been reported. It is also worth mentioning that one of the most promising approaches for the synthesis of NRI metamaterials at quasioptical or optical frequencies is based on fishnet structures, which can be implemented by means of micro/nanofabrication technologies (see Ref. [66] and references therein). A recent review of 3D photonic metamaterials can be found in Ref. [67].

The one-dimensional LH structure reported in Ref. [12] consists of a combination of metallic posts (or wires) and SRRs (Fig. 3.4). The structure supports backward waves in a narrow band above SRR resonance (where the effective permittivity and permeability are simultaneously negative) for incident radiation polarized with the electric field parallel to the wires and the magnetic field axial to the SRRs. The metallic wires and SRRs are responsible for the negative effective permittivity and permeability, respectively.

[14] Although the structure reported in Ref. [12] is a bulk LH medium, it is actually a one-dimensional metamaterial, since backward wave propagation occurs only in one direction of propagation and for a specific polarization conditions.

FIGURE 3.4 First synthesized one-dimensional bulk LH metamaterial. Photo courtesy by D.R. Smith.

3.2.4.1 Negative Effective Permittivity Media: Wire Media Let us first consider a square lattice of metallic wires without SRRs for incident plane waves polarized with the electric field parallel to the wires (Fig. 3.5a). Let l and r be the distance between wires and wire radius, respectively. This system can be modeled by a TEM transmission line loaded with metallic posts, as depicted in Figure 3.5b, which in turn can be described by the lumped-element circuit model of Figure 3.5c, where the per-section series inductance and shunt capacitance of the TEM line are $L_s = \mu_o l$ and $C_s = \varepsilon_o l$, respectively, and the shunt inductance of the metallic wires is [4, 28]

$$L = \frac{\mu_o l}{2\pi} \ln\left(\frac{l}{r}\right) \tag{3.18}$$

The dispersion of this periodic structure can be inferred by means of (2.33), with $Z_s(\omega) = j\omega L_s$ and $Z_p(\omega) = Y_p^{-1}(\omega) = [j\omega C_s + 1/(j\omega L)]^{-1}$, giving:

$$\cos(kl) = 1 - \frac{1}{2} L_s C_s \omega^2 \left(1 - \frac{1}{L C_s \omega^2}\right) \tag{3.19}$$

In the long wavelength limit ($kl \to 0$), $\cos(kl) \cong 1 - (kl)^2/2$, and the wavenumber (or phase constant) is found to be

$$k^2 = k_o^2 \left(1 - \frac{2\pi}{\mu_o \varepsilon_o l^2 \omega^2 \ln(l/r)}\right) \tag{3.20}$$

where $k_o^2 = \mu_o \varepsilon_o \omega^2$ is the free space wavenumber. The dispersion relation given by (3.20) is identical to that of an ideal plasma, provided the cutoff frequency is

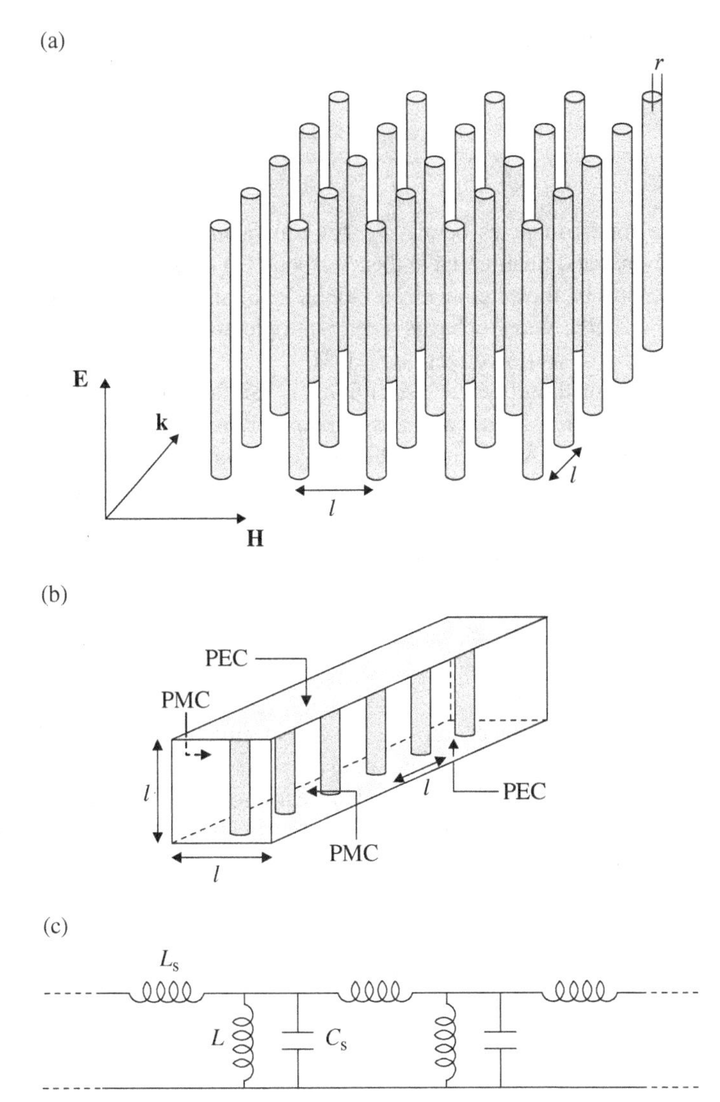

FIGURE 3.5 Artificial wire medium exhibiting negative effective permittivity for incident radiation with the electric field parallel to the wires (a), equivalent structure consisting of a post-loaded TEM transmission line (b), and equivalent circuit model of the transmission line (c). PMC and PEC stand for perfect magnetic and perfect electric conductor, respectively.

substituted by the plasma frequency. Thus the system of wires, for the considered polarization conditions, behaves as an ideal plasma, with plasma frequency:

$$\omega_p^2 = \frac{2\pi}{\mu_o \varepsilon_o l^2 \ln(l/r)} \tag{3.21}$$

and effective permittivity given by

$$\varepsilon_{\text{eff}} = \varepsilon_{\text{o}} \left(1 - \frac{\omega_{\text{p}}^2}{\omega^2} \right) \tag{3.22}$$

being negative for frequencies below ω_{p}. An equivalent expression to (3.21) was derived by Solymar and Shamonina in their textbook [21], by considering the current induced in the wires by the external electric field. It has also been demonstrated that a cubic mesh of metallic wires behaves as an isotropic artificial plasma for all wave polarizations and propagation directions [4].

The expression of the effective permittivity is said to correspond to the Drude model, which is a particular case of the well-known Lorentz model of materials that results when the restoring force is negligible. To gain insight on this, let us assume that an electric field is applied to a certain material. The motion of the electrons is derived from the Newton's second law, giving for the most general case the equation of a forced and damped oscillator, that is,

$$m\frac{d^2\vec{x}}{dt^2} = e\vec{E} - k_{\text{r}}\vec{x} - m\Gamma\frac{d\vec{x}}{dt} \tag{3.23}$$

where e and m are the electron charge and mass, respectively, $\vec{E}$ is the applied electric field, k_{r} is the proportionality constant between the restoring force and displacement (it is assumed that the attractive force of bounded electrons in atoms has a linear dependence on $\vec{x}$), and Γ is a damping factor that accounts for losses. The solution of (3.23), assuming harmonic type time dependence, is

$$\vec{x} = \frac{e}{m\left(\omega_{\text{o}}^2 - \omega^2 + j\Gamma\omega\right)}\vec{E} \tag{3.24}$$

where $\omega_{\text{o}}^2 = k/m$. If N is the density of electrons, the polarization induced by $\vec{E}$ is simply $\vec{P} = Ne\ \vec{x}$.[15] Since the polarization is related to the external field by the electric susceptibility, χ_{e}, that is, $\vec{P} = \varepsilon_{\text{o}}\chi_{\text{e}}\ \vec{E}$, it follows that the permittivity of the medium $\varepsilon = \varepsilon_{\text{o}}(1 + \chi_{\text{e}})$ is (Lorentz model):

$$\varepsilon = \varepsilon_{\text{o}} \left(1 + \frac{\omega_{\text{p}}^2}{\left(\omega_{\text{o}}^2 - \omega^2 + j\Gamma\omega\right)} \right) \tag{3.25}$$

[15] The polarization is defined as the density of electric dipole moments. Since the electric dipole moment of two point charges separated a distance $\vec{x}$ is $\vec{p} = e\ \vec{x}$, the aforementioned expression for the polarization is obtained.

with

$$\omega_p^2 = \frac{Ne^2}{\varepsilon_o m} \tag{3.26}$$

For free electrons in metals, the restoring force is zero ($\omega_o = 0$), and the Drude model results, that is,

$$\varepsilon = \varepsilon_o \left(1 - \frac{\omega_p^2}{(\omega^2 - j\Gamma\omega)}\right) \tag{3.27}$$

where the real part of the dielectric permittivity is found to be negative for frequencies satisfying $\omega^2 < {\omega_p}^2 - \Gamma^2$. If losses are neglected ($\Gamma = 0$), expression (3.22) results, and the dielectric permittivity is negative below the plasma frequency. It should be mentioned, however, that losses are always present in metals, and therefore (3.27) must be applied. At low frequencies, the real part of the permittivity is negative, but the imaginary part is very large and becomes dominant. Under these conditions, the conduction current dominates over the displacement current, and dispersion is dictated by (1.78). From a physical viewpoint, in the low-frequency regime, electron motion in metals is dictated by continuous dissipative collisions and the plasma-like behavior is not manifested. Conversely, near the plasma frequency, the amplitude of the collective electron oscillations generated by the applied electric field is small as compared to the electron mean free path, and the metal exhibits the plasma-like behavior. Notice that the main difference between the Drude and Lorentz models concerns the bandwidth of the frequency region where the real part of the dielectric permittivity is negative. The Lorentz model predicts a negative dielectric permittivity in a narrow band just above the resonance frequency ω_o.

In ordinary dielectric materials, electrons are bounded to atoms and the restoring force and damping dominate over the acceleration of the charges. Under these conditions, the solution of (3.23) gives a dielectric permittivity of the form

$$\varepsilon = \varepsilon_o \left(1 + \frac{\omega_p^2}{(\omega_o^2 + j\Gamma\omega)}\right) \tag{3.28}$$

corresponding to the Debye model of materials.

Notice that, in the previous analysis, the dielectric permittivity has been considered a scalar and the considered material has been assumed to be isotropic. However, many materials exhibit anisotropy (or even bianisotropy—a concept introduced later), and the permittivity is a tensor, rather than a scalar. Indeed, the one-dimensional wire media discussed before (described by the post-loaded TEM line of Fig. 3.5) and many other artificial media are highly anisotropic. In particular, the analyzed wire structure is an ENG medium only for incident radiation polarized with the electric field parallel to the wires.

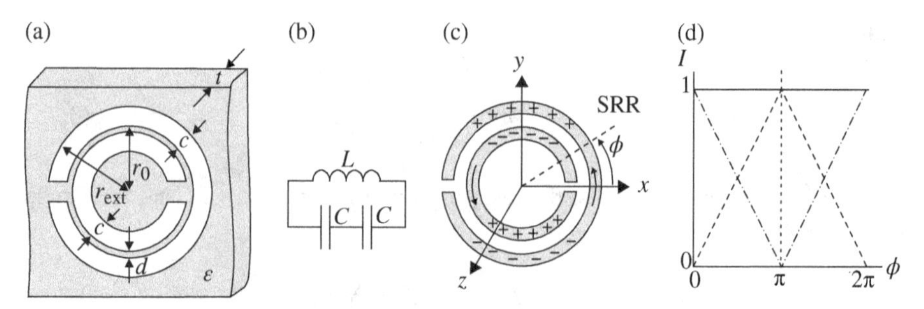

FIGURE 3.6 Topology of the SRR and relevant dimensions (a), equivalent circuit model (b), distribution of charges at the fundamental resonance (c), and plot of the angular dependence of the currents for the inner (dashed line) and outer (dash-dotted line) ring (d). The current flows in the same direction in both rings for the fundamental resonance, and the sum of the currents of both rings is uniform along the whole circumference [4]. Notice that the symmetry plane (*x*-*z* plane) is an electric wall. The metallization in (a) is indicated in white. Reprinted with permission from Ref. [4]; copyright 2008 John Wiley.

*3.2.4.2 **Negative Effective Permeability Media: SRRs*** Unlike dielectric materials, magnetic materials (i.e., materials exhibiting magnetic activity) are less common. In particular, except ferrites at certain frequencies, negative permeability materials (MNG) are not available in nature. The implementation of artificial media with negative permeability is of interest not only for the synthesis of LH media but also because the magnetic response of most materials tails off in the GHz range. In principle, a set of conducting rings (all of them with the axis oriented in the same direction) could be a good candidate for the realization of a negative permeability medium for EM radiation polarized with the magnetic field parallel to the rings axis. The time-varying magnetic field induces circulating currents in the rings, which in turn generate a secondary magnetic flux opposite to that created by the external field. The magnetic polarizability of a conducting ring is thus negative, providing a negative dipole moment, and hence a negative magnetic susceptibility for the set of rings. However, this diamagnetic behavior is not strong enough to generate negative permeability (a detailed justification is given in Ref. [4]).

The magnetic polarizability of a closed metallic ring can be enhanced by loading the loop with a capacitor [68], or by any other means providing a resonant behavior. In particular, Pendry *et al.* [30] proposed the topology depicted in Figure 3.6a, known as split-ring resonator (SRR). The SRR can be implemented in fully planar technology, yet achieving small electrical sizes[16] on account of the distributed capacitance of the particle. Above the resonance frequency of the SRR, a strong diamagnetic behavior is expected for the external magnetic field polarized in the axial direction; hence, an

[16] The electrical size of the SRR is usually given by its diameter expressed in terms of the free space wavelength at resonance. Alternatively, if the SRR loads a transmission line (as will be later considered), the electrical size can be expressed in terms of the guided wavelength at resonance.

array of SRRs is expected to exhibit a negative effective permeability in a certain band of frequencies, under these polarization conditions. A detailed demonstration of this is given in Refs. [4, 68]. Nevertheless, a simplified analysis is provided in the next paragraphs for completeness.

When a time-varying magnetic field is applied to the axial direction (z-direction), current loops are induced in the rings, and current flows from one ring to the other through the distributed (edge) capacitances between the rings. As long as both rings are very close one each other, the edge capacitance is significant, resulting in considerable coupling between both rings. Under these conditions, a quasistatic analysis applies, and the SRR can be modeled by the circuit depicted in Figure 3.6b, where L is the SRR self-inductance and C is the edge capacitance corresponding to each SRR half. This capacitance is $C = \pi r_o C_{pul}$, where r_o is the mean radius of the SRR ($r_o = r_{ext} - c - d/2$), and C_{pul} is the per-unit-length capacitance along the slot between the rings. The self-inductance can be calculated as the inductance of a single closed loop with identical width to that of the individual rings (c), and mean radius (r_o) [69]. Quasianalytical expressions for L and C_{pul} are given in Ref. [4] (in particular, C_{pul} is inferred from formulas extracted from Ref. [70]). The fundamental resonance frequency of the SRR is thus:

$$\omega_o = \sqrt{\frac{2}{LC}} = \sqrt{\frac{2}{L\pi r_o C_{pul}}} \tag{3.29}$$

Let us assume that the SRR is excited by an external axial magnetic field that generates a magnetic flux, $\Phi_{ext} = \pi r_o^2 B_z^{ext}$, across the SRR. The induced current in the SRR can be expressed as follows:

$$I = -j\omega \Phi_{ext} \left(\frac{2}{j\omega C} + j\omega L \right)^{-1} \tag{3.30}$$

Since the induced dipolar magnetic moment is given by $m_z = \pi r_o^2 I$, and it is related to the magnetic polarizability by $m_z = \alpha B_z^{ext}$, it follows that the magnetic polarizability is[17]

$$\alpha = \frac{\pi^2 r_o^4}{L} \left(\frac{\omega_o^2}{\omega^2} - 1 \right)^{-1} \tag{3.31}$$

[17] In Ref. [4], the magnetic polarizability given by expression (3.31) is designated as α_{zz}^{mm}, where the double subindex zz indicates that it is due to a force directed along the z-axis, and produces an effect along the same axis, and the double superindex mm indicates that the external force is magnetic, and that the effect is also magnetic. The reason for such nomenclature is that the SRR exhibits further polarizabilities, including a resonant-type electric polarizability along the y-axis, and resonant cross-polarizabilities. Such polarizabilities are not considered in this book (the expressions for such polarizabilities can be found in Ref. [4]).

and it experiences a strong variation from highly positive to highly negative values at the resonance frequency of the SRR.

In the previous analysis, losses have been neglected. However, a more realistic picture of the SRR must include ohmic losses. It can be easily demonstrated that the polarizability of the SRR, including the effects of losses through the AC resistance, R, of the SRR, is [4]

$$\alpha = \frac{\pi^2 r_o^4}{L}\left(\frac{\omega^2}{\omega_o^2 - \omega^2 + j\omega\Gamma}\right) \tag{3.32}$$

where $\Gamma = R/L$.

Once the magnetic polarizability has been inferred, the magnetic susceptibility is calculated, to a first-order approximation,[18] as $\chi_m = \mu_o \alpha / a^3$, and the effective permeability is [4]:

$$\mu_{eff} = \mu_o(1 + \chi_m) = \mu_o\left(1 + \frac{\mu_o \pi^2 r_o^4}{a^3 L}\left(\frac{\omega^2}{\omega_o^2 - \omega^2 + j\omega\Gamma}\right)\right) \tag{3.33}$$

where a cubic lattice SRR structure with unit cell volume given by a^3 has been considered (Fig. 3.7). Expression (3.33) can be written in a more compact form as follows:

$$\mu_{eff} = \mu_o\left(1 - \frac{F\omega^2}{\omega^2 - \omega_o^2 - j\omega\Gamma}\right) \tag{3.34}$$

where F is a dimensionless factor less than 1[19] given by

$$F = \frac{\mu_o \pi^2 r_o^4}{a^3 L} \tag{3.35}$$

If we assume that losses are very small, the magnetic permeability is negative in the range $\omega_o < \omega < \omega_o/(1 - F)^{1/2}$, that is, in a narrow band. The expression for the effective permeability given by (3.34), corresponding to the Lorentz model, is identical to the one reported by Shamonina and Solymar [21], and very similar to the one reported by Pendry *et al.* [30].

Notice that (3.34) gives the effective magnetic permeability of the structure of Figure 3.7 in the z-direction. Actually, the structure of Figure 3.7 is bianisotropic, that is, it exhibits cross polarizabilities, and the constitutive relationships are characterized by electric, magnetic, and magnetoelectric susceptibility tensors [4]. However, for the

[18] This approximation ignores the coupling between adjacent SRRs (more accurate expressions are given in Refs. [4, 21]).

[19] The reason is that $a^2 > \pi r_o^2$ and L is significantly larger than $\mu_o r_o$ [69] for reasonable values of c and r_o (a justification of this for a cylindrical ring can be found in Ref. [4]).

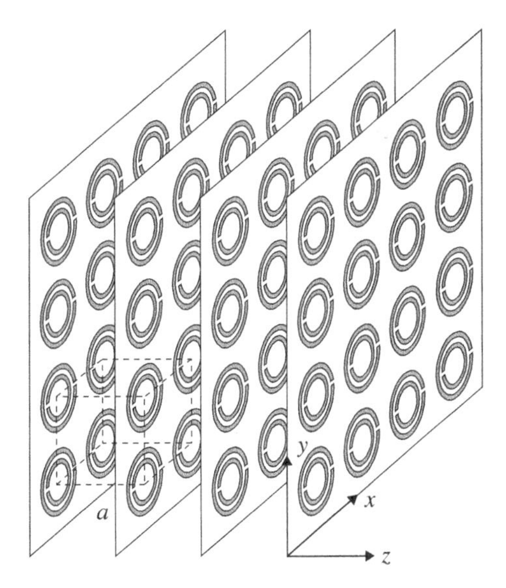

FIGURE 3.7 MNG artificial medium made of a cubic lattice of SRRs for incident radiation with the magnetic field polarized in the z-direction. Reprinted with permission from Ref. [4]; copyright 2008 John Wiley.

purposes of this book, the previous analysis suffices. Nevertheless, it is worth mentioning that the SRR exhibits also an electric dipole moment in the plane of the particle (y-direction), as can be seen from the distribution of charges in the rings (Fig. 3.6c). This means that the SRR can be driven by a time-varying electric field applied in the y-direction as well. In other words, the magnetic and the electric dipole moments of the particle can be induced by either an electric or a magnetic field (applied in the convenient direction). The fact that the SRR exhibits magnetoelectric coupling has implications for the implementation and modeling of metamaterial transmission lines based on these particles, as will be discussed later.

3.2.4.3 Combining SRRs and Metallic Wires: One-Dimensional LH Medium

The first one-dimensional LH artificial structure, designed by Smith and co-workers (Fig. 3.4), was implemented by combining the wire and SRR media discussed in the previous sections [12]. The superposition hypothesis that the combination of an ENG and an MNG medium gives an LH medium was assumed. However, such hypothesis is not in general justified, and the implementation of an LH medium by combining SRRs and wires (or posts) requires the consideration of the possible interactions between the constitutive elements. In particular, in the medium reported by Smith (Fig. 3.4), caution was taken to avoid the potential influence of the SRRs on the inductance of the posts by alternating the planes containing the wires and the SRRs (further details on the superposition principle are given in Refs. [4, 71, 72]).

The structure of Figure 3.4 was illuminated with a plane wave polarized with the electric and magnetic field directed toward the post and SRR axis, respectively, and a bandpass response was obtained in a narrow band, just above the SRR resonance frequency (Fig. 3.8a). It was also demonstrated that by removing the metallic posts, a stop band behavior results (Fig. 3.8a). Such stop band was attributed to the negative effective permeability of the SRR medium. However, by incorporating the posts, the

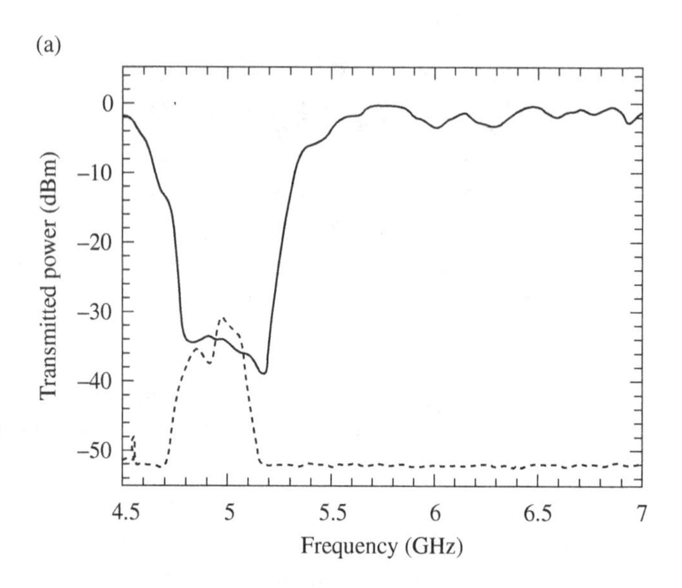

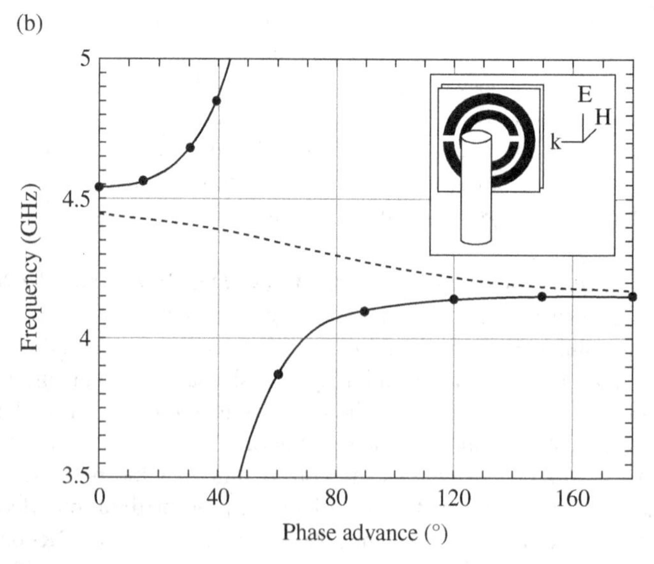

FIGURE 3.8 (a) Transmitted power for the structure of Figure 3.4 (dashed line) and for the structure without metallic posts (solid line); (b) dispersion relation. The polarization conditions are explained in the text, and the details of the experiment are given in Ref. [12]. Reprinted with permission from Ref. [12]; copyright 2000 American Physical Society.

SRRs are embedded in a negative permittivity medium with plasma frequency above the SRR resonance frequency (12 GHz approximately). Therefore, the composite medium exhibits LH wave propagation in the region of negative permeability, where the effective permittivity is also negative. Outside that band, the structure is an opaque ENG medium and the transmitted power is attenuated to below the noise floor. The small power transmission in the LH pass band can be mainly attributed to a significant mismatch between the composite medium and air. The LH behavior of the composite SRR-wire structure was confirmed through the dispersion relation, inferred from a commercial EM mode solver [12]. The results, depicted in Figure 3.8b, indicate that the phase and group velocities have different sign in the pass band region (dashed line between 4.15 GHz and 4.45 GHz), as one expects in a LH medium. The dispersion relation corresponding to the SRR medium without wires, also included in Figure 3.8b as bold line, points out that wave propagation in the allowed regions is forward, and that the stop band (associated to the negative effective permeability) roughly coincides with the LH band of the composite SRR-wire medium.

It is noticeable that an LH medium was also achieved by introducing a row of SRRs inside a rectangular metallic waveguide (along its middle E-plane) designed to exhibit its cutoff frequency above the SRR fundamental resonance frequency [73]. In such structure, the negative permittivity (for the fundamental TE_{10} mode) is provided by the cutoff waveguide, whereas the negative permeability is due to the SRRs. Indeed the dispersion relation of such medium is formally identical to that of the SRR-wire media, which can be inferred from (3.6) replacing ε and μ with $\varepsilon_{\mathrm{eff}}$ and μ_{eff}, respectively. To end this subsection, let us point out that the equivalence between a rectangular waveguide and a dielectric plasma, valid for evanescent TE modes, has a dual counterpart for magnetic plasmas and evanescent TM modes. Namely, an MNG medium can be simulated by means of TM modes below their cutoff frequencies. Using this concept, a DNG medium, consisting of a square waveguide filled with a transverse two-dimensional array of thin metallic wires, for evanescent TM modes, has been proposed [74].

3.3 ELECTRICALLY SMALL RESONATORS FOR METAMATERIALS AND MICROWAVE CIRCUIT DESIGN

The aim of this section is to analyze several resonator topologies for the synthesis of effective media metamaterials, for the implementation of metamaterial transmission lines, and for the design of RF/microwave components based on such resonators or on such artificial transmission lines. The list is limited to those resonators that are used in the following subsections and chapters of this book. All of them exhibit peculiarities that make them useful for certain applications, or represent some advantages over the original SRR. Indeed, with the exception of the folded stepped impedance resonators (SIRs), the other resonators included in the next subsections are derived from (or inspired by) the original SRR topology proposed by Pendry *et al.* [30]. The classification of these resonators can be made according to several criteria. In this book, the considered electrically small resonators are divided into metallic and complementary (or slot) resonators (the latter are derived from the former by applying duality considerations).

3.3.1 Metallic Resonators

It was justified in the previous section the need to load a metallic loop with a capacitance in order to achieve a resonant behavior able to produce a strong diamagnetic behavior above resonance. However, the SRR of Figure 3.6 is not the unique topology providing a negative magnetic polarizability. Further topologies are analyzed and discussed in this subsection. Moreover, the concept of open particle will be introduced (as will be seen later, these open particles can be driven by means of a voltage or current source and are also useful for the implementation of metamaterial transmission lines).

3.3.1.1 The Non-Bianisotropic SRR (NB-SRR) The topology of the NB-SRR [75] is depicted in Figure 3.9. The equivalent circuit model and magnetic polarizability are those of the SRR (provided the same dimensions and substrate are considered). However, the distribution of charges in the NB-SRR differs from that of the SRR (Fig. 3.9). Unlike the SRR, a net electric dipole in the y-direction is not induced by an axial magnetic field in the NB-SRR. This means that cross polarizabilities are not present in this particle, which is said to be non-bianisotropic. This resonator can be of interest in certain applications to avoid mixed (i.e., simultaneous electric and magnetic) coupling.

3.3.1.2 The Broadside-Coupled SRR (BC-SRR) Like the NB-SRR, the BC-SRR (Fig. 3.10) exhibits inversion symmetry and is non-bianisotropic [69]. The drawback of this particle as compared to the NB-SRR, or SRR, is that two metal levels are required for its implementation. However, since the rings are etched face-to-face at both sides of the substrate, the distributed capacitance between the rings can be substantially enhanced by merely reducing the substrate thickness (or by increasing its dielectric constant). Therefore, this particle can potentially be electrically much smaller than the SRR or the NB-SRR.

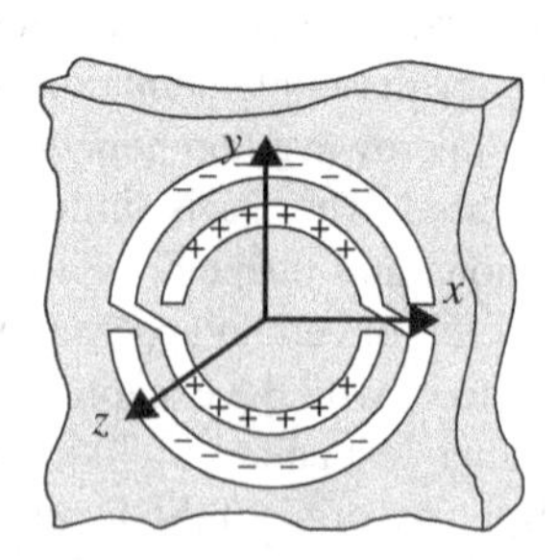

FIGURE 3.9 Topology of the NB-SRR and distribution of charges at the fundamental resonance. The metallization is indicated in white.

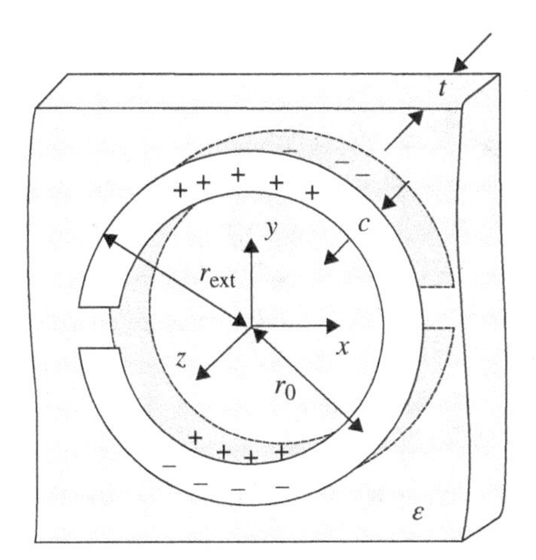

FIGURE 3.10 Topology of the BC-SRR and distribution of charges at the fundamental resonance. The metallization is indicated in white. Reprinted with permission from Ref. [4]; copyright 2008 John Wiley.

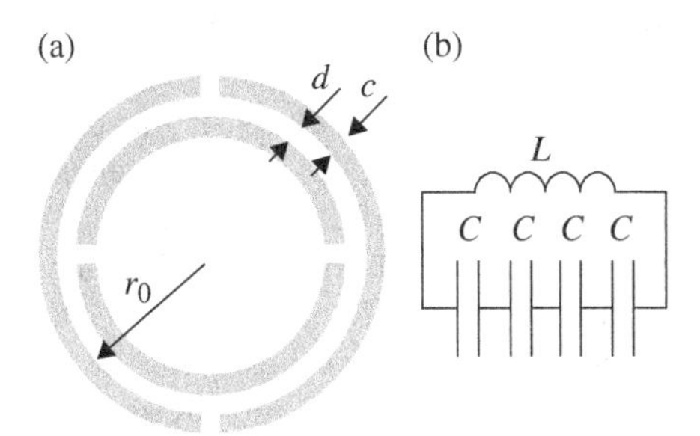

FIGURE 3.11 Topology of the DS-SRR and relevant dimensions (a) and equivalent circuit model (b).

3.3.1.3 ***The Double-Slit SRR (DS-SRR)*** The DS-SRR is derived from the original topology of the SRR by adding two cuts in each ring (Fig. 3.11) [75]. The resulting particle is also non-bianisotropic. It is electrically larger than the SRR since the capacitance of the particle is given by the series connection of the four capacitances corresponding to each quadrant. Such capacitances are $C = \pi r_o C_{pul}/2$, and therefore the total particle capacitance is four times smaller than that of the SRR. Since the inductance is identical to that of the SRR, it follows that the resonance frequency of the DS-SRR is twice the resonance frequency of the SRR (provided the same dimensions, shape, and substrate are considered). Although the particle is electrically larger, the symmetry of this particle is useful for the development of isotropic MNG media [4]. Since the ratio *L/C* is larger than the corresponding ratio in an SRR, the DS-SRR can also be useful for bandwidth broadening in transmission lines loaded with these particles (this aspect will be discussed in Chapter 6, in reference to common-mode suppressed differential lines).

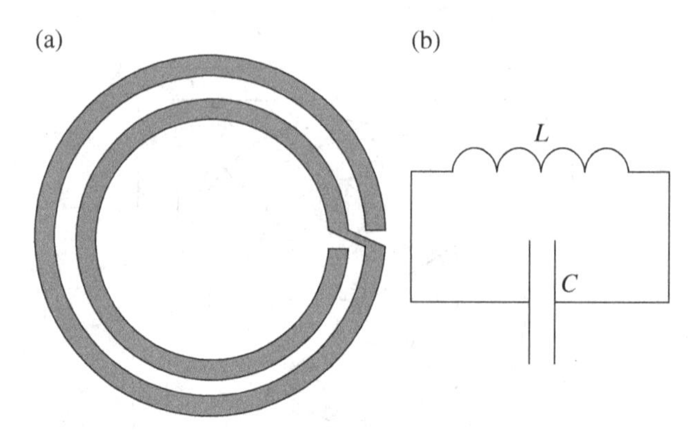

FIGURE 3.12 Topology of a two-turn SR (a) and equivalent circuit model (b).

3.3.1.4 The Spiral Resonator (SR) Spirals are well-known particles that have been largely used for the implementation of inductances in planar technology. However, the parasitic (distributed) capacitance between the turns actually provides a resonant behavior to the particle. Therefore, SRs exhibit a strong diamagnetic behavior above resonance and can be used for the synthesis of MNG media as well. The topology of a two-turn spiral resonator (SR) is depicted in Figure 3.12. Although the two-turn SR is older than the SRR, the elements of the equivalent circuit can be straightforwardly inferred from the analysis of the SRR [76]. Indeed, in terms of the equivalent circuit model, the single difference between these particles concerns the capacitance. In the SR, the edge capacitance is the one corresponding to the whole circumference, and thus given by $C = 2\pi r_o C_{pul}$. Since the inductance is identical to that of the SRR, the SR resonance frequency is half the resonance frequency of the SRR. In other words, the SR is electrically smaller than the SRR by a factor of two.

3.3.1.5 The Folded SIR As was pointed out in Chapter 1, half wavelength resonators are common building blocks in microwave circuit and filter design. Such elements are distributed resonators and hence are electrically large. However, dimensions can be reduced by narrowing and widening the central and external sections of the particle, respectively [77]. Further size reduction can be achieved by folding the structure, giving rise to the folded SIR (Fig. 3.13). The particle can be driven by a z-oriented magnetic field, inducing a strong magnetic dipole in that direction and an electric dipole in the plane of the particle (y-direction). Therefore, like the SRR, the folded SIR exhibits bianisotropy. In the folded SIR, the inductance of the particle is roughly the inductance of the narrow strip and the capacitance is the edge capacitance between the patches. However, coupled to transmission lines (specifically CPWs), a broadside coupling capacitance between the line and the particle arises, giving rise to very small resonance frequencies. This is of interest for size reduction as will be discussed later (see Section 4.3.2.3).

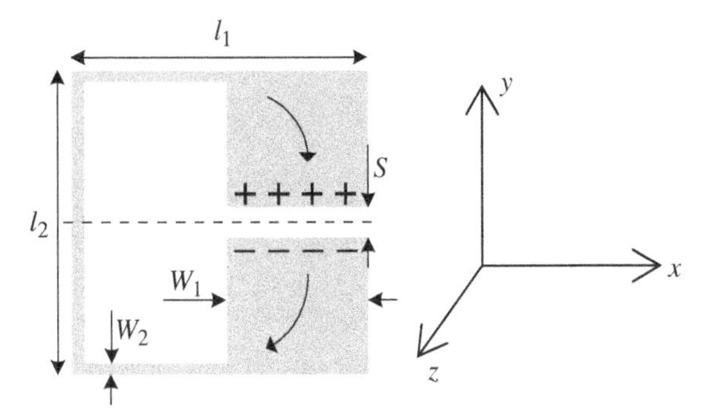

FIGURE 3.13 Topology of the folded SIR and relevant dimensions. The distribution of charges at the fundamental resonance is indicated, the symmetry plane being an electric wall. The depicted topology is square shaped, but it can be rectangular or circular as well.

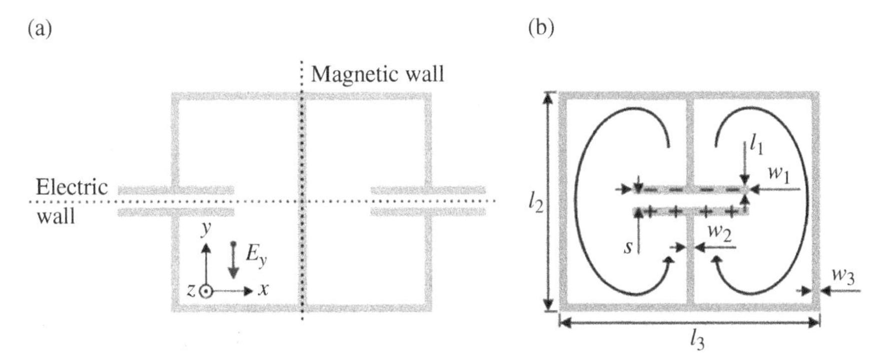

FIGURE 3.14 Topologies of the ELC resonator and relevant dimensions. (a) Loaded with capacitors in the external branches and (b) loaded with a capacitor in the central branch. The distribution of charges and current flow at the fundamental resonance (case b) are indicated.

3.3.1.6 The Electric LC Resonator (ELC) The ELC resonator was proposed in Ref. [78] as an alternative to the metallic wires for the implementation of ENG media with small absolute values of permittivity [4]. This is a bisymmetric resonator consisting of a pair of contacting capacitively loaded metallic loops (Fig. 3.14 depicts the two alternative topologies).[20] The application of a z-oriented time-varying magnetic field may induce a magnetic dipole moment in the particle. However, the induced currents will not flow through the central branch, since this would give opposite (canceling) magnetic dipole moments in each loop. This means that a uniform axial magnetic field is not able to excite the particle at the first (fundamental) resonance.[21] At the

[20] Obviously, circularly shaped ELCs are also possible.

[21] All the considered resonators exhibit higher-order (dynamic) resonances. Such resonances are analyzed in depth in Ref. [79]. However, the electrical size of the particles is large at the second- and higher-order resonances. Therefore, the homogeneity requirement for the implementation of effective media metamaterials is

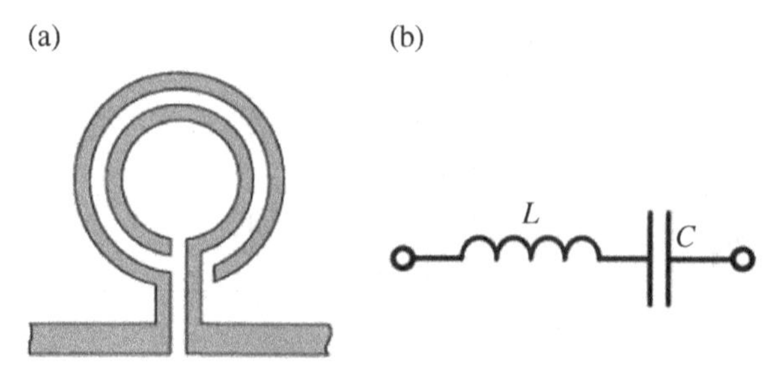

FIGURE 3.15 Topology of the OSRR (a) and equivalent circuit model (b).

fundamental resonance, current flows through the central branch, the horizontal symmetry plane is an electric wall, and an electric dipole in the y-direction appears. This means that the particle can be excited at the fundamental resonance by means of a uniform time-varying electric field applied in the y-direction, but not by a uniform z-oriented magnetic field. However, such electric field is not able to induce a magnetic dipole moment in the z-direction, and therefore the ELC resonator is non-bianisotropic. Notice that the vertical symmetry plane is a magnetic wall at the fundamental resonance. The presence of two planes of symmetry (an electric wall and a magnetic wall) is useful for the implementation of microwave components (sensors and bar codes) based on symmetry properties, as will be studied later (see Section 4.3.2.2).

3.3.1.7 The Open Split-Ring Resonator (OSRR) Unlike the previous resonators, the open split-ring resonator (OSRR) is an open resonator that can be driven by a voltage or current source applied to its terminals. The topology of the OSRR, firstly proposed by Martel *et al.* [80], can be derived from the two-turn SR topology by simply opening the metallic strip connecting the inner and outer rings of the spiral and providing access terminals to both rings (Fig. 3.15). This means that the equivalent circuit model of the OSRR can be derived from that of the two-turn SR by simply opening it. Namely, the OSRR can be modeled as an open series resonator where the inductance and capacitance are those of the two-turn SR. Thus, this particle is electrically smaller than the SRR by a factor of two.

3.3.2 Applying Duality: Complementary Resonators

In this subsection, complementary resonators are introduced by applying complementary considerations and duality [81–83] to the previous metallic particles. With the exception of the BC-SRR, the considered metallic resonators have their dual, or complementary, counterparts. The complementary particles are derived from the metallic ones by simply etching the corresponding topologies in a metallic screen. In other

---

not justified at these frequencies. Although the second resonance may be useful for certain microwave applications, most RF/microwave components based on these resonators or implemented by means of resonant-type metamaterial transmission lines exploit the fundamental resonance. For this reason, the analysis of the higher-order resonances is not included in this book.

words, they are negative images, and their EM behavior is roughly dual. Specifically, a negative effective permittivity can be expected for any medium comprising complementary particles of those exhibiting negative effective permeability. The following analysis is restricted to the complementary split-ring resonator (CSRR), first proposed in Refs. [84, 85], since for the other complementary particles the elements of the circuit model can be derived from those of the metallic particles in a similar way. However, the complementary version of the OSRR is also analyzed since further explanations are required for their full comprehension.

3.3.2.1 Complementary Split-Ring Resonator (CSRR) As it is well known, the complementary of a planar metallic structure is obtained by replacing the metal parts of the original structure with apertures, and the apertures with metal plates [81]. Due to symmetry considerations, it can be demonstrated that if the thickness of the metal plate is zero, and its conductivity is infinity (perfect electric conductor), then the apertures behave as perfect magnetic conductors. In that case, the original structure and its complementary are effectively dual; and if the field $\vec{F} = \left(\vec{E}, \vec{H}\right)$ is a solution for the original structure, its dual $\vec{F}'$ defined by

$$\vec{F}' = \left(\vec{E}', \vec{H}'\right) = \left(-\sqrt{\frac{\mu}{\varepsilon}} \cdot \vec{H}, \sqrt{\frac{\varepsilon}{\mu}} \cdot \vec{E}\right) \tag{3.36}$$

is the solution for the complementary structure. Thus, under these ideal conditions, a perfectly dual behavior is expected for the complementary screen of the SRR. Thus, whereas the SRR can be mainly considered as a resonant magnetic dipole that can be excited by an axial magnetic field, the CSRR (Fig. 3.16) essentially behaves as an electric dipole (with the same frequency of resonance) that can be excited by an axial electric field [82, 83]. The cross-polarization effects present in the SRR, discussed before, are also present in the CSRR. Thus, CSRRs exhibit a resonant magnetic polarizability along the y-axis (see Fig. 3.16); and therefore, its fundamental resonance can also be excited by an external magnetic field applied along this direction [10].

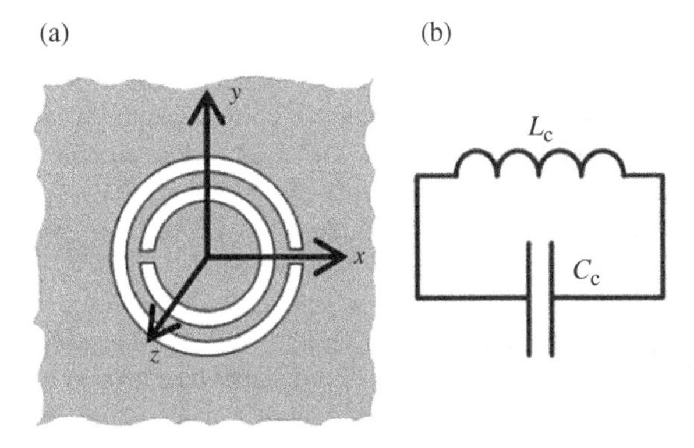

FIGURE 3.16 Topology of the CSRR (a) and equivalent circuit model (b).

Rigorously speaking, due to the continuity of the tangential/normal components of the electric/magnetic fields in the CSRR plane, $\vec{F}'$ in Equation 3.36 is the solution on one side of this plane, and $-\vec{F}'$ is the solution on the other side. With regard to the CSRR polarizabilities, this means that the sign of such polarizabilities changes from one side to another. Therefore, the net electric and magnetic dipoles on the CSRR must vanish, a result that can also be deduced from the fact that electric currents confined in a plane cannot produce any net normal/tangential electric/magnetic polarization. However, when the CSRR is seen from one side, the aforementioned effective polarizabilities arise.

The circuit model of the CSRR is depicted in Figure 3.16b, where L_c and C_c are the inductance and capacitance of the particle, respectively. The sub-index "c" denotes that these reactive elements are those corresponding to the CSRR. Since these parameters are related to the reactive elements of the circuit model of the SRR, let us redefine these elements as L_s and C_s, where the subindex indicates that these parameters are referred to the SRR. Moreover, let us define C_o as $C_o = 2\pi r_o C_{pul}$ (so that the capacitance C of the SRR model of Figure 3.6 is $C = C_o/2$ and the whole capacitance of the SRR is $C_s = C_o/4$). For the CSRR, the capacitance C_c is calculated as the capacitance of a disk of radius $r_o - c/2$ surrounded by a metal plane at a distance c of its edge. The inductance L_c is given by the parallel combination of the two inductances connecting the inner disk to the outer metallic region of the CSRR. Each inductance is given by $L_o/2$, where $L_o = 2\pi r_o L_{pul}$, and L_{pul} is the per-unit length inductance of the CPWs connecting the inner disk to the surrounding metallic region. For infinitely thin perfect conducting screens, and in the absence of any dielectric substrate, it directly follows from duality that the parameters of the circuit models for the SRRs and the CSRRs are related by $C_c = 4(\varepsilon_o/\mu_o)L_s$ and $C_o = 4(\varepsilon_o/\mu_o)L_o$ [4, 10]. Therefore, under such ideal conditions, the resonance frequencies of the SRR and CSRR are identical, as is expected from duality. In practice, the conductor has a finite width (and losses), and a substrate is needed, which means that the resonance frequencies of the metal and complementary particles are "roughly" the same (but not identical). An exhaustive analysis of the effects of substrate thickness and dielectric constant on the divergence between the resonance frequencies of SRRs and CSRRs is reported in Refs. [4, 10]. The formula providing the capacitance C_c of the CSRR when a dielectric substrate is present was derived in Ref. [10], whereas the value of L_o can be inferred from the design formulas giving the per-unit-length inductance of a CPW [70].

The previous analysis can be extended to the complementary version of the aforementioned particles, with exception of the BC-SRR. These complementary particles are designated canonically as non-bianisotropic complementary SRR (NB-CSRR), complementary SR (CSR), double-slit complementary SRR (DS-CSRR), folded complementary SIR (CSIR), and open complementary SRR (OCSRR). The complementary counterpart of the ELC resonator is a magnetically driven resonator (the magnetic field must be applied in the plane of the particle), and for this reason the particle is called magnetic LC (MLC) resonator. The complementary version of the OSRR, that is, the OCSRR, can also be analyzed from duality considerations. However, the fact that the OCSRR is an open particle needs some attention. Moreover, since this particle

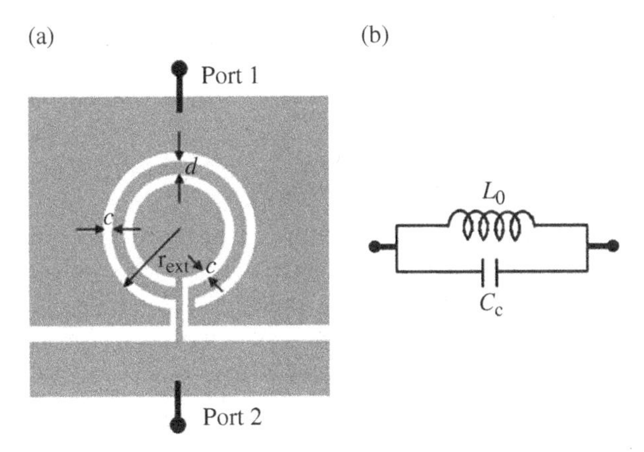

FIGURE 3.17 Topology of the OCSRR (a) and equivalent circuit model (b).

will be part of several circuits discussed later, the OCSRR is briefly analyzed in the next subsection.

3.3.2.2 Open Complementary Split-Ring Resonator (OCSRR) The OCSRR was first proposed in Ref. [86] (Fig. 3.17). This is the complementary version of the OSRR, but it can also be viewed as an open CSRR. However, notice that in the OCSRR, the connecting terminals (or ports) are located at both sides of the central slot. Between these ports, there is an electric short through the metal between the inner and outer slot rings forming the particle, but there is also capacitive connection through the capacitances across the slots. Thus, according to this, the circuit model of the particle is an open parallel resonant tank, as Figure 3.17b illustrates. The inductance of the particle is L_o, whereas the capacitance is identical to the capacitance of the CSRR, C_c. Since the inductance of the CSRR is $L_o/4$, it is expected that the resonance frequency of the OCSRR is one half the resonance frequency of the CSRR. Notice that from duality considerations, it follows that the OSRR and OCSRR have very similar resonance frequencies. As will be seen later, the combination of OSRRs and OCSRRs is very useful for the design of metamaterial transmission lines and many circuits based on them.

3.4 CANONICAL MODELS OF METAMATERIAL TRANSMISSION LINES

In this section, the canonical models for the implementation of metalines are introduced. By canonical model, we refer to the simplest circuit model that mimics an LH metamaterial (i.e., the dual transmission line model), and to a family of circuits that exhibit backward and forward wave propagation, depending on the frequency band, and are useful for microwave circuit design on the basis of dispersion and impedance engineering. Let us begin with the dual transmission line model. The following subsections are dedicated to those models able to exhibit backward and forward wave behavior, obtained by increasing the number of reactive elements. In these models, losses are not considered.

3.4.1 The Dual Transmission Line Concept

The concept of backward waves in RF/microwave engineering is an old concept [31]. Backward waves can be generated by feeding a ladder network with series-connected capacitors and shunt-connected inductors alternating, as shown in Figure 3.18a. This network is the dual version of the equivalent circuit model of a lossless transmission line (Fig. 3.18b); and for this reason, this ladder network is designated as dual transmission line. The dual transmission line exhibits backward (or LH) wave propagation, and wave propagation characteristics can be controlled by the value of the reactive elements. However, the potential of these lines by themselves is very limited due to the fact that the unit cell (see the T-circuit in Fig. 3.18c) contains only two reactive elements. Nevertheless, let us study the dual transmission line in detail and compare it with the circuit model of an ordinary lossless transmission line. Afterward, the CRLH transmission line concept will be introduced.

The analysis of the propagation characteristics of the dual and conventional transmission line models can be realized by means of the transfer matrix approach (see Chapter 2). Applying expressions (2.23), with A given by (2.28a), and (2.30) to the circuit of Figure 3.18a, the phase constant and characteristic (or Bloch) impedance of the dual transmission line are found to be

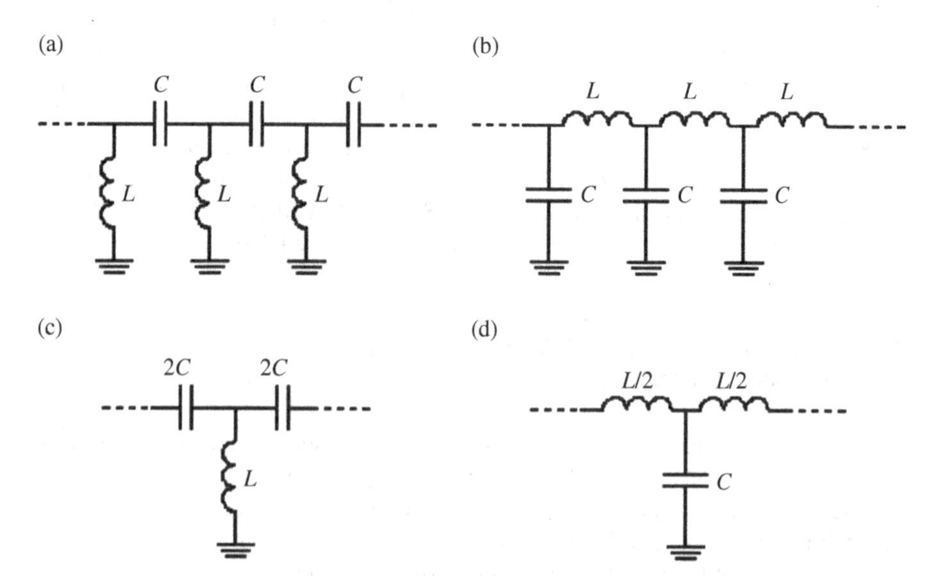

FIGURE 3.18 Equivalent circuit model of a backward (a) and forward (b) transmission line. The T-circuit models of the basic cell structures are also indicated in (c) and (d). Reprinted with permission from Ref. [4]; copyright 2008 John Wiley.

$$\cos\beta_L l = 1 - \frac{1}{2LC\omega^2} \tag{3.37}$$

$$Z_{BL} = \sqrt{\frac{L}{C}\left(1 - \frac{\omega_{cL}^2}{\omega^2}\right)} \tag{3.38}$$

where l is the period, $Z_s = -j/\omega C$, $Z_p = j\omega L$, and $\omega_{cL} = 1/2(LC)^{1/2}$ is an angular cutoff frequency (propagation below ω_{cL} is not allowed since the Bloch impedance is imaginary). In (3.37) and (3.38), the subscript "L" indicates that the phase constant and characteristic impedance are those of the dual, or LH, transmission line. For the model of a conventional (forward) transmission line, application of (2.23) and (2.30) gives

$$\cos\beta_R l = 1 - \frac{LC}{2}\omega^2 \tag{3.39}$$

$$Z_{BR} = \sqrt{\frac{L}{C}\left(1 - \frac{\omega^2}{\omega_{cR}^2}\right)} \tag{3.40}$$

where $\omega_{cR} = 2/(LC)^{1/2}$ is the angular cutoff frequency for the forward transmission line model of Figure 3.18b, and the subscript "R" is indicative of forward or RH wave propagation in the allowed regions of this line.[22]

The dispersion diagram and the dependence of Z_B on frequency for the networks of Figure 3.18 are depicted in Figure 3.19. Transmission is limited to those frequency intervals that make the phase constant and the characteristic impedance to be real numbers. It is worth mentioning that frequency dispersion is present in both structures. Even though the circuit of Figure 3.18b models an ideal lossless ordinary (forward) transmission line, where dispersion is absent, actually this circuit is only valid for frequencies satisfying $\omega << \omega_{cR}$, that is, in the long wavelength limit (corresponding to those frequencies where wavelength for guided waves satisfies, $\lambda_g >> l$). To correctly model an ideal lossless transmission line at higher frequencies, we simply need to reduce the period of the structure, and accordingly the per-section inductance and capacitance of the line, L and C, with the result of a higher cut-off frequency. Thus, the circuit of Figure 3.18b can properly describe ideal transmission lines without dispersion. To this end we simply need to select the period such that the long wavelength limit approximation holds. Under this approximation, expressions (3.39) and (3.40) lead us to expressions (1.6) and (1.9), and the phase and group velocities are given by (1.10) and (1.11), and they are positive and constant, as expected.

Conversely, the backward wave structure of Figure 3.18a is dispersive even in the long wavelength limit. In order to identify this structure as an effective medium (i.e., one-dimensional metamaterial), operation under this approximation is required. However, regardless of the operating frequency within the transmission band, the structure supports backward waves and, for this reason, it is an LH transmission line. In order to

[22] Notice that (3.40) is identical to (2.34).

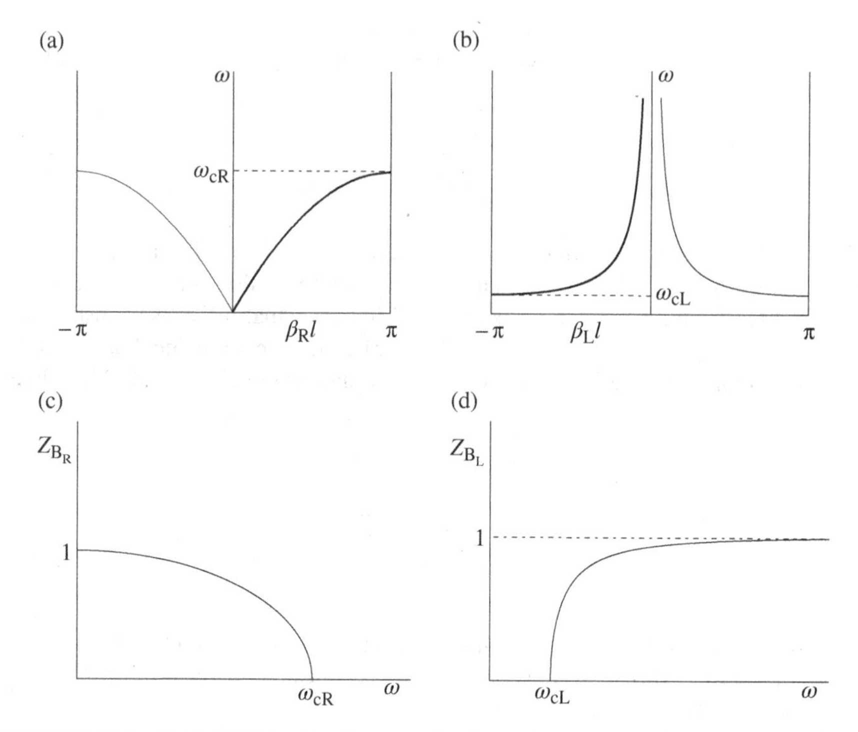

FIGURE 3.19 Typical dispersion diagram of a forward (a) and backward (b) transmission line model. The dependence of the normalized Bloch impedance with frequency is shown in (c) and (d) for the forward and backward lines, respectively. Reprinted with permission from Ref. [4]; copyright 2008 John Wiley.

properly identify effective constitutive parameters, μ_{eff} and $\varepsilon_{\mathrm{eff}}$, the long wavelength approximation is necessary. However, as has been previously discussed, operation under this approximation is not a due for microwave circuit design. Nevertheless, for coherence and simplicity, the phase constant, the characteristic impedance, as well as the phase and group velocities are derived under the long wavelength limit ($\omega >> \omega_{\mathrm{cL}}$). The following results are obtained:

$$\beta_{\mathrm{L}} l = -\frac{1}{\omega\sqrt{LC}} \tag{3.41}$$

$$Z_{\mathrm{BL}} = \sqrt{\frac{L}{C}} \equiv Z_{\mathrm{lw}} \tag{3.42}$$

$$v_{\mathrm{pL}} = \frac{\omega}{\beta_{\mathrm{L}}} = -\omega^2 l\sqrt{LC} < 0 \tag{3.43}$$

$$v_{gL} = \left(\frac{\partial \beta_L}{\partial \omega}\right)^{-1} = +\omega^2 l\sqrt{LC} > 0 \tag{3.44}$$

and the phase and group velocities have opposite signs. From a mathematical point of view, in the dispersion diagrams for the forward and backward lines depicted in Figure 3.19, and obtained from expressions (3.37) and (3.39), the sign of β can be either positive or negative. This ambiguity comes from the two possible directions of energy flow, namely from left to right or vice versa. If we adopt the usual convention of energy flow from left to right, then the sign of the phase constant is determined by choosing that portion of the curves that provide a positive group velocity (bold lines in Fig. 3.19a and b), as it was pointed out in Section 2.2. From this, it is clear that for the forward transmission line, β and v_p are both positive, whereas these magnitudes are negative for the backward transmission line. In both cases, v_g is positive as one expects on account of the co-directionality between power flow and group velocity.[23]

In the backward and forward transmission lines depicted in Figure 3.18, it is possible to identify effective constitutive parameters. To this end, we should take into account that TEM wave propagation in planar transmission lines and plane wave propagation in source-free, linear, isotropic, and homogeneous dielectrics are described by identical equations (telegrapher equations) provided the following mapping holds (Appendix A):

$$Z_s'(\omega) = j\omega\mu_{\text{eff}} \tag{3.45}$$

$$Y_p'(\omega) = j\omega\varepsilon_{\text{eff}} \tag{3.46}$$

where Z_s' and Y_p' are the per-unit length series impedance and shunt admittance of the considered line. Thus, for the forward line the effective permittivity and permeability are constant and given by[24]

$$\varepsilon_{\text{eff}} = \frac{C}{l} \tag{3.47}$$

$$\mu_{\text{eff}} = \frac{L}{l} \tag{3.48}$$

[23] As pointed out in Section (6.4.2), not always the group velocity is restricted to positive values smaller than c (the speed of light in vacuum). Superluminal (higher than c) and negative group velocities are possible in real structures. However, these abnormal velocities are not in contradiction with Einstein causality since the group velocity is not the velocity of information transfer.

[24] These expressions are equivalent to (A.22a) and (A.22b). Notice that the effective permeability and permittivity depend on the transverse geometry of the considered line and cannot be identified with the permittivity and permeability of the dielectric material of such line. This is true even for purely TEM transmission lines.

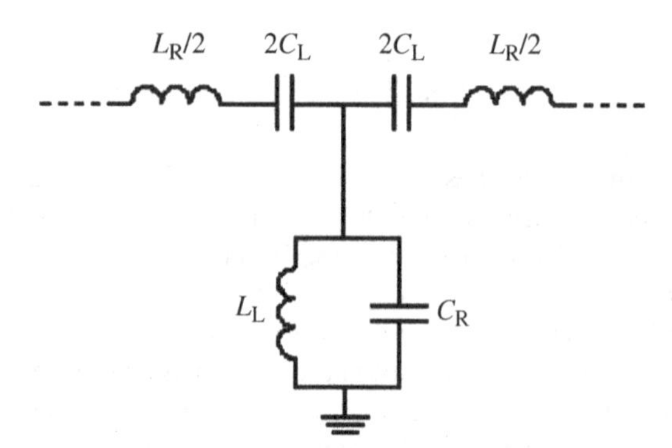

FIGURE 3.20 Equivalent circuit model (basic cell) of the CRLH transmission line.

whereas for the backward transmission line, the constitutive parameters are

$$\varepsilon_{\text{eff}} = -\frac{1}{\omega^2 Ll} \tag{3.49}$$

$$\mu_{\text{eff}} = -\frac{1}{\omega^2 Cl} \tag{3.50}$$

and they are both negative, a sufficient condition to obtain LH wave propagation.

3.4.2 The CRLH Transmission Line

From a practical point of view, in order to implement a dual (or backward) transmission line, a host line (microstrip or CPW, among others) is required. The host line introduces thus parasitic elements that in general may not be negligible, and they should be taken into account to accurately describe the propagation characteristics of the lines.[25] These structures may exhibit LH or RH wave propagation, depending on the frequency interval; and for this reason, they have been termed as CRLH transmission lines [11]. The equivalent circuit model (unit cell) of these structures is depicted in Figure 3.20. For clarity, we have renamed the reactive elements of the dual transmission line as C_L and L_L, while C_R and L_R correspond to the per-section capacitance and inductance of the host line. By using (2.23) and (2.30), the dispersion relation as well as the characteristic impedance of the CRLH transmission line can be inferred, namely

$$\cos\beta l = 1 - \frac{\omega^2}{2\omega_R^2}\left(1 - \frac{\omega_s^2}{\omega^2}\right)\left(1 - \frac{\omega_p^2}{\omega^2}\right) \tag{3.51}$$

[25] The parameters of the host line are parasitics for the synthesis of a purely LH line. However, these reactive elements play a key role in most practical CRLH lines, where they are typically tailored in order to achieve certain line specifications.

$$Z_B = \sqrt{\frac{L_R}{C_R}\frac{\left(1-(\omega_s^2/\omega^2)\right)}{\left(1-\left(\omega_p^2/\omega^2\right)\right)} - \frac{L_R^2\omega^2}{4}\left(1-\frac{\omega_s^2}{\omega^2}\right)^2} \quad (3.52)$$

where the following variables

$$\omega_R = \frac{1}{\sqrt{L_R C_R}} \quad (3.53)$$

$$\omega_L = \frac{1}{\sqrt{L_L C_L}} \quad (3.54)$$

and the series and shunt resonance frequencies

$$\omega_s = \frac{1}{\sqrt{L_R C_L}} \quad (3.55)$$

$$\omega_p = \frac{1}{\sqrt{L_L C_R}} \quad (3.56)$$

have been introduced to simplify the mathematical formulas. Expressions 3.51 and 3.52 are depicted in Figure 3.21. Two propagating regions, separated by a gap at the spectral origin can be distinguished. In the lowest frequency region, the parameters of the dual transmission line, C_L and L_L, are dominant and wave propagation is backward. This situation is reversed above the stop band, where the parasitic reactances of the host line make the structure to behave as a RH line. Indeed, at high frequencies the CRLH transmission line tends to behave as a purely right handed (PRH) line.

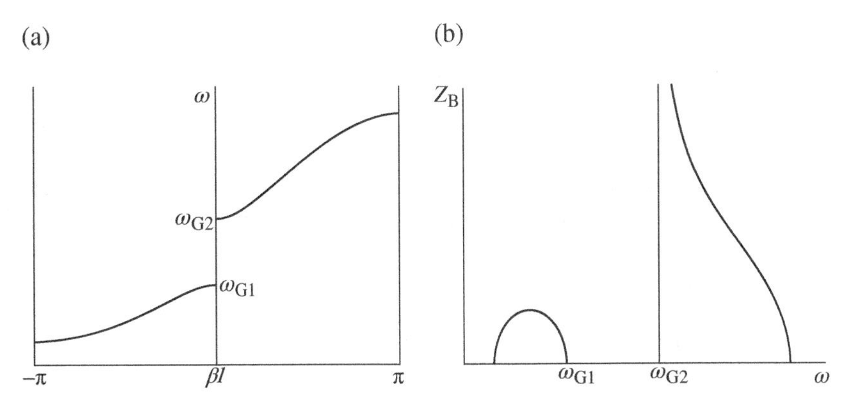

FIGURE 3.21 Typical dispersion diagram (a) and variation of Bloch impedance with frequency (b) in a CRLH transmission line model. In this example, $\omega_s < \omega_p$. Reprinted with permission from Ref. [4]; copyright 2008 John Wiley.

Conversely, in the lower limit of the first allowed band, the CRLH structure exhibits the characteristics of a purely left handed (PLH)—or dual—transmission line. The gap limits are given by the frequencies satisfying.

$$\omega_{G1} = \min(\omega_s, \omega_p) \tag{3.57}$$

$$\omega_{G2} = \max(\omega_s, \omega_p) \tag{3.58}$$

In the long wavelength limit, expression (3.51) rewrites as follows:

$$\beta = \frac{s(\omega)}{l}\sqrt{\frac{\omega^2}{\omega_R^2}\left(1-\frac{\omega_s^2}{\omega^2}\right)\left(1-\frac{\omega_p^2}{\omega^2}\right)} \tag{3.59}$$

where $s(\omega)$ is the following sign function:

$$s(\omega) = \begin{cases} -1 & \text{if} \quad \omega < \min(\omega_s, \omega_p) \\ +1 & \text{if} \quad \omega > \max(\omega_s, \omega_p) \end{cases} \tag{3.60}$$

From the phase constant (Eq. 3.59), the phase and group velocities can be easily inferred, these velocities being of opposite sign in the LH band and both being positive in the RH band.

With regard to the constitutive parameters for the CRLH transmission line, they can be inferred as previously indicated, that is,

$$\varepsilon_{\text{eff}} = \frac{C_R}{l} - \frac{1}{\omega^2 L_L l} \tag{3.61}$$

$$\mu_{\text{eff}} = \frac{L_R}{l} - \frac{1}{\omega^2 C_L l} \tag{3.62}$$

and they can be positive or negative, depending on the frequency range.

One particular case of interest is the so-called balanced CRLH line, which corresponds to the situation where the series and shunt resonances are identical, namely $\omega_s = \omega_p = \omega_o$ [3]. In this case, there is not a forbidden band between the LH and RH allowed bands; in other words, the change between backward and forward wave behaviour is continuous. Concerning the characteristic impedance, it reaches its maximum at ω_o (the transition frequency),[26] where

[26] Do not confuse the transition frequency of a balanced CRLH line with the fundamental resonance frequency of the SRR, introduced before and described by the same symbol.

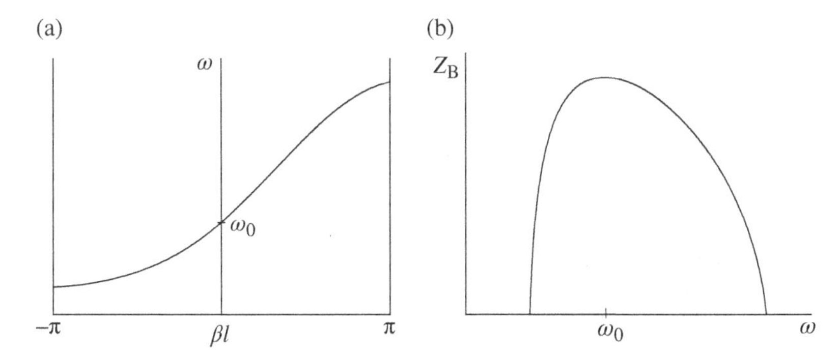

FIGURE 3.22 Typical dispersion diagram (a) and variation of Bloch impedance with frequency (b) in a balanced CRLH transmission line. Reprinted with permission from Ref. [4]; copyright 2008 John Wiley.

$$Z_B = \sqrt{\frac{L_R}{C_R}} = \sqrt{\frac{L_L}{C_L}} \tag{3.63}$$

and the characteristic impedance decreases as frequency increases or decreases from ω_o. At the limits of the allowed propagation interval (which contains both the LH and RH frequency bands), the characteristic impedance nulls, and beyond these limits, it takes imaginary values. Figure 3.22 depicts the dispersion diagram and the variation of line impedance with frequency. By contrast, the characteristic impedance dependence on frequency is more complicated for the unbalanced case (see Fig. 3.21). If we consider that $\omega_s < \omega_p$ (which is the situation considered in Fig. 3.21), the characteristic impedance is zero at the lower limit of the LH interval, it increases as frequency increases, reaches a maximum, and then decreases towards zero, at ω_s, the upper limit of the LH band. In the RH band, the impedance exhibits a pole at ω_p (the lower frequency limit), it decreases as frequency increases, and it nulls at the higher cutoff frequency of the RH band. However, the behavior changes if $\omega_s > \omega_p$; in this case, the impedance goes from zero up to infinity in the LH pass band, and it nulls at the edges of the RH band.

From the point of view of the characteristic impedance, the balanced CRLH transmission line is interesting because in the vicinity of the transition frequency, Z_B is not very dependent on frequency, and it allows for broadband matching, as compared to the unbalanced case, where the characteristic impedance is much more sensitive to frequency. On the other hand, the balanced case exhibits another important difference as compared to the unbalanced CRLH transmission line. Namely, at the transition frequency, where the phase velocity exhibits a pole, the phase shift[27] is zero, and the group velocity is different than zero. In other words, at this frequency, wave propagation is possible ($v_g \neq 0$) with $\beta = 0$. The phase origin of the balanced CRLH line

[27] Notice that by phase shift, we refer to the phase of the transmission coefficient, which is of opposite sign than the phase constant.

takes place thus at ω_o. When frequency is decreased below ω_o, the phase becomes positive (β negative) and it increases progressively; conversely, when frequency increases from ω_o, the phase increases in magnitude but with negative sign (as occurs in conventional transmission lines). Concerning the behaviour of wavelength for guided waves, λ_g, in the balanced CRLH transmission line, this reaches its maximum (infinity) at ω_o, and λ_g decreases as frequency increases or decreases from ω_o. Indeed, for the unbalanced case, the guided wavelength also increases when frequency approaches the spectral gap. However, at the edges of the gap the group velocity is zero, the line impedance takes extreme values and signal propagation is not allowed.

3.4.3 Other CRLH Transmission Lines

A CRLH transmission line can be achieved from various combinations of inductances and capacitances. Let us review the canonical models providing such composite behavior.

3.4.3.1 The Dual CRLH (D-CRLH) Transmission Line The D-CRLH line is obtained by simply interchanging the series LC resonator of the series branch with the parallel LC resonator of the shunt branch in the circuit of Figure 3.20. The result is the T-circuit depicted in Figure 3.23 [87]. This circuit exhibits forward wave propagation at low frequencies, and backward wave propagation at high frequencies, and it is opaque in a certain frequency band even under balanced conditions (i.e., identical series and shunt resonances). The dispersion diagram and the characteristic impedance for this line model, inferred from (2.23) and (2.30), are

$$\cos\beta l = 1 - \frac{\omega_L^2}{2\omega^2}\left(1 - \frac{\omega_s^2}{\omega^2}\right)^{-1}\left(1 - \frac{\omega_p^2}{\omega^2}\right)^{-1} \tag{3.64}$$

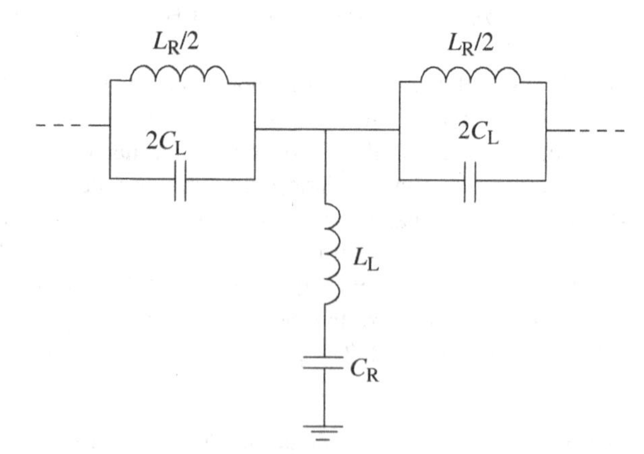

FIGURE 3.23 Equivalent circuit model (basic cell) of the D-CRLH transmission line.

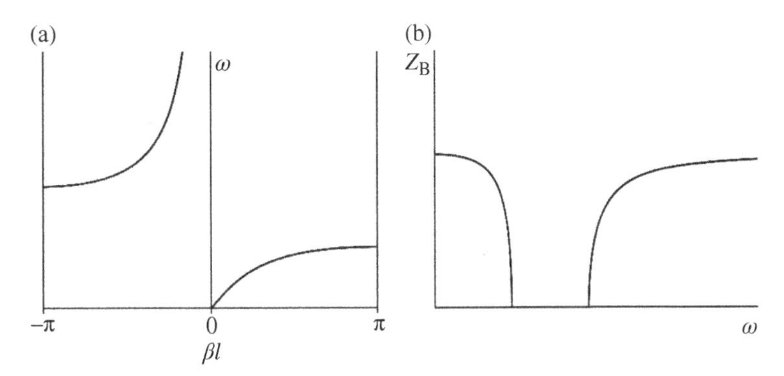

FIGURE 3.24 Typical dispersion diagram (a) and variation of Bloch impedance with frequency (b) in a balanced D-CRLH transmission line.

$$Z_B = \sqrt{\frac{L_L}{C_L}\frac{\left(1-\left(\omega_p^2/\omega^2\right)\right)}{\left(1-\left(\omega_s^2/\omega^2\right)\right)} - \frac{1}{4C_L^2\omega^2}\frac{1}{\left(1-\left(\omega_s^2/\omega^2\right)\right)^2}} \tag{3.65}$$

and typical diagrams are depicted in Figure 3.24. In (3.64) and (3.65), ω_L, ω_s, and ω_p are given by expressions (3.54)–(3.56).

3.4.3.2 Higher-Order CRLH and D-CRLH Transmission Lines Both the CRLH and the D-CRLH line models exhibit an LH band and an RH band. The number of bands can be arbitrarily extended by increasing the number of reactive elements in the series and shunt branches [88–91]. Indeed, the circuits of Figures 3.20 and 3.23 can be considered order-2 models, where the order index indicates the number of reactive elements of the series and shunt branches, and also the number of bands.[28] Figure 3.25 depicts order-3 and order-4 CRLH and D-CRLH transmission line models and the corresponding dispersion diagrams. The LH and RH bands alternate. However, the first transmission band is LH for the CRLH lines, whereas it is forward for the D-CRLH line models. The order-3 D-CRLH model of Figure 3.25 has been designated as double Lorentz metamaterial transmission line since the effective permittivity and permeability exhibit a Lorentz-type behavior [92]. The order-4 CRLH model of Figure 3.25 has been called extended CRLH (E-CRLH) transmission line [88]. As the order increases, the number of possible combinations of inductances and capacitances in the series and shunt branches providing a CRLH or a D-CRLH behavior also increases. In order to achieve a CRLH transmission line behavior, the reactance of the series branch and the susceptance of the shunt branch must be

[28] The order of a CRLH line should not be confused with the order of a filter. Notice that the unit cells of the CRLH (Fig. 3.20) and D-CRLH (Fig. 3.23) lines are third-order bandpass and bandstop filters, respectively. If several unit cells are cascaded, the resulting structure is a higher order periodic filter and an order-2 CRLH (or D-CRLH) line with multiple cells. CRLH lines can be applied to the implementation of bandpass filters, but it is sometimes convenient to sacrifice periodicity to achieve good filter performance.

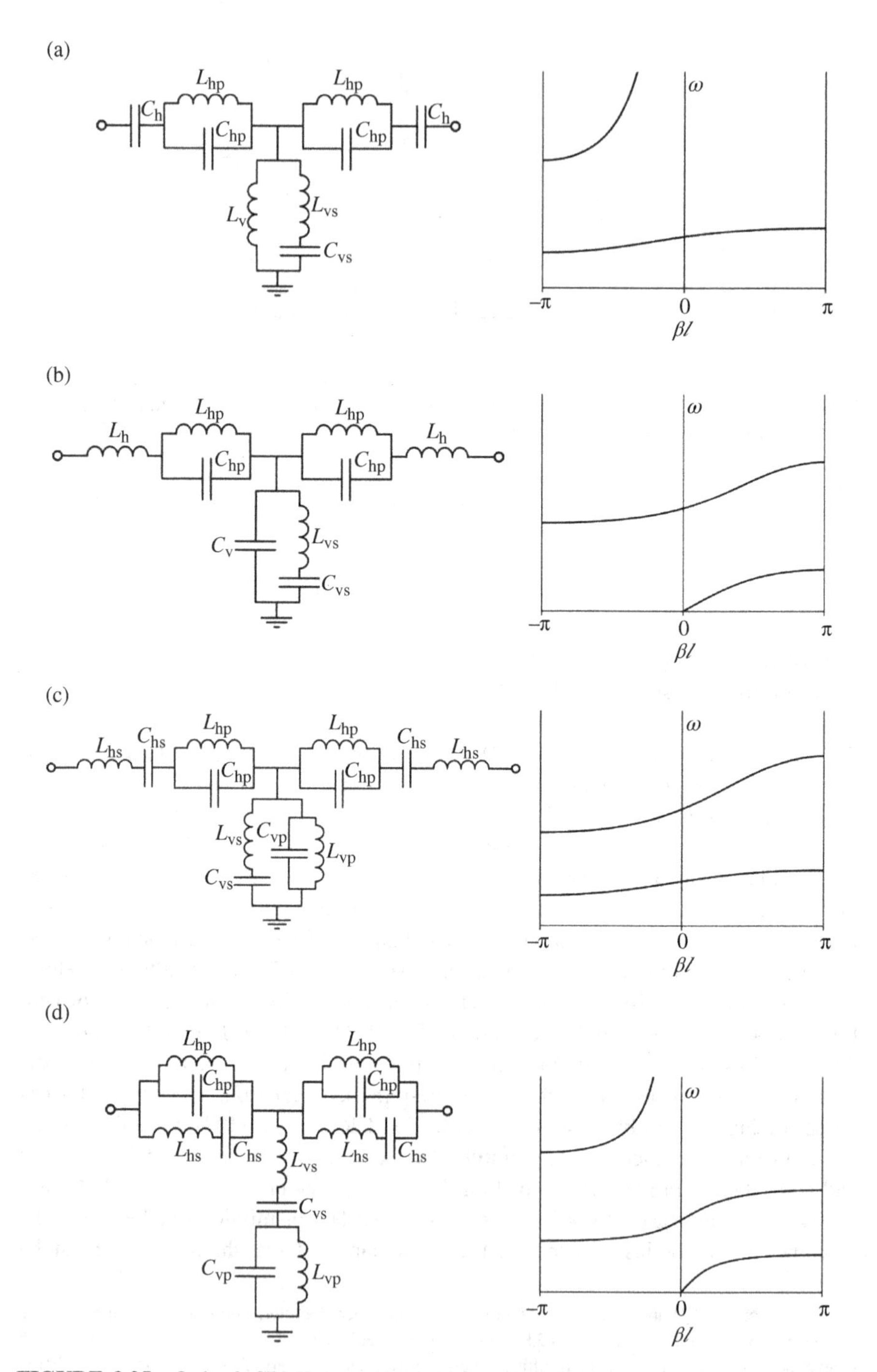

FIGURE 3.25 Order-3 CRLH (a), order-3 D-CRLH (b), order-4 CRLH (c), and order-4 D-CRLH (d) transmission line models (T-circuit) and the corresponding dispersion diagrams.

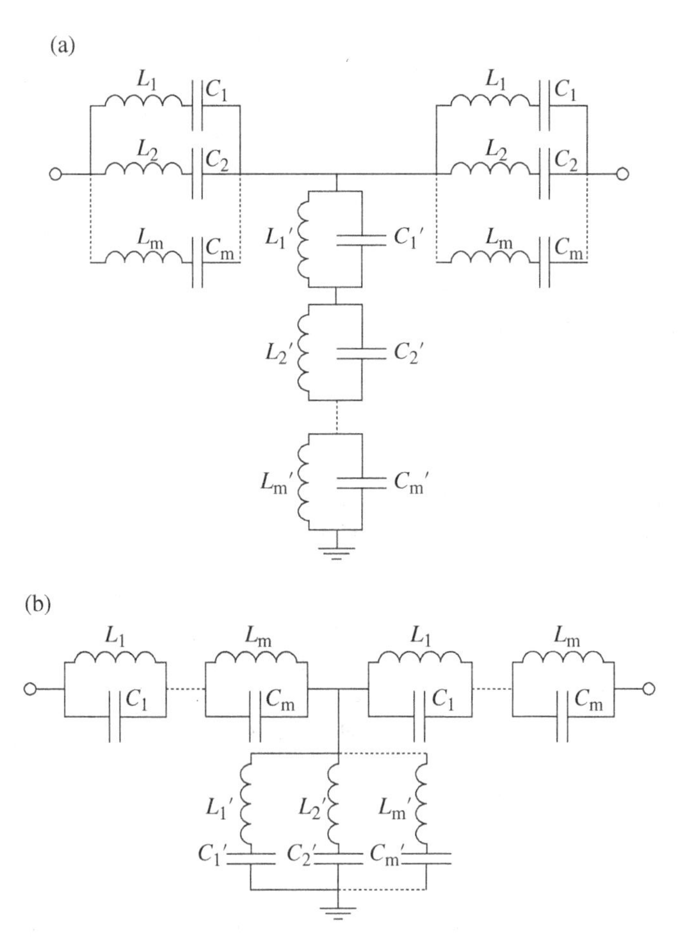

FIGURE 3.26 Generalized CRLH (a) and D-CRLH (b) transmission line models.

both negative in the DC limit (low frequencies). Conversely, the sign must be positive for a D-CRLH line. From the Foster reactance theorem [93], the sign of the reactance or susceptance slope for a network composed of inductances and capacitances is always positive. This means that the poles and zeros alternate. Therefore, to obtain a CRLH or a D-CRLH behavior, it suffices to place the zeros and poles of the reactance (series branch) at the same frequencies than the zeros and poles of the susceptance (shunt branch). It has been demonstrated that any network giving a CRLH behavior of even order, and containing the same number of reactive elements in the series and shunt branches, can be transformed to the generalized network depicted in Figure 3.26a (the generalized D-CRLH line model is depicted in Fig. 3.26b) [90].

In the previous paragraph, it has been considered that the number of reactive elements of the series and shunt branch is identical. However, this condition is not necessary in order to achieve a CRLH behavior with multiple bands. For instance, a CRLH transmission line can be achieved by simply using a series LC resonator in the series branch, and a complex network of capacitors and inductors in the shunt branch. The necessary (although not sufficient) condition to achieve an LH/RH band is a negative/positive reactance (series branch) and susceptance (shunt branch). Ultimately, the transmission bands and the forward or backward nature of wave propagation in those bands are dictated by the dispersion diagram, derived from (2.23). Examples of metamaterial transmission lines described by T-circuit unit cells with different number of reactive elements in the series and shunt branches will be reported later.

To end this section, let us highlight that there is another class of CRLH or D-CRLH transmission lines: those based on the lattice network and bridged-T topologies. Such transmission line models and their physical realizations are discussed in Chapter 6, focused on advanced topics.

3.5 IMPLEMENTATION OF METAMATERIAL TRANSMISSION LINES AND LUMPED-ELEMENT EQUIVALENT CIRCUIT MODELS

This section is devoted to the physical implementations of order-2 CRLH metamaterial transmission lines. The implementation of higher-order lines will be discussed in Chapter 4 (applications of metalines), specifically in the section focused on the design of multiband components, where quad-band devices based on order-4 CRLH lines are reported. There are two main physical implementations of order-2 CRLH transmission lines: (1) the CL-loaded approach and (2) the resonant-type approach.[29] Let us review them in the next subsections.

3.5.1 CL-Loaded Approach

Order-2 CRLH transmission lines can be implemented by loading a host line with series capacitors and shunt inductors. In the model of the CRLH transmission line of Figure 3.20, the host line is described by the per-section line inductance (L_R) and capacitance (C_R), whereas the loading elements are accounted for by the inductance L_L and the capacitance C_L. The host line can be any type of planar transmission line, such as microstrip, strip line, or CPW transmission line, among others. For the reactive elements (series capacitor and shunt inductor), two possibilities arise, namely the use of lumped (surface mount technology—SMT) elements, or semilumped (i.e., electrically small) planar components. The advantage of lumped elements is their size. However, lumped elements can operate only in a limited frequency range (typically below 5–6 GHz) due to parasitic effects that cause self-resonance to appear. Moreover, it is difficult to achieve the required electrical characteristics by using lumped

[29] Actually, resonant-type CRLH metalines based on SRRs or CSRRs, to be discussed later in this chapter, are described by circuit models (unit cells) with two reactive elements in one branch and three reactive elements in the other branch. Therefore, strictly speaking, these lines cannot be considered to be order-2 CRLH lines. However, these lines exhibit two frequency bands (one forward and the other backward). In this sense, we can include these SRR- and CSRR-based lines among the order-2 CRLH lines.

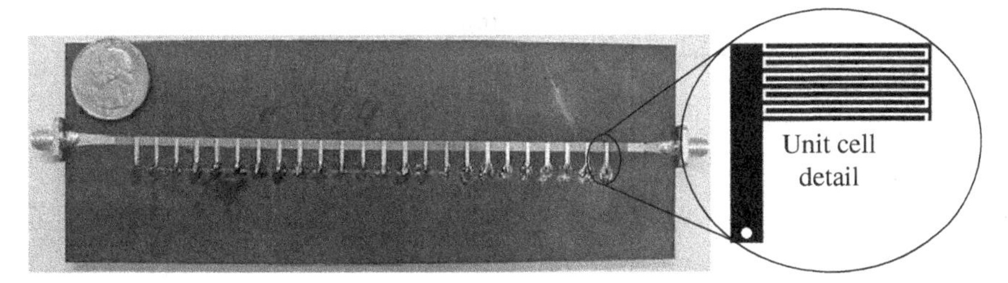

FIGURE 3.27 Microstrip CRLH transmission line implemented by means of shunt-connected grounded stubs and interdigital capacitors. Reprinted with permission from Ref. [3]; copyright 2006 by John Wiley.

elements since their values are restricted to those provided by the manufacturers. Finally, the use of lumped elements is more expensive, soldering is necessary (it causes additional losses), and it goes against the full integration of microwave components. Hence, semilumped elements are preferred.

Two main host lines have been considered for the synthesis of CRLH transmission lines: microstrip lines and CPWs. CL-loaded CRLH microstrip lines were proposed by Caloz *et al.* [94] in 2002, and subsequently used in many applications. The implementation of a CRLH structure by means of a CPW configuration was due to the group of Eleftheriades [15], who used the structure to generate backward leaky waves. The microstrip structure proposed by Caloz *et al.* consists of a periodic arrangement of series interdigital capacitors alternated with grounded (through metallic vias) stubs, which act as shunt connected inductors (Fig. 3.27). The interdigital capacitors are described by the series capacitance C_L in the circuit of Figure 3.20, whereas L_L models the grounded stubs. The other reactive elements C_R and L_R correspond to the line capacitance and inductance, respectively. The synthesis procedure and parameter extraction of these lines is not given in this book (see Ref. [3] for details). Such parameter extraction and synthesis methods will be exhaustively considered for resonant-type CRLH transmission lines, where the author's own experience is provided. Nevertheless, the frequency responses (measured, simulated through EM solvers and by means of circuit simulation) of a nine-cell balanced structure and a seven-cell unbalanced design are depicted in Figure 3.28 [3]. The transition frequency in the balanced structure is $f_o = 3.9$ GHz. This structure does not exhibit a stop band between the LH and the RH band; however, this stop band is present in the unbalanced structure. Alternatively, CRLH microstrip transmission lines can be implemented by replacing the shunt stubs with via holes in a mushroom type configuration similar to the high impedance surfaces proposed by Sievenpiper *et al.* [95] in one dimension. A two-dimensional structure, where caps are used to enhance the capacitive coupling between adjacent patches, was proposed in Ref. [58].

In CPW configuration, the shunt inductors can be implemented by means of connecting strips between the central strip and ground planes. The series capacitors can be implemented through interdigital geometries, or by means of series gaps. The latter are simpler, but the achievable capacitance values are much smaller. The Eleftheriades

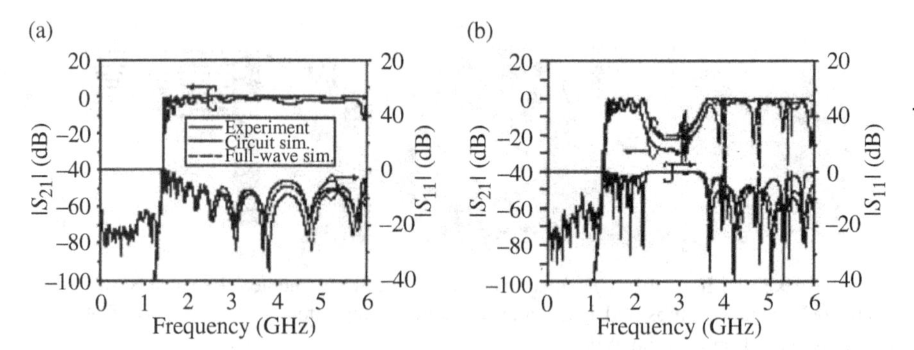

FIGURE 3.28 Typical frequency responses for a nine-cell balanced (a) and seven-cell unbalanced (b) CRLH transmission lines of the type depicted in Figure 3.27. Reprinted with permission from Ref. [3]; copyright 2006 by John Wiley.

Group at the University of Toronto (Canada) presented a realization of a CL-loaded CPW by using series gaps and shunt strips (see Fig. 3.29) [15]. This line was used as a backward leaky wave antenna (LWA), where the operating frequency was set in the fast wave region of the LH band. CRLH-based LWAs will be studied in Chapter 4.

To end this subsection, let us point out that CRLH transmission lines in multilayer vertical architectures have been reported [96], including realizations in low temperature co-fired ceramic (LTCC) technology [97]. The penalty of these vertically stacked CRLH lines is the need of multiple metal and dielectric layers. However, this approach has two inherent advantages: (1) small size and (2) the possibility of bandwidth enhancement, since it is possible to achieve significant series capacitances (C_L) and shunt inductances (L_L). To understand this later aspect, let us consider a balanced CRLH transmission line described by the model of Figure 3.20. From (3.51), considering $\omega_s = \omega_p = \omega_o$, it follows that wave propagation is delimited by the following angular frequencies:

$$\omega_{CL} = \omega_R \left(\sqrt{1 + \frac{\omega_o^2}{\omega_R^2}} - 1 \right) \tag{3.66a}$$

$$\omega_{CR} = \omega_R \left(\sqrt{1 + \frac{\omega_o^2}{\omega_R^2}} + 1 \right) \tag{3.66b}$$

Thus, bandwidth increases by increasing ω_R (i.e., $\Delta\omega = \omega_{CR} - \omega_{CL} = 2\omega_R$). This means that to achieve a wide CRLH band, L_R and C_R must be small, which leads to large values of C_L and L_L in order to set the transition frequency, ω_o, to the required value. From the point of view of filter theory, notice that the model of Figure 3.20 is also the canonical circuit of an order-3 bandpass filter. To achieve a wide filter bandwidth, it is necessary to use wideband series and shunt resonators, and this leads us to

(a)

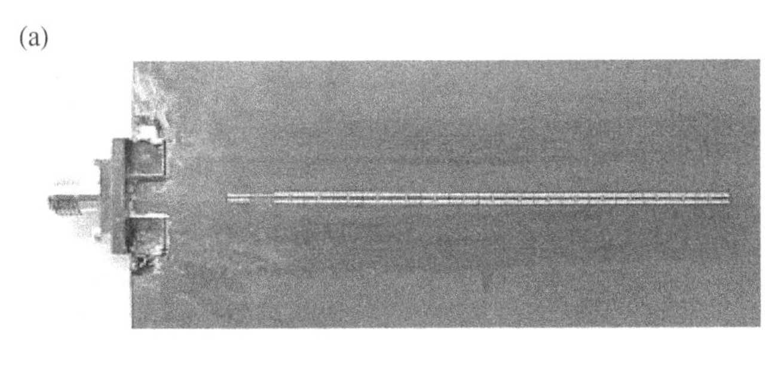

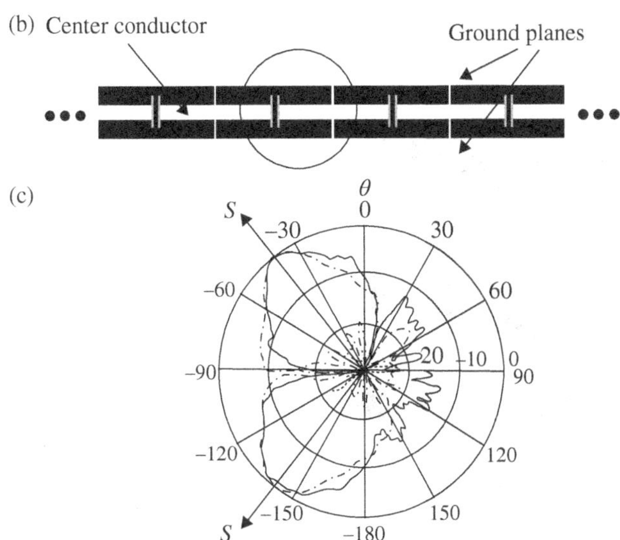

FIGURE 3.29 CPW CRLH transmission line leaky wave antenna (a). The detail of the structure and the radiation diagram are depicted in (b) and (c), respectively. Reprinted with permission from Ref. [15]. Copyright 2002, American Institute of Physics.

the same conclusion with regard to the values of C_L and L_L. The quasibalanced CRLH line reported in Ref. [97] (see Fig. 3.30) exhibits a bandwidth of 136%, whereas the unbalanced structure reported in Ref. [96] exhibits an LH band covering the range 0.2–0.6 GHz, that is, remarkable bandwidths as compared to those achievable with single-layered structures. The contributions of the Group of I. Vendik (St. Petersburg Electrotechnical University) to the design of compact broadband devices using LTCC technology and LH and RH transmission line sections are also remarkable (see, e.g., Ref. [98]).

To summarize, CRLH transmission lines consisting of a host line loaded with series capacitors and shunt inductors are described by the circuit model depicted in Figure 3.20 to a good approximation. In the next subsection, devoted to the resonant-type approach, the models of the reported CRLH lines present some modifications to the canonical model of 3.20, and will be discussed in detail.

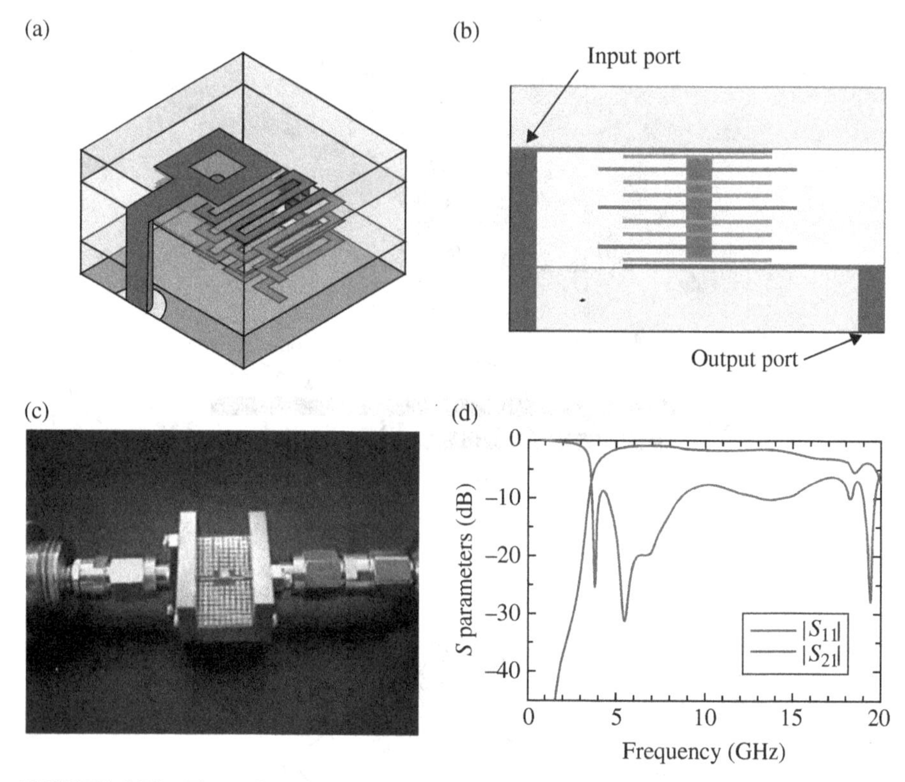

FIGURE 3.30 Three-dimensional view (a), side view (b), fabricated prototype (c), and frequency response (d) of a vertically stacked CRLH transmission line. Reprinted with permission from Ref. [97]. Copyright 2011 European Microwave Association.

3.5.2 Resonant-Type Approach

By resonant-type metamaterial transmission lines, we refer to the set of artificial lines where a host line is loaded with SRRs or other electrically small resonators (CSRRs, OSRRs, OCSRRs, etc.).[30] Notice, however, that the term given to this approach does not mean that the CL-loaded CRLH lines reviewed before are nonresonant (the canonical models of the CRLH lines presented in Section 3.4 clearly include series and parallel LC resonators). It has been argued that metamaterials based on split rings are narrow band and lossy due to the resonant nature of the constitutive elements (SRRs). Certainly, the first metamaterial structures described in Section 3.2, consisting of combinations of SRRs and metallic wires conveniently oriented, exhibit LH wave propagation in a narrow band. However, from this fact we cannot conclude that metamaterial transmission lines based on the resonant-type approach are intrinsically narrow band and lossy. Depending on the realizations, the LH band may be narrow, as

[30] Exception to this are the so-called hybrid lines, which can be considered to be a combination of the CL-loaded and resonant-type approach, and will be introduced in Section 3.5.3.

compared to CL-loaded lines; however, resonant-type balanced CRLH metalines can be designed to exhibit wide bandwidths, as will be shown later. Ultimately, the transmission bandwidths are dictated by the relations between the reactive elements of the circuit models. The achievable L/C ratios in SRR or CSRR based lines are limited, and these lines exhibit typically narrow LH bands. However, by combining open resonant particles (OSRRs and OCSRRs), CRLH lines with very wide bandwidths can be obtained, as will be demonstrated later. Very wideband bandpass filters, or moderate bandwidth bandpass filters with small insertion losses, based on CSRRs, have been reported as well. The applications of resonant-type metamaterial transmission lines will be studied in Chapter 4. Let us now discuss in detail the main approaches for their implementations and the specific circuit models.

3.5.2.1 Transmission Lines based on SRRs The first resonant-type metamaterial transmission line, based on SRRs [8], was designed as an attempt to mimic the first reported LH medium (depicted in Fig. 3.4) in planar technology. The considered host line was a CPW transmission line, loaded with pairs of SRRs (etched on the back substrate side) and shunt connected strips (Fig. 3.31). From the point of view of the effective constitutive parameters, the negative permeability is due to the SRRs, whereas the shunt-connected strips emulate the metallic wires of the wire medium, and therefore are responsible for the negative effective permittivity. In the structure of Figure 3.31, the pairs of SRRs are placed with their centers roughly aligned with the slots of the CPW. With this configuration, the magnetic flow lines generated by the host CPW can mostly "penetrate" the SRRs, and the particles can be magnetically excited. Hence, the CPW loaded only with SRRs behaves as a one-dimensional planar MNG medium in a narrow band just above the first SRR resonance frequency. Notice that in the structure of Figure 3.31, the SRRs are etched with their slits aligned in the direction orthogonal to the line axis and the shunt strips are allocated in the symmetry plane of the SRRs. This configuration avoids interactions between SRRs and strips [4]. Moreover, with this SRR orientation, the electric coupling between the CPW and the SRRs is cancelled, and the circuit model of the unit cell is simplified (the effects of particle rotation on the circuit model of the unit cell will be discussed later). Obviously, to implement an LH line, it is necessary that the cutoff frequency (plasma frequency) of the strip-loaded CPW is higher than the resonance frequency of the SRR, and this can be controlled by the distance between adjacent strips (period), and by the strip width, as discussed in Refs. [4, 8]. Notice that a CPW loaded only with strips can be described by the circuit model depicted in Figure 3.5, where L_s and C_s are the per-section line inductance and capacitance, and L is the inductance of the pair of parallel strips. The cutoff frequency is given by

$$\omega_c = \frac{1}{\sqrt{LC_s}} \tag{3.67}$$

Using the inductance value of the wire medium discussed in subsection 3.2.4.1 (expression 3.18) and $C_s = \varepsilon_0 l$, the cutoff frequency (3.67) is identical to the plasma frequency given by (3.21). Therefore, there is a clear link between the one-dimensional wire medium and the strip-loaded CPW.

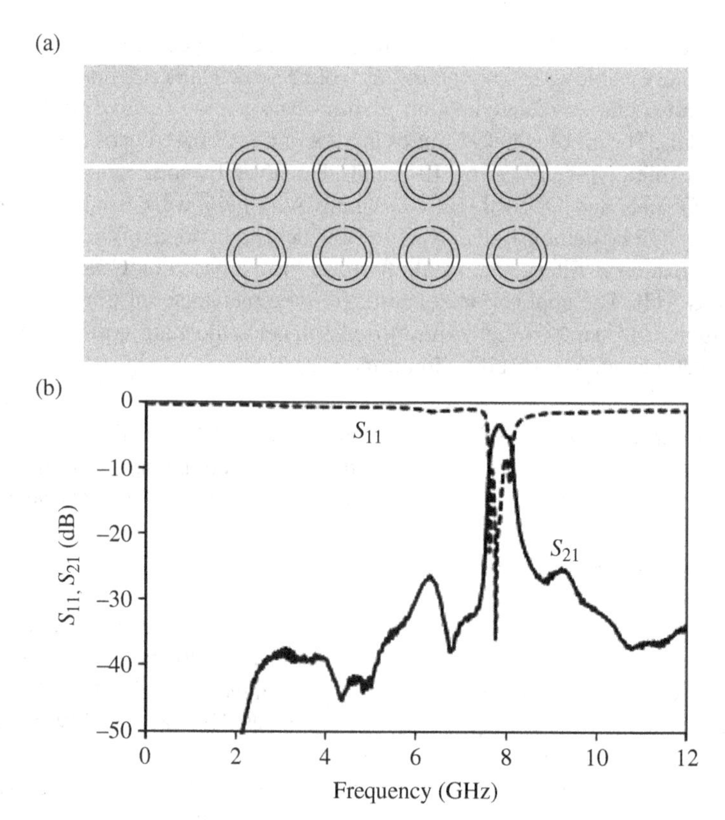

FIGURE 3.31 Layout (a) and measured frequency response (b) of a typical SRR/strip-loaded CPW. SRRs, etched on the back side of the substrate, are depicted in black, whereas the upper metal level is depicted in gray. Ring dimensions were determined following Refs [4, 99] to obtain a resonance frequency of $f_o = 7.7$ GHz, namely rings width and separation are $c = d =$ 0.2 mm, and the radius of the inner ring is $r = 1.3$ mm. The distance l between adjacent rings is 5 mm. Lateral CPW dimensions were calculated to obtain a 50 Ω characteristic impedance. Strip width is 0.2 mm. The considered substrate is the *Arlon 250-LX-0193-43-11* with $\varepsilon_r = 2.43$ and thickness $h = 0.49$ mm. Reprinted with permission from Ref. [8]. Copyright 2003, American Institute of Physics.

The frequency response of the SRR/strip-loaded CPW of Figure 3.31 is depicted in Figure 3.31b. A narrow transmission band, centered at roughly 8 GHz, can be appreciated. This pass band is an LH band as was demonstrated from full-wave EM simulations (not shown), and supported by the dispersion relation of the proposed lumped-element equivalent circuit model of the unit cell [4, 8]. The structure of Figure 3.31 is strongly unbalanced, but the geometry can be modified in order to narrow the gap present between the LH and the RH band (not present in the frequency range shown in Fig. 3.31b). Figure 3.32 depicts the SRR-loaded CPW without shunt strips and its frequency response. A stop band behavior, attributed to the negative effective permeability provided by the SRRs, is obtained in this case. Notice that

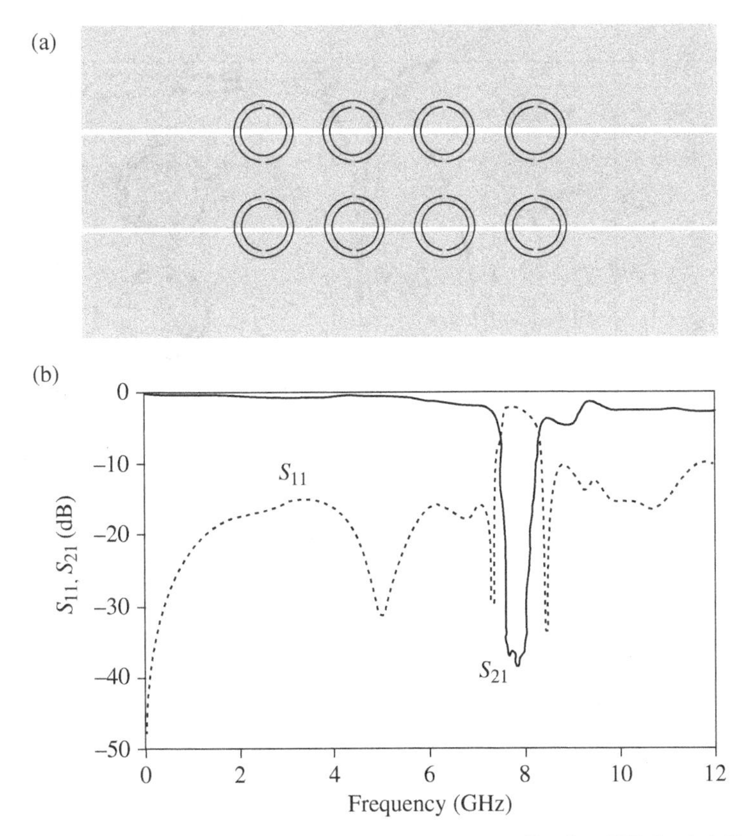

FIGURE 3.32 Layout (a) and measured frequency response (b) of an SRR-loaded CPW. The dimensions and substrate parameters are those of Figure 3.31. Reprinted with permission from Ref. [8]. Copyright 2003, American Institute of Physics.

the results of Figure 3.31b and 3.32b are conceptually similar to those reported in Figure 3.8a, as expected by virtue of the equivalence between the SRR/wire bulk media and SRR/strip-loaded CPWs.

Due to the presence of the SRRs, a transmission zero in the frequency response of both structures (with and without shunt strips) must be present. However, such transmission zero can be obscured by the noise floor level in multiple-cell structures. For this reason, in order to visualize the position of the transmission zero, it is convenient to obtain the frequency response in single-cell structures. Thus, Figure 3.33 depicts the insertion and return loss of a single-cell SRR-loaded CPW and SRR/strip-loaded CPW (both structures are depicted in the insets) [100]. As can be seen, the position of the transmission zeros (indicated in the figures) is different in both structures (the transmission zero frequency, f_z, is lower for the SRR/strip-loaded CPW). This shift in the transmission zero frequency is not obvious, and, indeed, in the first reported equivalent circuit models of the unit cell of the SRR-loaded and SRR/strip-loaded CPWs, such effect was ignored [8]. In other words, the formerly proposed models

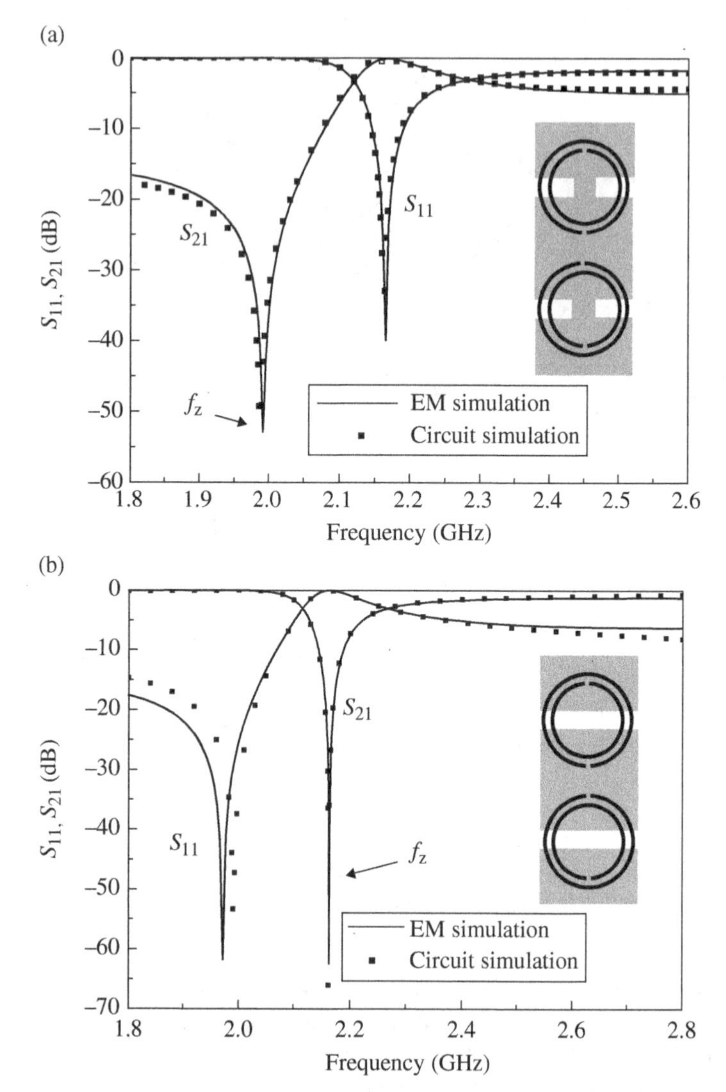

FIGURE 3.33 Simulated (through the *Agilent Momentum* commercial software) frequency responses of the unit cell structures shown in the insets. (a) CPW loaded with SRRs and shunt-connected strips and (b) CPW only loaded with SRRs. The response obtained from circuit simulation of the equivalent circuit model with extracted parameters (see Table 3.1) is also included. Dimensions are: the central strip width $W_c = 8$ mm, the width of the slots $G = 1.43$ mm, length of the line $D = 8$ mm, and the shunt strip width $w_s = 1.8$ mm. In all cases, the dimensions for the SRRs are: outer ring width $c_{out} = 0.364$ mm, inner ring width $c_{inn} = 0.366$ mm, distance between the rings $d = 0.24$ mm, and internal radius $r = 2.691$ mm. The considered substrate is *Rogers RO3010* with dielectric constant $\varepsilon_r = 10.2$ and thickness $h = 1.27$ mm. Reprinted with permission from Ref. [100]; copyright 2008 by Springer.

for these structures were not able to explain this different position of f_z. For this reason, the models were revised and improved in Ref. [101], and these latter models are those that are reported in detail in the next paragraph. Nevertheless, the first models of SRR-loaded and SRR-/strip-loaded CPW transmission lines are depicted in Figure 3.34 for completeness and for better comprehension of their limitations. Notice that the magnetic wall concept was applied to these models. L and C account for the line inductance and capacitance, respectively, C_s and L_s model the SRR, M is the mutual inductive coupling between the line and the SRRs (providing a mutual inductance of 2 M after application of the magnetic wall—see the caption of Fig. 3.34), and L_p is the inductance of the shunt strips. The π-models of Figure 3.34a and b can be transformed to those depicted in Figure 3.34c and d, through the indicated transformations (see details in Ref. [4, 8]). For the SRR-/strip-loaded CPW, the model (Fig. 3.34c) is very similar to that of the

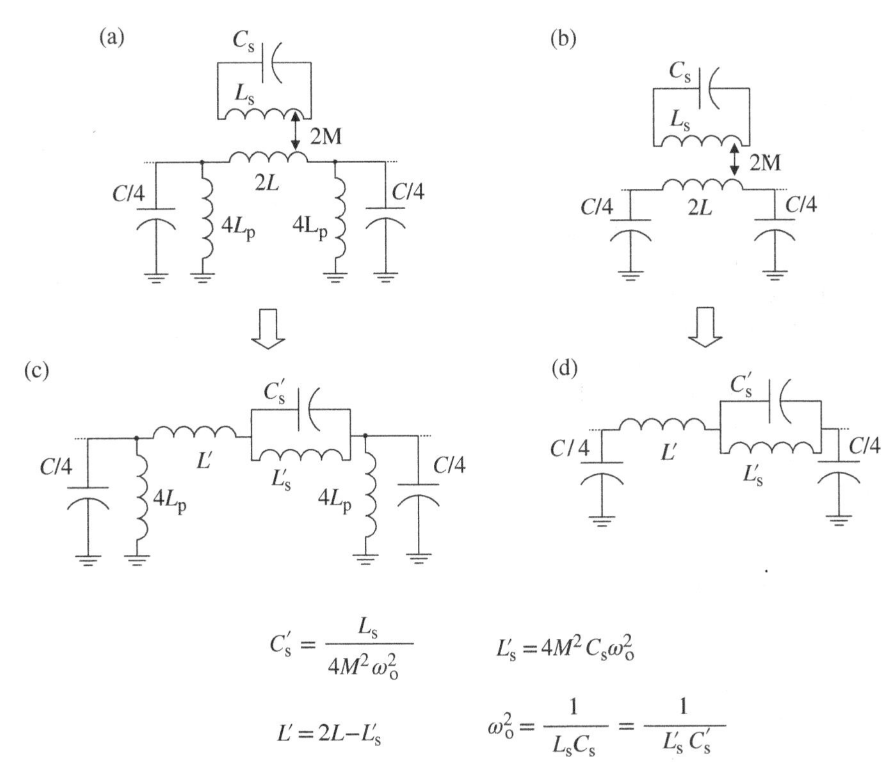

FIGURE 3.34 Lumped-element equivalent circuit for the basic cell of the SRR/strip-loaded CPW (a) and SRR-loaded CPW (b). These circuits can be transformed to those depicted in (c) and (d), according to the indicated transformations. In Refs. [4, 8], the inductance L' of the transformed models was approximated by $L' = 2L$, and the mutual inductance between the line and the SRR after application of the magnetic wall was considered to be M. This explains the difference between the indicated transformations and those in Refs. [4, 8]. However, in this book, the mutual inductance in (a) and (b) appears as 2 M for coherence with the definition of M in the text and with the circuit of Figure 3.35.

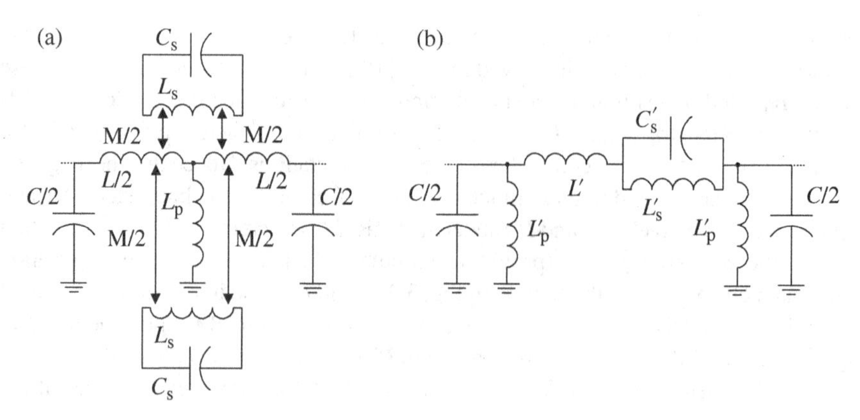

FIGURE 3.35 Improved circuit model for the basic cell of the SRR/strip-loaded CPW structure (a) and transformed model (b).

canonical order-2 CRLH transmission line model (Fig. 3.20).[31] The difference is the presence of the inductance L'_s, which is responsible for the transmission zero (given by the resonance frequency of the LC tank formed by L'_s and C'_s). However, the model of Figure 3.34d, corresponding to the SRR-loaded CPW has exactly the same series branch. Hence, according to these formerly proposed models, the transmission zeros of both structures should be present at the same frequency, which is not the case. Indeed, as will be seen later, the models of Figure 3.34c and d are formally correct. The difference is that the elements of the series branches of these models are different.

The improved model of the SRR-/strip-loaded CPW is depicted in Figure 3.35 [101]. Following Ref. [101], the magnetic wall concept has not been considered since this is not actually necessary. In this model, the reactive parameters have the same interpretation as in the models of Figure 3.34. However, the inductance of the shunt inductive strips, L_p, is now located between the two inductances ($L/2$) that model each line section, to the left and right of the position of the shunt strips. This improved model reflects the location of the inductive strips, as was reported in Ref. [102]. The model is neither a π-circuit nor a T-circuit. Thus the transmission zero frequency cannot be directly obtained from it. Due to symmetry considerations and reciprocity, the admittance matrix of the circuit of Figure 3.35a (which is a two-port) must satisfy $Y_{12} = Y_{21}$ and $Y_{11} = Y_{22}$. From these matrix elements, the equivalent π-circuit model can be obtained according to [103]:

$$Z_s(\omega) = -(Y_{21})^{-1} \tag{3.68a}$$

$$Z_p(\omega) = (Y_{11} + Y_{21})^{-1} \tag{3.68b}$$

[31] Notice that the unit cell of the canonical order-2 CRLH line of Figure 3.20 is a T-circuit. However, this line can also be obtained by cascading π-circuits comprising a series resonator L_R-C_L in the series branch and a parallel resonator $2L_L$-$C_R/2$ in the shunt branches. Such π-circuit is formally identical to the circuit of Figure 3.34(c) with the exception of the inductance L'_s.

Y_{21} is inferred by grounding the port 1 and obtaining the ratio between the current at port 1 and the applied voltage at port 2. Y_{11} is simply the input admittance of the two-port, seen from port 1, with a short-circuit at port 2. After a straightforward but tedious calculation, the elements of the admittance matrix are obtained, and by applying (3.68), we finally obtain

$$Z_s(\omega) = j\omega\left(2+\frac{L}{2L_p}\right)\left[\frac{L}{2}+M^2\frac{1+(L/4L_p)}{L_s((\omega_o^2/\omega^2)-1)-(M^2/2L_p)}\right] \tag{3.69a}$$

$$Z_p(\omega) = j\omega\left(2L_p+\frac{L}{2}\right) \tag{3.69b}$$

with $\omega_o = (L_s C_s)^{-1/2}$. Expression (3.69a) can be rewritten as follows:

$$Z_s(\omega) = j\omega\left[\left(2+\frac{L}{2L_p}\right)\frac{L}{2}-L'_s+\frac{L'_s}{1-L'_s C'_s \omega^2}\right] \tag{3.70}$$

with

$$L'_s = 2M^2 C_s \omega_o^2 \frac{(1+(L/4L_p))^2}{1+(M^2/2L_p L_s)} \tag{3.71}$$

$$C'_s = \frac{L_s}{2M^2\omega_o^2}\left(\frac{1+(M^2/2L_p L_s)}{1+(L/4L_p)}\right)^2 \tag{3.72}$$

These results indicate that the improved circuit model of the unit cell of the SRR-/strip-loaded CPW (Fig. 3.35a) can be formally expressed as the π-circuit of Figure 3.34c, but with modified parameters (Fig. 3.35b). These modified parameters are related to the parameters of the circuit of Figure 3.35a, according to (3.71), (3.72) and

$$L' = \left(2+\frac{L}{2L_p}\right)\frac{L}{2}-L'_s \tag{3.73}$$

$$L'_p = 2L_p+\frac{L}{2} \tag{3.74}$$

The transmission zero angular frequency, ω_z, for the circuit of Figure 3.35 is no longer given by the resonance frequency of the SRRs, ω_o, but it is smaller, that is, $\omega_z \le \omega_o$.

It is interesting to mention that the angular frequency where $\phi = \beta l = 0$, ω_s, obtained by forcing $Z_s(\omega) = 0$, that is,

$$\omega_s = \frac{1}{\sqrt{C_s(L_s - 2(M^2/L))}} \tag{3.75}$$

does not depend on the shunt inductance despite that $Z_s(\omega)$ is a function of L_p. This explains that ω_s is identical for both the SRR-loaded and SRR/strip-loaded CPW structures of Figure 3.33 (see Fig. 3.36; [100]).

According to this analysis, the previous reported circuit model of SRR/strip-loaded CPWs (Fig. 3.34c) is formally correct. The weakness relies on the physical interpretation of the elements of that model. These elements do not have any physical meaning (except C). However, they are related to the elements of the circuit model of Figure 3.35a, which describe the different components of the SRR/strip-loaded CPW.

The parameters of the circuit models for the structures of Figure 3.33 were extracted from the lossless EM simulations of their frequency responses, according to the procedure described in Appendix G [104] (see Table 3.1). The circuit of Figure 3.35b was considered for the SRR/strip-loaded CPW, whereas for the structure without shunt strips, the circuit model that results by forcing $L_p = \infty$ was considered. The circuit simulation of both circuits is also depicted in Figure 3.33 to ease the

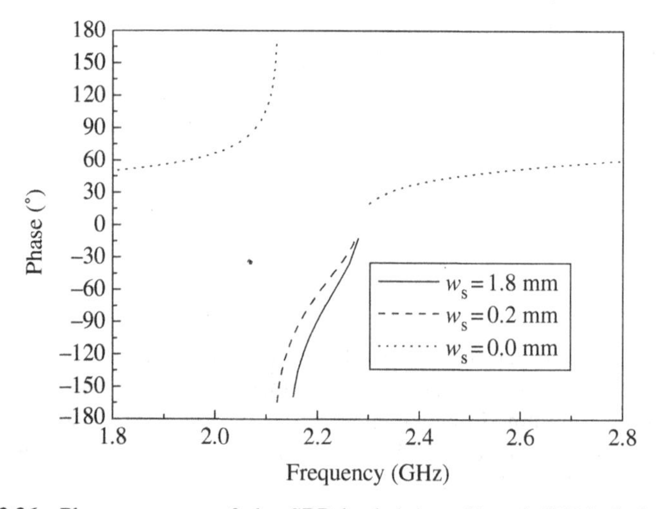

FIGURE 3.36 Phase response of the SRR-loaded ($w_s = 0$) and SRR/strip-loaded ($w_s =$ 1.8 mm) CPW structures depicted in Figure 3.33. The phase response of an identical structure with strip widths of $w_s = 0.2$ mm is also included. The phase nulls at identical frequency in all the considered cases. Reprinted with permission from Ref. [100]; copyright 2008 by Springer.

TABLE 3.1 Extracted element parameters for the structures of Figure 3.33

	C (pF)	L' (nH)	L'_p (nH)	C'_s (pF)	L'_s (nH)
With shunt strips	2.07	4.91	1.38	3.90	1.68
Without shunt strips	2.07	1.91	—	23.01	0.24

TABLE 3.2 Parameters for the circuit shown in Figure 3.35a, obtained from Table 3.1

	C (pF)	L (nH)	M (nH)	L_p (nH)	C_s (pF)	L_s (nH)
With shunt strips	2.07	1.94	1.23	0.20	0.31	17.16
Without shunt strips	2.07	2.15	1.45	—	0.31	17.57

comparison with the EM simulation. The agreement is good, thus pointing out the validity of the model.

However, to further validate the proposed circuit model of SRR/strip-loaded CPWs, Equations (3.71)–(3.74) were inverted in order to obtain the parameters of the model of Figure 3.35a (from the results shown in Table 3.1), and these coincide in reasonable agreement with those parameters inferred in the structure without shunt strips (see Table 3.2), and obtained through the transformations indicated in Figure 3.34 (or (3.71)–(3.74) with $L_p = \infty$).[32] This means that the presence of the shunt strips does not affect the parameters of the structure, hence having a clear physical interpretation.

Interestingly, application of (3.45) and (3.46) to the model of Figure 3.35b with the element values given by (3.71)–(3.74) gives an effective permittivity and permeability of the form of expressions (3.22) and (3.34), respectively. Namely, the impedance of the series branch, given by (3.70), can be expressed as follows:

$$Z_s(\omega) = j\omega\left[\left(2+\frac{L}{2L_p}\right)\frac{L}{2}+\frac{L_s'^2 C_s'\omega^2}{\left(1-L_s'C_s'\omega^2\right)}\right] \tag{3.76}$$

and it can be simplified to

$$Z_s(\omega) = j\omega\left(2+\frac{L}{2L_p}\right)\frac{L}{2}\left[1+\frac{2L_s'}{L\left(2+\left(L/2L_p\right)\right)}\frac{\omega^2}{\left(\omega_o'^2-\omega^2\right)}\right] \tag{3.77}$$

where $\omega_o' = \omega_z = \left(L_s'\ C_s'\right)^{-1/2}$ is the transmission zero frequency. Using (3.45), the effective permeability is

$$\mu_{\text{eff}}(\omega) = \left(2+\frac{L}{2L_p}\right)\frac{L}{2l}\left[1-\frac{2L_s'}{L\left(2+\left(L/2L_p\right)\right)}\frac{\omega^2}{\left(\omega^2-\omega_o'^2\right)}\right] \tag{3.78}$$

[32] It is very easy to demonstrate that if the magnetic wall concept in the circuit of Figure 3.34 is not applied, the resulting transformations are identical to those given by (3.71)–(3.74) with $L_p = \infty$. However, the model of a SRR/strip-loaded CPW of Figure 3.34a is not correct. As discussed in the text, the model of Figure 3.35a must be considered to describe these lines.

and introducing (3.71) in (3.78), we obtain

$$\mu_{\text{eff}}(\omega)=\left(2+\frac{L}{2L_{\text{p}}}\right)\frac{L}{2l}\left[1-\frac{2M^2C_{\text{s}}\omega_{\text{o}}^2(1+(L/4L_{\text{p}}))}{L(1+(M^2/2L_{\text{p}}L_{\text{s}}))}\frac{\omega^2}{(\omega^2-\omega_{\text{o}}'^2)}\right] \tag{3.79}$$

The mutual inductance is a fraction of the line inductance, that is, $M=L\cdot f$. Therefore, (3.79) can be written in the following form:

$$\mu_{\text{eff}}(\omega)=\left(2+\frac{L}{2L_{\text{p}}}\right)\frac{L}{2l}\left[1-\frac{2Lf^2}{L_{\text{s}}}\frac{\left(1+\dfrac{L}{4L_{\text{p}}}\right)}{\left(1+\dfrac{M^2}{2L_{\text{p}}L_{\text{s}}}\right)}\frac{\omega^2}{(\omega^2-\omega_{\text{o}}'^2)}\right] \tag{3.80}$$

which is formally identical to expression (3.34). Indeed, if we neglect the presence of the strips ($L_{\text{p}}\to\infty$), and consider that $L/l=\mu_{\text{eff}}(0)$, the effective permeability can be expressed as follows:

$$\mu_{\text{eff}}(\omega)=\mu_{\text{eff}}(0)\left[1-2\frac{\mu_{\text{eff}}(0)lf^2}{L_{\text{s}}}\frac{\omega^2}{(\omega^2-\omega_{\text{o}}'^2)}\right] \tag{3.81}$$

Notice that losses have been neglected, and the inductance of the SRR, L_{s}, appears as L in (3.34). Thus, the single difference between (3.81) and (3.34) concerns the factor F and the DC effective permeability, different than μ_{o}. In fact, $\mu_{\text{eff}}(0)$ is the effective permeability of the host line (CPW) without the presence of the SRRs, defined as the per-unit-length inductance of the line, according to (3.45).

Concerning the permittivity, application of (3.46) to the parallel resonator of the shunt branch of the circuit of Figure 3.35b, directly gives a permittivity of the form (3.22), that is,

$$\varepsilon_{\text{eff}}=\frac{C}{l}\left(1-\frac{\omega_{\text{p}}^2}{\omega^2}\right) \tag{3.82}$$

with $\omega_{\text{p}}=(L_{\text{p}}'\,C/2)^{-1/2}$. In (3.82), C/l is the effective permittivity of the structure at high frequencies ($\omega >> \omega_{\text{p}}$), or the effective permittivity of the CPW host line without the presence of the shunt strips. Thus, the equivalence between the SRR and wire medium of Figure 3.4 and the SRR/strip-loaded CPW has been demonstrated. Indeed, such equivalence was pointed out by Eleftheriades *et al.* [105] and by Solymar and Shamonina in their textbook [21].

Once the circuit model of the SRR/strip-loaded CPW has been expressed as a π-circuit, the dispersion relation and the characteristic impedance can be easily obtained. Application of (2.23) to the circuit of Figure 3.35b gives

$$\cos(\beta l) = 1 - \frac{1}{2} L' C \omega^2 \left(1 - \frac{\omega_p^2}{\omega^2}\right)\left(1 - \frac{1}{L' C_s' \omega^2 \left(1 - \omega_o'^2/\omega^2\right)}\right) \tag{3.83}$$

Notice that (3.83) can be reduced to (3.51) if $L_s' \to \infty$ (or $\omega_o' = 0$).[33] For a symmetric π-circuit, the characteristic impedance can be expressed as follows:

$$Z_B(\omega) = \sqrt{\frac{Z_s(\omega) Z_p(\omega)/2}{1 + \left(Z_s(\omega)/2Z_p(\omega)\right)}} \tag{3.84}$$

where Z_s and Z_p are the series and shunt impedances of the π-circuit, respectively. Therefore, the characteristic impedance is given by

$$Z_B(\omega) = \sqrt{\frac{L'}{C}\left(1 - \frac{\omega_p^2}{\omega^2}\right)^{-1}\left[\left(1 - \frac{1}{L' C_s' \omega^2 \left(1 - \left(\omega_o'^2/\omega^2\right)\right)}\right)^{-1} - \frac{C L' \omega^2}{4}\left(1 - \frac{\omega_p^2}{\omega^2}\right)\right]^{-1}} \tag{3.85}$$

In the limit $L_s' \to \infty$ (or $\omega_o' = 0$), expression (3.85) does not coincide with (3.52). The reason is that the characteristic impedance of a transmission line is different if it is described either by a cascade of symmetric π- or T-circuit unit cells. However, the dispersion relation is identical for both unit cells.

3.5.2.2 Transmission Lines based on CSRRs Since the CSRR is the negative image of the SRR and it can be driven (not exclusively) by an axial time-varying electric field, CSRR-based lines have been mainly implemented in microstrip technology by etching the CSRRs in the ground plane, beneath the conductor strip. In that region, there is a strong vertical electric field generated by the line, and the CSRR can be easily excited. The former artificial lines based on CSRRs were reported in Refs. [84, 85]. In Ref. [84], a microstrip line was loaded only with CSRRs, giving rise to a one-dimensional ENG medium, whereas the line reported in Ref. [85] was loaded with CSRR and series gaps. The series gaps provide the required negative permeability to achieve an LH medium. Figure 3.37 depicts the CSRR/gap-loaded microstrip structure reported in Ref. [85] and its frequency response, where a bandpass behavior in the LH band is obtained. For comparison purposes, the frequency response without the presence of the gaps is included, where a stop band functionality, attributed to the negative effective permittivity in the vicinity of the CSRR resonance frequency, is visible. Actually, the CSRR/gap-loaded lines exhibit a CRLH behavior. However, the RH band is beyond the frequency range shown in Figure 3.37.

[33] The expressions are identical if we identify the elements of the shunt and series branches (except L_s') of the circuit of Figure 3.35b with those of Figure 3.20.

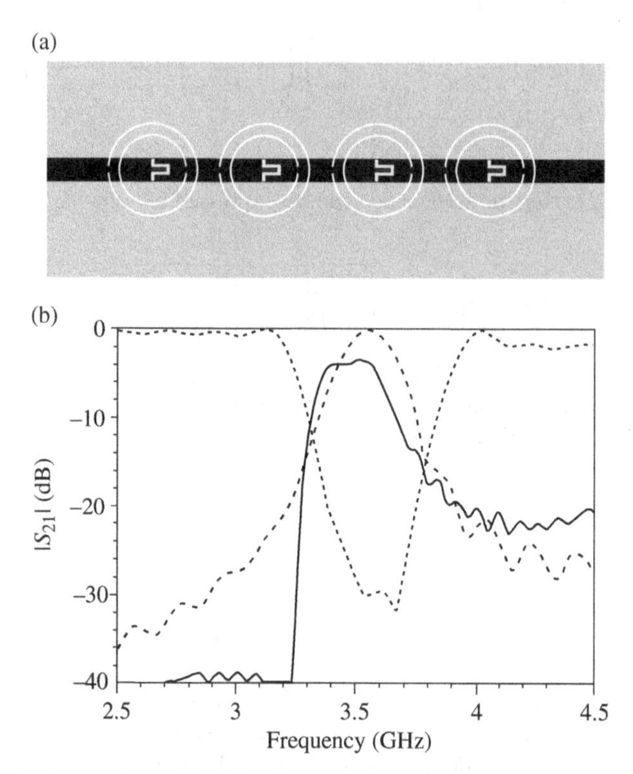

FIGURE 3.37 Layout (a) and insertion loss (b) of a CSRR/gap-loaded microstrip line. The EM simulation of the structure is represented by dashed line, whereas the measured response is represented by solid line. The simulated response without series gaps is also included and represented by dotted line. Substrate parameters are dielectric constant $\varepsilon_r = 10.2$ and thickness $h = 1.27$ mm. CSRR dimensions are $r_{ext} = 2.5$ mm and $c = d = 0.3$ mm. The strip width is $W = 1.2$ mm. Reprinted with permission from Ref. [85]. Copyright 2004, American Physical Society.

The former lumped-element equivalent circuit models of CSRR-loaded and CSRR/gap-loaded microstrip lines (unit cells) were reported in Ref. [10], and are depicted in Figure 3.38. According to Ref. [10], L and C are the per-section inductance and capacitance of the line, L_c and C_c model the CSRR, and C_g accounts for the capacitance of the series gaps. According to this interpretation of the circuit elements, the transmission zero, given by that frequency that nulls the shunt branch, that is,

$$f_z = \frac{1}{2\pi\sqrt{L_c(C + C_c)}} \tag{3.86}$$

should be independent of the characteristics of the series gap. However, it was found that the position of the transmission zero varies with gap dimensions and takes different values in CSRR-loaded microstrip lines with and without gaps (with identical line and CSRR dimensions) [100]. Notice that this behavior is similar to that of

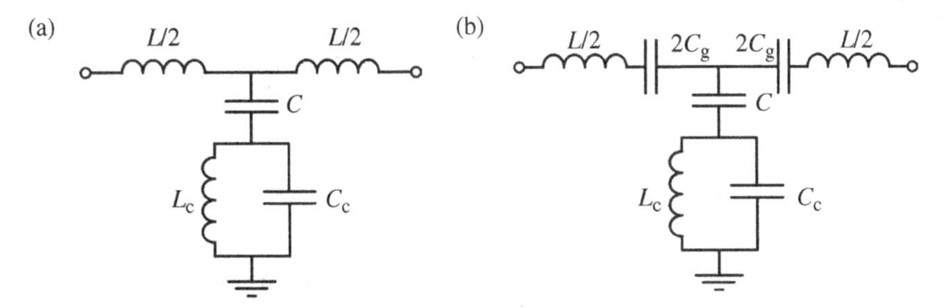

FIGURE 3.38 Formerly reported lumped-element equivalent circuits for the basic cell of the CSRR-loaded microstrip (a) and CSRR/gap-loaded microstrip (b) transmission lines.

SRR-loaded CPWs caused by variations in shunt-connected strip dimensions. Therefore, the models of the CSRR-based lines must be revised to account for the variation of the transmission zero frequency with gap dimensions.

Figure 3.39 depicts the layouts and frequency responses of two CSRR-loaded microstrip lines: one with series gap and the other one without it. The variation experienced by the transmission zero frequency is significant. The parameters of the models of Figure 3.38 were extracted (see Appendix G), and they are depicted in Table 3.3. Notice that the substantial variation in f_z is due to the strong variation of C. However, this cannot be attributed to the effects of the fringing capacitance of the gap.

There is no doubt that the models of Figure 3.38 accurately describe the EM response of the CSRR-based lines. However, the interpretation of the parameters as formerly reported in Ref. [10] is not correct. Neither C_g is actually the capacitance of the series gap, nor C is the line capacitance. A more realistic model of the structure is that depicted in Figure 3.40, where C_L is the line capacitance, C_f is the fringing capacitance of the gap and C_s is the series capacitance of the gap [106]. Obviously, from π-T transformation, the circuit model of Figure 3.38b, which is the previously reported model of microstrip lines loaded with CSRRs and series gaps, is obtained; but the values of C_g and C do not actually have a physical interpretation. Indeed, C_g and C can be expressed in terms of C_s and $C_{par} = C_f + C_L$ according to

$$C_g = 2C_s + C_{par} \tag{3.87}$$

$$C = \frac{C_{par}\left(2C_s + C_{par}\right)}{C_s} \tag{3.88}$$

Inspection of (3.88) indicates that as C_s decreases (gap distance increases), C increases. Indeed C can be made very large if C_s is sufficiently small. It is also apparent from (3.87) that despite the fact that the gap distance is increased, it is not expected that C_g experiences a significant reduction. The reason is that for small values of C_s, C_g is dominated by the line capacitance and the fringing capacitance (C_{par}). Such behavior of C and C_g was corroborated by considering several CSRR-loaded structures with identical geometry except in the gap distance [107]. This analysis reveals

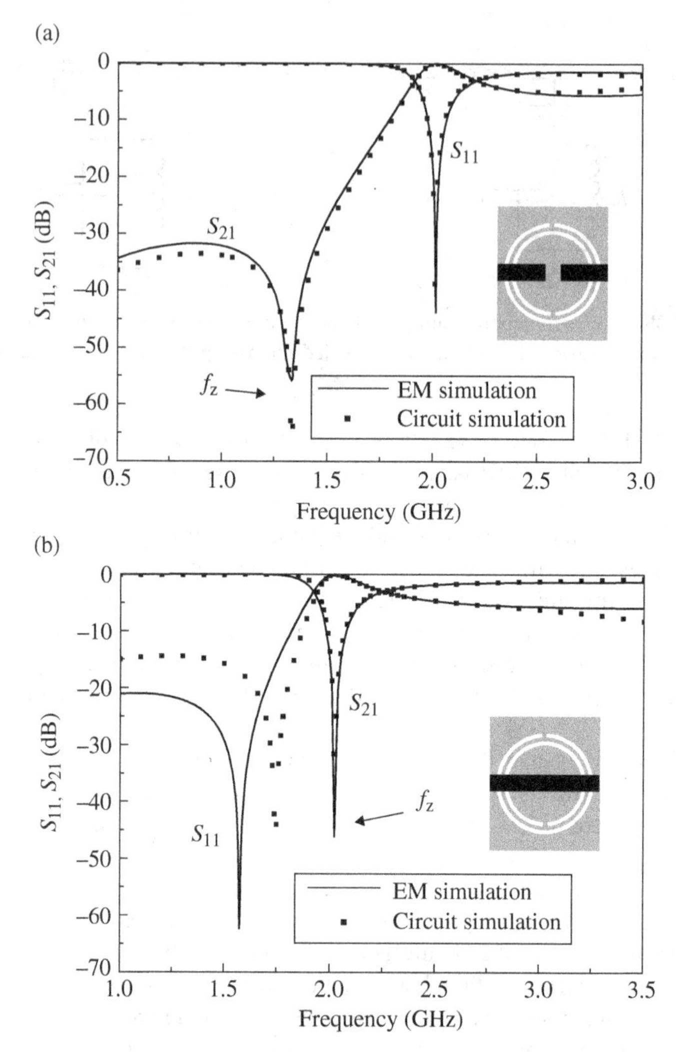

FIGURE 3.39 Simulated (through the *Agilent Momentum* commercial software) frequency responses of the unit cell structures shown in the insets. (a) Microstrip line loaded with CSRRs and series gaps and (b) microstrip line only loaded with CSRRs. The response obtained from circuit simulation of the equivalent model with extracted parameters is also included. Dimensions are: the strip line width $W_m = 1.15$ mm, the length $D = 8$ mm, and the gap width $w_g = 0.16$ mm. CSRRs dimensions: are outer ring width $c_{out} = 0.364$ mm, inner ring width $c_{inn} = 0.366$ mm, distance between the rings $d = 0.24$ mm, and internal radius $r =$ 2.691 mm. The considered substrate is *Rogers RO3010* with dielectric constant $\varepsilon_r = 10.2$ and thickness $h = 1.27$ mm. Reprinted with permission from Ref. [100]; copyright 2008 by Springer.

TABLE 3.3 Extracted element parameters for the structures of Figure 3.39

	C (pF)	L (nH)	C_g (pF)	C_c (pF)	L_c (nH)
With series gap	6.43	5.78	0.31	3.10	1.49
Without series gap	0.95	5.78	—	2.81	1.65

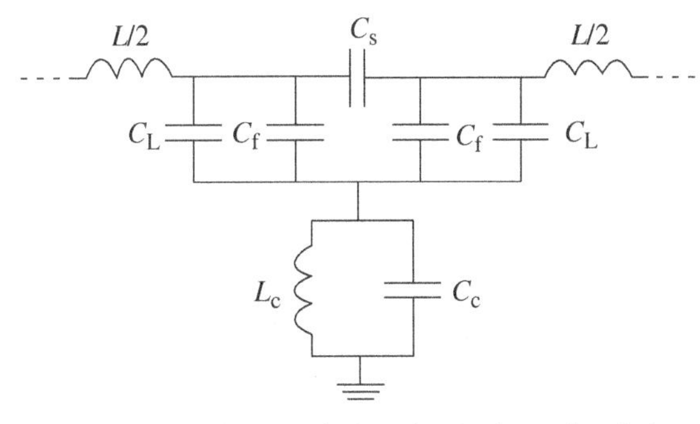

FIGURE 3.40 Improved circuit model for the basic cell of the CSRR/gap-loaded microstrip line.

that C can be enhanced by merely increasing the gap distance, without the penalty of a small C_g value (since C_f and C_L do also contribute to this capacitance—see expression (3.87)). These results are of interest because they reveal that it is possible to obtain high values of C (by decreasing C_s), regardless of the substrate thickness. These high values of C are typically necessary to drive the transmission zero frequency to small values.[34] Notice that in the limit $C \to \infty$, the model of Figure 3.38b is identical to the model of an order-2 CRLH transmission line.

The dispersion relation and the characteristic impedance of CSRR/gap-loaded lines is given by (2.33) and (2.30), that is,

$$\cos\beta l = 1 + \frac{(C/2C_g)\left(1-(\omega^2/\omega_s^2)\right)\left(1-\left(\omega^2/\omega_p^2\right)\right)}{\left(1-\left(\omega^2/\omega_p^2\right)-CL_c\omega^2\right)} \tag{3.89}$$

$$Z_B = \sqrt{\frac{L}{C_c}\frac{\left(1-(\omega_s^2/\omega^2)\right)}{\left(1-\left(\omega_p^2/\omega^2\right)\right)} - \frac{L^2\omega^2}{4}\left(1-\frac{\omega_s^2}{\omega^2}\right)^2 + \frac{L}{C}\left(1-\frac{\omega_s^2}{\omega^2}\right)} \tag{3.90}$$

Notice that the characteristic impedance is similar to that corresponding to the order-2 CRLH transmission line model (expression 3.52). The difference is the presence of the third term in the square root of (3.90). However, this term vanishes if $C \to \infty$, since in

[34] Nevertheless, in certain applications, it might be convenient to allocate the transmission zero close to the first (LH) band, in order to reject undesired frequencies.

this case the models of Figures 3.20 and 3.38b are identical. In the limit $C \to \infty$, (3.89) coincides with (3.51).

Both SRR/strip-loaded CPW and CSRR/gap-loaded microstrip lines exhibit a CRLH behavior. These lines can be balanced by merely forcing the series and shunt resonance frequencies to be identical. In particular, for CSRR/gap-loaded lines, the balanced condition is expressed as follows:

$$\omega_s = \frac{1}{\sqrt{LC_g}} = \frac{1}{\sqrt{L_c C_c}} = \omega_p = \omega_o \tag{3.91}$$

where ω_o is the transition frequency.[35] Under the balance condition, the characteristic impedance is given by

$$Z_B = \sqrt{\frac{L}{C_c} - \frac{L^2\omega^2}{4}\left(1 - \frac{\omega_o^2}{\omega^2}\right)^2 + \frac{L}{C}\left(1 - \frac{\omega_o^2}{\omega^2}\right)} \tag{3.92}$$

The frequency dependence of the characteristic impedance in a balanced CSRR/gap-loaded microstrip line is similar to that of a canonical order-2 CRLH line (Fig. 3.22b). Namely, the characteristic impedance is real in the allowed band (including the LH and the RH bands), it goes to zero at the extremes of the band, and it is imaginary in the forbidden bands. However, the maximum value of the characteristic impedance in CSRR/gap-loaded microstrip lines appears at a frequency slightly above the transition frequency, as can be appreciated in Figure 3.41.

In order to balance a CSRR-based line, relatively large values of C_g, or C_s, are typically necessary. Interdigital capacitors are thus an alternative to series gaps to implement the series capacitances. Figure 3.42 depicts the layout of a balanced unit cell, as well as the dispersion diagram and frequency response [108]. Actually, the structure is slightly unbalanced (see the dispersion diagram inferred from measurement), but the frequency response does not exhibit an appreciable gap between the LH and the RH bands (i.e., the transition between both bands is continuous). The agreement between the EM simulation (or measurement) and the circuit (electrical) simulation is good up to frequencies above the CSRR resonance (transition frequency). At sufficiently high frequencies (above 2 GHz), the discrepancy between the circuit and EM simulation is due to the fact that the circuit model (Fig. 3.38b) is no longer valid.

It is interesting to mention that there are clear similarities between the CSRR-based microstrip lines and SRR-based CPWs. Both lines exhibit a CRLH behavior, and the transmission zero frequency depends on the gap (for CSRR-based lines) or shunt strip (for SRR-based lines) geometry. With regard to the effective constitutive parameters, the permeability and permittivity follow the Lorentz and Drude models, respectively, for SRR-based lines, and the opposite models for CSRR-based lines. Indeed, the models of Figures 3.35b and 3.38b are circuit duals, in the sense that the series impedance

[35] Notice that in Section 3.5.2.1, ω_o was used to designate the intrinsic resonance frequency of the SRR.

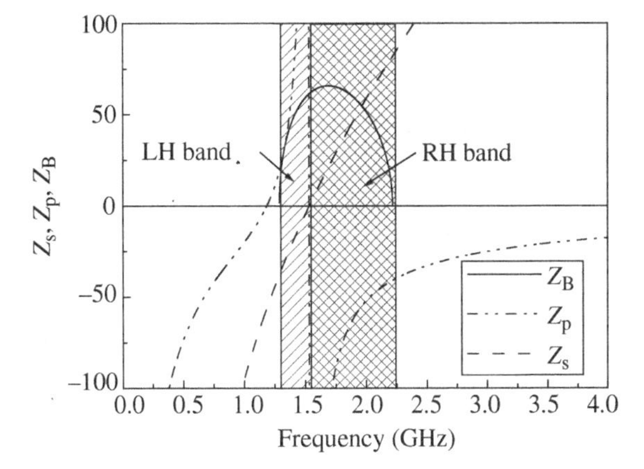

FIGURE 3.41 Representation of the series, Z_s, shunt, Z_p and characteristic impedance, Z_B, for a CRLH transmission line corresponding to the model of a balanced CSRR-based structure. The depicted values of Z_s and Z_p are actually the reactances. The transition frequency was set to roughly $f_o = 1.5$ GHz. Reprinted with permission from Ref. [108]; copyright 2007 IEEE.

of one circuit is inversely proportional to the shunt impedance of the other, as occurs in physical dual structures [83].[36] Although SRR-based CPWs and CSRR-based microstrip lines are not dual structures, they can be considered roughly duals [109].

3.5.2.3 Inter-Resonator Coupling: Effects and Modeling In the circuit models of the SRR- and CSRR-based lines of the preceding subsections the coupling between adjacent resonators has been neglected. This approximation can be made as long as the resonators are sufficiently separated and circular geometries are used. However, in SRR- or CSRR-based lines with closely located square or rectangular-shaped resonators, the effects of coupling cannot be neglected, and the models must be modified to account for inter-resonator coupling. As will be shown, inter-resonator coupling broadens the stop band (in single negative structures) or the pass band (in CRLH lines). Let us see in the next paragraphs the reason for such bandwidth enhancement, and the rich phenomenology associated to the lines with inter-resonator coupling, where multimode propagation, including the presence of complex modes,[37] arises [116]. The study is focused on CSRR-loaded microstrip lines only (the analysis of SRR-loaded CPWs is similar and is reported in Ref. [117]).

[36] According to Ref. [83], an impedance Z_1 representing an element in a planar circuit is inversely proportional to the impedance Z_2 of the corresponding element in its physical dual.

[37] Complex waves are modes that may appear in lossless structures that have complex propagation constants (despite the absence of losses) [110]. These modes appear as conjugate pairs in reciprocal lossless structures, and they carry power in opposite directions so that if the two modes are excited with the same amplitude, they do not carry net power [111]. Complex waves have been found in several structures, such as dielectrically loaded waveguides [112], finlines [113], shielded microstrip structures [114], and, more recently, in shielded mushroom-type Sievenpiper structures [115], among others.

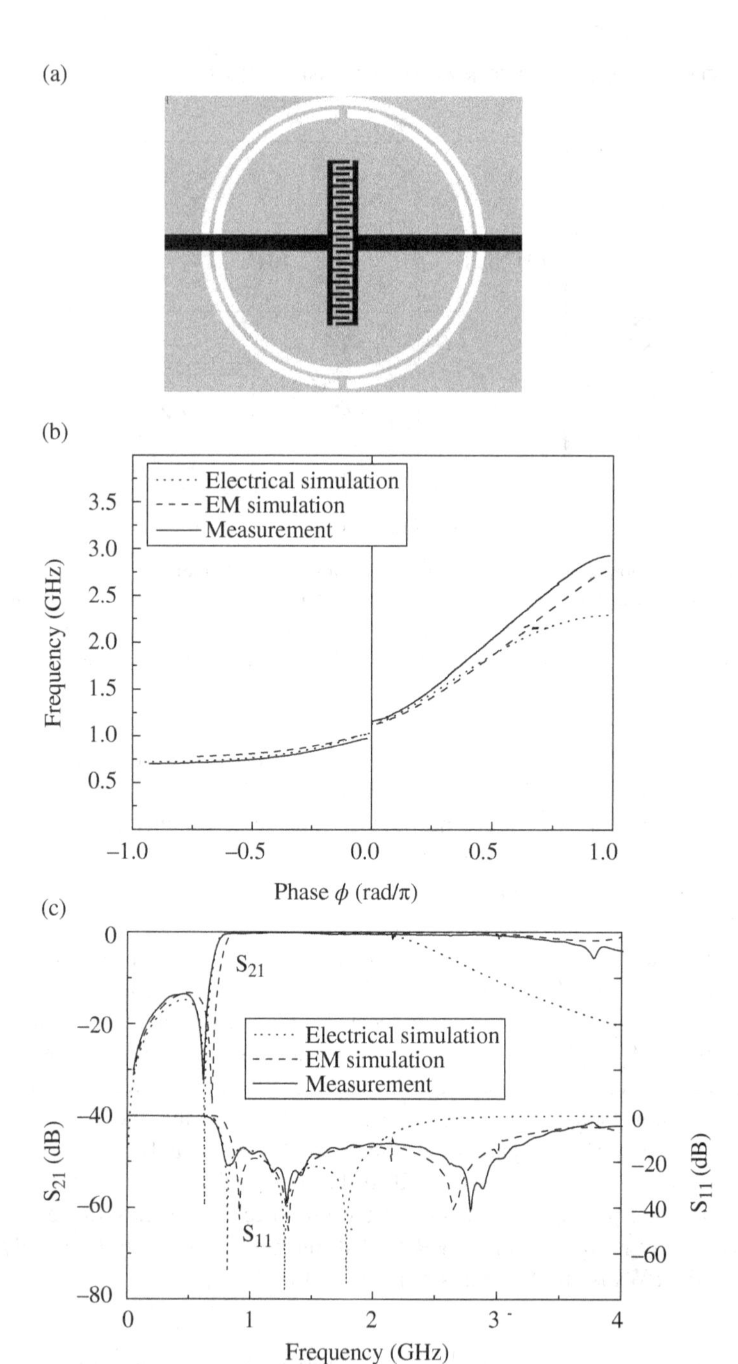

FIGURE 3.42 Balanced CRLH cell based on a microstrip line loaded with a CSRR and an interdigital capacitor (a), dispersion diagram (b), and frequency response (c). The structure was implemented in the *Rogers RO3010* substrate with dielectric constant $\varepsilon_r = 10.2$ and thickness $h = 1.27$ mm. Dimensions are: line width $W = 0.8$ mm, external radius of the outer ring $r = 7.3$ mm, rings width $c = 0.4$ mm, and ring separation $d = 0.2$ mm; the interdigital capacitor, formed by 28 fingers separated 0.16 mm, was used to achieve the required capacitance value. Reprinted with permission from Ref. [108]; copyright 2007 IEEE.

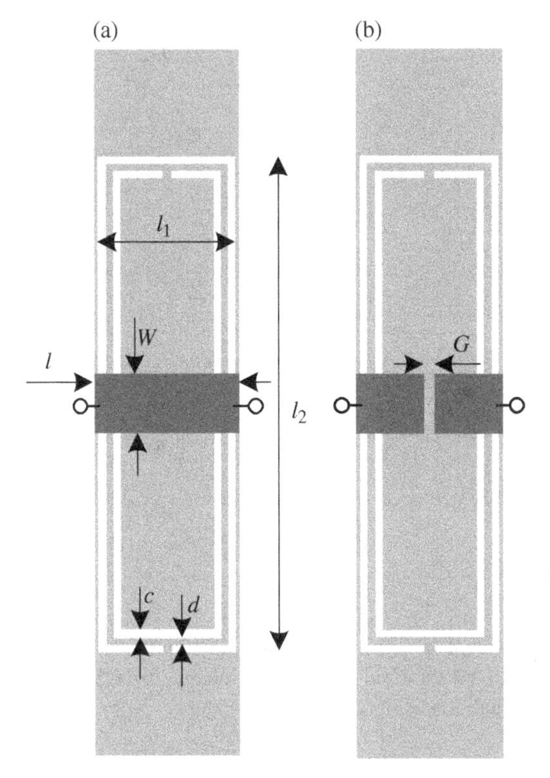

FIGURE 3.43 Unit cells of a CSRR-loaded (a) and CSRR/gap-loaded (b) microstrip line with rectangular-shaped resonators. The ground plane is depicted in light gray. Dimensions are: $W = 1.17$ mm, $l = 3$ mm, $c = d = 0.15$ mm, $l_1 = 2.8$ mm, $l_2 = 9.8$ mm, and $G = 0.2$ mm. The considered substrate is the Rogers *RO3010* with thickness $h = 1.27$ mm, dielectric constant $\varepsilon_r = 10.2$, and loss tangent $\tan\delta = 0.0023$.

Let us consider CSRR-loaded microstrip lines with rectangular CSRRs. To favor inter-resonator coupling the longer side of the CSRR is orthogonal to the line axis, as depicted in Figure 3.43. The lumped-element circuit models of these structures, including electric coupling between adjacent resonators, are depicted in Figure 3.44 (the first neighbor approximation for inter-resonator coupling is considered and losses are neglected). Notice that the circuit models of these CSRR- and CSRR/gap-loaded lines are derived from those depicted in Figure 3.38 by simply adding the coupling capacitances C_R. The resulting models are thus four-port networks that can be cascaded to describe multi-section CSRR- or CSRR/gap-loaded lines.

The dispersion characteristics of these CSRR-loaded lines can be inferred from Bloch mode theory applied to the circuits of Figure 3.44. These circuits are multiport networks, and thus we can appeal to multiconductor line theory [118] in order to

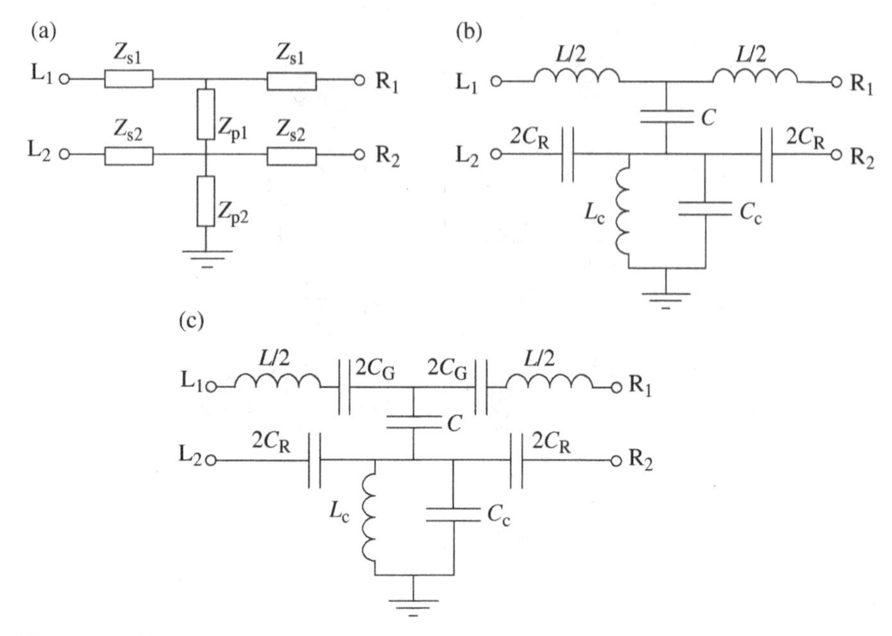

FIGURE 3.44 Equivalent circuit model of the structures of Figure 3.43, including inter-resonator coupling. (a) Generalized impedance model, (b) CSRR-loaded microstrip line model, and (c) CSRR/gap-loaded microstrip line model.

obtain the eigenmodes of the structure, and hence the propagation constants. Let us denote V_{L1}, V_{L2}, I_{L1}, and I_{L2} as the voltages and currents at the ports (1 and 2) of the left-hand side (subscript L) of the unit cell, and V_{R1}, V_{R2}, I_{R1}, and I_{R2} the variables at the right-hand side ports. The variables at both sides of the network are linked through a generalized order-4 transfer matrix, according to

$$\begin{pmatrix} \boldsymbol{V}_{\mathrm{L}} \\ \boldsymbol{I}_{\mathrm{L}} \end{pmatrix} = \begin{pmatrix} \boldsymbol{A} & \boldsymbol{B} \\ \boldsymbol{C} & \boldsymbol{D} \end{pmatrix} \begin{pmatrix} \boldsymbol{V}_{\mathrm{R}} \\ \boldsymbol{I}_{\mathrm{R}} \end{pmatrix} \tag{3.93}$$

where $\boldsymbol{V}_{\mathrm{L}}$, $\boldsymbol{I}_{\mathrm{L}}$, $\boldsymbol{V}_{\mathrm{R}}$, and $\boldsymbol{I}_{\mathrm{R}}$ are column vectors composed of the pair of port variables, and $\boldsymbol{A}$, $\boldsymbol{B}$, $\boldsymbol{C}$, and $\boldsymbol{D}$ are order-2 matrices. The dispersion relation is obtained from the eigenmodes of the system (3.93), that is

$$\det\begin{pmatrix} \boldsymbol{A}-e^{\gamma l}\cdot\boldsymbol{I} & \boldsymbol{B} \\ \boldsymbol{C} & \boldsymbol{D}-e^{\gamma l}\cdot\boldsymbol{I} \end{pmatrix} = 0 \tag{3.94}$$

where $\boldsymbol{I}$ is the identity matrix, the phase-shift factor $e^{\gamma l}$ is the eigenvalue, $\gamma = \alpha + j\beta$ is the complex propagation constant, and l is the unit cell length. For reciprocal, lossless and symmetric networks, the eigenvalues can be simplified to the solutions of [119, 120]:

$$\det(\boldsymbol{A} - \cosh(\gamma l)\cdot\boldsymbol{I}) = 0 \tag{3.95}$$

which gives

$$\cosh(\gamma l) = \frac{1}{2}\left(A_{11} + A_{22} \pm \sqrt{(A_{11} - A_{22})^2 + 4A_{12}A_{21}}\right) \tag{3.96}$$

where the elements of the $\boldsymbol{A}$ matrix (inferred from the network of Fig. 3.44a) are

$$\mathbf{A} = \mathbf{D}^t = \begin{pmatrix} 1 + \dfrac{Z_{s1}}{Z_{p1}} & -\dfrac{Z_{s1}}{Z_{p1}} \\ -\dfrac{Z_{s2}}{Z_{p1}} & 1 + \dfrac{Z_{s2}}{Z_{p1}} + \dfrac{Z_{s2}}{Z_{p2}} \end{pmatrix} \tag{3.97}$$

Since the network is lossless, the elements of $\boldsymbol{A}$ (A_{ij}) are real numbers. Hence, if the radicand of the square root in (3.96) is positive, the propagation constant is either purely real ($\alpha \neq 0, \beta = 0$) or purely imaginary ($\alpha = 0, \beta \neq 0$), corresponding to evanescent or propagating modes, respectively. However, if the radicand in (3.96) is negative, the two solutions are of the form $\gamma = \alpha \pm j\beta$, corresponding to complex modes. In order to obtain the frequency band that supports complex modes, the radicand in (3.96) is forced to be negative, that is:

$$\left(\frac{Z_{s1} - Z_{s2}}{Z_{p1}} - \frac{Z_{s2}}{Z_{p2}}\right)^2 + 4\frac{Z_{s1}Z_{s2}}{Z_{p1}^2} < 0 \tag{3.98}$$

According to (3.98) a necessary (although not sufficient) condition to have complex modes is an opposite sign for the reactances of Z_{s1} and Z_{s2}.

Let us first evaluate the dispersion relation of the CSRR-loaded line of Figure 3.43a, for which the pair of modal propagation constants given by expression (3.96) is depicted in Figure 3.45. In the first allowed band, there is a region with bivalued propagation constant; one (forward) corresponding to transmission-line type propagation, and the other (backward) related to electroinductive waves[38] [121]. Then, a region with conjugate modes (complex modes) appears, followed by a region with $\beta = 0$ and $\alpha \neq 0$ for both modes (evanescent waves). Finally, a forward wave

[38] Electroinductive waves (EIWs) may appear in chains of coupled CSRRs and are backward waves. They are similar to the magnetoinductive waves (MIWs) that may arise in chains of magnetically coupled coplanar SRRs [21]. EIW and MIW transmission lines will be studied in detail in Chapter 6.

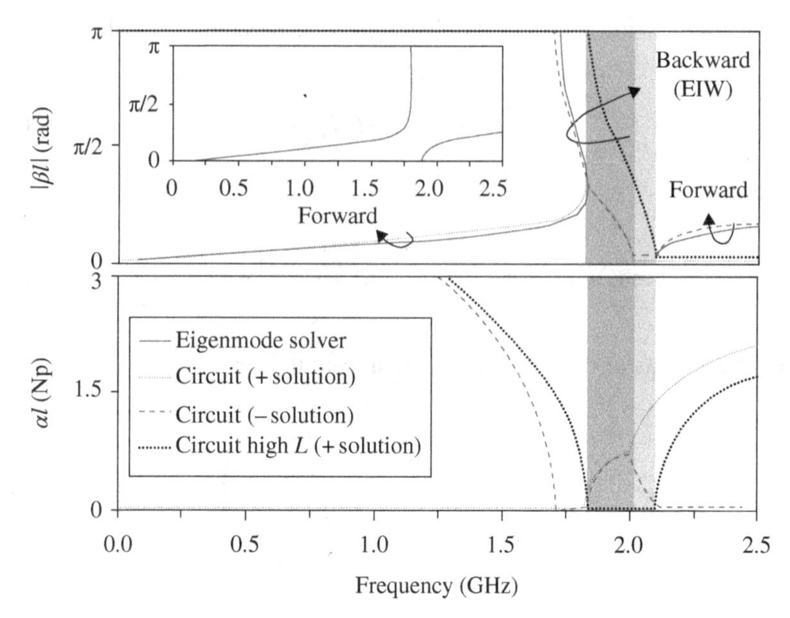

FIGURE 3.45 Dispersion relation for the structure of Figure 3.43a and for its equivalent circuit model of Figure 3.44b. The circuit parameters are: $L = 3.05$ nH, $C = 0.55$ pF, $L_c = 2.03$ nH, $C_c = 2.84$ pF, and $C_R = 0.21$ pF. The dispersion relation for the structure of Figure 3.43a with $l = 4.8$ mm is depicted in the inset. Notice that the eigenmode solver is not able to provide the complex modes (dark gray region) since these modes do not carry net power. The evanescent modes region is indicated in light gray. Reprinted with permission from Ref. [116]; copyright 2012 IEEE.

transmission band emerges again. It is apparent that the stop band characteristics in the complex wave region can be interpreted as a result of the antiparallelism between the transmission-line type (forward) and the electroinductive (backward) modes. To demonstrate the electroinductive nature of one of the modes in the upper region of the first allowed band, a large value of series inductance ($L \rightarrow \infty$, corresponding to an extremely narrow line) was considered, and the corresponding dispersion relation was inferred and depicted in Figure 3.45. According to this result, the nature of the aforementioned electroinductive waves is clear (the frequency shift is simply due to the fact that for the high-L case, the electroinductive waves propagate entirely through the inter-resonator capacitances).

The dispersion relation of a periodic structure composed of the cascaded unit cells of Figure 3.43a obtained by means of the eigenmode solver of *CST Microwave Studio* is also plotted in Figure 3.45. The agreement with the dispersion relation inferred from the circuit model is very good. The eigenmode solver was also used to obtain the dispersion relation that results after increasing the unit cell size to $l = 4.8$ mm (which corresponds to an increase in inter-resonator distance from 0.2 to 2 mm). For this inter-resonator distance (see the inset of Fig. 3.45), the electroinductive and complex modes are not present, and the fractional bandwidth of the stop band

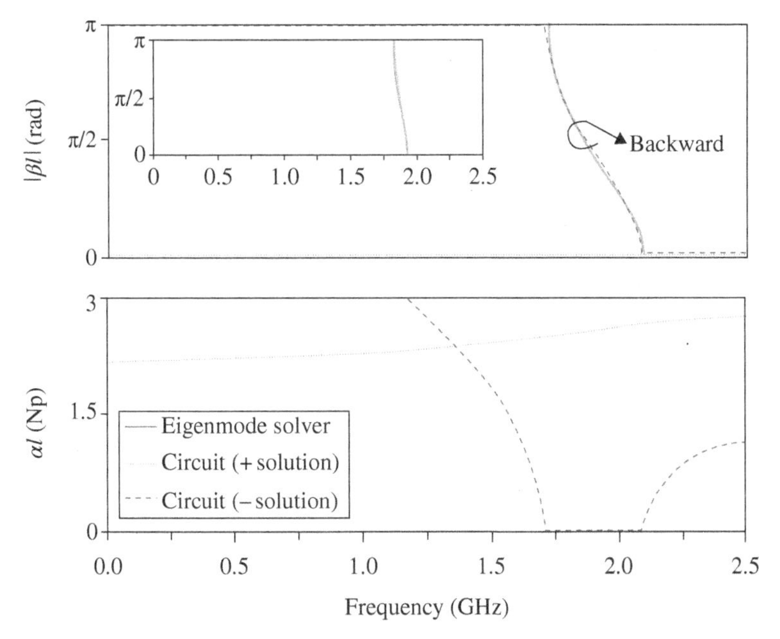

FIGURE 3.46 Dispersion relation for the structure of Figure 3.43b and for its equivalent circuit model of Figure 3.44c. The circuit parameters are: $L = 3.05$ nH, $C = 1.4$ pF, $L_c = 2.02$ nH, $C_c = 2.83$ pF, $C_G = 0.2$ pF, and $C_R = 0.23$ pF. The dispersion relation for the structure of Figure 3.43b with $l = 4.8$ mm is depicted in the inset. Reprinted with permission from Ref. [116]; copyright 2012 IEEE.

is reduced from 13.3 to 5.5%. These results indicate that, by reducing the distance between adjacent resonators, the forbidden band is enhanced due to the presence of complex modes, which do not carry net power and expand the stop band beyond the evanescent modes region.

The dispersion diagram for the structure of Figure 3.43b is depicted in Figure 3.46. As expected, a backward wave transmission band is obtained (the structure is unbalanced and the forward wave transmission band is above the depicted frequency range). The bandwidth in the first LH band broadens as inter-resonator coupling increases since electroinductive waves aid backward wave propagation.

The effects of inter-resonator coupling on bandwidth broadening are illustrated by comparing the frequency response of two order-9 CSRR-loaded stop band structures. One of them was fabricated by cascading the unit cells of Figure 3.43a, whereas the other one was fabricated with a larger unit cell size ($l = 4.8$ mm) in order to minimize inter-resonator coupling. The fabricated structures as well as their frequency responses are depicted in Figure 3.47. The agreement with the responses that result from the circuit models with the indicated parameters is good, and it is clear that bandwidth is enhanced by reducing the distance between adjacent resonators. Obviously, the stop band for the structure of Figure 3.47a coincides with the stop band predicted by the eigenmode analysis to a good approximation.

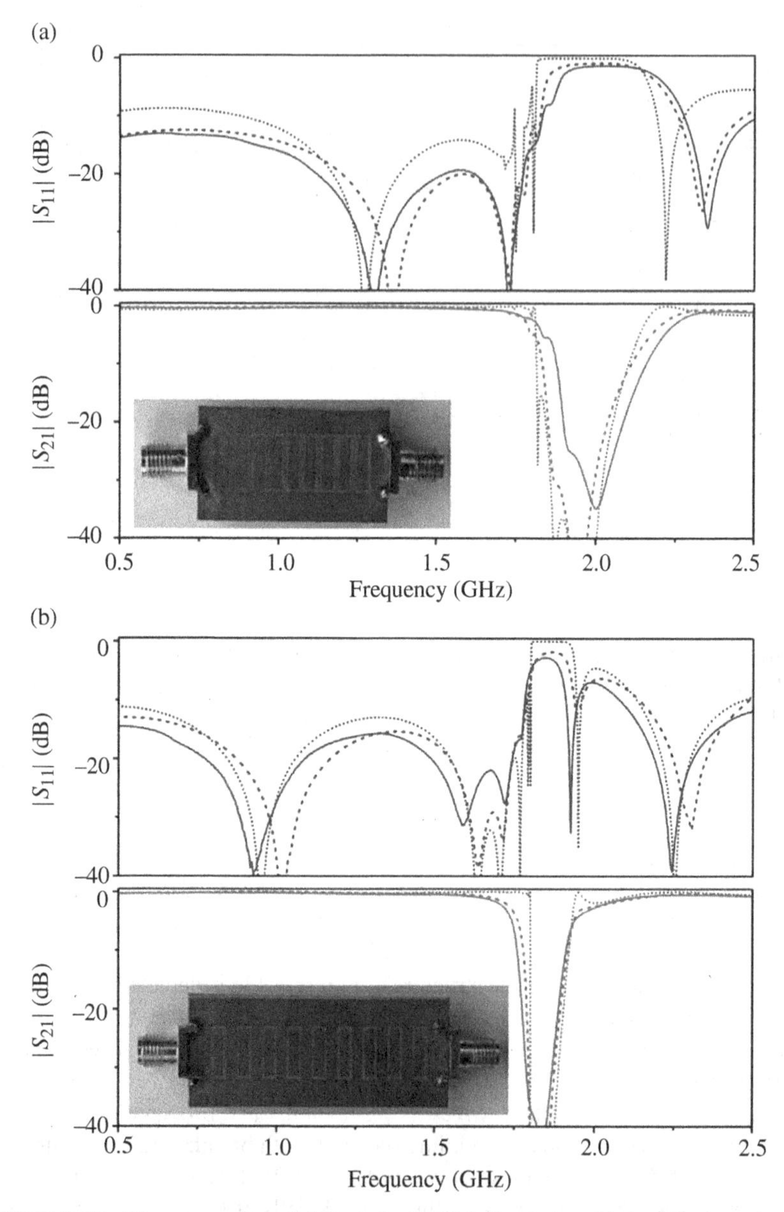

FIGURE 3.47 Measurement (solid line), lossy EM simulation (dashed line), and circuit simulation (dotted line) of the transmission and reflection coefficients of the CSRR-loaded structures shown in the insets (only the bottom face is shown). (a) $l = 3$ mm, the circuit parameters being those indicated in Figure 3.45; (b) $l = 4.8$ mm, the circuit parameters being those indicated in Figure 3.45, except $C_R = 0.017$ pF and $C_c = 3.2$ pF. To better fit the measurement in (b), a transmission line section of length 1.8 mm was cascaded between consecutive Z_{s1} in the circuit model. Reprinted with permission from Ref. [116]; copyright 2012 IEEE.

3.5.2.4 Effects of SRR and CSRR Orientation: Mixed Coupling The lumped-element equivalent circuit models of the SRR- and CSRR-based lines reported in the previous subsections are symmetric, that is, they are identical seen from both ports. Therefore, they must describe symmetric structures. Actually, the models of Figures 3.34, 3.35, 3.38, 3.40, and 3.44 are not valid for arbitrary orientation of the SRRs or CSRRs.[39] These models accurately describe the unit cells for SRRs or CSRRs oriented with the slits etched in a plane orthogonal to the line axis, corresponding to symmetric unit cells ($\phi = 0°$ orientation). Under this SRR/CSRR orientation, the coupling between the line and the resonator is purely magnetic/electric. Though SRRs/CSRRs exhibit cross-coupling effects, they cannot be electrically/magnetically excited for this orientation since there is not a net component of the electric/magnetic field generated by the line in the direction orthogonal to the slits plane. However, if the SRRs/CSRRs are rotated ($\phi \neq 0°$), the structures are no longer symmetric, and mixed coupling (electric and magnetic) between the line and the resonators arises. Such mixed coupling must be included in the circuit models for an accurate description of the SRR- or CSRR-based lines under arbitrary particle orientation [122].

Figure 3.48 depicts the frequency response of an SRR-loaded CPW with the SRRs oriented with an angle of $\phi = 90°$ (i.e., with the symmetry planes of the SRRs parallel to the symmetry plane of the line). Notice that the asymmetry leads to substantially different phase of the reflection coefficients S_{11} and S_{22} (even though $|S_{11}| = |S_{22}|$ and the transmission coefficients satisfy $S_{12} = S_{21}$, as results from unitarity and reciprocity). The same phenomenology applies to a CSRR-loaded microstrip line (see Fig. 3.49). To gain insight on the effects of SRR or CSRR rotation on the frequency response of the structures, let us consider the non-bianisotropic SRR (NB-SRR) and the non-bianisotropic CSRR (NB-CSRR). These particles do not exhibit cross polarization since they have inversion symmetry with regard to their center, and this prevents the appearance of the electric and magnetic dipole for the SRR and CSRR, respectively, in the plane of the particles. According to this, it is expected that CPW and microstrip lines loaded with NB-SRRs and NB-CSRRs, respectively, are insensitive to particle rotation. This was verified from EM simulation by considering two different rotation angles (Figs. 3.50 and 3.51). As can be seen, the frequency response does not change appreciably by rotating the NB-SRRs and NB-CSRRs. Moreover, the responses of the SRR- and CSRR-loaded lines for $\phi = 0°$ present the same pattern (unlike for $\phi = 90°$) as those of the lines loaded with NB-SRRs and NB-CSRRs.

The previous results support the hypothesis that the dependence of the EM behavior of the SRR- and CSRR-loaded lines on particle orientation is related to cross-polarization effects. Therefore, in order to accurately describe the lines with arbitrarily oriented SRRs or CSRRs, it is necessary to include both electric and magnetic coupling in the models. Thus, the circuit model for SRR-loaded CPWs is shown in

[39] Notice that in the CSRR/gap-loaded microstrip line of Figure 3.37, the slits are aligned with the line axis. Strictly speaking, the model of Figure 3.38b is not valid for this CSRR orientation since mixed coupling is not considered in that model. However, the electric coupling is dominant, and the structure can be modeled by the circuit of Figure 3.38b to a first-order approximation.

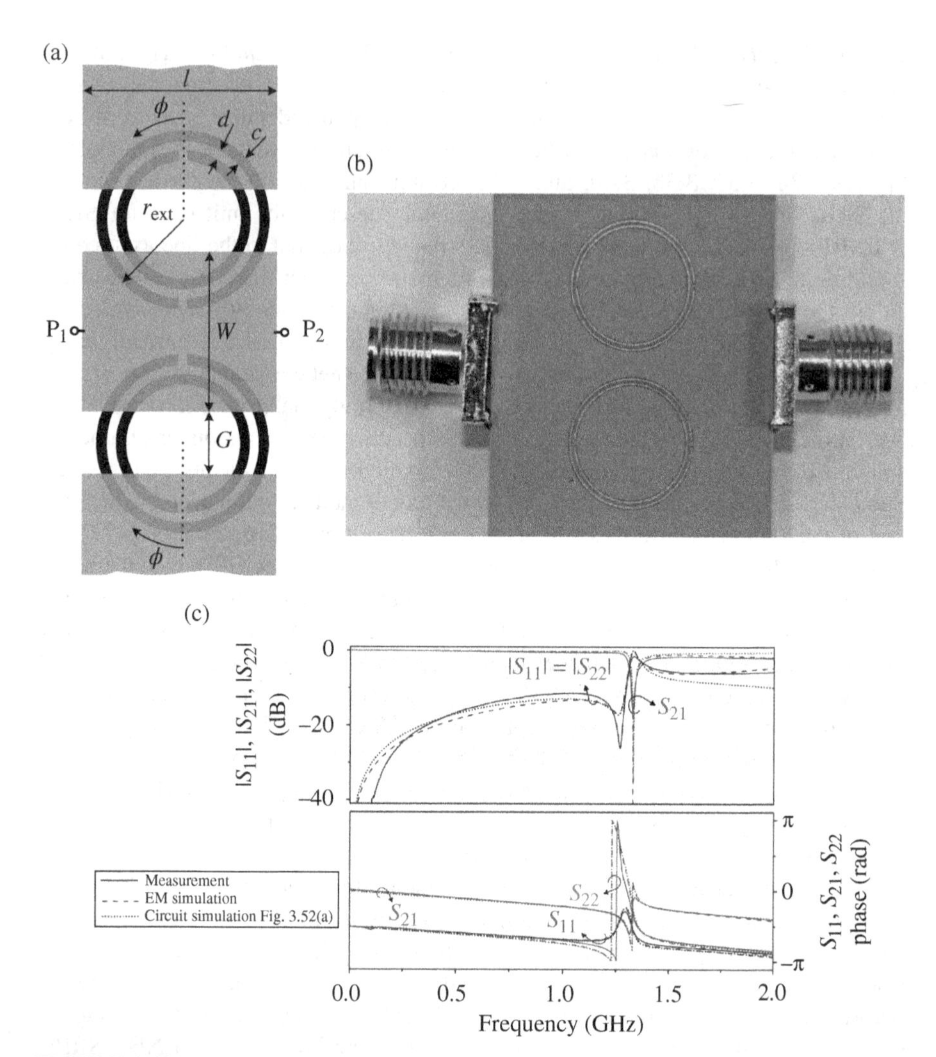

FIGURE 3.48 SRR-loaded CPW with arbitrary SRR orientation (a), fabricated SRR-loaded CPW structure with SRRs rotated $\phi = 90°$ (b), and frequency response (c). The substrate is the *Rogers RO3010* with dielectric constant $\varepsilon_r = 11.2$ and thickness $h = 1.27$ mm. Dimensions are: $W = 10.4$ mm, $G = 1.6$ mm, $l = 10.4$ mm, $r_{ext} = 5$ mm, and $c = d = 0.2$ mm. Reprinted with permission from Ref. [122]; copyright 2013 IEEE.

Figure 3.52a [122]. The circuit parameters are those in reference to Figure 3.34b, plus the coupling capacitance, C_a, that takes into account the electric coupling between the line and the SRRs (it is assumed that C_a depends on ϕ, being $C_a = 0$ if $\phi = 0°$). Notice that the asymmetry in the circuit model (necessary to explain that $S_{11} \neq S_{22}$) comes from the magnetic coupling (the dot convention for the mutual inductance is used, where the currents entering dot-marked terminals produce additive magnetic fluxes), although this is not manifested if electric coupling is not present.

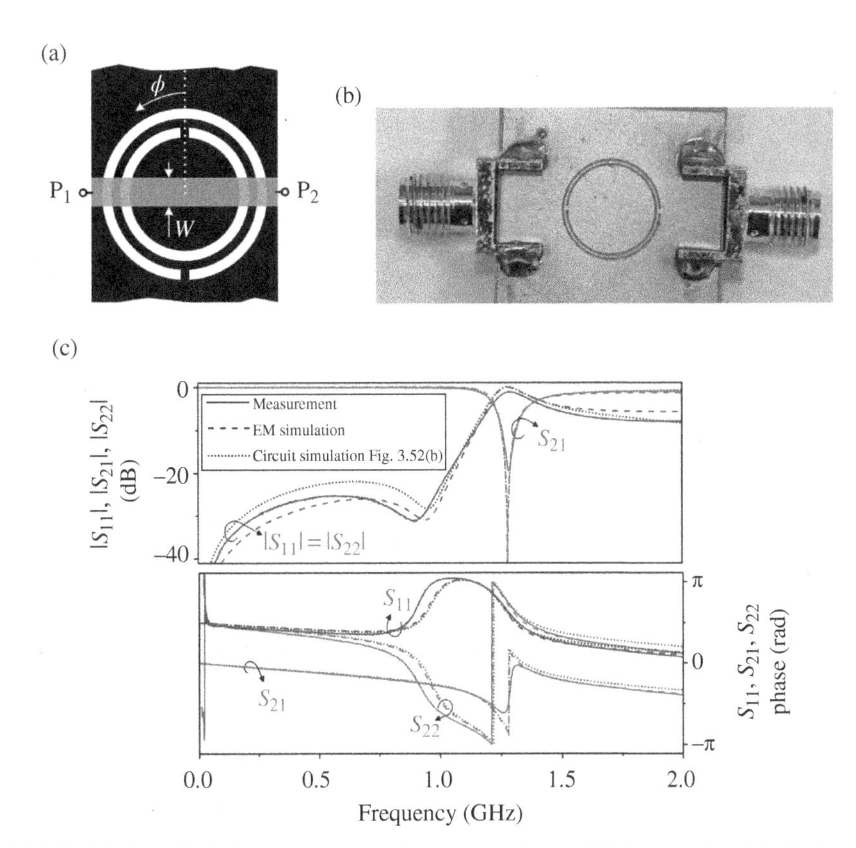

FIGURE 3.49 CSRR-loaded microstrip line with arbitrary CSRR orientation (a), fabricated CSRR-loaded microstrip line structure with the CSRR rotated $\phi = 90°$ (b), and frequency response (c). The substrate is the *Rogers RO3010* with dielectric constant $\varepsilon_r = 11.2$ and thickness $h = 1.27$ mm. Dimensions are: $W = 1.15$ mm, $l = 10.4$ mm, $r_{ext} = 5$ mm, and $c = d = 0.2$ mm. Reprinted with permission from Ref. [122]; copyright 2013 IEEE.

The parameters of the circuit model corresponding to the SRR-loaded line of Figure 3.48 were extracted. To this end, it was first considered the structure with SRRs oriented with $\phi = 0°$. By these means, the circuit parameters were inferred according to the procedure explained in Appendix G ($L' = 2.5$ nH, $C = 6$ pF, $L'_s = 0.4$ nH, $C'_s = 38.8$ pF, and $L = 1.45$ nH). Then, $L_s - C_s$, M, and C_a were adjusted by curve fitting the circuit simulation to the EM response, such that the resulting $L_s - C_s$ values were similar to the analytical SRR parameters given by Refs. [4, 69] ($L_s = 27.8$ nH and $C_s = 0.6$ pF). The resulting circuit parameters were found to be $L = 1.5$ nH, $C = 6.1$ pF, $L_s = 28.2$ nH, $C_s = 0.43$ pF, $M = 1.35$ nH, and $C_a = 0.08$ pF. The agreement between the circuit, EM simulation and measured response is good (see Fig. 3.48).

Figure 3.52b depicts the equivalent circuit model of CSRR-loaded microstrip lines that includes electric and magnetic coupling [122]. The circuit parameters are the same as in Figure 3.38a, plus the mutual inductance, M, that depends on ϕ ($M = 0$ if $\phi = 0°$),

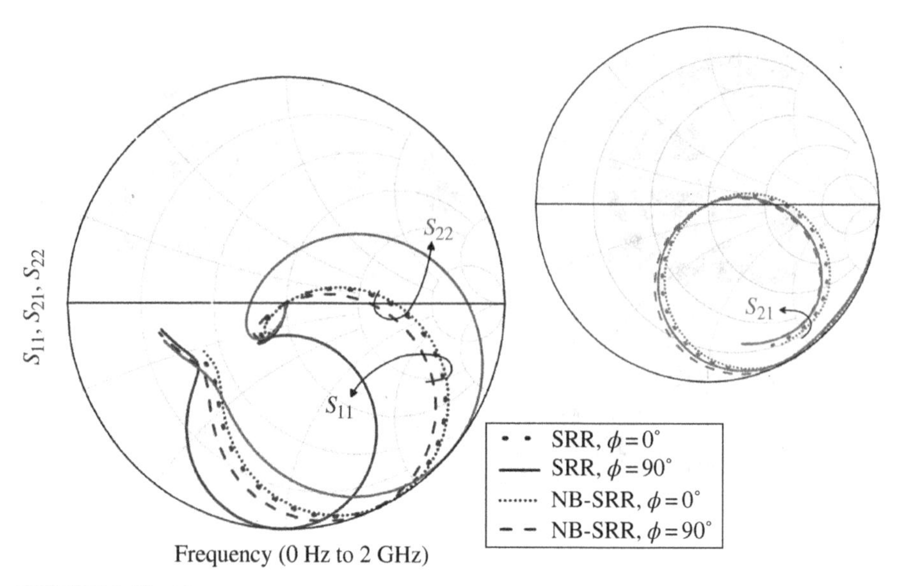

FIGURE 3.50 Frequency response in Smith chart of a CPW loaded with a pair of SRRs or NB-SRRs for two different rotation angles. The considered substrate and dimensions are those of Figure 3.48. Reprinted with permission from Ref. [122]; copyright 2013 IEEE.

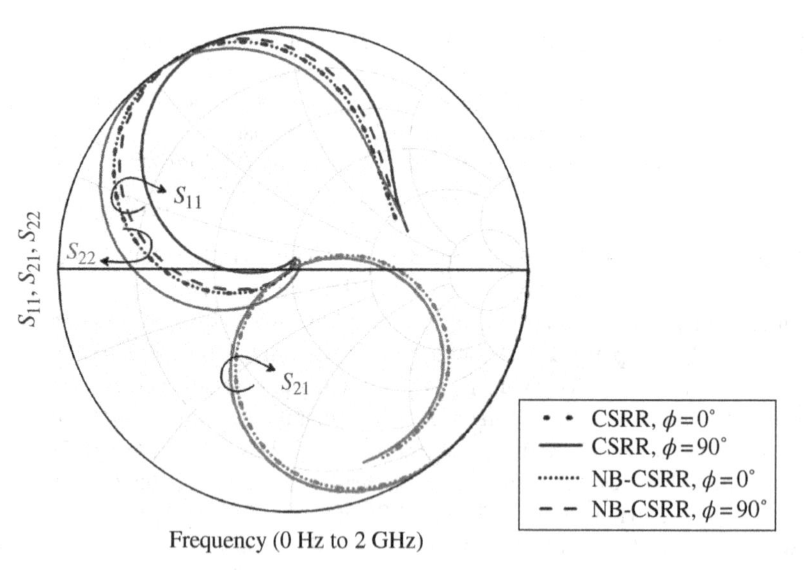

FIGURE 3.51 Frequency response in Smith chart of a microstrip line loaded with a CSRR or an NB-CSRR for two different rotation angles. The considered substrate and dimensions are those of Figure 3.49. Reprinted with permission from Ref. [122]; copyright 2013 IEEE.

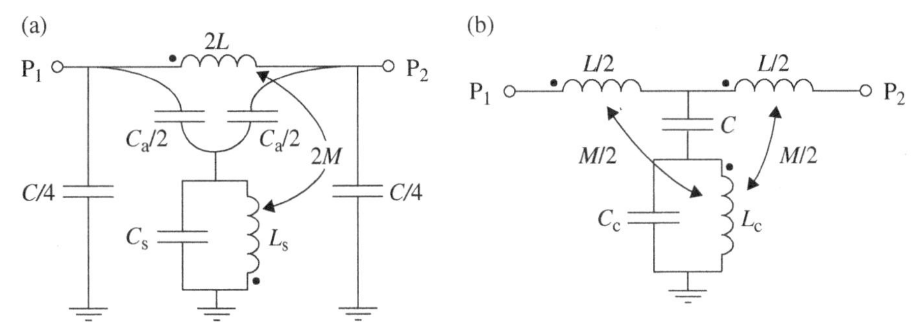

FIGURE 3.52 Equivalent circuit model (unit cell) of (a) a CPW loaded with arbitrarily oriented SRRs and (b) a microstrip line loaded with arbitrarily oriented CSRRs. The magnetic wall concept is applied in (a).

and accounts for the magnetic coupling between the line and the CSRR. The parameters of the circuit model corresponding to the CSRR-loaded line of Figure 3.49 were extracted. Similar to the SRR-loaded CPW, the parameters for $\phi = 0°$ (all except M) were first extracted according to the method reported in Appendix G ($L = 7.87$ nH, $C = 1.62$ pF, $L_c = 2.94$ nH, and $C_c = 3.9$ pF). Finally, M was obtained by curve fitting, where a slight optimization of the other parameters was required. The resulting circuit elements were found to be $L = 7$ nH, $C = 1.7$ pF, $L_c = 2.74$ nH, $C_c = 3.95$ pF, and $M = 0.27$ nH. As Figure 3.49 reveals, the circuit simulation accurately describes the EM simulation and measured response.

Notice that the structures of Figures 3.48 and 3.49 are single negative structures; that is, they do not include shunt strips (for the SRR-loaded CPWs) or series gaps (for the CSRR-loaded microstrip line). The models for SRR/strip-loaded CPWs and CSRR/gap-loaded microstrip lines with arbitrarily oriented SRRs or CSRRs are out of the scope of this book since these models are too complicated for design purposes. Thus, in general, the SRR- or CSRR-based lines of interest for circuit design are those with the symmetry plane of the resonator orthogonal to the line axis, where mixed coupling can be neglected. Exceptions to this are microwave sensors based on the symmetry properties of SRR- or CSRR-loaded lines, or differential lines with common mode suppression, as will be shown later. Nevertheless, structures with the symmetry plane of the SRRs, or CSRRs, parallel to the line axis can be described to a first-order approximation by neglecting mixed coupling, since this simplifies the design process [123].

3.5.2.5 Transmission Lines based on OSRRs and OCSRRs According to the circuit models of the OSRR and OCSRR, reported in Section 3.3, it follows that by cascading series connected OSRRs and shunt connected OCSRRs alternating, CRLH lines described by the canonical circuit model of Figure 3.20 can potentially be implemented. The combination of OSRRs and OCSRRs has been demonstrated to be a powerful approach for the implementation of resonant type CRLH transmission lines [124, 125]. As will be seen, these artificial lines can be designed to exhibit very broad bands,

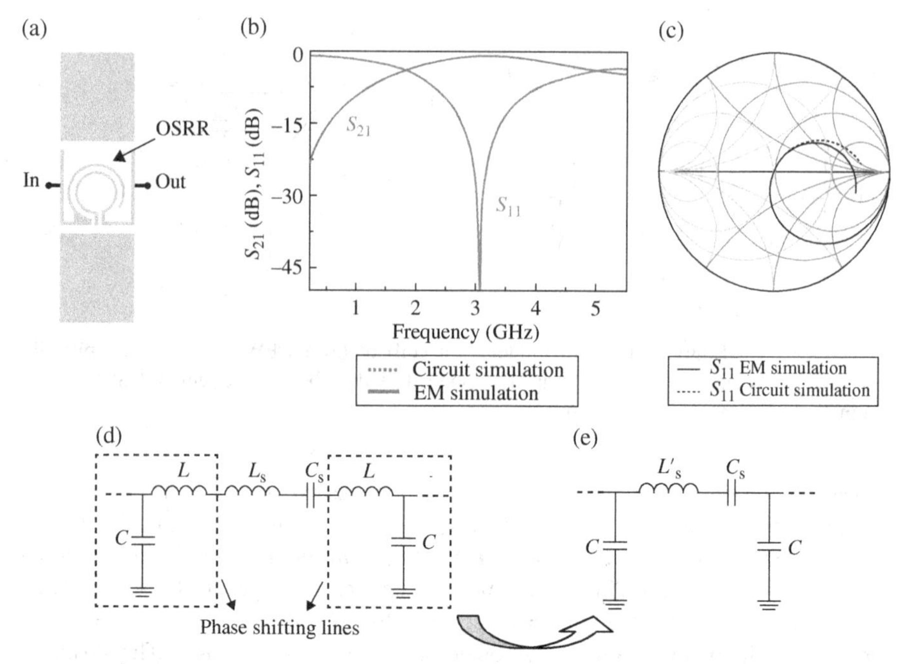

FIGURE 3.53 Topology (a), frequency response (magnitude) (b), representation of the reflection coefficient in the Smith Chart (c), and circuit model (d) and (e) of a typical OSRR-loaded CPW. This structure was implemented on the *Rogers RO3010* substrate with thickness $h = 0.254$ mm and measured dielectric constant $\varepsilon_r = 11.2$. Dimensions are $W = 5$ mm, $G = 0.55$ mm, $r_{ext} = 1.6$ mm, and $c = d = 0.2$ mm. The element values of the equivalent circuit inferred by parameter extraction are $C = 0.189$ pF, $L'_s = 5.55$ nH and $C_s = 0.58$ pF. This figure has been reproduced from Ref. [125] with permission from the author.

and the lumped-element equivalent circuit model is able to accurately describe the structures up to frequencies beyond the second (forward) transmission band.[40] The implementation of OSRR/OCSRR CRLH lines requires a host line, typically a CPW or a microstrip transmission line. The presence of the host line introduces parasitics in the circuit models of the particles. Therefore, the circuit model of CRLH lines based on the combination of OSRRs and OCSRR is not exactly the one of Figure 3.20. Let us now present these models for OSRRs and OCSRRs in CPW technology.

Figure 3.53 shows a typical topology of a series-connected OSRR in CPW technology, as well as the frequency response and the reflection coefficient, S_{11}, in the Smith chart. The fact that S_{11} does not lie in the unit resistance circle indicates that the model of the structure cannot merely be a series resonant tank. The presence of the host line introduces some phase shift that must be taken into account for an

[40] By contrast, the models of the SRR- or CSRR-based CRLH lines provide an accurate description of the lines in the first (LH) transmission band, but they tend to fail in the second (RH) transmission band, as Figure 3.42 illustrates.

accurate description of the OSRR-loaded CPW. Thus, in order to properly model the structure, additional elements to account for the phase shift must be introduced at both sides of the resonant tank describing the OSRR, that is, phase shifting lines. However, such phase shifting lines can be modeled through series inductances and shunt capacitances, as depicted in Figure 3.53. Notice that the model can be simplified as depicted in the figure, where C is a parasitic capacitance of the model. The circuit simulation of the model, also depicted in Figure 3.53, reveals that the agreement with the EM simulation is very good.

Figure 3.54 shows the topology of an arbitrary shunt connected-OCSRR pair in CPW technology. In order to prevent the slot mode, the different ground plane regions are connected through strips (etched in the ground plane) and vias. The strip that connects the ground planes at both sides of the central strip is optional and can be useful to enhance the shunt capacitance to ground, providing more degrees of freedom. The frequency response and S_{11} represented in the Smith chart are also included

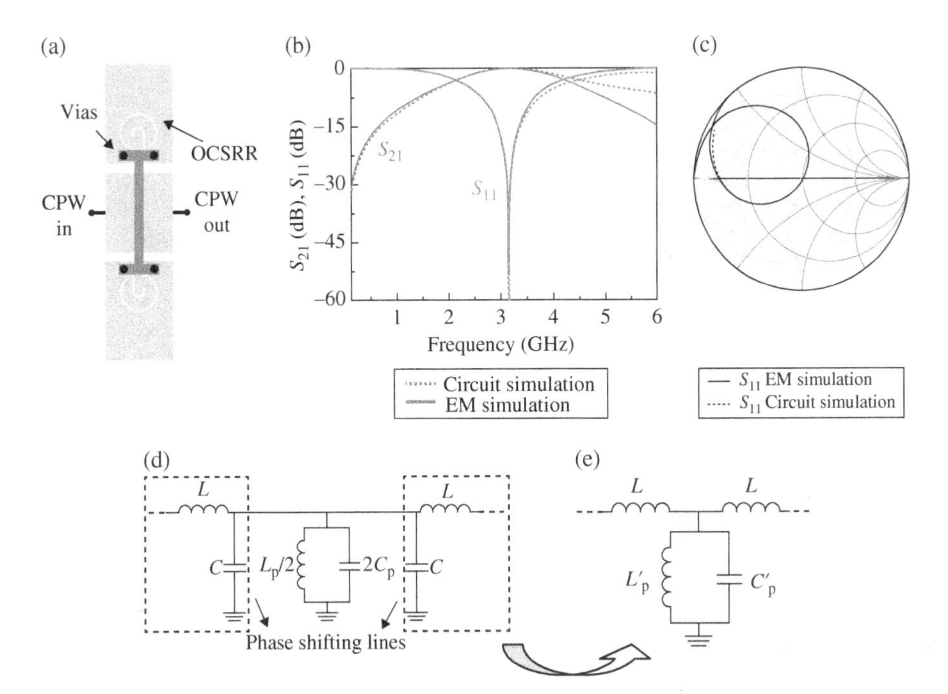

FIGURE 3.54 Topology (a), frequency response (magnitude) (b), representation of the reflection coefficient in the Smith Chart (c), and circuit model (d) and (e) corresponding to typical CPW transmission line loaded with a pair of shunt connected OCSRRs. Dimensions are $r_{ext} = 1.2$ mm, $c = 0.2$ mm, $d = 0.6$ mm, $W = 5$ mm, and $G = 0.55$ mm. The width of the strip connecting the different ground planes is 0.55 mm. The structure was implemented on the *Rogers RO3010* substrate with thickness $h = 0.254$ mm and measured dielectric constant $\varepsilon_r = 11.2$. The element values inferred by parameter extraction are $L = 0.32$ pF, $L'_p = 0.983$ nH and $C'_p = 2.85$ pF. This figure has been reproduced from Ref. [125] with permission from the author.

in Figure 3.54. The phenomenology is similar to that of the OSRR-loaded CPW. In this case, the trace of S_{11} deviates from the unit conductance circle, indicating that some phase shift is introduced by the host line, and preventing that the whole structure can be described by means of a shunt connected parallel resonant tank. The circuit model, and the simplified circuit model, are both depicted in Figure 3.54, where the parasitic element of the model is in this case the inductance L. The comparison between the circuit and EM simulation (Fig. 3.54) is also good, although some discrepancy in the insertion loss above 4 GHz can be appreciated. Interestingly, the broadband EM response of the structure of the OCSRR-loaded CPW of Figure 3.54 exhibits a transmission zero in the vicinity of 7.5 GHz. This transmission zero is caused by the inductance associated to the strip present between the central strip of the CPW and the inner metallic region of the OCSRRs. By taking into account this inductance (L_{sh}) in the model (see Fig. 3.55) [126], the agreement between the

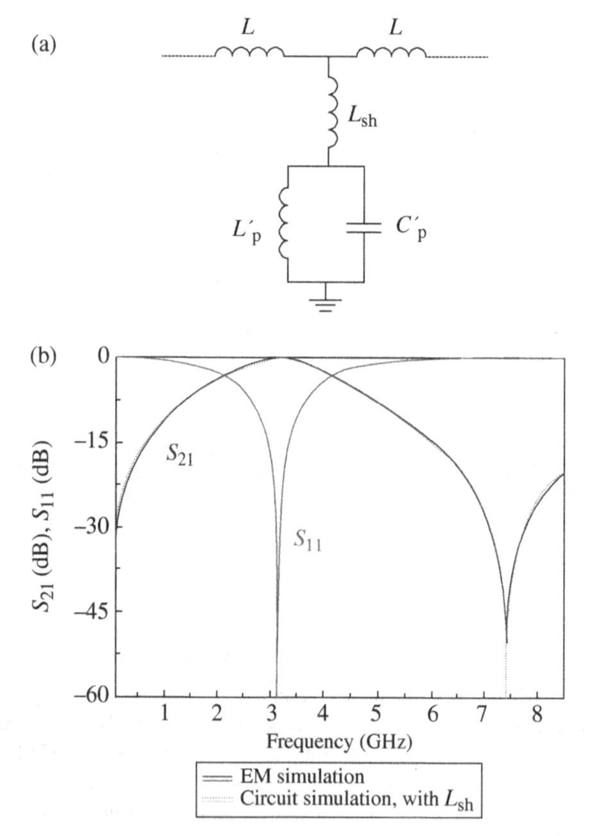

FIGURE 3.55 Wideband circuit model (a) and frequency response (magnitude) (b) of the pair of shunt connected OCSRRs shown in Figure 3.54a. The element values inferred by parameter extraction are $L = 0.345$ pF, $L'_p = 0.94$ nH, $C'_p = 2.98$ pF, and $L_{sh} = 0.185$ nH. Reprinted with permission from Ref. [126]; copyright 2011 IEEE.

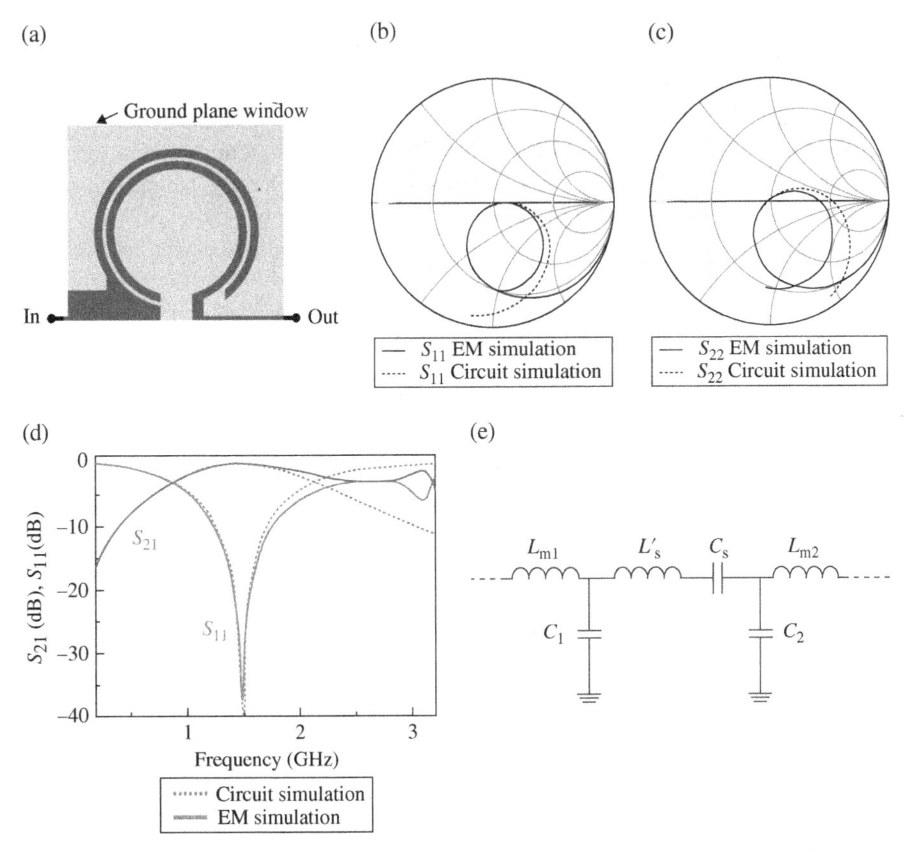

FIGURE 3.56 Topology (a), representation of S_{11} (b) and S_{22} (c) in the Smith chart, frequency response (magnitude) (d) and circuit model (e) of a typical series connected OSRR in microstrip technology with $r_{ext} = 4$ mm, $c = 0.4$ mm, and $d = 0.2$ mm. The considered substrate is the *Rogers RO3010* with thickness $h = 0.254$ mm and dielectric constant $\varepsilon_r = 10.2$. The element values are $L'_s = 12.93$ nH, $C_s = 1.33$ pF, $C_1 = 1.38$ pF, $C_2 = 0.98$ pF, $L_{m1} = 0.8$ nH, and $L_{m2} = 0$ nH. This figure has been reproduced from Ref. [125] with permission from the author.

frequency responses (circuit and EM simulation) is excellent, as Figure 3.55 reveals.[41] The model of Figure 3.55 is useful to accurately describe the frequency response of broadband filters (as will be seen in Chapter 4). However, in many applications the model of Figure 3.54 suffices.

In microstrip technology, series-connected OSRRs and shunt-connected OCSRRs exhibit different behavior. It has been found that OSRR-loaded microstrip lines (Fig. 3.56) cannot be properly described by the model of Figure 3.53. Indeed, the reflection coefficients, S_{11} and S_{22}, are quite different, as Figure 3.56 illustrates. This

[41] Depending on the distance between the two arms of the OSRR hooks, a capacitive effect may appear in the structure of Figure 3.53. This capacitance is the dual counterpart of the inductance L_{sh}, and to account for its effects in the model of Figure 3.53 it should be placed in parallel to the tank formed by L'_s and C_s.

suggests that the model must be asymmetric, and it has been found that the circuit model of Figure 3.56 provides a reasonable (although not very accurate) description of the OSRR-loaded microstrip line (look at the comparisons between the circuit and EM simulations depicted in Fig. 3.56).[42] However, the fact that the model is more complex does not mean that OSRR-loaded microstrip line sections cannot be used as part of CRLH lines or circuits based on them. Conversely, shunt-connected OCSRRs in microstrip technology (Fig. 3.57) behave as purely parallel LC resonators (the trace of S_{11} lies in the unit conductance circle), and the agreement between the circuit and EM simulation is excellent.

According to the models of OSRR- and OCSRR-loaded CPWs, the circuit models of the T-type or π-type CRLH lines based on OSRRs and OCSRRs are depicted in Figure 3.58 (the inductance L_{sh} has been omitted for simplicity). Notice that if the values of the parasitic elements L and C are close to zero, the models can be approximated by the canonical T- or π-type circuit models of a CRLH line (see Fig. 3.20). Indeed, the parasitic effects can be minimized by choosing appropriate geometries for the resonators. Thus, if the terminals of the OCSRR section are shortened in the topology of Figure 3.58b, L vanishes, and the structure can be described by the canonical π-type circuit model of the CRLH line since the parasitic capacitance C of the OSRR is added to the shunt capacitance of the OCSRR [125]. Similarly, for the structure of Figure 3.58a, if the slots of the CPW line are expanded, C is very small, and the parasitic inductance L of the OCSRR is absorbed by the inductance of the OSRR, resulting in the canonical T-type circuit model of the CRLH transmission line. The structure of Figure 3.58b is a quasibalanced CRLH line, where the effects of the parasitic elements are very small. The dispersion diagram and the frequency response of this structure are depicted in Figure 3.59. Notice that the series and shunt resonance frequencies (look at the element values in the caption of Fig. 3.59) of the structure of Figure 3.58b are not identical. The reason is that in order to balance the structures described by the models of Figure 3.58 (with the presence of parasitics) the condition of equal resonance frequencies for the series and shunt branch no longer holds. To balance the structures described by the models of Figure 3.58, it is necessary to force the series reactance and shunt susceptance of the equivalent T or π circuit model to zero at the desired transition frequency. For the circuit of Figure 3.58c, corresponding to the structure of 3.58a, the series and shunt impedances of the equivalent T-circuit model are given by [125]:

$$Z_s(\omega) = -j\frac{\left[\left(1-\omega^2 L_s' C_s\right) + L\omega^2\left(CL_s' C_s\omega^2 - C - C_s\right)\right]}{\omega\left[C_s + C\omega^2\left(CL\left(\omega^2 L_s' C_s - 1\right) - 2C_s L - L_s' C_s + \frac{1}{\omega^2}\right)\right]} \tag{3.99a}$$

$$Z_p(\omega) = -j\frac{\omega L_p' C_s^2}{\left(1-\omega^2 L_p' C_p'\right)C_1^2 + CC_2\omega^2\left(CC_2 LL_2\omega^2 - 2C_1 L_1\right)} \tag{3.99b}$$

[42] Since the number of reactive elements of the OSRR-loaded microstrip line model is too high, such elements have been determined by curve fitting, rather than by parameter extraction.

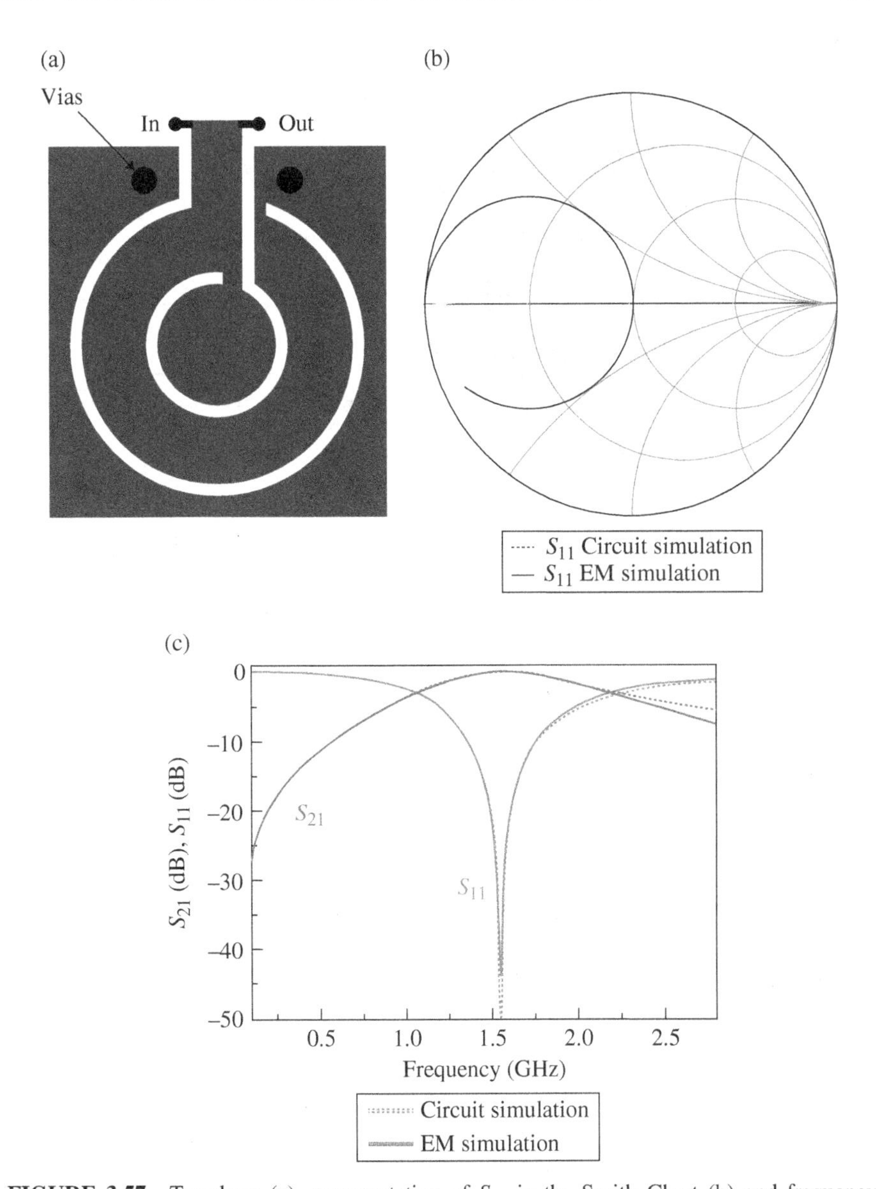

FIGURE 3.57 Topology (a), representation of S_{11} in the Smith Chart (b) and frequency response (magnitude) (c) of a typical shunt-connected OCSRR in microstrip technology with $r_{\text{ext}} = 2.7$ mm, $c = 0.2$ mm, and $d = 1.2$ mm. The considered substrate is the *Rogers RO3010* with thickness $h = 0.254$ mm and dielectric constant $\varepsilon_r = 10.2$. The element values of the shunt resonator in reference of Figure 3.17b are $L_o = 2.02$ nH and $C_c = 5.25$ pF. This figure has been reproduced from Ref. [125] with permission from the author.

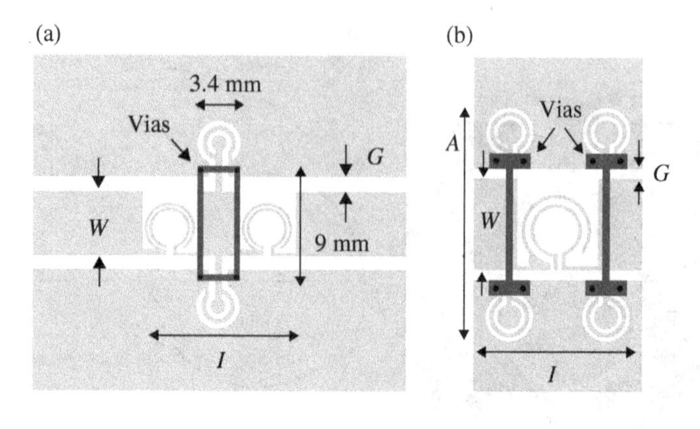

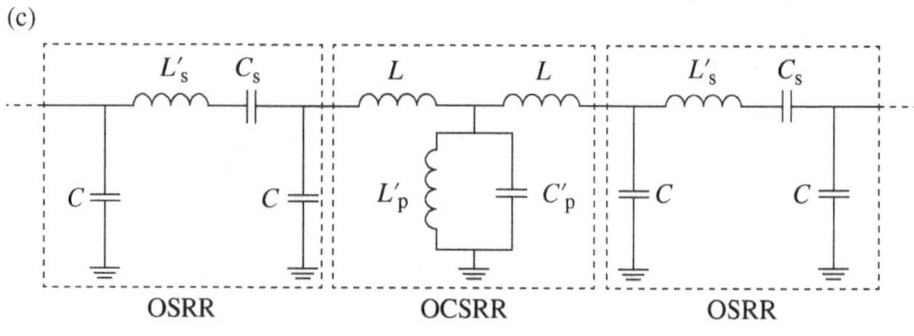

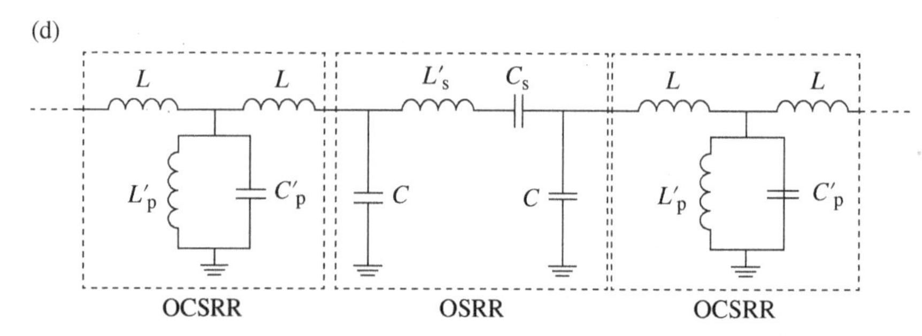

FIGURE 3.58 Typical CPW topologies to implement the T- (a) or π- (b) circuit model (unit cell) of a CRLH transmission line based on a combination of OSRRs and OCSRRs. The corresponding circuit models of the structures of (a) and (b) are depicted in (c) and (d), respectively. This figure has been reproduced from Ref. [125] with permission from the author.

with

$$C_1 = C_s + C\left(1 - \frac{\omega^2}{\omega_s^2}\right) \tag{3.100a}$$

$$C_2 = 2C_s + C\left(1 - \frac{\omega^2}{\omega_s^2}\right) \tag{3.100b}$$

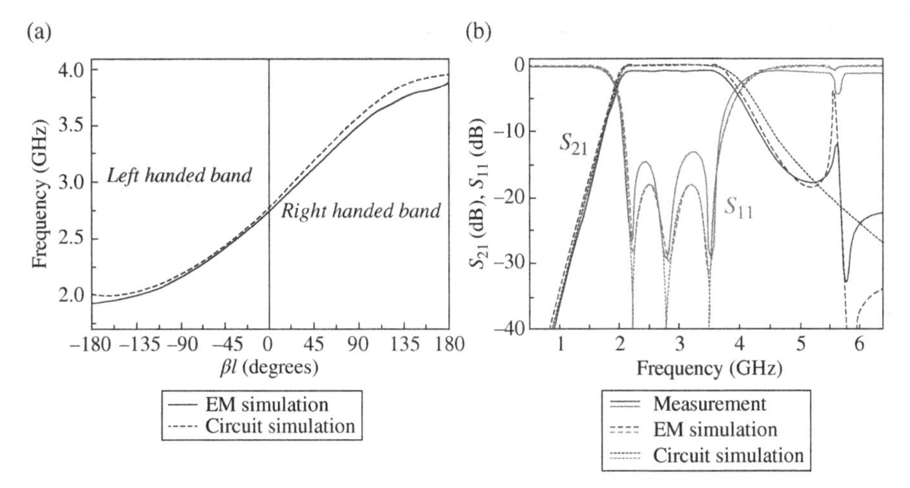

FIGURE 3.59 Dispersion diagram (a) and frequency response (b) of the CPW quasibalanced CRLH line based on open resonators shown in Figure 3.58b. The considered substrate is the *Rogers RO3010* with thickness $h = 0.254$ mm and measured dielectric constant $\varepsilon_r = 11.2$. Dimensions (in reference to Fig. 3.58b) are $l = 7.4$ mm, $A = 11.5$ mm, $W = 4.5$ mm, and $G = 0.52$ mm; for the OCSRRs, $r_{ext} = 1.2$ mm, $c = 0.2$ mm, and $d = 0.2$ mm. For the OSRRs, $r_{ext} = 1.8$ mm, $c = 0.2$ mm, and $d = 0.3$ mm. The values of the equivalent circuit in reference to Figure 3.58d are $C = 0.19$ pF, $L = 0.05$ nH, $C_s = 0.515$ pF, $L'_s = 6.16$ nH, $C'_p = 1.91$ pF, and $L'_p = 1.55$ nH. This figure has been reproduced from Ref. [125] with permission from the author.

$$L_1 = L'_p + L\left(1 - \frac{\omega^2}{\omega_p^2}\right) \quad (3.100c)$$

$$L_2 = 2L'_p + L\left(1 - \frac{\omega^2}{\omega_p^2}\right) \quad (3.100d)$$

$$\omega_s = \frac{1}{\sqrt{L'_s C_s}} \quad (3.100e)$$

$$\omega_p = \frac{1}{\sqrt{L'_p C'_p}} \quad (3.100f)$$

Thus, (3.99) can be used for design purposes, including the synthesis of balanced lines. As will be seen in Chapter 4, CRLH lines based on OSRRs and OCSRRs are useful for the design of dual-band components and broadband bandpass filters.

3.5.2.6 Synthesis Techniques This section has been focused on the resonant type approach of metamaterial transmission lines, where the main topologies and circuit models have been presented and discussed. The author and his research group have dedicated a significant effort in recent years to the automated synthesis of these artificial lines on the basis of the so-called aggressive space mapping (ASM) optimization. This synthesis technique, based on quasi-Newton type iteration, is applied to automatically find the layout of the structure (unit cell) from the elements of the equivalent circuit model (which are assumed to be determined from circuit specifications). ASM optimization is important for the design of microwave circuits based on these artificial lines. However, the general formulation of ASM, as well as its application to the implementation of practical synthesis tools is very specific and requires significant effort for understanding. Therefore, this part is left for Appendix H (recommended to those readers willing to design microwave circuits based on resonant type metamaterial transmission lines).

3.5.3 The Hybrid Approach

In 2006, metamaterial transmission lines based on a combination of CSRRs, series gaps, and shunt-connected stubs were presented for the first time [127]. These artificial lines were called hybrid lines [128] because they are implemented by using the reactive elements of the CL-loaded approach (series gaps and shunt connected stubs) and the resonant type approach (CSRRs). Since they include more reactive elements than CL-loaded lines and resonant type metalines, hybrid lines exhibit a richer phenomenology. Indeed, these lines were proposed for the implementation of narrow (and moderate) band bandpass filters exhibiting backward wave propagation in the transmission band and a transmission zero above it (with the intention of improving the stop band response of the filters) [127]. The typical topology and circuit model of the unit cell of a hybrid line are depicted in Figure 3.60. The interpretation of the

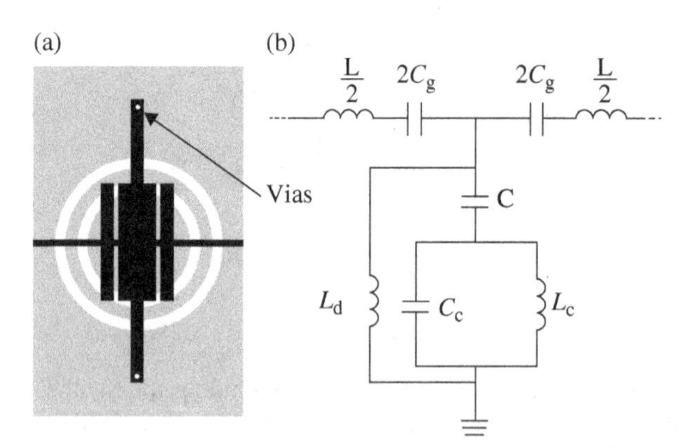

FIGURE 3.60 Typical topology of the unit cell of a metaline based on the hybrid approach (a) and lumped-element equivalent circuit model (b). In (a), the upper metal layer is depicted in black, whereas the lower metal layer (ground plane), where the CSRR is etched, is depicted in gray.

different model parameters is similar to that given for CSRR-based lines, except that now we have an additional element, the shunt inductance, L_d, that accounts for the pair of grounded (through vias) stubs.

Due to the presence of shunt stubs, hybrid lines exhibit three transmission bands, rather than two, plus one transmission zero, which is given by (3.86). As discussed before, wave propagation is possible in those frequency regions where the phase constant is a real number. The dispersion characteristics can easily be inferred from the circuit model numerically, or by means of a circuit simulator. Depending on the relative positions of the relevant frequencies of the structure, two types of dispersion result, where the differences refer to the nature, LH or RH, of the different transmission bands [9]. The two possibilities are (1) LH–LH–RH and (2) LH–RH–RH. Hence, the difference is that the central (second) band can be either LH or RH (the first and third band always exhibit LH and RH wave propagation, respectively). If $f_{ser} < f_z$ (with $f_{ser} = (LC_g)^{-1/2}/2\pi$), then the transmission bands are LH–RH–RH; if $f_{ser} > f_z$, wave transmission in the bands follows the sequence LH–LH–RH. Of particular interest is the balanced case, where two of the bands merge. Let us denote the two poles of the shunt branch of the circuit model as f_{p1} and f_{p2}, with $f_{p1} < f_{p2}$. If (i) $f_{p1} = f_{ser}$, the first (LH) and the second (RH) bands merge, and the third (RH) band is isolated; if (ii) $f_{p2} = f_{ser}$, the second (LH) and the third (RH) band merge, and the first (LH) band is isolated. The former situation is interesting for wide band filter design because it is possible to achieve broad bands with a transmission zero above these bands [129].

The dispersion and characteristic impedance corresponding to the particular hybrid structure shown in Figure 3.61a are depicted in Figure 3.61b and c. This structure is un-balanced with $f_{ser} > f_z$, thus, the sequence of the bands is LH–LH–RH. The agreement between experimental data, EM simulation and circuit simulation is good. The parameters of the circuit model were extracted following a parameter extraction method similar to that reported in Appendix G.

Although hybrid lines exhibit three transmission bands, these lines cannot be considered to be order-3 CRLH transmission lines. The reason is that the number of elements in the series branch of the T-circuit model is limited to two (L and C_g). Therefore, it is not possible to obtain certain functionalities achievable with order-3 CRLH transmission lines. For instance, it is not possible to implement tri-band components by means of hybrid lines of the type shown in Figures 3.60 or 3.61. As will be later shown (Chapter 4), to implement tri-band components it is necessary to force the characteristic impedance and phase shift of the constituent artificial lines of such tri-band circuits to the required values at three operating frequencies, and this is only possible if at least three elements for both the series and shunt branches are present. Hence hybrid lines are essentially useful to achieve certain functionality in two bands (or in a single band). However, the presence of the shunt inductive stub enhances design flexibility. This is the main advantage of hybrid lines over SRR- or CSRR-loaded lines. This design flexibility has been exploited in the design of bandpass filters, where superior performance (i.e., improved bandwidth and stop-band) with respect to filters based on SRR- or CSRR-loaded lines has been achieved [127–129] (however this aspect will be discussed in Chapter 4).

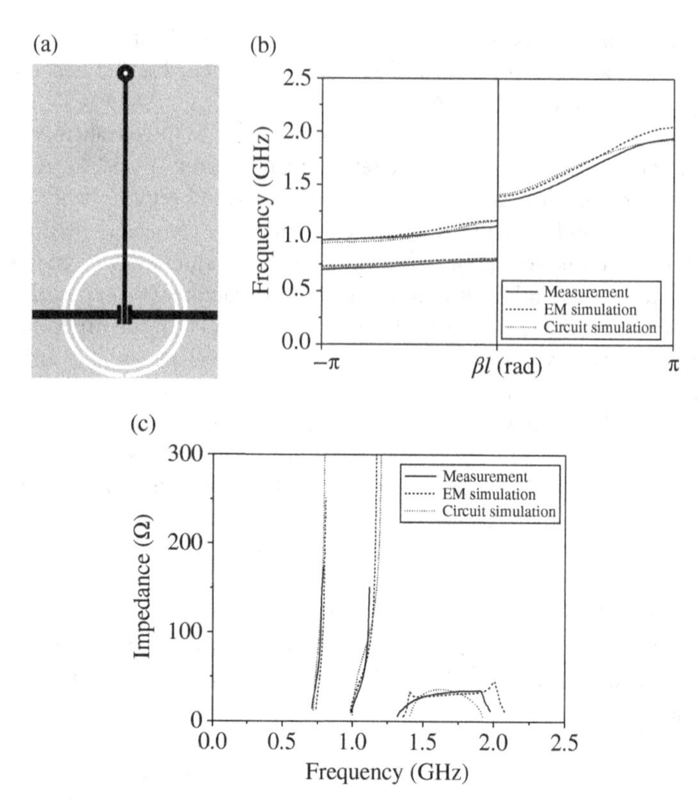

FIGURE 3.61 Topology (a), dispersion diagram (b) and characteristic impedance (c) for a hybrid microstrip structure, inferred from EM simulation (dashed line), from measurement (solid line) and from circuit simulation (dotted line). The considered substrate is the *Rogers RO3010* with thickness $h = 0.635$ mm and dielectric constant $\varepsilon_r = 10.2$. Dimensions are: host line width $W = 1$ mm, line length $l = 21.97$ mm, gaps separation $s_g = 0.2$ mm, gap width $W_g = 2.17$ mm, stub length $l_s = 26.23$ mm, and stub width $W_s = 0.44$ mm; for the CSRR, external radius $r_{ext} = 7.48$ mm, ring width $c = 0.43$ mm, and ring separation $d = 0.54$ mm. The extracted element values, used for the circuit simulation, are $L = 18.65$ nH, $C_g = 0.68$ pF, $C = 5.35$ pF, $C_c = 42.82$ pF, $L_c = 0.67$ nH, and $L_d = 4.77$ nH. Reprinted with permission from Ref. [9]; copyright 2011 IEEE.

REFERENCES

1. D. Schurig, J. J. Mock, B. J. Justice, S. A. Cummer, J. B. Pendry, A. F. Starr, and D. R. Smith, "Metamaterial electromagnetic cloak at microwave frequencies," *Science*, vol. **314**, pp. 977–980, 2006.
2. G. V. Eleftheriades and K. G. Balmain, Ed., *Negative-Refraction Metamaterials: Fundamental Principles and Applications*, Wiley-Interscience, Hoboken, NJ, 2005.
3. C. Caloz and T. Itoh, *Electromagnetic Metamaterials: Transmission Line Theory and Microwave Applications*, John Wiley & Sons, Hoboken, NJ, 2006.
4. R. Marqués, F. Martín, and M. Sorolla, *Metamaterials with Negative Parameters: Theory, Design and Microwave Applications*, Wiley-Interscience, Hoboken, NJ, 2007.

5. A. K. Iyer and G. V. Eleftheriades, "Negative refractive index metamaterials supporting 2-D waves," *IEEE-MTT Int. Microw. Symp. Dig.*, Seattle, WA, June 2002, pp. 412–415.
6. A. A. Oliner, "A periodic-structure negative-refractive-index medium without resonant elements," *URSI Dig. IEEE-AP-S USNC/URSI Natl. Radio Sci. Meet.*, San Antonio, TX, June 2002, p. 41.
7. C. Caloz and T. Itoh, "Application of the transmission line theory of left-handed (LH) materials to the realization of a microstrip LH transmission line," *Proc. IEEE-AP-S USNC/URSI National Radio Science Meeting*, vol. 2, San Antonio, TX, June 2002, pp. 412–415.
8. F. Martín, F. Falcone, J. Bonache, R. Marqués, and M. Sorolla, "Split ring resonator based left handed coplanar waveguide," *Appl. Phys. Lett.*, vol. **83**, pp. 4652–4654, 2003.
9. M. Durán-Sindreu, A. Vélez, G. Sisó, J. Selga, P. Vélez, J. Bonache, and F. Martín, "Recent advances in metamaterial transmission lines based on split rings," *Proc. IEEE*, vol. **99**, pp. 1701–1710, 2011.
10. J. D. Baena, J. Bonache, F. Martín, R. Marqués, F. Falcone, T. Lopetegi, M. A. G. Laso, J. García, I. Gil, M. Flores-Portillo, and M. Sorolla, "Equivalent circuit models for split ring resonators and complementary split rings resonators coupled to planar transmission lines," *IEEE Trans. Microw. Theory Tech.*, vol. **53**, pp. 1451–1461, 2005.
11. C. Caloz and T. Itoh, "Novel microwave devices and structures based on the transmission line approach of metamaterials," *IEEE-MTT Int. Microw. Symp. Dig.*, vol. **1**, Philadelphia, PA, June 2003, pp. 195–198.
12. D. R. Smith, W. J. Padilla, D. C. Vier, S. C. Nemat-Nasser, and S. Schultz, "Composite medium with simultaneously negative permeability and permittivity," *Phys. Rev. Lett.*, vol. **84**, pp. 4184–4187, 2000.
13. R. A. Shelby, D. R. Smith, and S. Schultz, "Experimental verification of a negative index of refraction," *Science*, vol. **292**, pp. 77–79, 2001.
14. R. A. Shelby, D. R. Smith, S. C. Nemat-Nasser, and S. Schultz, "Microwave transmission through a two-dimensional, isotropic, left-handed metamaterial," *Appl. Phys. Lett.*, vol. **78**, pp. 489–491, 2001.
15. A. Grbic and G. V. Eleftheriades, "Experimental verification of backward wave radiation from a negative refractive index metamaterial," *J. Appl. Phys.*, vol. **92**, pp. 5930–5935, 2002.
16. V. G. Veselago, "The electrodynamics of substances with simultaneously negative values of ε and μ," *Sov. Phys. Usp.*, vol. **10**, pp. 509–514, 1968.
17. N. Engheta and R. W. Ziolkowski, *Metamaterials: Physics and Engineering Explorations*, Wiley-Interscience, Hoboken, NJ, 2006.
18. A. K. Sarychev and V. M. Shalaev, *Electrodynamics of Metamaterials*, World Scientific Publishing Co., Singapore, 2007.
19. P. Markos and C. M. Sokoulis, *Wave Propagation: From Electrons to Photonic Crystals and Left-Handed Materials*, Princeton University Press, Princeton, NJ, 2008.
20. F. Capolino, Ed., *Metamaterials Handbook*, CRC Press, Boca Raton, FL 2009.
21. L. Solymar and E. Shamonina, *Waves in Metamaterials*, Oxford University Press, Oxford, 2009.
22. T. J. Cui, D. R. Smith, and R. Liu, Eds., *Metamaterials: Theory, Design and Applications*, Springer, New York, 2010.

23. R. W. Ziolkowski and E. Heynman, "Wave propagation in media having negative permeability and permittivity," *Phys. Rev. E*, vol. **64**, paper 056625, 2001.
24. V. G. Veselago, "Formulating Fermat's principle for light travelling in negative refraction materials," *Phys. Usp.*, vol. **45**, pp. 1097–1099, 2002.
25. I. V. Lindell, S. A. Tretyakov, K. I. Nikoskinen, and S. Ilvonen, "BW-media with negative parameters, capable of supporting backward waves," *Microw. Opt. Technol. Lett.*, vol. **31**, pp. 129–133, 2001.
26. A. Sihvola, "Electromagnetic emergence in metamaterials," in *Advances in Electromagnetics of Complex Media and Metamaterials*, S. Zouhdi, A. Sihvola, and M. Arsalane, Eds., NATO Science Series: II: Mathematics, Physics, and Chemistry, vol. **89**, pp. 1–17, Kluwer Academic Publishers, Dordrecht, 2003.
27. W. Rotman, "Plasma simulation by artificial dielectrics and parallel-plate media," *IRE Trans. Antennas Propag.*, vol.**10**, pp. 82–95, 1962.
28. J. B. Pendry, A. J. Holden, W. J. Stewart, and I. Youngs, "Extremely low frequency plasmons in metallic mesostructures," *Phys. Rev. Lett.*, vol. **76**, pp. 4773–4776, 1996.
29. J. B. Pendry, A. J. Holden, D. J. Robbins, and W. J. Stewart, "Low frequency plasmons in thin-wire structures," *J. Phys. Condens. Matter.*, vol. **10**, pp. 4785–4810, 1998.
30. J. B. Pendry, A. J. Holden, D. J. Robbins, and W. J. Stewart, "Magnetism from conductors and enhanced nonlinear phenomena," *IEEE Trans. Microw. Theory Tech.*, vol. **47**, pp. 2075–2084, 1999.
31. S. Ramo, J. R. Whinnery, and T. Van-Duzer, *Fields and Waves in Communication Electronics*, Wiley, Hoboken, NJ, 1965, and 2nd (1984) and 3rd (1994) Editions.
32. M. Mrozowski, *Guided Electromagnetic Waves. Properties and Analysis*, Research Studies Press, Taunton and Wiley, New York, 1997.
33. J. A. Kong, *Electromagnetic Wave Theory*, EMW Publishing, Cambridge, MA, 2000.
34. R. F. Harrington, *Time-harmonic Electromagnetic Fields*, McGraw-Hill, New York, 1961.
35. L. D. Landau, E. M. Lifshitz, and L. P. Pitaevskii, *Electrodynamics of Continuous Media*, Pergamon, New York, 1984.
36. J. D. Jackson, *Classical Electrodynamics*, 3rd Edition, Wiley, New York, 1999.
37. D. R. Smith, D. Schurig, and J. B. Pendry, "Negative refraction of modulated electromagnetic waves," *Appl. Phys. Lett.*, vol. **81**, pp. 2713–2715, 2002.
38. S. Foteinopoulou, E. N. Economou, and C. M. Soukoulis, "Refraction in media with a negative refractive index," *Phys. Rev. Lett.*, vol. **90**, paper 107402, 2003.
39. C. G. Parazzoli, R. B. McGregor, K. Li, B. E. C. Kontenbah, and M. Tlienian, "Experimental verification and simulation of negative index of refraction using Snell's law," *Phys. Rev. Lett.*, vol. **90**, paper 107401, 2003.
40. K. Li, S. J. McLean, R. B. Gregor, C. G. Parazzoli, and M. H. Tanielian, "Free-space focused-beam characterization of left-handed materials," *Appl. Phys. Lett.*, vol. **82**, pp. 2535–2537, 2003.
41. A. A. Houck, J. B. Brock, and I. L. Chuang, "Experimental observations of a left-handed material that obeys Snell's law," *Phys. Rev. Lett.*, vol. **90**, paper 137401, 2003.
42. D. Felbacq and A. Moreau, "Direct evidence of negative refraction at media with negative ε and μ," *J. Opt. A Pure Appl. Opt.*, vol. **5**, pp. L9–L11, 2003.
43. J. B. Pendry, "Negative refraction makes perfect lens," *Phys. Rev. Lett.*, vol. **85**, pp. 3966–3969, 2000.

44. A. Grbic, R. Merlin, E. M. Thomas, and M. F. Imani, "Near-field plates: metamaterial surfaces/ arrays for subwavelength focusing and probing," *Proc. IEEE*, vol. **99**, pp. 1806–1815, 2011.
45. J. Lu, T. M. Gregorczyc, Y. Zhang, J. Pacheco, B. Wu, J. A. Kong, and M. Chen, "Cerenkov radiation in materials with negative permittivity and permeability," *Opt. Express*, vol. **11**, pp. 723–734, 2003.
46. P. R. Berman, "Goos-Hächen shift in negatively refractive media," *Phys. Rev. E*, vol. **66**, paper 067603, 2002.
47. J. A. Kong, B. I. Wu, and Y. Zhang, "Lateral displacement of a Gaussian beam reflected from a grounded slab with negative permittivity and permeability," *Appl. Phys. Lett.*, vol. **80**, pp. 2084–2086, 2002.
48. R. W. Ziolkowski, "Pulsed and CW Gaussian beam interactions with double negative metamaterial slabs," *Opt. Express*, vol. **11**, pp. 662–681, 2003.
49. A. Lakhtakia, "On plane wave remittances and Goos-Hänchen shifts of planar slabs with negative real permittivity and permeability," *Electromagnetics*, vol. **23**, pp. 71–75, 2002.
50. I. V. Shadrivov, A. A. Zharov, and Y. S. Kivshar, "Giant Goos-Hänchen effect at the refraction from left-handed metamaterials," *Appl. Phys. Lett.*, vol. **83**, pp. 2713–2715, 2003.
51. C. Luo, S. G. Johnson, J. D. Joannopoulos, and J. B. Pendry, "All-angle negative refraction without negative effective index," *Phys. Rev. B*, vol. **65**, paper 201104(R), 2002.
52. M. Notomi, "Theory of light propagation in strongly modulated photonic crystals: refraction like behavior in the vicinity of the photonic band gap," *Phys. Rev. B*, vol. **62**, pp. 10696–10705, 2000.
53. E. Cubukcu, K. Aydin, E. Ozbay, S. Foteinopoulou, and C. M. Soukoulis, "Electromagnetic waves: negative refraction by photonic crystals," *Nature*, vol. **423**, pp. 604–605, 2003.
54. N. B. Kundtz, D. R. Smith, and J. B. Pendry, "Electromagnetic design with transformation optics," *Proc. IEEE*, vol. **99**, pp. 1122–1633, 2011.
55. R. Marques, L. Jelinek, M. J. Freire, J. D. Baena, and M. Lapine, "Bulk metamaterials made of resonant rings," *Proc. IEEE*, vol. **99**, pp. 1660–1668, 2009.
56. D. O. Guney, T. Koschny, and C. M. Soukoulis, "Intra-connected three-dimensionally isotropic bulk negative index photonic metamaterial," *Opt. Express*, vol. **18**, pp. 12348–12353, 2010.
57. A. Grbic and G. V. Eleftheriades, "Overcoming the diffraction limit with a planar left-handed transmission-line lens," *Phys. Rev. Lett.*, vol. **92**, paper 117403, 2004.
58. A. Sanada, C. Caloz, and T. Itoh, "Planar distributed structures with negative refractive index," *IEEE Trans. Microw. Theory Tech.*, vol. **52**, pp. 1252–1263, 2004.
59. F. Elek and G. V. Eleftheriades, "A two-dimensional uniplanar transmission-line metamaterial with a negative index of refraction," *New J. Phys.*, vol. **7**, p. 163, 2005.
60. A. Grbic and G. V. Eleftheriades, "A 3-D negative-refractive-index transmission-line medium," *IEEE AP-S/URSI Int. Symp.*, Washington, DC, July 2005.
61. A. Grbic and G. V. Eleftheriades, "An isotropic three-dimensional negative-refractive-index transmission-line metamaterial," *J. Appl. Phys.*, vol. **98**, paper 043106, 2005.
62. P. Alitalo, S. Maslovski, and S. Tretyakov, "Three-dimensional isotropic perfect lens based on LC-loaded transmission lines," *J. Appl. Phys.*, vol. **99**, paper 064912, 2006.
63. P. Alitalo, S. Maslovski, and S. Tretyakov, "Experimental verification of the key properties of a three-dimensional isotropic transmission-line superlens," *J. Appl. Phys.*, vol. **99**, paper 124910, 2006.

64. M. Zedler, C. Caloz, and P. Russer, "A 3-D isotropic left-handed metamaterial based on the rotated transmission-line matrix (TLM) scheme," *IEEE Trans. Microw. Theory Tech.*, vol **55**, pp. 2930–2941, 2007.
65. A. K. Iyer and G. V. Eleftheriades, "A three-dimensional isotropic transmission-line metamaterial topology for free-space excitation," *Appl. Phys. Lett.*, vol. **92**, paper 261106, 2008.
66. C. Garcia-Meca, J. Hurtado, J. Martí, A. Martínez, W. Dickson, and A. V. Zayats, "Low-loss multilayered metamaterial exhibiting a negative index of refraction at visible wavelengths," *Phys. Rev. Lett.*, vol. **106**, paper 067402, 2011.
67. C. M. Soukoulis and M. Wegener, "Past achievements and future challenges in the development of three-dimensional photonic metamaterials," *Nat. Photonics*, vol. **5**, pp. 523–530, 2011.
68. S. A. Shelkunoff and H. T. Friis, *Antennas: Theory and Practice*, 3rd Edition, Wiley, New York, 1966.
69. R. Marques, F. Medina, and R. Rafii-El-Idrissi, "Role of bi-anisotropy in negative permeability and left handed metamaterials," *Phys. Rev. B*, vol. **65**, paper 144441, 2002.
70. I. Bahl and P. Bhartia, *Microwave Solid State Circuit Design*, Wiley, New York, 1988.
71. A. L. Pokrovsky and A. L. Efros, "Electrodynamics of photonic crystals and the problem of left-handed metarials," *Phys. Rev. Lett.*, vol. **89**, paper 093901, 2002.
72. R. Marques and D. R. Smith, "Comments to Electrodynamics of photonic crystals and the problem of left-handed metamaterials," *Phys. Rev. Lett.*, vol. **92**, paper 059401, 2004.
73. R. Marqués, J. Martel, F. Mesa, and F. Medina, "Left-handed-media simulation and transmission of EM waves in subwavelength split-ring-resonator-loaded metallic waveguides," *Phys. Rev. Lett.*, vol. **89**, paper 183901, 2002.
74. J. Esteban, C. Camacho-Peñalosa, J. E. Page, T. M. Martin-Guerrero, and E. Marquez-Segura, "Simulation of negative permittivity and negative permeability by means of evanescent waveguide modes—theory and experiment," *IEEE Trans. Microw. Theory Tech.*, vol. **53**, pp. 1506–1514, 2005.
75. R. Marqués, J. D. Baena, J. Martel, F. Medina, F. Falcone, M. Sorolla, and F. Martin, "Novel small resonant electromagnetic particles for metamaterial and filter design," *Proc. ICEAA'03*, Torino, Italy, September 8–12, 2003, pp. 439–442.
76. J. D. Baena. R. Marqués, F. Medina, and J. Martel, "Artificial magnetic metamaterial design by using spiral resonators," *Phys. Rev. B*, vol. **69**, paper 014402, 2004.
77. M. Makimoto, S. Yamashita, "Compact bandpass filters using stepped impedance resonators," *Proc. IEEE*, vol. **67**, pp. 16–19, 1979.
78. D. Schurig, J. J. Mock, and D. R. Smith, "Electric-field-coupled resonators for negative permittivity metamaterials," *Appl. Phys. Lett.*, vol. **88**, paper 041109, 2006.
79. J. García-García, F. Martín, J. D. Baena, R. Marqués, and L. Jelinek, "On the resonances and polarizabilities of split ring resonators," *J. Appl. Phys.*, vol. **98**, paper 033103, 2005.
80. J. Martel, R. Marqués, F. Falcone, J. D. Baena, F. Medina, F. Martín and M. Sorolla, "A new LC series element for compact band pass filter design," *IEEE Microw. Wireless Compon. Lett.*, vol. **14**, pp. 210–212, 2004.
81. H. G. Booker, "Slot aerials and their relation to complementary wire aerials (Babinet's principle)," *J. IEE*, vol. **93**, pt. III–A, no. 4, pp. 620–626, 1946.
82. G. A. Deschamps, "Impedance properties of complementary multiterminal planar structures," *IRE Trans. Antennas Propag.*, vol. **AP–7**, pp. 371–378, 1959.

83. W. J. Getsinger, "Circuit duals on planar transmission media," *IEEE MTT-S Int. Microw. Symp. Dig.*, Boston, MA, May–June 1983, pp. 154–156.

84. F. Falcone, T. Lopetegi, J. D. Baena, R. Marqués, F. Martín, and M. Sorolla, "Effective negative-ε stop-band microstrip lines based on complementary split ring resonators," *IEEE Microw. Wireless Compon. Lett.*, vol. **14**, pp. 280–282, 2004.

85. F. Falcone, T. Lopetegi, M. A. G. Laso, J. D. Baena, J. Bonache, R. Marqués, F. Martín, and M. Sorolla, "Babinet principle applied to the design of metasurfaces and metamaterials," *Phys. Rev. Lett.*, vol. **93**, paper 197401, 2004.

86. A. Velez, F. Aznar, J. Bonache, M. C. Velázquez-Ahumada, J. Martel, and F. Martín, "Open complementary split ring resonators (OCSRRs) and their application to wideband CPW band pass filters," *IEEE Microw. Wireless Compon. Lett.*, vol. **19**, pp. 197–199, 2009.

87. C. Caloz, "Dual composite right/left-handed (D-CRLH) transmission line metamaterial," *IEEE Microw. Wireless Compon. Lett.*, vol. **16**, pp. 585–587, 2006.

88. A. Rennings, S. Otto, J. Mosig, C. Caloz, and I. Wolff, "Extended composite right/left-handed metamaterial and its application as quadband quarter-wavelength transmission line," *Proc. Asia-Pacific Microw. Conf. (APMC)*, Yokohama, Japan, December 2006, pp. 1405–1408.

89. G. V. Eleftheriades, "A generalized negative-refractive-index transmission-line (NRL-TL) metamaterial for dual-band and quad-band applications," *IEEE Microw. Wireless Compon. Lett.*, vol. **17**, pp. 415–417, 2007.

90. G Sisó, M. Gil, J. Bonache, and F. Martín, "Generalized model for multiband metamaterial transmission lines," *IEEE Microw. Wireless Compon. Lett.*, vol. **18**, pp. 728–730, 2008.

91. C. Camacho-Peñalosa, T. M. Martín-Guerrero, J. Esteban, and J.E. Page, "Derivation and general properties of artificial lossless balanced composite right/left-handed transmission lines of arbitrary order," *Prog. Electromagn. Res. B*, vol. **13**, pp. 151–169, 2009.

92. A. Rennings, T. Liebig, C. Caloz, and I. Wolff, "Double-Lorentz transmission line metamaterial and its application to tri-band devices," *IEEE MTT-S Int. Microw. Symp., Dig.*, Honolulu, HI, June 2007, pp. 1427–1430.

93. R. A. Foster, "A reactance theorem," *Bell Syst. Tech. J.*, vol. **3**, pp. 259–267, 1924.

94. C. Caloz, H. Okabe, T. Iwai, and T. Itoh, "Transmission line approach of left handed materials," *USNC/URSI Natl. Radio Sci. Meet.*, San Antonio, TX, vol. **1**, June 2002, p. 39.

95. D. Sievenpiper, L. Zhang, R. F. J. Broas, N. G. Alexopolous, and E. Yablonovitch, "High impedance electromagnetic surfaces with a forbidden frequency band," *IEEE Trans. Microw. Theory Tech.*, vol. **47**, pp. 2059–2074, 1999.

96. Y. Horii, C. Caloz, T. Itoh, "Super-compact multilayered left-handed transmission line and diplexer application," *IEEE Trans. Microw. Theory Tech.*, vol. **53**, pp. 1527–1534, 2005.

97. Y. Horii, N. Inoue, T. Kawakami, and T. Kaneko, "Super-compact LTCC-based multilayered CRLH transmission lines for UWB applications," *Proc. 41st Eur. Microw. Conf.*, Manchester, UK, October 2011.

98. P. Turalchuk, I. Munina, P. Kapitanova, D. Kholodnyak, D. Stoepel, S. Humbla, J. Mueller, M. A. Hein, and I. Vendik, "Broadband small-size LTCC directional couplers," *Proc. 40 Eur. Microw. Conf.*, Paris, France, September 2010.

99. R. Marqués, F. Mesa, J. Martel, and F. Medina, "Comparative analysis of edge and broadside coupled split ring resonators for metamaterial design. Theory and experiment," *IEEE Trans. Antennas Propag.*, vol. **51**, pp. 2572–2582, 2003.

100. F. Aznar, M. Gil, J. Bonache, and F. Martín, "Modelling metamaterial transmission lines: a review and recent developments," *Opto-Electron. Rev.*, vol. **16**, pp. 226–236, 2008.

101. F. Aznar, J. Bonache, and F. Martín, "Improved circuit model for left handed lines loaded with split ring resonators," *Appl. Phys. Lett.*, vol. **92**, paper 043512, 2008.

102. L. Roglá, J. Carbonell, and V. E. Boria, "Study of equivalent circuits for open-ring and split-ring resonators in coplanar waveguide technology," *IET Microw. Antennas Propag.*, vol. **1**, pp. 170–176, 2007.

103. D. M. Pozar, *Microwave Engineering*, Addison Wesley, New York, 1990.

104. F. Aznar, M. Gil, J. Bonache, J. D. Baena, L. Jelinek, R. Marqués, and F. Martín, "Characterization of miniaturized metamaterial resonators coupled to planar transmission lines," *J. Appl. Phys.*, vol. **104**, paper 114501, 2008.

105. G. V. Eleftheriades, O. Siddiqui, and A. Iyer, "Transmission line models for negative refractive index media and associated implementations without excess resonators," *IEEE Microw. Wireless Compon. Lett.*, vol. **13**, pp. 51–53, 2003.

106. J. Bonache, M. Gil, O. García-Abad, and F. Martín, "Parametric analysis of microstrip lines loaded with complementary split ring resonators," *Microw. Opt. Technol. Lett.*, vol. **50**, pp. 2093–2096, 2008.

107. F. Aznar, M. Gil, J. Bonache, and F. Martín, "Revising the equivalent circuit models of resonant-type metamaterial transmission lines," *IEEE MTT-S Int. Microw. Symp. Dig.*, Atlanta, GA, June 2008.

108. M. Gil, J. Bonache, J. Selga, J. García-García, and F. Martín, "Broadband resonant type metamaterial transmission lines," *IEEE Microw. Wireless Compon. Lett.*, vol. **17**, pp. 97–99, 2007.

109. F. Aznar, M. Gil, J. Bonache, and F. Martín, "SRR- and CSRR-loaded metamaterial transmission lines: a comparison to the light of duality," *2nd Int. Congr. Adv. Electromagn. Mater. Microw. Opt. (Metamaterials'08)*, Pamplona, Spain, September 2008.

110. T. Tamir and A. A. Oliner, "Guided complex waves," *Proc. IEEE*, vol. **110**, pp. 310–34, 1963.

111. R. Islam and G. V. Eleftheriades, "On the independence of the excitation of complex modes in isotropic structures," *IEEE Trans. Antennas Propag.*, vol. **58**, pp. 1567–1578, 2010.

112. P. J. B. Clarricoats and K. R. Slinn, "Complex modes of propagation in dielectric loaded circular waveguide," *Electron. Lett.*, vol. **1**, pp. 145–146, 1965.

113. A. S. Omar and K. F. Schünemann, "The effect of complex modes at finline discontinuities," *IEEE Trans. Microw. Theory Tech.*, vol. **34**, pp. 1508–1514, 1986.

114. W. Huang and T. Itoh, "Complex modes in lossless shielded microstrip lines," *IEEE Trans. Microw. Theory Tech.*, vol. **36**, pp. 163–164, 1988.

115. F. Elek and G. V. Eleftheriades, "Dispersion analysis of the shielded Sievenpiper structure using multiconductor transmission line theory," *IEEE Microw. Wireless Compon. Lett.*, vol. **14**, pp. 434–436, 2004.

116. J. Naqui, M. Durán-Sindreu, A. Fernández-Prieto, F. Mesa, F. Medina, and F. Martín, "Multimode propagation and complex waves in CSRR-based transmission line metamaterials," *IEEE Antennas Wireless Propag. Lett.*, vol. **11**, pp. 1024–1027, 2012.

117. J. Naqui, A. Fernández-Prieto, F. Mesa, F. Medina, and F. Martín, "Effects of inter-resonator coupling in split ring resonator (SRR) loaded metamaterial transmission lines," *J. Appl. Phys.*, vol. **115**, paper 194903, 2014.

118. R. Mongia, I. Bahl, and P. Barthia, *RF and Microwave Coupled Line Circuits*, Artech House, Norwood, MA, 1999.

119. J. Shekel, "Matrix analysis of multi-terminal transducers," *Proc. IRE*, vol. **42**, pp. 840–847, 1954.

120. R. Islam, M. Zedler, and G. V. Eleftheriades, "Modal analysis and wave propagation in finite 2D transmission-line metamaterials," *IEEE Trans. Antennas Propag.*, vol. **59**, pp. 1562–1570, 2011.

121. M. Beruete, F. Falcone, M. J. Freire, R. Marqués, and J. D. Baena, "Electroinductive waves in chains of complementary metamaterial elements," *Appl. Phys. Lett.*, vol. **88**, paper 083503, 2006.

122. J. Naqui, M. Durán-Sindreu, and F. Martín, "Modeling split ring resonator (SRR) and complementary split ring resonator (CSRR) loaded transmission lines exhibiting cross polarization effects," *IEEE Ant. Wireless Propag. Lett.*, vol. **12**, pp. 178–181, 2013.

123. J. Naqui, A. Fernández-Prieto, M. Durán-Sindreu, F. Mesa, J. Martel, F. Medina, and F. Martín, "Common mode suppression in microstrip differential lines by means of complementary split ring resonators: theory and applications," *IEEE Trans. Microw. Theory Tech.*, vol. **60**, pp. 3023–3034, 2012.

124. M. Durán-Sindreu, A. Vélez, F. Aznar, G. Sisó, J. Bonache, and F. Martín, "Application of open split ring resonators and open complementary split ring resonators to the synthesis of artificial transmission lines and microwave passive components," *IEEE Trans. Microw. Theory Tech.*, vol. **57**, pp. 3395–3403, 2009.

125. M. Durán-Sindreu, *Miniaturization of planar microwave components based on semi-lumped elements and artificial transmission lines: application to multi-band devices and filters*, PhD dissertation, Universitat Autònoma de Barcelona, Barcelona, Spain, 2011.

126. M. Durán-Sindreu, P. Vélez, J. Bonache, and F. Martín, "Broadband microwave filters based on open split ring resonators (OSRRs) and open complementary split ring resonators (OCSRRs): improved models and design optimization," *Radioengineering*, vol. **20**, pp. 775–783, 2011.

127. J. Bonache, I. Gil, J. García-García, and F. Martín, "Novel microstrip band pass filters based on complementary split rings resonators," *IEEE Trans. Microw. Theory Tech.*, vol. **54**, pp. 265–271, 2006.

128. J. Bonache, M. Gil, I. Gil, J. García-García, and F. Martín, "Limitations and solutions of resonant-type metamaterial transmission lines for filter applications: the hybrid approach," *IEEE MTT-S Int. Microw. Symp. Dig.*, San Francisco, CA, June 2006, pp. 939–942.

129. M. Gil, J. Bonache, J. García-García, J. Martel, and F. Martín, "Composite right/left handed (CRLH) metamaterial transmission lines based on complementary split rings resonators (CSRRs) and their applications to very wide band and compact filter design," *IEEE Trans. Microw. Theory Tech.*, vol. **55**, pp. 1296–1304, 2007.

4

METAMATERIAL TRANSMISSION LINES: RF/MICROWAVE APPLICATIONS

4.1 INTRODUCTION

This chapter is entirely focused on the RF/microwave applications of the artificial lines considered in Chapter 3, that is, composite right-/left-handed (CRLH) transmission lines, and also on the applications of transmission lines with metamaterial loading, that is, transmission lines loaded with electrically small resonators, formerly used for the implementation of effective media metamaterials.

The applications of CRLH transmission lines are mainly based on the controllability of the dispersion and characteristic impedance of the lines. Thus, impedance and dispersion engineering, which represents a genuine metamaterial-related design approach, allow us to implement RF/microwave devices with superior performance (as compared to devices based on ordinary lines and stubs), or with novel functionalities. Moreover, since the unit cells of metamaterial transmission lines are electrically small at the frequencies of interest, device dimensions can be made very small as compared to the typical sizes of ordinary distributed circuits. The implementation of dual-band and multiband devices, where the characteristic impedance and electrical length of the constitutive artificial lines must be set to certain values at the operating frequencies, is a clear application (among many others) where impedance and dispersion engineering apply. On the other hand, as anticipated in the introductory section of Chapter 3, the functionality of transmission lines with metamaterial loading is mainly based on particle resonance, rather than on dispersion and impedance engineering. Examples of applications of transmission lines with metamaterial loading are the multiband printed dipole and monopole antennas based on the concept of trap antennas, where

Artificial Transmission Lines for RF and Microwave Applications, First Edition. Ferran Martín.

the multiband functionality is achieved by loading the dipole, or the monopole, with SRRs or with other electrically small resonators.

The RF/microwave applications considered in this chapter are grouped into two categories: applications of CRLH transmission lines, mainly, although not exclusively, based on impedance and dispersion engineering, and applications of transmission lines with metamaterial loading. The reported applications of this chapter are necessarily influenced by the own experience of the author. Thus, most of these applications have been proposed by the author's Group, although there are many others proposed by other active researchers in the field. Nevertheless, it is impossible to cover all the applications of metamaterial transmission lines reported in the literature in a single chapter. Hence, many applications not considered here are quoted in the reference list. The author also recommends several reference books on the topic [1–6], as complementary and excellent lectures to the present book.

4.2 APPLICATIONS OF CRLH TRANSMISSION LINES

Dispersion and impedance engineering is a design process that consists of tailoring the dispersion and characteristic impedance of metamaterial transmission lines in order to achieve certain functionalities, which are normally not achievable by means of conventional lines. Dispersion and impedance engineering can be applied to the design of enhanced bandwidth components, multiband components, filters and diplexers, leaky wave antennas (LWAs), couplers, distributed amplifiers and mixers, and sensors, among others. Let us now study the principles behind these applications or the advantages of using CRLH lines and some illustrative results.

4.2.1 Enhanced Bandwidth Components

Metamaterial transmission lines can be applied to the design of microwave components with a broad operation bandwidth as compared to those bandwidths achievable in devices implemented by means of ordinary transmission lines and stubs. We will start this subsection by studying the principles behind bandwidth enhancement and by discussing their fundamental limitations, and then several enhanced bandwidth components will be reported as illustrative examples.

4.2.1.1 Principle and Limitations The operative bandwidth of microwave components is given by that frequency interval where the required characteristics are satisfied within certain limits. In distributed circuits, bandwidth is limited by the phase shift experienced by transmission lines and stubs when frequency is varied from the nominal operating value. In a conventional transmission line of length l, the electrical length, or phase shift,[1] of the line at a certain angular frequency ω_o is given by

[1] In this discussion, by phase shift we refer to the electrical length of the line, and hence it has the same sign as the phase constant β. The phase shift is thus positive for ordinary lines, and it is negative/positive in the left/right handed transmission bands of CRLH metamaterial transmission lines.

$$\phi_o = \beta l = \frac{l}{v_p}\omega_o \tag{4.1}$$

where v_p is the phase velocity of the line. According to the previous comment, bandwidth is intimately related to the derivative of ϕ with frequency, also known as group delay. From this, it follows that the shorter the line is, the broader the bandwidth becomes. In other words, bandwidth is inversely proportional to the required phase of the line, which is dictated by the design. This means that the operative bandwidth is not an easily controllable parameter in conventional distributed circuits.[2] The reason is the limited number of degrees of freedom of conventional transmission lines. However, in metamaterial transmission lines, the loading elements provide additional parameters, and certain control over the phase response is possible. Namely, one expects that the required phase does not dictate the bandwidth. The key question is: Is it possible to improve the bandwidth (or decrease the group delay) of conventional lines by means of metamaterial transmission lines regardless of the required phase shift? Let us discuss now this aspect (a detailed analysis is given in Ref. [7]).

Assuming that for a certain transmission line the required phase and characteristic impedance at the operating frequency are ϕ_o and Z_o (where Z_o is considered to be a real number), expressions (2.30) and (2.33) can be inverted, and the series and shunt impedances must take the following values at the design frequency (a T-circuit model—single stage—has been assumed for the artificial transmission line under consideration):[3]

$$Z_s = Z_o\sqrt{\frac{\cos\phi_o - 1}{\cos\phi_o + 1}} \tag{4.2}$$

$$Z_p = \frac{Z_o}{\sqrt{\cos^2\phi_o - 1}} \tag{4.3}$$

The derivative of the phase shift with frequency is given by

$$\left.\frac{d\phi}{d\omega}\right|_{\omega_o} = \frac{1}{\sin\phi}\frac{1}{\chi_p^2}\left(-\chi_p\left.\frac{d\chi_s}{d\omega}\right|_{\omega_o} + \chi_s\left.\frac{d\chi_p}{d\omega}\right|_{\omega_o}\right) \tag{4.4}$$

where χ_s and χ_p are the reactances of the series and shunt branch of the T-circuit, respectively ($Z_s = j\chi_s$ and $Z_p = j\chi_p$). By introducing (4.2) and (4.3) in (4.4), and after

[2] Nevertheless, bandwidth in distributed circuits can be enhanced by increasing the number of elements. An example is the multisection branch line coupler, which can be found in many general textbooks devoted to microwave engineering.

[3] To obtain (4.2) and (4.3), the impedance of the series branch of the T-model (basic cell) has been considered to be Z_s, rather than $Z_s/2$.

some simple calculation, the derivative of the phase at the operating frequency is found to be:

$$\left.\frac{d\phi}{d\omega}\right|_{\omega_o}=\frac{1}{Z_o}\left(\left.\frac{d\chi_s}{d\omega}\right|_{\omega_o}-(\cos\phi_o-1)\left.\frac{d\chi_p}{d\omega}\right|_{\omega_o}\right) \tag{4.5}$$

The two right-hand side terms in (4.5) are effectively positive since according to the Foster reactance theorem [8], the derivative of the reactance (or susceptance) of a passive and lossless network with frequency is always positive, and $\cos\phi_o \leq 1$. To enhance bandwidth, we must force the derivative of ϕ with frequency to be as small as possible. Thus, according to (4.5), the frequency derivatives of the series and shunt reactances must be both as small as possible. From the Foster theorem, it follows that the optimum metamaterial transmission line model to minimize expression (4.5) is the canonical dual transmission line (see the appendix of Ref. [7]). Hence, it is convenient to calculate expression (4.5) for this particular case. By introducing

$$\chi_s=-\frac{1}{2C_L\omega} \tag{4.6}$$

$$\chi_p=L_L\omega \tag{4.7}$$

in (4.5), one obtains, after some straightforward calculation:

$$\left.\frac{d\phi}{d\omega}\right|_{\omega_o}=\frac{2}{\omega_o}\sqrt{\frac{1-\cos\phi_o}{1+\cos\phi_o}}=\frac{2}{\omega_o}\tan\frac{|\phi_o|}{2} \tag{4.8}$$

Thus, for the optimum transmission line model (in terms of the operative bandwidth) the derivative of phase with frequency does not depend on the required characteristic impedance. It is simply determined by the operating frequency and phase.

It is important to take into account that expression (4.8) is the derivative of the phase with frequency corresponding to a single-unit cell. If a number N of cells is used to obtain the nominal phase shift, ϕ_o, then the phase shift of either cell is ϕ_o/N, and expression (4.8) rewrites as follows:

$$\left.\frac{d\phi}{d\omega}\right|_{\omega_o}=\frac{2}{\omega_o}N\tan\frac{|\phi_o|}{2N} \tag{4.9}$$

A simple analysis of expression (4.9) reveals that $d\phi/d\omega$ decreases as N increases, namely,

$$\frac{2}{\omega_o}N\tan\frac{|\phi_o|}{2N}>\frac{2}{\omega_o}(N+1)\tan\frac{|\phi_o|}{2(N+1)} \tag{4.10}$$

for $N = 1, 2\ldots, \infty$. Therefore, in terms of bandwidth, the optimum solution is an N-stage artificial transmission line with $N \to \infty$. In this case, (4.9) can be expressed as follows:

$$\left.\frac{d\phi}{d\omega}\right|_{\omega_o} = \frac{|\phi_o|}{\omega_o} \tag{4.11}$$

According to (4.11), the derivative of phase with frequency for a dual transmission line in the considered limiting case ($N \to \infty$) is identical to that of a conventional line with identical phase (but different sign) at the same frequency. Therefore, from this result we may conclude that in those applications where the sign of the phase shift is irrelevant (e.g., in 90° impedance inverters and many microwave components based on them [9]), it is not possible to enhance the bandwidth by using artificial lines.[4] If the number of cells is limited to a finite number, the derivative of phase with frequency increases, and the operative bandwidth is degraded, as compared to that of a conventional transmission line.

Bandwidth improvement can be obtained if the sign of the phase is relevant. In this case, we have to compare the dual transmission line with phase ϕ_o (ϕ_o being negative) with a conventional transmission line with equivalent phase shift, that is, $2\pi + \phi_o$. From this comparison, it is obvious (from 4.11) that as long as the required phase shift is higher than π (i.e., $\pi < 2\pi + \phi_o < 2\pi$), the dual transmission line exhibits smaller phase dependence with frequency, and hence it exhibits a better solution in terms of bandwidth. If the number of stages is limited to a finite number, then the limiting phase above which the artificial lines exhibit a better phase response (lower derivative) is no longer π. For a single-stage dual transmission line, such limiting phase can be inferred by simply forcing (4.8) to be $(2\pi + \phi_o)/\omega_o$. It gives $\phi_o = -0.7\pi$, or a positive phase (conventional line of 1.3π). It means that in applications where the sign of the phase shift is fundamental, by using a single-stage dual transmission line bandwidth can be improved if the required (positive) phase shift is comprised between 1.3π and 2π. Let us consider, for instance, a required phase shift of 1.5π, equivalent to $-\pi/2$, to be achieved with an N-stage dual transmission line at the angular frequency ω_o. If N is high enough, the derivative of the phase shift with frequency can be approximated by (4.11),

$$\left.\frac{d\phi}{d\omega}\right|_{\omega_o} = \frac{\pi}{2\omega_o} \tag{4.12}$$

whereas the derivative of the phase with frequency for the ordinary line is

$$\left.\frac{d\phi}{d\omega}\right|_{\omega_o} = \frac{3\pi}{2\omega_o} \tag{4.13}$$

[4] However, device size can be substantially reduced by virtue of the small electrical size of the unit cell of LH or CRLH lines, as can be seen in the power dividers reported in Ref. [9].

Therefore, it is clear that the bandwidth is superior by using a dual transmission line. It should be noted that if the number of stages, N, of the dual transmission line is high, the long wavelength approximation can be applied, and the characteristic impedance is roughly constant with frequency (i.e., 3.38 can be approximated by 3.42). However, if the number of stages is small, or if only one stage is considered, the characteristic impedance will no longer be constant with frequency in the vicinity of the operating frequency. Under these conditions, the functionality of the device may also be limited by the variation of the characteristic impedance.

As highlighted earlier, the optimum metamaterial transmission line for bandwidth enhancement is the dual transmission line. Thus, the phase shifts (π for the infinite stage structure and 1.3π for the one-stage transmission line) below which the dual transmission line is not able to provide an improved bandwidth are fundamental limits. Namely, with CRLH transmission lines (either CL-loaded or resonant type) this limits are even lower. However, despite this, bandwidth enhancement is possible since many microwave components are based on phase differences between transmission lines. To enhance bandwidth, it is necessary that the dispersion characteristics exhibit similar slopes in the vicinity of the operating frequency, and this can be achieved through the considered artificial CRLH lines, as will be shown in the next subsection (see Fig. 4.1). On the other hand, in certain applications, transmission lines with electrical lengths higher than π are required. Under these conditions, the ordinary lines can be replaced with metamaterial transmission lines, with the corresponding bandwidth improvement, as anticipated above and as it will be shown at the beginning of the next subsection.

4.2.1.2 *Illustrative Examples*

Broadband Series Power Divider The first considered example is a series power divider. Such devices can be of interest to feed an array of antennas, where in-phase signals at either radiating element may be required. Therefore, series power dividers

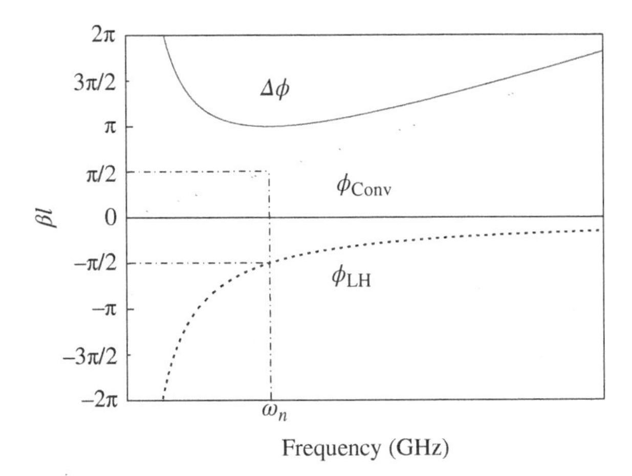

FIGURE 4.1 Illustration of bandwidth enhancement by using a conventional and an LH transmission line. At the design frequency, ω_n, the phase difference, $\Delta\phi$, between the two lines is π, and this difference is roughly preserved in a wide band.

are implemented by means of transmission lines with in-phase input and output signals. If ordinary lines are considered, the required phase shift is 2π at the operating frequency.[5] Alternatively, a balanced CRLH line operating at the transition frequency can be designed to actually achieve zero-phase shift at that frequency. The group delay for the conventional line is $\tau_C = 2\pi/\omega_o$. Therefore, to enhance bandwidth by means of the CRLH line, it is necessary that the group delay at the transition frequency satisfies $\tau_{CRLH} < \tau_C$. In reference to CL-loaded balanced CRLH lines, the phase shift in the vicinity of the transition frequency can be inferred by forcing $\omega_s = \omega_p = \omega_o$ in Equation 3.59. It gives

$$\phi = \beta l = \frac{N\omega}{\omega_R}\left(1 - \frac{\omega_o^2}{\omega^2}\right) \tag{4.14}$$

where we have added N to account for the number of stages (not necessarily one). The group delay at ω_o is simply:

$$\tau_{CRLH} = \left.\frac{d\phi}{d\omega}\right|_{\omega_o} = \frac{2N}{\omega_R} \tag{4.15}$$

By forcing (4.15) to be smaller than τ_C, the following inequality results:

$$\frac{N}{\omega_R} < \frac{\pi}{\omega_o} \tag{4.16}$$

which leads to

$$N\sqrt{\frac{L_R}{C_L}} < \pi Z_o \tag{4.17}$$

with

$$Z_o = \sqrt{\frac{L_R}{C_R}} = \sqrt{\frac{L_L}{C_L}} \tag{4.18}$$

From (4.17), it follows that to obtain a significant bandwidth enhancement, L_R must be small and C_L must be high. This leads also to a small value of C_R and a high value of L_L since (4.18) must be satisfied. Consequently, ω_R must be high and ω_L must be small (see expressions (3.53) and (3.54)).

As an alternative to CRLH lines, enhanced bandwidth series power dividers can be implemented by replacing the 2π ordinary lines with artificial lines consisting of purely left-handed (PLH) line sections alternating with conventional line sections providing exactly the same phase shift but with opposite sign (i.e., a 0° phase shifting

[5] Transmission lines with a phase of $2\pi n$ (n being a positive integer) can also be considered, but bandwidth is optimized by considering $n = 1$ for the reasons explained in Section 4.2.1.1.

line). Let us consider that the 0° line is implemented by combining a conventional line with phase shift $\phi_{CONV} = \phi_C$ at the design frequency, with a PLH line with $\phi_{LH} = -\phi_C$ at that frequency (so that $\phi = \phi_{CONV} + \phi_{LH} = 0$). The derivative of the line phase with frequency at ω_o is

$$\left.\frac{d\phi}{d\omega}\right|_{\omega_o} = \frac{\phi_C}{\omega_o} + \frac{2}{\omega_o}\tan\frac{\phi_C}{2} \tag{4.19}$$

From (4.19), it follows that to maximize the bandwidth it is convenient to choose ϕ_C as small as possible. From this, one concludes that the conventional line must be as short as possible, and C_L and L_L (see expression 3.37) as high as possible. Notice that these conditions are equivalent to those inferred in the previous paragraph in reference to CRLH lines. In practice, ϕ_C is limited since excessive large values for C_L and L_L should be avoided, and a minimum size for the conventional line sections is required.

Following the previous idea, bandwidth improvement was experimentally demonstrated in a 1:4 series power divider operative at 1.92 GHz, where 0° phase shifting lines were used [10]. In-phase signals at the input of either divider branch were achieved through the designed 0° metamaterial transmission lines, placed between the different output branches of the divider. As compared to the conventional device, the metamaterial-based power divider exhibits an improved bandwidth. The conventional and the metamaterial-based series power dividers are compared in Figure 4.2. It is worth mentioning that a significant area reduction is achieved in the divider based on metalines, since these lines are substantially shorter than the conventional 2π transmission lines. This example demonstrates that in applications where line phases larger than π are required, bandwidth can be enhanced by replacing these lines with CRLH lines.[6]

Broadband Rat-Race Hybrid Coupler The second illustrative example is a broadband rat-race hybrid coupler where bandwidth enhancement is obtained by engineering the dispersion of the constitutive transmission lines, with an eye toward achieving similar phase slopes at the design frequencies, following the principle illustrated in Figure 4.1. By using this concept, a rat-race hybrid coupler where the phase balance for the Δ and Σ ports exhibits broader bandwidth, as compared to the conventional counterpart, was demonstrated by Okabe *et al.* [11]. The topology of the conventional rat-race hybrid coupler is depicted in Figure 4.3a. It is essentially a four-port device consisting of a 1.5λ ring structure (where λ is the guided wavelength at the design frequency, ω_o) with the ports equally spaced in the upper half of the ring. In the coupler reported in Ref. [11], the +270° (0.75λ) line section was replaced with a –90° LH line designed to provide the required characteristic impedance (70.7 Ω) and phase (–90°) at the operating frequency. The LH line was implemented by loading a host line with lumped inductors and capacitors in shunt and series connection, respectively. The performance and size of the device is good, with a wide bandwidth for the coupling coefficient and phase balance (see Ref. [11] for more details).

[6] Since the combination of a PLH line and an ordinary line provides a backward and a forward transmission band, the structure can be considered to belong to the category of CRLH lines.

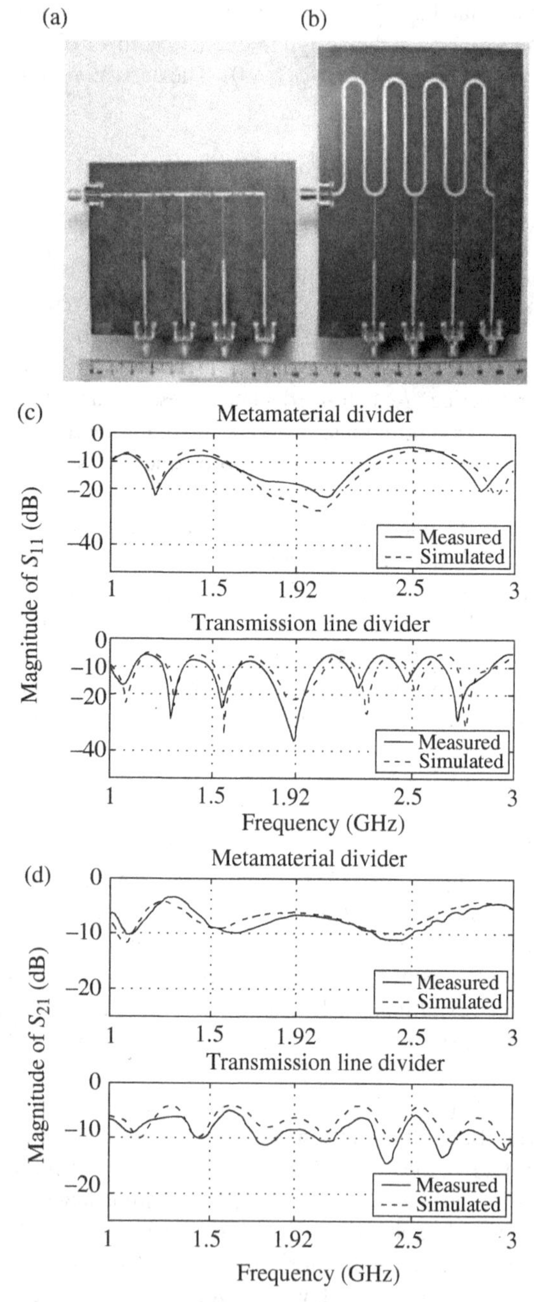

FIGURE 4.2 Metamaterial-based (a) and conventional (b) 1:4 series power dividers with their corresponding measured and simulated frequency responses (only S_{11} (c) and S_{21} (d) are depicted). Reprinted with permission from Ref. [10]; copyright 2005 IEEE.

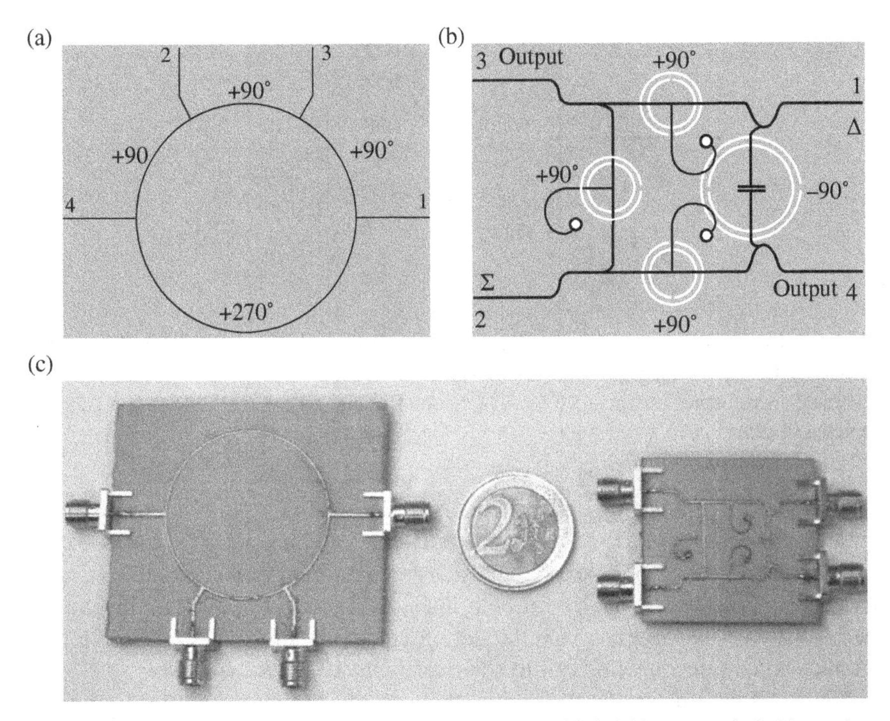

FIGURE 4.3 Layout of the conventional (a) and fully artificial (b) rat-race hybrid couplers. A comparative photograph of both devices can be seen in (c). The devices were fabricated on the *Rogers RO3010* substrate with dielectric constant $\varepsilon_r = 10.2$ and thickness $h = 635\ \mu m$. The active area (excluding access lines) of the CSRR-based hybrid coupler is $3.62\ cm^2$, whereas the conventional one occupies an area of $10.33\ cm^2$. Reprinted with permission from Ref. [12], copyright 2007 IEEE.

Alternatively, the metamaterial rat-race hybrid coupler can be implemented by means of the resonant type approach [12]. In this case, not only the 270° line is replaced with an artificial –90° LH line but also the three +90° (right handed – RH) transmission lines are implemented as artificial lines based on CSRRs (the characteristic impedance of these artificial lines is set to 70.7 Ω). Such artificial RH lines are implemented by combining CSRRs with shunt stubs, as Figure 4.4 illustrates. As discussed in Ref. [13], by means of this fully artificial structure (all the lines are artificial) a further controllability on the frequency dependence of the phase of the lines can be achieved, as compared to the structure with only one artificial line (i.e., the LH line providing –90° phase shift, in substitution of the +270° conventional line). Indeed, it is not possible to achieve similar phase slopes between the conventional +90° lines and the –90° LH line in the vicinity of the design frequencies [13] for the structure with only one artificial line. The +90° and –90° CSRR-based lines were optimized in order to exhibit similar phase slopes at the design frequency (notice that expressions 4.4 and 4.5 are valid for the structure of Fig. 4.4). Nevertheless, it should be mentioned that the characteristic impedance of these lines is frequency dependent,

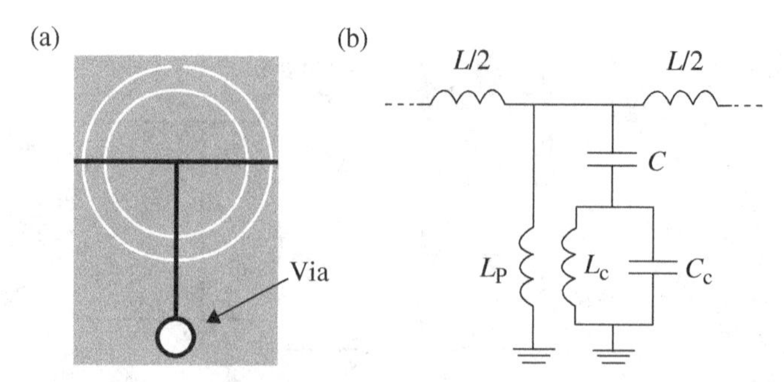

FIGURE 4.4 Typical topology (a) and circuit model (b) of the CSRR-based RH line unit cell. L_p models the shunt stub, grounded by means of a via, whereas the other circuit parameters were described in reference to Figure 3.38. The ground plane, where the CSRRs are etched, is depicted in gray.

and therefore its frequency derivative must be minimized as much as possible for bandwidth optimization. The design equations for both lines are (4.2) and (4.3), subjected to equal phase slopes (expression 4.5) and minimum variation of the characteristic impedance with frequency for both lines. In practice, however, optimization at the layout level is required since the design equations are complex, involve many parameters, and they are difficult to solve analytically.

Figure 4.5 depicts the phase difference between the designed artificial lines, where it can be appreciated that it experiences less variation with frequency as compared with the phase difference of conventional lines or with the phase difference resulting by using a conventional line and a LH line (the design frequency is 1.6 GHz). The layout of the fully artificial rat-race hybrid coupler is also depicted in Figure 4.3 (also in this figure are included the photographs of the conventional and the metamaterial rat-race hybrid couplers, in order to appreciate the achievable size reduction by using metamaterial transmission lines). The simulated and measured impedance matching, coupling and isolation for both structures are depicted in Figure 4.6. In Figure 4.7, the phase balance for the Σ and Δ ports, namely $\phi(S_{42}) - \phi(S_{32})$ and $\phi(S_{41}) - \phi(S_{31})$, respectively, are depicted. In terms of flatness, the phase balance of the fully artificial rat-race exhibits superior performance. Power splitting for both the Δ and the Σ ports exhibits similar characteristics to those measured in the conventional implementation. A comparison between the fully artificial and the conventional rat-race hybrid coupler in terms of bandwidth is represented in Table 4.1. Isolation is comparable in both couplers, although matching is better in the conventional implementation. However, as has been mentioned, phase balance is substantially improved by using metamaterial transmission lines.

The design of the fully artificial rat-race is compatible with planar technology since no lumped elements are used, and size reduction by a factor of 3 is achieved as compared to the conventional implementation. There are other means to achieve size reduction in rat-race hybrid couplers. For instance, small-size devices have been achieved by using folded lines [14], periodically loaded lines [15], synthetic meander lines [16], and stepped impedance resonators [17]. However, in general, these

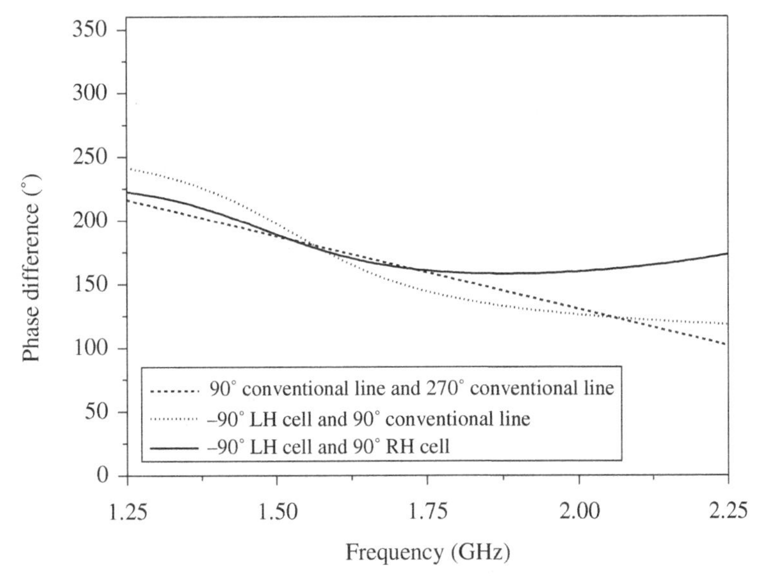

FIGURE 4.5 Representation of the phase difference $\Delta\phi_{S21}$ between the conventional +90° and +270° lines (dashed line), the conventional +90° and LH –90° lines (dotted line) and the artificial RH +90° and LH –90° lines (solid line). The phase difference is roughly 180° at the design frequency (1.6 GHz). Reprinted with permission from Ref. [13], copyright 2008 John Wiley.

techniques do not provide a comparable performance, in terms of bandwidth, as compared to the devices based on metamaterial transmission lines [11, 12]. An exception to this is the rat-race hybrid presented in Ref. [18], where the 270° branch is replaced by a –90° branch realized by a +90° microstrip line and a CPW phase inverter. However the design requires a CPW-to-microstrip transition and fabrication is more complex.

The technique for bandwidth enhancement based on metamaterial transmission lines has been applied to other microwave components, such as phase shifters [19, 20], Wilkinson baluns [21], and T-junction power splitters with balanced [22] or quadrature phase [23] outputs over a broad band, among others.

4.2.2 Dual-Band and Multiband Components

Dual-band components exhibit certain functionality at two arbitrary frequencies (the concept can be extended to tri-band, quad-band and, in general, to n-band components by satisfying the required functionality at three, four, or n controllable frequencies, respectively).[7] Conventional distributed microwave components typically exhibit a

[7] These microwave components exhibiting multiple band functionality at controllable frequencies (i.e., those dictated by design requirements) are designated as multiband devices. By contrast, devices operative at a single frequency are called single-band or mono-band components.

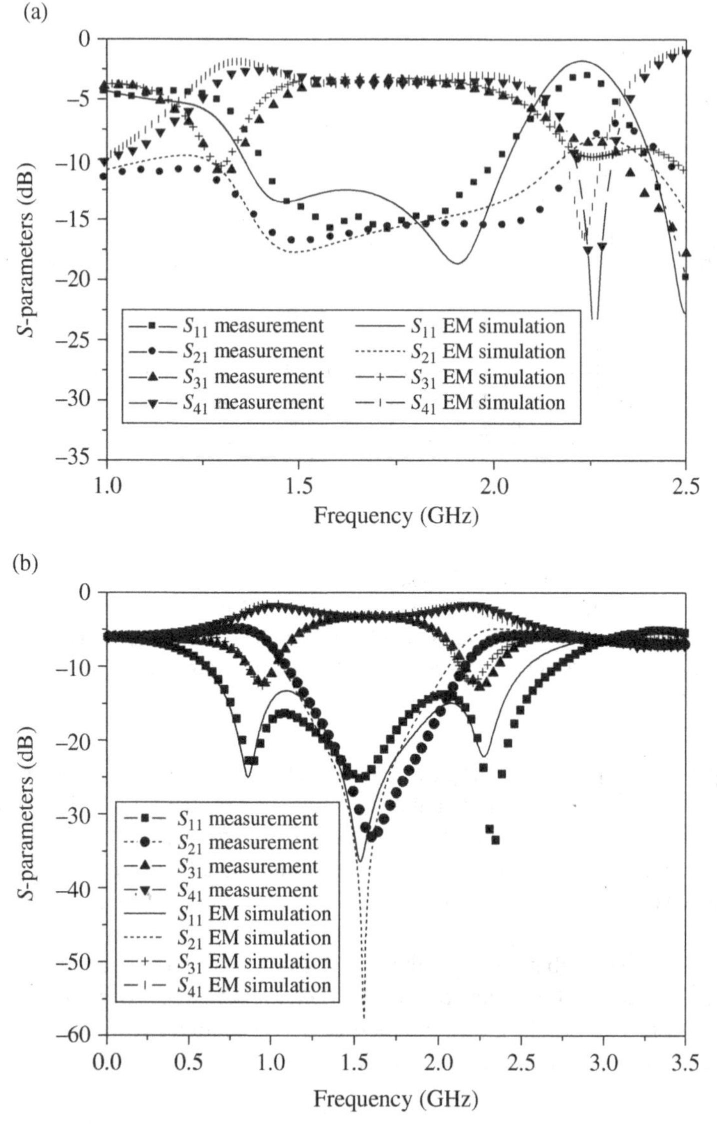

FIGURE 4.6 Impedance matching (S_{11}), coupling, (S_{31}, S_{41}) and isolation (S_{21}) for the CSRR-based hybrid coupler (a) and conventional coupler (b). Reprinted with permission from Ref. [12]; copyright 2007 IEEE.

periodic response and hence exhibit the required functionality at the design frequency and at its odd harmonics. However, such devices cannot be considered multiband components since the operating frequencies cannot be set to those values corresponding to system requirements. They are thus single-band components, that is, their functionality can only be satisfied at a single-design frequency.

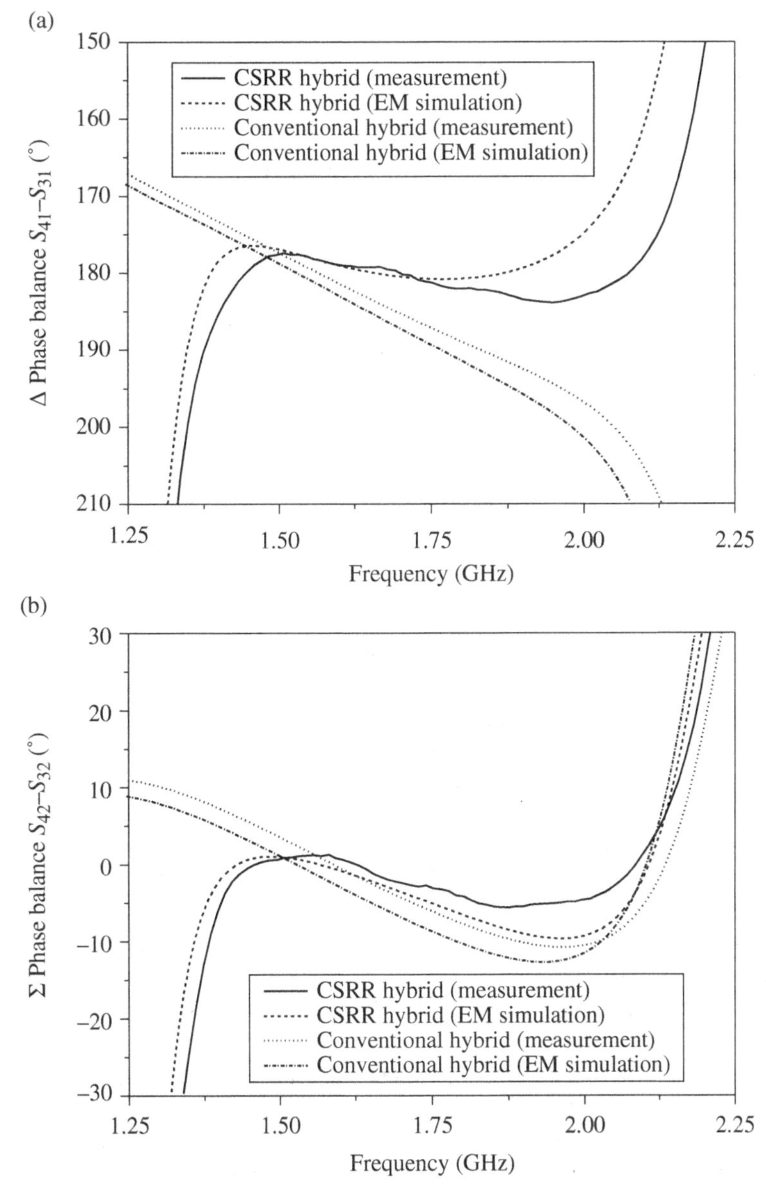

FIGURE 4.7 Phase balance for the Δ (a) and Σ (b) ports. Reprinted with permission from Ref. [12], copyright 2007 IEEE.

4.2.2.1 Principle for Dual-Band and Multiband Operation The larger number of free parameters of CRLH transmission lines (as compared to conventional lines) is useful for the design of dual-band and multiband components through dispersion and impedance engineering. To achieve dual-band operation, it is necessary to satisfy

TABLE 4.1 Bandwidth characteristics in the conventional and artificial rat-race couplers

	Conventional		Fully artificial	
	Σ Input	Δ Input	Σ Input	Δ Input
Output (dB)	**3.10 ± 0.25**	**3.10 ± 0.25**	**3.5 ± 0.25**	**3.38 ± 0.25**
Range (GHz)	1.38–1.70	1.40–1.72	1.55–1.89	1.54–1.89
Bandwidth (%)	20	20	20	20
Phase balance (degrees)	**0 ± 5**	**180 ± 5**	**0 ± 5**	**180 ± 5**
Range (GHz)	1.40–1.65	1.40–1.65	1.40–2.10	1.40–2.10
Bandwidth (%)	16	16	40	40
Isolation (dB)	**< −15**		**< −15**	
Range (GHz)	1.20–1.90		1.35–2.10	
Bandwidth (%)	45		44	
Return loss (dB)	**< −10**		**< −10**	
Range (GHz)	1.13–1.99	0.59–2.50	1.41–2.26	1.40–2.01
Bandwidth (%)	55	123	46	36

Bold indicates the main characteristics of the rat-race couplers.

the phase and impedance requirements (for the transmission lines and stubs of the circuit) at the two design frequencies, f_1 and f_2. Let us consider that the required phases are ϕ_1 and ϕ_2 at the design frequencies. As can be seen in Figure 4.8, in general, such phase values cannot simultaneously be satisfied by means of ordinary lines. However, by using artificial lines, we can tailor the dispersion diagram to set the phases to the required values at the two operating frequencies (i.e., Fig. 4.8 illustrates). One important aspect is that, contrary to conventional lines, metamaterial transmission lines exhibit frequency-dependent characteristic impedance. Thus, such artificial lines must be designed to satisfy also the impedance requirements at the design frequencies. For multiband functionality, the idea is to increase the number of elements of the artificial lines in order to enhance the degrees of freedom, and thus be able to satisfy the phase and impedance requirements at more than two frequencies, as will be shown later.

4.2.2.2 Main Approaches for Dual-Band Device Design and Illustrative Examples Practical dual-band components have been implemented either by using artificial lines that combine ordinary (RH) lines with PLH lines[8] made of SMT components, or by means of fully planar CRLH transmission lines.

Dual-Band Devices Based on a Combination of Ordinary and PLH Lines: Application to a Dual-Band Branch Line Hybrid Coupler For the implementation of dual-band metamaterial transmission lines, one approach is to cascade multiple

[8] Although PLH lines do not exist in practice, as discussed in Chapter 3, we intentionally use this terminology to emphasize the fact that the considered SMT-based LH lines roughly behave as PLH lines in the frequency region of interest. This means that the parameters of the host line can be neglected in such region.

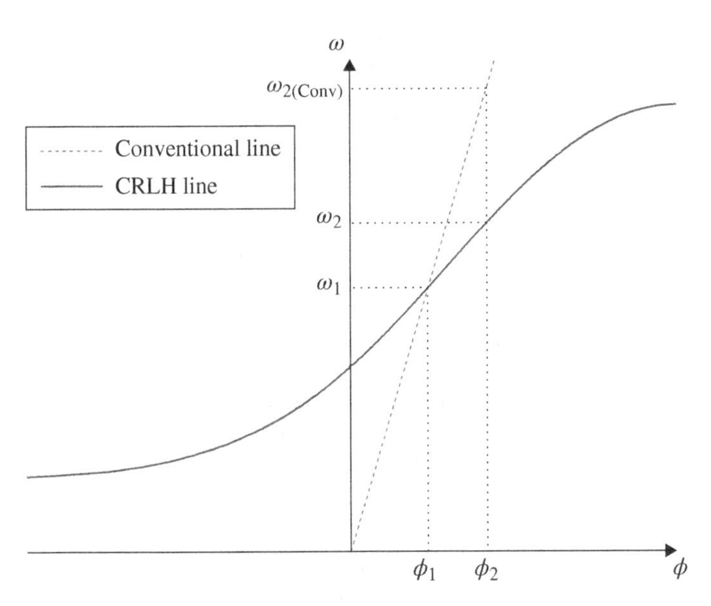

FIGURE 4.8 Illustration of the single-band operation of conventional lines, as compared to the possibility of designing dual-band metamaterial transmission lines, with the required phase at arbitrary frequencies.

(N) effectively homogeneous unit cells. For instance, a balanced line in the vicinity of the transition frequency exhibits roughly constant (i.e., frequency independent) characteristic impedance. Thus, by setting this impedance to the required value and adjusting the slope of the dispersion diagram so that the phases of the unit cells are ϕ_1/N and ϕ_2/N at the design frequencies, dual-band operation can be achieved. Practical dual-band components have indeed been implemented by cascading conventional lines (with positive phase shift) and backward (i.e., PLH) lines implemented by means of lumped elements [2]. As long as the phase shift per unit cell in the backward lines is small, the phase can be approximated by

$$\phi_{\text{CRLH}} = \phi_{\text{R}} - \frac{N}{\omega\sqrt{L_{\text{L}}C_{\text{L}}}} \tag{4.20}$$

where the subindex *CRLH* denotes the phase of the composite structure, N is the number of unit cells of the backward wave transmission line, and $\phi_{\text{R}} = \beta_{\text{R}} l = (\omega/v_{\text{p}})l$, where v_{p} is the phase velocity of the conventional line and l its length. By forcing the phase of the composite structure to be $\phi_{\text{CRLH}} = \phi_1$ at $\omega_1 = 2\pi f_1$ and $\phi_{\text{CRLH}} = \phi_2$ at $\omega_2 = 2\pi f_2$ (with $\omega_2 > \omega_1$), and the characteristic impedance of both lines to Z_{o}, the length of the line and the reactive elements of the PLH line are found to be[9]

[9] Since the PLH line is considered to satisfy the homogeneity condition, the characteristic impedance of this line is simply given by $Z_{\text{o}} = (L_{\text{L}}/C_{\text{L}})^{1/2}$.

$$l = \frac{v_{\mathrm{p}}(\phi_2\omega_2 - \phi_1\omega_1)}{\omega_2^2 - \omega_1^2} \tag{4.21}$$

$$C_{\mathrm{L}} = \frac{N}{Z_{\mathrm{o}}} \frac{\omega_2^2 - \omega_1^2}{\omega_1\omega_2(\phi_2\omega_1 - \phi_1\omega_2)} \tag{4.22}$$

$$L_{\mathrm{L}} = Z_{\mathrm{o}} N \frac{\omega_2^2 - \omega_1^2}{\omega_1\omega_2(\phi_2\omega_1 - \phi_1\omega_2)} \tag{4.23}$$

and the transverse dimensions of the conventional line must be adjusted to provide the required characteristic impedance, Z_{o}. Notice that, depending on the required phases, the conventional line length and/or the elements of the PLH line, L_{L} and C_{L}, may be negative. For instance, if $\phi_1 = \phi_2 > 0$, l is positive, but L_{L} and C_{L} take negative values. Although this negative element values cannot be implemented (unless non-Foster active components are used, as discussed in the last chapter of this book), from a mathematical viewpoint the negative phase of the PLH line reverses its sign and for this reason a mathematical solution exists for negative values of L_{L} and C_{L}. In order to have positive values of L_{L} and C_{L}, the following condition must be satisfied:

$$\phi_2\omega_1 - \phi_1\omega_2 > 0 \tag{4.24}$$

Notice that this condition forces $\phi_2 > \phi_1$, and hence it leads also to a positive value of l.

If both phases are positive, it is necessary that $\phi_2/\phi_1 > \omega_2/\omega_1$ in order to satisfy (4.24). Thus, the relative values of the operating frequencies dictate the minimum ratio between the (positive) phases at these frequencies. Let us consider, for instance, that $\phi_1 = +90°$ and $\phi_2 = +270°$. This is an interesting case because such phases are equivalent (hence providing the same functionality) for many applications (i.e., the implementation of impedance inverters). According to (4.24), the ratio of the operating frequencies is limited to 3. If $\phi_2 > 0$ and $\phi_1 < 0$, expression (4.24) is always satisfied, and a solution always exists regardless of the values of ω_1 and ω_2 (with $\omega_2 > \omega_1$). Let us consider, for instance, that $\phi_2 = -\phi_1 = 90°$. Introducing these phase values in (4.21)–(4.23), the following results are obtained:

$$l = \frac{\pi v_{\mathrm{p}}}{2(\omega_2 - \omega_1)} \tag{4.25}$$

$$C_{\mathrm{L}} = \frac{2N}{\pi Z_{\mathrm{o}}} \frac{(\omega_2 - \omega_1)}{\omega_1\omega_2} \tag{4.26}$$

$$L_{\mathrm{L}} = \frac{2NZ_{\mathrm{o}}}{\pi} \frac{(\omega_2 - \omega_1)}{\omega_1\omega_2} \tag{4.27}$$

Thus, expressions (4.25)–(4.27) are simple design equations for the implementation of dual-band impedance inverters exploiting the LH and the RH band of the CRLH artificial transmission line.

Following the previous strategy, several dual-band components have been implemented. For instance, a dual-band Wilkinson power divider is reported in Ref. [2], where the phases of the two impedance inverters of the divider are set to $\phi_1 = -90°$ and $\phi_2 = +90°$ at the operating frequencies $f_1 = 1$ GHz and $f_2 = 3.1$ GHz, respectively (Eqs. 4.25–4.27 apply in this case). As an illustrative result of the dual-band approach based on the combination of ordinary and PLH line sections, a dual-band branch line hybrid coupler is reported here [24] (the lumped elements are implemented by means of SMT chip components). The design frequencies are $f_1 = 0.93$ GHz and $f_2 = 1.78$ GHz, and the electrical lengths of the four coupler branches are set to $\phi_1 = +90°$ and $\phi_2 = +270°$ at those frequencies. The design equations are thus given by (4.21)–(4.23), with the required values of the phases. The fabricated prototype device is depicted in Figure 4.9, where matching, coupling, and isolation are also depicted. The characterization results reveal that the dual-band functionality is satisfied at the desired frequencies, with near −3 dB transmission between the input port and the coupled and through ports. Other devices based on this approach are reported in Ref. [25–27], including an active device (i.e., a dual-band Class-E power amplifier [27]).

Dual-Band Devices based on Fully Planar CRLH Lines: Application to Power Dividers and Branch Line Hybrid Couplers Although the size of the dual-band components considered in the previous subsection is moderately small, the use of SMT components may represent a severe limitation in certain applications, and inaccuracies may appear due to the limited number of available reactive element values. The use of lumped elements can be avoided by implementing the CRLH lines with semilumped (i.e., fully planar) components. Typically, the size of semilumped elements is larger than the size of lumped components. For this reason, dual-band devices based on semilumped CRLH lines are typically implemented by means of a single unit cell.

Let us consider the implementation of dual-band ±90° transmission lines by means of fully planar CRLH lines based on the canonical T-circuit model of Figure 3.20. The line must be designed so that the first and second operating frequencies lie in the LH and RH bands, respectively. According to this, the electrical length and characteristic impedance must be set to $\phi = \phi_1 = -90°$ and $Z_o = Z_1$ at the lower frequency, and $\phi = \phi_2 = +90°$ and $Z_o = Z_2$ at the upper frequency (typically $Z_1 = Z_2$, but at the moment this condition is relaxed). By forcing the above phases and characteristic impedances at the design frequencies, the following conditions result:

$$\chi_s(\omega_1) = -Z_1 \quad (4.28a)$$

$$\chi_p(\omega_1) = +Z_1 \quad (4.28b)$$

$$\chi_s(\omega_2) = +Z_2 \quad (4.28c)$$

$$\chi_p(\omega_2) = -Z_2 \quad (4.28d)$$

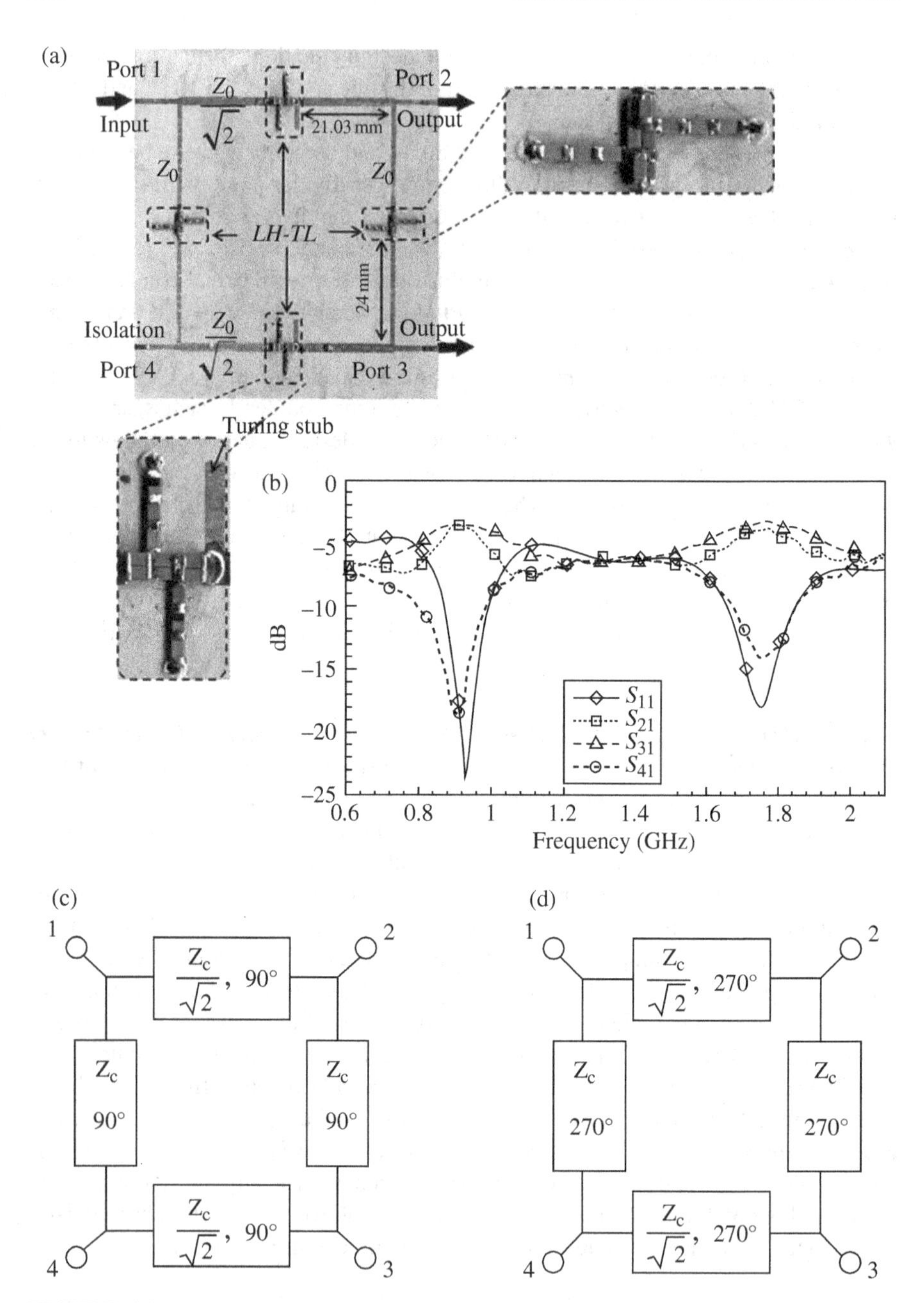

FIGURE 4.9 Dual-band branch line coupler (a), measured frequency response (b), and schematics of the coupler for the two operating frequencies f_1 (c) and f_2 (d). Reprinted with permission from Ref. [24]; copyright 2004 IEEE.

as it is derived from (2.30) and (2.33), χ_s and χ_p being the series and shunt reactance, respectively, of the T-circuit model of the unit cell. Since the circuit model has four reactive elements, they are univocally determined from Equations 4.28, that is,

$$L_R = \frac{2(Z_1\omega_1 + Z_2\omega_2)}{\omega_2^2 - \omega_1^2} \tag{4.29a}$$

$$C_L = \frac{\omega_2^2 - \omega_1^2}{2\omega_1\omega_2(Z_1\omega_2 + Z_2\omega_1)} \tag{4.29b}$$

$$L_L = \frac{(\omega_2^2 - \omega_1^2)Z_1Z_2}{\omega_1\omega_2(Z_1\omega_1 + Z_2\omega_2)} \tag{4.29c}$$

$$C_R = \frac{Z_1\omega_2 + Z_2\omega_1}{(\omega_2^2 - \omega_1^2)Z_1Z_2} \tag{4.29d}$$

It is interesting to mention that the series, ω_s, and shunt, ω_p, resonance frequencies satisfy the following condition:

$$\omega_s\omega_p = \omega_1\omega_2 \tag{4.30}$$

In general, the elements given by (4.29) do not correspond to a balanced CRLH unit cell. However, if the characteristic impedances at both design frequencies are forced to take the same value ($Z_1 = Z_2$),[10] the resulting line is balanced and

$$\omega_s = \omega_p = \sqrt{\omega_1\omega_2} \tag{4.31}$$

In this case ($Z_1 = Z_2 = Z_o$), the elements are simplified to

$$L_R = \frac{2Z_o}{\omega_2 - \omega_1} \tag{4.32a}$$

$$C_L = \frac{\omega_2 - \omega_1}{2\omega_1\omega_2 Z_o} \tag{4.32b}$$

[10] In most dual-band applications, the devices must exhibit the same functionality at the design frequencies and, therefore, $Z_1 = Z_2$. Nevertheless, since the considered artificial lines allow us to implement dual-band impedance inverters with different impedance value at each frequency, the analysis of the more general case with $Z_1 \neq Z_2$ is also of interest.

$$L_L = \frac{(\omega_2 - \omega_1) Z_o}{\omega_1 \omega_2} \tag{4.32c}$$

$$C_R = \frac{1}{(\omega_2 - \omega_1) Z_o} \tag{4.32d}$$

and (4.32) can be used for design purposes.

In practice, fully planar dual-band components based on dual-band impedance inverters can be implemented by means of OSRR/OCSRR loaded lines. Certainly, the circuit model of these lines (see Fig. 3.58) is not exactly the canonical circuit model of a CRLH line, due to the presence of the parasitic elements (L and C). Nevertheless, the previous analysis can be applied for design purposes. To clarify the design process, let us consider the implementation of a dual-band impedance inverter functional at f_1 = 2.4 GHz and f_2 = 3.75 GHz, with Z_o = 35.35 Ω and electrical length of $\phi_1 = -90°$ at f_1 and $\phi_2 = +90°$ at f_2, (this inverter will be applied to the design of a power divider) [28]. According to the circuit model of Figure 3.58c, the design equations are those given by (4.28) with $Z_1 = Z_2 = Z_o = 35.35$ Ω and χ_s and χ_p given by (3.99). However, there are six unknowns (the six reactive elements of the circuit of Fig. 3.58c). The procedure to determine the element values is as follows: in a first step, L and C (the parasitics in the model of Fig. 3.58c) are neglected, and the other four element values are inferred from the four equations (4.32) (with $L'_p = L_L$, $C'_p = C_R$, $L'_s = L_R/2$, and $C_s = 2C_L$). Then a layout for the OSRR and OCSRR stages is generated so that the extracted parameters for the resonators (see Appendix G) are identical to those inferred in the first step. This gives also the parasitic values, which are introduced in (3.99). Finally, the other element values (L'_p, C'_p, L'_s, and C_s) are recalculated in order to satisfy (4.28). Obviously, the layout must also be readjusted in order to be in coherence with the recalculated reactive elements (the parasitics do not experience a significant variation). By considering the *Rogers RO3010* substrate with thickness $h = 0.635$ mm and dielectric constant $\varepsilon_r = 10.2$; and by applying the previous procedure, the elements of the circuit model of Figure 3.58c were found to be $C = 0.2$ pF, $L = 0.25$nH, $C_s = 0.66$ pF, L'_s = 3.74 nH, C'_p = 2.99 pF, and L'_p = 0.83 nH. The layout of the dual-band impedance inverter is depicted in Figure 4.10, with the circuit and electromagnetic (EM) simulation of the phase and characteristic impedance. These results reveal that the required characteristics are satisfied. By cascading a 50 Ω input (access) line and two 50 Ω output lines, the dual-band power splitter results. The photograph of this device and the simulated and measured power splitting and matching are depicted in Figure 4.11. The required functionality at the two operating frequencies is clearly achieved.

Let us now consider the implementation of dual-band ±90° transmission lines by means of CSRR/gap-loaded CRLH microstrip lines. The lumped-element equivalent circuit model of the unit cell of the CSRR/gap-loaded line (depicted in Fig. 3.38b) is described by means of five independent parameters, whereas the phase and impedance requirements represent only four conditions (expressions 4.28). Thus, the circuit parameters of the unit cell cannot be univocally determined, unless an additional condition is imposed. If the structure is forced to be balanced, the series and shunt resonance

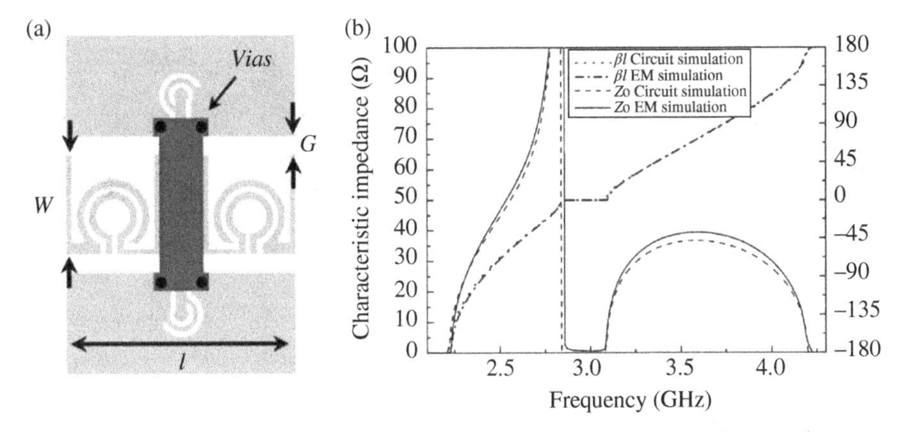

FIGURE 4.10 Layout (a) and circuit and EM simulation (b) of the OSRR/OCSRR-based dual-band impedance inverter. Dimensions are $l = 9$ mm, $W = 4$ mm, and $G = 0.74$ mm. For the OCSRRs $r_{ext} = 0.9$ mm, $c = 0.2$ mm, and $d = 0.2$ mm. For the OSRRs, $r_{ext} = 1.5$ mm, $c = 0.3$ mm, and $d = 0.2$ mm. The wide metallic strip in the back substrate side was added in order to enhance the shunt capacitance of the OCSRR stage, as required to achieve the electrical characteristics of the device. Reprinted with permission from Ref. [28]; copyright 2009 IEEE.

frequencies must be identical. This additional condition leads to the following element values:

$$L = \frac{2(Z_1\omega_1 + Z_2\omega_2)}{\omega_2^2 - \omega_1^2} \tag{4.33a}$$

$$C_g = \frac{\omega_2^2 - \omega_1^2}{2\omega_1\omega_2(Z_1\omega_2 + Z_2\omega_1)} \tag{4.33b}$$

$$C = \frac{Z_1\omega_2 + Z_2\omega_1}{(Z_2^2 - Z_1^2)\omega_1\omega_2} \tag{4.33c}$$

$$C_c = \frac{Z_1\omega_2 + Z_2\omega_1}{(\omega_2^2 - \omega_1^2)Z_1Z_2} \tag{4.33d}$$

$$L_c = \frac{(Z_1\omega_1 + Z_2\omega_2)(\omega_2^2 - \omega_1^2)Z_1Z_2}{\omega_1\omega_2(Z_1\omega_2 + Z_2\omega_1)^2} \tag{4.33e}$$

where Z_1 and Z_2 are the inverter impedances at the operating frequencies. Inspection of Equations 4.33 reveals that L, C_g, C_c, and L_c are always positive (notice that $\omega_2 > \omega_1$). However, C may be negative (for $Z_1 > Z_2$), or $C = \infty$ (for a dual-band impedance inverter with identical impedances $Z_1 = Z_2$). If $Z_1 > Z_2$ ($C < 0$), the inverter cannot be synthesized by means of a balanced cell described by the model of Figure 3.38b.

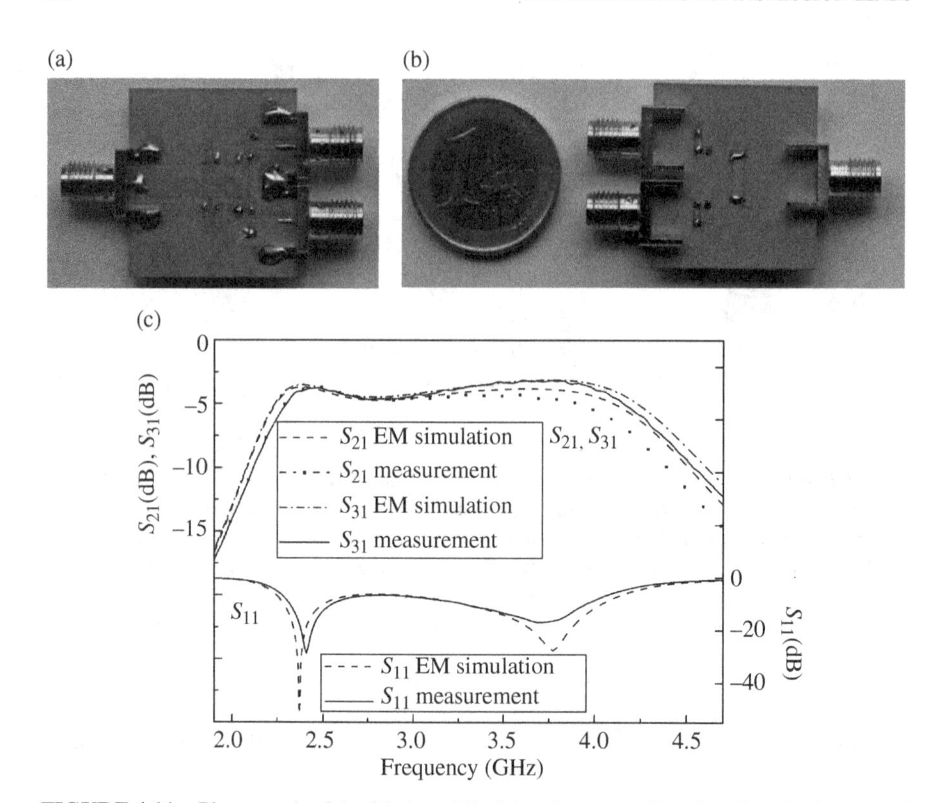

FIGURE 4.11 Photograph of the fabricated dual-band power splitter based on the inverter of Figure 4.10. (a) Top, (b) bottom, and (c) frequency response. Reprinted with permission from Ref. [28]; copyright 2009 IEEE.

If $Z_1 = Z_2 = Z_o$, as most applications require, the inverter can be synthesized by balancing the line, but substituting the capacitance C with an electrical short. In practice, this is not possible by using the cell topology shown in Figure 3.38b. Notice, however, that this case corresponds to the balanced CRLH line based on the canonical circuit model of Figure 3.20, discussed in the previous paragraph. In other words, expressions (4.33) with $Z_1 = Z_2 = Z_o$ provide element values identical to those given by (4.32). Hence, in order to implement dual-band quarter wavelength impedance inverters with identical impedances (at the frequencies of interest) by means of CSRR/gap-loaded CRLH unit cells, such cells must be designed without the restriction of being balanced. Nevertheless, the circuit parameters of the series branch, L and C_g are still univocally determined (and given by expressions 4.33a and 4.33b) under the unbalance condition.[11] The elements C, L_c, and C_c must be chosen in order to satisfy (4.28b) and (4.28d). This gives certain flexibility in the determination of such elements that

[11] Notice that according to (4.28), the elements of the series (L, C_g) and shunt (C, L_c, C_c) branches are independently determined. Thus, introduction of an additional element (the coupling capacitance C) in the shunt branch does not modify the equations giving the elements of the series branch.

can be of interest to ease their physical implementation. However, as discussed in Chapter 3, CSRR/gap-loaded lines exhibit a transmission zero to the left of the first (LH) band (expression 3.86). By considering the position of this transmission zero as an additional condition to be satisfied by the elements of the shunt branch, such elements are univocally determined and given by[12]

$$\frac{L_c\omega_1}{1-L_cC_c\omega_1^2}-\frac{1}{C\omega_1}=Z_o \tag{4.34a}$$

$$\frac{L_c\omega_2}{1-L_cC_c\omega_2^2}-\frac{1}{C\omega_2}=-Z_o \tag{4.34b}$$

$$\frac{1}{\sqrt{L_c(C+C_c)}}=\omega_z \tag{4.34c}$$

where $\omega_z = 2\pi f_z$ is the angular frequency of the transmission zero.

Depending on the specific values of the characteristic impedance of the inverter, the design frequencies (including the transmission zero frequency) and the parameters of the substrate, it may be difficult to synthesize the element values with an implementable layout. Let us consider this situation by means of a specific example and the proposed solution [29]. The objective is again to design a dual-band power divider by means of a $Z_o = 35.35\ \Omega$ quarter wavelength impedance inverter. The operating frequencies are set to to $f_1 = 0.9$ GHz and $f_2 = 1.8$ GHz and the transmission zero to $f_z = 0.5$ GHz. The resulting element values are $L = 12.5$ nH, $C = 24.9$ pF, $C_g = 1.25$ pF, $L_c = 3.38$ nH, and $C_c = 5.10$ pF. With the element values of L, C, and C_g, the convergence region in the L_c–C_c plane was obtained according to the method reported in Appendix H,[13] and the target values of L_c and C_c do not belong to such region (the considered substrate is the *Rogers RO3010*, with thickness $h = 0.635$ mm and dielectric constant $\varepsilon_r = 10.2$). This means that it is not possible to implement the dual-band impedance inverter in that substrate by merely considering the CSRR/gap-loaded line (some element values are too extreme). However, it is expected that by cascading transmission line sections at both sides of the CSRR/gap-loaded line, the element values of the cell are relaxed, and a solution within the convergence region arises. Therefore, two identical 35.35 Ω transmission line sections are cascaded at both sides of the CSRR/gap-loaded line. The width of these line sections is 1.127 mm, corresponding to the indicated characteristic impedance in the considered substrate. Notice that by cascading such 35.35 Ω lines, the electrical length at the operating frequencies is the sum of the electrical lengths of the lines plus the electrical length

[12] Isolation of C, L_c, and C_c from expressions (4.34) is cumbersome and gives huge analytical expressions to be included.

[13] Actually, the method given in Appendix H is for CSRR-loaded lines, but the procedure to determine the convergence region in the L_c–C_c subspace for CSRR/gap-loaded lines is very similar (see Ref. [29] for further details).

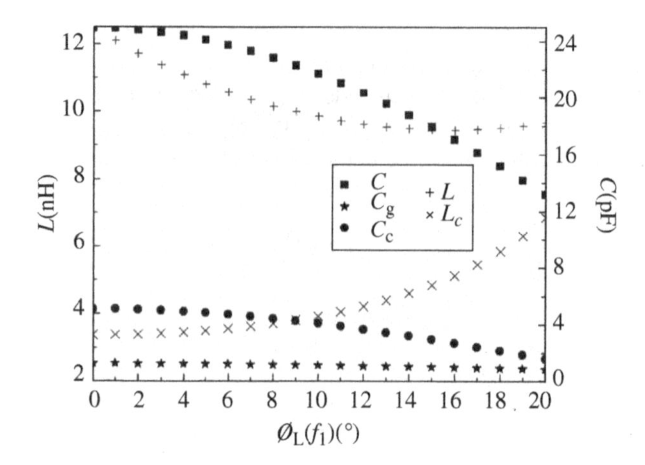

FIGURE 4.12 Dependence of the element values of the CSRR/gap-loaded line with the phase of each cascaded transmission line section at f_1. Reprinted with permission from Ref. [29]; copyright 2013 IEEE.

of the CSRR-based cell. Thus, the phase condition that must satisfy the CSRR/gap-loaded line can be expressed as $\phi_1 = -90° - 2\phi_L(f_1)$ and $\phi_2 = +90° - 2\phi_L(f_2)$, where ϕ_1 and ϕ_2 are the electrical lengths of the CSRR-based cell at the design frequencies f_1 and f_2, and ϕ_L is the phase introduced by the line at the indicated frequency.

A parametric analysis, consisting of obtaining the element values of the CSRR-based cell for different values of the length of the cascaded transmission line sections (and hence $\phi_L(f_1)$ and $\phi_L(f_2)$), was made. The results are depicted in Figure 4.12. It can be observed that for small values of $\phi_L(f_1)$, L and C are too large for being implemented. Large L means a small value of the strip width, W, of the host line, but this is not compatible with a large C value. On the other hand, the values of L_c and C_c without cascaded line sections, that is, $\phi_L(f_1) = 0°$, give extreme values of d and c, that is, large value of d and small value of c. However, by increasing $\phi_L(f_1)$ (or the length of the cascaded lines), the variation of the elements of the central CSRR-gap-based cell goes in the correct direction for their implementation. Specifically, a pair of transmission line sections with $\phi_L(f_1) = 15°$ was considered. With this phase, the required electrical lengths for the CSRR-based cell at the operating frequencies are $\phi_1 = -120°$ and $\phi_2 = +30°$. The element values providing these phases and the required characteristic impedance are $L = 9.45$ nH and $C = 17.9$ pF, $C_g = 1.01$ pF, $L_c = 4.85$ nH, and $C_c = 2.95$ pF, and these values lead us to an implementable layout.[14] The reason of choosing this phase shift for the lines is that it gives the minimum value of L (see Fig. 4.12) and a reasonable small value of C, with L_c not so small and C_c not so large. Notice that C_g does not experience significant variations with $\phi_L(f_1)$.

[14] These element values were derived from (2.30) and (2.33) by forcing the phases and characteristic impedance to the required values at the operating frequencies, and by means of (3.86). Expressions (4.34) only apply if $\phi_1 = -90°$ and $\phi_2 = +90°$.

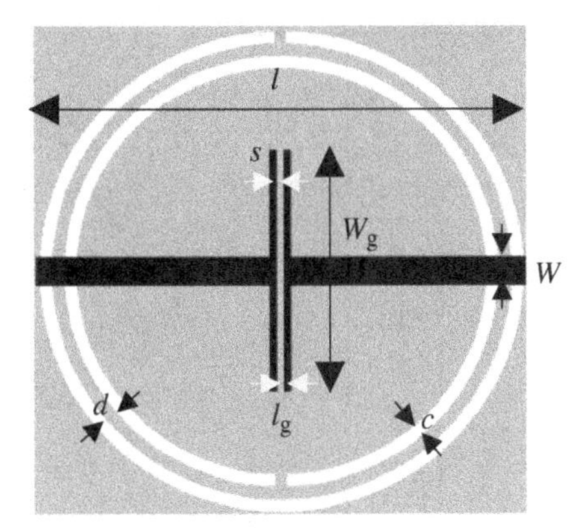

FIGURE 4.13 CSRR/gap-loaded microstrip line with T-shaped gap, and relevant dimensions.

The layout of the CSRR/gap-loaded line was inferred from the ASM-based synthesis method reported in Appendix H, where a T-shaped geometry for the gap was considered.[15] The geometrical parameters of the synthesized structure are (see Fig. 4.13) $l = 14.42$ mm, $W = 0.87$ mm, $c = 0.34$ mm, $d = 0.40$ mm, $W_g = 7.13$ mm, and convergence (with a relative error of 0.012—see Appendix H for more details) was obtained after six iterations. The comparison of the electrical length and characteristic impedance inferred from EM simulation of the synthesized impedance inverter (the CSRR-based cell plus the cascaded 35.35 Ω transmission line sections), and the ones inferred from circuit simulation are shown in Figure 4.14. The agreement is excellent in the LH region, where the model describes the structure to a very good approximation, and progressively degrades as frequency increases, as it is well known and expected. Nevertheless, the phase shift and the characteristic impedance at f_2 are reasonably close to the nominal values, and hence we do expect that the functionality of the power divider at f_2 is preserved. Figure 4.15 depicts the schematic and photograph of the divider (fabricated after cascading two output 50 Ω access lines for connector soldering) and the frequency response. It can be appreciated that optimum matching occurs at f_1 and slightly below f_2, for the reasons explained. Nevertheless, the functionality of the power divider covers both design frequencies. The discrepancy between the measured response and the target is not due to a failure of the ASM-based synthesis method, but it is due to the fact that the circuit model of the CSRR/gap-loaded line does not accurately describe the structure at high frequencies, including part of the RH band (notice that this does not occur with OSRR/OCSRR-based CRLH lines, where the model accurately describes the structure up to frequencies including the second –RH–band).

Since the use of the synthesis method tends to underestimate the upper matching frequency, one possibility to improve the power divider response is to design it by increasing the upper operating frequency with regard to the nominal value. Alternatively, optimization at the layout level can be carried out. Various additional examples

[15] This geometry may be used in applications where a relatively large gap capacitance is required.

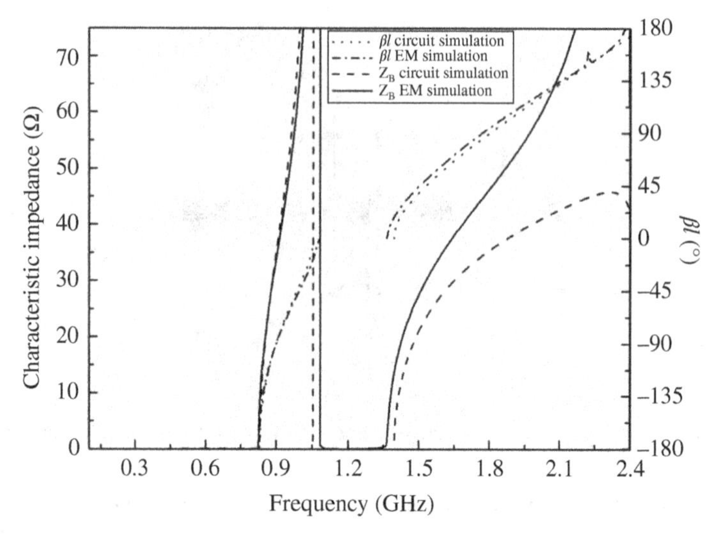

FIGURE 4.14 Comparison between the EM and circuit simulation corresponding to the characteristic impedance and electrical length of the designed dual-band impedance inverter described in the text. Reprinted with permission from Ref. [29]; copyright 2013 IEEE.

of power dividers based on the CSRR/gap-loaded lines can be found in Ref. [30]. In one of those dividers, a complementary spiral resonator (CSR) was used in order to reduce device size.

In both dual-band dividers, depicted in Figures 4.11 and 4.15, the upper band exhibits a wider bandwidth as compared to the first one. This is due to the softer variation of phase and characteristic impedance with frequency at the second design frequency. If both dividers are compared, the matching level and insertion loss at the design frequencies are of the same order. The OSRR/OCSRR-based divider exhibits broader bands, but splitters implemented with CSRRs exhibit a transmission zero that can be interesting for filtering purposes. Concerning size, despite that OSRRs and OCSRRs are electrically smaller than SRRs and CSRRs (this aspect was discussed in Chapter 3), the unit cell of the CSRR/gap-loaded CRLH line requires only one resonator, whereas three resonator stages are needed for the CRLH line based on the combination of OSRRs and OCSRRs. For this reason, dual-band components based on quarter wavelength impedance inverters implemented by means of CSRRs are typically electrically smaller as compared to those based on OSRR/OCSRR-based lines. On the other hand, the circuit model of OSRR/OCSRR provides very accurate results in a broader bandwidth as compared to the model of CSRR/gap-loaded lines. This eases circuit design for OSRR/OCSRR based CRLH lines. Nevertheless, a synthesis method has been developed to aid the design process in devices based on CSRRs, as detailed in Appendix H.

A dual-band branch line hybrid coupler based on CSRR/gap-loaded microstrip lines acting as quarter wavelength transmission lines was reported in Ref. [31].

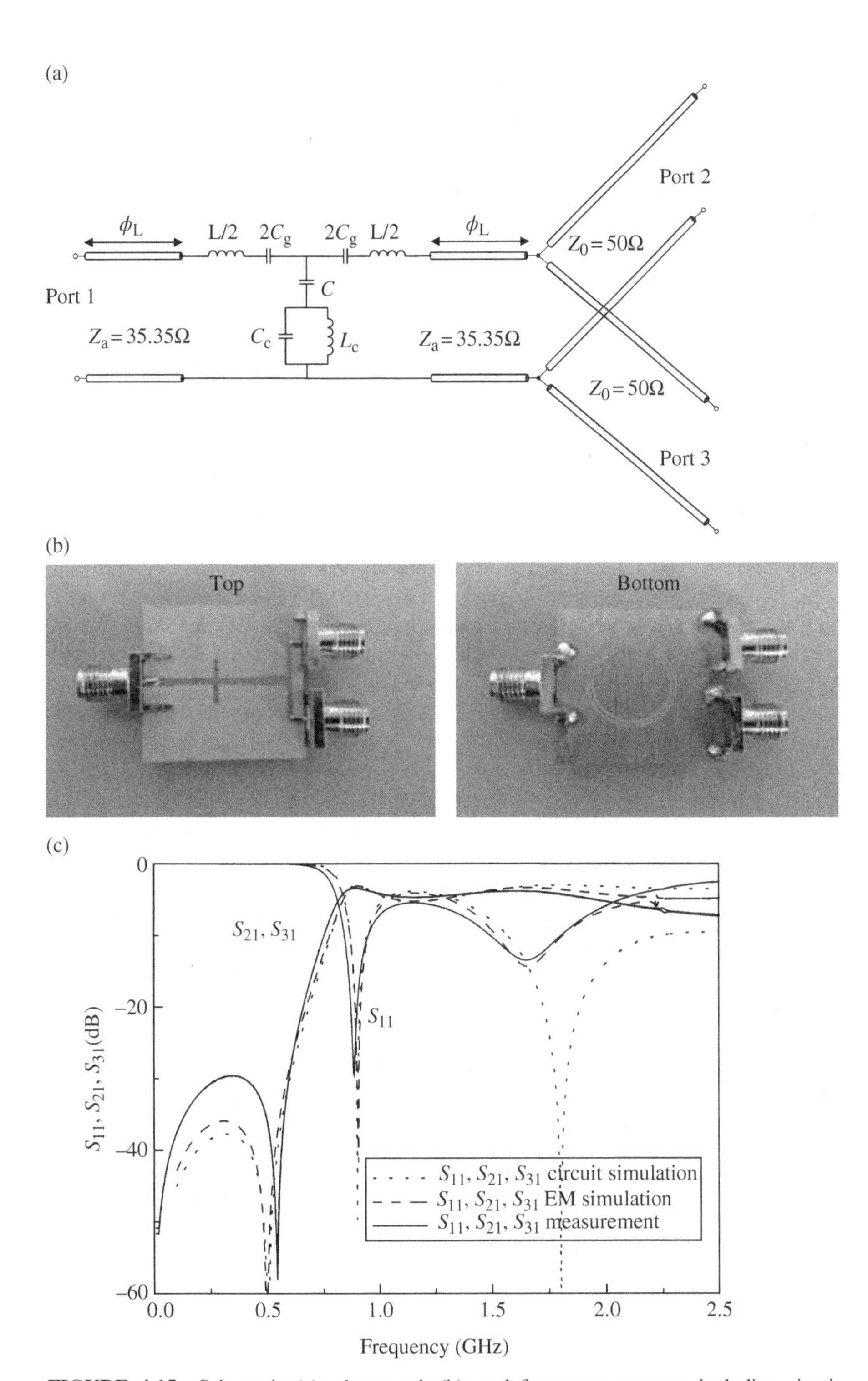

FIGURE 4.15 Schematic (a); photograph (b); and frequency response including circuit simulation, EM simulation, and measurement (c) of the designed and fabricated dual-band power divider. Reprinted with permission from Ref. [29]; copyright 2013 IEEE.

It is depicted in Figure 4.16 for completeness (the details of the design are given in Ref. [31]). The measured results show the dual-band performance around 0.90 GHz and 1.81 GHz (minimum return loss). The return loss, coupling, transmission, and isolation are 25.8 dB, 4.19 dB, 4.31 dB, and 21.5 dB at the first band, and 30.5 dB, 4.43 dB, 4.72 dB, and 17.69 dB at the second band. The measured phase balance is −93° and +92° at 0.90 GHz and 1.81 GHz, respectively. In terms of bandwidth and coupling at both operating frequencies, the device depicted in Figure 4.9 exhibits better performance. However, the coupler of Figure 4.16 is fully planar and electrically smaller (the size of the square is $0.19\lambda \times 0.20\lambda$, where λ is the guided wavelength at f_1).

Dual-Band Diplexer based on Branch Line Hybrid Couplers Although filters and diplexers will be considered in Section 4.2.3, an interesting application of dual-band hybrid couplers is reported here: the realization of a dual-band microwave diplexer operative at the mobile GSM frequency bands (GSM-850, DCS-1800) [32]. The schematic diagram of the dual-band diplexer is sketched in Figure 4.17 (the diplexer principle for single-band operation is reported in Ref. [33]). The port 1 of hybrid A is the transmitter (Tx) port, whereas ports 2 and 3 of the hybrid B are the receiver (Rx) and antenna ports, respectively. For dual-band operation, the hybrids and the band-stop filters must be designed to operate at the desired frequency bands. Indeed, the hybrids must cover both the transmitter and receiver frequencies at both bands, whereas the band-stop filters must be designed to efficiently reject the Rx signal and transmit the Tx signal for both GSM bands. Filters A and B have been designed to provide the central rejection frequencies at 0.89 and 1.73 GHz, respectively, that is the receiver signal frequencies. Transmitter frequencies are centered at 0.83 GHz for the lower band, and at 1.80 GHz for the upper band.

From the scattering matrix of the quadrature hybrid coupler (assuming that the pairs of isolated ports are ports 1–4 and 2–3), and considering that ports 1 and 4 of hybrid B are terminated by loads with identical reflection coefficient (the reflection coefficients of the two bandstop filters, Γ), it follows that the transmission from the antenna port to the receiver port is given by

$$S_{\text{antenna}-\text{Rx}} = j\Gamma \tag{4.35}$$

From (4.35), it is clear that if $|\Gamma| = 1$, the received signal at the antenna will be transmitted to the Rx port without any loss. This requires band-stop filters with high rejection and very low loss in the reflected signals at the two receiver band frequencies. From the analysis of the scattering matrix of the hybrids, and assuming that the filters only provide a certain phase shift at the transmit frequencies (which will be different for each frequency band), it follows that the power transmitted from the Tx port to the Rx port is null (isolated ports), whereas the injected power in the Tx port is totally transmitted to the antenna port according to:

$$S_{\text{Tx}-\text{antenna}} = e^{-j\left(\theta_{1,2}-\pi/2\right)} \tag{4.36}$$

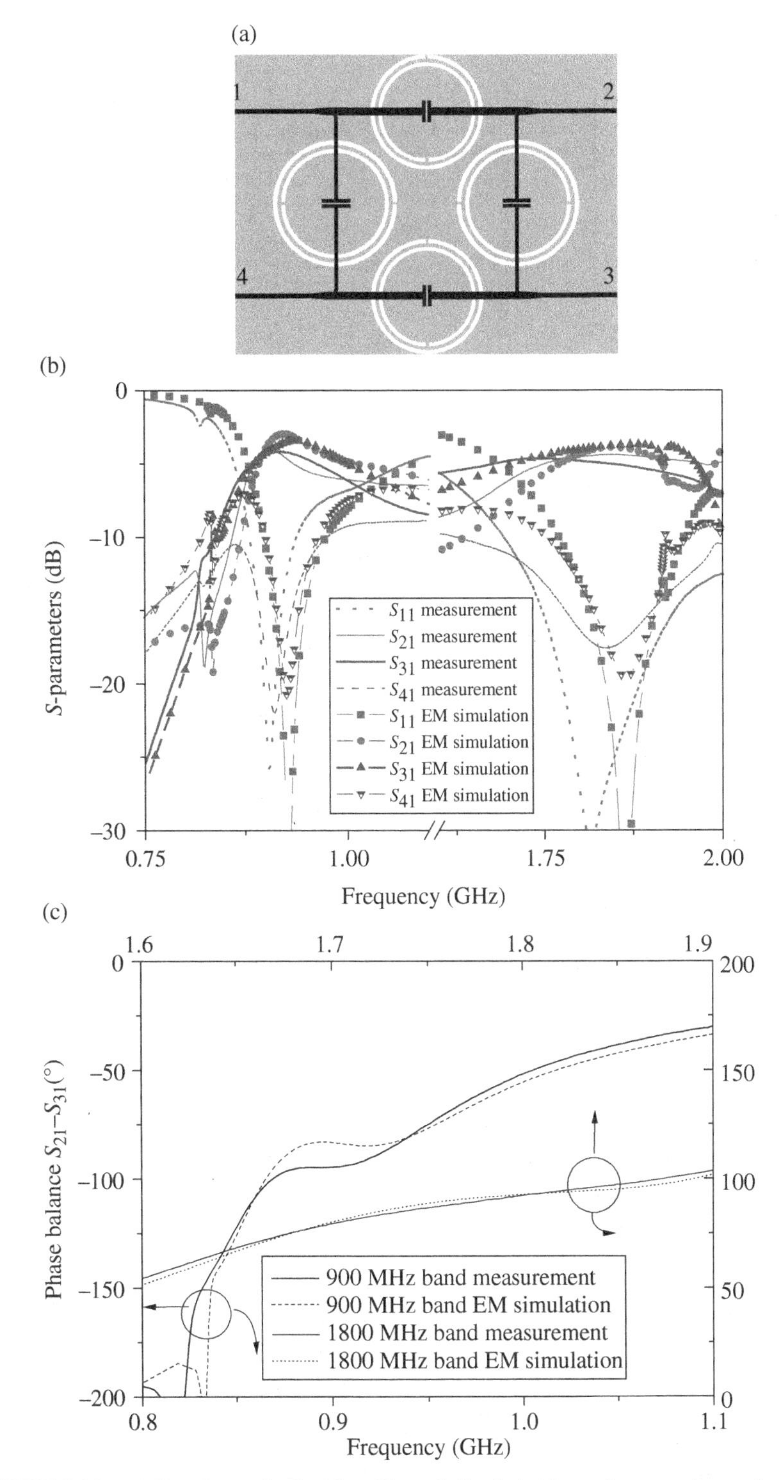

FIGURE 4.16 (a) Topology of a dual-band branch line hybrid coupler operating at the GSM frequencies $f_1 = 0.9$ GHz and $f_2 = 1.8$ GHz; (b) simulated and measured power splitting (S_{21}, S_{31}), matching (S_{11}) and isolation (S_{41}); and (c) simulated and measured phase response. For

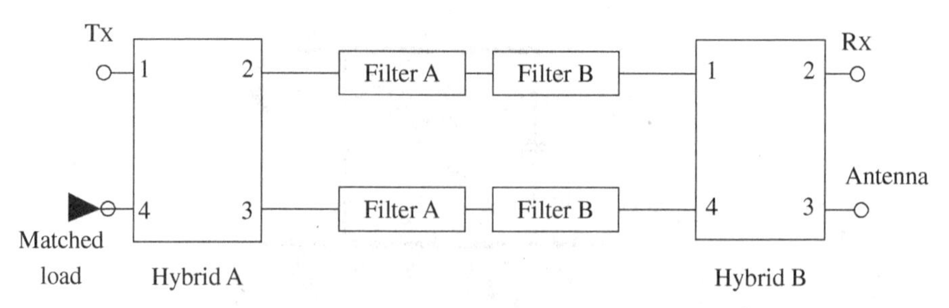

FIGURE 4.17 Schematic diagram of the dual-band diplexer.

It is clear that the signals at the antenna and Tx ports are identical except by certain phase shift, which depends on the phase shift introduced by the filters at the transmit frequencies of the two operating bands, θ_1 and θ_2.

The design of the dual-band branch-line hybrid couplers (implemented by means of CSRR/gap loaded lines) is described in Ref. [31]. The resulting devices are similar to that depicted in Figure 4.16. The dual-band band-stop filters were implemented by means of spiral resonators (SRs) coupled to the microstrip line connecting both hybrid couplers. To achieve dual-band operation, a pair of SRs was designed to provide rejection at the reception frequency of the lower band, whereas a second pair was tuned at the reception frequency of the upper band. The rejection filters could have been implemented by coupling other resonators to the line, but spirals provide small size and narrow band, as is required in this application (due to the proximity of the transmit frequencies). A photograph of the fabricated diplexer, including the dual-band hybrid couplers, the filters and the access lines, is depicted in Figure 4.18a.

The simulated and measured transmission coefficients between the different ports of the diplexer are depicted in Figure 4.18b and c. The simulated and experimental data are in good agreement, laying the receiver and transmitter frequencies within the desired GSM frequency bands. Concerning dimensions, the side length of the hybrids is as small as $\lambda/7$, λ being the guided wavelength at the lower frequency band (which is the limiting one in terms of size). Thanks to the small size of the hybrids, plus the miniature spirals used as rejection filters, the overall dimensions of the proposed prototype dual-band diplexer are small.

FIGURE 4.16 (*Continued*) the 35.35 Ω impedance inverters (lines between ports 1–2 and 3–4), the dimensions are internal radius $r = 7.53$ mm, ring width $c = 0.42$ mm, separation between rings $d = 0.55$ mm, width of the host line $W = 1.00$ mm, gap separation $s = 0.27$ mm, and gap width $W_g = 2.77$ mm. For the 50 Ω impedance inverters (lines between ports 1–4 and 2–3) $r = 8.27$ mm, $c = 0.46$ mm, $d = 0.60$ mm, $W = 0.46$ mm, $s = 0.27$ mm, and $W_g = 3.77$ mm. The device was fabricated on the *Rogers RO3010* substrate with thickness $h = 0.635$ mm and dielectric constant $\varepsilon_r = 10.2$. Reprinted with permission from Ref. [31]; copyright 2008 IEEE.

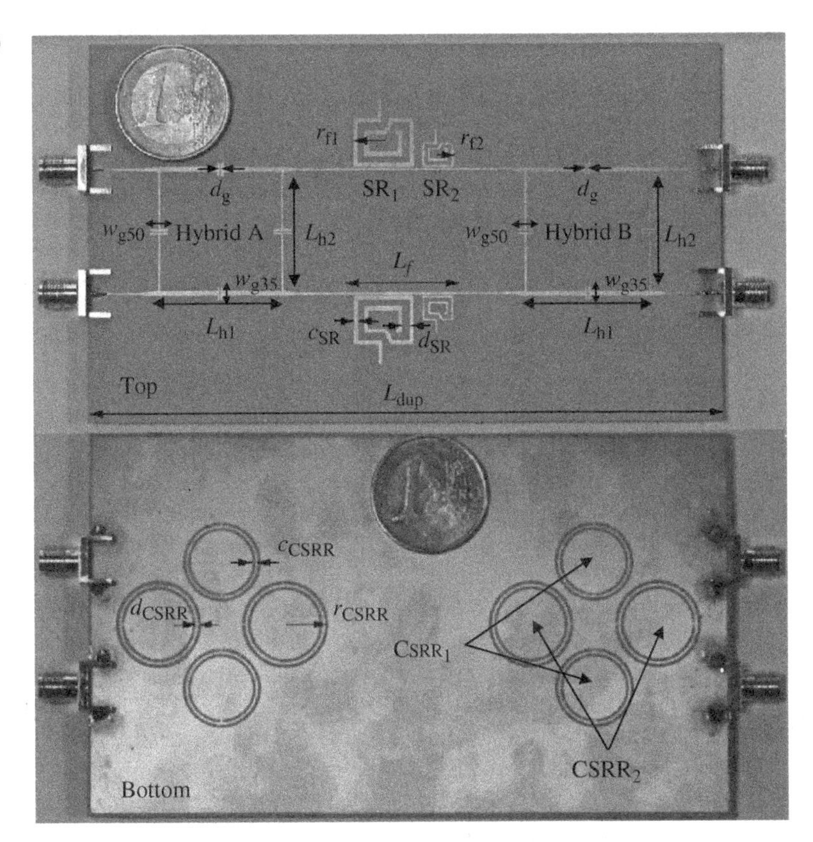

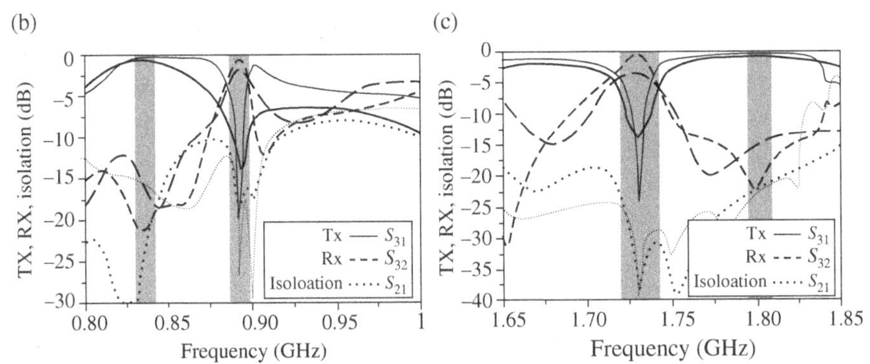

FIGURE 4.18 Photograph of the fabricated diplexer (a) and comparison between simulated (thin line) and measured (thick line) transmission characteristics for the fabricated diplexer at the lower (b) and upper (c) GSM bands. Dimensions are $L_{dup} = 124$ mm, $L_f = 21.44$ mm, $r_{CSRR1} = 7.4$ mm, $r_{CSRR2} = 7.9$ mm (external radius of the CSRR), $c_{CSRR1} = c_{CSRR2} = 0.5$ mm (width of the rings), $d_{CSRR1} = d_{CSRR2} = 0.5$ mm (distance between rings), $r_{f1} = 6$ mm, $r_{f2} = 3$ mm, $c_{SR1} = 0.9$ mm, $c_{SR2} = 0.5$ mm (width of the spirals), $d_{SR1} = 1.1$ mm, and $d_{SR2} = 0.6$ mm (distance between loops of the spiral). The gap separation is $d_g = 0.27$ mm and gap widths are $w_{35} = 2.77$ mm and $w_{50} = 3.77$ mm. Length (L_{h1}, L_{h2}) of the artificial lines constituting the hybrids are 18 mm and 17.6 mm for the 35 Ω impedance and 50 Ω impedance lines, respectively. The considered substrate is the *Rogers RO3010* substrate with measured dielectric constant $\varepsilon_r = 11.2$ and thickness $h = 635$ μm. Reprinted with permission from Ref. [32]; copyright 2011 Springer.

4.2.2.3 Quad-Band Devices based on Extended CRLH Transmission Lines The order-4 CRLH transmission line T-circuit model depicted in Figure 3.25c (also known as extended CRLH—E-CRLH—line) is useful for the implementation of quad-band components based on quarter wavelength transmission lines. For multiband functionality Equations 4.28 can be generalized as

$$\chi_s(\omega_n) = (-1)^n Z_A \tag{4.37a}$$

$$\chi_p(\omega_n) = (-1)^{n+1} Z_A \tag{4.37b}$$

where n is the frequency index (i.e., $n = 1, 2, 3, 4$ for a quad-band device). In (4.37), it has been assumed that the characteristic impedance of the quarter wavelength transmission lines is identical at all the operating frequencies and given by Z_A. For quad-band functionality, expressions (4.37) provide eight equations that univocally determine the eight independent reactive elements of the circuit model of Figure 3.25c.

For the synthesis of the extended CRLH line model of Figure 3.25c in fully planar technology, both CPW [34] and microstrip [35] host lines have been considered. Let us consider the implementation of a quad-band impedance inverter at the GSM ($f_1 = 0.9$ GHz, $f_4 = 1.8$ GHz) and GPS ($f_2 = 1.17$ GHz, $f_3 = 1.57$ GHz) frequency bands in CPW technology by using OSRRs and OCSRRs. The characteristic impedance is forced to be $Z_A = 35.35\ \Omega$ at these frequencies in order to subsequently implement a quad-band Y-junction power divider. The electrical length of the structure is set to −90° (LH bands) at the odd frequencies and to +90° (RH bands) at the even frequencies. The values of the circuit model corresponding to Figure 3.25c that satisfy these conditions are $L_{hs} = 22.458$ nH, $C_{hs} = 0.766$ pF, $L_{hp} = 0.883$ nH, $C_{hp} = 14.06$ pF, $L_{vs} = 8.785$ nH, $C_{vs} = 1.414$ pF, $L_{vp} = 1.915$ nH, and $C_{vp} = 8.986$ pF. With these line parameters and frequencies, the resulting reactive element values are reasonable for their implementation by means of OSRRs or OCSRRs, with exception of the parallel resonators of the series branch, that are realized by parallel connecting a capacitive patch and a meander inductor (the topology of the final quad-band impedance inverter is depicted in Fig. 4.19a).

To a first-order approximation, the structure of Figure 4.19a can be described by the canonical order-4 E-CRLH model shown in Figure 3.25c. However, for an accurate description, line parasitics must be taken into account, as it was done in Section "Dual-Band Devices Based on Fully Planar CRLH Lines" for the description of dual-band impedance inverters implemented by means of OSRRs and OCSRRs. Thus, a more accurate circuit model of the quad-band CRLH line of Figure 4.19a is depicted in Figure 4.19b. For the determination of the layout, a procedure similar to that reported in Section "Dual-Band Devices Based on Fully Planar CRLH Lines" in reference to dual-band CRLH lines is applied. Once the element values of the canonical circuit model (Fig. 3.25c) are known, the layout of each section is determined so that the element parameters (inferred from curve fitting) of the resonators coincide with those of the E-CRLH model of Figure 3.25c. This procedure also provides the values of the parasitic elements.

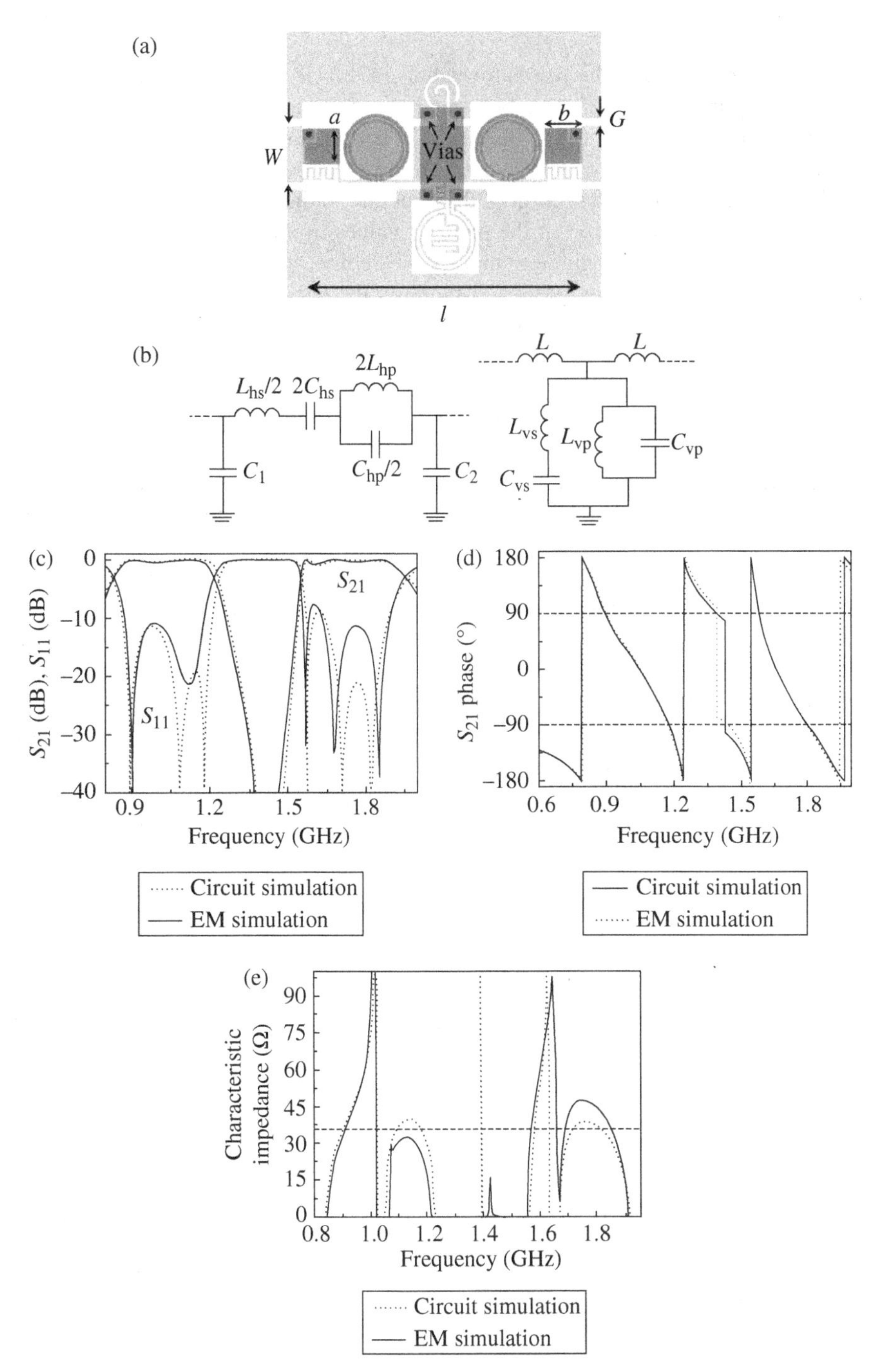

FIGURE 4.19 Topology (a), circuit models of the series and shunt branches including parasitics (b), magnitude of the frequency response (c), phase response (d), and characteristic impedance (e) of the designed quad-band CPW impedance inverter. The dimensions are

The next step is to tune the resonator values in the circuit model of Figure 4.19b until the required values of impedance and phase at the operating frequencies are obtained (this is simple and fast because this is done at the circuit level). Finally, the topology of the different resonators is modified at the layout level in order to fit the EM simulation of each section of the structure to the circuit simulation. This is simple because the variation of the element values in the tuning procedure of the previous step gives us the guide to modify the geometry of the different resonators (tuning the layout does not affect substantially the effects of the parasitics). With this procedure, the layout of the whole structure that provides the required values of characteristic impedance and phase at the operating frequencies is directly obtained (additional tuning or optimization is not required). The layout of the quad-band impedance inverter obtained by this procedure is the one depicted in Figure 4.19a. The electrical size of the device is $0.105\lambda \times 0.074\lambda$, λ being the guided wavelength at the lower frequency band (900 MHz), and the considered substrate is the *Rogers RO3010* with thickness $h = 0.254$ mm, and measured dielectric constant $\varepsilon_r = 11.2$.

The simulated S-parameters of the designed impedance inverter, using port impedances of 35.35 Ω, are depicted in Figure 4.19c. Both the EM (without losses) and circuit simulation are included in the figure. The simulated phase response and characteristic impedance of the designed quad-band impedance inverter are depicted in Figure 4.19d and e, respectively. The agreement between the EM and circuit simulation is good, and the required functionality of the impedance inverter is achieved at the whole operating frequencies to a good approximation. This validates the model of the quad-band CRLH structure implemented by means of OSRRs and OCSRR in CPW technology that includes parasitic elements (Fig. 4.19b).

In order to implement a Y-junction power divider, two 50 Ω access lines are cascaded to the output port of the impedance inverter and a 50 Ω access line is cascaded to the input port. The fabricated quad-band power divider is shown in Figure 4.20. The simulated and measured power division and matching, also depicted in the figure, indicate that the desired functionality at the design frequencies is achieved, with a matching level better than −10 dB at the whole frequencies.

FIGURE 4.19 (*Continued*) $W = 4.5$ mm, $G = 0.524$ mm, and $b = 3$ mm. The width and separation of all the meanders is of 0.2 mm, $a = 2.8$ mm, and $l = 22.5$ mm; for the series connected OSRRs, $r_{ext} = 2.6$ mm, $c = 0.2$ mm, and $d = 0.15$ mm; for the shunt connected OSRR, $r_{ext} = 2.5$ mm, $c = 0.2$ mm, and $d = 0.2$ mm; for the OCSRR, $r_{ext} = 1.4$ mm, $c = 0.2$ mm, and $d = 0.6$ mm. r_{ext}, c, and d are the external radius, width, and separation of the rings, respectively. The element values corresponding to the circuit model shown in (b) are $L'_{hs} =$ 21 nH, $C_{hs} = 0.775$ pF, $L_{hp} = 0.88$ nH, $C_{hp} = 13.96$ pF, $L_{vs} = 8.8$ nH, $C_{vs} = 1.41$ pF, $L_{vp} = 1.95$ nH, $C_{vp} = 8.3$ pF, $C_1 = 0.41$ pF, $C_2 = 0.22$ pF, and $L = 0.4$ nH. Notice that the phase of S_{21} exhibits opposite sign to the electrical length. Reprinted with permission from Ref. [34]; copyright 2010 IEEE.

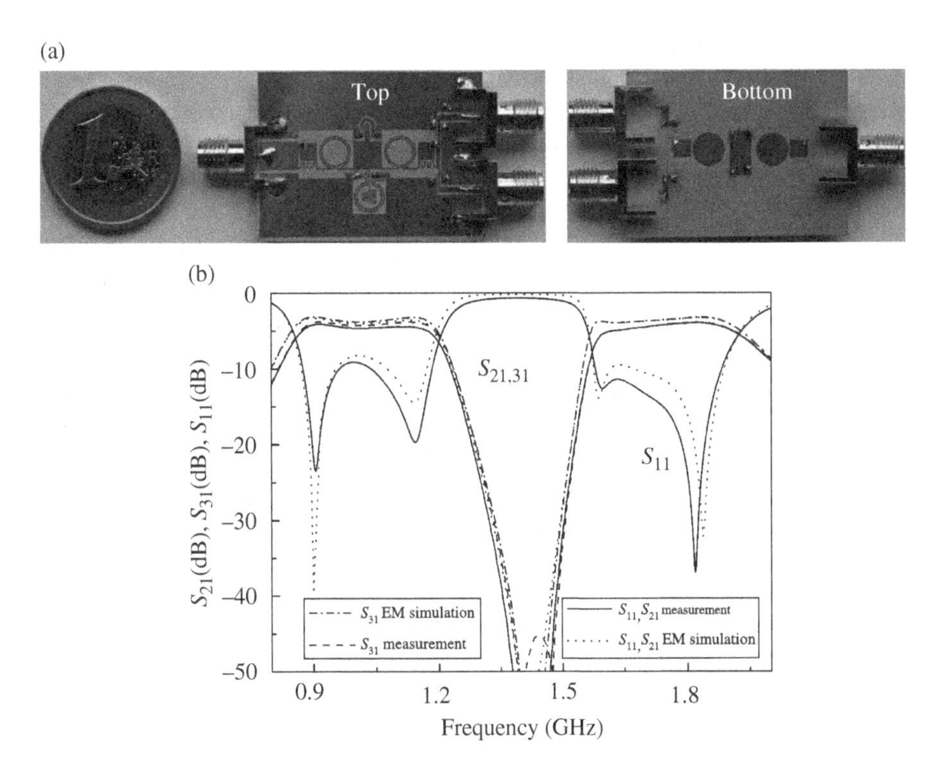

FIGURE 4.20 Photograph of the fabricated quad-band Y-junction power divider implemented by means of the inverter of Figure 4.19 (a), and simulated and measured power division and matching (b). Reprinted with permission from Ref. [34]; copyright 2010 IEEE.

The implementation of fully planar quad-band CRLH transmission lines in microstrip technology is reported in Ref. [35], and applied to the design of power splitters and branch-line hybrid couplers. However, in microstrip technology the series-connected OSRRs can no longer be described by a simple model (as it was shown in Chapter 3). Moreover, the typical required values of resonator inductances and capacitances further limit the implementation of such resonators by means of OSRRs and OCSRRs, as compared to CPW technology. Thus, the microstrip quad-band CRLH lines of Ref. [35] are implemented through a combination of OCSRRs (for the parallel resonator of the shunt branch), meander inductors, patch capacitances and interdigital capacitances. The quad-band power divider reported in Ref. [35] exhibits comparable performance to that of Figure 4.20. The quad-band branch line hybrid coupler also reported in Ref. [35] is a complex structure with tiny details at the layout level. Thus, losses degrade somehow device performance, especially at the third operating frequency, where the insertion loss is substantially higher than the ideal value (3 dB). Nevertheless, the measured matching, isolation, and phase balance are reasonable considering the complexity of the circuit layout (see Ref. [35] for further details).

Fully planar order-4 E-CRLH microstrip lines were also reported in Ref. [36], but the agreement between the circuit model and the EM response of the designed structure is very limited. On the other hand, quad-band Wilkinson power dividers have been implemented in microstrip technology by combining semilumped and lumped elements [37]. It is also remarkable that order-4 CRLH lines can also be applied to the implementation of triband components. However, in this case there is an excess of elements, and such triband components are preferably implemented by means of order-3 CRLH transmission lines, such as the double-Lorentz metamaterial transmission line reported in Ref. [38].

To end this subsection, it is important to clarify that dual-band and multiband components are not exclusive products of CRLH transmission lines. There are many other approaches that either exploit transmission line loading (this also enhances the number of parameters of the constitutive transmission lines) or make use of complex structures combining transmission line and stubs in order to satisfy the required functionality at several frequencies simultaneously. In many of the reported multiband devices, the performance is very satisfactory, but device dimensions are usually relatively large, as compared to the typical dimensions resulting by applying the CRLH transmission line concept. As illustrative examples of dual-band components based on other approaches, the author suggests the papers [39, 40] and references therein (a complete list would be too large for inclusion in this book).

4.2.3 Filters and Diplexers

Metamaterial transmission lines exhibit pass bands and stop bands, and therefore they are filtering structures by nature. The purpose of this subsection is to present some of the main approaches for the implementation of different type of filters (narrowband, wideband, etc.) and filter responses (stopband, bandpass, lowpass, and highpass), including those strategies that lead to small filter size and spurious suppression. Several examples will be reported, including narrow and wideband filters, microwave diplexers, and multifunctional filters, among others. It is impossible to provide a wide overview of metamaterial-based filters in this subsection. Therefore, an extensive list of references on this topic is provided. This includes a review paper [41], where some of the approaches reported here are succinctly considered. The topic of tunable and reconfigurable filters will be studied in Chapter 5, devoted to reconfigurable, tunable, and nonlinear artificial transmission lines.

4.2.3.1 Stopband Filters based on SRR- and CSRR-Loaded Lines[16] Stopband filters are of interest in RF/microwave engineering for the suppression of undesired responses or interfering signals. As it was studied in Chapter 3, SRR-loaded CPWs and CSRR-loaded microstrip lines are single negative artificial transmission lines

[16] The filters reported in this subsection are indeed based on transmission lines with metamaterial loading. Namely, the functionality of these filters is based on a resonance phenomenon (rather than on dispersion and impedance engineering), and the considered structures are not based on CRLH lines. However, such filters are included here (rather than in Section 4.3) for thematic coherence. This also applies to Section 4.2.3.2.

able to inhibit signal propagation in the vicinity of particle resonance. In periodic structures, strong rejection can be achieved if a sufficient number of cascaded resonators is considered. However, the typical rejection bandwidths of these filtering structures are small due to the typical quality factors of SRRs and CSRRs. Bandwidth can be enhanced by cascading tightly coupled SRRs or CSRRs, as it was discussed in Chapter 3. However this approach is not very effective since the coupling level is limited in practice. To overcome this limitation, we may take advantage of the high reflectivity of the considered particles, and cascade several resonators tuned at slightly different frequencies over the frequency band of interest. The resulting non-periodic structures may provide wide stop bands if the number of cascaded resonators is high enough.

Frequency tuning can be achieved either by scaling up (or down) the dimensions of the SRRs (or CSRRs) or by increasing (or decreasing) their radius. This approach has been driven to practice in microstrip and CPW technology using CSRRs and SRRs [42–44]. As illustrative examples, two stopband filters are reported. The first one is a multiple tuned SRR-loaded CPW stopband filter with five SRR pairs (Fig. 4.21). Tuning was implemented by increasing the inner radius of the SRRs in 0.05 mm step increments and leaving unaltered c and d (as compared to the SRR geometry of Fig 3.31). The simulated and measured frequency responses (also depicted in Fig. 4.21) show that the rejected band broadens toward lower frequencies. This is expected since an increment of r has the effect of decreasing the resonant frequency of SRRs. The second example is a multiple tuned CSRR-loaded microstrip line (Fig. 4.22), where rectangular-shaped resonators are etched in the ground plane (they can also be etched in the central strip [45]). The measured frequency response of this structure, which is also depicted in Figure 4.22, reveals that rejection above 20 dB can be achieved within a 25% fractional bandwidth.

4.2.3.2 Spurious Suppression in Distributed Filters The filtering structures of the previous subsection can be applied to the elimination of spurious bands in distributed filters. Such undesired bands can seriously degrade filter performance, and may be critical in several applications where high rejection bandwidths are required. For instance, in parallel-coupled line bandpass filters, the first spurious band appears at twice the filter central frequency as consequence of the different phase velocities of the even and odd modes of the coupled-line sections. The rejection of spurious bands has been a subject of interest in microwave engineering for years. Specifically, in coupled line filters, several approaches based on modified structures, aimed to obtain equal modal phase velocities, have been proposed as a means to improve the out-of-band filter performance [46–49]. It was also reported in Chapter 2 that EBG-based structures embedded in the coupled line sections efficiently suppress the spurious bands in coupled line bandpass filters. This approach is very effective and can be applied to many other filter types (see, e.g., the filter depicted in Fig. 2.36). However, EBG structures scale with frequency and may be excessively large for the suppression of spurious bands in compact planar filters.

An alternative approach for the elimination of spurious frequency bands in distributed filters is the use of SRRs or CSRRs. Their small electrical size, rejection

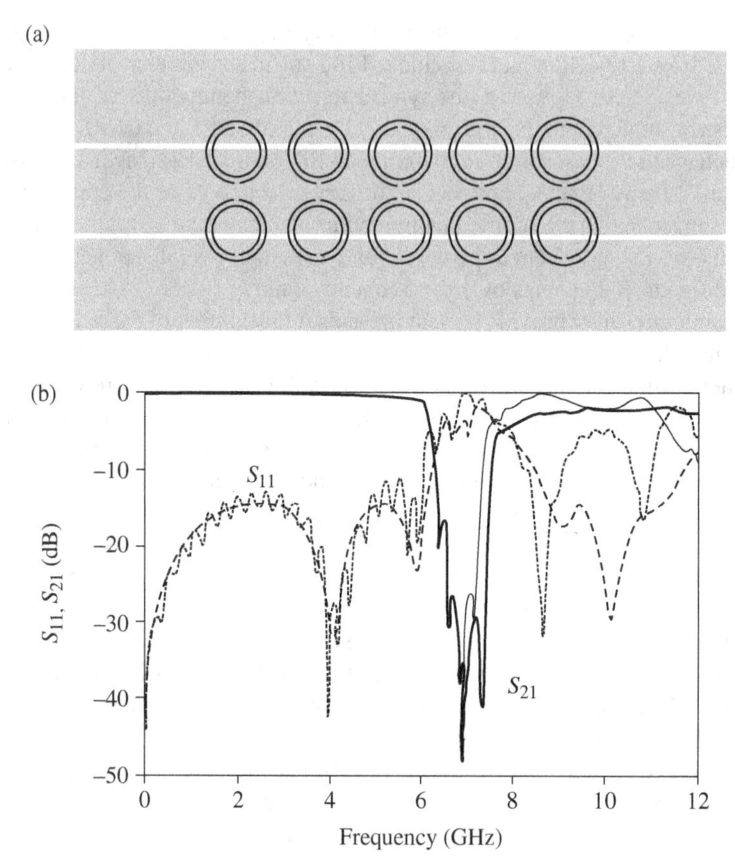

FIGURE 4.21 Layout (a), simulated (thin line) and measured (bold line) frequency responses (b) of the multituned SRR-based CPW stopband filter. The length of the active region is 4 cm. The dimensions of the smallest SRRs are $c = d = 0.2$ mm and internal radius $r = 1.3$ mm. For the other SRRs, r was incremented in 0.05 mm steps, whilst c and d were left unaltered. The structure was fabricated on the *Arlon* 250-LX-0193-43-11 substrate ($\varepsilon_r = 2.43$, thickness $h =$ 0.49 mm). Reprinted with permission from Ref. [42]; copyright 2003 IEEE.

efficiency, and versatility (stopband filters based on these particles have been demonstrated in both microstrip and CPW technology [42–44]) make these resonant elements very attractive to achieve effective spurious suppression with minimum area increase. As occurs with EBGs, the main advantage of the approach is the fact that SRRs and CSRRs can be embedded within the filter active region. Namely, it is not necessary to cascade the SRRs or the CSRRs at the input or output access lines (as it was done in the first reported structure using this concept [50]). This represents an optimum design in terms of area saving. To illustrate the efficiency of SRRs and CSRRs for the elimination of spurious bands and its versatility, a parallel coupled line bandpass filter implemented in microstrip technology is reported [44]. The layout of the device, a third-order Butterworth bandpass filter with a central frequency

(a)

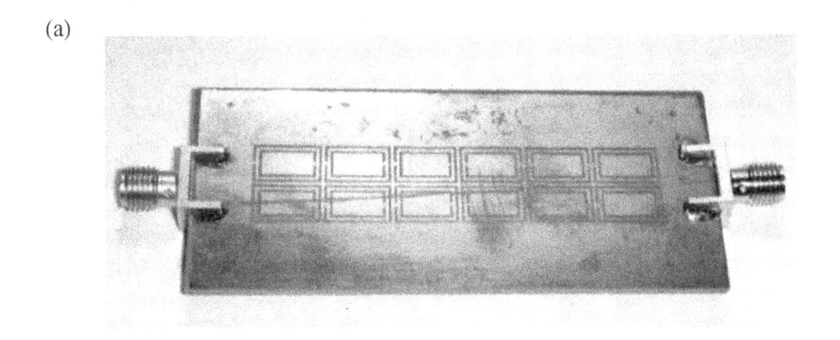

(b)

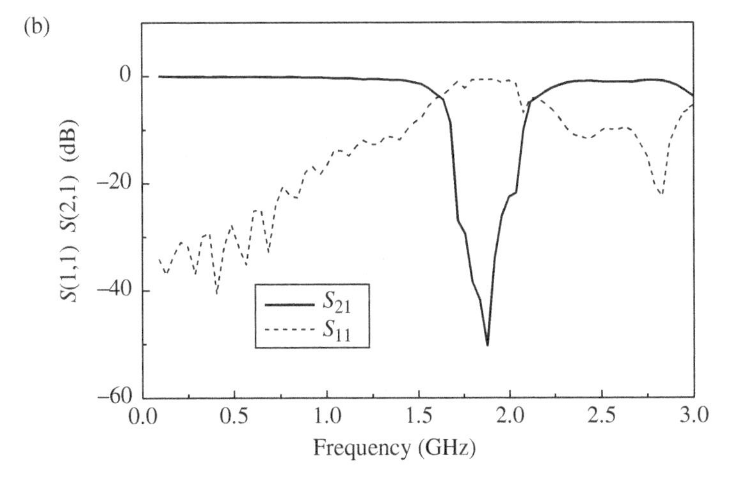

FIGURE 4.22 Microstrip line with square-shaped CSRR etched in the ground plane (a) and measured frequency response (b). $c = d = 0.4$ mm while the external edges of the CSRRs were tuned in the vicinity of 9.4 mm × 5.8 mm. The structure was fabricated on the *Rogers RO3010* substrate ($\varepsilon_r = 10.2$, $h = 1.27$ mm). Reprinted with permission from Ref. [44]; copyright 2005 IEEE.

of $f_o = 2.4$ GHz and 10% fractional bandwidth, is depicted in Figure 4.23. Rectangular SRRs are etched adjacent to the coupled lines in order to reject the first and second spurious bands. The multiple tuning procedure explained in the previous subsection was used. The smaller rings (etched in the central stages) are responsible for the rejection of the second spurious band, whereas the first undesired band is rejected by the action of the larger SRRs, which are allocated in the first and forth filter stages. The simulated and measured frequency responses of this device are also represented in Figure 4.23. In comparison to the conventional filter, the first spurious band is rejected with attenuation levels near 40 dB, while insertion losses in the second band are clearly above 20 dB. This efficiency can be attributed to the significant number of SRRs pairs distributed along the device, which is possible by virtue of their small electrical size. It is also worth mentioning that the use of square or rectangular-shaped SRRs enhances the coupling between line and rings, and this allows for further

(a)

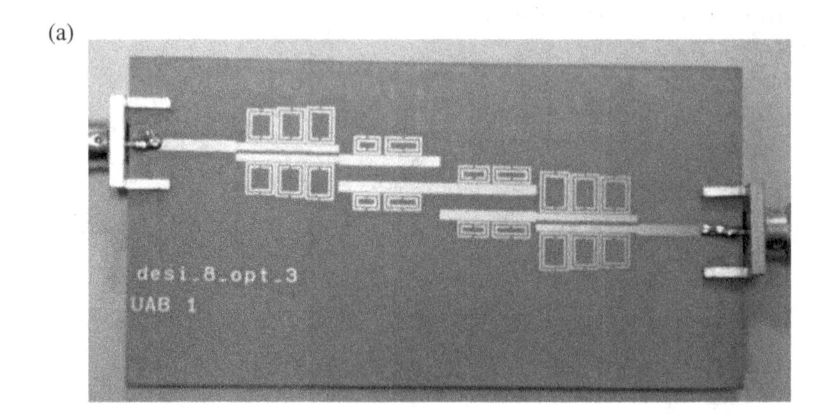

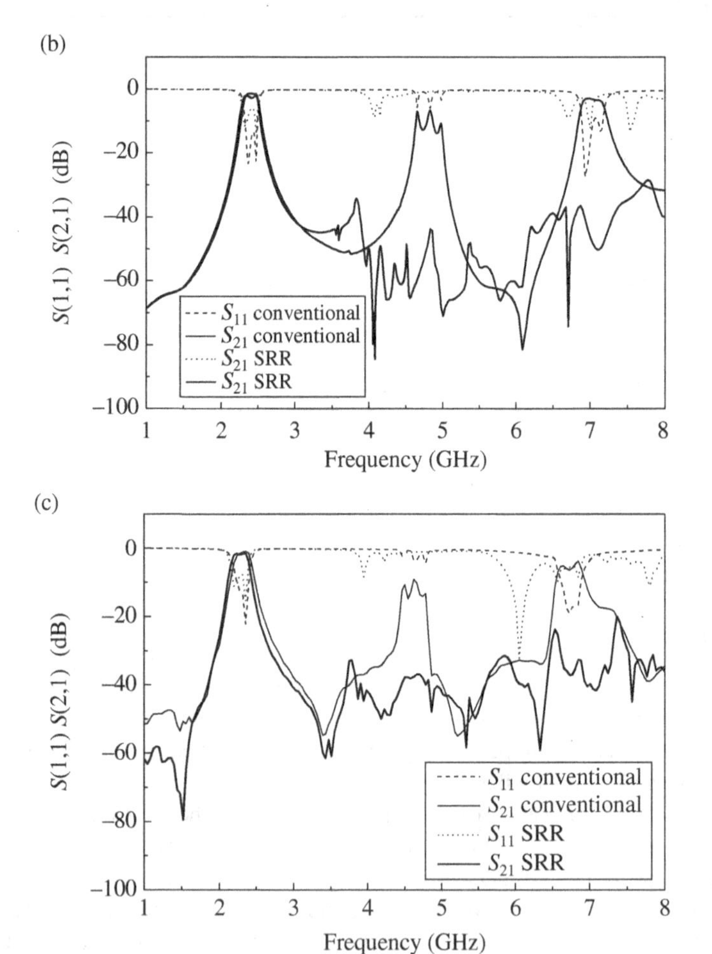

FIGURE 4.23 (a) Coupled-line microstrip bandpass filter loaded with SRRs for spurious suppression, (b) simulated frequency response compared to that obtained in the device without SRRs, and (c) measured frequency responses. The device was fabricated on the *Rogers RO3010* substrate (thickness $h = 1.27$ mm, dielectric constant $\varepsilon_r = 10.2$). Reprinted with permission from Ref. [44]; copyright 2005 IEEE.

rejection levels, as compared to circular SRRs. This strategy has also been applied (using CSRRs) to the elimination of spurious bands in CPW coupled line bandpass filters [44] and stepped impedance lowpass filters [51].

4.2.3.3 Narrow Band Bandpass Filters and Diplexers Based on Alternate Right-/Left-Handed Unit Cells The SRR/strip-loaded CPW transmission lines and the CSRR/gap-loaded microstrip CRLH transmission lines discussed in Chapter 3 can be used for the implementation of compact bandpass filters. By balancing these lines, filters with wide bands result, as will be discussed later. Conversely, if these structures are unbalanced the first, and narrow, LH band can in principle be exploited for narrow band filtering applications. The presence of a transmission zero to the left of the LH band provides good frequency selectivity at the lower edge of the transmission band. However, the frequency selectivity at the upper stop band is poor. To overcome this limitation, one strategy is to combine the LH structures based on SRRs or CSRRs with RH cells (also based on SRRs or CSRRs), the latter exhibiting a transmission zero above the RH band. Obviously, it is necessary to tune the cells in order to exhibit the same transmission band. This approach has been demonstrated to be useful for the implementation of very compact narrow band bandpass filters and diplexers in CPW and microstrip technology [52–56].

In CPW technology, an RH transmission band with a transmission zero to the right of this band can be achieved by combing the pairs of SRRs with series gaps. The equivalent circuit model of such cell is depicted in Figure 4.24a. According to this model, signal transmission occurs in that frequency region where the reactance of the series branch is positive (inductive), and this takes place in a narrow band before the resonance frequency of the SRRs (transmission zero frequency).

In microstrip technology, narrow RH transmission bands are expected in structures where the CSRRs are combined with shunt connected inductors. According to the model of this RH cell (Fig. 4.24b), the shunt reactance must be negative (capacitive) in the transmission bands, and this occurs to the left of the transmission zero frequency.

As illustrative examples of this alternate right-/left-handed (ARLH) approach, a narrow band CPW bandpass filter at S-band and a microstrip diplexer are reported.

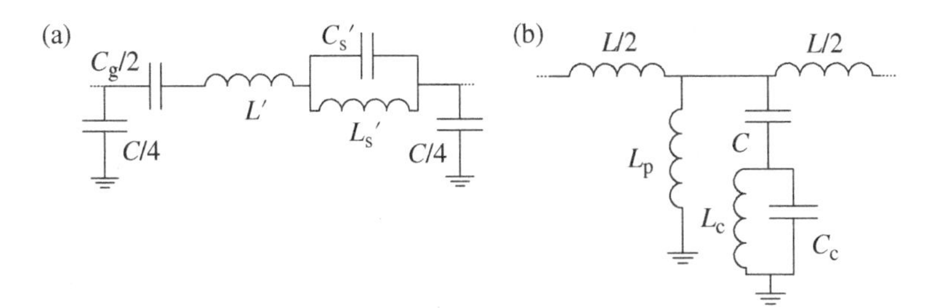

FIGURE 4.24 Lumped-element equivalent circuit for the SRR/gap-loaded CPW (a) and CSRR/strip-loaded microstrip line. C_g is the capacitance of the series gap and L_p is the inductance of the grounded strip. In (a), the magnetic wall concept is applied.

The S-band filter [55], inspired by the first implementation of the ARLH concept [52], was designed by cascading two unit cells: an LH SRR/strip and an RH SRR/gap combination (Fig. 4.25). The central filter frequency is $f_o = 2.4$ GHz. Filter length (active region) is as small as 22.5 mm, that is, five times smaller than the wavelength at f_o. The frequency response of the filter exhibits good frequency selectivity by virtue of the transmission zeros present at both sides of the pass band. The measured average

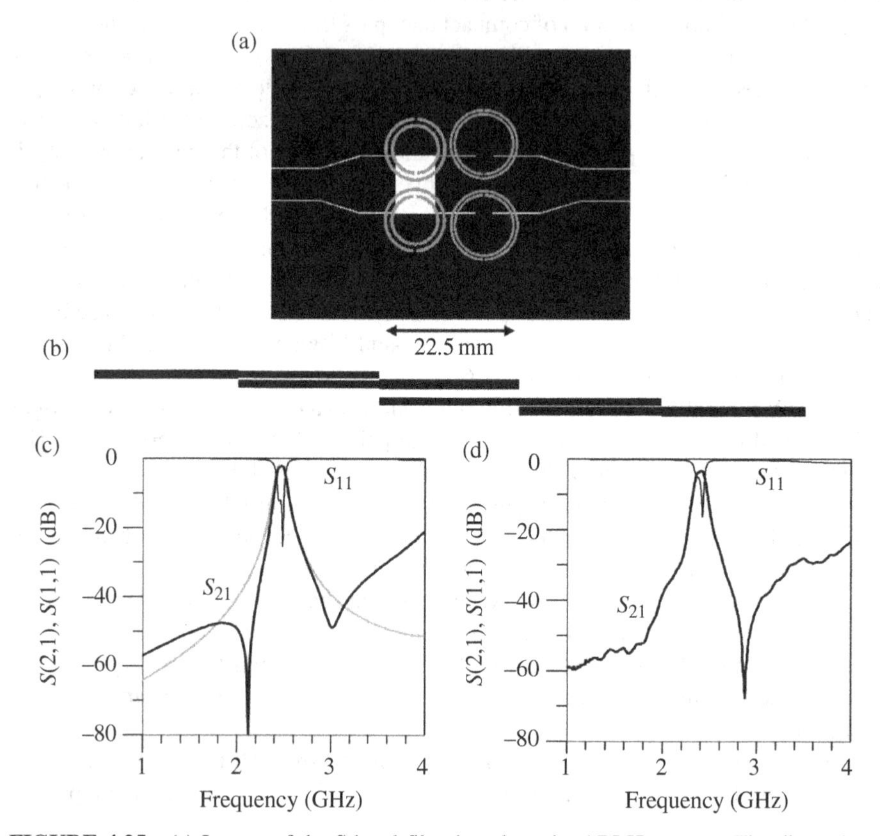

FIGURE 4.25 (a) Layout of the S-band filter based on the ARLH concept. The dimensions of the SRRs pairs corresponding to the strip and gap stages are $r = 5.2$ mm, $c = 0.44$ mm, $d = 0.22$ mm, and $r = 4.1$ mm, $c = d = 0.55$ mm, respectively. Strip and gap dimensions have been set to $w_w = 2.8$ mm and $l_g = 6.6$ mm, respectively. The host CPW is a 50 Ω line with a wide strip ($W = 10$mm) in order to accommodate the rings. Since the separation between ground planes is wider than the dimensions of the connectors in use, it was necessary to cascade taper transitions at the input and output ports to allow for connectors insertion. The length of the active device region is 22.5 mm. The considered substrate is the *Arlon 250-LX-0193-43-11* with dielectric constant $\varepsilon_r = 2.43$, and thickness $h = 0.49$ mm. (b) Layout of a conventional microstrip parallel-coupled line bandpass filter with comparable performance, drawn to scale. (c) Simulated frequency response for the S-band filter compared to that of the coupled line bandpass filter (insertion loss in gray line). (d) Measured frequency response of the S-band filter. Reprinted with permission from Ref. [55]; copyright 2005 John Wiley.

slope of the upper and lower transition bands is 150 dB/GHz and 125 dB/GHz, respectively. A conventional coupled line bandpass filter (in microstrip technology) exhibiting similar central frequency, bandwidth, and selectivity is also depicted in Figure 4.25 for comparison purposes. The active region of the coupled line bandpass filter is substantially longer (63.6 mm). Thus, the ARLH transmission line concept has been revealed to be an efficient approach for size reduction in narrow band bandpass filter. Limitations of this approach are the difficulty in controlling bandwidth and selectivity, and the sensitivity of the filter response with SRR dimensions (it was found that variations of the SRR geometry due to fabrication related tolerances degrade in-band losses).

For the implementation of the microwave diplexer an Rx and a Tx filter is required. The configuration of the considered diplexer is shown in Figure 4.26, where the Rx and Tx filters are cascaded at the output ports of a Y-junction. The main relevant parameters representative of diplexer performance are in-band looses for the Tx and Rx channels (which should be as small as possible) and Rx/Tx isolation (which should be high to avoid interfering signals between the Rx and Tx channels). A prototype device operative in the 2.4–3.0 GHz frequency band is depicted in Figure 4.27a [56]. The Tx and Rx filters (implemented by means of CSRR-based ARLH microstrip lines[17]) have been designed to provide pass bands centered at 2.4 and 3.0 GHz, respectively, with absolute bandwidths of 0.25 GHz (namely 10.3% and 8% fractional bandwidths for transmission and reception). Figure 4.27b depicts the measured transmission coefficient for the Tx and Rx filters (i.e., S_{21} and S_{31}, respectively), as well as the measured Rx/Tx isolation (S_{32}). In-band losses lower than 2 dB are measured for each filter, while return losses (also represented in Fig. 4.27b) are better than 10 dB. The frequency response of the filters is quite symmetric and the measured isolation between ports 2 and 3 is in the vicinity of 40 dB. Remarkable are also the dimensions of the diplexer (see the region indicated in Fig. 4.27a), which are as small as 29.8 mm × 16.3 mm (namely $0.63\lambda \times 0.34\lambda$, λ being the signal wavelength at the Tx frequency) thanks to the compact resonators employed (more details on diplexer design are given in Ref. [56]).

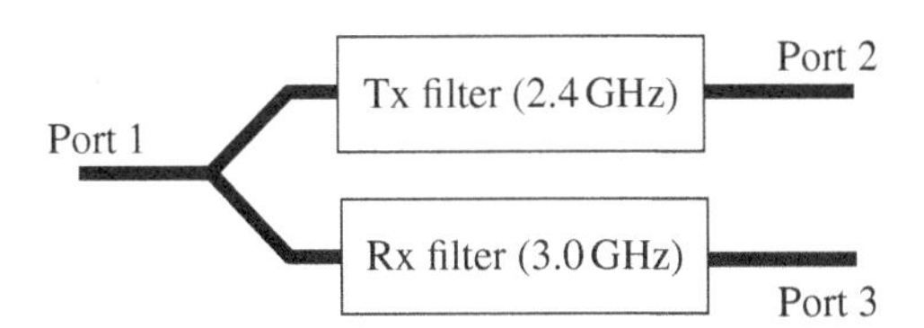

FIGURE 4.26 Structure of the diplexer.

[17] Notice that in the RH sections of the Rx and Tx filters, two CSRRs are used. The reason is the moderate bandwidth required for each filter. To achieve such bandwidths, it is necessary to implement unit cells also with moderate bandwidths. For the forward sections, this is more easily achieved by using two CSRRs, rather than only one, as it was pointed out in Ref. [53].

(a)

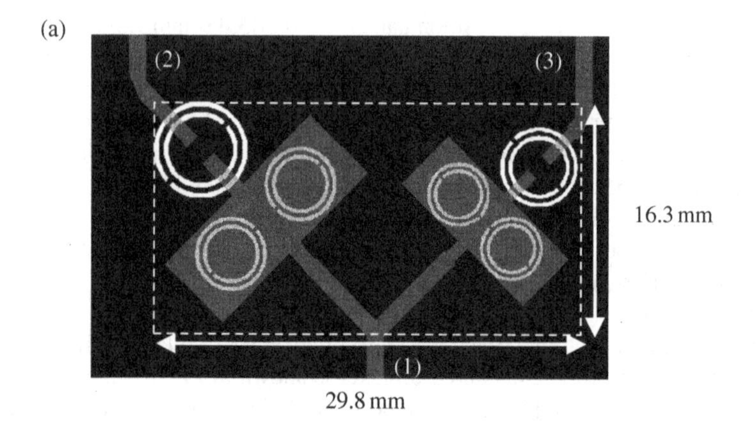

(b)

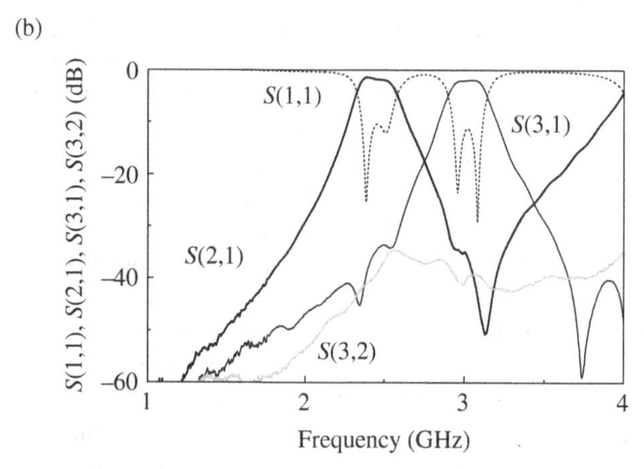

FIGURE 4.27 (a) Topology of the fabricated ARLH microstrip diplexer and (b) measured frequency response. The upper metal level is depicted in gray, whereas the lower metal is drawn in black. Gap spacing is 1.33 mm and 1.67 mm for the Rx and Tx filters respectively, whereas shunt strip dimensions are 4.54 mm × 11.9 mm and 5.70 mm × 14.48 mm. External CSRR radius are 2.18 mm, 2.73 mm, 2.56 mm, and 3.27 mm for the Rx (first stage), Rx (second stage), Tx (first stage), and Tx (second stage), respectively. The shunt strips are grounded through vias. The structure was implemented on the *Rogers RO3010* substrate (thickness $h = 1.27$ mm, dielectric constant $\varepsilon_r = 10.2$). Reprinted with permission from Ref. [56]; copyright 2005 IET.

4.2.3.4 Compact Bandpass Filters based on the Hybrid Approach Although the filters based on the ARLH transmission line concept are electrically small and provide moderate or narrow bands with symmetric responses and high selectivity, the controllability of the filter response is not simple. In this subsection, a methodology for the design of moderate and narrow band bandpass filters subjected to specifications (with the possibility to synthesize standard filter responses such as Butterworth and

Chebyshev[18]) is presented [57]. Such filters are based on the hybrid model of metamaterial transmission lines presented in Chapter 3. Since these structures can be designed to exhibit a transmission zero above the first LH band, the out-of-band performance of the filters can be improved through the suppression of undesired harmonic bands.

Design Methodology The microstrip filters proposed in this subsection are planar structures that can be modeled by the circuit of Figure 4.28a, which consists of a cascade of admittance inverters (with normalized admittance $\bar{J} = 1$) alternating with shunt connected resonators tuned at the central frequency of the filter band, f_o [33, 58]. This circuit is inferred from the lowpass filter prototype by the well-known frequency and element transformations that can be found in any microwave textbook [59] or in books devoted to microwave filters [33, 58].[19] By properly designing the shunt resonators, the synthesis of standard frequency approximations is potentially possible. Actually, the transformation from the lowpass filter prototype leads to the structure of Figure 4.28b, with parallel *LC* resonant tanks. Thus, as long as resonator's admittances fit to those of the *LC* tanks (which depend on the *L* and *C* values inferred from circuit transformation), the targeted approximation (Butterworth, Chebyshev, etc.) can be achieved in the vicinity of resonance.

The elemental cell and circuit model of the proposed filters is depicted in Figure 3.60. Following Ref. [57], in the present study the series inductance of the line, *L*, is neglected since it is assumed that the region of interest lies in that frequency band where the series reactance is capacitive. The design methodology of the filters is a two-step process: first, the electrical parameters of the equivalent circuit model for each filter section (provided the filter is not periodic) are determined from specifications (central frequency, fractional bandwidth and filter order); then, the layout of the

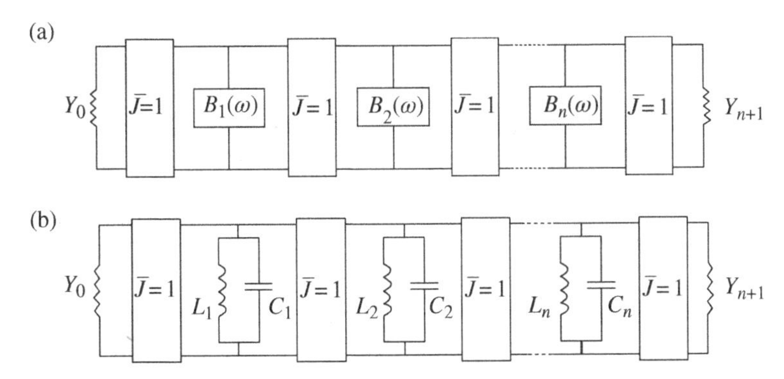

FIGURE 4.28 Generalized bandpass filter network with admittance inverters and shunt resonators (a). In (b), the resonators are *LC* resonant tanks.

[18] For that purpose, periodicity must be generally sacrificed.

[19] Alternatively, bandpass filters can be synthesized by alternating impedance inverters and series-connected resonators.

structure is synthesized. Let us consider the first aspect. In the circuit of Figure 4.28b, the filter central frequency, f_o, is determined by the resonance frequency of the shunt *LC* tanks, which nulls their admittance. However, this is not exactly the case for the filter implemented by cascading the elemental cells depicted in Figure 3.60, since resonators and admittance inverters are intermixed. Specifically, neither f_o coincides with the intrinsic resonance of CSRRs, nor it nulls the admittance of the shunt branch, $Y_p(\omega) = Z_p^{-1}(\omega)$. Nevertheless, at f_o the phase shift and transmission coefficient between the input and output ports of the basic cell should be $\phi = -90°$ (corresponding to backward wave propagation) and $|S_{21}| = 1$, respectively. This means that at f_o the characteristic impedance, Z_B, should coincide with the reference impedance of the ports, which is usually set to $Z_o = 50\,\Omega$. Thus, the series and shunt reactances of the T-circuit model of Figure 3.60 must satisfy: $Z_s(\omega_o) = -jZ_o$ and $Z_p(\omega_o) = jZ_o$. Actually, the dual solution ($Z_s(\omega_o) = jZ_o$ and $Z_p(\omega_o) = -jZ_o$) satisfies also the previous conditions on phase shift and impedance matching, but this solution is not compatible with the circuit of Figure 3.60 if L is neglected (i.e., with LH wave propagation). The above condition for Z_s and Z_p does not univocally determine the element values of the shunt impedance. These values are also determined by the 3 dB bandwidth of the resonators,

$$\Delta = \frac{\omega_2 - \omega_1}{\omega_o} \tag{4.38}$$

(where ω_o, ω_1, and ω_2 are the central—angular—frequency and 3 dB frequencies, respectively), and by the transmission zero, which occurs at that frequency where the shunt impedance reduces to zero (expression 3.86)

For a *LC* parallel resonant tank, with inductance and capacitance L_{eq} and C_{eq}, respectively, we have

$$\Delta = \frac{2}{Z_o}\sqrt{\frac{L_{eq}}{C_{eq}}} \tag{4.39}$$

If we consider the filter structure of Figure 4.28b, where the L and C values come from the lowpass filter prototype by frequency and element transformation according to [33]

$$C_{eq} = \left[\frac{1}{\text{FWB}\cdot\omega_o\cdot Z_o}\right] g_i \tag{4.40}$$

$$L_{eq} = \frac{1}{\omega_o^2 C_{eq}} \tag{4.41}$$

and (4.40) and (4.41) are introduced in (4.39), the following expression for resonator bandwidths results:

$$\Delta_i = \frac{2\text{FBW}}{g_i} \tag{4.42}$$

where g_i's are the element values of the lowpass filter prototype (which depend on the specific filter approximation and can be inferred from published tables) and Δ_i is the 3 dB bandwidth of the resonators. According to expression (4.42), Δ_i is proportional to the fractional bandwidth, FBW, and hence it is determined from the required filter bandwidth. Obviously, Δ_i is also dependent on g_i and it is therefore determined by the type of response and order. Once Δ_i are known, we can force the frequencies ω_1 and ω_2 to be equidistant from the central filter frequency. At these frequencies, under the assumption that $Z_s(\omega)$ does not substantially vary along the pass band, the shunt impedance of the unit cell satisfies: $Z_p(\omega_1) = jZ_o/2$ and $Z_p(\omega_2) = \infty$. These conditions plus $Z_p(\omega_o) = jZ_o$ give

$$\frac{L_d L_c \omega_1^3 (C + C_c) - L_d \omega_1}{L_c \omega_1^2 (C + C_c) - C\omega_1^2 L_d (L_c C_c \omega_1^2 - 1) - 1} = \frac{Z_o}{2} \tag{4.43}$$

$$L_c \omega_2^2 (C + C_c) - C\omega_2^2 L_d (L_c C_c \omega_2^2 - 1) - 1 = 0 \tag{4.44}$$

$$\frac{L_d L_c \omega_o^3 (C + C_c) - L_d \omega_o}{L_c \omega_o^2 (C + C_c) - C\omega_o^2 L_d (L_c C_c \omega_o^2 - 1) - 1} = Z_o \tag{4.45}$$

The previous approximation (which is valid for narrow and moderate bandwidths) leads us to simple analytical expressions (Eqs. 4.43 and 4.44). If this approximation is not applied, then the conditions arising from the 3 dB frequencies are not mathematically simple. Solution of Eqs. (3.86) and (4.43)–(4.45) leads us to the parameters of the shunt reactance, while the series capacitance is given by

$$C_g = \frac{1}{2Z_o \omega_o} \tag{4.46}$$

The criterion to set the transmission zero frequency, f_z, obeys to a compromise between the need to obtain a sharp transition in the upper band edge, and the convenience to far the spurious responses as much as possible, and thus optimize the out-of-band performance of the filter. Namely, a narrow spurious band above the pass band of interest arises. By adjusting f_z, this spurious can be minimized, and the filter frequency response can be significantly improved. Since the position of the spurious is not known a priori, the transmission zero frequency is tentatively set to $f_z = 2f_o$.[20] From the previous equations, the parameters of Figure 3.60b are inferred (except L,

[20] There is not a physical reason that justifies this choice. However, in practice, the spurious band lies in the vicinity of the first harmonic of the filter central frequency. This has been corroborated from several examples.

which is neglected, as mentioned before), and the topology of each filter cell (Fig. 3.60a) can be obtained.

For the synthesis of the layout, the method reported in Appendix H cannot be applied (this is of application in CSRR/gap-loaded lines but not in hybrid unit cells). Nevertheless, the model described in Ref. [60] can be used to obtain an initial guess for CSRR dimensions (the validity of the model is subjected to conditions that do not exactly apply in the considered filter cells [60]). The coupling capacitance, C, can be adjusted by partially removing the metal delimited by the CSRR contour. The length and width of the grounded metal strips (stubs) can be determined from independent full-wave EM simulations carried out in microstrip transmission lines loaded with these elements, or by using standard formulas [59]. A similar procedure can be used to determine the geometry of the series capacitances. From this initial layout, the frequency response of the complete filter is then simulated by means of a full wave EM solver, thus being visible the position of the spurious band. To eliminate this band, f_z is adjusted, and the model parameters are then recalculated. To determine the final layout, cell dimensions should be adjusted (starting from the seeding topology) in order to fit the electrical response obtained from the latter model parameters as much as possible. In practice, this is simple since cell bandwidth is mainly controlled by L_d and C (provided f_z and f_o are distant enough). Therefore, stub dimensions and the etched area inside the CSRRs can be adjusted to match to the required bandwidth, while the transmission zero frequency can be tailored by scaling CSRR dimensions. The parameter extraction method for CSRR-based lines, conveniently modified, can also be of help for the synthesis of the circuit layout.

Illustrative Examples Let us now consider the design of a bandpass filter with moderate bandwidth following this approach. In this example, a periodic structure is considered, although periodicity is not a necessary condition. The specifications are order 3, central frequency $f_o = 1$ GHz and fractional bandwidth FBW = 10%. This filter does not obey any standard approximation; therefore, we cannot directly determine g_i from table values, as usual. Obviously, due to periodicity, all resonators must have the same Δ and, accordingly, the same g. To obtain the value of g, an order-3 lowpass filter prototype with identical element values has been considered, and it has been forced to exhibit the 3 dB cutoff at the normalized $\omega = 1$ rad/s angular frequency. From this, it is obtained $g = 1.521$; hence, the 3 dB bandwidth of the resonators, Δ, is perfectly determined. From the previously explained procedure, the element values of the equivalent circuit model were obtained. They are identical for the three filter cells and are depicted in Table 4.2. Layout dimensions for the basic cell are indicated in

TABLE 4.2 Element values of the equivalent circuit model for the filter of Figure 4.30

C_g(pF)	L_d(nH)	C(pF)	C_c(pF)	L_c(nH)
1.59	1.33	12.33	21.7	0.23

Figure 4.29 (the parameters of the *Rogers RO3010* substrate with thickness $h = 1.27$ mm and dielectric constant $\varepsilon_r = 10.2$ have been used).

The EM simulation of the frequency response (amplitude and phase) of the single-cell structure is represented in Figure 4.29 and compared to that obtained from the circuit simulation (insertion loss only) of the equivalent circuit model (with the parameters indicated in Table 4.2). Reasonable agreement between the circuit and EM simulations is obtained. The simulated and measured insertion and return loss for the fabricated three-stage filter are depicted in Figure 4.30. Thanks to the transmission zero, the frequency response is spurious free up to approximately $3f_o$. In-band insertion and return losses are good (i.e., $IL < 1.5$ dB and $RL > 17$ dB) and frequency selectivity at both band edges is high, with near symmetric transition bands. The measured fractional bandwidth is FBW = 8%, which coincides to a good approximation with the nominal value. Cell dimensions (length) are small as compared to signal wavelength at f_o (i.e., $l_c \sim \lambda/7$). Further miniaturization can be achieved with the penalty of critical (smaller) dimensions being closer to the limits imposed by the fabrication technology (typically ≈ 100 μm). To avoid problems related to fabrication tolerances, critical dimensions substantially larger than 100 μm were considered in this design, with the result of a moderately small cell size. In Figure 4.30b, the simulated frequency response (insertion loss) obtained on a conventional microstrip coupled line bandpass filter with similar performance is also included (layout comparison is shown in the inset of Fig 4.30b). As compared to the conventional response, where a spurious band is present at $2f_o$, the CSRR-based prototype device filter exhibits near 40 dB rejection at that frequency, and device length, excluding access lines, is 2.4 times shorter. This prototype device, published in Ref. [57], was the first LH transmission line based on CSRRs (hybrid approach) employed for the design of a bandpass filter following a methodology able to provide the element values of the equivalent circuit model. Another interesting example of the approach is the design of an order-3 Chebyshev (i.e., nonperiodic) bandpass filter with 9% fractional bandwidth, 0.3 dB ripple, centered at $f_o = 2.5$ GHz (see Ref. [57] for details). Following this design methodology, an ultracompact bandpass filter at K-band, implemented in MCM-D technology, was reported in Ref. [61].

Alternatively to the unit cell of Figure 3.60 (hybrid cell), the shunt resonators (composed of the CSRRs plus the shunt inductive strip) may be coupled through quarter wavelength transmission lines or by means of inductive elements. The design approach, summarized in Ref. [4], is similar to that reported before and has been applied to the design of bandpass filters with moderate bandwidths [62–64]. It is also possible to achieve wide band filter responses by considering balanced hybrid cells, as discussed in Ref. [65, 66]. Figure 4.31 depicts a bandpass filter based on a periodic structure consisting of four balanced hybrid cells, and subjected to the following specifications: bandwidth covering the 4–6 GHz range or wider, at least 80 dB rejection at 2 GHz, in-band insertion losses lower than 1 dB, and group-delay variation smaller than 1 ns. The simulated frequency response and group delay satisfy these requirements, although due to the critical dimensions of some geometry parameters (and fabrication related tolerances), the measured response exhibits higher in-band losses. It is

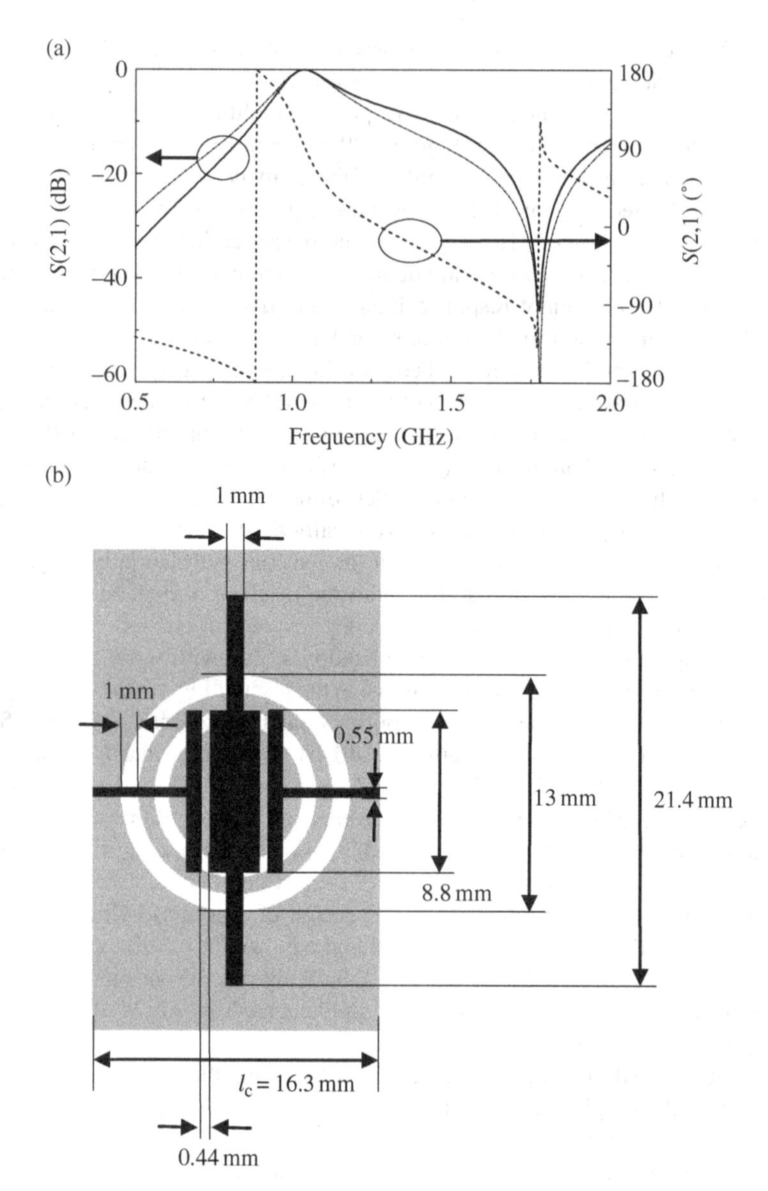

FIGURE 4.29 Simulated frequency response (amplitude and phase) corresponding to the basic filter cell of the designed filter, depicted in (b). The insertion loss obtained by circuit simulation of the equivalent circuit model is also depicted (thin line). The positive phase (+90°) of S_{21} (negative electrical length), obtained at f_o, clearly points out the LH nature of the structure. Reprinted with permission from Ref. [57]; copyright 2006 IEEE.

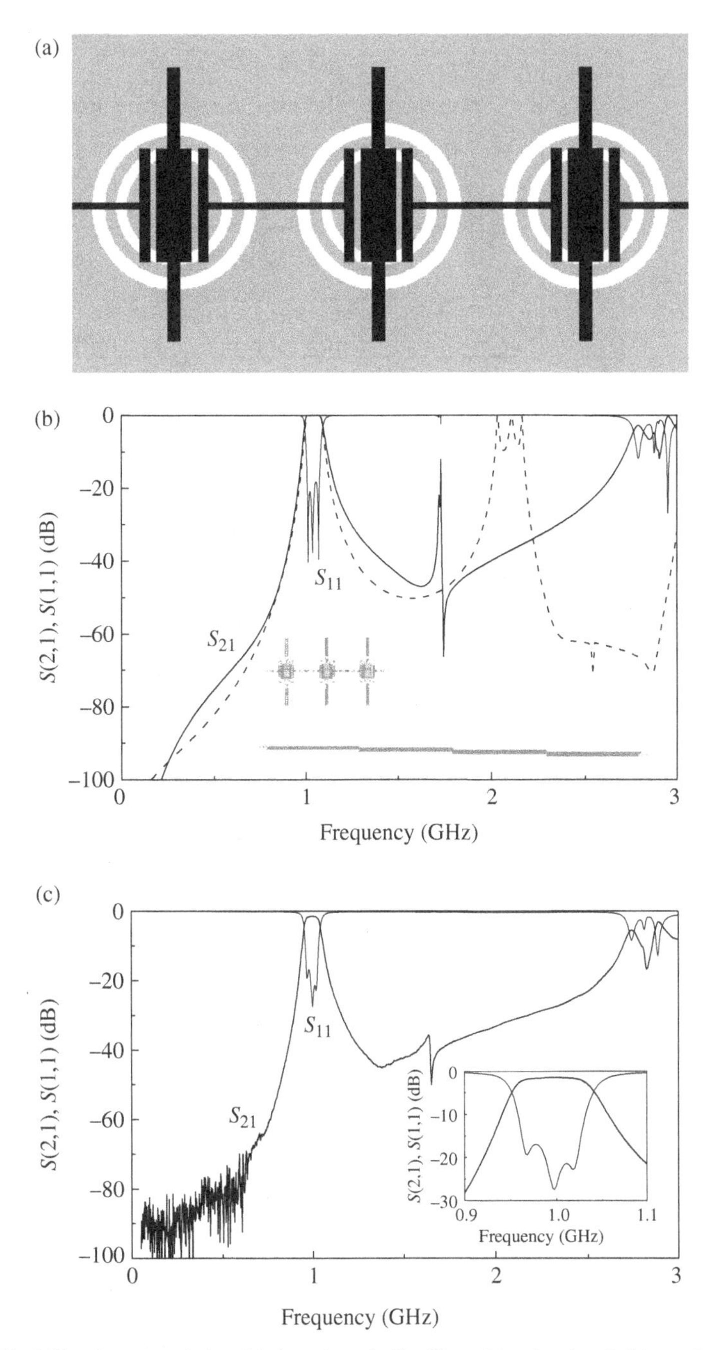

FIGURE 4.30 Layout of the fabricated periodic filter (a), simulated (b), and measured (c) insertion and return loss. Total device length excluding access lines is 4.56 cm. In (b), the simulated insertion loss for a conventional order-3 coupled line filter with similar performance is depicted (dashed line). Reprinted with permission from Ref. [57]; copyright 2006 IEEE.

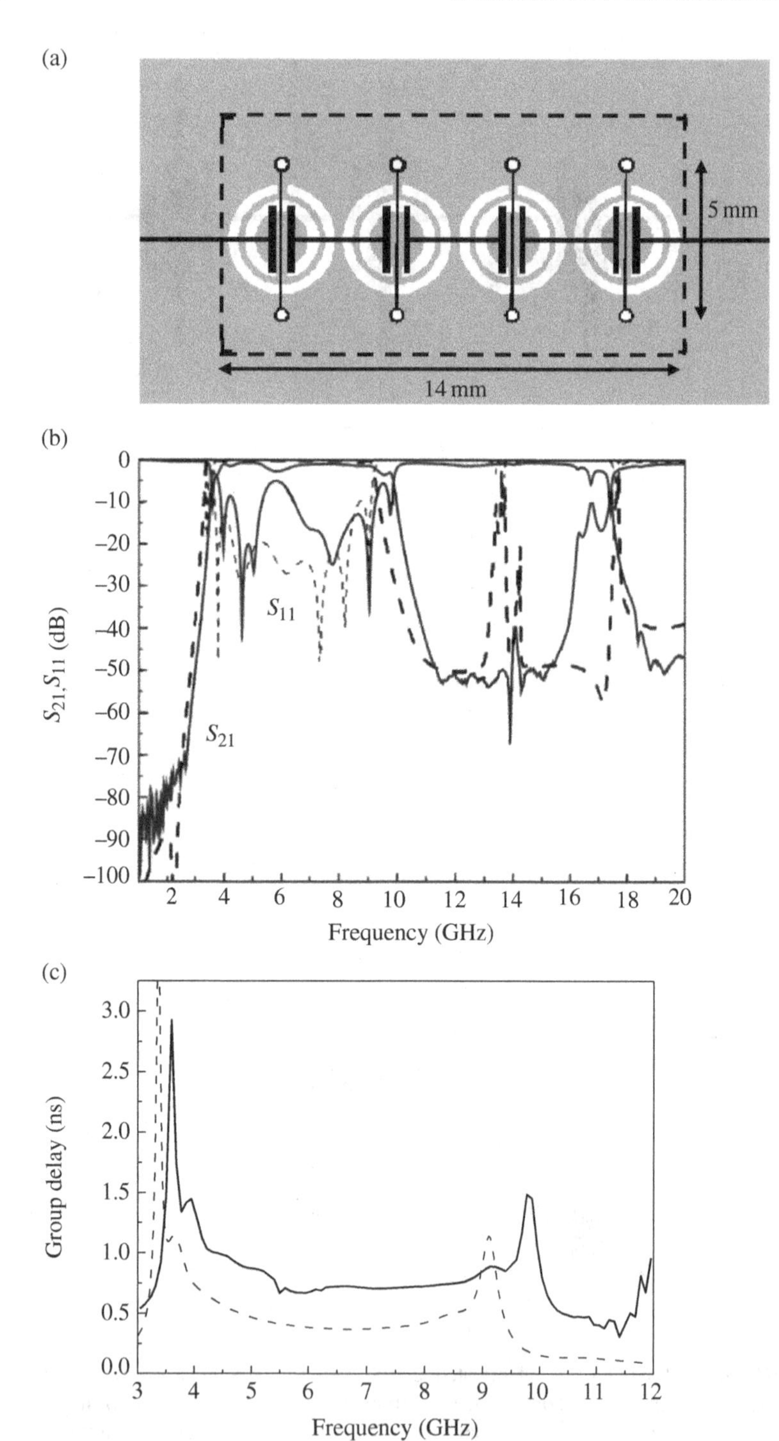

FIGURE 4.31 Layout of the wide band filter based on balanced hybrid cells (a), simulated (dashed line) and measured (solid line) frequency response (b), and simulated (dashed line) and

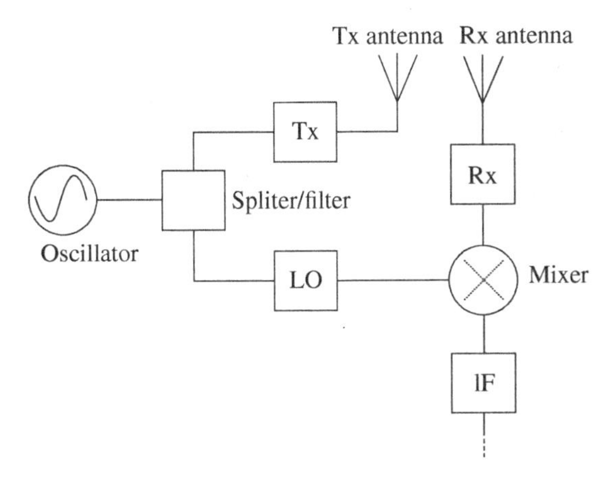

FIGURE 4.32 Schematic of the front-end of a FMCW-GPR system that uses one transmitter (Tx) antenna and one receiver (Rx) antenna.

remarkable that the filter area is as small as 0.7 cm^2. This size points out the potential of CSRR filters based on the hybrid cell for device miniaturization.

The last example of this subsection is a multifunctional filter, namely, a wideband filter with power splitting capability, of interest, for instance, for the front-end miniaturization of frequency-modulated continuous-wave ground penetrating radars (FMCW-GPRs) [67, 68]. The schematic of such system is depicted in Figure 4.32. The input signal to the power divider is modulated so that its frequency varies gradually. The power divider splits the signal between the transmitter antenna and the local oscillator of the receiver mixer, so that the radar signal is mixed with the signal reflected from the target object. Its depth can be determined from the intermediate (IF) signal frequency according to standard formulas [59]. To improve resolution and to obtain high penetration depths, it is convenient to deal with moderately low carrier frequencies (operation at the L-band) and wide radar frequency sweep. Thus, power splitters with significant fractional bandwidth are necessary. Such power dividers must also be able to reject the first harmonic of the radar signal within the whole frequency sweep. For this reason, filters are also needed (notice that harmonic rejection at a single frequency, reported in several papers [69, 70], does not suffice in the present application). In order to save space, a new power divider concept, with filtering capability, was proposed in Ref. [67] (this concept is also included in the review

---

FIGURE 4.31 (*Continued*) measured (solid line) group delay (c). The metallic parts are depicted in black in the top layer, and in gray in the bottom layer. The dashed rectangle has an area of 1 cm^2. Dimensions are line width $W = 0.126$ mm, external radius of the outer rings $r = 1.68$ mm, rings width $c = 0.32$ mm, and rings separation $d = 0.19$ mm; inductor width is 0.10 mm and the distance between the metals forming the gap is 0.4 mm. The considered substrate is the *Rogers RO3010* with dielectric constant $\varepsilon_r = 10.2$ and thickness $h = 0.127$ mm. Reprinted with permission from Ref. [65], copyright 2007 IEEE.

paper [68]), and reported here as an illustrative example of application of the hybrid approach of metamaterial transmission lines. The following system requirements were considered: frequency sweep 1—2 GHz; at least 20 dB rejection for the first harmonic; return losses for the splitter/filter better than 10 dB; insertion losses for the splitter/filter below 5 dB. Since the first harmonics (lower frequencies) are very close to the upper frequency of the modulated radar signal, the whole bandwidth was covered by means of two splitter/filters: one designed to scan the 1–1.5 GHz band; the other covering the 1.4—2 GHz frequency band. The area of both devices was required to be smaller than 12 cm^2.

The considered strategy for the implementation of the filters/splitters was to add an inductive transformer to the output stage of a bandpass filter. By this means, the load of the device (corresponding to the parallel connection of the two terminated output ports, that is, $R = 25\,\Omega$) can be driven to the required value to achieve input matching ($R_{eq} = 50\,\Omega$), and the filtering and splitting action are combined in a single multifunctional device. According to this strategy, the device is indeed a bandpass filter with modified output stage. The inductive transformer equations are (Fig. 4.33):

$$R_{eq} = \left(1 + \frac{L_1}{L_2}\right)^2 \cdot R \tag{4.47}$$

$$L_{eq} = L_1 + L_2 \tag{4.48}$$

Thus, it is possible to adjust the values of L_1 and L_2 to obtain the required value of the impedance R_{eq}, that is $R_{eq} = 50\,\Omega$. We report here the implementation of the splitter/filter covering the band 1–1.5 GHz. To this end, a fourth-order filter was considered, where three filter stages were implemented by means of a balanced hybrid cell (similar to that of Fig. 4.31), and the additional stage was implemented by parallel connecting an inductive stub (acting as inductive transformer) and a series resonator (implemented by series connecting a narrow inductive strip, L_3, and a capacitive patch, C_1), as depicted in Figure 4.34. This later stage, with the presence of the inductive stub, is required to include the inductive transformer, which cannot be integrated within the hybrid cell. Essentially, this fourth stage is a shunt connected parallel resonator. However, the additional inductance L_3 series connected to the capacitor C_1 provides an additional transmission zero, of interest to improve the filter stop band response.

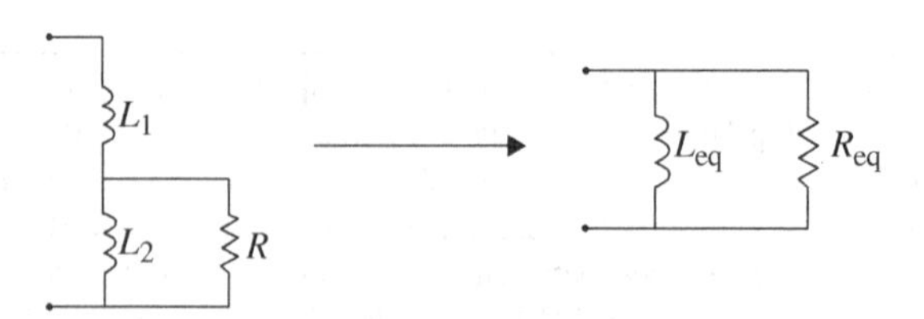

FIGURE 4.33 Inductive transformer.

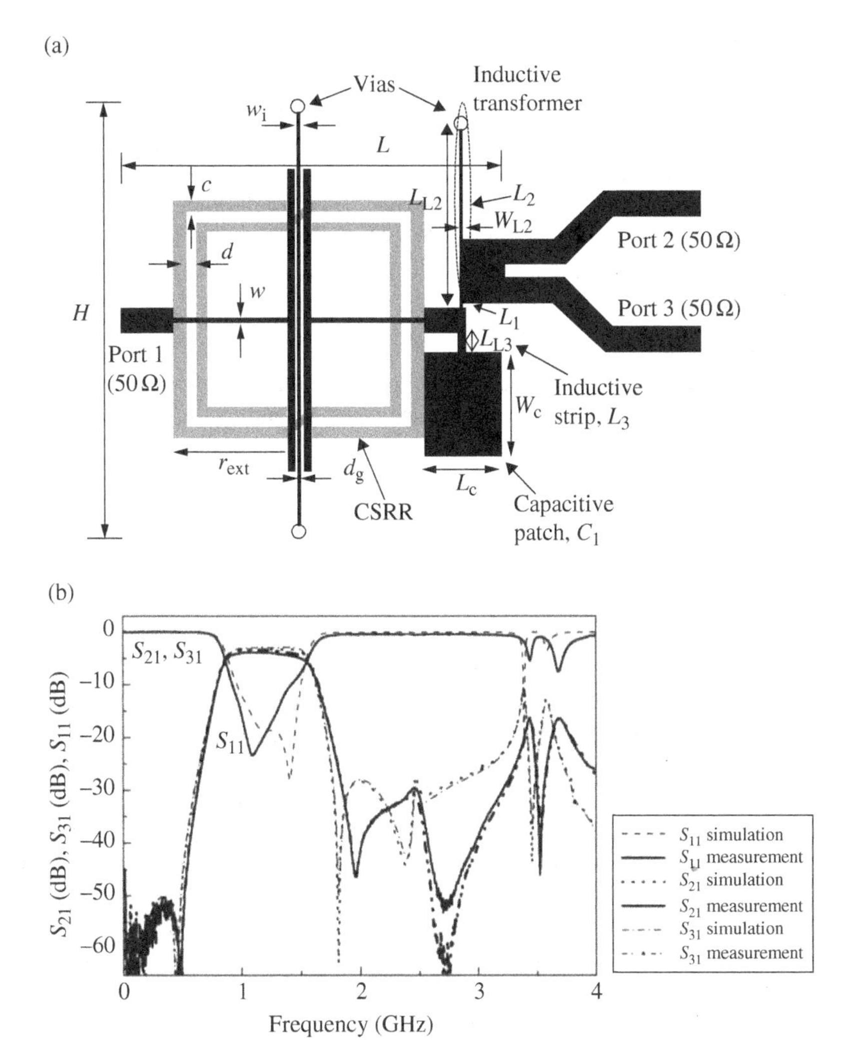

FIGURE 4.34 Topology of the designed filter/splitter (a), and EM simulation and measurement of the power splitting (S_{21}, S_{31}) and matching (S_{11}) (b). Dimensions are $H = 32.6$ mm, $L = 27.3$ mm, $W_l = 0.2$ mm, $c = 0.79$ mm, $d = 0.79$ mm, $w = 0.16$ mm, $r_{ext} = 8.12$ mm, $d_g = 0.26$ mm, $L_c = 5.7$ mm, $W_c = 7.7$ mm, $L_{L2} = 12.96$ mm, $W_{L2} = 0.2$ mm, and $L_{L3} = 1.53$ mm. The device was implemented on the *Rogers RO4003* substrate with thickness $h = 0.81$ mm and dielectric constant $\varepsilon_r = 3.55$. Notice that for better comprehension, in this figure the CSRR, etched in the back substrate side (ground plane), is depicted in gray (the back side metallization is not depicted). Black color corresponds to the upper metal level. Reprinted with permission from Ref. [68]; copyright 2011 IEEE.

The topology of the proposed filter/splitter is depicted in Figure 4.34. For design purposes, the equivalent circuit model of the hybrid cell, as well as the elements describing the inductive transformer and the shunt connected series resonator, were considered. However, the structure is complex and optimization at layout level was necessary. The EM simulation and measurement of the frequency response of the filter/splitter are depicted in Figure 4.34b. The specifications given earlier were fulfilled, with measured insertion losses in the center frequency of the pass band ($S_{21} = 3.8$ dB and $S_{31} = 3.1$ dB) close to the ideal value (3 dB) of a splitter, and a device area of 8.9 cm^2.

4.2.3.5 Highpass Filters Based on Balanced CRLH Lines In microwave engineering, the difference between a highpass and a bandpass filter is subtle since highpass filters implemented in planar technology do not exhibit an unlimited bandwidth above the cutoff frequency. Thus, microwave highpass filters are actually bandpass filters with a very wide band and a controllable cutoff frequency at the lower band edge. These filters can be implemented by cascading balanced CSRR/gap-loaded lines with unit cells similar to that depicted in Figure 3.42, by virtue of the wide band and transmission zero that such balanced cells exhibit. Rejection below cutoff can be controlled by the number of cells, as illustrated in Figure 4.35. This figure depicts the layout of a fabricated three-cell highpass filter [66]. The filter exhibits a measured rejection better than 50 dB below the cutoff frequency, and in-band insertion and return losses are good up to roughly 3 GHz.

The second reported prototype is a highpass filter implemented by means of a balanced SRR/strip-loaded CPW with three pairs of SRRs (Fig. 4.36) [71]. In this device, designed to exhibit the transmission zero frequency at $f_z = 1.4$ GHz, the measured rejection in the stop band is better than 30 dB, and insertion loss in the transmission band is good ($IL < 2$ dB between 1.40 and 2.48 GHz). Notice that the topology of the host CPW was modified (as compared to the typical topologies of SRR/strip-loaded CPWs) in order to balance the structure.

Bandpass filters with ultra wideband (UWB) responses can be designed by introducing additional elements to the balanced CSRR/gap-loaded lines able to generate a controllable upper edge. Such structures have been applied to the design of UWB bandpass filters covering the mask supplied by the Federal Communications Commission (FCC) [41, 72].

4.2.3.6 Wideband Filters Based on OSRRs and OCSRRs The open particles, OSRRs and OCSRRs, are useful for the design of bandpass filters with wide fractional bandwidths and small dimensions. Two main strategies have been reported for the implementation of wideband filters based on these particles. One of them consists of using series connected OSRRs or shunt connected OCSRRs coupled through impedance inverters; the other strategy combines series OSRRs and shunt OCSRRs in configurations similar to those corresponding to CRLH lines based on these open resonant particles (reported in Section 3.5.2.5).

(a)

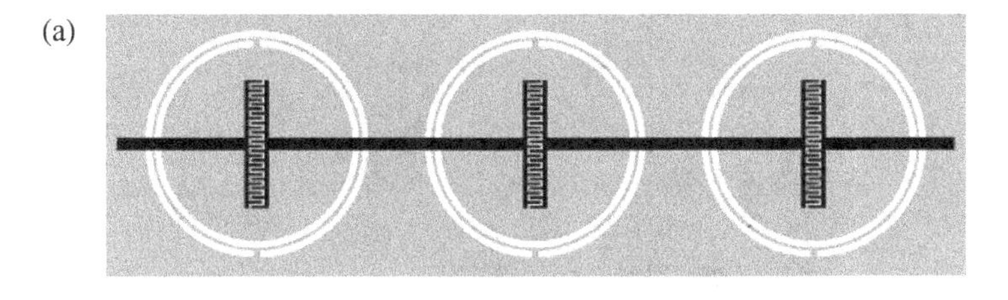

(b)

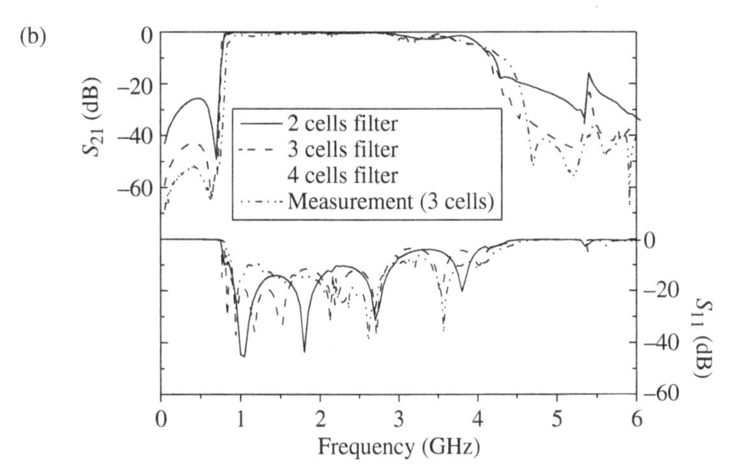

FIGURE 4.35 Layout of the microstrip highpass filter formed by three balanced CRLH CSRR-based cells (a) and measured frequency response (b). In (b), the simulated insertion and return losses for a two-, three-, and four-stage device filters are also depicted. The metallic parts are depicted in black in the top layer, and in gray in the bottom layer. Dimensions are total length $l = 55$ mm, line width $W = 0.8$ mm, external radius of the outer rings $r = 7.3$ mm, rings width $c = 0.4$ mm, and rings separation $d = 0.2$ mm; the interdigital capacitors are formed by 28 fingers separated 0.16 mm. The substrate is the *Rogers RO3010* with thickness $h = 1.27$ mm and dielectric constant $\varepsilon_r = 10.2$. Reprinted with permission from Ref. [66]; copyright 2007 IEEE.

Bandpass Filters Based on OSRRs or OCSRRs Coupled through Impedance Inverters The idea of using OSRRs for the implementation of wideband bandpass filters was first reported in Ref. [73], and the systematic approach for the design of filters where such resonators are coupled through impedance inverters was reported in Ref. [74]. The relatively large capacitance values of OSRRs are the responsible for the wideband response of these particles, and hence for the broad achievable fractional bandwidths. Filter design is based on the generalized bandpass filter network depicted in Figure 4.37, where series LC resonators are coupled through impedance inverters [33].[21] As it is pointed out in Ref. [33], there are two possibilities for filter design.

[21] This network is an alternative to the network depicted in Figure 4.28 for the design of bandpass filters. However, the approach for filter design considered in this subsection is the standard approach reported in Ref. [33], and hence different than the one reported in Section 4.2.3.4.

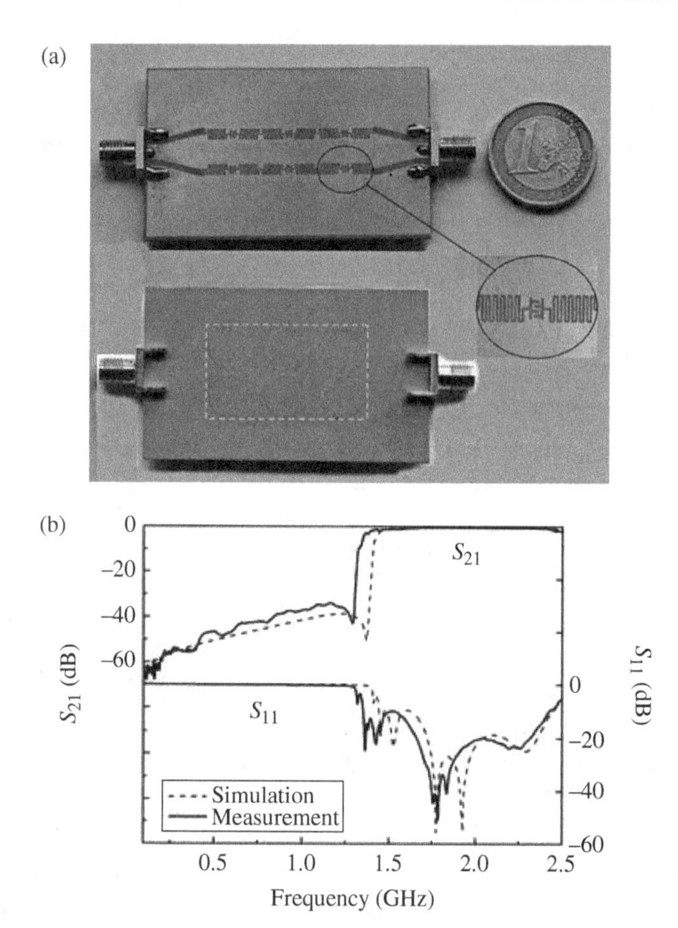

FIGURE 4.36 Photograph of the CPW highpass filter implemented by means of three CRLH SRR-based cells (a) and frequency response (b). Dimensions are external radius of the SRRs $r_{ext} = 4.68$ mm, width of the rings $c = 0.223$ mm, and rings separation $d = 0.237$ mm. The shunt inductance is implemented by means of a meander with width $w_m = 0.2$ mm and separation distance $s_m = 0.2$ mm. The gap distance between the central strip and the ground planes is $g = 0.16$ mm. The substrate is the *Rogers RO3010* with thickness $h = 1.27$ mm and measured dielectric constant $\varepsilon_r = 11.2$. Reprinted with permission from Ref. [71]; copyright 2010 IET.

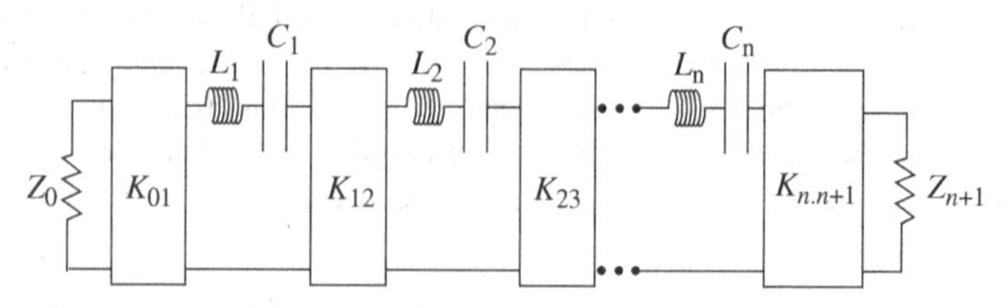

FIGURE 4.37 Generalized bandpass filter network with impedance inverters and series LC resonators.

In the first case, all the inverters are forced to have the same normalized impedance, $\overline{K}_{ij}=1$ (i.e., the inverters are quarter wavelength transmission lines with the usual characteristic impedance $Z_o = 50\,\Omega$). In this case, the design algorithm provides different values of L_i and C_i for each filter stage. The second option, considered in Ref. [73, 74],[22] employs the same resonator for all filter stages (therefore, the values of L_i and C_i are identical in the different filter stages) and different values of K_{ij} for the impedance inverters.

As an example, an order-7 microstrip bandpass filter with central frequency $f_o = 5$ GHz, 40% fractional bandwidth and 0.01 dB ripple is reported [74] (the considered substrate is the *Rogers RO3010* with dielectric constant $\varepsilon_r = 10.2$ and thickness $h = 0.254$ mm). The geometry of the OSRR was chosen to provide a series inductance and capacitance of $L = 2.99$ nH and $C = 0.338$ pF (the effects of the parasitics are accounted for in the impedance inverters). With these reactive values, the impedances of the inverters that result by applying the formulas given in Ref. [33] (un-normalized values) are: $K_{01} = K_{78} = 48.6\,\Omega$, $K_{12} = K_{67} = 35.7\,\Omega$, $K_{23} = K_{56} = 24.1\,\Omega$, and $K_{34} = K_{45} = 22.2\,\Omega$. The determination of the length and width of the inverters was done through optimization, by forcing 90° phase shift for all filter stages at f_o. The filter layout and the frequency response are depicted in Figure 4.38. In-band insertion losses are very small (<1 dB), whereas the measured filter bandwidth is 36%, slightly below the nominal value as expected on account of the limited functionality of the impedance inverters.[23] Optimization of filter size by meandering the inverters is reported in Ref. [74] through two additional examples.

Wideband bandpass filters based on OCSRRs and implemented in CPW technology were reported for the first time in Ref. [75] (the design methodology is similar to that corresponding to the OSRR-based filters detailed above). Notice that the wideband filters considered in this subsection are not based or inspired on artificial transmission lines. They are simply implemented by means of the open versions of the SRR or CSRR (i.e., OSRRs and OCSRRs). Nevertheless, the inclusion of these filters in this book obeys to the fact that these filters were conceived before the OSRR/OCSRR-based CRLH lines (discussed in Section 3.5.2.5), and they were in fact the precursor of these artificial lines and the filters based on them, to be discussed in the next subsection.

Bandpass Filters Based on CRLH Lines Implemented by Means of OSRRs and OCSRRs The CRLH unit cell based on the combination of OSRRs and OCSRRs, used in Section "Dual-Band Devices Based on Fully Planar CRLH Lines" for the

[22] The reason for this choice is that the synthesis of transmission lines with a given characteristic impedance is more simple than the synthesis of OSRRs with given values of L and C.

[23] It is well known that wideband filters based on impedance inverters implemented by means of quarter-wavelength transmission lines experience bandwidth degradation as compared to the nominal value. The reason is that in the extremes of the transmission band, the phase of the line may be distant from the 90° requirement for inverter functionality.

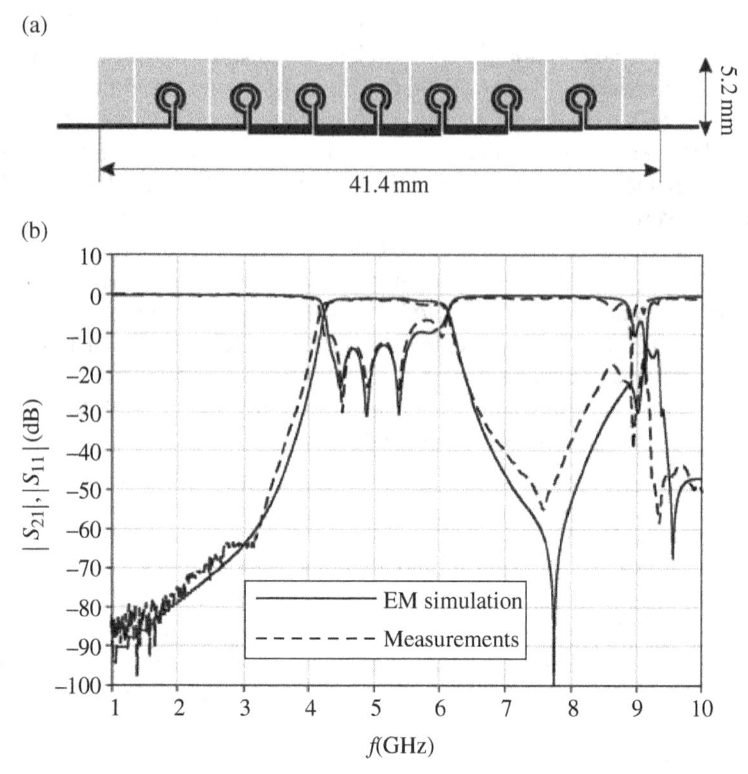

FIGURE 4.38 Layout (a) and frequency response (b) of the filter based on OSRR coupled through impedance inverters. OSRR dimensions are $r_{ext} = 1.1$ mm and $c = d = 0.2$ mm. The gray region indicates the windows in the ground plane. Reprinted with permission from Ref. [74]; copyright 2007 IEEE.

implementation of dual-band components, is described by a model that is also the canonical circuit of an order-3 bandpass filter.[24] Therefore, such structures can be used as building blocks for wideband bandpass filter design. By sacrificing periodicity, it is possible to implement higher-order filters subjected to standard responses, such as Chebyshev or Butterworth, among others. Filter synthesis follows the same approach considered for the synthesis of dual-band impedance inverters and the dual-band components based on them. Therefore, it will not be repeated in this subsection. Let us recall, however, that synthesis requires tuning the element values of the circuit model with the presence of parasitics in order to achieve the required response. If the parasitics are small, and for low-order filters, this tuning can be done manually. However, if the filter order and/or the parasitic element values increase, the manual

[24] Notice that by neglecting the parasitics, the circuits of Figure 3.58c, or d, are the circuit models of order-3 bandpass filters and also the circuit models of order-2 CRLH lines, where the line order indicates the number of elements of the series and shunt branch of the CRLH line, as discussed in Chapter 3. Therefore, caution must be taken to avoid confusion between filter order and line order (for CRLH lines).

tuning of the structure can be complex and time-consuming. For this reason, an automatic optimization routine that forces the frequency positions at the N matching points (N being the filter order) of the Chebyshev response (which is the considered response in all the examples) can be implemented in a commercial circuit simulator, such as the *Agilent ADS* [76]. Thus, from the matching frequencies and ripple, given by the standard Chebyshev response, the optimum circuit elements (considering the parasitics set to certain values) can be obtained. This procedure allows us to obtain filter responses at the circuit level, with the presence of parasitics, which are very close to the required Chebyshev response. The resulting circuits are slightly unbalanced in order to preserve the equal-ripple magnitude response. One important point for filter design, especially for wide band filters, is the inclusion of the inductance L_{sh} in the circuit model of the OCSRR stages (see Fig. 3.55). This typically gives higher rejection at the upper stop band, as compared to the Chebyshev response, and the agreement with the EM simulation and measurement is expected to be more accurate.

The synthesis technique was applied to the design of various filters with different order and bandwidth [28, 68, 76–78], including order-7 filters and fractional bandwidths in the vicinity of 100%. As an example, Figure 4.39 depicts an order-3 Chebyshev bandpass filter with central frequency $f_o = 2$ GHz, 0.15 dB ripple and 70% fractional bandwidth, and its frequency response as well. In the narrow band response,[25] where the lossless EM simulation and the circuit simulation with and without the inclusion of L_{sh} are considered, it can be appreciated the excellent agreement between the circuit and EM simulation when L_{sh} is included. The wideband response indicates that the measured filter response is in good agreement with the circuit simulation and with the ideal Chebyshev response. It is also remarkable that filter size is as small as $0.13\lambda \times 0.15\lambda$, where λ is the guided wavelength at the filter central frequency. In summary, the combination of OSRRs and OCSRRs is useful for the design of filters requiring wide transmission bands and small size. The synthesis procedure is simple, and the circuit models accurately describe the filter responses.

Dual-Band Bandpass Filters Based on the E-CRLH Transmission Line Concept It is also possible to design dual-band bandpass filters by means of the order-4 E-CRLH structures of the type depicted in Figure 4.19. This dual-band functionality can be obtained from the lowpass filter prototype through convenient transformations. In a first step, a single-band bandpass response results from the well known transformation [58]:

$$\frac{\Omega}{\Omega_c} = \frac{1}{w}\left(\frac{\omega}{\omega_0} - \frac{\omega_0}{\omega}\right) \tag{4.49}$$

Ω and ω being the angular frequency of the lowpass and bandpass filter, w and ω_0 the fractional bandwidth and central frequency of the bandpass filter and Ω_c the angular cutoff frequency of the lowpass filter. Then, a second transformation is applied to the single-band bandpass filter to obtain dual-band behavior [79]:

[25] By narrow band response, we mean the insertion and return loss restricted to the bandpass region (that can be arbitrarily broad).

$$\omega = \frac{\omega_0}{w'}\left(\frac{\omega'}{\omega_0} - \frac{\omega_0}{\omega'}\right) \tag{4.50a}$$

$$w' = \left(\frac{\omega_2 - \omega_1}{\omega_0}\right) \tag{4.50b}$$

$$\omega_0 = \sqrt{\omega_1 \omega_2} \tag{4.50c}$$

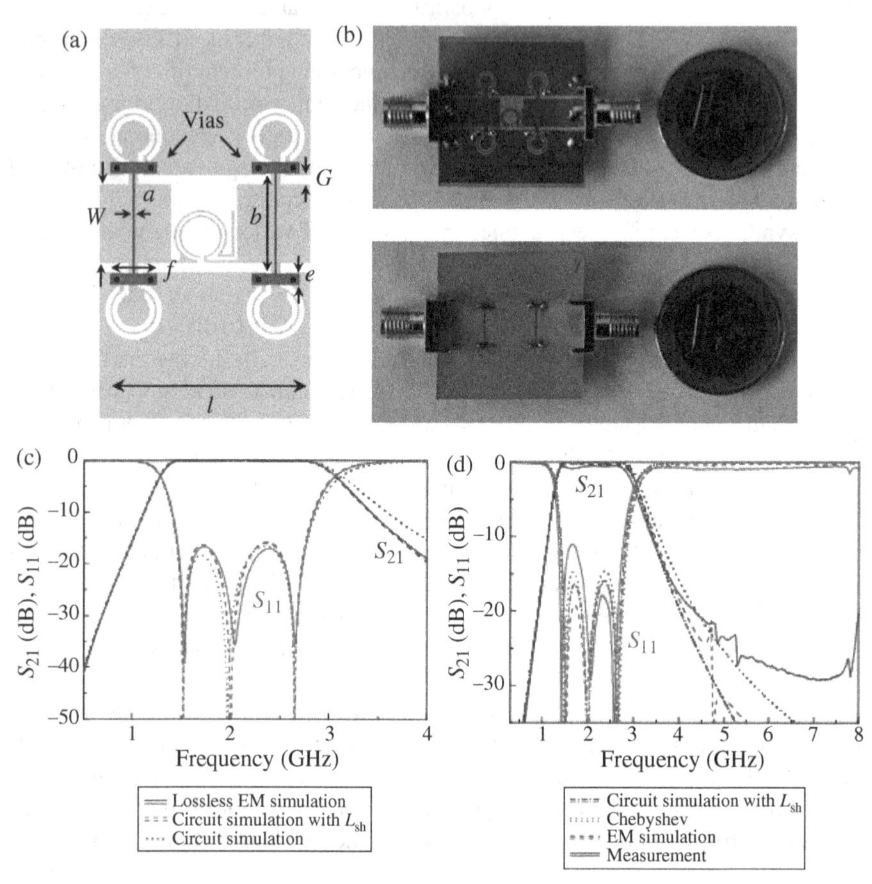

FIGURE 4.39 Topology (a), photograph (b), narrow band response (c), and wideband response (d) of the designed third-order filter with 70% fractional bandwidth. The considered substrate is the *Rogers RO3010* with thickness $h = 0.254$ mm and measured dielectric constant $\varepsilon_r = 11.2$. Dimensions are $l = 13.38$ mm, $W = 5$ mm, $G = 0.547$ mm, $a = 0.16$ mm, $b = 6.27$ mm, $e = 0.73$ mm, and $f = 3$ mm. For the OCSRR, $r_{ext} = 1.8$ mm, $c = 0.3$ mm, and $d = 0.16$ mm. For the OSRR, $r_{ext} = 1.7$ mm, and $c = d = 0.16$ mm. The values of the circuit simulation considering parasitics are (in reference to Figure 3.58d): $C = 0.188$pF, $L = 0.62$nH, $C_s = 0.85$pF, $L'_s = 6.3$nH, $C'_p = 2.43$pF, and $L'_p = 2.7$nH. The modified values of the OCSRR considering the wideband model with the additional parasitic element L_{sh} are (in reference to Figure 3.55a): $L = 0.62$ nH, $C'_p = 2.54$ pF, $L'_p = 2.56$ nH, and $L_{sh} = 0.15$ nH. Reprinted with permission from Ref. [68]; copyright 2011 IEEE.

ω' being the angular frequency of the dual-band bandpass filter and ω_1, ω_2 the angular central frequencies of the first and second filter bands. Once these transformations are applied to the lowpass filter prototype, the circuit of Figure 3.25c is obtained, provided identical fractional bandwidths for the two bands, as well as the narrow band approximation (i.e., a fractional bandwidth, *FBW*, for each band less than 10%), are considered [79]. As reported in Ref. [79], the element values of the circuit of Figure 3.25c are obtained from a set of equations dependent on the lowpass filter prototype coefficients (g_i), Ω_c and on the relevant design parameters, namely, the angular central frequencies (ω_1, ω_2) and the *FBW* (these equations are not reproduced here for simplicity).

Using this procedure, a dual-band bandpass filter that covers the GSM (0.9–1.8 GHz) and the *L*1 (1575.42 MHz) – *L*5 (1176.45 MHz) civil GPS frequency bands is reported [35]. An order-3 Chebyshev response with 0.01 dB ripple, central frequencies of $f_{c1} = 1.02$ GHz and $f_{c2} = 1.68$ GHz, and 20% fractional bandwidth was considered. Even though the narrow band approximation is not satisfied, the filter response is in good agreement with the ideal (single-band) Chebyshev response (only small deviations in the transition bands are observed). From the aforementioned filter specifications, the values of the circuit elements (Fig. 3.25c) are $L_{hs} = 19.05$ nH, $C_{hs} = 0.79$ pF, $L_{hp} = 1.16$ nH, $C_{hp} = 12.69$ pF, $L_{vs} = 10.29$ nH, $C_{vs} = 1.43$ pF, $L_{vp} = 2.58$ nH, and $C_{vp} = 5.71$ pF.

Using the design methodology applied to the quad-band inverters, the filter layout was obtained (this is depicted in Figure 4.40 together with the photograph of the fabricated prototype). Filter optimization taking parasitics into account was done by tuning the reactive elements at the circuit level. Specifically, the values that are changed (in reference of Fig. 4.19b) are $L_{hs} = 18.26$ nH, $C_{vp} = 4.4$ pF, $C_1 = 0.33$ pF, $C_2 = 0.51$ pF, and $L = 0.45$ nH.

The frequency response and group delay of the filter, including measurement, EM simulation, circuit simulation of the accurate circuit model of Figure 4.19b and the dual-band Chebyshev response (circuit of Fig. 3.25c), are also depicted in Figure 4.40. As can be seen, good agreement between the different responses is obtained, with measured return losses better than 20 dB and a rejection level better than 20 dB up to 4.6 GHz ($3.5f_o$), f_o given by expression 4.50c. Moreover, the size of the filter is 19.5 mm × 13.5 mm, which corresponds to $0.22\lambda_g \times 0.15\lambda_g$, λ_g being the guided wavelength at f_o.

4.2.3.7 Elliptic Lowpass Filters Based on OCSRRs The elliptic lowpass filters considered in this subsection are actually structures based on transmission lines with metamaterial loading (OCSRRs). Namely, they are neither based on the CRLH transmission line concept nor on the hybrid approach or on other structures that can be used as building blocks of metamaterial transmission lines. Nevertheless, the proposed OCSRR-based elliptic lowpass filters are introduced here for thematic coherence and because the procedure for filter design is not very different to that considered for the design of bandpass filters based on the combination of OSRRs and OCSRRs.

The reported OCSRR-based elliptic-function lowpass filters are described by the network of Figure 4.41 and implemented in coplanar waveguide (CPW) technology.

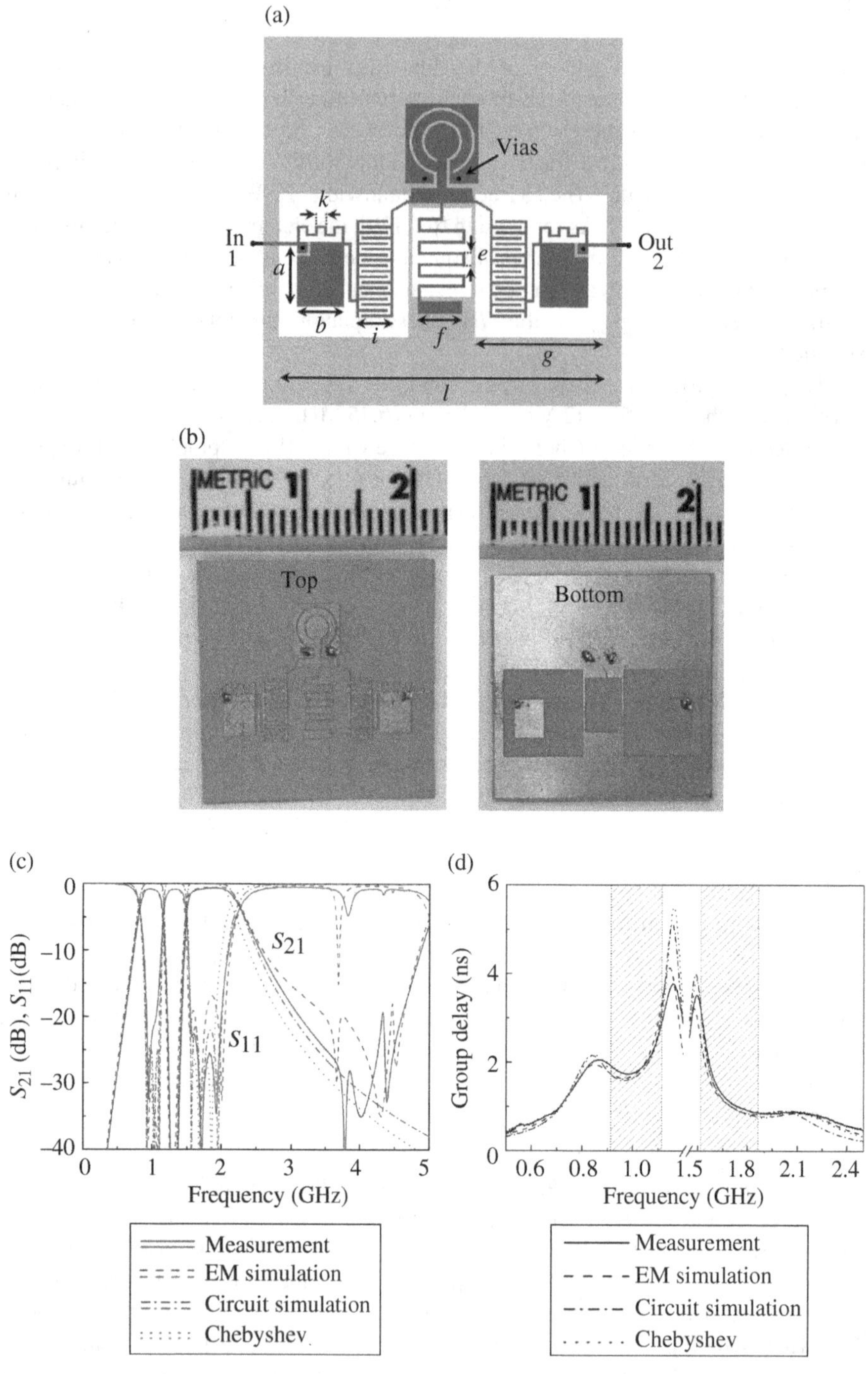

FIGURE 4.40 Layout (a), photograph (b), frequency response (c), and group delay (d) of the dual-band bandpass filter. Dimensions are $a = 3.52$ mm, $b = 2.85$ mm, $f = 2.64$ mm,

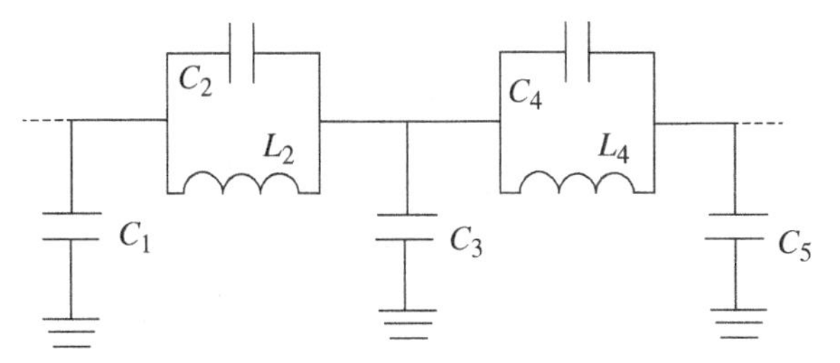

FIGURE 4.41 Low-pass elliptic-function prototype filter. The circuit corresponds to a fifth-order prototype.

The series connected parallel resonators are implemented by means of OCSRRs, whereas the shunt capacitances are implemented by etching metallic patches in the back side of the substrate connected to the ground planes through vias. The device (presented in Ref. [80]) is an order-5 elliptic-function lowpass filter with a pass band ripple of $L_{Ar} = 0.1$ dB, a cutoff frequency of $f_c = 1$ GHz and an equal-ripple stop band starting frequency of 1.4085 GHz (with stop band ripple of $L_{As} = 39.59$ dB) [33]. The element values corresponding to this elliptic filter response are (referred to the circuit of Fig. 4.41): $C_1 = 3.20$ pF, $C_2 = 0.58$ pF, $L_2 = 9.44$ nH, $C_3 = 5.02$ pF, $C_4 = 1.73$ pF, $L_4 = 6.87$ nH, $C_5 = 2.41$ pF. Filter implementation was realized by means of two series connected OCSRRs (etched in the central strip) cascaded between electrically short (to avoid the parasitic effects of the series inductance) transmission line sections with a metallic patch etched in the back substrate side and connected to the ground planes by means of vias. These patches are necessary to obtain the high required values of the shunt capacitors.

The dimensions of the metallic patches are inferred from the EM simulation of the S-parameters of the short CPW transmission line section with the metallic patch below it. From these results, the shunt (capacitive) reactance of the π-circuit of such transmission line section can be obtained through standard formulas linking the S-parameters and Z-parameters (see Appendix C). Then, the capacitance is extracted and compared to the nominal value. The procedure is repeated (by modifying the patch dimensions) until good agreement results. In practice, the length, l_P, of the patch was modified, while its width was set to a fixed value ($w_P = 15.84$ mm) [80].

FIGURE 4.40 (*Continued*) $e = 0.49$ mm, $g = 7.71$ mm, $i = 2.19$ mm, and $l = 19.4$ mm. The OCSRR dimensions are $r_{ext} = 1.95$ mm, $c = 0.16$ mm, and $d = 0.6$ mm, where r_{ext}, c, and d are the external radius, width, and separation of the rings. All the meanders have a width of 0.16 mm. The interdigital capacitors have a width and separation between fingers of 0.16 mm. The radius of the vias is of 0.15 mm. The considered substrate is the *Rogers RO3010* with thickness $h = 0.254$ mm and measured dielectric constant $\varepsilon_r = 10.5$. Reprinted with permission from Ref. [35]; copyright 2010 IEEE.

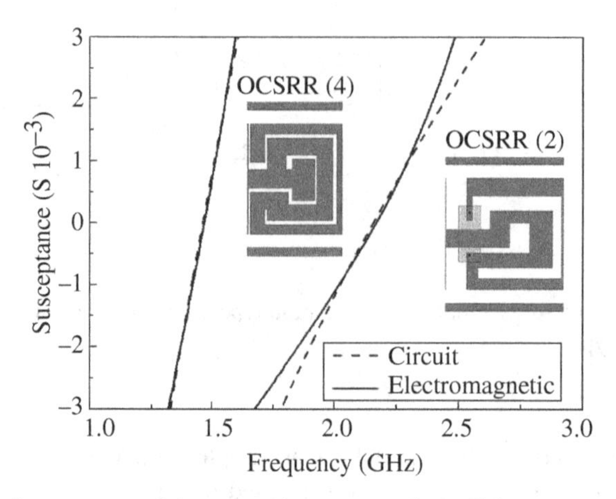

FIGURE 4.42 Susceptance of the two series connected parallel resonators of the designed elliptic-function lowpass filter. The susceptance inferred by means of EM simulation was obtained from the optimized layout topologies included in the figure. The *Rogers RO3010* substrate with measured dielectric constant $\varepsilon_r = 11.5$ and thickness $h = 0.635$ mm was considered. Reprinted with permission from Ref. [80]; copyright 2009 IEEE.

The dimensions of the OCSRRs are determined by forcing their resonance frequencies to the required values and by forcing the susceptance slopes to be identical to those of the lumped element resonators (of the elliptic-function model) at resonance. Depending on the values of the resonator inductance and capacitance, it is necessary to substantially modify the topology of the OCSRR in order to obtain the desired values of the elements. Thus, the formulas that link the geometry of the OCSRR and the element values are only used as reference. Optimization is required to provide a good description of the resonators in the vicinity of resonance. Figure 4.42 depicts resonator's susceptance for the two parallel resonators of the filter of Figure 4.41 in the vicinity of resonance. The susceptances that have been inferred from EM simulation of the optimized OCSRR topologies (shown in the inset) are also depicted in this figure. As required, there is good agreement between the two pair of curves in the vicinity of resonance.

The layout and photograph of the fabricated lowpass filter are depicted in Figure 4.43. The device was fabricated on the *Rogers RO3010* substrate with measured dielectric constant $\varepsilon_r = 11.5$ and thickness $h = 0.635$ mm. Besides the patch capacitances, there is an additional strip on the back side of the substrate that connects (through metallic vias), the external metallic regions of OCSRR (2) (the one which exhibits the upper resonance frequency). With this strategy, a parasitic slot mode that is otherwise generated in this resonator is suppressed, as revealed from the simulated magnetic currents (not shown). This parasitic mode obscures the first resonance frequency of the OCSRR (2). Thus, by etching such additional metallic strip, the stop band filter response is improved, and the two transmission zeros can be clearly identified. The simulated and measured frequency responses of the device are also shown in Figure 4.43. The circuit simulation of the model of Figure 4.41 with the above

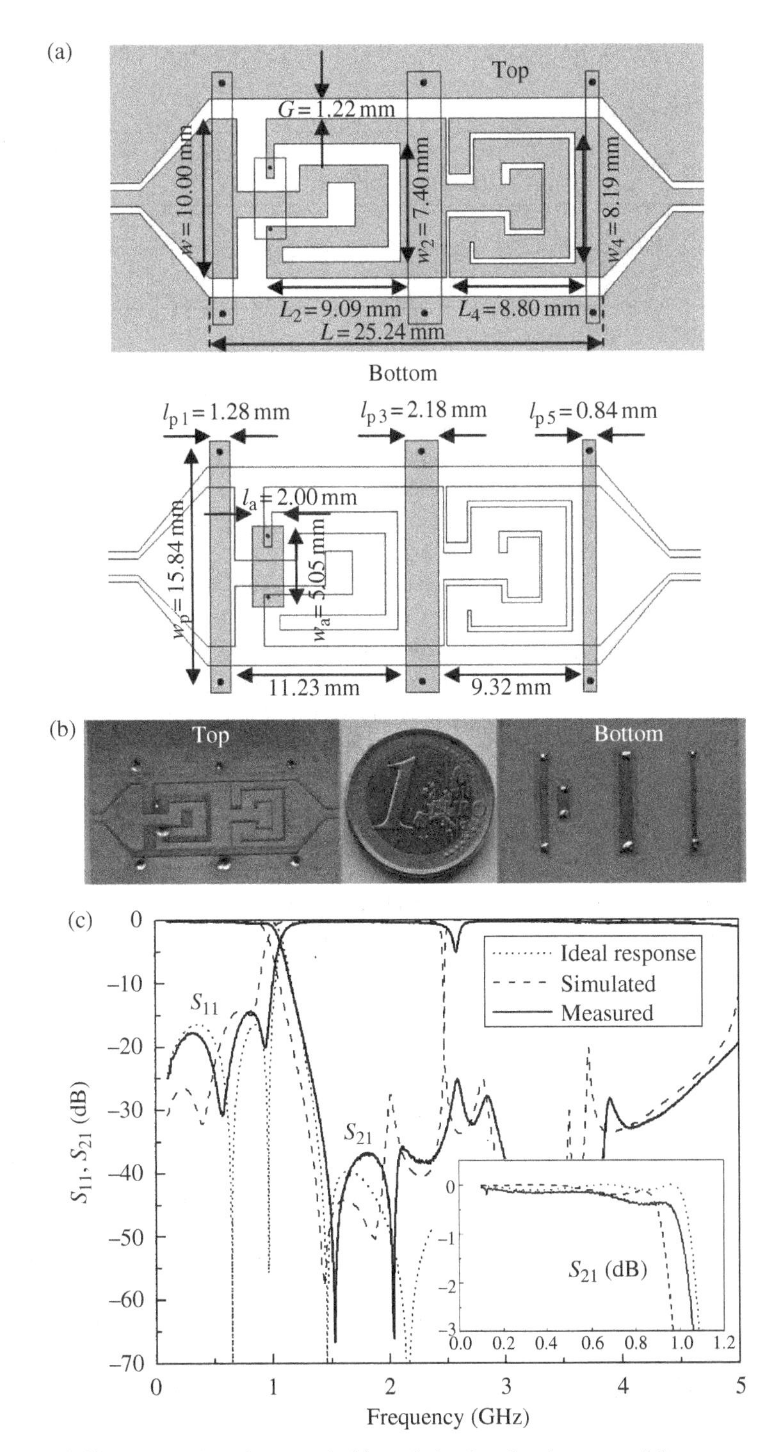

FIGURE 4.43 Layout (a), photograph (b), and simulated and measured frequency response (c) of the fabricated OCSRR-based lowpass filter. The ideal elliptic-function response is also depicted. The detail of the in-band insertion loss is depicted in the inset. Reprinted with permission from Ref. [80]; copyright 2009 IEEE.

TABLE 4.3 Comparison of split-ring elliptic lowpass filters

Reference	Pass-band IL (dB)	Pass-band RL (dB)	Stop-band rejection (up to $2f_c$) (dB)	Electrical size
[89]	<1	N.A.	>35	$0.52\lambda \times 0.27\lambda$
[90]	<0.7	>9	>32	$1.2\lambda \times 0.26\lambda$
Figure 4.43	<0.4	>14	>35	$0.13\lambda \times 0.086\lambda$

indicated element values (corresponding to the ideal elliptic-function response) is also depicted the figure. The agreement between experimental data and the elliptic function response is good up to approximately 2 GHz. At higher frequencies, the mismatch is due to the fact that the lumped element models of the different filter elements (OCSRRs and shunt capacitors) fail. Measured insertion losses in the pass band differ in less than 0.3 dB to the ideal response, and return losses are better than 14 dB. The frequency response exhibits a sharp cutoff, with a stop band rejection better than 25 dB up to roughly 4.5 GHz. It is also remarkable that filter length is small, that is, 2.5 cm, which corresponds to 0.13λ, where λ is the guided wavelength at the cutoff frequency. This combination of size (relative to the filter cutoff frequency) and performance (especially for which concerns the width of the stop band) is not easily found in other elliptic lowpass filters (see for instance [81–88]). Moreover, the electrical size of this lowpass filter is much smaller than the size of other lowpass filters based on split rings, such as CSRRs [89, 90]. The dimensions and performance of these latter filters and the filter of Figure 4.43 are compared in Table 4.3.

In summary, OCSRRs provide a good solution for the implementation of compact elliptic function lowpass filters in CPW technology. OCSRR-based lowpass filters in microstrip technology are reported in Ref. [91]. Although these filters are very selective and spurious free up to high frequencies, they are merely designed by cascading identical unit cells and therefore they do not give standard filter responses.

4.2.4 Leaky Wave Antennas (LWA)

Leaky wave radiation in periodic structures was succinctly considered in Chapter 2.[26] Leaky wave antennas (LWAs) can be defined as guiding structures with the wave radiating, or leaking, along the structure. The topic of LWAs has been in steady development for many years, and there are several excellent sources that are recommended to the interested reader on the topic [92–96]. Nevertheless, the field has been activated in recent years, especially for planar LWAs, due to the advent of metamaterials and metamaterial transmission lines [97–99]. In this subsection, the objective is to point out the potential of CRLH transmission lines for the implementation of planar LWAs,

[26] In this subsection, there are general antenna concepts not covered in any appendix to avoid an excessively long book. The author recommends general textbooks on antennas to those readers not familiarized with the topic. Nevertheless, there are several references on the specific topic of LWAs in this subsection.

and to highlight some advantages and properties of such CRLH-based antennas over other LWAs.

Traditionally, one-dimensional LWAs have been divided in two categories: uniform and periodic. In uniform LWAs, the guiding structure is uniform along the direction of propagation, and the structure supports fast waves with respect to free space, so that the complex wavenumber of the leaky mode ($k_n = \beta - j\alpha$) has a phase constant satisfying $0 < \beta < k$, k being the free space wavenumber. Typically, uniform LWAs only radiate in the forward direction since $\beta > 0$, and broadside radiation is not possible in conventional uniform LWAs. Moreover, uniform LWAs, such as microstrip antennas [100–102], typically require complex feeding structures in order to excite the required leaky mode and prevent the excitation of the dominant (slow-wave) microstrip mode.

Periodic LWAs are based on guiding structures with slow waves as compared to free space (i.e., $\beta > k$), where leaky wave radiation is achieved by introducing periodicity along the propagation direction. As studied in Chapter 2, due to periodicity the fields in these periodic structures are characterized by space harmonics with wavenumbers given by (2.6), and the structure may radiate if the leaky wave condition ($|\beta_n| < k$) is satisfied for any of the space harmonics (see Fig. 2.2).[27] In periodic LWAs, the fundamental mode does not radiate. Periodic LWAs can either radiate in the forward or backward direction since the phase constant of the $n = -1$ space harmonic can either be positive or negative. This is a fundamental difference of periodic LWAs, as compared to uniform antennas. However, in general, the presence of gaps at periodic frequencies prevents broadside radiation as a leaky wave structure [103, 104]. In practice, continuous frequency scanning from backward to forward direction is not possible due to the lack of radiation over the frequency gaps.[28] Periodic bidirectional LWAs (fed in the center to create a bidirectional leaky wave) can, however, radiate at broadside as an LWA by operating slightly away from the broadside frequency, so that $\beta_{-1} > 0$ or $\beta_{-1} < 0$. In this case, a bidirectional beam is created in which the beam from each half of the structure is slightly deviated from broadside. However, the combination of the two beams results in a symmetrical broadside beam.

Contrarily to resonant antennas, where dimensions scale with the operating frequency, the length of the leaky wave structure in LWAs (uniform or periodic) is related to the antenna directivity. Indeed, one important feature of LWAs is that they can achieve a high directivity with a simple structure, without the need for a complicated and costly feeding network as typically used in phased arrays. Directivity increases with the length of the LWA provided the leakage factor α is small enough so that the guided power is not completely leaked along the structure (i.e., it is partially absorbed by a matched load). Under these conditions, the radiation aperture is given

[27] In the analysis carried out in Section 2.2 to obtain the leaky wave condition, the attenuation constant α was neglected. Hence $k_n = \beta_n$, and for that reason the leaky wave condition was expressed as $|k_n| < k$.

[28] The stop band that occurs when the beam is scanned at broadside is known in the literature as open stopband region. In it, the radiated power drops substantially. However, there are strategies to effectively close this region and achieve leaky wave radiation at broadside in periodic structures [103, 104].

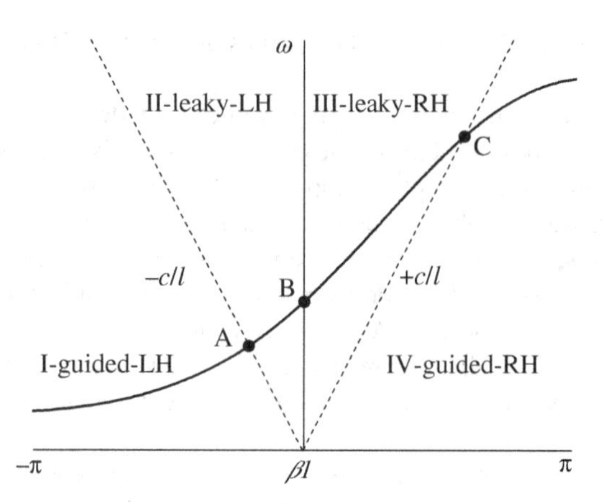

FIGURE 4.44 Typical dispersion diagram of a balanced CRLH transmission line and indication of the four main regions (enumerated I, II, III, and IV) and the relevant points separating such regions. A, B, and C points correspond to backfire, broadside, and endfire leaky wave radiation, respectively.

by the length of the LWA, and hence a high directivity is obtained for long LWA structures.

As an alternative to periodic and uniform leaky wave structures, LWAs can be implemented by means of balanced CRLH transmission lines [97]. Indeed, balanced CRLH lines are periodic structures by nature. However, in such lines, the periodicity is much smaller than the wavelength and does not play an active role in the radiation. Balanced CRLH-based LWAs radiate from the fundamental ($n = 0$) space harmonic, and essentially they behave as uniform leaky wave structures in the sense that they are homogeneous one-dimensional media.[29] Therefore, CRLH-based LWAs may be considered to belong to the category of uniform LWAs despite the fact that they are structurally periodic. To distinguish them from conventional uniform LWAs, CRLH LWAs can be referred to as quasi-uniform LWAs [96].

According to the typical dispersion of a CRLH transmission line (reproduced in Fig. 4.44), it is clear that there are regions where the dispersion curve lies in the fast wave region. Therefore, open CRLH transmission lines can be used as LWAs. Although the balance condition is not strictly necessary, by balancing the line it is possible to achieve continuous frequency scanning from backward to forward directions, including broadside radiation at the transition frequency, ω_o. Thus, balanced CRLH LWAs radiate from the fundamental mode and exhibit backfire-to-endfire capability. This is an important feature of CRLH LWAs, which is not achievable with their conventional uniform or periodic counterparts (although, as mentioned before, there are strategies to mitigate the effects of the open stop bands

[29] Although it was stated in Chapter 3 that, in general, periodicity and homogeneity are not necessary for the implementation of metamaterial transmission line-based circuits, CRLH-based LWAs are implemented by homogeneous periodic structures.

in periodic LWAs). The fact that CRLH LWAs operate at the fundamental mode is important since the feeding structures do not need to be complex (a simple transmission line suffices).

The first implementation of a CRLH LWA exploiting the backward-to-forward radiation capability was presented in Ref. [97], and consisted of a 24-unit cell microstrip line periodic-loaded with series interdigital capacitors and shunt-connected grounded stubs. The structure is the one depicted in Figure 3.27 (substrate parameters and dimensions are given in Ref. [97]). The phase and attenuation constants of this structure, as well as the radiation patterns at different frequencies, are depicted in Figure 4.45. For this design, the LH and RH guided-wave regions, designated as I and IV, respectively, in the figure, extend to frequencies below 3.1 GHz and above 6.3 GHz. Region II is the backward radiated region (with $|\beta| < k_o$ and $\beta < 0$), extending from 3.1 up to 3.9 GHz, and region III is the forward radiated region (with $|\beta| < k_o$ and $\beta > 0$), extending from 3.9 up to 6.3 GHz. The radiation patterns at the considered frequencies are consistent with the dispersion characteristics of the LWA, and confirm the frequency scanning capability of the designed antenna. At this point, it is important to mention that backward leaky wave radiation in metamaterial transmission lines was demonstrated almost simultaneously in Ref. [98] (see Chapter 3), by considering a CPW transmission line loaded with series capacitive gaps and shunt connected (inductive) strips. However, in such antenna, backward-to-forward frequency scanning capability was not demonstrated. The same authors reported a forward CPW LWA consisting of a similar structure, but only loaded with series capacitive gaps [99]. It was argued in Refs. [98, 99] that the radiating elements in the reported CPW-based LWAs are the capacitive gaps, which cause a backward [98] or forward [99] transverse magnetic wave front. Conversely, the pairs of inductive strips support opposite (antiparallel) currents, causing cancelation in the far field (E-plane) due to line symmetry and even excitation of the fundamental CPW mode.

CPW transmission lines loaded with pairs of SRRs and shunt-connected inductive strips do also exhibit fast wave regions for the fundamental ($n = 0$) mode. However, such structures are inefficient to produce leaky wave radiation from the fundamental mode due to far field cancelation caused by opposite currents. That is, due to line symmetry and even mode excitation (CPW mode), the symmetry plane of the structure is a magnetic wall, and the magnetic currents in the slots are antiparallel.[30] In order to achieve efficient leaky wave radiation in SRR-based CPW transmission lines, a strong asymmetry can be forced by removing the set of SRRs present at one side of the symmetry plane of the line. This allows reducing far field cancelation from the

[30] The far-field cancellation effect due to antiparallel magnetic currents can be explained by considering a magnetic current (that of a single slot) parallel and close to a perfect magnetic conductor (PMC), located at the symmetry plane of the CPW transmission line. The PMC generates an antiparallel image current, causing far field cancellation. The effect is similar to the antiparallel electric (image) current generated by a perfect electric conductor (PEC), preventing the radiation from an electric dipole, if it is parallel oriented and closely located to the PEC.

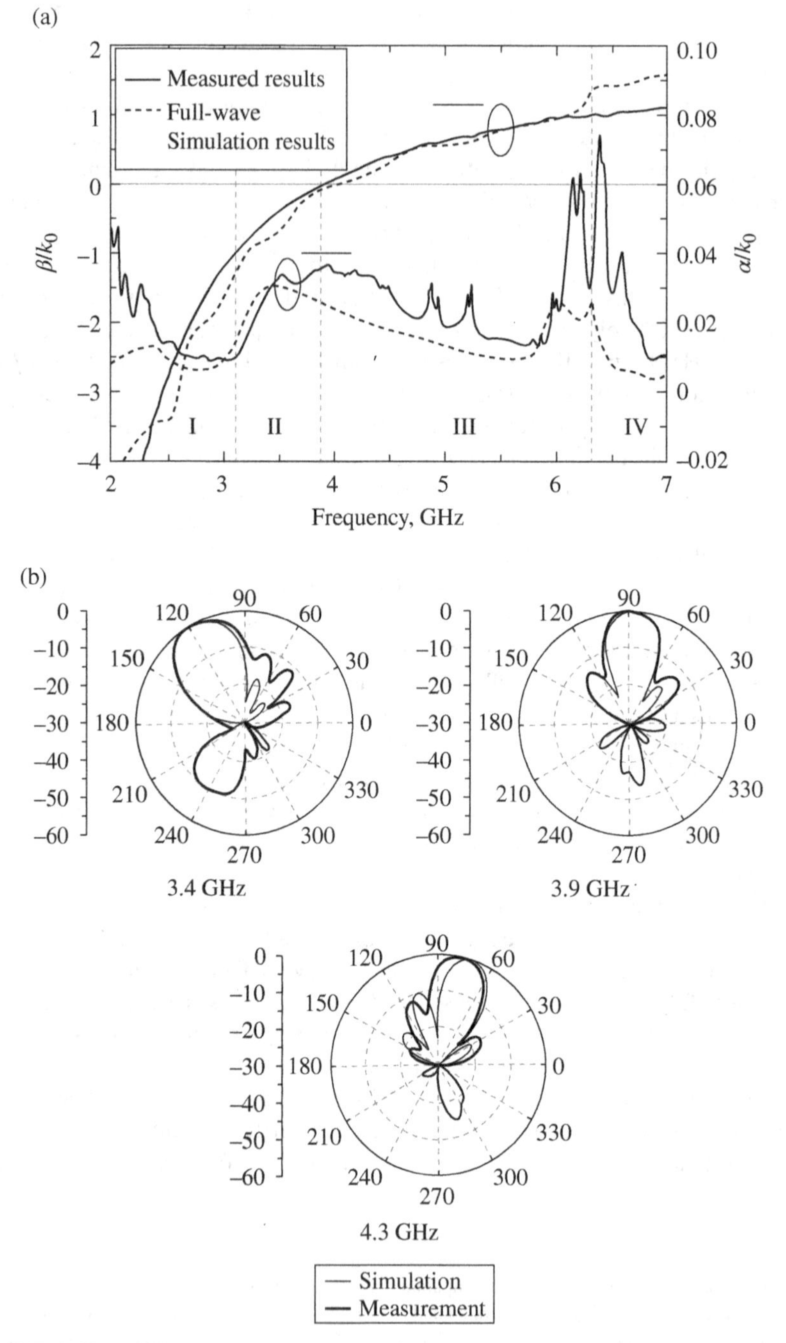

FIGURE 4.45 (a) Dispersion characteristics of the structure of Figure 3.27; (b) radiation patterns (E-plane) for the indicated frequencies. Reprinted with permission from Ref. [97]; copyright 2002 IET.

magnetic current balance between both slots, providing the excitation of the slot line mode, which results in an important contribution to increase leaky wave radiation [105].

Alternatively to the previous strategy, efficient leaky wave radiation can be achieved by loading a slot line with SRRs and shunt inductive strips [106]. With such a structure, it is possible to scan the radiation angle with frequency by balancing the line. The design of the structure starts by conceiving a balanced SRR/strip-loaded CPW with sufficiently distant slots (wide central strip) to minimize the field coupling between the slots. Under these conditions, the CPW can be viewed as a pair of two independent slot lines sharing the same substrate, with antiparallel slot magnetic currents. Therefore, it is expected that a symmetric CPW with a wide central strip exhibits similar propagation characteristics than the slot line resulting by removing one half of the CPW structure. With this in mind, a balanced SRR/strip-loaded CPW with the transition frequency at $f_o = 2.5$ GHz was designed in Ref. [106]. The phase constant inferred from EM simulation of the structure (unit cell), depicted in Figure 4.46, is roughly the same as the phase constant corresponding to the slot line that results by removing one half of the CPW transmission line (the fabricated structure is depicted in Fig. 4.47a). Unlike the phase constant, the analysis of a single-unit cell is inaccurate for the determination of the attenuation constant α of leaky periodic structures. This is due to mutual coupling and edge effects. However, as the number of the considered unit cells increases, edge effects become less significant and the obtained attenuation constant is more accurate [107]. Therefore, in order to determine the required number of unit cells, N, the attenuation constant was computed by simulating an N-element two-port structure, using the *Agilent Momentum* commercial software. Figure 4.47b shows the normalized attenuation constant α/k_0 at the transition frequency. It can be observed that more than 15 elements are required to achieve

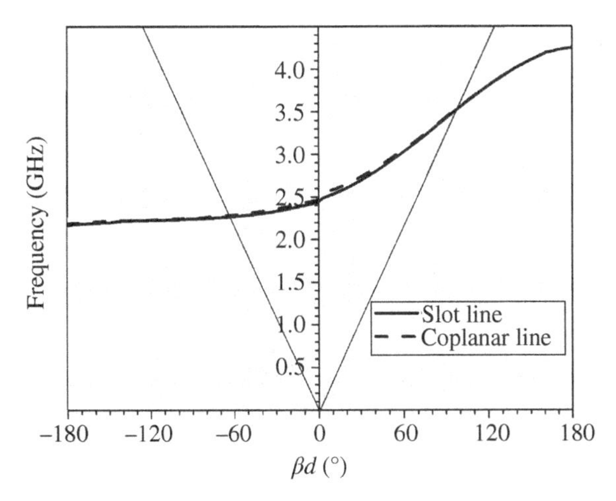

FIGURE 4.46 Dispersion diagram of the SRR/strip-loaded CPW unit cell (dashed line) and SRR/strip-loaded slot line unit cell (solid line) inferred from EM simulation.

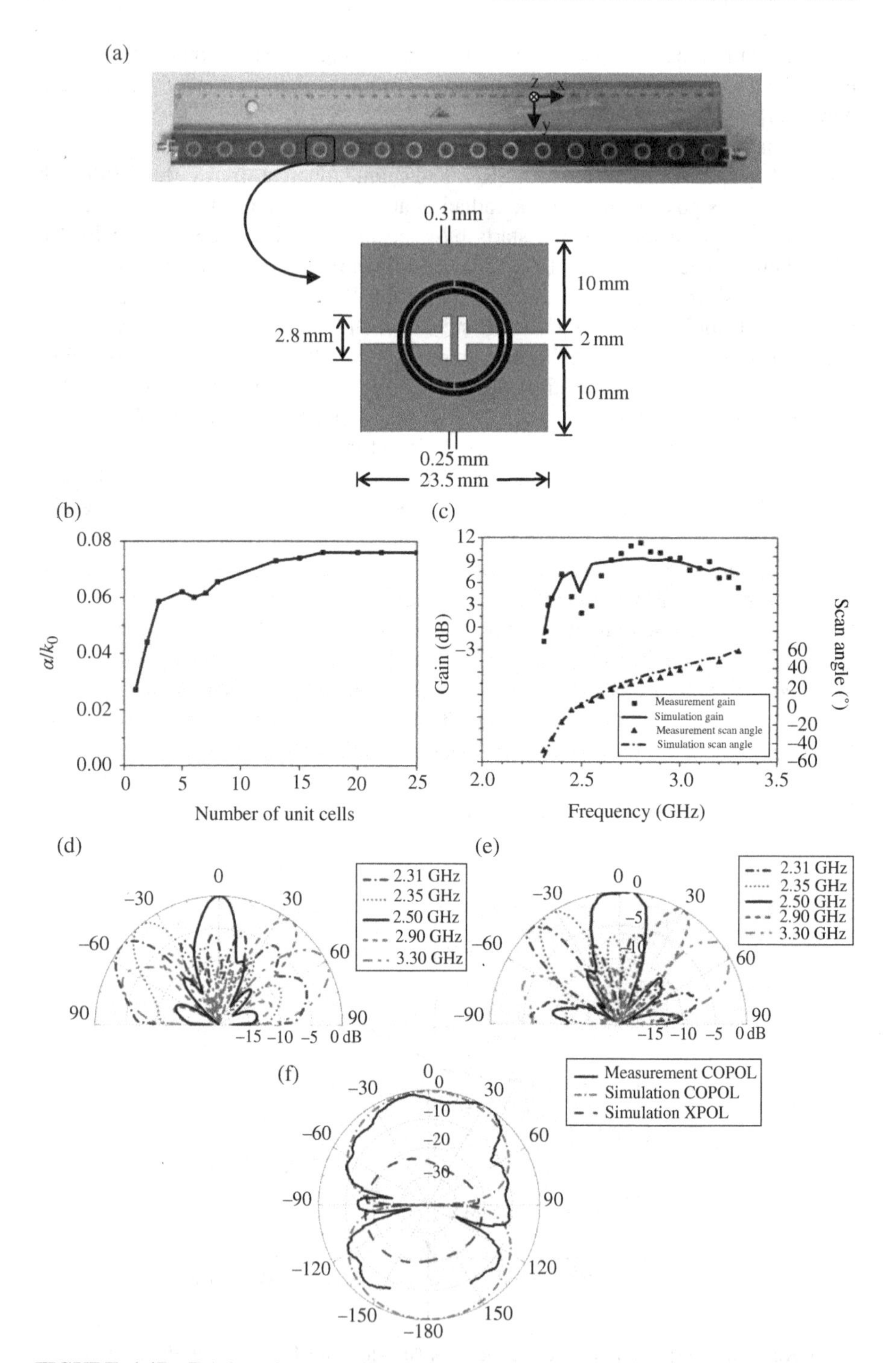

FIGURE 4.47 Fabricated 17-cell SRR/strip-loaded slot LWA and antenna performance. (a) Photograph and relevant dimensions of the unit cell, (b) attenuation constant versus the

convergence to a value of $\alpha/k_0 = 0.076$. This value was used to obtain the length of the LWA from the expression for the power flow along a lossy periodic structure with periodicity d

$$P_n = P_o e^{-2\alpha nd} \tag{4.51}$$

where P_0 is the power delivered to the LWA, P_n is the power at the nth terminal of the periodic structure and d is the unit cell length. For 95% of the power dissipated before reaching the antenna termination and using the converged value of $\alpha/k_0 = 0.076$, the required number of unit cells is $N = 17$.

According to simulation, the characteristic impedance of the loaded line at the transition frequency is $Z_o = 150\,\Omega$. This impedance was matched to the $50\,\Omega$ connector using a semilumped microstrip matching network, connected to the slot line by means of a via hole. The measured and simulated gains are depicted in Figure 4.47c, with a maximum measured gain of 7.1 dB and 11.3 dB for the LH and RH bands, respectively. There is a relatively good agreement between measured and simulated data. The angle of the maximum beam θ_m as a function of frequency is also plotted in Figure 4.47c, showing good agreement between measurement and simulation. The measured backward to forward scanning range is from −50° to +60° while maintaining an acceptable gain level. The simulated and measured normalized radiation patterns are shown in Figure 4.47d and e. As it can be appreciated, backward, broadside and forward leaky wave radiation is obtained as frequency increases from the LH to the RH frequency bands. The simulated and measured orthogonal pattern at 2.5 GHz is depicted in Figure 4.47f. It can be observed that while the radiated beam is directive in the xz-plane (see Fig. 4.47d and e), it is fat in the perpendicular direction (yz-plane), thus resulting in a fan beam.

In microstrip technology, LWAs loaded with CSRRs have been recently demonstrated [108]. There are many other CRLH-based LWAs reported in the recent literature. It is not possible to cover all of them in a general textbook like the present one. However, let us mention to finish this subsection that besides microstrip, CPW and slot-line technology, CRLH LWAs have been also demonstrated using CPSs [109] and SIW technology [110, 111] (these antennas will be analyzed in the last chapter, where a section is devoted to SIW technology). Electrically scanned (tunable) LWAs, with fixed frequency but variable radiation angle [112, 113], and dual-band LWAs based on E-CRLH lines have also been reported [111, 114]. Finally, active CRLH LWAs consisting of CRLH line sections interconnected through amplifiers have been

FIGURE 4.47 (*Continued*) number of unit cells at 2.5 GHz, (c) simulated and measured antenna gain and scanning angle, (d) simulated normalized radiation patterns at the longitudinal plane, (e) measured normalized radiation patterns at the longitudinal plane, and (f) simulated and measured orthogonal plane radiation pattern of the proposed LWA at 2.5 GHz. The antenna was fabricated on the *Arlon Cuclad 250XL* substrate with dielectric constant $\varepsilon_r = 2.43$, loss tangent $\tan\delta = 0.002$, and thickness $h = 0.49$ mm. Reprinted with permission from Ref. [106]; copyright 2013 IEEE.

presented in Ref. [115]. Since the amplifiers regenerate the power progressively leaked out of the structure in the radiation process, the effective aperture of the antenna is increased, and hence the directivity and gain are enhanced.

4.2.5 Active Circuits

The CRLH transmission lines or some of the considered CRLH-based passive circuits studied in the previous subsections can be combined with active elements such as diodes and transistors, or with amplifiers, for the implementation of RF/microwave circuits with enhanced performance or functionality. Thus, for instance, dual-band balanced mixers implemented by means of dual-band CRLH-based hybrid couplers [116], dual-band class-E power amplifiers with CRLH-based impedance matching networks [117], and active LWAs [115], among others, have been reported. In this subsection, the aim is to demonstrate the potential of CRLH transmission lines for the implementation of active microwave circuits, on the basis of two types of circuits: distributed amplifiers with enhanced functionalities and dual-band recursive active filters.

4.2.5.1 Distributed Amplifiers Distributed amplification was proposed as a means of improving the gain-bandwidth product [118]. Essentially, distributed amplifiers consist of a pair of actively coupled transmission lines, the gate line, and the drain line. Active coupling is achieved through the distributed transconductance of a set of transistors (normally field-effect transistors or FETs) with their gates and drains connected to the gate and drain lines, respectively. By injecting the input signal to any of the ports of the gate line (the other one being matched), the signal in the drain line is amplified by virtue of the distributed transconductance [119]. This results in a certain gain between the input and the so-called forward and reverse output ports.[31] Since the fifties of the past century, there have been many notable contributions to the field of distributed amplifiers. Among them, those amplifiers that first used FETs [120, 121], the first monolithic implementations based on AsGa-FETs (achieving bandwidths beyond 10 GHz) [122–124], and the dual-fed concept to improve device gain Ref. [125], are worth mentioning. More recently, several works have been focused on improving the amplifier bandwidth and efficiency [126, 127], and notable achievements have been reported in the field of distributed mixers [128, 129], formerly reported in Ref. [130, 131].

In all the previous reported distributed amplifiers and mixers, forward (RH) gate and drain transmission lines are used. The first structure that made use of LH transmission lines was presented by Eccleston in 2005 [132]. However, the aim in that work was to reduce device dimensions, rather than using the unusual dispersion characteristics of LH lines. The combination of distributed amplification with the nonlinear dispersive behavior of CRLH transmission lines was first pointed out in Ref. [133]. This work opened the path toward the design of distributed amplifiers with a radically new

[31] The forward and reverse output ports are those ports of the drain line located in the opposite and in the same extreme, respectively, than the input (gate line) port.

intrinsic behavior. Thanks to the controllability of the dispersion in the gate and drain lines, it is possible, for instance, to achieve maximum gain at arbitrary frequencies in the forward and reverse ports, and thus achieve a dual-band/diplexing action, avoiding the use of additional costly components. In other words, the introduction of CRLH lines in distributed amplifiers, rather than improving the performance of the conventional counterparts, has generated novel and interesting functionalities for these devices. Distributed amplifiers based on metamaterial transmission lines constitute a genuine example of devices designed through dispersion and impedance engineering. Such devices have been designated as metadistributed amplifiers to highlight the fact that they are based on metamaterial structures (CRLH transmission lines). CRLH transmission lines can also be introduced in distributed mixers (meta-distributed mixers) [134], but the analysis of these devices is out of the scope of this book.

To understand the high potential of CRLH transmission lines to the implementation of distributed amplifiers with advanced functionalities, it is necessary to briefly review the principles behind distributed amplification and to derive some useful expressions that link the amplifier gain to the main device parameters, including the phase constants of the drain and gate lines. Such expressions will then be analyzed to point out several amplifier functionalities, related to the dependence of the gain in the forward and reverse output ports with frequency. Finally, an illustrative implementation and its characterization will be reported.

Principle of Distributed Amplification Let us introduce the principle of distributed amplification by considering a lossless, unilateral, and continuous model for the elemental cell of the distributed amplifier. It consists of two lossless transmission lines with active unilateral coupling through a distributed transconductance, as depicted in Figure 4.48a [120, 135]. In this basic cell, $\beta_g(\omega)$ and $\beta_d(\omega)$ are the phase constants of the gate and drain lines, respectively, and Z_g and Z_d the corresponding characteristic impedances. The per-unit-length transconductance, g_m, provides a current in the drain line that depends on the voltage on the gate line, $V_{gs}(z)$. By considering the input port of the distributed amplifier to be one of the ports of the gate line, the ratio of the output power in the forward port (see Fig. 4.48b), P_o^{fwd}, to the input power, P_i, is called forward gain, whereas the reverse gain is the ratio of the power collected in the reverse port, P_o^{rev}, to the input power, that is,

$$G_{fwd} = \frac{P_o^{fwd}}{P_i} \tag{4.52a}$$

$$G_{rev} = \frac{P_o^{rev}}{P_i} \tag{4.52b}$$

Let us consider that all the ports of the amplifier are matched, so that no reflected waves are generated in the structure. Under these conditions, the output current in the forward port is given by the contribution of all the source currents along the structure, namely

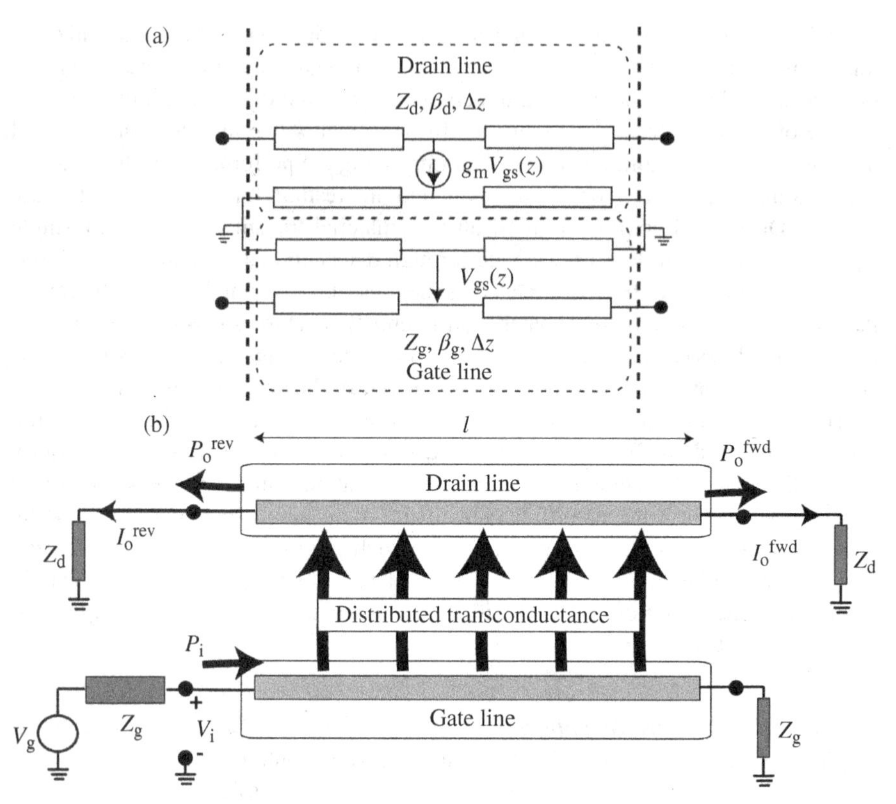

FIGURE 4.48 Basic cell (a) and scheme (b) of the continuous unilateral distributed amplifier. Reprinted from Ref. [135], with permission from the author.

$$I_o^{fwd} = -\int_0^l V_{gs}(z) g_m e^{-j\beta_d(\omega)(l-z)} dz = -\int_0^l V_i e^{-j\beta_g(\omega)z} g_m e^{-j\beta_d(\omega)(l-z)} dz \tag{4.53}$$

where V_i is the input voltage. The integral in (4.53) gives[32]

$$I_o^{fwd} = -V_i g_m l \cdot \mathrm{sinc}\left(\left(\beta_g(\omega) - \beta_d(\omega)\right) \frac{l}{2} \right) \cdot e^{-j\left(\beta_g(\omega) + \beta_d(\omega)\right)\frac{l}{2}} \tag{4.54}$$

From (4.54) the power gain is found to be

$$G_{fwd} = \frac{\left|I_o^{fwd}\right|^2}{4\left|V_i\right|^2} Z_g Z_d = \frac{g_m^2 l^2 Z_g Z_d}{4} \mathrm{sinc}^2\left(\left(\beta_g(\omega) - \beta_d(\omega)\right) \frac{l}{2} \right) \tag{4.55}$$

[32] The sinc function is defined as $\mathrm{sinc}(x) = \sin(x)/x$.

The power gain in the reverse port is derived by simply changing the signs in the exponential function of (4.53), that is,

$$I_{\mathrm{o}}^{\mathrm{rev}} = -\int_{\mathrm{o}}^{l} V_{\mathrm{gs}}(z) g_{\mathrm{m}} e^{-j\beta_{\mathrm{d}}(\omega)(z-l)} dz \tag{4.56}$$

This gives

$$G_{\mathrm{rev}} = \frac{|I_{\mathrm{o}}^{\mathrm{rev}}|^2}{4|V_{\mathrm{i}}|^2} Z_{\mathrm{g}} Z_{\mathrm{d}} = \frac{g_{\mathrm{m}}^2 l^2 Z_{\mathrm{g}} Z_{\mathrm{d}}}{4} \operatorname{sinc}^2\left((\beta_{\mathrm{g}}(\omega) + \beta_{\mathrm{d}}(\omega)) \frac{l}{2} \right) \tag{4.57}$$

According to this continuous model, the frequency behavior of the distributed amplifier is exclusively determined by the phase constants of the gate and drain lines. The maximum power gain for the forward and reverse ports is given by

$$G_{\mathrm{MAX}} = \frac{g_{\mathrm{m}}^2 l^2 Z_{\mathrm{g}} Z_{\mathrm{d}}}{4} \tag{4.58}$$

and it depends on the values of g_{m}, l, Z_{g}, and Z_{d}. Notice that, according to expressions (4.55) and (4.57), the maximum power gain for the forward port is obtained for those frequencies satisfying $\beta_{\mathrm{g}}(\omega) = \beta_{\mathrm{d}}(\omega)$, whereas for the reverse port the maximum power (i.e., maximum gain) is collected when $\beta_{\mathrm{g}}(\omega) = -\beta_{\mathrm{d}}(\omega)$. In distributed amplifiers implemented by means of forward transmission lines (conventional distributed amplifiers), the latter condition ($\beta_{\mathrm{g}}(\omega) = -\beta_{\mathrm{d}}(\omega)$) can only be satisfied at DC, where $\beta_{\mathrm{g}}(\omega) = \beta_{\mathrm{d}}(\omega) = 0$. Hence, the maximum power gain for the reverse port occurs at DC, and the power at the reverse port exhibits a lowpass behavior. If the gate and drain lines are identical (i.e., they have the same phase constant), the gain in the forward port takes the value given by (4.58) at all frequencies. Thus, distributed amplifiers ideally exhibit an infinite gain-bandwidth product (forward port). In practice, this of course does not occur since discretization due to the presence of transistors at periodic positions in the lines limits amplifier bandwidth. Bandwidth is also limited by potentially slight differences in the phase constants of the gate and drain lines. The ideal power gains (forward and reverse ports) of a continuous distributed amplifier that result by considering conventional gate and drain lines with identical (and linear) phase constant ($\beta_{\mathrm{g}}(\omega) = \beta_{\mathrm{d}}(\omega)$) are depicted in Figure 4.49a.

Dispersion Engineering in the Drain and Gate Lines To the light of the dependence of the forward and reverse power gains with the phase constants of the gate and drain lines, it is simple to deduce that by engineering such phase constants it is possible to obtain functionalities beyond those achievable with conventional distributed amplifiers. To start, let us consider that the gate and drain lines are replaced with PLH transmission lines. If the lines are identical and exhibit a dependence of the phase constant with frequency of the type given by expression (3.41), the forward gain is independent of frequency (since $\beta_{\mathrm{g}}(\omega) = \beta_{\mathrm{d}}(\omega)$) and given by (4.58). However, since the phase

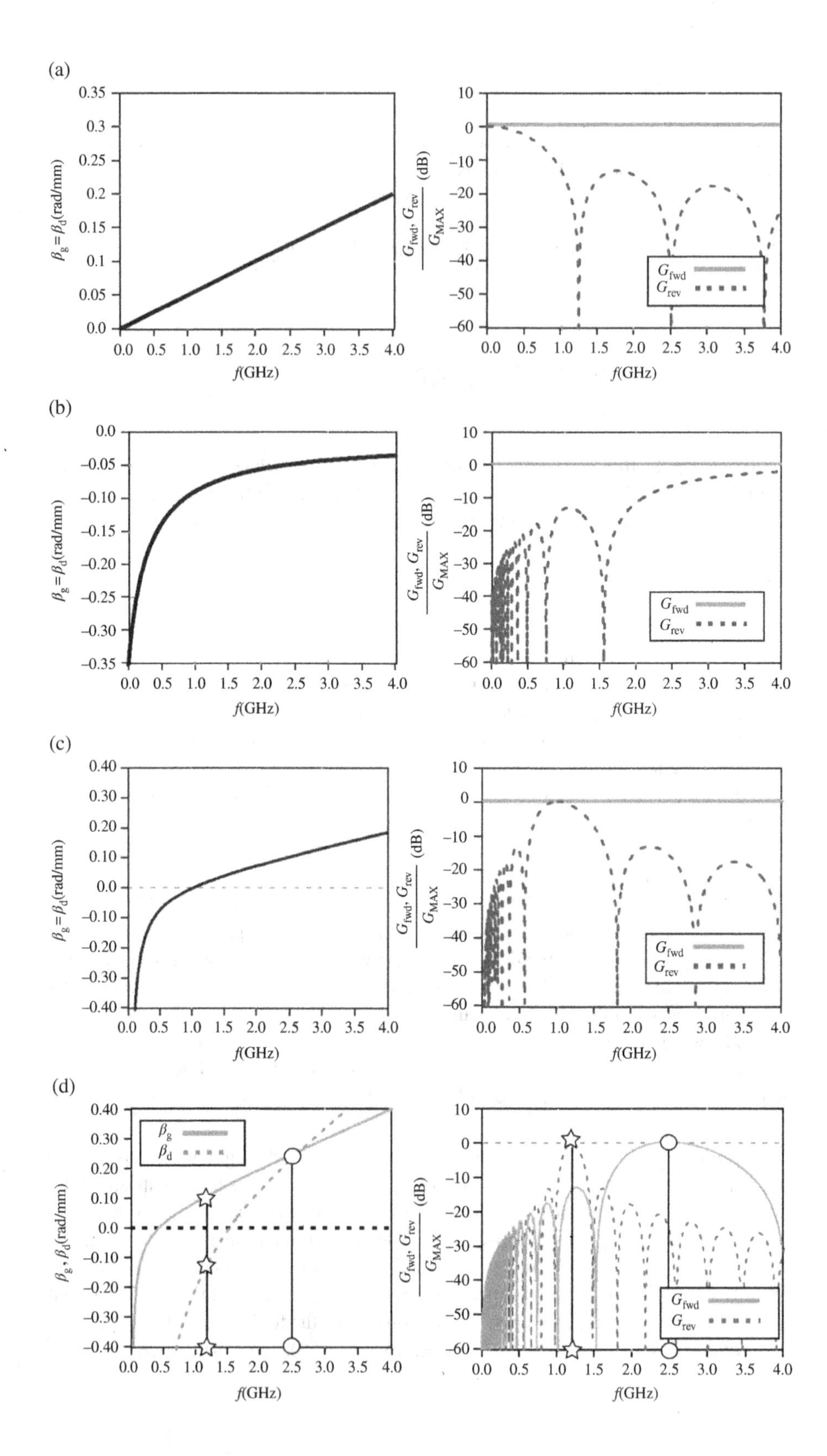
(a)
$\beta_g = \beta_d$(rad/mm)
f(GHz)
$\frac{G_{fwd}, G_{rev}}{G_{MAX}}$ (dB)
G_{fwd}
G_{rev}
(b)
$\beta_g = \beta_d$(rad/mm)
f(GHz)
$\frac{G_{fwd}, G_{rev}}{G_{MAX}}$ (dB)
G_{fwd}
G_{rev}
(c)
$\beta_g = \beta_d$(rad/mm)
f(GHz)
$\frac{G_{fwd}, G_{rev}}{G_{MAX}}$ (dB)
G_{fwd}
G_{rev}
(d)
β_g, β_d(rad/mm)
β_g
β_d
f(GHz)
$\frac{G_{fwd}, G_{rev}}{G_{MAX}}$ (dB)
G_{fwd}
G_{rev}

constant is null in the high-frequency limit, it follows that the reverse gain takes the maximum value in that limit, and hence it exhibits a high-pass behavior (this situation is illustrated in Fig. 4.49b). In practice, discretization limits the bandwidth of the forward gain since the phase constant is purely imaginary at low frequencies (expression 3.41 is valid in the long wavelength limit, that is, at high frequencies for a PLH line). On the other hand, a PLH line cannot be implemented in practice, as it was discussed in Chapter 3, with the result of a forward gain also decaying at sufficiently high frequencies.

Another interesting situation results by considering gate and drain lines with CRLH behavior. In the long wavelength limit (continuous limit), the phase constant of the CRLH transmission line is given by expression (3.59). By balancing the line ($\omega_s = \omega_p = \omega_o$) the phase constant is simplified to

$$\beta l = \frac{\omega}{\omega_R}\left(1 - \frac{\omega_o^2}{\omega^2}\right) \tag{4.59}$$

where ω_R was defined in Section (3.4.2) and ω_o is the so-called transition frequency. In this case, the reverse gain exhibits an intrinsic bandpass behavior with maximum gain at ω_o. This situation is illustrated in Figure 4.49c. Many other interesting situations arise by considering different gate and drain lines (i.e., with different CRLH-type phase constants). We will not cover all of them in the present book (the author recommends [135] for an in-depth analysis). However, let us point out the situation that results by considering the phase constants for the gate and drain lines indicated in Figure 4.49d (the corresponding normalized forward and reverse gains are also depicted in the figure). In view of this figure, both the forward and reverse power gains exhibit an intrinsic bandpass behavior. The important aspect is that by tailoring the dispersion of the gate and drain lines, it is possible to independently control the frequency where the forward and reverse gains are maximized. Therefore, such distributed amplifier exhibits a dual-band/diplexing amplifying functionality that can be of interest in numerous applications. As pointed out in Ref. [135], such diplexing capability can also be achieved through a conventional gate line and a CRLH drain line with properly engineered phase constants.

The frequency dependence of the forward and reverse gains depicted in Figure 4.49 has been inferred by introducing the considered phase constants ($\beta_g(\omega)$ and $\beta_d(\omega)$) in expressions (4.55) and (4.57). This provides the intrinsic (ideal) behavior of the continuous distributed or metadistributed amplifiers. The analysis of the discrete model

FIGURE 4.49 Phase constants of the drain and gate lines and the corresponding forward and reverse power gains for an ideal continuous unilateral distributed amplifier. (a) Forward gate and drain lines with identical phase constants, (b) backward gate and drain lines with identical phase constants, (c) CRLH gate and drain lines with identical phase constants, and (d) CRLH gate and drain lines with different phase constants. Reprinted from Ref. [135], with permission from the author.

can be found in Ref. [119, 135], and it will not be reproduced here. Nevertheless, the expressions of the forward and reverse gains for a distributed (or metadistributed) amplifier consisting of two transmission lines actively coupled through N FETs with transconductance g_m are

$$G_{\mathrm{fwd}} = \frac{(g_m N)^2 Z_g(\omega) Z_d(\omega)}{4} \frac{\sin^2\left[\frac{N}{2}(\beta_g(\omega) - \beta_d(\omega))\right]}{N^2 \sin^2\left[\frac{1}{2}(\beta_g(\omega) - \beta_d(\omega))\right]} \tag{4.60}$$

$$G_{\mathrm{rev}} = \frac{(g_m N)^2 Z_g(\omega) Z_d(\omega)}{4} \frac{\sin^2\left[\frac{N}{2}(\beta_g(\omega) + \beta_d(\omega))\right]}{N^2 \sin^2\left[\frac{1}{2}(\beta_g(\omega) + \beta_d(\omega))\right]} \tag{4.61}$$

Due to the similarities between (4.60), (4.61) and (4.55), (4.57), the main features relative to the continuous (ideal) distributed or metadistributed amplifiers pointed out before can be extended to the discrete structure, with exception of bandwidth, as has been discussed before (notice that the explicit dependence of the characteristic impedances of the gate and drain lines with frequency has been included in (4.60) and (4.61)). Indeed, the continuous model can be used to analyze the discrete distributed amplifier provided the frequency regions of interest are far away from the cutoff frequencies and close to the frequencies where the forward and reverse gains are maximized [135].

An Illustrative Example: Dual-Band/Diplexer Meta-Distributed Amplifier A dual-band/diplexer metadistributed amplifier, based on two actively coupled CRLH transmission lines, is reported as an illustrative example [136]. The amplifier was designed to exhibit maximum forward and reverse gains at 440 MHz and 640 MHz, respectively, with a bandwidth of approximately 100 MHz in both cases, and characteristic impedances for the drain and gate lines of 50 Ω. The CRLH lines were implemented by means of lumped elements. The element values can be inferred from the required phase constants of the CRLH lines, which in turn are determined by expressions (4.60) and (4.61). However, as pointed out in Ref. [136], it is necessary to take into account the input and output FET capacitances, which are parallel connected to the shunt capacitance of the CRLH lines (the considered active devices are *Infineon CFY-30* FET transistors). On the other hand, the shunt inductor of the CRLH lines is a short-circuit at DC, and therefore it prevents the active device from being properly biased. This problem was circumvented in Ref. [136] by series connecting a capacitor with this inductor. This capacitor allows for device biasing, but it must exhibit a very low impedance at all the significant frequencies, being effectively a short circuit for those frequencies. To achieve this, 100 pF capacitors were used in the design. Figure 4.50a depicts the scheme of the amplifier, where the dual-band/diplexing functionality is represented, and the fabricated device is shown in Figure 4.50b. The measured forward and reverse gains, depicted in Figure 4.50c and obtained for the bias

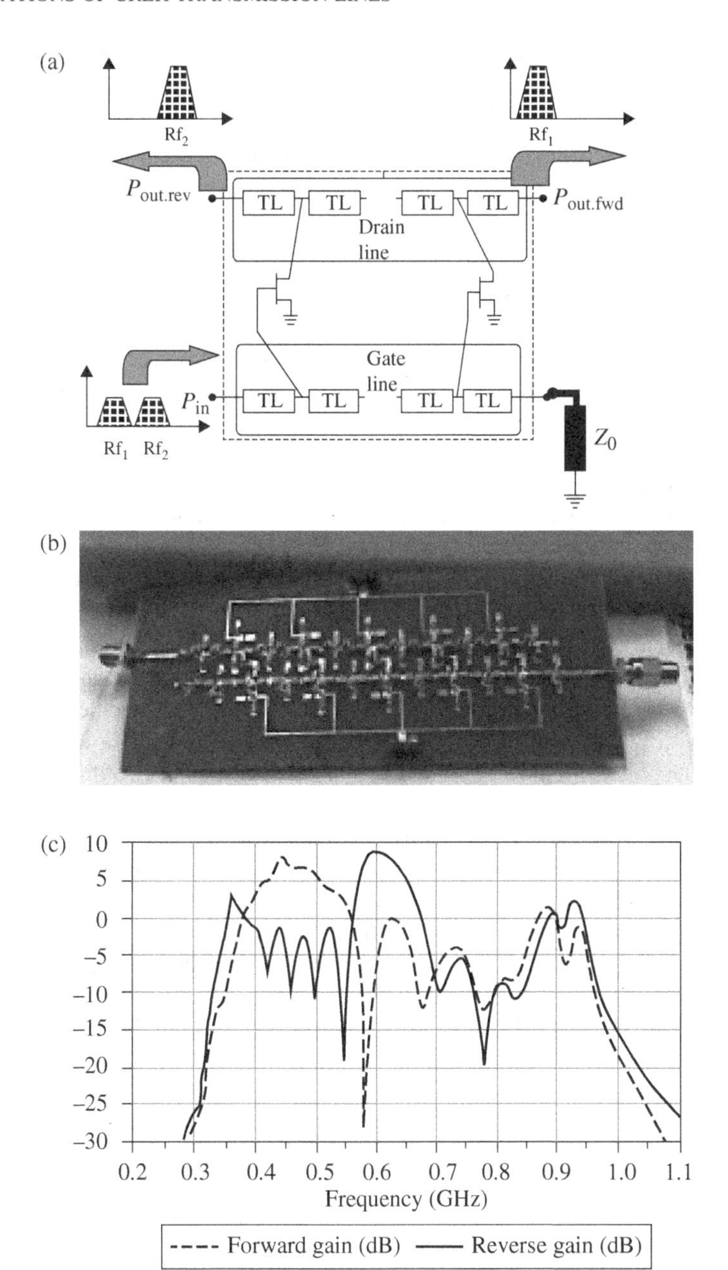

FIGURE 4.50 Scheme showing the dual-band/diplexer functionality of the designed distributed amplifier (a), photograph (b), and measured forward and reverse power gains (c). Reprinted with permission from Ref. [136]; copyright 2007 EuMA.

point defined by $V_{DS} = 3.5$ V and $V_{GS} = -0.7$ V, reveal that the required functionality is achieved. Noise measurements (not depicted here) reveal that the forward and reverse noise figures are similar to those achieved in conventional distributed amplifiers [136].

Other representative works of the recent progress in metadistributed amplifiers and mixers are given in Refs. [137–142]. The list includes distributed amplifiers integrated with CRLH LWAs [138, 141], and the first monolithic metadistributed amplifier implemented in GaN technology operative at C-band [142].

4.2.5.2 Dual-Band Recursive Active Filters Recursive active filters are a subclass of active filters based on distributed components and a feedback scheme, where the frequency selective response is generated by combining two signals: the main path signal component and a component properly weighted and delayed that constitutes the feedback [143]. The design of the filter is based on providing a certain phase response (phase delay) to the feedback section at the center frequency of the pass band. Such phase delay can be achieved by means of a transmission line section. Since the phase delay is a periodic function, it follows that recursive active filters implemented with conventional lines actually exhibit a periodic response. However, since the higher-order bands are centered at the harmonics of the fundamental frequency (i.e., not controllable), such filters are single-band devices. However, it is possible to extend the functionality of recursive active filters by replacing the conventional lines in the feedback section with CRLH transmission lines, in order to achieve a frequency response with two transmission bands (dual-band filter) [144]. In order to produce a dual-band response, the phase condition should be fulfilled at two different controllable frequencies (an inherent property of CRLH lines, as it was discussed in Section 4.2.2). Thus, the purpose of this subsection is to demonstrate that CRLH lines can be applied to the implementation of dual-band recursive active filters.

The typical schematic of a recursive active filter is depicted in Figure 4.51. The filter uses power dividers and combiners that can be implemented by means of branch line couplers with a matched load in one of the ports, as depicted in Figure 4.51. The weighting factor is achieved by means of the amplifier, whereas the phase delay in the feedback signal is obtained thanks to the transmission line in the feedback branch.

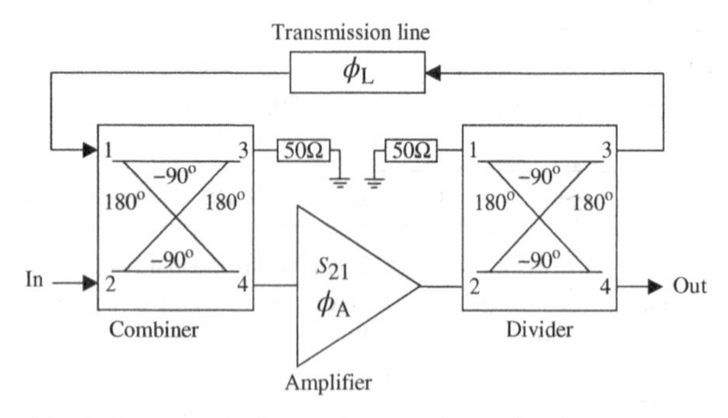

FIGURE 4.51 Bloch diagram of a first-order recursive active filter that makes use of branch line hybrid couplers for the power divider and combiner stages.

According to this scheme, the central frequencies of the filter pass bands are given by those angular frequencies, ω_o, satisfying the following phase condition (loop phase):

$$\phi_A(\omega_o) + \phi_D(\omega_o) + \phi_L(\omega_o) + \phi_C(\omega_o) = n \cdot 2\pi \tag{4.62}$$

where ϕ_A, ϕ_D, ϕ_L, and ϕ_C are the phase responses of the amplifier, power divider, transmission line section and power combiner, respectively, n being an integer number. Notice that, according to (4.62), the feedback signal and the input power signal are combined constructively (i.e., with null relative phase shift) at any of the filter-operating frequencies. Assuming that the dividers and combiners produce a −3 dB insertion loss and perfect matching between blocks, the filter transmission coefficient at ω_o is given by [144]

$$S_{21,F}(\omega_o) = \frac{1}{2} \frac{|S_{21,A}|}{|1 - 0.5 \cdot |S_{21,A}||} \tag{4.63}$$

where $S_{21,A}$ is the amplifier gain. Depending on the amplifier gain the whole structure may be unstable. This aspect is discussed in Refs. [144, 145] where it is concluded that first-order recursive filter topologies with −3 dB couplers are unconditionally unstable when using amplifier gain values over 6 dB, and potentially unstable or unconditionally stable below that value.

Although important, the stability issue is out of the scope of this subsection, focused on the potential of CRLH lines as delay lines for achieving dual-band functionality in the recursive active filters. If we assume that the phase delays given by the power combiner and divider take fixed values at the operating frequencies, expression (4.62) can be reduced to an expression only dependent on the phases of the transmission line section and the amplifier. The phase delay of a branch-line coupler is π; thus, the phase condition in (4.62) can be rewritten as follows:[33]

$$\phi_A(\omega_i) + \phi_L(\omega_i) = n \cdot 2\pi \tag{4.64}$$

where the index $i = 1, 2$ is used to designate the two desired operating frequencies of the dual-band recursive active filter. Expression (4.64) represents actually a pair of equations that have a solution provided the phase response of the considered transmission line has a nonlinear dispersion. Thus, it is obvious that by using CRLH lines in the feedback branch it is possible to achieve the dual-band functionality. As an illustrative example, Figure 4.52a depicts the dual-band recursive active filter reported in Ref. [144]. The filter response is also depicted in the figure, where the dual-band behavior can be appreciated. In Figure 4.52d the phase responses of the amplifier and transmission line stages are depicted; it can be seen that (4.64), with $n = 0$, is fulfilled at the design frequencies (0.8 GHz and 1.7 GHz). Higher-order dual-band and tunable recursive active filters are reported in Refs. [146, 147].

[33] It is assumed that the branch line coupler is also a dual-band device.

4.2.6 Sensors

Metamaterial transmission lines are also of interest for the implementation of microwave sensors with advanced properties. In microwave sensors, the sensing principle is based on the effects caused by the measured variable (temperature, humidity, pressure, position, dielectric loading, etc.) on the characteristics (phase, magnitude, resonance frequency, delay, etc.) of a feeding microwave signal.[34] For instance, a transmission line loaded with an SRR exhibits a notch in the transmission coefficient

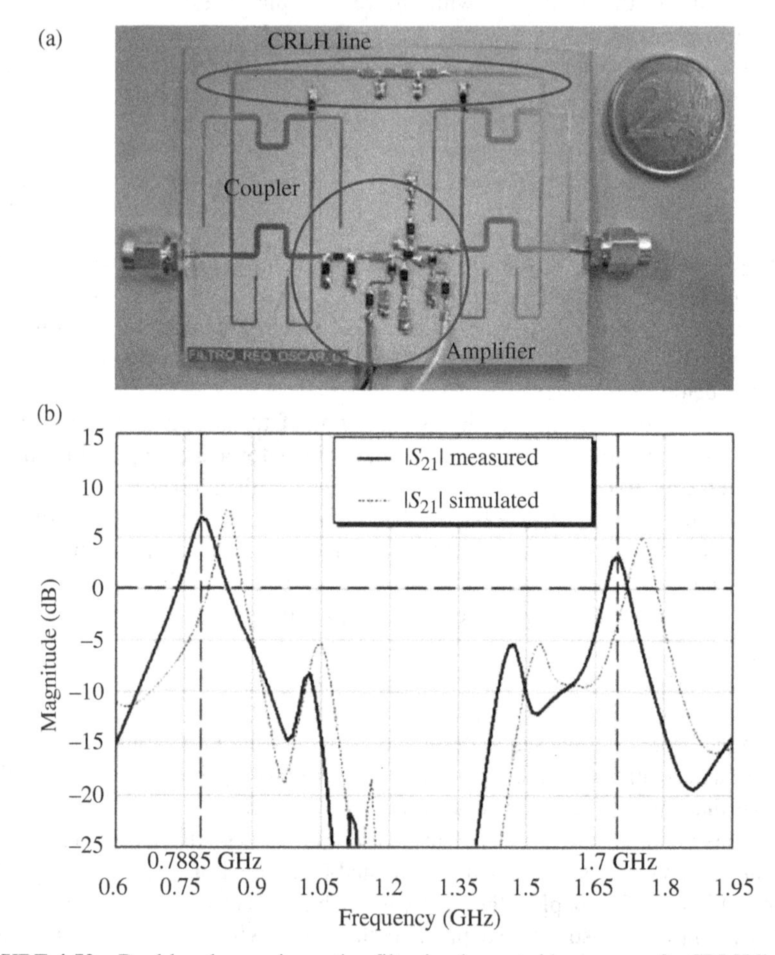

FIGURE 4.52 Dual-band recursive active filter implemented by means of a CRLH line and dual-band couplers: (a) photograph of the prototype, (b) insertion loss, (c) return loss, and (d) simulated phase response of the gain stage and feedback line. Reprinted with permission from Ref. [144]; copyright 2009 IEEE.

[34] In microwave sensors, the feeding signal can be provided by the source thorough direct (wire) connection or wirelessly. In the latter case, the devices are designated as wireless microwave sensors.

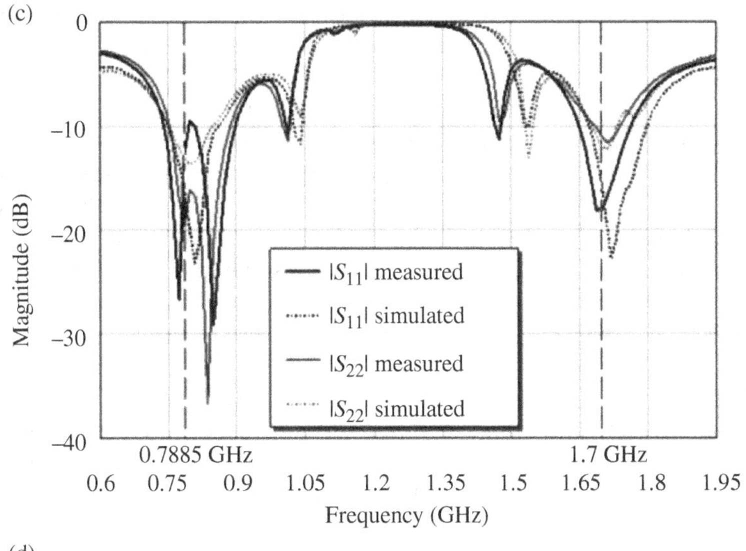

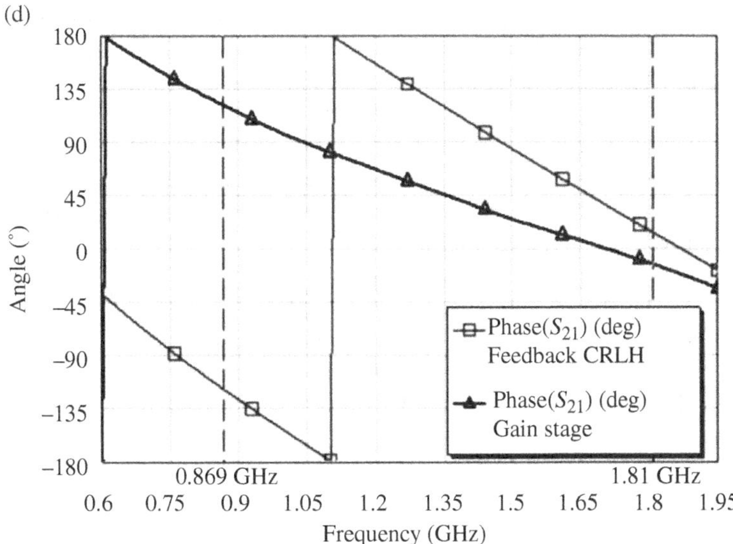

FIGURE 4.52 (*Continued*)

at the fundamental SRR resonance frequency (and at higher-order resonances). Thus, any variable able to modify the resonance frequency of the SRR can be easily sensed. Indeed, SRR-loaded lines can be used for sensing purposes, but these microwave sensors are considered to belong to the category of lines with metamaterial loading, and will be studied later in this chapter.

CRLH lines can be engineered in order to implement sensors with small size and high sensitivity (the author recommends the paper [148] and references therein to those readers willing to penetrate more deeply in the topic of CRLH transmission line-based sensors—from now on CRLH-sensors to simplify the terminology). Typically, CRLH-sensors use capacitive elements as transducers. Hence, in combination with functional layers (e.g., for the monitoring of chemical or biological substances), with specific mechanical setups such as movable capacitor plates, or with tunable materials, such CRLH-sensors can be useful to measure very different variables. CRLH-sensors can be subdivided into two groups: spatially integrating sensors and spatially resolved sensors [148]. The first group refers to sensors where the output signal is determined by a change of the overall (average) transmission line properties caused by changes within the line capacitors as transducer elements. Spatially resolved CRLH sensors have the same structure, but each transducer element is read out individually, which leads to an output vector instead of a single measured value.

As an example of a spatially integrating CRLH-sensor, we report here a mass flow sensor first presented in Ref. [149], able to detect velocities in a conveyor belt system. The sensor consists of a 10-cell CRLH resonator coupled to the input ports by means of SMD capacitors. Actually, the CRLH line is composed of series (patch) capacitances and shunt inductances, as depicted in Figure 4.53. However, parasitics are unavoidable and therefore the line is actually a CRLH line. The sensor is fed by a microwave signal tuned to the desired resonance of the sensor, and, simultaneously, such signal is fed into a phase comparator. The time-dependent dielectric load of the sensor modulates the phase of the resonator output signal. This phase is compared with the reference phase originated directly from the source. The difference provides a measure of the dielectric loading of the resonator [149]. For velocity measurement of single particles passing on the measurement section, the time-dependent phase shift describes a sinusoidal wave. From the time difference between two adjacent phase minima, the velocity can be inferred.

Another interesting sensing device reported by the same authors is the CRLH transmission line-based differential sensor [150]. This sensor, depicted in Figure 4.54, utilizes two identical CRLH lines and two power dividers/combiners. The microwave signal is divided into two equal parts (by means of a Wilkinson power divider) and fed into the two CRLH lines. The feeding lines of the upper artificial line are one half wavelength longer than the ones of the lower line, so that the signals at the end of both lines exhibit a phase difference of 180°. One of these CRLH lines acts as the reference, whereas the other one is detuned by dielectric loading. The altered signal is subtracted (due to the presence of the combiner) from the unaltered reference to form the output signal. The resulting signal to noise ratio for this configuration has been found to be roughly 60 dB, allowing a high measurement dynamic range [150] (see Fig. 4.54b). In this sensor, the variable to be evaluated is simply the magnitude of S_{21} at the output port, rather than the phase. However, such magnitude is intimately related to the phase variation in the active line, caused by the presence of a dielectric load. Thus, the sensor principle is based on a power measurement, rather than on a phase measurement. Another advantage of this setup is that the sensitivity can be linearly increased with the number of used cells. These devices can be applied in medical diagnosis or as biosensors for the detection of specific substances.

Examples of spatially resolved sensors are reported in Ref. [148] and references therein. Among them, wireless temperature sensors with identification [151] and level

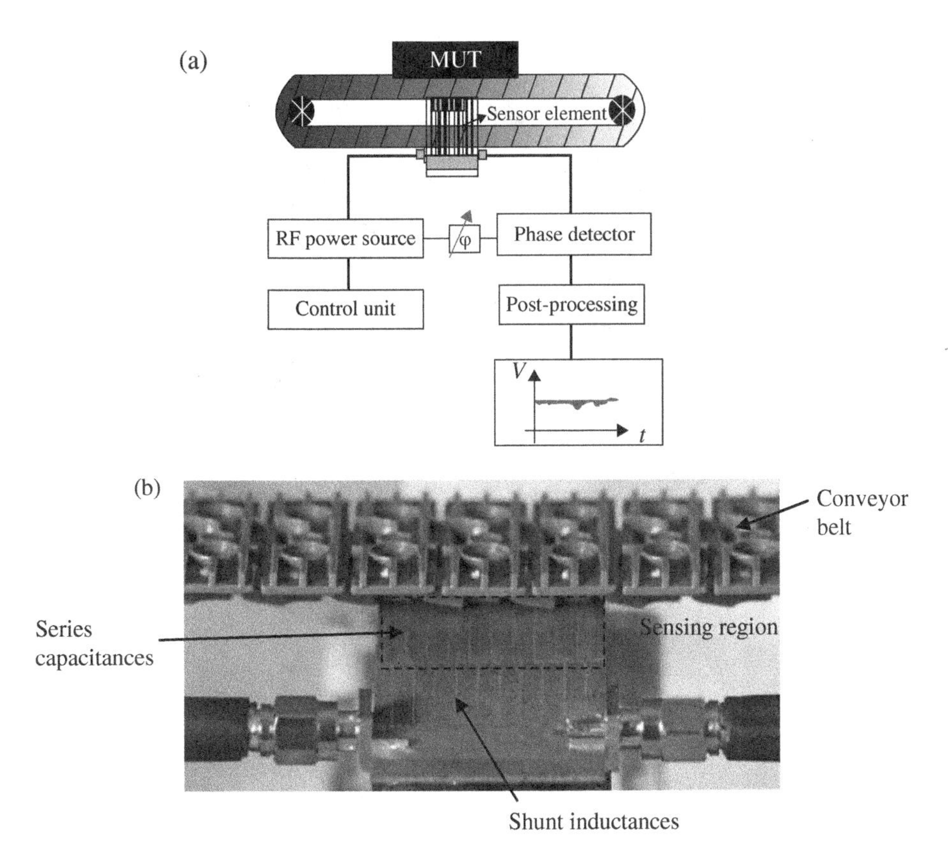

FIGURE 4.53 Bloch diagram of the mass flow sensor setup (a) and photograph of the sensor under the conveyor belt (b). Reprinted with permission from Ref. [147]; copyright 2012 IEEE.

sensors [152] constitute good examples of the potential of CRLH transmission lines for sensing purposes.

4.3 TRANSMISSION LINES WITH METAMATERIAL LOADING AND APPLICATIONS

By loading a transmission line with electrically small resonators, such as those formerly used for the implementation of metamaterials (SRRs, CSRRs) or other related resonators (see Section 3.3), multiband planar antennas, sensors, radiofrequency bar codes, and other RF/microwave devices can be implemented.[35] Let us review some of these applications in the next subsections.

[35] Indeed, planar notch and stopband filters consisting of transmission lines loaded with SRRs, CSRRs, and so on, could have been included in this section (notice that the design of these filters is not based on dispersion/impedance engineering). However, the author has opted for the inclusion of SRR-based notch and stopband filters in a single subsection devoted to filters (Section 4.2.3) for coherence. Nevertheless, tunable filters and other advanced filters will be studied in the next chapters.

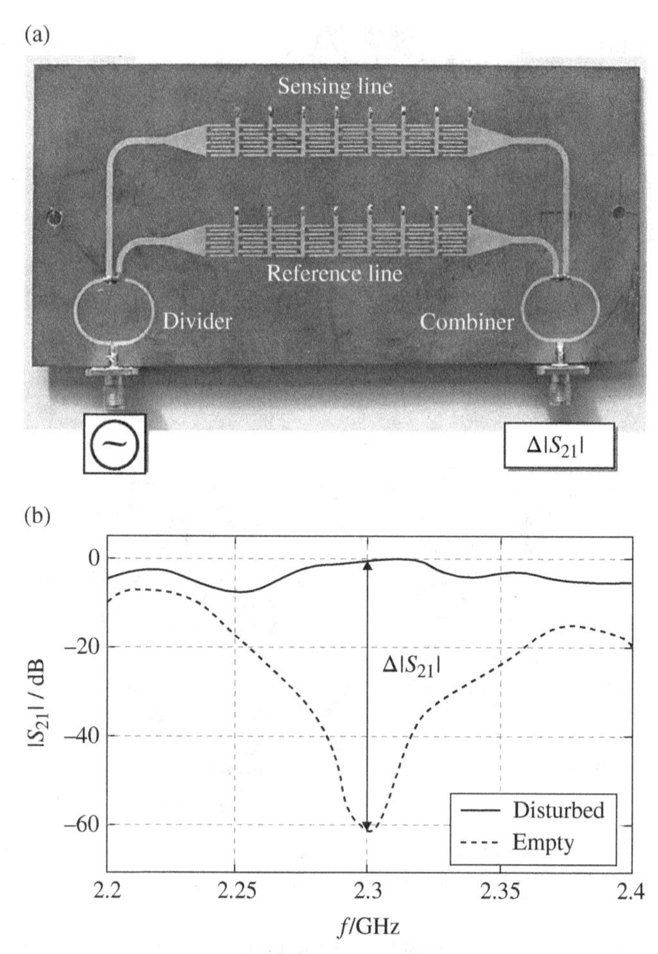

FIGURE 4.54 Photograph of the proposed CRLH transmission line-based differential sensor (a) and comparison between the transmission coefficients for the empty case (unloaded sensing line) and for a small dielectric disturbance applied to the sensing line (b). Reprinted with permission from Ref. [150]; copyright 2009 IEEE.

4.3.1 Multiband Planar Antennas

There are several approaches for the implementation of multiband antennas in planar technology. Here we report two strategies: one based on the concept of trap antennas and the other based on a perturbation method, of special interest in ultra-high-frequency radiofrequency identification (UHF-RFID) for the implementation of tags able to operate in two of the regulated UHF-RFID bands.

4.3.1.1 Multiband Printed Dipole and Monopole Antennas In this subsection, the focus is on printed dipole and monopole antennas with metamaterial loading

[153–157]. The idea of this technique is based on loading a conventional printed antenna with a set of resonant particles. In Ref. [154], it is shown that a dual-band antenna is achieved by coupling a set of SRRs to a printed dipole. Using this approach, the benefits of printed antennas are kept while dual-band antennas are achieved by using a simple design technique. The SRRs produce open circuits in their positions at the resonance frequencies. Hence the antenna resonance is achieved not only when the effective length of the dipole arms is $\lambda/4$ (λ being the guided wavelength), but also when the different locations of the SRR are $\lambda/4$ from the antenna feeding point (the SRRs must be tuned at these frequencies). By using SRRs, narrow bandwidths are reported in Ref. [154] for the bands associated with the SRR loading. An open circuit in the dipole arms can also be obtained by means of series connected OCSRRs. By using these particles, broadband responses can be achieved (as compared to SRRs), due to the relative values of capacitance and inductance in OCSRRs.

To illustrate the potential of OCSRR-loaded printed antennas, a dual-band dipole [156] and a tri-band monopole [157] are reported. The layout of the printed dipole (an antipodal structure) and the details of the OCSRR are depicted in Figure 4.55 (the photograph of the antenna, compared to the conventional mono-band dipole, is also shown in Fig. 4.55). The parameters of each dipole strip are the length L and the width W (which must be engineered to optimize matching). This antipodal configuration was chosen in Ref. [156] because it avoids the use of a balun to feed the antenna. The dimensions of the feeding line are the length L_f and the width W_f. An OCSRR is connected in series to each dipole strip at a distance d_{OCSRR} from the center of the antenna. The different dimensions of the proposed antenna were optimized to simultaneously operate at the L1-GPS frequency (1.575 GHz) and the WiFi band of 2.4–2.48 GHz, namely, $L = 22.00$ mm, $W = 2.5$ mm, $L_f = 25$ mm, $W_f = 1.15$ mm, $d_{OCSRR} = 17$ mm, $l_{ext} = 4$ mm, and $c = d = 0.3$ mm. The used substrate is the *Rogers RO3010* with $\varepsilon_r = 10.2$ and $h = 1.27$ mm.

In the case of the conventional unloaded antenna, there is only one series resonance at 2.1 GHz. This resonance corresponds to the fundamental mode of the dipole antenna. In the vicinity of resonance, the value of the real part of the antenna impedance is close to 50 Ω (not shown), which produces proper matching, as it can be appreciated in Figure 4.56. OCSRR loading introduces a parallel resonance in the proposed dual-band antenna input impedance. The addition of this parallel resonance has a double effect on the input impedance of the proposed dual-band antenna. The first one is a slight shift of the series resonance of the dipole antenna toward higher frequencies. In this case, this resonance is found at 2.45 GHz. The second effect is the appearance of an additional series resonance below the parallel resonance of the OCSRRs. This additional series resonance is found at 1.6 GHz. The real part of the input impedance is around 50 Ω in the vicinity of this additional series resonance. This allows achieving an additional band with proper matching in the proposed antenna (Fig. 4.56). As it can be seen, this additional frequency is below the fundamental frequency of the unloaded dipole antenna.

Considering $|S_{11}|$ below −10 dB, the first working band is centered at 1.56 GHz with 5% bandwidth and the second one is centered at 2.46 GHz with 9% bandwidth. These bandwidths represent a considerable improvement with respect to the

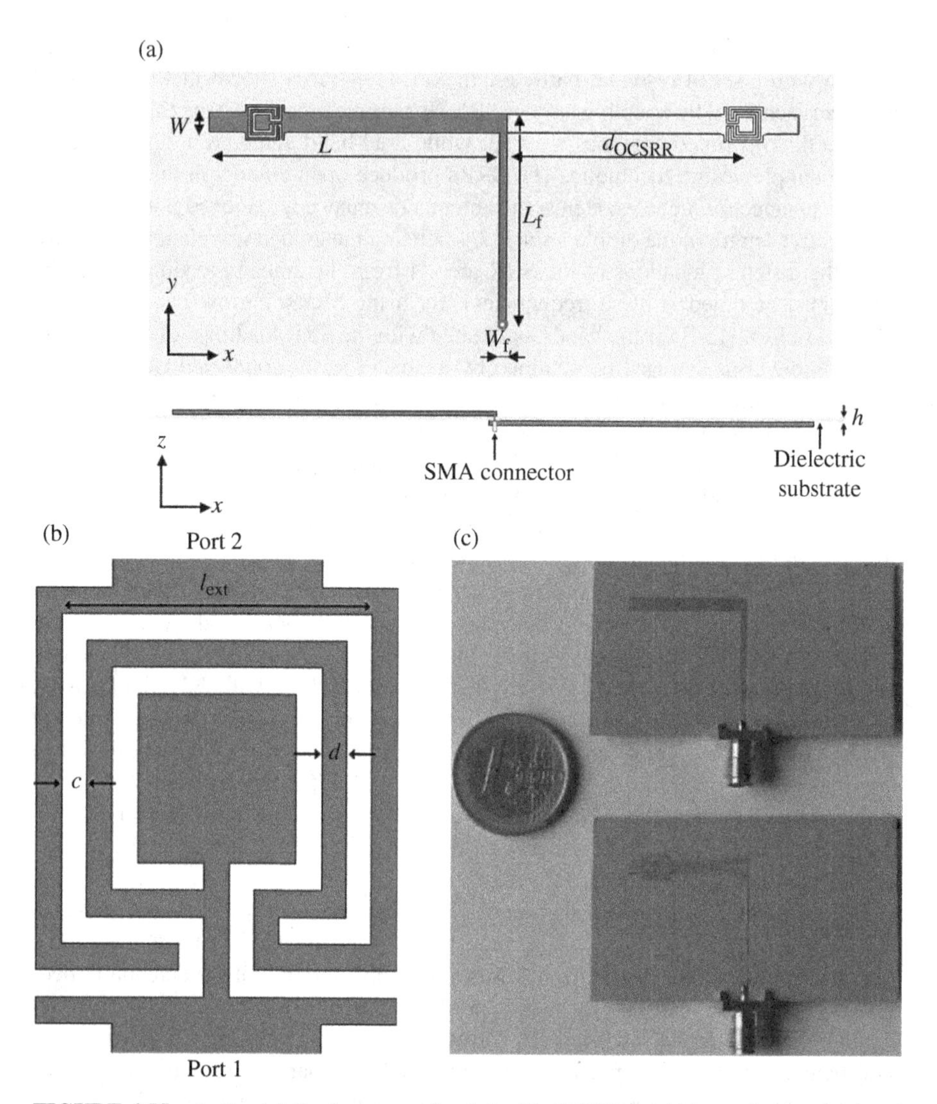

FIGURE 4.55 Antipodal dipole antenna loaded with OCSRRs. (a) Top and cross-sectional views; (b) detail of the OCSRR and relevant dimensions; and (c) comparative photographs of the unloaded (top) and loaded (bottom) printed dipole antenna. Reprinted with permission from Ref. [156]; copyright 2012 John Wiley.

implementations based on SRRs [154, 155]. The measured gain of the antenna (estimated in a TEM cell from the power received by the antenna and the incident field measured by a probe) is 0.85 dB at the GPS band (1.575 GHz) and 2 dB at the WiFi band (2.45 GHz). The radiation patterns inferred with *CST Microwave Studio* for the proposed dual-band antenna are shown in Figure 4.57. A dipolar-like radiation pattern is obtained at both working bands. The typical figure of eight is obtained

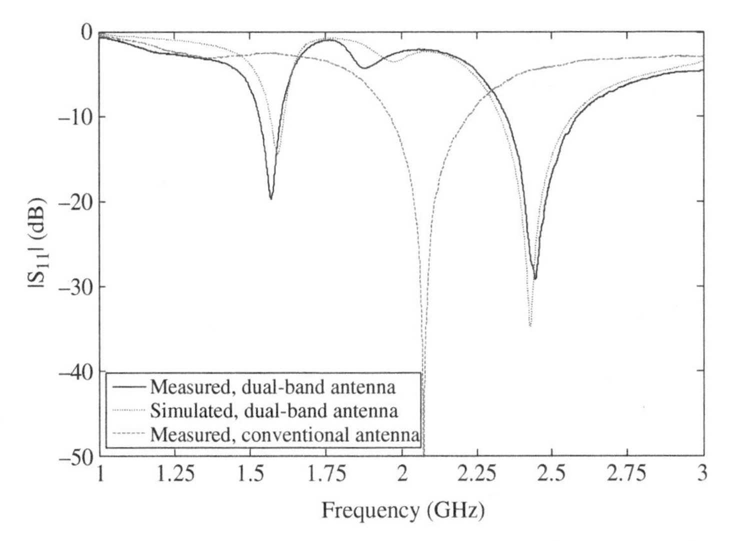

FIGURE 4.56 Measured and simulated reflection coefficient of dual-band printed dipole antenna compared to the unloaded antenna. Reprinted with permission from Ref. [156]; copyright 2012 John Wiley.

in the *xz*-plane and an omnidirectional pattern is obtained in the *yz*-plane at both frequencies (1.575 and 2.45 GHz). The only difference between both working bands is the cross-polarization level, which is around −10 dB in the first band while it is −20 dB in the second one.

OCSRR loading can also be applied to printed monopole antennas [157]. Figure 4.58 shows a photograph of two fabricated prototypes: a dual-band and a tri-band printed monopole antenna. The considered substrate in this case was the low-cost *FR4* ($\varepsilon_r = 4.5$ and $h = 1.5$ mm). The dual-band antenna covers the bands of 2.40–2.48 GHz (Bluetooth and WiFi) and 5.15–5.80 GHz (WiFi). The dimensions of the monopole are $L_m = 21$ mm and $W_m = 5.85$ mm. The parameters of the feeding line were set to obtain a 50 Ω CPW. Hence, $S = 2.44$ mm and $W = 0.30$ mm. The dimensions of each ground plane are $L_g = 16$ mm and $W_g = 13.48$ mm. The gap between the ground planes and the monopole is $g = 0.30$ mm. The OCSRR was placed at a distance $d_o = 12.50$ mm and its parameters are $l_{ext} = 2.30$ mm, $c = d = 0.25$ mm. The gap g_o was set to 0.50 mm.

The tri-band monopole antenna is an extension of the previous dual-band antenna, and it covers the previous bands and the IEEE 802.11y band of 3.65–3.70 GHz. According to the layout of the tri-band monopole (Fig. 4.58), the additional OCSRR was placed at a distance $d_{o2} = 18$ mm. Its dimensions are $l_{ext} = 2.70$ mm and $c = d = 0.25$ mm. This corresponds to a resonance frequency of 3.65 GHz. The gap g_{o2} was set to 0.40 mm. These values were optimized to only cover the desired bandwidth and not interfere with other systems. The other parameters of the antenna remain unchanged with respect to the dual-band design, except the length of the monopole which is

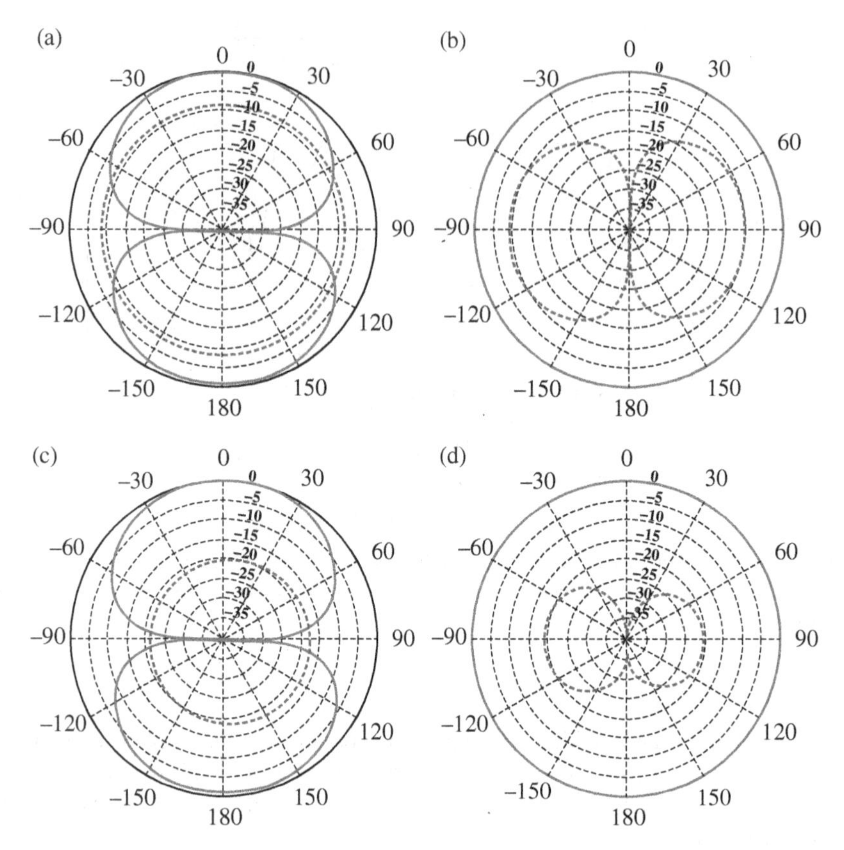

FIGURE 4.57 Simulated normalized radiation patterns for the proposed dual-band dipole antenna. Solid line: copolar component and dashed line: crosspolar component. (a) 1.575 GHz, *xz*-plane; (b) 1.575 GHz, *yz*-plane; (c) 2.45 GHz, *xz*-plane; and (d) 2.45 GHz, *yz*-plane. Reprinted with permission from Ref. [156]; copyright 2012 John Wiley.

reduced to $L_m = 19.75$ mm to compensate the inductive behavior of the OCSRRs below their resonance frequencies.

The simulated and measured reflection coefficients of the dual-band and tri-band monopoles are depicted in Figure 4.59. The dual-band printed monopole antenna exhibits good matching ($|S_{11}| < -10$ dB) from 2.29 GHz to 2.52 GHz at the lower frequency band. This corresponds to a 9.6% bandwidth. In the upper band, the antenna is well matched from 4.66 GHz to at least 7 GHz. Thus, the fabricated antenna satisfies the specifications of Bluetooth and WiFi (bands of 2.40–2.48 GHz and 5.15–5.80 GHz). The fabricated tri-band monopole antenna is well matched from 2.30 GHz to 2.52 GHz for the first band. Its reflection coefficient is below −10 dB between 3.56 GHz and 3.78 GHz for the second band, and between 5.06 GHz and 6.71 GHz for the third band. Hence, the fabricated prototype is well matched within the regulated bandwidths of Bluetooth and WiFi including IEEE 802.11y (3.65–3.70 GHz band).

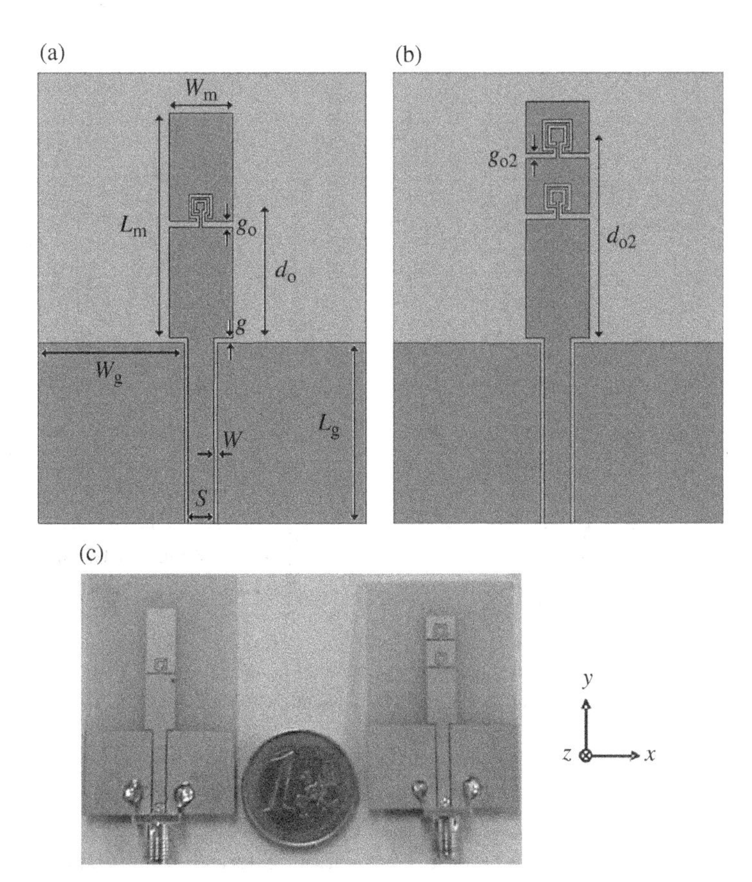

FIGURE 4.58 Sketch of the dual-band (a), and tri-band (b) printed monopole antenna loaded with OCSRRs, and respective photographs (c). Reprinted with permission from Ref. [157]; copyright 2011 IEEE.

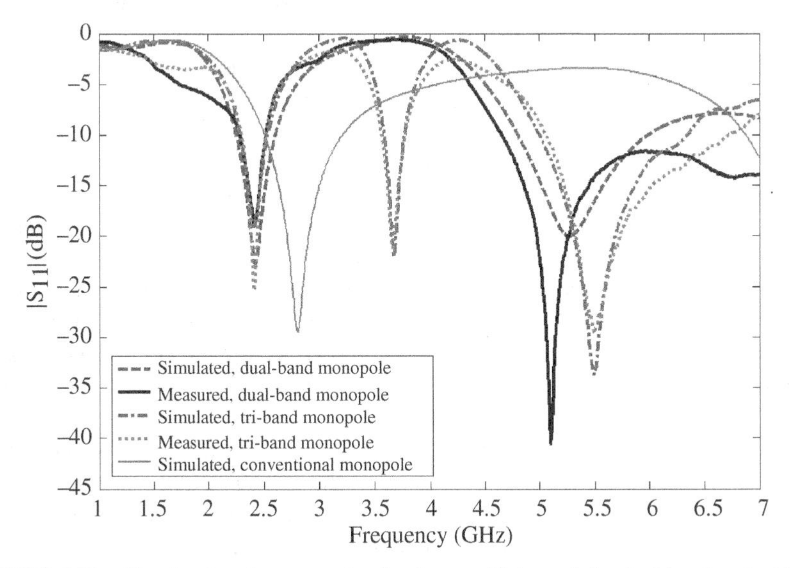

FIGURE 4.59 Simulated and measured reflection coefficient of the dual-band and tri-band printed monopole antennas. The simulated reflection coefficient of the conventional monopole antenna is also plotted. Reprinted with permission from Ref. [157]; copyright 2011 IEEE.

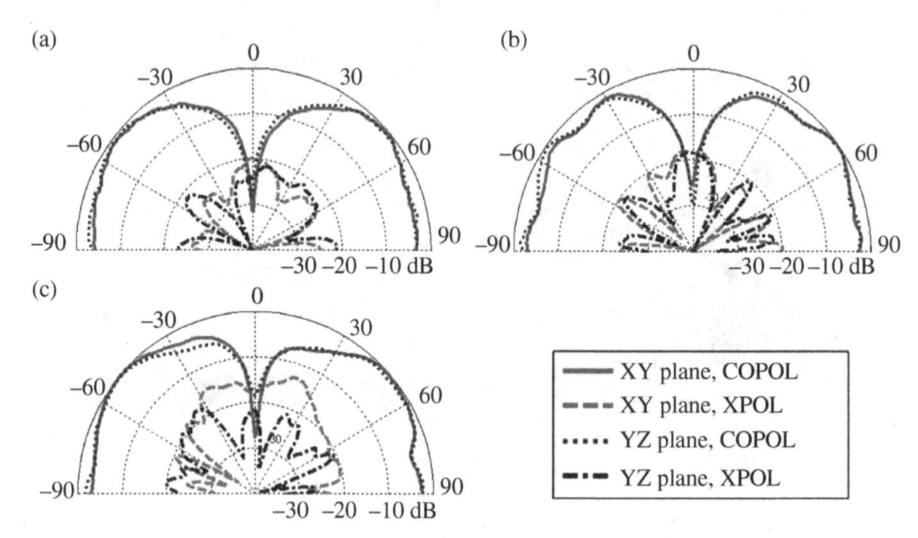

FIGURE 4.60 Measured radiation patterns of the tri-band printed monopole antenna. (a) 2.45 GHz, (b) 3.65 GHz, and (c) 5.40 GHz. Reprinted with permission from Ref. [157]; copyright 2011 IEEE.

The proposed antennas present monopolar radiation characteristics at all the bands. As an example, the normalized measured radiation patterns of the tri-band monopole antenna are shown in Figure 4.60. A monopolar radiation pattern is obtained at the three frequencies. The crosspolar component (XPOL) has low values (below −20 dB in all cases, except the *xy*-plane of the third frequency where is below −15 dB). The gains of this design are 1.4 dB, 1.2 dB, and 1.7 dB at the first, second, and third bands, respectively. These results are in good agreement with simulations, in which the radiation efficiency is 92%, 83%, and 94% and the overall efficiency is 91%, 82%, and 93% at the central frequency of each band.

4.3.1.2 Dual-Band UHF-RFID Tags Metamaterial loading is also interesting for the implementation of dual-band antennas with close frequency bands. This is the case of UHF-RFID, where the different regulated bands worldwide are contained in the spectral region comprised between 860 MHz and 960 MHz. Our aim in this subsection is to demonstrate that by loading tag antennas with electrically small resonators, it is possible to implement UHF-RFID tags operative at two of the regulated UHF-RFID bands. The dual-band functionality is achieved by applying a perturbation method similar to that reported in Ref. [158] for dual-band matching networks, properly adapted.

The key aspect in the implementation of long read-range UHF-RFID chip-based tags is to achieve conjugate matching between the antenna and the integrated circuit (or chip). The input impedance of the chip is provided by the manufacturer and varies with frequency. Therefore, the implementation of dual-band UHF-RFID tags means to design the antenna (and the matching network, if it is present) so that the chip "sees" its conjugate impedance at the required frequencies. This can be done by cascading a dual-band impedance matching network between the antenna and the chip, consisting of a transmission line loaded with a metamaterial resonator [158]. The resonator produces a perturbation in both the characteristic impedance and the phase constant of the transmission line, and conjugate matching at two frequencies can be obtained (the details are given in Ref. [158]). However, it is possible to directly actuate on the impedance of the antenna, by loading it with metamaterial resonant particles [159] (avoiding thus the matching circuitry). The principle is very similar to that reported in Ref. [158] for matching networks.

Following the earlier cited perturbation approach, several dual-band UHF-RFID tag antennas have been implemented. One prototype device consists of a meander line antenna (MLA) loaded with a spiral resonator (SR) [159]. The layout of the antenna, compared to the one without the SR is depicted in Figure 4.61. The dual-band antenna was designed to be operative at the European (867 MHz) and USA (915 MHz) UHF-RFID regulated bands. The antenna was designed on the *Rogers RO3010* with dielectric constant $\varepsilon_r = 10.2$ and thickness $h = 0.127$ mm. The considered RFID chip was the *SL31001* from *NXP Semiconductors*.

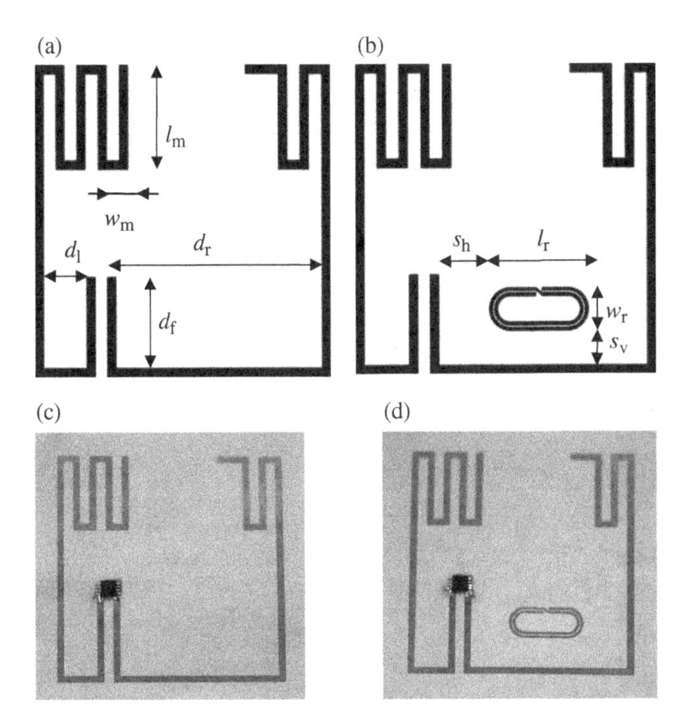

FIGURE 4.61 Layout of the unloaded (a) and SR-loaded (b) MLA, and relevant dimensions. The photographs of the fabricated prototypes are depicted in (c) and (d). The strip width of the SR is 0.5 mm, and the separation between strips is 0.3 mm. Reprinted with permission from Ref. [159]; copyright 2011 IEEE.

The impedance of the chip at the intermediate frequency (between 867 MHz and 915 MHz) is $Z_{chip} = 20 - j485\ \Omega$. This impedance was considered as reference impedance, so that the antenna was designed to roughly exhibit the conjugate impedance of the chip at the intermediate frequency (891 MHz), and then the perturbation (by means of the SR) was introduced in order to achieve conjugate matching at the required frequencies. The dimensions of the MLA are 48 mm × 48 mm, and the strip width is 1.4 mm. The other relevant dimensions (in reference to Fig. 4.61) are $l_m = 16.3$ mm, $w_m = 4.8$ mm, $d_l = 7.3$ mm, $d_r = 33.9$ mm, $d_f = 14.2$ mm, $l_r = 16.1$ mm, $w_r = 6.8$ mm, $s_h = 8.3$ mm, and $s_v = 5.4$ mm. The input impedance and matching of the dual-band SR-loaded MLA are depicted in Figure 4.62, where it can be appreciated that the dual-band functionality is achieved.

Both antennas were fabricated, and the read range, given by

$$r = \frac{\lambda}{4\pi}\sqrt{\frac{\mathrm{EIRP}\, G_r \tau}{P_{chip}}} \tag{4.65}$$

was measured through the experimental setup sketched in Figure 4.63. In (4.65), λ is the wavelength, *EIRP* (equivalent isotropic radiated power), determined by local

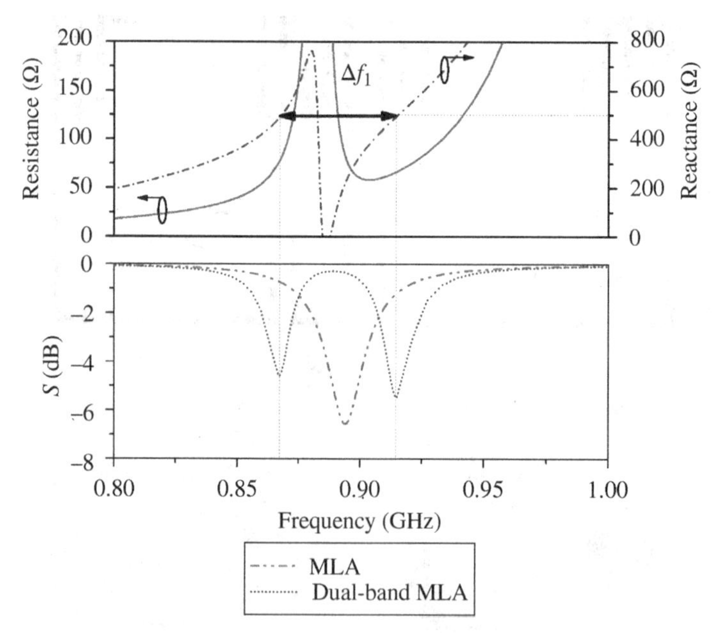

FIGURE 4.62 Input impedance of the dual-band MLA with coupled SR and power-wave reflection coefficient of the mono-band and dual-band meander line antenna. The frequencies where reactance matching is obtained, separated by Δf_1, are indicated and correspond to the resonance frequencies of the dual-band MLA. Reprinted with permission from Ref. [159]; copyright 2011 IEEE.

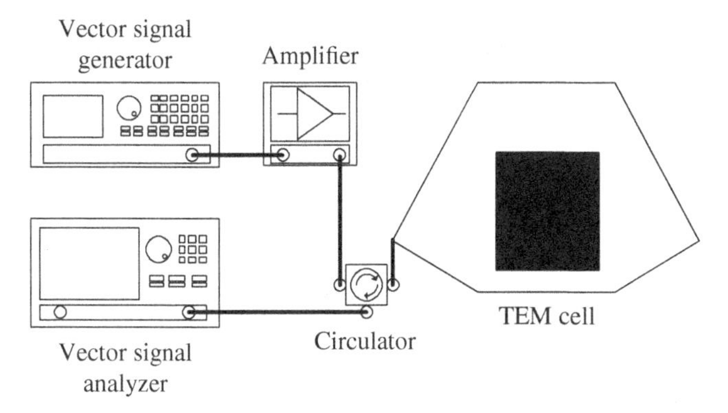

FIGURE 4.63 Bloch diagram of the experimental setup for UHF-RFID tag characterization.

country regulations, is the product of P_tG_t which are the transmission power and the transmission gain, respectively, P_{chip} is the minimum threshold power necessary to activate the RFID chip, G_r is the gain of the receiving tag antenna, and τ is the power transmission coefficient. The value of *EIRP* in Europe is 3.3 W, whereas in the United States it is 4 W. The power transmission coefficient was inferred from the simulation of the return loss of the antenna, using as port impedance that of the chip. The tag gain was also obtained from simulation. Using (4.65), the theoretical read range was calculated.

The experimental setup for read-range measurement consists of a *N5182A* vector signal generator which creates RFID frames. Such generator was connected to a TEM cell by means of a circulator. The tag under test was located inside the TEM cell, and it was excited by the frame created by the generator, generating a backscatter signal to a *N9020A* signal analyzer through the circulator. The RFID frame frequency was swept with a specific power. Once the operation frequencies were identified, the output power was decreased until the tag stopped working. Finally, a probe was placed into the TEM cell in order to determine the incident electric field intensity E_0 at each frequency, which is related to the received power by the chip according to

$$P_{chip} = SA_{ef}\tau = \frac{|E_0|^2}{2\eta}\frac{\lambda^2 G_r}{4\pi}\tau \tag{4.66}$$

where S is the incident power density, A_{ef} is the effective area of the tag antenna, and η is the free-space wave impedance ($120\pi\ \Omega$). The measured read range can be inferred by introducing (4.66) in (4.65), resulting

$$r = \frac{\sqrt{60\ \mathrm{EIRP}}}{E_0} \tag{4.67}$$

As indicated before, the *EIRP* in Europe is lower than the *EIRP* in the United States, so the read range in Europe is expected to be reduced for the same incident

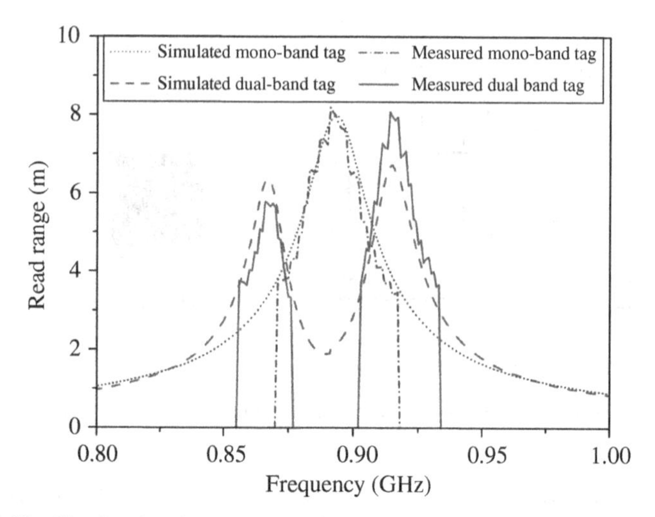

FIGURE 4.64 Simulated and measured read ranges of the mono-band and dual-band MLA RFID tags. Reprinted with permission from Ref. [159]; copyright 2011 IEEE.

electric field intensity. The simulated and measured read ranges of both tags are depicted in Figure 4.64. The read range obtained in the dual-band tag (at the frequencies of interest) is superior to that of the mono-band tag. The read range of the single-band MLA tag is roughly 4 m at the frequencies of interest, whereas almost 6 m and 8 m at the European and USA frequency bands, respectively, are achieved by means of the designed dual-band MLA.

The perturbation method based on loading the antenna with SRs can also be applied to other planar antennas, such as folded dipole antennas [160], for the implementation of dual-band UHF-RFID tags. It was also shown in Ref. [160] that detuning caused by the object or material to which the tag is attached can be corrected by tailoring the distance between the antenna and the SR (see Ref. [160] for more details).

4.3.2 Transmission Lines Loaded with Symmetric Resonators and Applications

Along this chapter, several microwave devices based on transmission lines loaded with symmetric resonators (SRRs, CSRRs, etc.) have been studied. However, the symmetry properties of transmission lines loaded with symmetric resonators have not been analyzed so far. Such symmetry properties are very interesting and can be exploited for the implementation of microwave sensors and RF bar codes. In both cases, the working principle is the truncation of symmetry. In the next subsection, this working principle is analyzed in detail, whereas the applications are left for the following subsections.

4.3.2.1 Symmetry Properties: Working Principle for Sensors and RF Bar Codes Let us consider a transmission line, such as a CPW, loaded with a single SRR, rather than with a pair of SRRs. In absence of loading elements, the symmetry plane of the line is a magnetic wall for the fundamental CPW mode. Let us now consider that the symmetry plane of the SRR, etched in the back substrate side of the CPW, is aligned with the symmetry plane of the line. Under these conditions, there is not a net axial magnetic field able to inductively drive the particle, and the SRR is not excited, that is, the line is transparent (see Fig. 4.65). However, if symmetry is truncated, for instance by laterally displacing or rotating the SRR, or by any other means causing asymmetry (i.e., dielectric loading), the resonator is excited, causing a notch in the transmission coefficient. The magnitude of this notch increases with the level of asymmetry. Therefore, the notch magnitude can provide a measure of the variable responsible for symmetry truncation (displacement, rotation angle, etc.) [161, 162]. Indeed, with perfectly aligned CPW and SRR symmetry planes, the resonator is not excited because such symmetry planes are of distinct EM nature (the symmetry plane of the SRR is an electric wall at the fundamental resonance). However, a slight misalignment between these symmetry planes suffices to drive the particle, causing a notch in the transmission coefficient. According to these words, any planar resonator exhibiting an electric wall at resonance can potentially be useful as sensing element loading a transmission line excited with an even mode (necessary to generate a magnetic wall in the symmetry plane of the line).

Symmetric resonators exhibiting a magnetic wall at resonance, such as the CSRR, can also be used for sensing purposes. However, in this case the host line must be chosen to exhibit an electric wall, rather than a magnetic wall, at its symmetry plane. Under these conditions, the symmetry planes of the line and resonator are also of

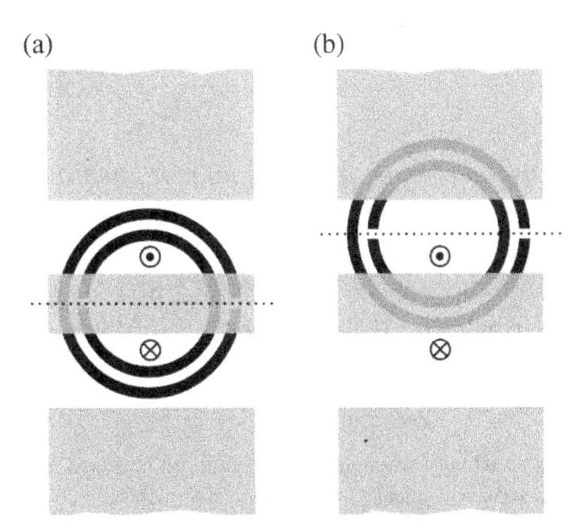

FIGURE 4.65 CPW loaded with an aligned (a) and misaligned (b) SRR. Due to symmetry, there is not a net axial magnetic field in (a) able to excite the SRR, contrary to the situation depicted in (b).

different EM nature, and the loaded line is transparent if such planes are perfectly aligned. For instance, a differential microstrip line loaded with a CSRR (etched in the ground plane) in a symmetric configuration is transparent for the differential mode.[36] However, the CSRR causes a strong notch for the common mode because for this mode the symmetry planes (line and CSRR) are both a magnetic wall (common-mode filters based on CSRRs will be discussed in the Section devoted to differential lines in the last chapter [163]). Obviously, the CSRR acts as a notch filter for the differential mode as well if symmetry is truncated.

Symmetric resonators can also be used for the implementation of RF bar codes, also known as spectral signatures codes [164].[37] In RF bar codes, a transmission line is loaded with multiple resonators, each tuned at a different frequency. Each resonance frequency corresponds to a different bit, and the logic '1' and '0' states are given by the presence or absence, respectively, of a transmission zero in the spectrum for that frequency. For the implementation of RF bar codes, symmetric resonators are not a due. Indeed, in Ref. [165] SRs are used as loading elements of a microstrip line based 35-bit bar code. Encoding can be achieved by etching (close to the microstrip line) only those SRs with resonance frequency corresponding to the required logic '1' states. The coupling between the line and the spiral inhibits transmission at the resonance frequency, with the result of as many notches as spirals. An alternative for line encoding is to etch the whole set of spirals close to the line and short-circuit those spirals with resonance frequency corresponding to the required "0" states. By short-circuiting the spiral, the corresponding resonance frequency is moved away, providing the same effect as removing the spiral.

By using symmetric resonators as loading elements in RF bar codes, writing the '1' state is as simple as truncating symmetry, for instance by rotating or laterally displacing the resonant element [164]. This opens the possibility to implement mechanically reconfigurable bar codes, or bar codes where encoding can be achieved by any other means of truncating symmetry. Moreover, since the level of asymmetry determines the notch magnitude, it is potentially possible to increase the stored information by simply providing more than two logic states to each resonant element.

4.3.2.2 Rotation, Displacement, and Alignment Sensors Most microwave sensors that make use of resonators as sensing elements are based on the variation of the resonance frequency, phase, or quality factor caused by the physical magnitude under measurement. Split rings, including SRRs, CSRRs or other related resonators, have been widely used as sensors [161, 162, 166–176]. Among them, there are several

[36] This configuration is roughly the dual of the CPW loaded with an SRR, and hence a similar behavior is expected.

[37] RF bar codes are chipless RFID tags, where the information can be stored in a set of resonant elements causing a specific spectral response. Typically, such resonant elements are coupled to a host transmission line, which is equipped with cross-polarized receiver and transmitter antennas for wireless identification. As compared to other chipless RFID tags based on time domain reflectometry (TDR) [164], RF bar codes need a relatively large bandwidth for high data capacity, since each resonant element is tuned at a different frequency and contributes with a single bit to the whole code.

implementations where a host line is loaded with such metamaterial resonators. For instance, sensors able to spatially resolve the dielectric properties of a material were achieved in Ref. [172] by loading a line with an array of SRRs tuned at different frequencies. In such sensors, a frequency shift of one individual resonant peak indicates the dielectric properties of the material under test and its location within the array. Two-dimensional displacement sensors consisting of a bended microstrip line loaded with a pair of triangular and orthogonally oriented CSRRs (etched on the ground plane of an independent and movable substrate) were also proposed [173]. The working principle is based on detuning the CSRR by a change in the line-to-CSRR coupling capacitance. The triangular shape of the CSRR was considered in order to enhance the sensitivity of the sensor. Moreover, two orthogonally polarized receiving and transmitting antennas were cascaded to the input and output ports of the host line in order to provide a wireless connection to the reader. The scheme of the wireless link and the photograph of the designed displacement sensor are depicted in Figure 4.66a and b, respectively. Figure 4.66c shows the variation of the frequency response that is obtained by laterally displacing one of the rings with regard to the line axis.

Alternatively to the frequency shift caused by the variable under measurement, symmetry truncation can be used for sensing and detection purposes in transmission lines loaded with symmetric resonant particles. One advantage of this strategy is a major robustness in front of changing ambient conditions. Such environmental variations may cause frequency detuning, but have little effect on the notch magnitude. Moreover, notice that sensors based on symmetry properties can be used as detectors, able to "detect" the lack of symmetry. For instance, SRR-loaded CPWs can be used as alignment sensors, where the lack of alignment between two surfaces is detected by the presence of a notch in the transmission spectrum. Under perfect alignment, such notch is not present, regardless of the environmental conditions. The first sensors based on the symmetry properties of SRR loaded lines were reported in Ref. [161]. Both one-dimensional displacement and rotation sensors based on rectangular-shaped SRRs and CPWs were reported. The rectangular shape of the SRRs was chosen in order to optimize the sensitivity. Namely, if circularly shaped SRR with an internal diameter much larger than the distance between the CPW slots is considered, the magnetic field lines within the SRR region do not exactly cancel if symmetry is broken, but the effects of asymmetry are not significant. Conversely, if the SRR does not extend beyond the slots of the CPW, the effects of the asymmetry on the frequency response of the structure (notch magnitude) are more pronounced. Indeed, it was found that the magnitude of the notch increases by etching the SRR just below the central strip of the CPW [161]. Another important parameter regarding sensitivity is the vertical distance between the CPW metallic layer and the resonator. Sensitivity increases by reducing such distance because the coupling between the line and the resonator also increases. Thus, by using rectangular SRRs, it was found in Ref. [161] that high sensitive sensors with good linearity and dynamic range can be implemented for both linear displacement and rotation.

In Ref. [174], a proof of concept of a two-dimensional displacement sensor was proposed by combining the symmetry properties of the one-dimensional displacement

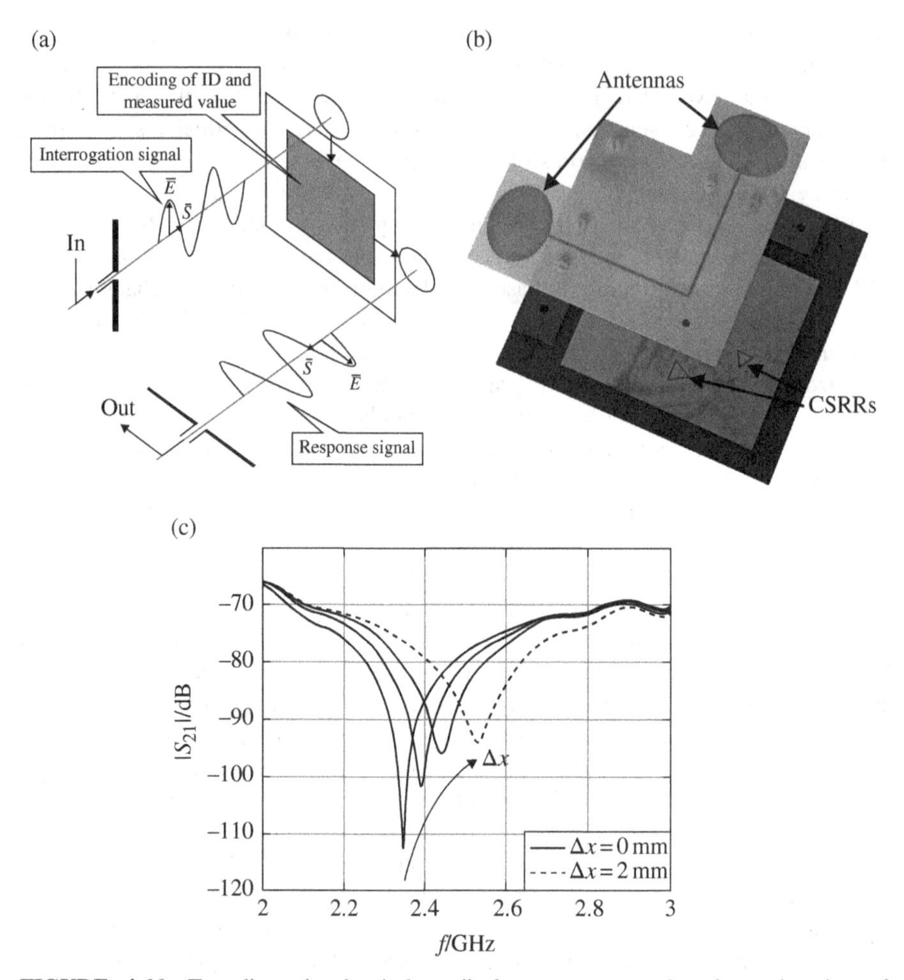

FIGURE 4.66 Two-dimensional wireless displacement sensor based on detuning of triangular shaped CSRRs. (a) Scheme showing the polarization decoupling of interrogation and response signals, (b) photograph, and (c) frequency response for various displacements. For better visualization in (b), the movable sensitive plate was rotated 180°. Reprinted with permission from Ref. [173]; copyright 2011 EuMA.

sensor of Ref. [161], and the bending scheme proposed in Ref. [173].[38] Like in Ref. [173], a right-angle bended CPW loaded with a pair of rectangular-shaped SRRs tuned at different frequencies and orthogonally oriented was proposed in Ref. [174]. The structure with the SRRs aligned with the line axis is depicted in Figure 4.67. This structure is able to detect lack of alignment and linear displacement in two

[38] In this proof of concept, rather than etching the SRRs in a movable substrate, these resonators were etched in the back substrate side, and different samples with different displacements were considered for characterization.

(a)

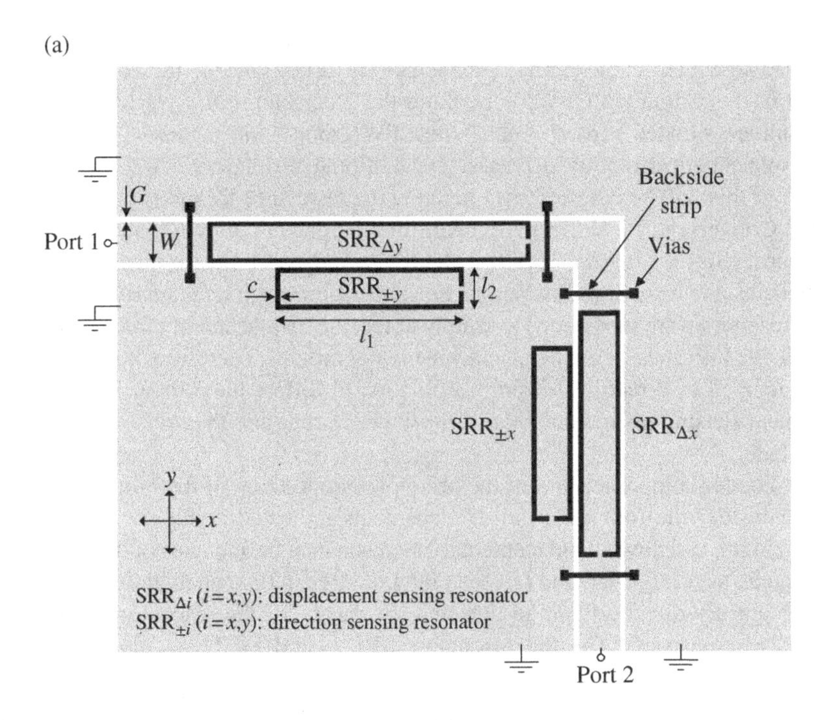

(b) (c)

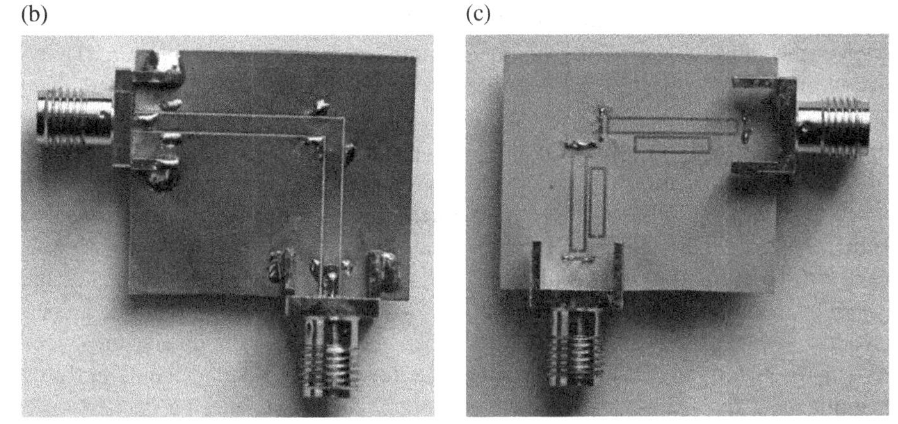

FIGURE 4.67 Layout (a) and top (b) and bottom (c) photographs of the two-dimensional displacement sensing device for the aligned position (i.e., the CPW and the displacement sensing SRRs are aligned). The CPW strip and slot widths are $W = 1.67$ mm and $G = 0.2$ mm, respectively; the vias have a 0.2 mm radius; and the narrow strips between vias have a width of 0.2 mm. The dimensions of the SRRs are $l_1(\text{SRR}_{\Delta x}) = 9.95$ mm, $l_1(\text{SRR}_{\pm x}) = 7.05$ mm, $l_1(\text{SRR}_{\Delta y}) = 13.4$ mm, $l_1(\text{SRR}_{\pm y}) = 7.8$ mm, $l_2 = 1.67$ mm, and $c = 0.2$ mm. The vias and backside strip metals are used to connect the CPW ground plane regions, and thus prevent the appearance of the parasitic slot mode. The considered substrate is the *Rogers RO3010* with dielectric constant $\varepsilon_r = 10.2$, thickness $h = 127$ μm, and loss tangent $\tan\delta = 0.0023$. Reprinted with permission from Ref. [174]; copyright 2012 MDPI.

dimensions. In order to distinguish the displacement direction, two additional resonators (also tuned at different frequencies) are used. One of these resonators is etched in the x-oriented CPW section and the other one in the y-oriented section, and both are situated beneath one of the CPW ground planes, near a CPW slot. If the displacement direction drives such additional resonators toward the slot of the CPW, this will be detected by a notch at the resonance frequency of these resonators. Conversely, by shifting the SRRs in the opposite direction such notch will not appear. Since it is necessary that the four required SRRs are tuned at different frequencies, the resonator dimensions must be different. Notice that these two additional resonators are introduced to simply detect the displacement direction (they do not provide information on the displacement magnitude). Therefore, these SRRs can be designated as direction sensing resonators, to differentiate them from the displacement sensing resonators, that is, those which measure the linear displacement magnitude.

For a better comprehension of the principle of operation of the proposed sensor, let us consider the four different displacements indicated in Figure 4.68, that is, right, left, up, and down displacements. The resonance frequencies of the four SRRs are denoted as $f_{\Delta y}, f_{\Delta x}, f_{\pm y}$, and $f_{\pm x}$ (see Fig. 4.68). It can be seen that displacement in the $\pm x$- and $\pm y$-direction can be detected (by means of the resonators $SRR_{\pm x}$ and $SRR_{\pm y}$) and measured (by the resonators $SRR_{\Delta x}$ and $SRR_{\Delta y}$). Any other linear displacement is a combination of the previous ones, and hence it can also be detected and measured. In order to validate the proposed approach, positive and negative displacement in the $x = y$ direction (diagonal shift) was considered. Figure 4.69 depicts the dependence of the notch magnitude (simulated and measured) with displacement. As expected, for positive displacements, the $SRR_{\pm x}$ and $SRR_{\pm y}$ are activated as it is manifested by a clear increase in the notch at $f_{\pm x}$ and $f_{\pm y}$, whereas the specified −3 dB threshold level is not exceeded for negative displacements (indicating that the shift is in the negative direction). The dependence of the notch magnitude for $f_{\Delta x}$ is similar and roughly linear in both directions, with a measured value of approximately −20 dB for $\Delta x = \pm 0.3$ mm, which is indicative of a significant sensitivity of roughly 65 dB/mm (average value). The same occurs for the dependence of the notch magnitude for $f_{\Delta y}$. However, the notch magnitude associated to a displacement sensing resonator depends not only on the displacement, but also on resonator dimensions. This causes that, for the same displacement, the notch magnitude of the y-axis displacement sensing resonator produces a deeper notch than that of the x-axis.

Rotation sensors with enhanced dynamic range and linearity (as compared to those reported in Ref. [161]) can be implemented by coupling circular resonators to a non-uniform (circularly shaped) CPW. Specifically, a circular ELC resonator was proposed as sensing element in Ref. [175] (see Fig. 4.70). As it was pointed out in Section 3.3.1.6, the ELC is a bisymmetric resonator with two orthogonal symmetry planes, one being a magnetic wall and the other one an electric wall at the fundamental resonance. If the electric wall is perfectly aligned with the symmetry plane of the line (a magnetic wall), the line is transparent. However, by rotating the particle 90°, the two magnetic walls (the one of the line and the

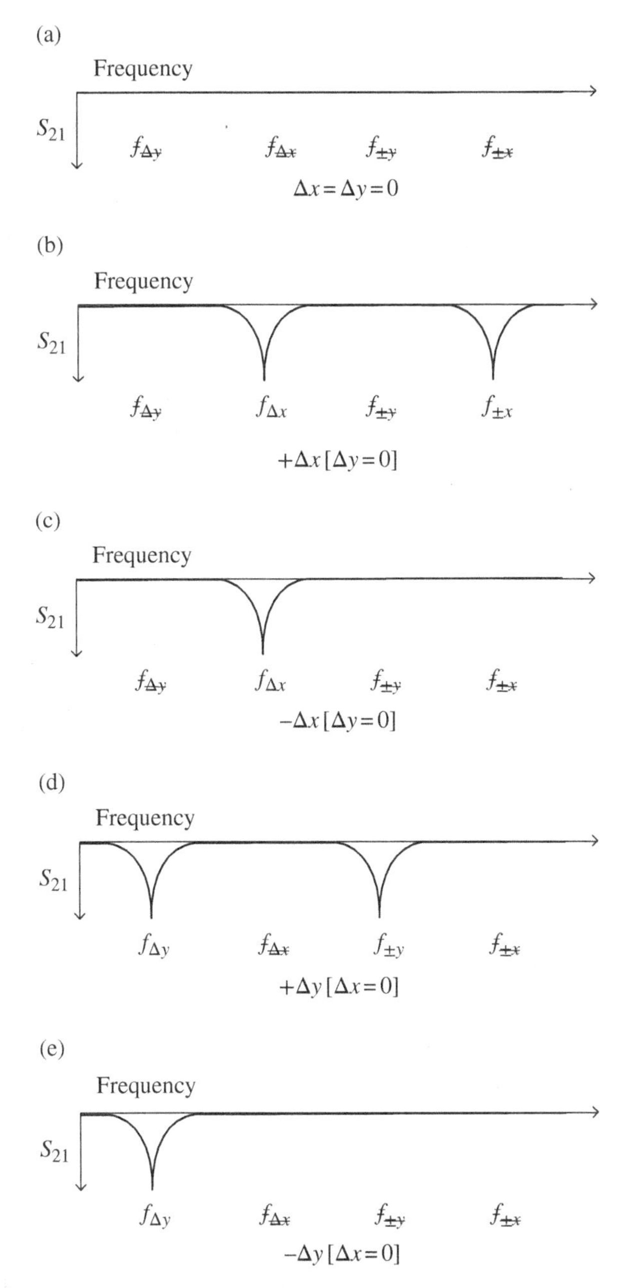

FIGURE 4.68 Scheme indicating the primitive shifting operations (a–e) and the resulting transmission coefficient S_{21}. A notch is indicative of an SRR excitation. A linear displacement in the x- and y-orientation is indicated as Δx and Δy, respectively, relative to the aligned position (i.e., $\Delta x = \Delta y = 0$). Reprinted with permission from Ref. [174]; copyright 2012 MDPI.

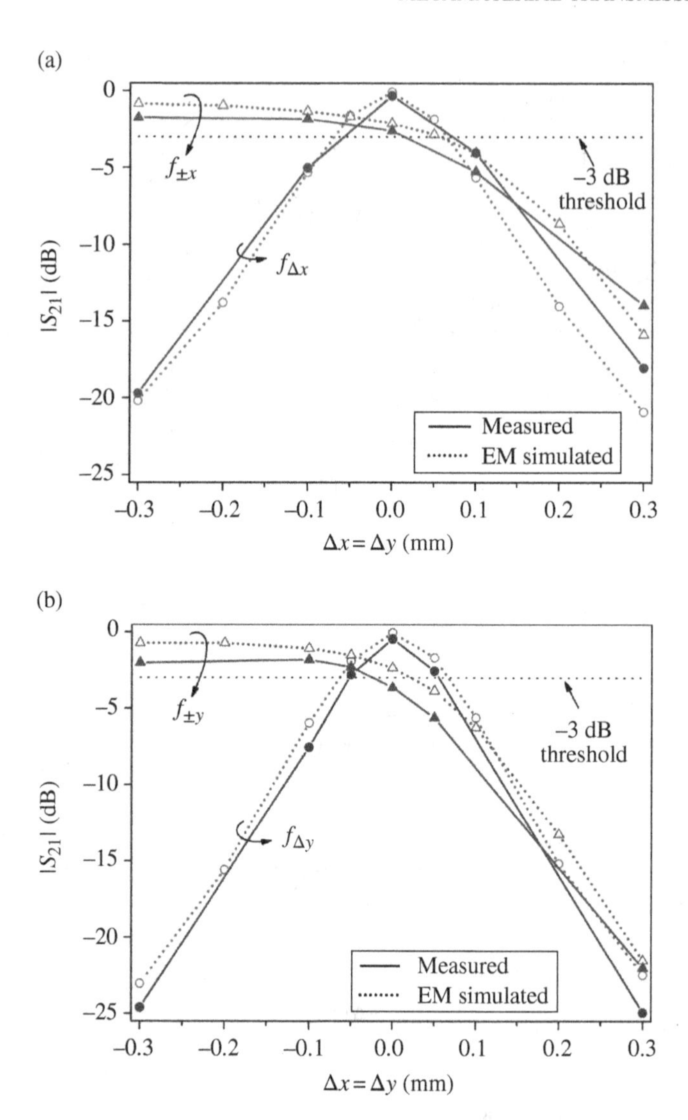

FIGURE 4.69 Notch magnitude of the transmission coefficient S_{21} at the indicated frequencies for $x = y$ displacement; results for (a) x- and (b) y-axis position sensing. Reprinted with permission from Ref. [174]; copyright 2012 MDPI.

one of the ELC) become aligned, and a strong notch in the transmission coefficient is expected. This explains that the dynamic range can be improved as compared to the rotation sensor based on a rectangular-shaped SRR. Linearity is improved thanks to optimization of resonator and line shape (the key aspect is that the external contour of the resonator roughly coincides with the slots of the nonuniform CPW host line).

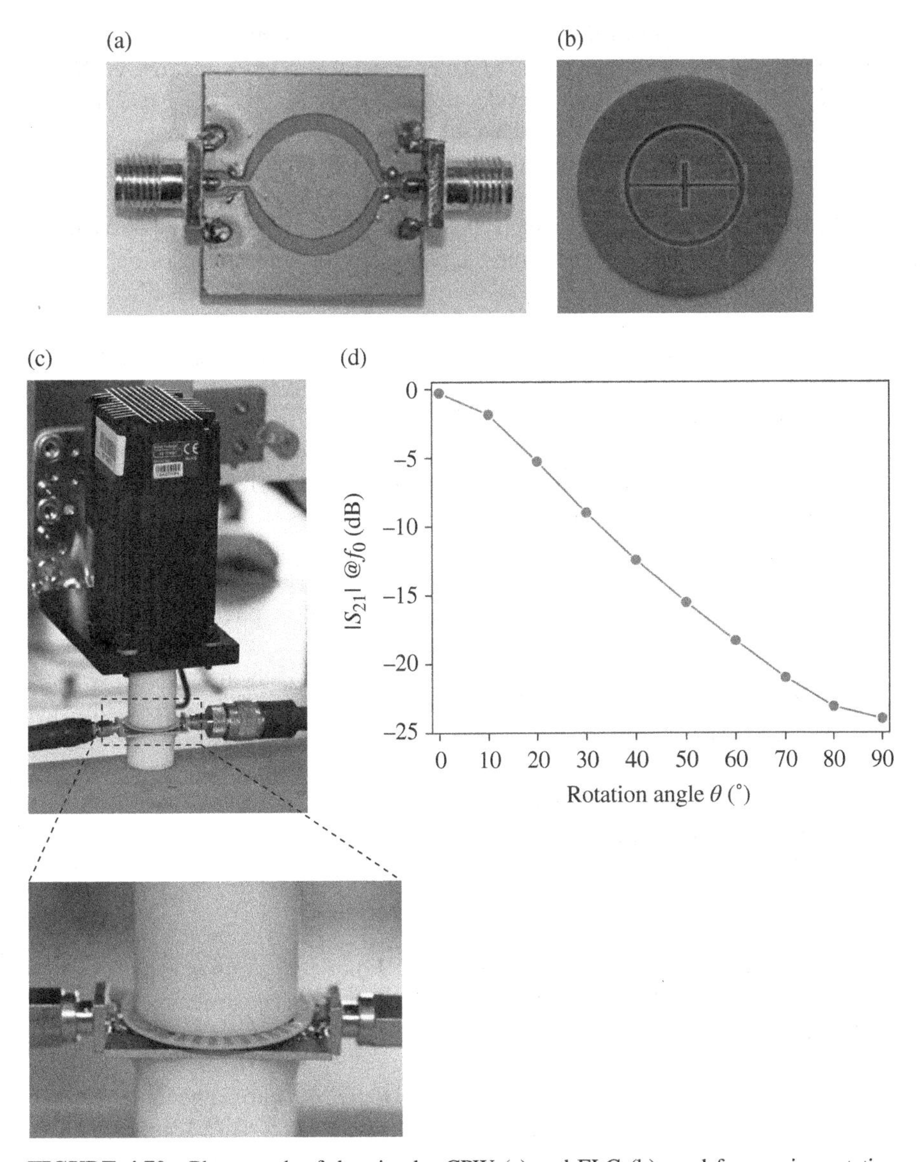

FIGURE 4.70 Photograph of the circular CPW (a) and ELC (b) used for sensing rotation angles through the setup shown in (c). The notch magnitude as a function of the rotation angle is depicted in (d). The ELC is attached to a rotating cylinder made of Teflon. The CPW and the ELC were etched on the *Rogers RO3010* substrate with thickness $h = 1.27$ mm, dielectric constant $\varepsilon_r = 11.2$, and loss tangent $\tan\delta = 0.0023$. Dimensions are: for the CPW, W and G are tapered such that the characteristic impedance is 50 Ω; for the ELC resonator, the diameter is 16.6 mm. Vias and backside strips are used to eliminate the slot mode.

The proof of concept of this rotation sensor was demonstrated in Ref. [175] by considering two different scenarios: (1) different CPW-loaded structures, each with the ELC etched in the back substrate side and rotated a different angle and (2) a single CPW with the ELC etched on a movable substrate. An experimental setup, where the ELC resonator is attached to a cylinder made of Teflon which can rotate through the action a step motor, was then implemented in order to perform angle and angular velocity measurements in rotating elements (Fig. 4.70). The notch depth in the transmission coefficient as a function of the rotation angle for the structure of Figure 4.70 is also depicted in the figure. Good linearity, dynamic range, and sensitivity can be appreciated.

The setup of Figure 4.70 was used for the measurement of angular velocities. The idea is very simple: the CPW transmission line is fed by a harmonic signal (carrier) tuned at the notch frequency of the ELC resonator.[39] At the output port of the CPW the amplitude of the carrier signal is modulated by the effect of the rotating ELC, that is, the amplitude is a maximum each time the electric wall of the ELC is aligned with the symmetry plane of the line (this occurs twice per cycle). Therefore, by rectifying the modulated signal through an envelope detector, the distance between adjacent peaks gives the rotation speed. In order to avoid unwanted reflections caused by the envelope detector, a circulator is cascaded between the output port of the CPW transmission line and the input of the envelope detector, as depicted in Figure 4.71a. The typical waveform obtained through an oscilloscope in the envelope detector, for a rotation speed of 10 cycles per second, is depicted in Figure 4.71b. The pronounced peaks allow for the measurement of angular velocities with high accuracy. Since the carrier frequency is very high, it is obvious that very high angular velocities can be measured with this approach. Further details on the design of this rotation speed sensor are given in Ref. [177]; an alternative version of the rotation speed sensor in microstrip technology is reported in Ref. [178].

4.3.2.3 RF Bar Codes As an example of a RF bar code implemented by symmetric resonators, a 3-bit structure is reported here. The structure is a CPW loaded with folded SIRs. Such resonators exhibit an electric wall in its symmetry plane at the fundamental resonance. By aligning them with the line, the structure is transparent for the corresponding resonance (logic '0' state). The logic '1' state is achieved by laterally displacing the folded SIR. Folded SIRs loading a CPW are interesting since they are electrically small at resonance. The reason is that line-to-resonator coupling is dominated by the large capacitance between the SIR and the line. The basic cell (for the symmetric configuration) and the circuit model are depicted in Figure 4.72 [162]. Coupling between the line and resonator is represented by the capacitances C_{sci} and C_{sgi} ($i = 1, 2$). Such capacitances depend on the relative position between the SIR and the CPW. Thus, any symmetry truncation can be modeled by adjusting these capacitances. If the structure

[39] Actually, the notch frequency slightly depends on the orientation (angle) of the ELC resonator. Therefore, the carrier frequency is chosen as the average value.

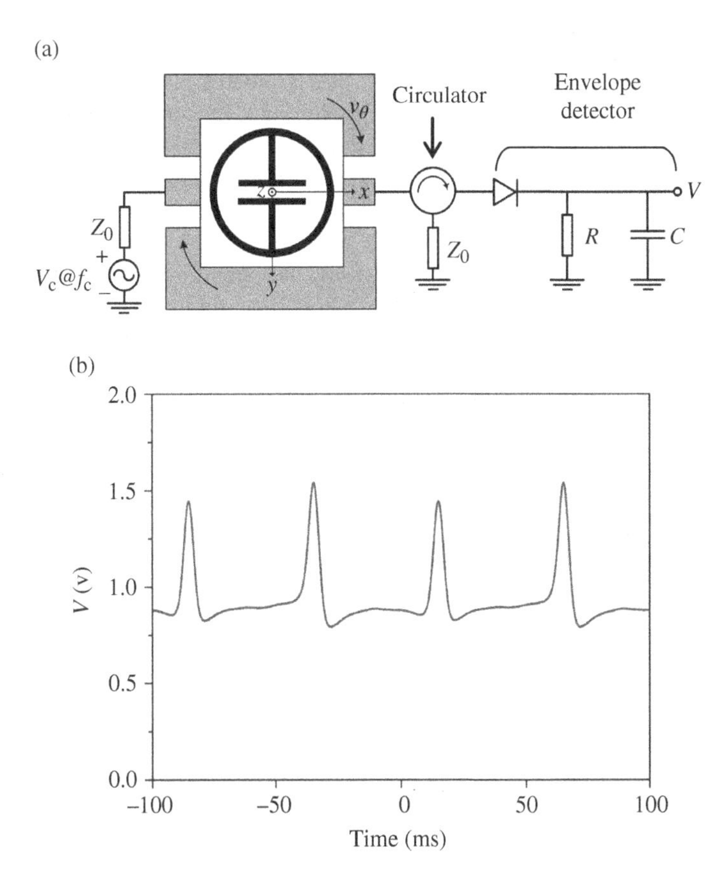

FIGURE 4.71 Schematic of the proposed system for measuring angular velocities in time domain, including the sensing line, circulator and envelope detector (a) and output waveform for a rotating speed of 10 cycles per second (b).

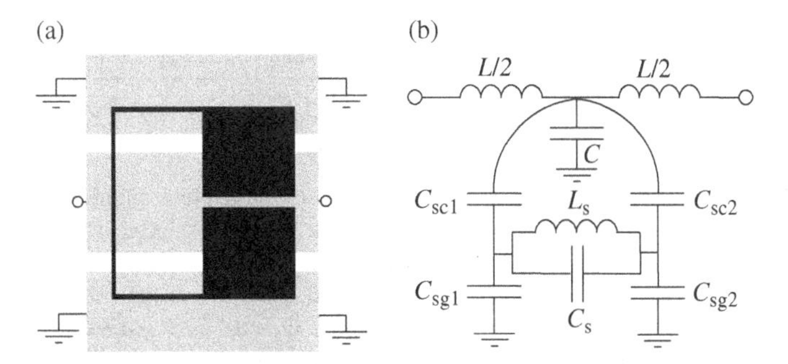

FIGURE 4.72 Typical layout of a CPW loaded with a folded SIR (a) and lumped-element equivalent circuit model (b). The metallization in the back substrate side is indicated in black.

is symmetric (i.e., $C_{sc1} = C_{sc2}$ and $C_{sg1} = C_{sg2}$), the resonator L_s–C_s is opened due to the presence of the magnetic wall at the symmetry plane, and the resulting model is the one of a conventional transmission line. By contrast, if the symmetry is broken, the capacitances are unbalanced and a transmission zero appears. In the simplest case in which $C_{sc1} = C_{sg2} = 0$ (or $C_{sc2} = C_{sg1} = 0$), the model is equivalent to that of a transmission line loaded with a grounded series resonator (the capacitance C_s can be neglected). The model parameters were extracted for the two cases indicated in Figure 4.73. In both cases, the circuit simulations are in good agreement with the EM simulations. For comparison purposes, the response of a CPW loaded with an SRR of the same dimensions is also included. It can be appreciated the interest of folded SIRs in this application, not only because the particles are electrically smaller, but also because the balance between notch depth and width is good. Narrow notches occupy less bandwidth, but at the expense of small notch depth due to losses.

A 3-bit RF bar code consisting of a CPW loaded with three different folded SIRs with the code '111' is depicted in Figure 4.74a [162]. In order to prevent the presence of the slot mode, the ground planes are connected through backside strips and vias. In view of the transmission coefficients of this structure and those corresponding to other bit configurations (Fig. 4.74b), the principle is validated. Although somehow

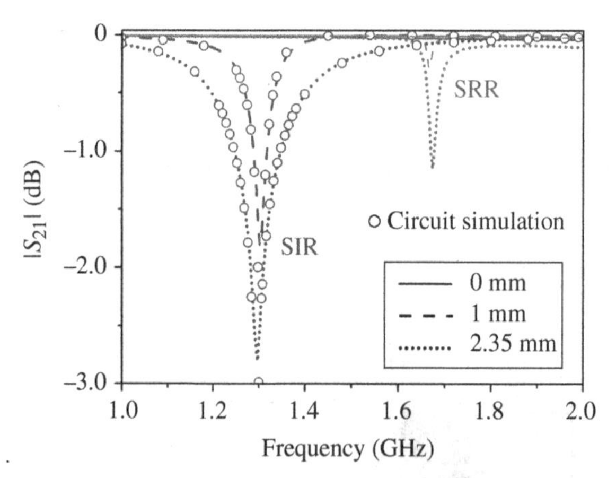

FIGURE 4.73 Simulated transmission coefficient of the structure of Figure 4.72a and those corresponding to the same structure with the resonators laterally displaced. The substrate is the *Rogers RO3010* with dielectric constant $\varepsilon_r = 10.2$, thickness $h = 0.635$ mm, and $\tan\delta = 0.0023$. CPW dimensions are $L_{CPW} = 9.6$ mm, $W = 4$ mm, and $G = 0.7$ mm; SIR dimensions are (in reference to Fig. 3.13): $l_1 = l_2 = 7.6$ mm, $W_1 = 3.8$ mm, $W_2 = 0.2$ mm, and $S = 0.4$ mm. Circuit parameters are $L = 2.4$ nH, $C = 0.97$ pF, $L_s = 13.8$ nH, $C_s = 0$pF; for 1 mm, $C_{sc1} = 0.57$ pF, $C_{sc2} = 1.8$ pF, $C_{sg1} = 1.6$ pF, and $C_{sg2} = 0.4$ pF; for 2.35 mm, $C_{sc1} = 0$ pF, and $C_{sc2} = 2.2$ pF, $C_{sg1} = 2.2$ pF, and $C_{sg2} = 0$pF. Reprinted with permission from Ref. [162]; copyright 2012 IEEE.

(a)

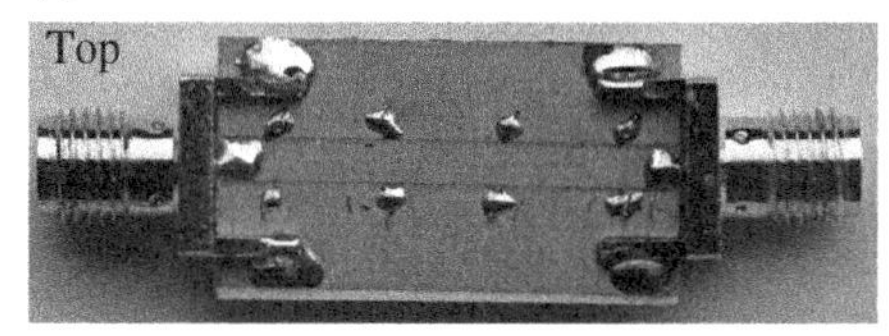

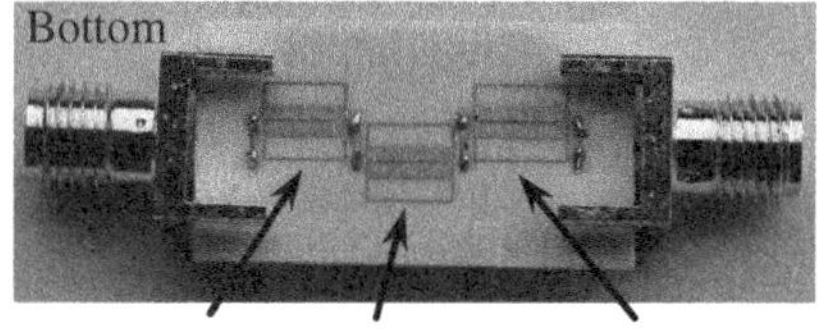

(b)

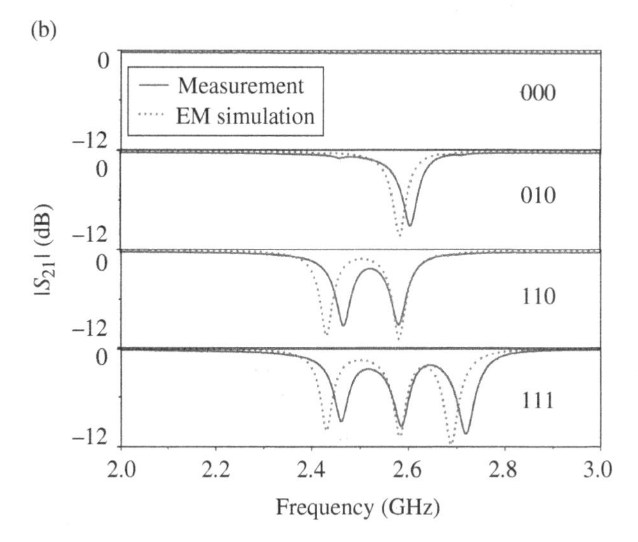

FIGURE 4.74 Fabricated CPW with three folded SIRs corresponding to the binary code '111' of a 3-bit RF bar code (a) and simulated and measured transmission coefficient of the structure corresponding to the indicated codes (b). The substrate is the *Rogers RO4003C* with $\varepsilon_r = 3.55$, $h = 0.8128$ mm, and $\tan\delta = 0.0021$. CPW dimensions are $L_{CPW} = 23.9$ mm, $W = 2.16$ mm, and $G = 0.15$ mm; SIR dimensions are (in reference to Fig. 3.13): $W_1 = 0.825$ mm, $W_2 = 0.15$ mm, $S = 0.15$ mm, $l_2 = 4.47$ mm, and l_1 is shown in the photograph. Reprinted with permission from Ref. [162]; copyright 2012 IEEE.

degraded, the bar code functionality (not shown) is still preserved if the backside strip and vias are removed.

RF-bar codes implemented with symmetric resonators in microstrip technology (thus avoiding the use of vias) have been also reported [179].

REFERENCES

1. G. V. Eleftheriades and K. G. Balmain, Ed., *Negative-Refraction Metamaterials: Fundamental Principles and Applications*, Wiley-Interscience, Hoboken, NJ, 2005.
2. C. Caloz and T. Itoh, *Electromagnetic Metamaterials: Transmission Line Theory and Microwave Applications*, John Wiley & Sons, Hoboken, NJ, 2006.

3. N. Engheta and R. W. Ziolkowski, *Metamaterials: Physics and Engineering Explorations*, Wiley-Interscience, Hoboken, NJ, 2006.
4. R. Marqués, F. Martín, and M. Sorolla, *Metamaterials with Negative Parameters: Theory, Design and Microwave Applications*, Wiley-Interscience, Hoboken, NJ, 2007.
5. F. Capolino, Ed., *Metamaterials Handbook*, CRC Press, Boca Raton, FL, 2009.
6. L. Solymar and E. Shamonina, *Waves in Metamaterials*, Oxford University Press, Oxford, 2009.
7. G. Sisó, M. Gil, J. Bonache, and F. Martín, "On the dispersion characteristics of metamaterial transmission lines," *J. Appl. Phys.*, vol. **102**, paper 074911, 2007.
8. R. A. Foster, "A reactance theorem," *Bell System Tech. J.*, vol. **3**, pp. 259–267,1924.
9. M. Gil, J. Bonache, I. Gil, J. García-García, and F. Martín, "Miniaturization of planar microwave circuits by using resonant-type left handed transmission lines," *IET Microw. Antennas Propag.*, vol. **1**, pp. 73–79, 2007.
10. M. A. Antoniades and G. V. Eleftheriades, "A broadband series power divider using zero-degree metamaterial phase shifting lines," *IEEE Microw. Wireless Compon. Lett.*, vol. **15**, pp. 808–810, 2005.
11. H. Okabe, C. Caloz, and T. Itoh, "A compact enhanced bandwidth hybrid ring using an artificial lumped element left handed transmission line section," *IEEE Trans. Microw. Theory Tech.*, vol. **52**, pp. 798–804, 2004.
12. G. Sisó, J. Bonache, M. Gil, J. García-García, and F. Martín, "Compact rat-race hybrid coupler implemented through artificial left handed and right handed lines," *IEEE MTT-S Int. Microwave Symp. Dig.*, Honolulu, HI, June 2007, pp. 25–28.
13. G. Sisó, J. Bonache, M. Gil, and F. Martín, "Application of resonant-type metamaterial transmission lines to the design of enhanced bandwidth components with compact dimensions," *Microw. Opt. Technol. Lett.*, vol. **50**, pp. 127–134, 2008.
14. R. K. Settaluri, G. Sundberg, A. Weisshaar, and V. K. Tripathi, "Compact folded line rat race hybrid couplers," *IEEE Microw. Guided Wave Lett.*, vol. **10**, pp. 61–63, 2000.
15. K. W. Eccleston and S. H. M. Ong, "Compact planar microstripline branch-line and rat race couplers," *IEEE Trans. Microw. Theory Tech.*, vol. **51**, pp. 2119–2125, 2003.
16. C.-C. Chen and C.-K. C. Tzuang, "Synthetic quasi-TEM meandered transmission lines for compacted microwave integrated circuits," *IEEE Trans. Microw. Theory Tech.*, vol. **52**, pp. 1637–1647, 2004.
17. J.-T. Kuo, J.-S. Wu, and Y.-C. Chiou, "Miniaturized rat race coupler with suppression of spurious passband," *IEEE Microw. Wireless Compon. Lett.*, vol. **17**, pp. 46–48, 2007.
18. T. T. Mo, Q. Xue, and C. H. Chan, "A broadband compact microstrip rat-race hybrid using a novel CPW inverter," *IEEE Trans. Microw. Theory Tech.*, vol. **55**, pp. 161–167, 2007.
19. I. B. Vendik, O. G. Vendik, D. V. Kholodnyak, E. V. Serebryakova, and P. V. Kapitanova, "Digital phase shifters based on right- and left-handed transmission lines," *Proc. Eur. Microw. Assoc.*, vol. **2**, pp. 30–37, 2006.
20. I. T. Nassar, A. A. Gheethan, T. M. Weller, and G. Mumcu, "A Miniature, broadband, non-dispersive phase shifter based on CRLH TL unit cells," *IEEE Int. Symp. Antennas Propag. (APS-URSI)*, Chicago, IL, 2012.
21. M. A. Antoniades and G. V. Eleftheriades, "A broadband Wilkinson balun using microstrip metamaterial lines," *IEEE Antennas Wireless Propag. Lett.*, vol. **4**, pp. 209–212, 2005.

22. S.-G. Mao and Y.-Z. Chueh, "Broadband composite right/left-handed coplanar waveguide power splitters with arbitrary phase responses and balun and antenna applications," *IEEE Trans. Antennas Propag.*, vol. **54**, pp. 243–250, 2006.
23. G. Sisó, M. Gil, J. Bonache, and F. Martín, "Application of metamaterial transmission lines to the design of quadrature phase shifters," *Electron. Lett.*, vol. **43**, pp. 1098–1100, 2007.
24. I.-H. Lin, M. DeVincentis, C. Caloz, and T. Itoh, "Arbitrary dual-band components using composite right/left-handed transmission lines," *IEEE Trans. Microw. Theory Tech.*, vol. **52**, pp. 1142–1149, 2004.
25. I.-H. Lin, K. M. K. H. Leong, C. Caloz, and T. Itoh, "Dual-band sub-harmonic quadrature mixer using composite right/left-handed transmission lines," *IEE Proc. Microw. Antennas Propag.*, vol. **153**, pp. 365–375, 2006.
26. C.-H. Tseng and T. Itoh, "Dual-band bandpass and bandstop filters using composite right/left-handed metamaterial transmission lines," *IEEE MTT-S Int. Microwave Symp. Dig.*, San Francisco, CA, June 2006, pp. 931–934.
27. S. H. Ji, C. S. Cho, J. W. Lee, and J. Kim, "Concurrent dual-band class-E power amplifier using composite right/left-handed transmission lines," *IEEE Trans. Microw. Theory Tech.*, vol. **55**, pp. 1341–1347, 2007.
28. M. Durán-Sindreu, A. Vélez, F. Aznar, G. Sisó, J. Bonache, and F. Martín, "Application of open split ring resonators and open complementary split ring resonators to the synthesis of artificial transmission lines and microwave passive components," *IEEE Trans. Microw. Theory Tech.*, vol. **57**, pp. 3395–3403, 2009.
29. J. Selga, A. Rodríguez, V. E. Boria, and F. Martín, "Synthesis of split rings based artificial transmission lines through a new two-step, fast converging, and robust aggressive space mapping (ASM) algorithm," *IEEE Trans. Microw. Theory Tech.*, vol. **61**, pp. 2295–2308, 2013.
30. G. Sisó, J. Bonache, and F. Martín, "Dual-band Y-junction power dividers implemented through artificial lines based on complementary resonators," *IEEE MTT-S Int. Microwave Symp. Dig.*, Atlanta, GA, June 2008, pp. 663–666.
31. J. Bonache, G. Sisó, M. Gil, A. Iniesta, J. García-Rincón, and F. Martín, "Application of composite right/left handed (CRLH) transmission lines based on complementary split ring resonators (CSRRs) to the design of dual band microwave components," *IEEE Microw. Wireless Compon. Lett.*, vol. **18**, pp. 524–526, 2008.
32. A. Vélez, G. Sisó, A. Campo, M. Durán-Sindreu, J. Bonache, and F. Martín, "Dual-band microwave duplexer based on spiral resonators (SR) and complementary split ring resonators (CSRRs)," *Appl. Phys. A Mater. Sci. Proc.*, vol. **103**, pp. 911–914, 2011.
33. J.-S. Hong and M. J. Lancaster, *Microstrip Filters for RF/Microwave Applications*, Wiley, Hoboken, NJ, 2001.
34. M. Durán-Sindreu, G. Sisó, J. Bonache, and F. Martín, "Fully planar implementation of generalized composite right/left handed transmission lines for quad-band applications," *IEEE-MTT-S Int. Microwave Symp. Dig.*, Anaheim, CA, May 2010.
35. M. Durán-Sindreu, G. Sisó, J. Bonache, and F. Martín, "Planar multi-band microwave components based on the generalized composite right/left handed transmission line concept," *IEEE Trans. Microw. Theory Tech.*, vol. **58**, pp. 3882–3891, 2010.
36. B. H. Chen, Y. N. Zhang, D. Wu, and K. Seo, "A novel composite right/left-handed transmission line for quad-band applications," *11th IEEE Singapore International Conference on Communication Systems (ICCS)*, Guangzhou, November 2008, pp. 617–620.

37. A. C. Papanastasiou, G. E. Georghiou, and G. V. Eleftheriades, "A quad-band wilkinson power divider using generalized NRI transmission lines," *IEEE Microw. Wireless Compon. Lett.*, vol. **18**, pp. 521–523, 2008.
38. A. Rennings, T. Liebig, C. Caloz, and I. Wolff, "Double-Lorentz transmission line metamaterials and its applications to tri-band devices," *IEEE MTT-S Int. Microwave Symp. Dig.*, Honolulu, HA, June 2007, pp. 1427–1430.
39. K.-K. M. Cheng and C. Law, "A novel approach to the design and implementation of dual-band power divider," *IEEE Trans. Microw. Theory Tech.*, vol. **56**, pp. 487–492, 2008.
40. K.-S. Chin, K.-M. Lin, Y.-H. Wei, T.-H. Tseng, and Y.-J. Yang, "Compact dual-band branch-line and rat-race couplers with stepped-impedance-stub lines," *IEEE Trans. Microw. Theory Tech.*, vol. **58**, pp. 1213–1221, 2010.
41. M. Gil, J. Bonache, and F. Martín, "Metamaterial filters: a review," *Metamaterials*, vol. **2**, pp. 186–197, 2008.
42. F. Martín, F. Falcone, J. Bonache, R. Marqués, and M. Sorolla, "Miniaturized CPW stop band filters based on multiple tuned split ring resonators," *IEEE Microw. Wireless Compon. Lett.*, vol. **13**, pp. 511–513, 2003.
43. J. García-García, J. Bonache, I. Gil, F. Martín, R. Marqués, F. Falcone, T. Lopetegi, M. A. G. Laso, and M. Sorolla, "Comparison of electromagnetic bandgap and split rings resonator microstrip lines as stop band structures," *Microw. Opt. Technol. Lett.*, vol. **44**, pp. 376–379, 2005.
44. J. García-García, F. Martín, F. Falcone, J. Bonache, J. D. Baena, I. Gil, E. Amat, T. Lopetegi, M. A. G. Laso, J. A. Marcotegui-Iturmendi, M. Sorolla, and R. Marqués, "Microwave filters with improved stop band based on sub-wavelength resonators," *IEEE Trans. Microw. Theory Tech.*, vol. **53**, pp. 1997–2006, 2005.
45. M. Gil, J. Bonache, and F. Martín, "Synthesis and applications of new left handed microstrip lines with complementary split rings resonators (CSRRs) etched in the signal strip," *IET Microw. Antennas Propag.*, vol. **2**, pp. 324–330, 2008.
46. C. Person, A. Sheta, J. Ocupes, and S. Toutain, "Design of high performance band pass filters by using multilayer thick film technology," *Proc. 24th Eur. Microw. Conf.*, Cannes, France, September 1994, vol. **1**, pp. 446–471.
47. J. T. Kuo and M. Jiang, "Suppression of spurious resonance for microstrip band pass filters via substrate suppression," *Asia Pac. Microw. Conf.*, Kyoto, Japan, 2002, pp. 497–500.
48. M. Le Roy, A. Perennec, S. Toutain, and L. C. Calvez, "Continously varying coupled transmission lines applied to design band pass filters," *Int. J. RF Microw. Comput. Aided Eng.*, vol. **12**, pp. 288–295, 2002.
49. J. T. Kuo and W. Hsu, "Parallel coupled microstrip filters with suppression of harmonic response," *IEEE Microw. Wireless Compon. Lett.*, vol. **12**, pp. 383–385, 2002.
50. J. García-García, F. Martín, F. Falcone, J. Bonache, I. Gil, T. Lopetegi, M. A. G. Laso, M. Sorolla, and R. Marqués, "Spurious passband suppression in microstrip coupled line band pass filters by means of split ring resonators," *IEEE Microw. Wireless Compon. Lett.*, vol. **14**, pp. 416–418, 2004.
51. J. García-García, J. Bonache, F. Falcone, J. D. Baena, F. Martín , I. Gil, T. Lopetegi, M. A. G. Laso, A. Marcotegui, R. Marqués, and M. Sorolla, "Stepped-impedance low pass filters with spurious passband suppression," *Electron. Lett.*, vol. **40**, pp. 881–883, 2004.

52. J. Bonache, F. Martín, F. Falcone, J. García, I. Gil, T. Lopetegi, M. A. G. Laso, R. Marqués, F. Medina, and M. Sorolla, "Super compact split ring resonators CPW band pass filters," *IEEE-MTT-S Int. Microw. Symp. Dig.*, Fort Worth, TX, June 2004, pp. 1483–1486.
53. J. Bonache, F. Martín, F. Falcone, J. D. Baena, T. Lopetegi, J. García-García, M. A. G. Laso, I. Gil, A. Marcotegui, R. Marqués, and M. Sorolla, "Application of complementary split rings resonators to the design of compact narrow band pass structures in microstrip technology," *Microw. Opt. Technol. Lett.*, vol. **46**, pp. 508–512, 2005.
54. P. Mondal, M. K. Mandal, A. Chaktabarty, and S. Sanyal, "Compact bandpass filters with wide controllable fractional bandwidth," *IEEE Microw. Wireless Compon. Lett.*, vol. **16**, pp. 540–542, 2006.
55. J. Bonache, F. Martín, F. Falcone, J. García-García, I. Gil, T. Lopetegi, M. A. G. Laso, R. Marqués, F. Medina, M. Sorolla, "Compact coplanar waveguide band pass filter at S-band," *Microw. Opt. Technol. Lett.*, vol. **46**, pp. 33–35, 2005.
56. J. Bonache, I. Gil, J. García-García, and F. Martín, "Complementary split rings resonator for microstrip diplexer design," *Electron. Lett.*, vol. **41**, pp. 810–811, 2005.
57. J. Bonache, I. Gil, J. García-García, and F. Martín, "Novel microstrip band pass filters based on complementary split rings resonators," *IEEE Trans. Microw. Theory Tech.*, vol. **54**, pp. 265–271, 2006.
58. G. L. Matthaei, L. Young, and E. M. T. Jones, *Microwave Filters, Impedance-Matching Networks, and Coupling Structures*, Artech House, Norwood, MA, 1980.
59. D. M. Pozar, *Microwave Engineering*, Addison Wesley, Reading, MA 1990.
60. J. D. Baena, J. Bonache, F. Martín, R. Marqués, F. Falcone, T. Lopetegi, M. A. G. Laso, J. García, I. Gil and M. Sorolla, "Equivalent circuit models for split ring resonators and complementary split rings resonators coupled to planar transmission lines," *IEEE Trans. Microw. Theory Tech.*, vol. **53**, pp. 1451–1461, 2005.
61. J. Bonache, G. Posada, G. Garchon, W. De Raedt, and F. Martín, "Compact ($<0.5 mm^2$) K-band metamaterial band pass filter in MCM-D technology," *Electron. Lett.*, vol. **43**, pp. 45–46, 2007.
62. J. Bonache, F. Martín, J. García-García, I. Gil, R. Marqués, and M. Sorolla, "Ultra wide band pass filtres (UWBPF) based on complementary split rings resonators," *Microw. Opt. Technol. Lett.*, vol. **46**, pp. 283–286, 2005.
63. J. Bonache, F. Martín, I. Gil, J. García-García, R. Marqués, and M. Sorolla, "Microstrip bandpass filters with wide bandwidth and compact dimensions," *Microw. Opt. Technol. Lett.*, vol. **46**, pp. 343–346, 2005.
64. J. Bonache, I. Gil, J. García-García, and F. Martín, "Complementary split rings resonators (CSRRs): towards the miniaturization of microwave device design," *J. Comput. Electron.*, vol. **5**, pp. 193–197, 2006.
65. J. Bonache, J. Martel, I. Gil, M. Gil, J. García-García, F. Martín, I. Cairó, and M. Ikeda, "Super compact ($<1 cm^2$) band pass filters with wide bandwidth and high selectivity at C-band," *Proc. Eur. Microw. Conf.*, Manchester, UK, September 2006, pp. 599–602.
66. M. Gil, J. Bonache, J. García-García, J. Martel, and F. Martín, "Composite right/left handed (CRLH) metamaterial transmission lines based on complementary split rings resonators (CSRRs) and their applications to very wide band and compact filter design," *IEEE Trans. Microw. Theory Tech.*, vol. **55**, pp. 1296–1304, 2007.

67. A. Vélez, P. Vélez, J. Bonache, and F. Martín, "Compact power dividers with filtering capability for ground penetrating radar (GPR) applications," *Microw. Opt. Technol. Lett.*, vol. **54**, pp. 608–611, 2012.
68. M. Durán-Sindreu, A. Vélez, G. Sisó, J. Selga, P. Vélez, J. Bonache, and F. Martín, "Recent advances in metamaterial transmission lines based on split rings," *Proc. IEEE*, vol. **99**, pp. 1701–1710, 2011.
69. D.-J. Woo, T.-K. Lee, "Suppression of harmonics in Wilkinson power divider using dual-band rejection by asymmetric DGS," *IEEE Trans. Microw. Theory Tech.*, vol. **53**, pp. 2139–2144, 2005.
70. C.-M. Lin, H.-H. Su, J.-C. Chiu, and Y.-H. Wang, "Wilkinson power divider using micro-strip EBG cells for the suppression of harmonics," *IEEE Microw. Wireless Compon. Lett.*, vol. **17**, pp. 700–702, 2007.
71. J. Selga, M. Gil, J. Bonache, and F. Martín, "Composite right/left handed coplanar waveguides loaded with split ring resonators and their application to high pass filters," *IET Microw. Antennas Propag.*, vol. **4**, pp. 822–827, 2010.
72. M. Gil, J. Bonache, J. García-García, and F. Martín, "Metamaterial filters with attenuation poles in the pass band for ultra wide band (UWB) applications," *Microw. Opt. Technol. Lett.*, vol. **49**, pp. 2909–2913, 2007.
73. J. Martel, R. Marqués, F. Falcone, J. D. Baena, F. Medina, F. Martín, and M. Sorolla, "A new LC series element for compact band pass filter design," *IEEE Microw. Wireless Compon. Lett.*, vol. **14**, pp. 210–212, 2004.
74. J. Martel, J. Bonache, R. Marqués, F. Martín, and F. Medina, "Design of wide-band semi-lumped bandpass filters using open split ring resonators," *IEEE Microw. Wireless Compon. Lett.*, vol. **17**, pp. 28–30, 2007.
75. A. Velez, F. Aznar, J. Bonache, M. C. Velázquez-Ahumada, J. Martel, and F. Martín, "Open complementary split ring resonators (OCSRRs) and their application to wideband CPW band pass filters," *IEEE Microw. Wireless Compon. Lett.*, vol. **19**, pp. 197–199, 2009.
76. M. Durán-Sindreu, *Miniaturization of planar microwave components based on semi-lumped elements and artificial transmission lines: application to multi-band devices and filters*, PhD dissertation, Universitat Autònoma de Barcelona, Barcelona, 2011.
77. M. Durán-Sindreu, P. Vélez, J. Bonache, and F. Martín, "High-order coplanar waveguide (CPW) filters implemented by means of open split ring resonators (OSRRs) and open complementary split ring resonators (OCSRRs)," *Metamaterials*, vol. **5**, pp. 51–55, 2011.
78. M. Durán-Sindreu, P. Vélez, J. Bonache, and F. Martín, "Broadband microwave filters based on open split ring resonators (OSRRs) and open complementary split ring resonators (OCSRRs): improved models and design optimization," *Radioengineering*, vol. **20**, pp. 775–783, 2011.
79. X. Guan, Z. Ma, P. Cai, Y. Kobayashi, T. Anada, and G. Hagiwara, "Synthesis of dual-band bandpass filters using successive frequency transformations and circuit conversions," *IEEE Microw. Wireless Compon. Lett.*, vol. **16**, pp. 110–112, 2006.
80. F. Aznar, A. Vélez, M. Durán-Sindreu, J. Bonache, and F. Martín, "Elliptic-function CPW low-pass filters implemented by means of open complementary split ring resonators (OCSRRs)," *IEEE Microw. Wireless Compon. Lett.*, vol. **19**, pp. 689–691, 2009.
81. F. Giannini, M. Salerno, and R. Sorrentino, "Effects of parasitics in lowpass elliptic filters for MIC," *Electron. Lett.*, vol. **18**, pp. 284–285, 1982.

82. F. Giannini, M. Salerno, and R. Sorrentino, "Design of low-pass elliptic filters by means of cascaded microstrip rectangular elements," *IEEE Trans. Microw. Theory Tech.*, vol. **MTT-30**, pp. 1348–1353, 1982.
83. L. H. Hsieh and K. Chang, "Compact elliptic-function low-pass filters using microstrip stepped-impedance hairpin resonators," *IEEE Trans. Microw. Theory Tech.*, vol. **51**, pp. 193–199, 2001.
84. N. Yang, Z. N. Chen, Y. Y. Wang, and M. Y. W. Chia, "An elliptic low-pass filter with shorted cross-over and broadside-coupled microstrip lines," *IEEE-MTT-S Int. Microw. Symp. Dig.*, Philadelphia, PA, June 2003, vol. **1**, pp. 535–538.
85. W. Tu and K. Chang, "Compact microstrip low-pass filter with sharp rejection," *IEEE Microw. Wireless Compon. Lett.*, vol. **15**, pp. 404–406, 2005.
86. W. Tu and K. Chang, "Microstrip elliptic-function low-pass filters using distributed elements or slotted ground structure," *IEEE Trans. Microw. Theory Tech.*, vol. **54**, pp. 3786–3792, 2006.
87. M. C. Velazquez-Ahumada, J. Martel, and F. Medina, "Design of compact low-pass elliptic filters using double-sided MIC technology," *IEEE Trans. Microw. Theory Tech.*, vol. **55**, pp. 121–127, 2007.
88. A. Balalem, A. R. Ali, J. Machac, and A. Omar, "Quasi-elliptic microstrip low-pass filters using an interdigital DGS slot," *IEEE Microw. Wireless Compon. Lett.*, vol. **17**, pp. 586–588, 2007.
89. B. Wu, B. Li, and C. Liang, "Design of lowpass filter using a novel split-ring resonator defected ground structure," *Microw. Opt. Technol. Lett.*, vol. **49**, pp. 288–291, 2007.
90. X. Chen, L. Weng, and X. Shi, "Novel complementary split ring resonator DGS for low-pass filter design," *Microw. Opt. Technol. Lett.*, vol. **51**, pp. 1748–1751, 2009.
91. F. Aznar, A. Vélez, J. Bonache, J. Menés, and F. Martín, "Compact low pass filters with very sharp transition bands based on open complementary split ring resonators (OCSRRs)," *Electron. Lett.*, vol. **45**, pp. 316–317, 2009.
92. A. A. Oliner, "Leaky-wave antennas," in *Antenna Engineering Handbook*, R. C. Johnson, Ed. McGraw-Hill, New York, pp. 10-1–10-59, 1993.
93. F. B. Gross, Ed., *Frontiers in Antennas: Next Generation Design & Engineering*, McGraw Hill, New York, 2011.
94. J. L. Volakis, Ed., *Antenna Engineering Handbook*, 4th Edition, McGraw Hill, London, 2007.
95. C. A. Balanis, Ed., *Modern Antenna Handbook*, Wiley, Hoboken, NJ, 2008.
96. D. R. Jackson, C. Caloz, and T. Itoh, "Leaky wave antennas," *Proc. IEEE*, vol. **100**, pp. 2194–2206, 2012.
97. L. Liu, C. Caloz, and T. Itoh, "Dominant mode leaky-wave antenna with backfire to endfire scanning capability," *Electron. Lett.*, vol **38**, pp. 1414–1416, 2002.
98. A. Grbic and G. V. Eleftheriades, "Experimental verification of backward wave radiation from a negative refractive index metamaterial," *J. Appl. Phys.*, vol. **92**, pp. 5930–5935, 2002.
99. A. Grbic and G. V. Eleftheriades, "Leaky CPW-based slot antenna arrays for millimeter-wave applications," *IEEE Trans. Antennas Propag.*, vol. **50**, pp. 1494–1504, 2002.
100. W. Menzel, "A new travelling-wave antenna in microstrip," *Arch. Elek. Ubertragung.*, vol. **33**, pp. 137–140, 1979.

101. A. Oliner and K. Lee, "Microstrip leaky wave strip antennas," *Proc. IEEE Int. Symp. Antennas Propag. Symp.*, Philadelphia, PA, June 1986, pp. 443–446.
102. Y. Lin, J. Sheen, and C. Tzuang, "Analysis and design of feeding structures for microstrip leaky wave antenna," *IEEE Trans. Microw. Theory Tech.*, vol. **44**, pp. 1540–1547, 1996.
103. P. Burghignoli, G. Lovat, and D. R. Jackson, "Analysis and optimization of leaky-wave radiation at broadside from a class of 1-D periodic structures," *IEEE Trans. Antennas Propag.*, vol. **54**, pp. 2593–2603, 2006.
104. S. Paulotto, P. Baccarelli, F. Frezza, and D. R. Jackson, "A novel technique for open-stopband suppression in 1-D periodic printed leaky-wave antennas," *IEEE Trans. Antennas Propag.*, vol. **57**, pp. 1894–1906, 2009.
105. I. Arnedo, J. Illescas, M. Flores, M. A. G. Laso, F. Falcone, J. Bonache, J. García-García, F. Martín, J. A. Marcotegui, R. Marques, and M. Sorolla, "Forward and backward leaky wave radiation in split-ring-resonator-based metamaterials," *IET Microw. Antennas Propag.*, vol. **1**, pp. 65–68, 2007.
106. G. Zamora, S. Zuffanelli, F. Paredes, F. J. Herraiz Martínez, F. Martín, and J. Bonache, "Fundamental mode leaky-wave antenna (LWA) using slot line and split-ring-resonator (SRR)-based metamaterials," *IEEE Antennas Wireless Propag. Lett.*, vol. **12**, pp. 1424–1427, 2013.
107. T. Kokkinos, C. D. Sarris, and G. V. Eleftheriades, "Periodic FDTD analysis of leaky-wave structures and applications to the analysis of negative-refractive-index leaky-wave antennas," *IEEE Trans. Microw. Theory Tech.*, vol. **54**, pp. 1619–1630, 2006.
108. S. Eggermont and I. Huynen, "Leaky wave radiation phenomena in metamaterial transmission line based on complementary split ring resonators," *Microw. Opt. Technol. Lett.*, vol. **53**, pp. 2025–2029, 2011.
109. M. A. Antoniades and G. V. Eleftheriades, "A CPS leaky-wave antenna with reduced beam squinting using NRI-TL metamaterials," *IEEE Trans. Antennas Propag.*, vol. **56**, pp. 708–721, 2008.
110. J. Machac and M. Polivka, "A dual band SIW leaky wave antenna," *IEEE-MTT Int. Microwave Symp. Dig.*, Montréal, Canada, June 2012.
111. M. Durán-Sindreu, J. Choi, J. Bonache, F. Martín, and T. Itoh, "Dual-band leaky wave antenna with filtering capability based on extended-composite right/left-handed transmission lines," *IEEE MTT-S Int. Microw. Symp. Dig.*, Seattle, WA, June 2013.
112. S. Lim, C. Caloz, and T. Itoh, "Electronically scanned composite right/left handed microstrip leaky-wave antenna," *IEEE Microw. Wireless Compon. Lett.*, vol. **14**, pp. 277–279, 2004.
113. S. Lim, C. Caloz, and T. Itoh, "Metamaterial-based electronically-controlled transmission line structure as a novel leaky-wave antenna with tunable angle and beamwidth," *IEEE Trans. Microw. Theory Tech.*, vol. **52**, pp. 2678–2690, 2004.
114. C. G. M. Ryan and G. V. Eleftheriades, "A dual-band leaky-wave antenna based on generalized negative-refractive-index transmission-lines," *IEEE Int. Symp. Antennas Propag. (APS-URSI)*, Toronto, Canada, July 2010.
115. F. P. Casares-Miranda, C. Camacho-Peñalosa, and C. Caloz, "High-gain active composite right/left-handed leaky-wave antenna," *IEEE Trans. Antennas Propag.*, vol. **54**, pp. 2292–2300, 2006.

116. P. de Paco, R. Villarino, G. Junkin, O. Menéndez, E. Corrales, and J. Parrón, "Dual-band mixer using composite right/left-handed transmission lines," *IEEE Microw. Wireless Compon. Lett.*, vol. **17**, pp. 607–609, 2007.

117. S. H. Ji, G. S. Hwang, C. S. Cho, J. W. Lee, and J. Kim, "836 MHz/1 .95GHz dual-band class-E power amplifier using composite right/left-handed transmission lines," *Proc. 36th Eur. Microw. Conf.*, Manchester, UK, September 2006, pp. 356–359.

118. E. L. Ginzton, W. R. Hewlett, J. H. Jasberg, and J. D. Noe, "Distributed amplification," *IRE*, vol. **36**, pp. 956–969, 1948.

119. T. Wong, *Fundamental of Distributed Amplification*, Artech House, Boston, MA, 1993.

120. G. W. McIver, "A travelling wave transistor," *IRE*, vol. **53**, pp. 1747–1748, 1965.

121. W. Jutzi, "A MESFET distributed amplifier with 2GHz bandwidth," *Proc. IEEE*, vol. **57**, pp. 1195–1196, 1969.

122. Y. Ayasli, J. L. Vorhaus, R. Mozzi, and L. Reynolds, "Monolithic GaAs travelling-wave amplifier," *Electron. Lett.*, vol. **17**, pp. 12–13, 1981.

123. J. A. Archer, F. A. Petz, H. P. Weidlich, "GaAs FET distributed amplifier," *Electron. Lett.*, vol. **17**, p. 433, 1981.

124. E. W. Strid, K. R. Gleason, "A DC-12 GHz monolithic GaAs FET distributed amplifier," *IEEE Trans. Microw. Theory Tech.*, vol. **30**, pp. 969–975, 1982.

125. C. S. Aitchison, N. Bukhari, C. Law, and N. Nazoa-Ruiz, "The dual-fed distributed amplifier," *IEEE MTT-S Int. Microwave Symp. Dig.*, New York, May 1988, vol. **2**, pp. 911–914.

126. P. N. Shastry and A. S. Ibrahim, "Design guidelines for a novel tapered drain line distributed power amplifier," *Proc. 36th Eur. Microw. Conf.*, Manchester, UK, September 2006, pp. 1274–1277.

127. J. Gassmann, P. Watson, L. Kehias, and G. Henry, "Wideband, high-efficiency GaN power amplifiers utilizing a non-uniform distributed topology," *IEEE MTT-S Int. Microwave Symp. Dig.*, Honolulu, HI, June 2007, vol. **1**, pp. 615–618.

128. P. N. Shastry and E. W. Cullerton, "A novel wideband GaAs FET source injected distributed mixer," *Proc. 36th Eur. Microw. Conf.*, Manchester, UK, September 2006, pp. 1533–1536.

129. F.-C. Chang, P.-S. Wu, M.-F. Lei, and H. Wang, "A 4–41 GHz singly balanced distributed mixer using GaAs pHEMT technology," *IEEE Microw. Wireless Compon. Lett.*, vol. **17**, pp. 136–138, 2007.

130. O. S. A. Tang and C. S. Atchison, "A practical microwave travelling-wave MESFET gate mixer," *IEEE MTT-S Int. Microwave Symp. Dig.*, Sant Louis, MO, June 1985, pp. 605–608.

131. O. S. A. Tang and C. S. Atchison, "Very wide-band microwave MESFET mixer using the distributed mixing principle," *IEEE Trans. Microw. Theory Tech.*, vol. **33**, pp. 1470–1478, 1985.

132. W. K. Eccleston, "Application of left-handed media in distributed amplifiers," *Microw. Opt. Tech. Lett.*, vol.**44**, pp. 527–530, 2005.

133. J. Mata-Contreras, T. M. Martín-Guerrero, and C. Camacho-Peñalosa, "Distributed amplifiers with composite left/right-handed transmission lines," *Microw. Opt. Technol. Lett.*, vol.**48**, pp. 609–613, 2006.

134. J. Mata-Contreras, C. Camacho-Peñalosa, and T. M. Martín-Guerrero, "Active distributed mixers based on composite right/left-handed transmission lines," *IEEE Trans. Microw. Theory Tech.*, vol. **57**, pp. 1091–1101, 2009.

135. J. Mata-Contreras, *Amplificadores y Mezcladores Distribuidos con Líneas de Transmision basadas en Metamateriales*, PhD dissertation, University of Málaga, Málaga, Spain, 2010.

136. J. Mata-Contreras, T. M. Martín-Guerrero, and C. Camacho-Peñalosa, "Experimental performance of a meta-distributed amplifier," *Proc. 37th Eur Microw Conf*, Munich, Germany, October 2007, pp. 743–746.

137. C. Xie, "Directional dual band distributed power amplifier with composite left/right-handed transmission lines," *IEEE MTT-S Int. Microw. Symp. Dig.*, Atlanta, GA, June 2008, pp. 1135–1138.

138. K. Mori and T. Itoh, "Distributed amplifier with CRLH-transmission line leaky wave antenna," *Proc. 38th Eur. Microw. Conf.*, Amsterdam, The Netherlands, October 2008, pp. 686–689.

139. J. Mata-Contreras, C. Camacho-Penalosa, and T. M. Martin-Guerrero, "Diplexing distributed amplifier with improved isolation," *Electron. Lett.*, vol. **47**, pp. 922–924, 2011.

140. J. Mata-Contreras, C. Camacho-Penalosa, and T. M. Martin-Guerrero, "Diplexing dual-gate FET distributed mixer," *Electron. Lett.*, vol. **47**, pp. 381–382, 2012.

141. C.-T. Michael-Wu, Y. Dong, J. S. Sun, and T. Itoh, "Ring-resonator-inspired power recycling scheme for gain-enhanced distributed amplifier-based CRLH-transmission line leaky wave antennas," *IEEE Trans. Microw. Theory Tech.*, vol. **60**, pp. 1027-1037, 2012.

142. J. Mata-Contreras, D. Palombini, T. M. Martín-Guerrero, A. Bentini, C. Camacho-Peñalosa, and E. Limiti, "Active GaN MMIC diplexer based on distributed amplification concept," *Microw. Opt. Technol. Lett.*, vol.**55**, pp. 1041–1045, 2013.

143. C. Rauscher, "Microwave active filters based on transversal and recursive principles," *IEEE Trans. Microw. Theory Tech.*, vol. **MTT-33**, pp. 1350–1360, 1985.

144. O. García-Pérez, A. García-Lampérez, V. González-Posadas, M. Salazar-Palma, and D. Segovia-Vargas, "Dual-band recursive active filters with composite right/left-handed transmission lines," *IEEE Trans. Microw. Theory Tech.*, vol. **57**, pp. 1180–1187, 2009.

145. L. Billonnet, B. Jarry, and P. Guillon, "Stability diagnosis of microwave recursive structures using the NDF methodology," *IEEE MTT-S Int. Microw. Symp. Dig.*, Orlando, FL, May 1995, vol. **3**, pp. 1419–1422.

146. O. García-Pérez, L. E. García-Muñoz, D. Segovia-Vargas, and V. González-posadas, "Multiple order dual-band active ring filters with composite right/left-handed cells," *Prog. Electromag. Res. (PIER)*, vol. **104**, pp. 201–219, 2010.

147. D. Segovia-Vargas, O. García-Pérez, V. González-Posadas, and F. Aznar-Ballesta, "Dual-band tunable recursive active filter," *IEEE Microw. Wireless Compon. Lett.*, vol. **21**, pp. 92–94, 2011.

148. M. Schüßler, C. Mandel, M. Puentes, and R. Jakoby, "Metamaterial inspired microwave sensors," *IEEE Microw. Mag.*, vol. **13**, pp. 57–67, 2012.

149. M. Puentes, B. Stelling, M. Schüßler, A. Penirschke, and R. Jakoby, "Planar sensor for permittivity and velocity detection based on metamaterial transmission line resonator," *Proc. 39th Eur. Microw. Conf.*, Rome, Italy, 2009, pp. 57–60.

150. C. Damm, M. Schüßler, M. Puentes, H. Maune, M. Maasch, and R. Jakoby, "Artificial transmission lines for high sensitive microwave sensors," *Proc. IEEE Sens. Conf.*, Christchurch, New Zealand, 2009, pp. 755–758.

151. C. Mandel, M. Schüßler, M. Maasch, and R. Jakoby, "A novel passive phase modulator based on LH delay lines for chipless microwave RFID applications," *Proc. IEEE MTT-S Int. Microw. Workshop Wireless Sens., Local Positioning RFID*, Cavtat, Croatia, 2009.

152. M. Schüßler, C. Mandel, M. Puentes, and R. Jakoby, "Capacitive level monitoring of layered fillings in vessels using composite right/left-handed transmission lines," *IEEE MTT-S Int. Microw. Symp. Dig.*, Baltimore, MD, June 2011.

153. J. Zhu and G. V. Eleftheriades, "Dual-band metamaterial-inspired small monopole antenna for WiFi applications," *Electron. Lett.*, vol. **45**, pp. 1104–1106, 2009.

154. F. J. Herraiz-Martínez, L. E. García-Muñoz, D. González-Ovejero, V. González-Posadas, and D. Segovia-Vargas, "Dual-frequency printed dipole loaded with split ring resonators," *IEEE Antennas Wireless Propag. Lett.* vol. **8**, pp. 137–140, 2009.

155. J. Montero-de-Paz, E. Ugarte-Muñoz, F. J. Herraiz-Martínez, V. González-Posadas, L. E. García-Muñoz, and D. Segovia-Vargas, "Multifrequency self-diplexed single patch antennas loaded with split ring resonators," *Prog. Electromag. Res.*, vol. **113**, pp. 47–66, 2011.

156. F. J. Herraiz-Martínez, F. Paredes, G. Zamora, F. Martín, and J. Bonache, "Dual-band printed dipole antenna loaded with open complementary split-ring resonators (OCSRRs) for wireless applications," *Microw. Opt. Technol. Lett.*, vol. **54**, pp. 1014–1017, 2012.

157. F. J. Herraiz-Martínez, G. Zamora, F. Paredes, F. Martín, and J. Bonache, "Multiband printed monopole antennas loaded with open complementary split ring resonators for PANs and WLANs," *IEEE Antennas Wireless Propag. Lett.*, vol. **10**, pp. 1528–1531, 2011.

158. F. Paredes, G. Zamora, J. Bonache, and F. Martín, "Dual-band impedance matching networks based on split ring resonators for applications in radiofrequency identification (RFID)," *IEEE Trans. Microw. Theory Tech.*, vol. **58**, pp. 1159–1166, 2010.

159. F. Paredes, G. Zamora, F. J. Herraiz-Martínez, F. Martín, and J. Bonache, "Dual-band UHF-RFID tags based on meander line antennas loaded with spiral resonators," *IEEE Antennas Wireless Propag. Lett.*, vol. **10**, pp. 768–771, 2011.

160. F. Paredes, G. Zamora, F. J. Herraiz-Martínez, F. Martín, and J. Bonache, "Dual-band RFID tags based on folded dipole antennas loaded with spiral resonators," *IEEE Int. Workshop Antennas Technol. Small Antennas Unconven. Appl. (IWAT 2012)*, Tucson, AZ, March 2012.

161. J. Naqui, M. Durán-Sindreu, and F. Martín, "Novel sensors based on the symmetry properties of split ring resonators (SRRs)," *Sensors*, vol **11**, pp. 7545–7553, 2011.

162. J. Naqui, M. Durán-Sindreu, and F. Martín, "On the symmetry properties of coplanar waveguides loaded with symmetric resonators: analysis and potential applications," *IEEE MTT-S Int. Microw. Symp. Dig.*, Montreal, Canada, June 2012.

163. J. Naqui, A. Fernández-Prieto, M. Durán-Sindreu, F. Mesa, J. Martel, F. Medina, and F. Martín, "Common mode suppression in microstrip differential lines by means of complementary split ring resonators: theory and applications," *IEEE Trans. Microw. Theory Tech.*, vol. **60**, pp. 3023–3034, 2012.

164. S. Preradovic, N. C. Karmakar, "Chipless RFID: bar code of the future," *IEEE Microw. Mag.*, vol. **11**, pp. 87–97, 2010.

165. S. Preradovic, I. Balbin, N. C. Karmakar, and G. F. Swiegers, "Multiresonator-based chipless RFID system for low-cost item tracking," *IEEE Trans. Microw. Theory Tech.*, vol. **57**, pp. 1411–1419, 2009.

166. T. Driscoll, G. O. Andreev, D. N. Basov, S. Palit, S. Y. Cho, N. M. Jokerst, and D. R. Smith, "Tuned permeability in terahertz split ring resonators for devices and sensors," *Appl. Phys. Lett.*, vol. **91**, paper 062511, 2007.

167. H.-J. Lee and J.-G. Yook, "Biosensing using split-ring resonators at microwave regime," *Appl. Phys. Lett.*, vol. **92**, paper 254103, 2008.

168. R. Melik, E. Unal, N. K. Perkgoz, C. Puttlitz, and H. V. Demir, "Metamaterial-based wireless strain sensors," *Appl. Phys. Lett.*, vol. **95**, paper 011106, 2009.

169. E. Cubukcu, S. Zhang, Y. S. Park, G. Bartal, and X. Zhang, "Split ring resonator sensors for infrared detection of single molecular monolayers," *Appl. Phys. Lett.*, vol. **95**, paper 043113, 2009.

170. X.-J. He, Y. Wang, J.-M. Wang, and T.-L. Gui, "Thin film sensor based tip-shaped splits ring resonator metamaterial for microwave applications," *Microsyst. Technol.*, vol. **16**, pp. 1735–1739, 2010.

171. E. Ekmekci and G. Turhan-Sayan, "Metamaterial sensor applications based on broadside-coupled SRR and V-shaped resonator structures," *Proc. Int. Symp. Antennas Propag.*, Jeju, South Korea, 2011, pp. 1170–1172.

172. M. Puentes, C. Weiss, M. Schüßler, and R. Jakoby, "Sensor array based on split ring resonators for analysis of organic tissues," *IEEE MTT-S Int. Microw. Symp. Dig.*, Baltimore, MD, June 2011.

173. C. Mandel, B. Kubina, M. Schüßler, and R. Jakoby, "Passive chipless wireless sensor for two-dimensional displacement measurement," *Proc. 41st Eur. Microw. Conf.*, Manchester, UK, 2011, pp. 79–82.

174. J. Naqui, M. Durán-Sindreu, and F. Martín, "Alignment and position sensors based on split ring resonators," *Sensors*, vol. **12**, pp. 11790–11797, 2012.

175. J. Naqui, M. Durán-Sindreu, and F. Martín, "Transmission lines loaded with bisymmetric resonators and applications," *Proc. IEEE MTT-S Int. Microw. Symp.*, Seattle, WA, June 2013.

176. A. Karami-Horestani, C. Fumeaux, S. F. Al-Sarawi, and D. Abbott, "Displacement sensor based on diamond-shaped tapered split ring resonator," *IEEE Sensors J.*, vol. **13**, pp. 1153–1160, 2013.

177. J. Naqui and F. Martín, "Transmission lines loaded with bisymmetric resonators and their application to angular displacement and velocity sensors," *IEEE Trans. Microw. Theory Tech.*, vol. **61**, pp. 4700–4713, 2013.

178. J. Naqui and F. Martín, "Angular displacement and velocity sensors based on electric-LC (ELC) loaded microstrip lines," *IEEE Sens. J.*, vol. **14**, pp. 939–940, 2014.

179. J. Naqui, M. Durán-Sindreu, and F. Martín, "Strategies for the implementation of sensors and RF barcodes based on transmission lines loaded with symmetric resonators," *21st Int. Conf. Appl. Electromag. Comm. (ICECom 2013)*, Dubrovnic, Croatia, October 2013.

5

RECONFIGURABLE, TUNABLE, AND NONLINEAR ARTIFICIAL TRANSMISSION LINES

5.1 INTRODUCTION

In this chapter, some materials, components, and technologies to implement reconfigurable and tunable artificial transmission lines are briefly reviewed and some applications of these lines are highlighted. In the considered structures, tuning is achieved through electronic actuation where, typically, changes in a capacitance, or local changes in the permittivity of a dielectric layer, are generated by the actuated voltage or applied field. These tunable or reconfigurable structures operate in the lineal regime. Conversely, the final part of the chapter is devoted to the topic of nonlinear transmission lines (NLTLs), where nonlinearity and dispersion combined give rise to soliton wave propagation.

The topics covered in this chapter could be the subject of an extensive study. Indeed, there are several textbooks entirely focused on some of the technologies considered to implement tunable components. It is not the purpose of this chapter to make an in-depth study of such technologies, but to discuss some ideas for the implementation of tunable and reconfigurable transmission lines based on their use, and to point out some relevant applications.

5.2 MATERIALS, COMPONENTS, AND TECHNOLOGIES TO IMPLEMENT TUNABLE DEVICES

The tuning components, materials, or technologies that will be considered for the implementation of reconfigurable artificial transmission lines in the following section,

Artificial Transmission Lines for RF and Microwave Applications, First Edition. Ferran Martín.

are briefly analyzed in the present section. The tuning principles, as well as their advantages and limitations are highlighted.

5.2.1 Varactor Diodes, Schottky Diodes, PIN Diodes, and Heterostructure Barrier Varactors

Semiconductor diodes are electronic devices well known since long time ago and typically (although not exclusively) consisting of a PN junction [1]. Usually, these devices are used as rectifiers and switches in many electronic circuits such as AC–DC converters, voltage multipliers, envelope detectors, and so on. By applying a forward voltage, the diodes exhibit a small DC resistance and current flows across the diode terminals. Conversely, by applying a reverse bias, the diode does not allow a DC current flow. In the reverse-biased state, diodes behave as voltage-dependent capacitors since the depletion region width varies (increases) with the applied voltage. Generally, the width of the depletion region is proportional to the square root of the applied voltage.[1] Since the capacitance is inversely proportional to the depletion region width, it follows that the capacitance is inversely proportional to the square root of applied voltage. Therefore, reverse-biased diodes can be used as voltage-controlled capacitors, and can be applied in various electronic circuits, such as voltage-controlled oscillators (VCOs), tunable filters, and so on. They can also be used in nonlinear regime for the implementation of frequency multipliers and mixers, among other components.

In their use as voltage-controlled capacitors, it is important to enhance the dependence of the capacitance of the diode with the applied voltage. Specific diodes manufactured to exploit this effect are usually called varactor diodes, varicap diodes, or tuning diodes. A figure of merit of varactor diodes is the range of variability of the capacitance with the applied bias; whereas in ordinary diodes, the focus is to minimize such capacitance. Varactor diodes are low-cost components, commercially available in surface-mount technology (SMT). With their use, it is possible to achieve significant tuning ranges since there are available components from the main semiconductor providers exhibiting very wide capacitance spans.

It is also worth mentioning that there are specific diodes for fast switching speeds (Schottky diodes). Schottky diodes are based on a metal–semiconductor junction (where the semiconductor is typically n-type doped). Fast switching in such diodes is related to the fact that, as compared to ordinary diodes (where the switching mechanism involves the removal or injection of minority carriers at both sides of the junction by diffusion and recombination), Schottky diodes do not involve minority carriers. Schottky diodes are majority carrier semiconductor devices. Therefore, no slow recombination of n- and p-type minority carriers is involved in the switching process, and Schottky diodes can switch much faster than ordinary PN rectifier diodes.

Another type of diodes frequently used in RF/microwave applications are those consisting of a wide and slightly doped (almost intrinsic) semiconductor layer

[1] This dependence is analytically achieved by considering an abrupt PN junction. In a more realistic junction, the width of the depletion region is proportional to the m-root of the reverse voltage, with $2 < m < 3$ [1].

sandwiched between a p-type and an n-type semiconductor region. Such semiconductor devices, called PIN diodes, exhibit a high breakdown voltage and a low level of junction capacitance by virtue of the presence of the wide intrinsic (i-type) layer. The depleted region in PIN diodes extends almost entirely in the whole intrinsic region. Under forward bias, the carrier concentration in the intrinsic region is much higher than the intrinsic-level carrier concentration. Due to this high injection level, the electric field extends deeply (almost the entire length) into the region and helps in speeding up the transport of charge carriers from the P to the N region. This results in faster operation of the diode, making it a suitable device for high-frequency applications.

Let us finalize this subsection by dedicating few words to the heterostructure barrier varactor (HBV), proposed by Kollberg and Rydberg [2]. Such semiconductor device exhibits a voltage-controlled capacitance, similar to a varactor diode. However, unlike a varactor diode, the HBV exhibits an antisymmetric current–voltage relationship and a symmetric capacitance–voltage relationship (Fig. 5.1). The device can be either forward or reverse biased giving the same capacitance values (for the same magnitude of the applied bias), and the maximum capacitance is achieved in the absence of bias. Obviously, this symmetric capacitance–voltage characteristic can only be achieved if the structure is symmetric. Thus, instead of using a metal–semiconductor junction (providing a triangular-shaped band structure) or a PN junction, the HBV active region is made by sandwiching a wide band gap semiconductor layer between two semiconductor layers with lower band gap material. The resulting heterostructure acts as a potential (rectangular-shaped) barrier for electrons, avoiding the flow of DC current under moderate and low applied voltages (the current at high voltages is due to

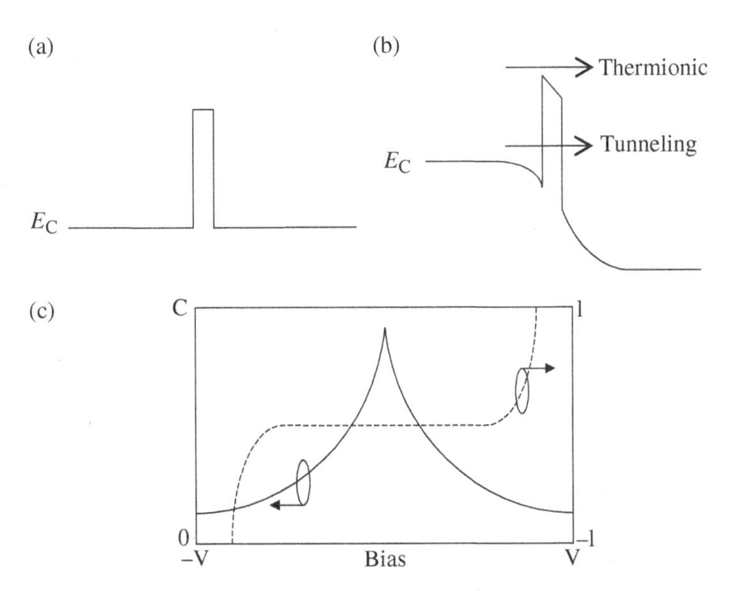

FIGURE 5.1 Conduction band diagram of an unbiased (a) and biased (b) HBV in the active (barrier) region, and typical capacitance–voltage and current–voltage diagrams (c). The current at high voltages may be due to thermionic emission or tunneling.

thermionic emission or tunneling). Basically, the barrier profile in equilibrium is symmetric around the middle of the barrier, and this symmetry is preserved if the heterostructure barrier is clad between two moderately doped semiconductor regions where the depleted region modulation takes place. Essentially, the structure exhibits a capacitive behavior where the extension of the depletion regions (essential to determine the voltage-dependent capacitance) is the same for forward and reverse conditions, and the capacitance–voltage characteristic is an even function. The symmetric capacitance–voltage characteristic of HBVs is not only of interest for biasing purposes in linear tunable components; several studies have also demonstrated the advantages of HBVs in nonlinear applications, particularly for efficiently rejecting the even-order harmonics in frequency multipliers [3] (this aspect will be considered in the last section of this chapter). In Section 5.3, the use of varactor diodes for the implementation of tunable metamaterial resonators and metamaterial transmission lines will be illustrated, and several examples of applications will be reported.

5.2.2 RF-MEMS

Microelectromechanical systems for radiofrequency applications (RF-MEMS) constitute a key enabling technology for telecommunication systems (see Ref. [4] and references therein). There are several areas where RF-MEMS may find application, but the focus here is the implementation of switches and variable capacitances for tunable/reconfigurable components. RF-MEMS can be classified depending on the actuation mechanism (electrostatic, thermal, magnetostatic, piezoelectric), configuration (fixed-fixed beam or cantilever beam type), and movement (continuous movement or switch), among others. The tunable/reconfigurable transmission lines and components that are described in the next section are based on electrostatically actuated RF-MEMS. Both configurations (fixed-fixed beam or cantilever) are considered, and RF-MEMS act as switchable components in most applications.

Figure 5.2 depicts the cross-sectional views of a nonactuated (up-state) and actuated (down-state) switchable cantilever type RF-MEMS. By application of a voltage, V,

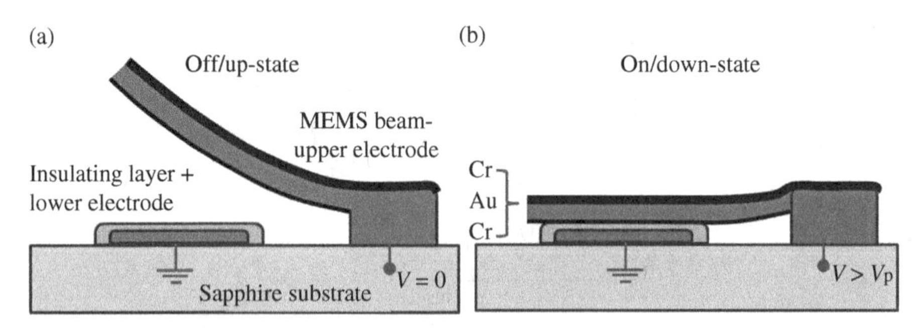

FIGURE 5.2 Cross sectional view of a cantilever-type beam RF-MEMS showing the two states: (a) nonactuated up-state and (b) actuated down-state when the applied voltage is higher than the MEMS pull-down voltage (V_p).

between the cantilever beam and the pull-down electrode, an electrostatic force is induced on the beam, and it deflects down as shown in Figure 5.2b. If the electrostatic force is larger than the mechanical restoring force, the switch stays in the down-state position. In order to estimate the required voltage to switch the RF-MEMS to the down-state, let us consider that the cantilever beam and the pull-down electrode are parallel oriented, so that their distance (beam height) is uniform in the whole area, regardless of the actuation voltage. Let us denote by g the distance between the beam and the pull-down electrode, g_o being the distance in the non-actuated state. The applied electrostatic force is given by

$$F_e = \frac{1}{2} V^2 \frac{dC(g)}{dg} \tag{5.1}$$

where the capacitance is[2]

$$C(g) = A \frac{\varepsilon_o}{g} \tag{5.2}$$

A being the effective area of the RF-MEMS capacitance. Equating the applied electrostatic force with the mechanical restoring force due to the stiffness of the beam, the following condition results [4]:

$$\frac{1}{2} \frac{\varepsilon_o A V^2}{g^2} = k(g_o - g) \tag{5.3}$$

where k is the spring constant. From (5.3), the necessary voltage to force a certain beam height, g, is

$$V = \sqrt{\frac{2kg^2(g_o - g)}{\varepsilon_o A}} \tag{5.4}$$

Notice that the right-hand side term in (5.4) exhibits a maximum for $g = (2/3)g_o$, and the corresponding voltage is

$$V_p = \sqrt{\frac{8kg_o^3}{27\varepsilon_o A}} \tag{5.5}$$

Thus, if the actuating voltage is $V > V_p$, there is not solution for (5.4), the system is unstable, and the beam and pull-down electrode collapse [4]. V_p is called pull-down voltage, and above that voltage the RF-MEMS is forced to be in the down-state. The

[2] Notice that the narrow insulating layer above the pull-down electrode has been neglected in the calculation of the capacitance $C(g)$.

instability is caused because for actuating voltages above V_p the restoring force is not able to compensate the electrostatic force. There is a positive feedback in the system, that is, as g decreases, the electrostatic force increases at a higher rate than the restoring force, and the beam finally collapses to the down-state. Cantilever beams generally exhibit smaller pull-down voltages, as compared to fixed-fixed beams, because they exhibit smaller spring constants [4].

In Section 5.3, several strategies to implement tunable metamaterial resonators using RF-MEMS are reported, including details relative to the fabrication process. As compared to varactor-based devices (useful for manufacturing tunable components in the L and S frequency bands), RF-MEMS can operate at much higher frequencies. Furthermore, the performance of varactor-based tunable microwave systems is generally limited by losses, power consumption, and nonlinearity. Conversely, the successful integration of RF-MEMS switches in electrically small metamaterial resonators is expected to be an enabling technology for many microwave applications (including the synthesis of metamaterial transmission lines and devices based on them), thanks to their low loss, near-zero power dissipation, compactness, high linearity on broad frequency bands, and very high capacitance ratios. RF-MEMS are still limited by reliability issues and moderate switching time, but progress has been done on these specific aspects [5–7]. The ability of this technology has been demonstrated over the past few years to provide an efficient solution to the tuning of microwave circuits [8, 9].

5.2.3 Ferroelectric Materials

Another possibility to implement tunable/reconfigurable RF/microwave components is to use materials with field-dependent constitutive parameters. Among them, ferroelectric materials are of special interest since they experience significant changes of the dielectric permittivity with an external electric field.[3] Hence, ferroelectric materials can be used for the implementation of variable capacitances, or as core substrates for the implementation of tunable resonators in planar technology [10].

Ferroelectric materials are able to exhibit spontaneous electric polarization in the absence of an external electric field (polar phase). Below the so-called Curie temperature, the spontaneous polarization can be reversed by an applied electric field, and the polarization is dependent not only on the instantaneous electric field but also on its history, yielding a hysteresis loop (materials exhibiting these properties are called ferroelectrics by analogy to ferromagnetic materials, which have spontaneous magnetization and also exhibit hysteresis loops). Above the Curie temperature, ferroelectric materials experience a phase transition, spontaneous polarization is not allowed, and they become paraelectric[4] (nonpolar phase). For microwave applications ferroelectric materials are typically used in the paraelectric phase since hysteresis is prevented [11]. This means that the operation temperature of the material must be above the

[3] The dielectric permittivity in ferroelectric materials also depends on temperature.

[4] Paraelectric materials exhibit a nonlinear dependence of the polarization with the electric field (or a field dependent permittivity).

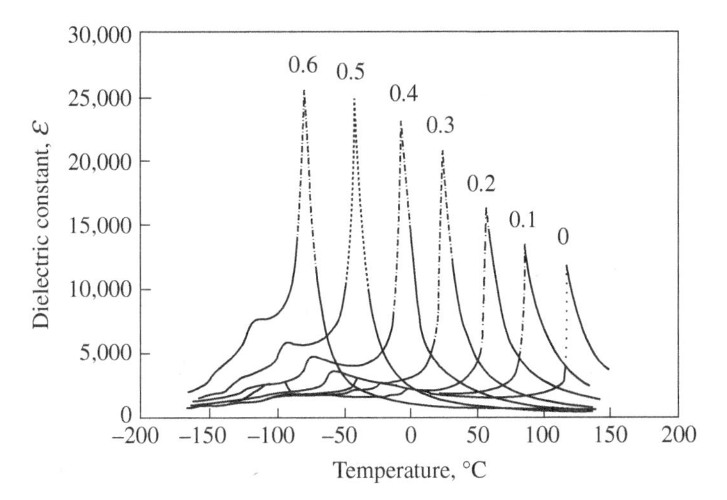

FIGURE 5.3 Dependence of the dielectric permittivity with temperature for several samples of BST with different values of the stoichiometic factor x. Reprinted with permission from Ref. [13]; copyright 2004 Elsevier.

Curie temperature. However, due to the fact that tunability (the variation of the permittivity with the external electric field) decreases as temperature increases from the Curie temperature, it is convenient to work just above the Curie point, where the permittivity reaches its maximum value [12]. Most ferroelectric materials are oxides with a perovskite crystalline structure, the most representative being the barium titanate ($BaTiO_3$). This material exhibits a Curie temperature of $T_c = 120°C$, that is, far above room temperature. Conversely, in strontium titanate ($SrTiO_3$) $T_c = -235°C$. This suggests that by combining such materials to form a mixed crystal of the type $Ba_xSr_{1-x}TiO_3$ (generally abbreviated as BST), the Curie temperature can be tailored with the stoichiometric factor x (see Fig. 5.3) [12, 13]. In particular, it has been found that for $x = 0.6$, the Curie temperature is $T_c = -2°C$. Therefore, $Ba_{0.6}Sr_{0.4}TiO_3$ exhibits paraelectric behavior at room temperature and is a good candidate material for the implementation of tunable components controlled by electric field.

BST exhibits high breakdown field strength (around 400 V/μm). This means that high biasing voltages can be applied for tuning, enhancing the achievable material tunability. Moreover, since the dielectric constant of BST can be set to various hundreds at room temperature,[5] film thicknesses as small as 2–6 μm are enough to implement tunable components. Typically thick (few microns) films of BST are screen printed on a carrier substrate (e.g., Al_2O_3) and the metallic layers are patterned on top of it. By virtue of the tunable permittivity of the BST layer, the capacitance between noncontacting (but closely separated) metallic strips (e.g., a capacitive gap) can be tailored by applying a voltage between the metallic terminals [14, 15]. This also applies to

[5] The dielectric permittivity is controlled by the stoichiometric factor x, but it is strongly influenced by the porosity of the BST material.

electrically small planar resonators, as will be discussed later. As compared to RF-MEMS, BST-based components are more robust, exhibit (in general) higher tunability ranges, but dielectric losses may limit device performance.

5.2.4 Liquid Crystals

Liquid crystals are a special class of materials that exhibit some properties of liquids and crystals simultaneously [16–18]. At the macroscopic level, liquid crystals behave like a liquid, being able to flow, move, and occupy arbitrarily shaped fixed volumes. At the microscopic level, liquid crystals have a rod-like molecule structure. Typically, the molecules are uniaxial, that is, they have one axis that is longer and preferred, with the other two being equivalent, but some liquid crystals are biaxial. The location of the particles is not fixed; but in the so-called nematic state, they are oriented along the main axis, maintaining long-range directional order (Fig. 5.4). Due to the high asymmetry of the liquid crystal molecules, the electromagnetic (EM) properties of these materials are expected to exhibit significant anisotropy. This is the case of the electric polarization, and the dielectric flux is related to the electric field by a permittivity tensor, rather than by a scalar. Nevertheless, if we consider the liquid crystal to be uniaxial, the permittivity tensor in the local molecule coordinate system (i.e., with axes parallel to the principal axes of the rod-shaped molecule) is diagonal, with parallel (to the main axis) and perpendicular (orthogonal to the main axis) permittivity denoted by $\varepsilon_{r\parallel}$ and $\varepsilon_{r\perp}$, respectively.

For tuning purposes in microwave circuits, nematic liquid crystals with controllable orientation of their molecules are of interest. Due to the anisotropy, the effective permittivity can be tuned by manipulating such orientation. In practice, the alignment of the main axis is usually achieved by application of a static electric or magnetic field.

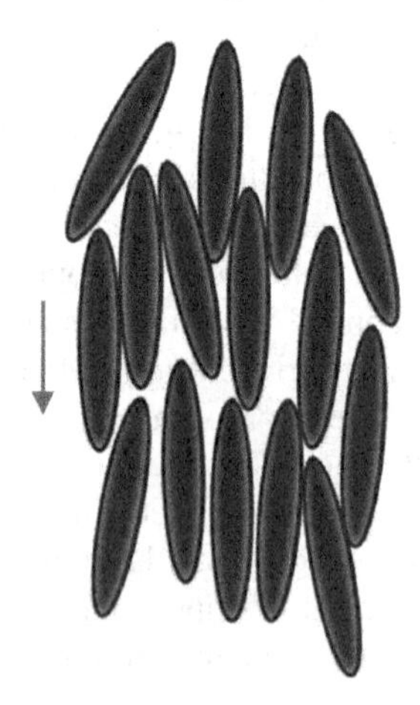

FIGURE 5.4 Scheme of the liquid crystal molecules in the nematic phase. The molecules do not necessarily all point in the same direction; however, the average of all the directors points in one direction (indicated).

Thus, in tunable microwave devices based on nematic liquid crystals, anisotropy plays a fundamental role and the tuning principle is completely different than those tuning principles studied in the previous subsections.

Liquid crystals typically exhibit small losses (at least as compared to ferroelectric materials), although the fact that they can flow makes the design/fabrication of liquid crystal-based devices more cumbersome (i.e., proper sealing is necessary). Other general advantages are high-frequency operation (up to hundreds of GHz), large tuning ranges, and fast switching times. This performance justifies the use of liquid crystals (despite the confinement requirements) in certain applications where these excellent properties must be satisfied simultaneously. Ignoring the inherent limitation (from the fabrication viewpoint) of dealing with a liquid, it is difficult to find a technology/material able to compete with liquid crystals for tuning purposes.

5.3 TUNABLE AND RECONFIGURABLE METAMATERIAL TRANSMISSION LINES AND APPLICATIONS

Several strategies for the implementation of tunable and reconfigurable metamaterial transmission lines, using the tunable components and materials presented in the previous section, are discussed in this section. The first part of the section is devoted to the implementation and applications of tunable lines based on SRRs and other related resonators. The second part focuses on the CL-loaded approach.

5.3.1 Tunable Resonant-Type Metamaterial Transmission Lines

The tuning principle in most metamaterial transmission lines implemented by means of split rings is based on the variation of the resonance frequency of the resonant elements through electronic actuation. This tuning can be achieved by loading the split rings (SRRs, OCSRRs, etc.,) with varactor diodes (or by means of other tuning diodes) or RF-MEMS. Alternatively, tunable resonant-type metamaterial transmission lines can be implemented by etching the resonant elements on top of tunable materials, such as ferroelectrics. Let us now discuss different approaches for the implementation of tunable split rings.

5.3.1.1 Varactor-Loaded Split Rings and Applications Probably, the simplest form to tune the resonance frequency of an SRR is by loading it with a voltage-controlled capacitor [19–21]. In practice, these capacitances can be implemented by means of tuning diodes, such as varactor diodes. The first SRR loaded with a varactor diode, designated as varactor-loaded split-ring resonator (VLSRR), was reported in Ref. [19]. The topology of that VLSRR [19] is depicted in Figure 5.5a. As compared to the SRR topology originally proposed by Pendry [22], the VLSRR reported in Ref. [19] was rectangular shaped, with a nonuniform distance d between the inner and outer ring, in order to accommodate the diode varactor between such rings. Moreover, a metal pad was added in the center of the particle to ease diode biasing. With this configuration, the EM behavior of the VLSRR does

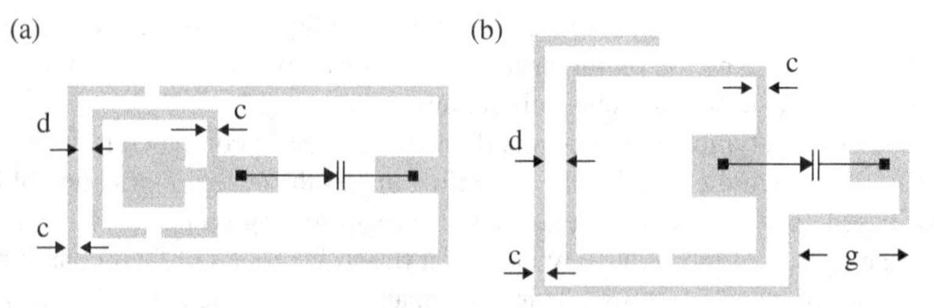

FIGURE 5.5 Topologies of the tunable VLSRR (a) and VLSRR with improved geometry (b). The relevant dimensions are indicated. Reprinted with permission from Ref. [20]; copyright 2006 IEEE.

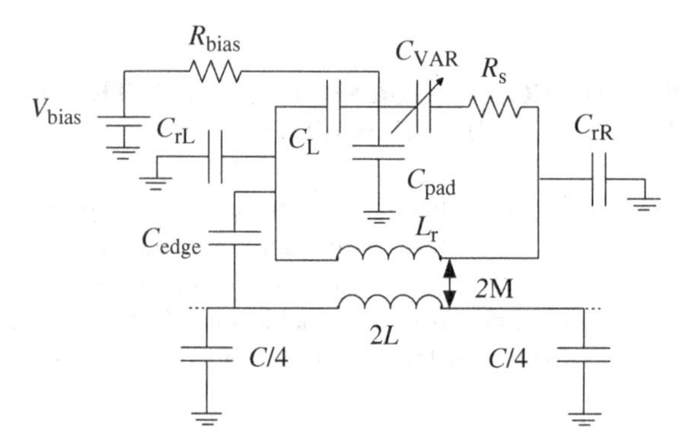

FIGURE 5.6 Lumped-element equivalent circuit model for the elemental cell of a biased VLSRR coupled to a microstrip transmission line. L and C are the per-section inductance and capacitance of the line. The magnetic wall concept has been used. Reprinted with permission from Ref. [20]; copyright 2006 IEEE.

not substantially differ from that of the SRR, except by the fact that the resonance frequency of the VLSRR can be tuned by virtue of the presence of the diode varactor (such varactor dominates over the edge capacitance of the right half of the structure). Due to the presence of the inner metallic pad, it is expected that particle excitation is mainly achieved through the magnetic field penetrating the inter-rings region, where the varactor is allocated, rather than by the axial magnetic field present within the inner ring of the particle (as in ordinary SRRs). An alternative, and simple, topology of the VLSRR is the one depicted in Figure 5.5b [20], where the right-hand arm of the outer ring has been shortened since no appreciable current flows through it (i.e., most of the electric current is absorbed by the varactor diode), and the metallic pad for diode soldering and polarization has been reduced. The result is a topology similar to that of a square-shaped SRR.

The lumped-element equivalent circuit model of a biased VLSRR coupled to a microstrip transmission line is depicted in Figure 5.6 [20]. The diode varactor is modeled by a variable capacitance, C_{VAR}, and a series resistance, R_s, which takes into account not only the intrinsic losses of the diode, R_D, but also the resistance associated to the varactor–metal junctions, R_{VM}. C_R (which is neglected due to the shunt connection of C_{VAR} to it) and C_L are the edge capacitances corresponding to the right and left halves, respectively, of the VLSRR, whilst L_r and C_{pad} model the equivalent inductance of the VLSRRs and the pad-to-ground capacitance, respectively. Diode biasing is applied through a variable voltage source, V_{bias}, which has an equivalent output resistance termed R_{bias}. Concerning line-to-VLSRRs coupling, we a priori assume that VLSRRs can be driven either by the axial magnetic field generated by the line (inductive coupling) or by the electric field present between the line and the external ring (capacitive coupling). Both couplings have been properly modeled by means of a mutual inductance, M, (magnetic coupling) and the edge capacitance between the line and the external ring, C_{edge} (electric coupling). The capacitances C_{rL} and C_{rR} are the rings-to-ground capacitances of the left and right halves, respectively.

The VLSRR model of Figure 5.6 is validated by comparing the frequency response of a fabricated two-stage VLSRR-loaded microstrip line with the frequency response obtained through circuit simulation of the corresponding two-stage equivalent circuit of the VLSRR-loaded line. The fabricated device is depicted in Figure 5.7a, where *BB833-Infineon Technologies* silicon tuning diodes have been used as nonlinear capacitances (the capacitance window for these diodes is 0.75–9 pF for varying voltages in the interval 0–30 V). The device, fabricated on the *Rogers RO3010* substrate with dielectric constant $\varepsilon_r = 10.2$, thickness $h = 1.27$ mm and $\tan \delta = 0.0023$, exhibits stopband functionality, similar to SRR-loaded lines and attributed to the negative effective permeability of the structure in the vicinity of the fundamental resonance. However, due to the presence of varactors, the central frequency can be electronically tuned (Fig. 5.7b). Except the mutual inductance, M, and the varactor–metal junction resistance, R_{VM}, which have been used as fitting parameters, the other element values have been estimated either through geometrical considerations or with the help of the commercial software *Agilent ADS* (see Ref. [20] for details). The frequency responses depicted in Figure 5.7c have been obtained by setting $M = 1.2$ nH and $R_{VM} = 3\ \Omega$ (the other parameters are indicated in the caption of Fig. 5.7), and are those that have optimally fitted to the experiment over the considered tuning interval. The good agreement between theory and experiment supports the validity of the model. Nevertheless, the model can be simplified since no appreciable differences in the circuit simulation result by switching the electric coupling off (this can be simply achieved by removing the coupling capacitance, C_{edge}, and the capacitance C_{pad}, as Fig. 5.8a illustrates). The circuit simulations considering only inductive coupling as the single driving mechanism of the VLSRRs (Fig. 5.8a) are depicted in Figure 5.8b. The fact that the frequency responses for the different diode polarizations are very similar to those depicted in Figure 5.7b and c points out that magnetic coupling is the dominant coupling mechanism. According to this model, signal can flow through the resistance R_{bias}. However, since C_{rL} and C_{rR} are small, it is a good approximation, in order to simplify the analysis of the circuit, to neglect these capacitances and consider R_{bias} as a polarization path for the diodes (i.e., irrelevant for signal analysis). With this approximation, the equivalent impedance of the series branch of the transmission-line section

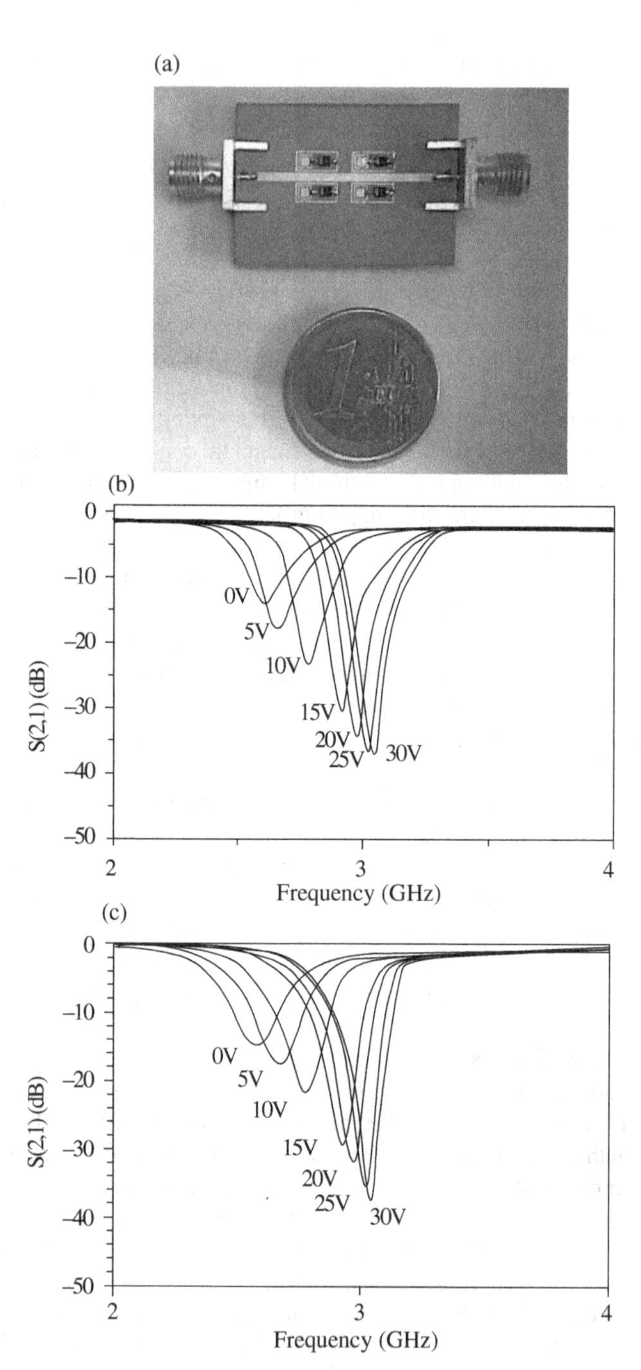

FIGURE 5.7 Photograph of a two-stage VLSRR-loaded microstrip line (a), measured insertion losses for different diode polarizations (b) and frequency responses obtained by circuit simulation of the equivalent circuit model shown in Fig. 5.6 (c). Relevant dimensions are $c = d = 0.2$ mm, separation between line and external rings is 0.2 mm and length and width of VLSRRs are 6.2 mm and 2.8 mm, respectively. Element values are $L_r = 3$ nH, $C_L = 1.4$ pF, $C_{edge} = 1$ fF, $C_{rL} = 0.1$ pF, $C_{pad} = 0.44$ pF, $C_{rR} = 0.89$ pF, $L = 3.3$ nH, $C = 1.3$ pF, $R_S = 4.8\ \Omega$, (i.e., $R_D = 1.8\ \Omega$ and $R_{VM} = 3.0\ \Omega$), $R_{bias} = 50\ \Omega$, $M = 1.2$ nH, and $1\ \text{pF} \leq C_{VAR} \leq 9$ pF. Reprinted with permission from Ref. [20]; copyright 2006 IEEE.

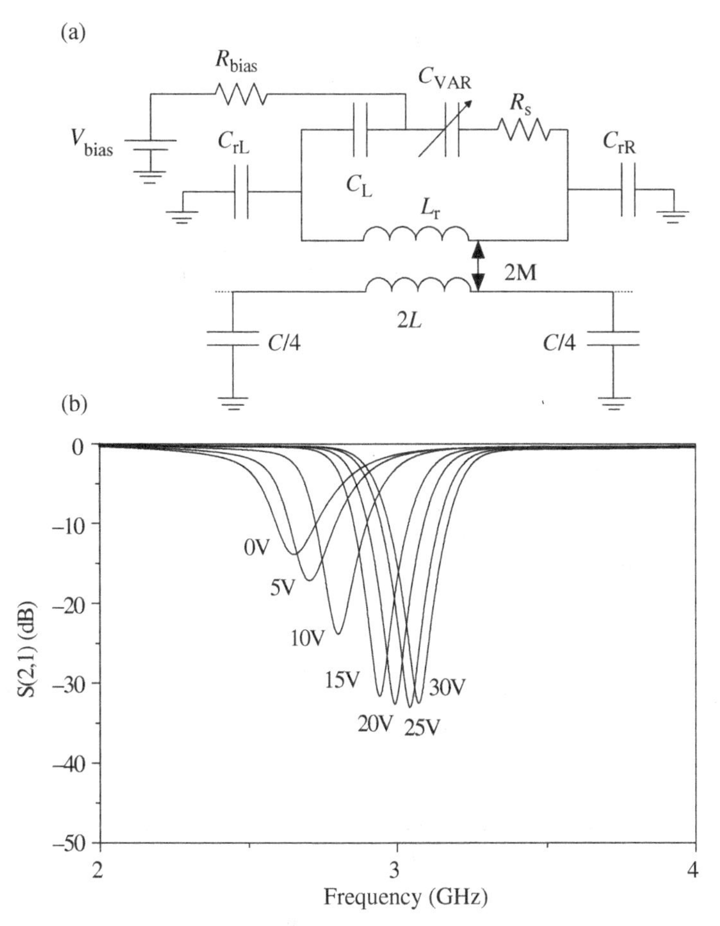

FIGURE 5.8 Equivalent circuit model of the VLSRR-loaded microstrip line with electric coupling removed (a), and circuit simulations for different diode polarizations (b). Reprinted with permission from Ref. [20]; copyright 2006 IEEE.

(including the effects of the resistance R_s) can be inferred. The result is the series connection of an inductor (with inductance $L' = 2L - L_s'$) and a parallel RLC tank, with element values given by[6]

$$L_s' = 4M^2 C_{eq} \omega_o^2 \tag{5.6a}$$

[6] These expressions are derived by applying the transformer equations to the circuit of Figure 5.8a with the approximations cited in the text. Notice also that L', L_s', and C_s' are identical to the expressions given in Figure 3.34, which are valid for a line loaded with SRRs only (i.e., without shunt-inductive elements).

$$C_s' = \frac{L_r}{4M^2\omega_o^2} \tag{5.6b}$$

$$R' = \frac{4M^2\omega_o^2}{R_s} \tag{5.6c}$$

where $\omega_o = (L_r C_{eq})^{-1/2}$. Expressions (5.6) explain the reduction in gap width and rejection enhancement when ω_o increases. Namely, the Q-factor of the RLC tank described by the elements of (5.6) is

$$Q = \omega_o R' C_s' = \omega_o \frac{L_r}{R_s} \tag{5.7}$$

and it increases with ω_o (R_s and L_r do not depend on frequency). Therefore, deeper and narrower stopbands are expected in the upper extreme of the tuning interval, as actually occurs. The agreement between measurements and circuit simulations of the equivalent-circuit model is good over the whole tuning interval. Therefore, the proposed equivalent-circuit model provides a good description of the VLSRR-loaded microstrip line.

Microstrip lines loaded with VLSRRs can be useful as tunable and reconfigurable stopband filters. It is possible to tune the filter central frequency and/or the filter bandwidth. As an illustrative example, Figure 5.9 depicts a four-stage stopband filter, where the VLSRR stages are all identical. By applying the same bias to the diodes, the central frequency is tuned (the bandwidth is slightly tuned as well, for the reasons explained before), as depicted in Figure 5.9b. However, we can also apply different bias voltages to the diodes in order to enhance filter bandwidth, as illustrated in Figure 5.9c.

By adding vias (acting as shunt connected inductors) or capacitive gaps (acting as series capacitors) to the VLSRR-loaded microstrip lines, the stopband is switched to a pass band [20]. Despite the fact that a tunable bandpass filter with roughly constant characteristics in the tuning interval was reported in Ref. [20] by adding gaps to the line, losses were found to be significant for most practical applications. To synthesize tunable bandpass filters with good performance, low-loss tuning elements, such as RF-MEMS, are necessary (the combination of RF-MEMS and SRRs, and their application to bandpass and bandstop filters, is discussed later).

CSRRs can also be combined with varactor diodes for the implementation of tunable components. The particle, first reported in Ref. [23], was called varactor-loaded complementary split-ring resonator (VLCSRR). Although CSRRs and SRRs are dual particles, the extension of the VLSRR concept to the VLCSRR is not obvious. The topology of the VLCSRR coupled to a microstrip transmission line, proposed in Ref. [23], is depicted in Figure 5.10. It consists of a CSRR etched in the ground plane, beneath the conductor strip, a varactor diode (with a variable capacitance C_{var}) and a lumped capacitance (C_L). The varactor diode and the lumped capacitance are series connected, and this combination is placed between the internal and external metallic

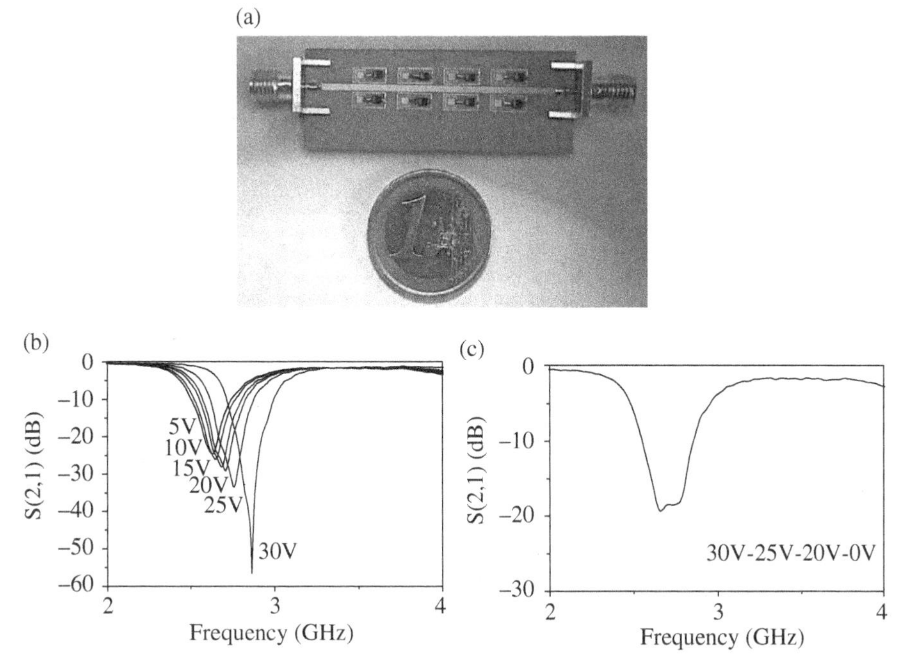

FIGURE 5.9 Fabricated four-stage VLSRR-loaded microstrip line (a), measured frequency response for identical bias voltages applied to the diode varactors (b), and measured transmission characteristics by applying the indicated voltages to the different VLSRR stages (c). Dimensions are identical to those of Figure 5.7a. Reprinted with permission from Ref. [20]; copyright 2006 IEEE.

regions of the CSRR, as can be seen in Figure 5.10. Actually, the varactor diode is soldered in the upper substrate side (see Fig. 5.10b), whereas the lumped capacitor is placed at the lower substrate side. The presence of the two indicated vias allows for the aforementioned configuration. Since the lumped-element equivalent circuit model of the CSRR is a parallel resonant tank, by adding a varactor diode between the internal and external metallic regions of the CSRR, the equivalent capacitance of the structure can be tuned. In practice, this can be achieved by placing the varactor diode on the top substrate side and connection to the bottom side through metallic vias. However, this is not so simple since the varactor diode must be biased and both diode ports are electrically connected with this arrangement. To solve this problem, the via 2 has been surrounded by a slot in the ground plane and a lumped capacitance has been placed between the ground plane and the via, as Figure 5.10a illustrates. It is clear that C_{var} and C_L are series connected. Since the equivalent capacitance of this combination is dominated by the smaller capacitance, it is convenient that C_L is at least of the same order than the maximum achievable capacitance of the varactor diode. According to these comments, the equivalent circuit model of the VLCSRR electrically coupled to the microstrip line is the one depicted in Figure 5.11, where

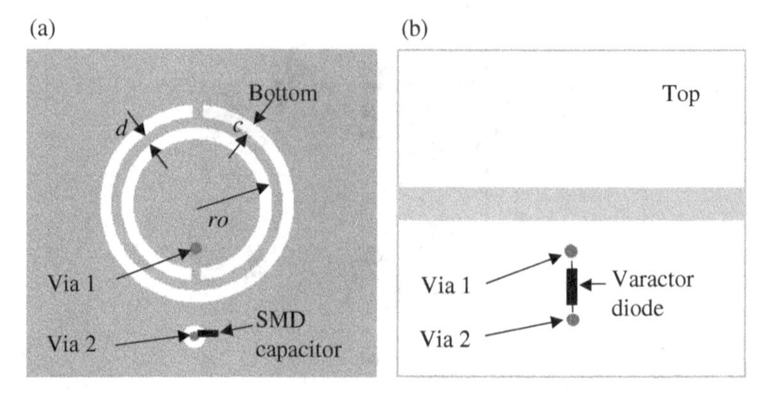

FIGURE 5.10 Topology of VLCSRR and relevant dimensions (a) bottom view and (b) top view. Reprinted with permission from Ref. [23]; copyright 2008 IEEE.

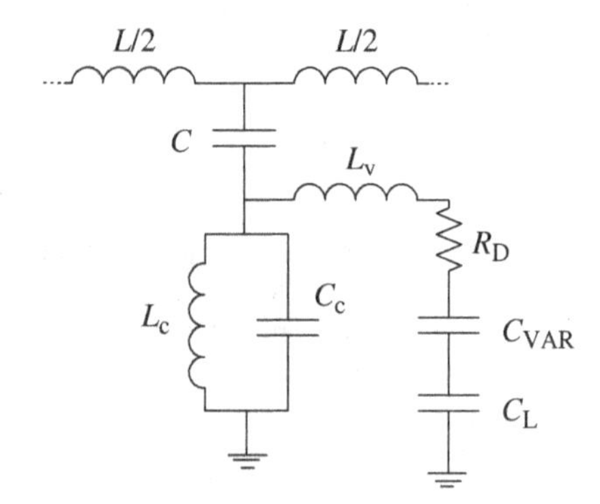

FIGURE 5.11 Lumped-element equivalent circuit model of the unit cell of a microstrip line loaded with VLCSRRs.

L_c and C_c are the inductance and capacitance of the CSRR, C is the coupling capacitance between the line and the VLCSRR, L is the line inductance, L_v is the inductance of the vias and R_D accounts for varactor losses. According to this model, to achieve significant tuning it is necessary that C_{var} dominates over C_c.

As it is, the structure of Figure 5.10 is expected to exhibit a stopband with tuning capability. This stopband is related to the presence of a transmission zero, f_z, (corresponding to that frequency that nulls the shunt reactance), where rejection is maximum. For a homogeneous structure comprising multiple unit cells, the stopband can also be interpreted as due to the high effective permittivity of the structure in the vicinity of the transmission zero, this being positive to the left of f_z and negative between f_z and the resonance frequency of the VLCSRRs.

The circuit model of Figure 5.11 is validated by comparing experimental data and circuit simulations. The fabricated device is a two-stage VLCSRR-loaded line (see Fig. 5.12) implemented on the *Rogers RO3010* substrate with thickness $h = 1.27$ mm

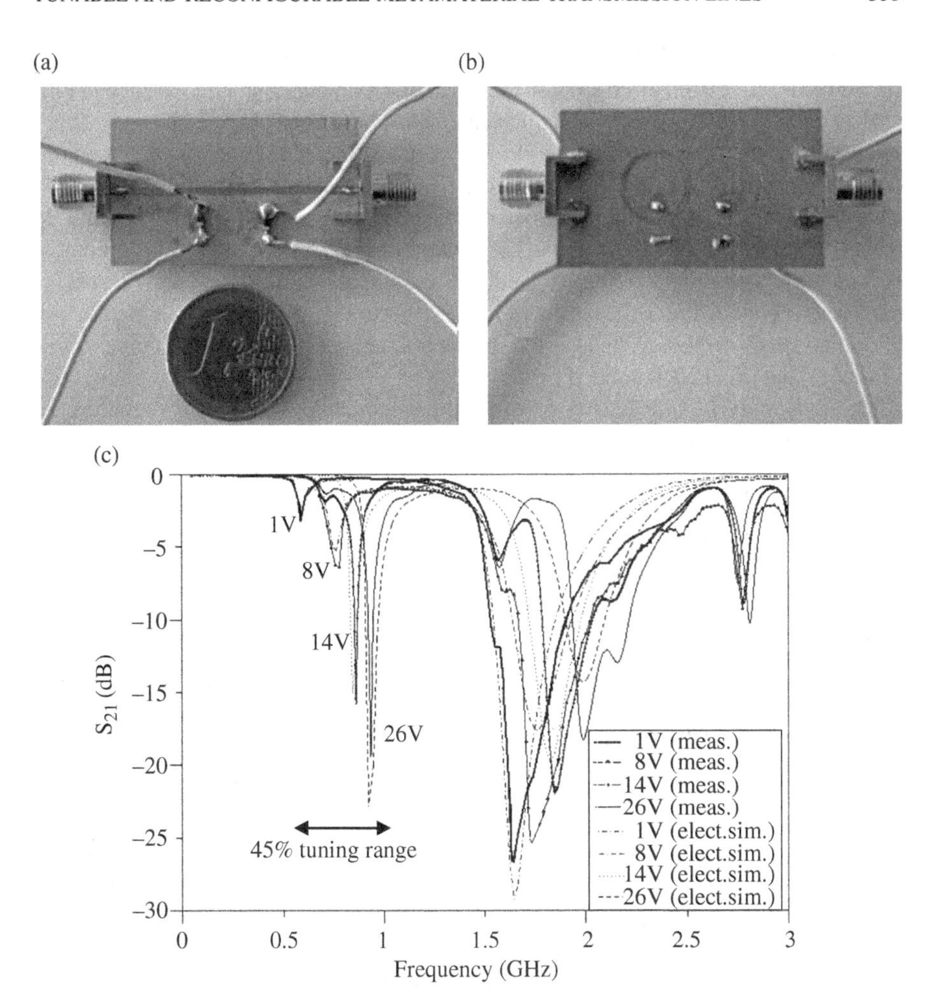

FIGURE 5.12 Photograph of the fabricated two-stage VLCSRR-loaded microstrip line. (a) Top view, (b) bottom view, and (c) measured and circuit simulation of the insertion loss for different bias polarizations. The region of interest is comprised between 0.5 GHz and 1 GHz, but the frequency response has been depicted up to 3 GHz to appreciate the validity of the model beyond that region. Circuit parameters are $L = 7.2$ nH, $C = 1.9$ pF, $C_c = 40$ pF, $L_c = 3.8$ nH, $R_D = 5.2\ \Omega$, and $L_v = 2.6$ nH. Reprinted with permission from Ref. [23]; copyright 2008 IEEE.

and dielectric constant $\varepsilon_r = 10.2$. In reference to Figure 5.10, dimensions are $c =$ 0.3 mm, $d = 0.18$ mm and $r_o = 5.2$ mm. The diameter of the vias is 0.2 mm. The host line is a 50 Ω line with a width of 1.15 mm. The diode varactors are the *BB535-Infineon Technologies* silicon tuning diodes. These devices exhibit a high capacitance ratio, namely, device capacitance is 20 pF and 2.1 pF at 1 V and 28 V reverse bias, respectively. Finally, 4.7 pF SMD-type lumped capacitances have been used for C_L. The measured transmission coefficients corresponding to different bias voltages are depicted in Figure 5.12b. Except for C_{var}, C_L, L_v, and R_D, the parameter values of the circuit model have been inferred from the parameter extraction method reported in Ref. [24], and presented in Appendix G. Such parameters have been obtained from the structures without the presence of the diode varactors and lumped capacitors and are indicated in the caption of Figure 5.12. The resistance R_D and the inductance of the vias L_v have been considered as fitting parameters. The best fit has been obtained by setting R_D and L_v to the values also indicated in the caption of Figure 5.12. The good agreement between the experimental curves and the circuit simulation in Figure 5.12c is remarkable. Notice that in the region of interest the tuning modifies the filter response, providing smaller rejection level as the central frequency decreases (a phenomenology similar to that of VLSRR-loaded lines).

By adding series capacitive gaps or shunt-inductive vias to the structure of Figure 5.10, the stopband behavior can be switched to a pass band. This possibility was demonstrated in Ref. [23], where a single cell VLCSRR-loaded line with a gap etched in the signal strip was fabricated and modeled by simply adding series capacitances to the line inductance, L, in the circuit of Figure 5.11. Although the agreement between the measured responses, for different voltages applied to the varactor, and the circuit simulations was found to be good, losses are important for using the structure as a tunable bandpass filter. This limitative aspect, related to diode losses, is identical to the encountered limitations of VLSRRs as tuning elements for bandpass filters (as indicated before). According to these comments and the results of this subsection, it is clear that VLSRR- and VLCSRR-loaded transmission lines exhibit very similar behavior.

The open counterparts of SRRs and CSRRs can also be combined with diode varactors for the implementation of tunable components [25]. Let us discuss the strategies for diode biasing in varactor-loaded open split-ring resonators (VLOSRRs) and varactor-loaded open complementary split-ring resonators (VLOCSRRs) in CPW technology. The VLOSRR is considered to be in shunt connection to the line (useful for the implementation of stopband filters [25]). The VLOCSRR, also in shunt connection, is useful for the implementation of tunable bandpass filters.

The typical topology of a shunt connected VLOSRR is depicted in Figure 5.13. The external ring is electrically connected to the central strip of the CPW transmission line, whereas the internal ring is electrically connected to ground. The varactor diode is connected between the inner ring and the metallic pad etched in the ground plane window close to the external ring. Diode polarization is achieved by applying a DC voltage to the metallic pad (in reference to the ground plane). To effectively short the upper pin of the diode to the external ring at the signal frequency, a large capacitance is required. This capacitance blocks the DC voltage, but acts as a bypass capacitor at microwave frequencies. Thus, with this topology we can electrically tune the resonance frequency of the shunt element. The equivalent circuit of the structure can be approximated by the model depicted in Figure 5.13b, where L and C are roughly the inductance and capacitance of the open split-ring resonator, and C_{VAR} is the variable capacitance of the varactor diode.

The typical topology of a shunt-connected VLOCSRR is depicted in Figure 5.14. In this case, the different metal regions of the structure are electrically shorted (similar

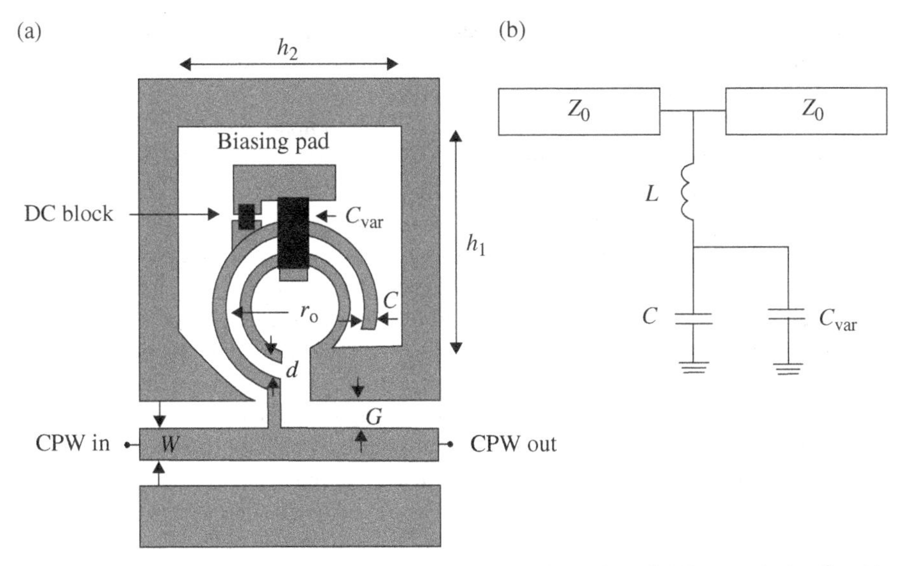

FIGURE 5.13 Typical layout of a shunt-connected VLOSRR in a CPW transmission line (a), and circuit model (b). Only one half of the CPW structure is shown. Reprinted with permission from Ref. [25]; copyright 2011 IET.

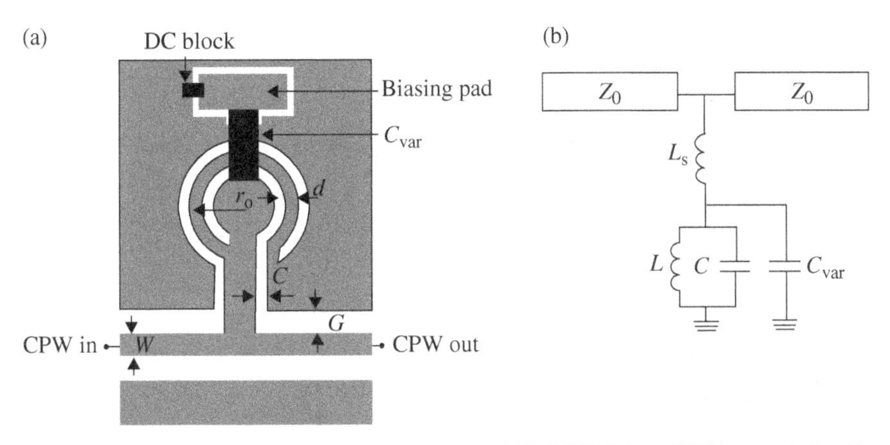

FIGURE 5.14 Typical layout of a shunt-connected VLOCSRR in a CPW transmission line (a) and circuit model (b). Only one half of the CPW structure is shown. Reprinted with permission from Ref. [25]; copyright 2011 IET.

to VLCSRRs). For this reason, we must etch an isolated bias pad, as shown in Figure 5.14. The varactor diode is soldered between the inner metallic region of the resonator and the bias pad. Again, a bypass capacitor is required in order to short the bias pad to the ground plane at microwave frequencies. The equivalent circuit model of the structure is that shown in Figure 5.14b, where L and C are roughly the inductance and capacitance of the open complementary split-ring resonator, C_{VAR} is the variable capacitance of the varactor diode and L_s represents the inductive strip between the resonator and the central line of the CPW.

Both equivalent circuit models are validated by comparison of the EM response (actually the co-simulated electric and EM response) and the electrical (circuit) response for different voltage levels applied to the diode varactor. The co-simulations can be inferred by means of the *Agilent Momemtum* EM simulator. To this end, the EM responses of the structures, by considering internal ports in those positions where the lumped elements are connected, are obtained. Then, these results are exported to the circuit simulator, and the lumped components (i.e., the capacitances of the varactors at the considered voltages) are added to get the final results. In practice, these co-simulations were carried out by considering commercial diode model libraries (those of the *Infineon BB833* varactor diodes) and different applied voltages. The circuit simulations were obtained by means of the circuit simulator included in *Agilent ADS*. This comparison is presented in Figures 5.15 and 5.16 for a CPW structure loaded with a single VLOSRR and VLOCSRR, respectively, and shows the excellent accuracy of the model to describe the behavior of the tunable cells (this good agreement is relevant for the design and characterization of tunable filters). For the structure of Figure 5.13, applied voltages between 0 V and 28 V were considered, corresponding to varactor capacitance values of 9.3 pF and 0.75 pF, respectively, for the used varactors (Fig. 5.15 depicts the responses for the indicated voltages). The inductance L and capacitance C of the circuit model (Fig. 5.13b) were inferred by curve fitting, with the result of $L = 6.3$ nH and $C = 0.75$ pF. These values are of the same order than

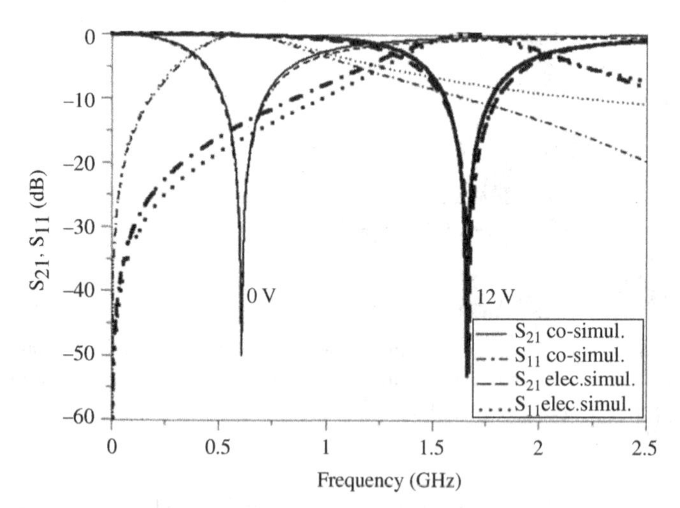

FIGURE 5.15 Comparison between circuit simulation and co-simulation (EM-electric) for different voltages applied to the VLOSRR of Figure 5.13. Dimensions (in reference to Fig. 5.13) are $W = 2.1$ mm, $G = 0.7$ mm, $r_0 = 2.29$ mm, $c = d = 0.38$ mm, $h_1 = 6.14$ mm, and $h_2 = 5.9$ mm. The circuit elements are $L = 6.3$ nH and $C = 0.75$ nH. Reprinted with permission from Ref. [25]; copyright 2011 IET.

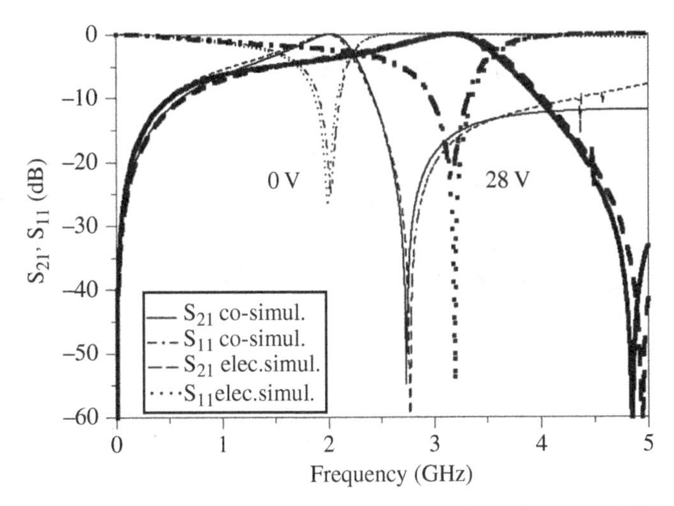

FIGURE 5.16 Comparison between circuit simulation and co-simulation (EM-electric) for different voltages applied to the VLOCSRR of Figure 5.14. Dimensions (in reference to Fig. 5.14) are $W = 1.68$ mm, $G = 3.7$ mm, $r_0 = 1.7$ mm, and $c = d = 0.3$ mm. The circuit elements are $L = 3.8$ nH, $C = 1.25$ nH and $L_s = 0.7$ nH. Reprinted with permission from Ref. [25]; copyright 2011 IET.

the values of capacitance and inductance of the isolated OSRR. For the structure of Figure 5.14, the same applied voltages were considered. The resulting circuit elements (of the circuit model of Fig. 5.14b), inferred by curve fitting, are in this case $L = 3.8$ nH, $C = 1.25$ pF and $L_s = 0.7$ nH.

Using the previous VLOSRRs- and VLOCSRR-loaded CPWs, tunable stopband and bandpass filters can be implemented. In this case, the filters were designed/implemented by coupling the resonators by means of admittance inverters [25], following the well-known approach reported in many textbooks devoted to microwave filters [26], and applied to the design of non-tunable bandpass and bandstop filters based on these open resonators [27, 28]. Notice, however, that tuning the resonance frequency of the particles would require tuning also the phase of the transmission line sections present between adjacent resonators, in order to achieve the admittance inversion functionality at the (tunable) filter central frequency. This strategy would require additional tuning elements and was discarded in the implementations reported in Ref. [25]. The designed tunable bandstop and bandpass filters, based on VLOSRRs and VLOCSRRs, respectively, are depicted in Figures 5.17 and 5.18, respectively (see Ref. [25] for further design details). The devices were fabricated on the *Rogers RO3010* substrate with dielectric constant $\varepsilon_r = 10.2$, thickness $h = 1.27$ mm and loss tangent $\tan\delta = 0.0023$. Relevant device dimensions are indicated in the caption of those figures. The tuning elements are the *Infineon BB833* varactor diodes (with a capacitance varying in the range 9.3–0.75 pF for voltages between 1 V and 28 V), and the bypass capacitors are the SMD *Phycom* capacitances with a value of 50 pF.

(a)

(b)

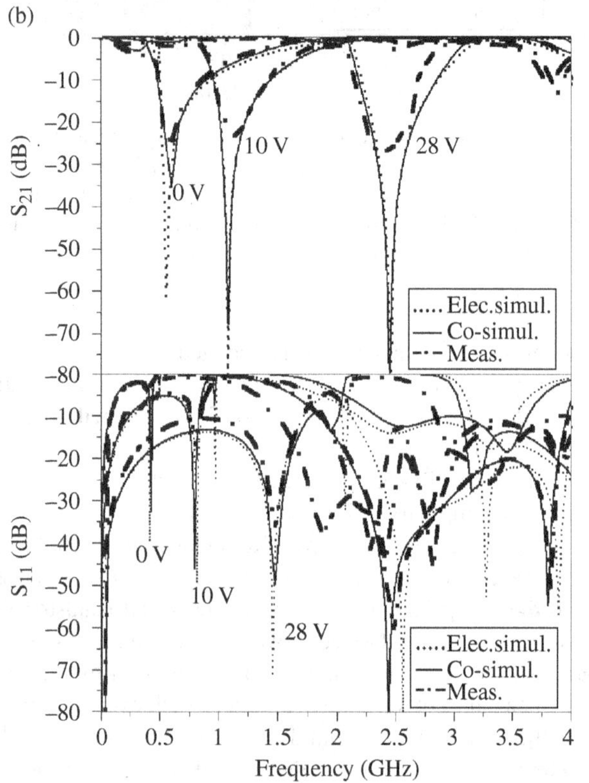

FIGURE 5.17 Photograph of the fabricated tunable band-stop filter (a) and comparison between co-simulated, pure electrically simulated and measured frequency responses for different applied voltages (b). Dimensions are $W = 2.1$ mm, $G = 0.7$ mm, $r_0 = 2.29$ mm, $c = d = 0.38$ mm, $h_1 = 6.14$ mm, and $h_2 = 5.9$ mm, and total device length is $l = 23$ mm. The different ground plane regions have been electrically connected through backside strips and vias to avoid the slot mode of the CPW structure. Reprinted with permission from Ref. [25]; copyright 2011 IET.

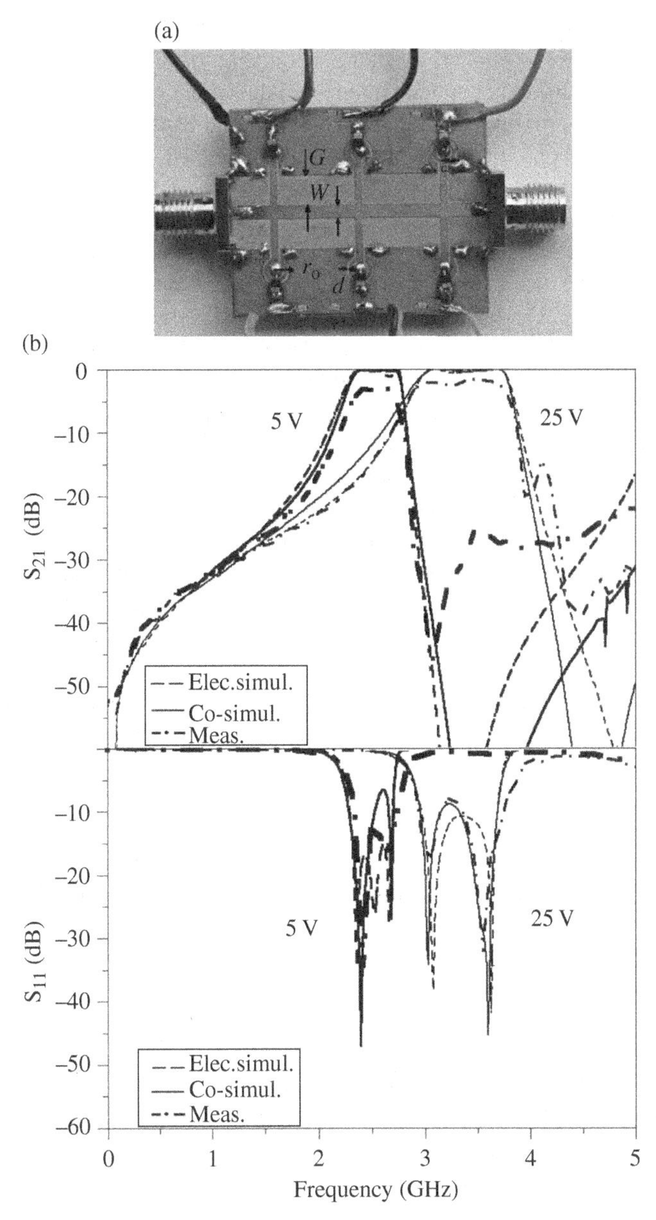

FIGURE 5.18 Photograph of the fabricated tunable bandpass filter (a) and comparison between co-simulated, pure electrically simulated and measured frequency response for different applied voltages (b). Dimensions are $W = 1.68$ mm, $G = 3.7$ mm, $r_0 = 1.7$ mm, $c = d = 0.3$ mm, and total device length is $l = 32$ mm. The different ground plane regions have been electrically connected through backside strips and vias to avoid the slot mode of the CPW structure. Reprinted with permission from Ref. [25]; copyright 2011 IET.

Prior to fabrication, the frequency response of the filters at different applied voltages was inferred from co-simulations (electric and EM). The results of the co-simulations (at different applied voltages) of the whole structures are also depicted in Figures 5.17 and 5.18 for the tunable stopband and bandpass filters, respectively. The pure electrical (circuit) simulations are also depicted in Figures 5.17 and 5.18, together with the measured frequency responses of the fabricated filters. The good agreement between the circuit simulations, the co-simulations and the measurements is remarkable, and further supports the validity of the reported models. The higher tuning range for the bandstop filter (as compared to that of the bandpass filter) is simply because the tunable capacitance clearly dominates over the resonator's capacitance in the considered OSRRs. In the OCSRRs of the band-pass filters, the tunable capacitance and resonator's capacitance are comparable and the tuning range is smaller. For the bandpass filter, in-band losses are present, but this level of losses may be acceptable for a tunable bandpass filter. Notice, however, that bandwidth is not controlled with the reported approach, and it decreases as the central frequency decreases, as Figure 5.18 illustrates.

Other tunable components based on varactor-loaded split rings have been reported. For instance, a tunable 1:2 power divider based on a VLCSRR-loaded line acting as impedance inverter (with 35.35 Ω impedance) is reported in Ref. [29]. To achieve the inverter functionality over the tuning interval, the structure is implemented with two additional tuning elements, acting as series connected variable capacitances. By this means, the line impedance and phase (90°) can be controlled over the tuning interval (nevertheless, the design procedure is complex and the author suggests paper [29] to the interested reader).

5.3.1.2 Tunable SRRs and CSRRs Based on RF-MEMS and Applications As anticipated in Section 5.2.2, RF-MEMS are very interesting components to implement tunable and reconfigurable split rings. As compared to varactor diodes, RF-MEMS can operate at higher frequencies and exhibit smaller levels of losses. The use of RF-MEMS for the implementation of tunable SRR-based metamaterials was reported in several papers (see, e.g., Refs [30, 31]). The purpose here is the implementation of transmission lines loaded with RF-MEMS-based tunable split rings. To this end, two main strategies have been considered: (1) split rings with RF-MEMS loading, where tuning is achieved through electrostatic actuation on a set of MEMS switches loading the resonators [32, 33], and (2) cantilever-type SRRs, where the arms of the SRR are deflectable [34]. Let us now discuss the principles behind these approaches, and some applications to tunable filters.

Split Rings with RF-MEMS Loading Concerning the first approach, both SRR [33] and CSRRs [32] loaded with RF-MEMS switches have been implemented. SRR tunability can be achieved by adding cantilever-type RF-MEMS switches, which are composed of one anchor and one movable beam suspended above an actuation electrode. The SRRs and the RF-MEMS can be combined following different configurations. In the configuration I (Fig. 5.19a), the external ring is used as DC ground electrode and is the anchor of the cantilever beams, while the internal ring,

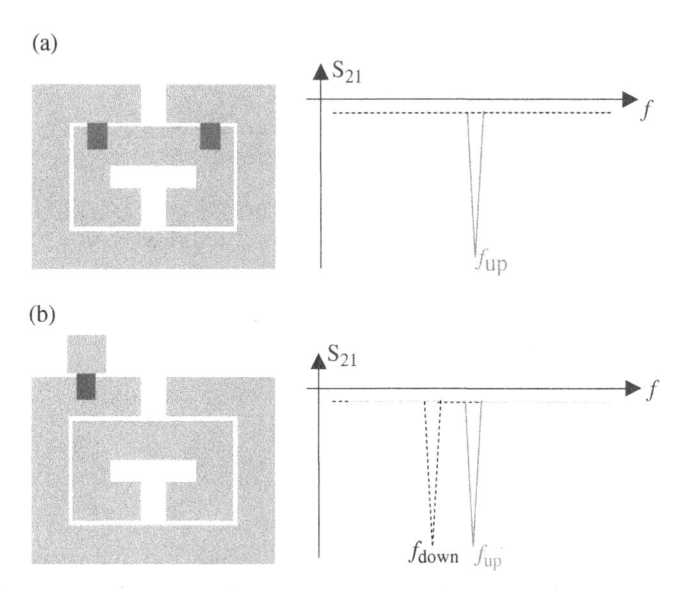

FIGURE 5.19 Illustration of cantilever-type RF-MEMS capacitive switches (represented by dark rectangles) loading a square SRR for configuration I (a) and II (b), as described in the text. The typical frequency responses of the tunable resonators coupled to a transmission line when all switches are at the up-state (solid lines) and when all switches are at the down-state (dotted lines), are also depicted. Reprinted with permission from Ref. [33]; copyright 2012 IOP Publishing.

under them and covered by a thin dielectric layer, acts as the DC actuation electrode. When the cantilever beams are at the up-state, the resulting capacitances formed with the internal ring are low. When they are actuated (down-state), the coupling between rings dramatically increases and this leads to a very large shift of the resonance frequency of the resonator. In the configuration II (Fig. 5.19b), the cantilever beam is anchored in a metallic patch and suspended over the outer ring. The switch actuation results in a resonance frequency decrement, and the frequency shift depends essentially on the metallic patch dimensions and location around the resonator.

The previous configurations can be applied to the implementation of compact tunable metamaterial-based filters at X-band, as will be shown in the illustrative examples. Extensive details on the fabrication of similar MEMS actuators are described elsewhere [35, 36]. Briefly, for the specific structures reported later, the actuation electrodes are realized by the thermal evaporation of a Cr/Au (60/1200 Å) thin layer on a 250 μm-thick Sapphire substrate (with dielectric constant $\varepsilon_r = 9.8$). They are covered by a 0.4 μm-thick Al_2O_3 dielectric layer deposited by plasma-enhanced chemical vapor deposition (PECVD). The alumina dielectric layer serves as electrical insulator between the lower electrode (outer ring of the SRR structure) and the MEMS cantilever beam (the upper moveable electrode, as shown in Fig. 5.2). It follows the lift-off of a 50 nm-thick doped Carbon layer, deposited by reactive laser ablation, to realize

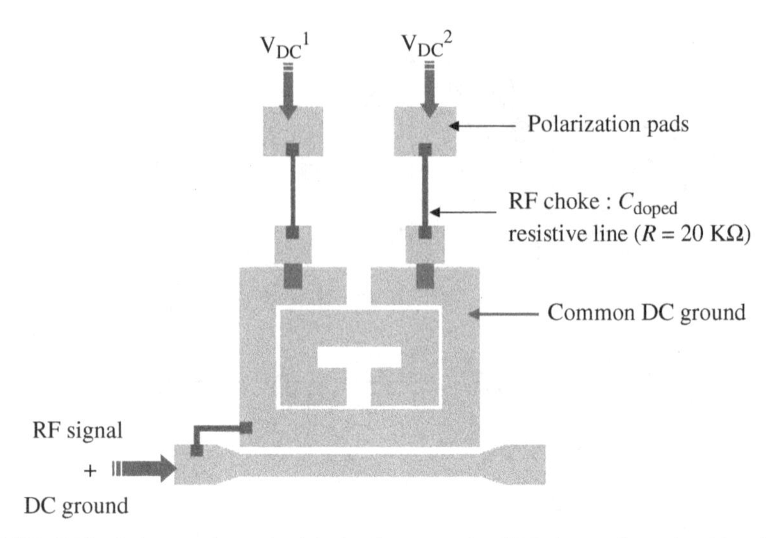

FIGURE 5.20 DC actuation principle in the case of a SRR in configuration II, with two cantilever beams independently actuated. Reprinted with permission from Ref. [33]; copyright 2012 IOP Publishing.

the 20 KΩ resistive lines (see Fig. 5.20 for the specific configuration II with the SRR coupled to a microstrip transmission line). The suspended parts of the structure (moveable cantilever beam) are defined by patterning a 0.5 μm-thick sacrificial polymethylglutarimide (PMGI) resist. The metallization is done using the Cr/Au seed layer which is gold-electroplated up to 1.5 μm. Next, a 90 Å Cr stress layer is deposited and patterned, in order to provide an appropriate stress gradient in the foldable areas. Finally, the device is realized and dried in a critical point drying system for avoiding stiction to the dielectric of the suspended structures. As illustrated in Figure 5.20, the structure integrates carbon-doped resistive lines and metallic polarization pads for the electrostatic actuation of the RF-MEMS switches.

On the basis of configuration I, stopband filters with electronically controllable number of poles can be implemented [33]. By this means, it is possible to tune the filter central frequency and the bandwidth. The idea is to couple multiple resonators, with slightly different resonance frequency, to the host line [37]. If the resonators are uncoupled, each resonator contributes with a filter pole (transmission zero) and bandwidth can be tailored. In the framework of this approach, it is clear that filter characteristics can be tuned by removing one or more resonators (and hence the corresponding poles). However, by using MEMS switches in combination with SRRs according to configuration I, the poles can be removed without the need for resonator removal. We simply need to actuate the MEMS, and the pole (or poles) of the corresponding SRR will be largely shifted. Following this idea, a four-pole reconfigurable bandstop filter, consisting of a 50 Ω microstrip transmission line loaded with four pairs of RF-MEMS-loaded SRRs, was designed and fabricated (Fig. 5.21) [33]. The difference between SRRs called A, B, C, and D is the side length H_i of the external ring (Fig. 5.21b), where $H_A = 1430$ μm, $H_B = 1475$ μm, $H_C = 1530$ μm, and

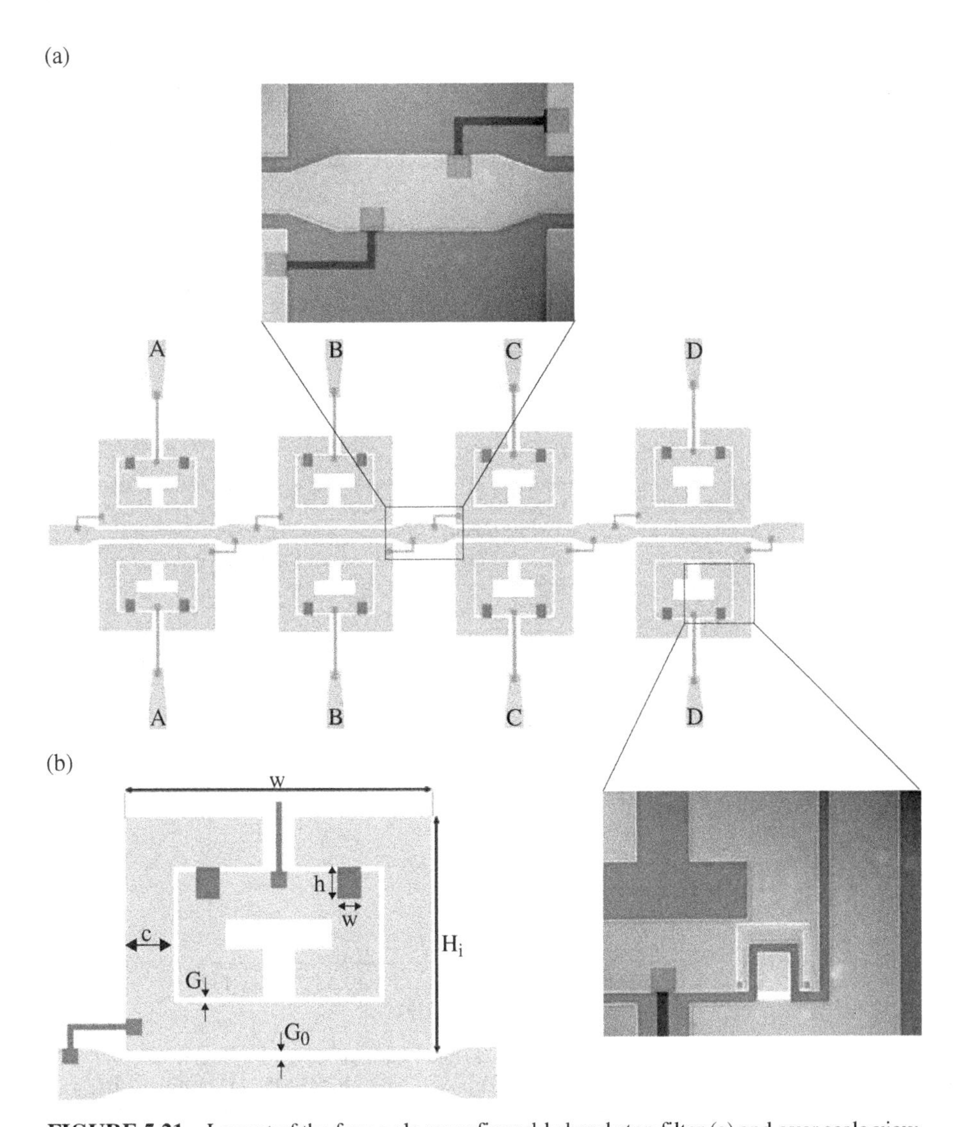

FIGURE 5.21 Layout of the four-pole reconfigurable band stop filter (a) and over scale view of one RF-MEMS-loaded SRR (b). The total size of the device is 6.4 × 14 mm^2. The dimensions of the cantilever-type MEMS are $h \times w = 200\ \mu\text{m} \times 150\ \mu\text{m}$. Width and distance between rings are $C = 300\ \mu\text{m}$ and $G = 30\ \mu\text{m}$. The gap between SRRs and the microstrip line is $G_0 = 50\ \mu\text{m}$. The side length of the SRRs in the longitudinal direction is $W = 1940\ \mu\text{m}$. Zoom photographs of the indicated parts of the fabricated device are also shown. Reprinted with permission from Ref. [33]; copyright 2012 IOP Publishing.

$H_D = 1580\ \mu\text{m}$. Without electrostatic actuation, this configuration provides a bandstop behavior with four poles corresponding to the resonance frequencies f_A (10.36 GHz), f_B (10.15 GHz), f_C (9.92 GHz), and f_D (9.73 GHz) of the SRRs of cells A, B, C, and D,

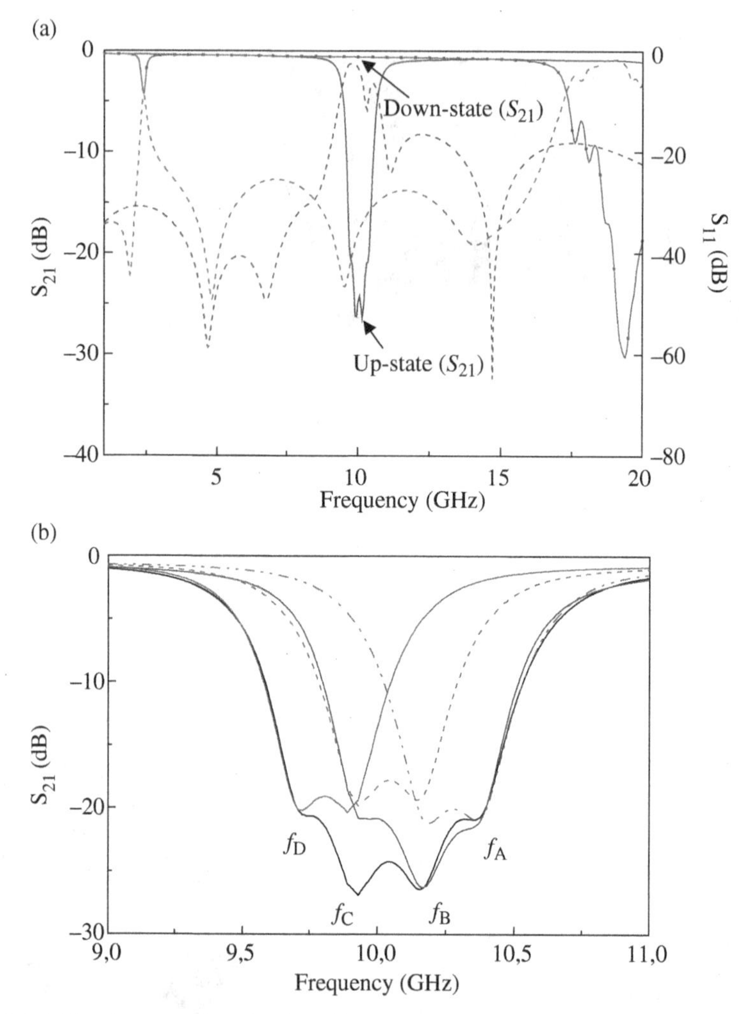

FIGURE 5.22 Simulated insertion (solid lines) and return (dashed lines) loss of the four-pole reconfigurable filter when all MEMS are at up-state and all MEMS are at down-state (a). Simulated responses of the device for different combinations of switches actuated (b). Reprinted with permission from Ref. [33]; copyright 2012 IOP Publishing.

respectively (Fig. 5.22). As shown in Figure 5.21a and b, the common DC ground signal is supplied to all external rings through the transmission line and resistive lines, while each internal ring acts as a DC-independent electrode. The reconfigurable filter, designed to operate at the X-band, was simulated by using the *Agilent Momentum* EM simulator. The ON/OFF RF-MEMS switches were designed to provide a ratio between up-state and down-state capacitances of 10, which leads to a shift of the resonance frequencies of the resonators from X-band to L-band. Owing to the actuation of switches and taking into account that both SRRs of one cell must always present the same resonance frequency, we obtain a 4-bit (called A, B, C, and D) reconfigurable filter. The simulated S-parameters of the device are displayed in Figure 5.22a. When all switches are at up-state, the rejection is higher than 20 dB in a 0.7 GHz range. When they are all at down-state, insertion losses are less than 1 dB and return losses higher than 20 dB in a range from 3 GHz to 16 GHz. The curves in Figure 5.22b present other simulated filter responses corresponding to different bit-combinations. By varying the number of switches actuated, we can digitally tune the filter bandwidth and central frequency. The measured insertion and return losses of the filter with all MEMS at up-state (non-actuated) are presented in Fig. 5.23a and compared with the full wave simulations. As expected, the filter exhibits a four pole rejection band around 10 GHz and the rejection is higher than 20 dB on a 0.72 GHz frequency range. There is good agreement between simulation and experiment, except that out of the stopband measured insertion losses are higher and return losses are lower than those predicted by the simulation. This is due to the connection between the transmission line of the filter and the two SMA connectors. Other measured filter responses corresponding to different combinations of switches simultaneously actuated with 60 V are depicted in Figure 5.23b. The number of poles of the stopband corresponds to the number of non-actuated switches. With these results, the digital reconfigurability principle is validated.

The RF-MEMS-loaded SRRs of configuration II can be applied to the design of tunable bandpass filter. The topology of the considered filter, depicted in Figure 5.24, consists of a pair of coupled SRRs fed by 50 Ω microstrip transmission lines. In the proposed configuration, the electric coupling between these two adjacent resonant elements results in the bandpass behavior of the filter [38]. Such configuration provides also transmission zeros, present at both sides of the pass band, which are relevant for frequency selectivity improvement and are caused by the feeding structure, as discussed in Ref. [39]. Since the central frequency of the filter depends directly on the resonance frequency of the SRRs, the position of the filter pass band can be tailored by tuning the SRRs. On the other hand, bandwidth is mainly controlled by the coupling level between SRRs, which is scarcely dependent on MEMS actuation. Therefore, the proposed tunable filters are specifically designed to tune the filter central frequency. For the untuned state (nonactuated MEMS), there is a systematic approach for the design of these type of filters. That is, given filter specifications (bandwidth, order, central frequency, and minimum in-band return losses), the inter-resonator coupling coefficients and external quality factors are determined; and from these values, inter-resonator distance and the position of the feeding lines are determined [26]. However, since the main relevant aspect of this section is to highlight novel tuning concepts for artificial transmission lines based on RF-MEMS switches, rather than pursuing a specific frequency response for the considered type of tunable filter, the inter-resonator distance has been tailored in order to obtain a filter response with a fractional bandwidth (nonactuated switches) of roughly 15%.[7] In

[7] Indeed, the filter depicted in Figure 5.24, rather than being based on a tunable artificial transmission line, it is implemented following the well-known coupled resonator's approach. Nevertheless, it constitutes a good example to illustrate the potential of RF-MEMS-loaded SRRs in the design of tunable components, and for this reason it has been included in this section.

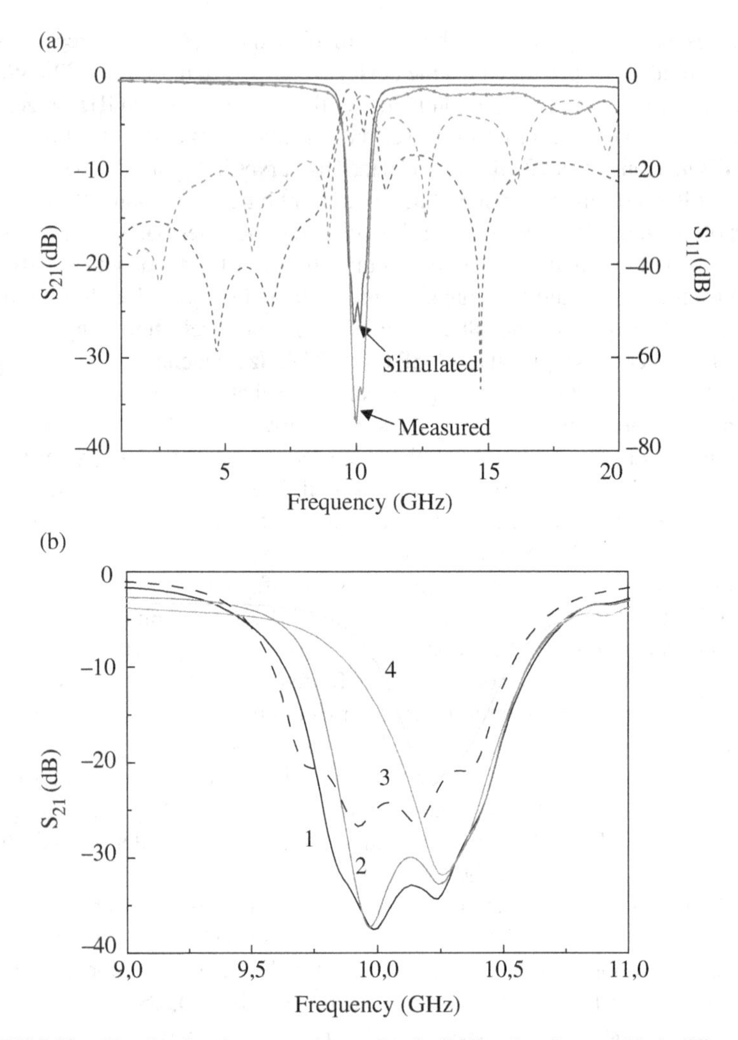

FIGURE 5.23 Simulated and measured insertion (solid lines) and return (dashed lines) loss of the four-pole reconfigurable stopband filter when all switches are at up-state (a). Simulated (dashed line) and measured (solid lines) responses of the device for different combinations of switches actuated (b). Measurements indicated as 1, 2, 3, and 4 correspond to bits ABCD set to "0000," "0001," "0011," and "1011," respectively, with "0" corresponding to MEMS at up-state (i.e., nonactuated) and "1" corresponding to MEMS at down-state. Notice that the bit combinations of (b) do not have direct correspondence to those of Figure 5.22b, except for the cases "0000," "0001," "0011." Reprinted with permission from Ref. [33]; copyright 2012 IOP Publishing.

order to improve the frequency selectivity of the filter, rather than considering a symmetric topology, a skew symmetric 0° feed configuration has been used, as reported in Ref. [39]. This produces one transmission zero at each side of the pass band, at those frequencies where the two signal paths between the feed point and the edges of the

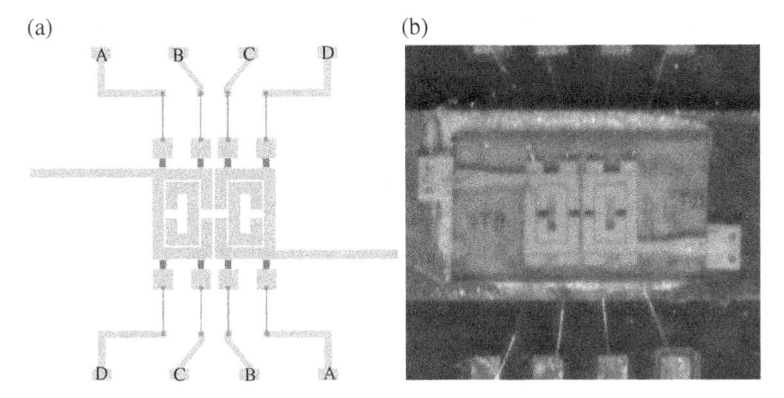

FIGURE 5.24 (a) Layout of the four-bit tunable bandpass filter. The geometrical parameters illustrated in Figure 5.21b are $h \times w = 200 \times 150\ \mu\text{m}$, $C = 300\ \mu\text{m}$, $G = 30\ \mu\text{m}$, and $W \times H = 2190\ \mu\text{m} \times 1490\ \mu\text{m}$. (b) Photograph of the fabricated device. Reprinted with permission from Ref. [33]; copyright 2012 IOP Publishing.

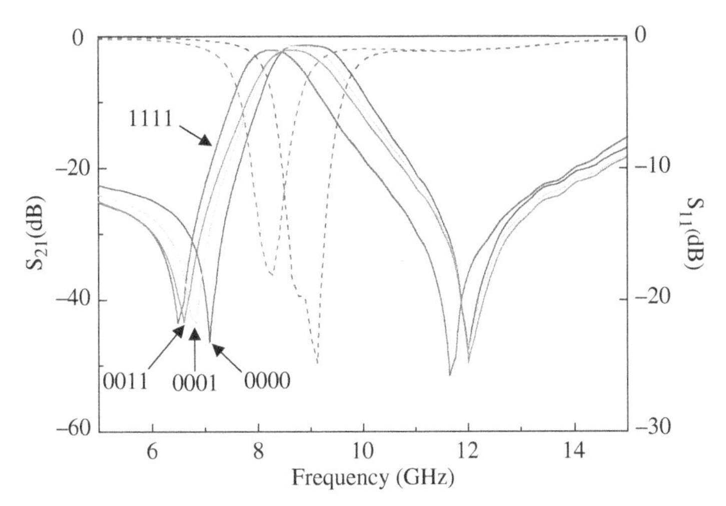

FIGURE 5.25 Measured insertion (solid line) and return (dashed line) loss of the designed digitally tunable bandpass filter for four different switching states. The switching states are "0000," "0001," "0011," and "1111," where the bit sequence is for the switches ABCD, and "0" and "1" stands for nonactuated and actuated, respectively. The return loss is only shown for the extreme states "0000" and "1111." Reprinted with permission from Ref. [33]; copyright 2012 IOP Publishing.

external ring are about one-quarter wavelength. The dimensions of the metallic patches to which the RF-MEMS are anchored determine the tuning range. The designed prototype was merely fabricated as a proof of concept; hence, the metallic patches were chosen with arbitrary dimensions. The measured frequency responses corresponding to four different states are depicted in Figure 5.25. The responses at the two extremes correspond to all switches ON and OFF. From this, the tuning range is found to be roughly 10%. Filter performance is good, with measured in-band insertion losses (central frequency) of 1.3 dB and 2 dB at the two extremes of the tuning range, and return losses better than 35 dB in both cases. It is found that MEMS actuation does also shift down the transmission zero frequencies.

Tunable CSRRs using fixed-fixed beam RF-MEMS have been reported in Ref. [32]. The CSRRs are etched in the central strip of a CPW, and the RF-MEMS are implemented on top of them, as Figure 5.26 illustrates. Through electronic actuation, the MEMS are bended down, modifying the capacitance of the CSRRs and hence the resonance frequency. Thus, this configuration is useful for the implementation of tunable band stop filters. The prototype reported in Ref. [32], depicted in Figure 5.27, exhibits a tuning range of 20% and operates at the Q-band (for the interested reader, the details of the fabrication process are given in [32]). The

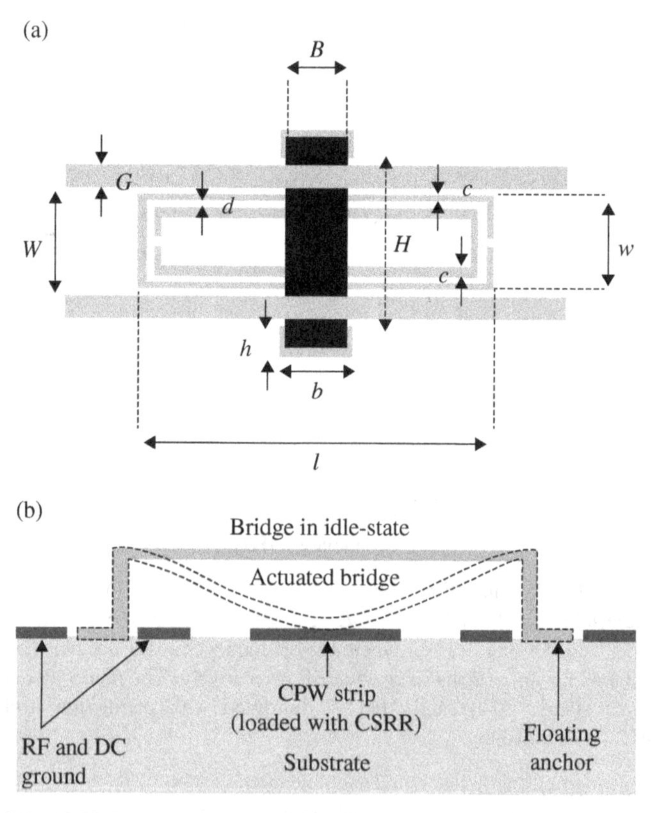

FIGURE 5.26 (a) Unit cell of the CSRR/RF-MEMS loaded CPW, with slot regions of the CPW depicted in gray, and relevant dimensions and (b) cross section of a CPW with a RF-MEMS bridge. The down-state corresponds to the application of an actuation voltage to the strip line of the CPW; in the up-state, no actuation voltage is applied. Reprinted with permission from Ref. [32]; copyright 2007 IET.

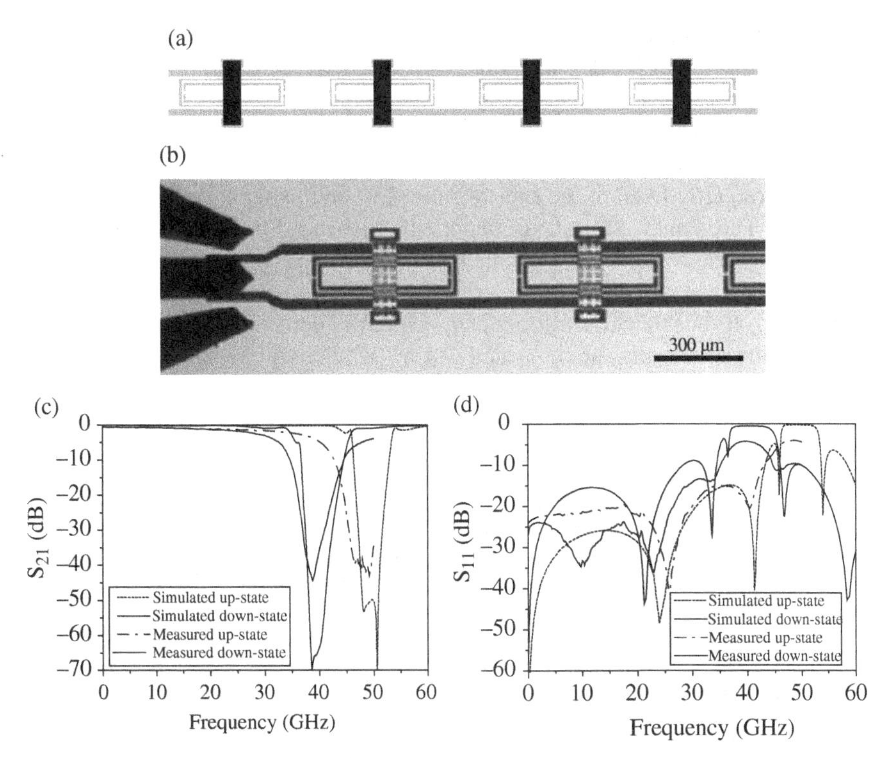

FIGURE 5.27 Layout of the fabricated tunable stopband filter (a), microphotograph of the first two stages of the filter, including RF probes (b), simulated and measured insertion losses (c), and simulated and measured return losses (d). The simulations were done by considering plate heights of 0.5 μm and 2 μm for the down- and up-state, respectively. Reprinted with permission from Ref. [32]; copyright 2007 IET.

dimensions of the CSRRs are (in reference to Fig. 5.26) $c = d = 10$ μm, $l = 480$ μm and $w = 130$ μm. CPW dimensions are: strip width $W = 150$ μm and slot width $G = 30$ μm. Finally, the geometry of the MEMS bridges is: $B = 80$ μm, $b = 100$ μm, $h = 40$ μm and $H = 290$ μm. The structure is a 4-stage periodic device where the distance between adjacent CSRRs is 220 μm. The simulated (by means of the *Agilent Momentum* by excluding losses) and measured S-parameters of the device are also depicted in Figure 5.27. As expected, the structure exhibits stopband behavior with tuning capability. The central frequency of the stopband is varied between 39 GHz and 48 GHz for corner actuation voltages of 17 V (down-state) and 0 V (up-state). This corresponds to a tuning range of roughly 20%. Measured rejection in the stopband is good ($IL > 40$ dB), whereas insertion losses in the allowed band are very small.

According to the results presented in this subsection, relative to SRRs and CSRRs loaded with RF-MEMS, it is clear that it is possible to implement tunable filters with very reasonable performance by using transmission lines loaded with these tunable metamaterial resonators. As compared to diode varactors, RF-MEMS not only decrease the level of losses (especially critical in bandpass filters) but also allow for the implementation of tunable filters at very high frequencies.

Cantilever-Type SRRs An alternative approach for the implementation of tunable resonators consists of using the RF-MEMS as part of the SRR [34]. The rings forming the SRRs are partly fixed to the substrate (anchor) and partly suspended (up-curved cantilever). Through electrostatic actuation, the suspended parts are deflected down, the distributed capacitance between the pair of coupled rings is modified, and hence the resonance frequency of the SRR can be electrically tuned. A typical top view of the cantilever-type tunable SRR (rectangular shaped) is depicted in Figure 5.28a. The movable parts of the rings are indicated in gray. Obviously, we can arbitrarily select

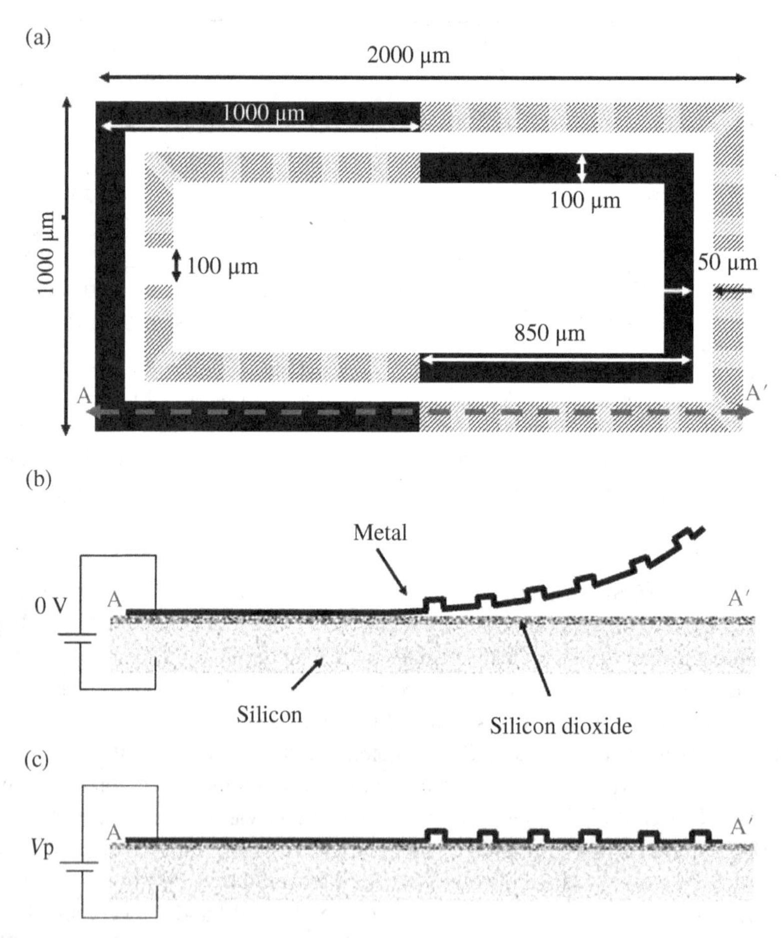

FIGURE 5.28 Tunable SRR based on cantilever-type RF-MEMS. (a) Top view with relevant dimensions. Black and gray parts correspond to anchors and suspended parts (including corrugations), respectively; (b) cross section in the up-state; (c) cross section in the down-state. Reprinted with permission from Ref. [34]; copyright 2011 IEEE.

the movable portion of each ring, which has direct influence on the tuning range. Figure 5.28b and c depict the cross-sectional view of the anchor and the cantilever, without (up-state) and with (down-state) electrostatic actuation, respectively. The tuning principle was validated in Ref. [34] by coupling the SRR of Figure 5.28 to a 50 Ω microstrip line (Fig. 5.29). Since we can independently actuate on both the internal and external ring of the tunable SRR, four different states arise. The measured transmission coefficients corresponding to the four states are also depicted in Figure 5.29. Without actuation (both cantilevers at up-state), the resonance frequency of the SRR is 13.42 GHz. It decreases to 11.45 GHz by actuating the outer ring or to 9.78 GHz by actuating the inner ring. The smaller resonance frequency (9.34 GHz) is that

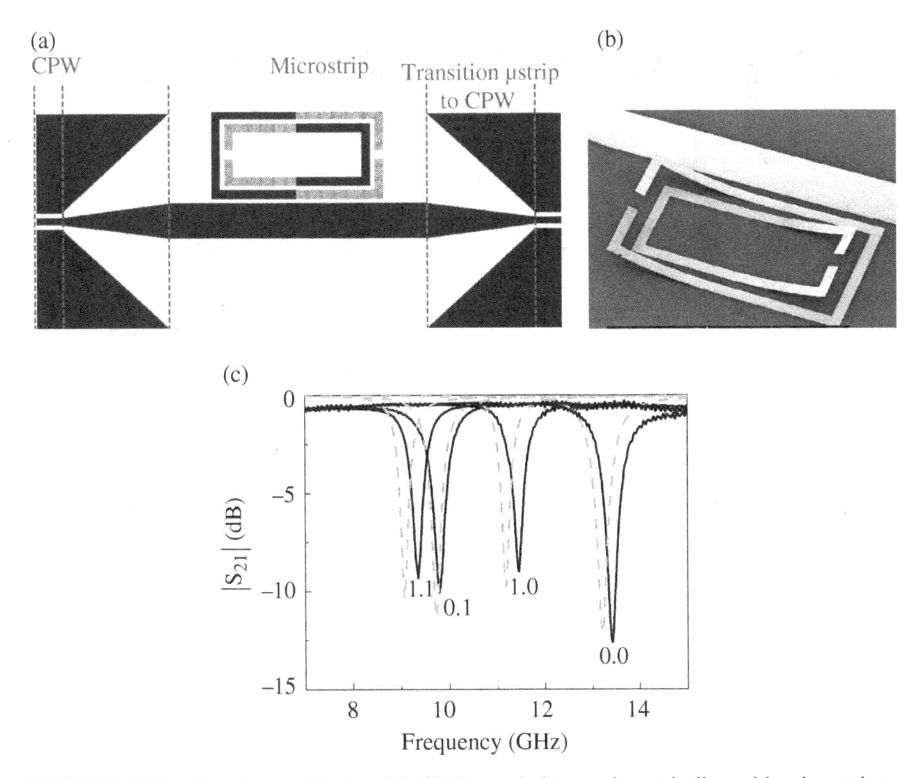

FIGURE 5.29 Topology of the tunable SRR coupled to a microstrip line with microstrip to coplanar waveguide transition (a), photograph of the nonactuated SRR (b), and measured (solid lines) and simulated (dashed lines) frequency response of the structure for the four different states (c). The separation between the SRR and the microstrip line is 50 μm, and the width of the microstrip line is 400 μm. The applied voltage for either ring actuation is 30 V. The state of the rings is indicated, where "1" (ring actuation) stands for down-state and "0" for up-state, and the first bit corresponds to the inner ring. Details on the fabrication process, substrate parameters and simulation are given in [34]. Reprinted with permission from Ref. [34]; copyright 2011 IEEE.

corresponding to the two rings in the down-state, as expected on account of the larger distributed capacitance between the two rings of the SRR.

By cascading the proposed MEMS-based SRRs in a microstrip transmission line, tunable stopband filters can be implemented (the rejection level can be controlled by the number of stages). Two prototype devices are depicted in Figure 5.30, where the

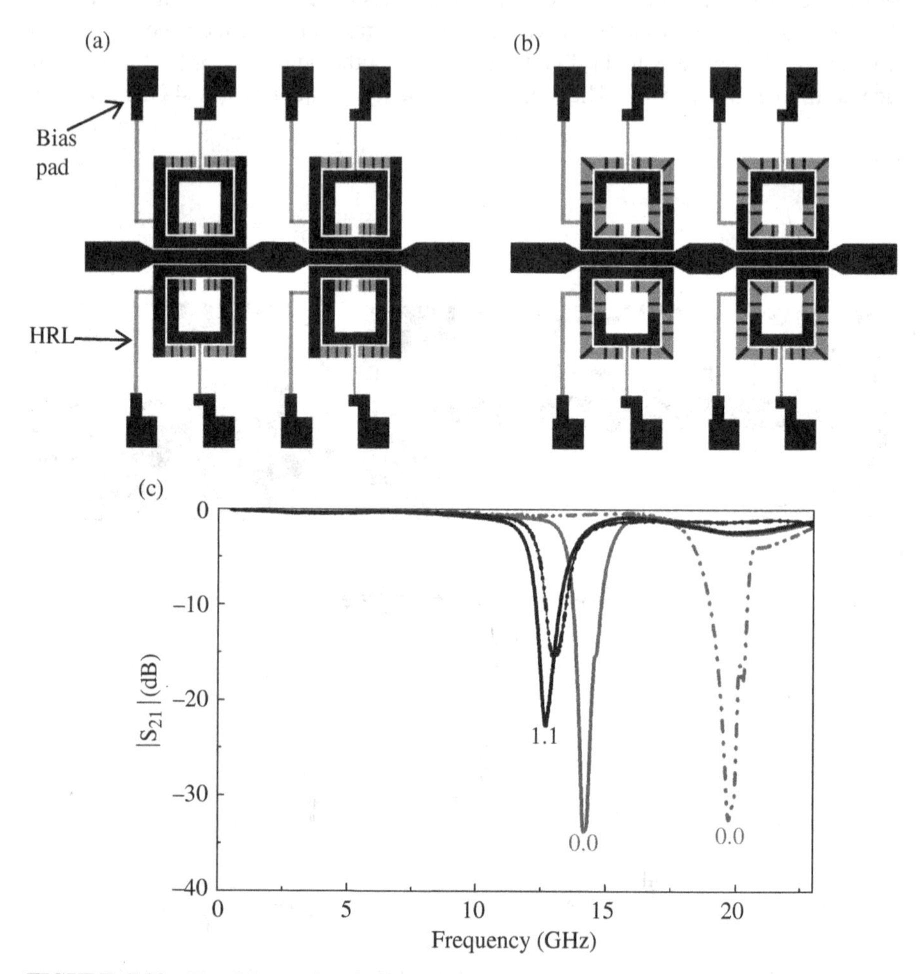

FIGURE 5.30 Tunable stopband filters based on square-shaped short (a) and long (b) cantilever-type SRRs, and measured transmission coefficients for the extreme switching states (c). SRR side length is 1200 μm, ring width 150 μm and ring separation 30 μm. The separation between the SRR and the microstrip line is 25 μm. The actuation voltages are applied to the rings through the bias pads and high resistive lines (HRLs). Solid lines correspond to the filter of (a) amd dash-dotted lines correspond to the filter of (b). As frequency decreases, rejection is reduced due to the degradation of the quality factor of the resonators (this aspect was discussed in Section 5.3.1.1). Actuation voltage is 30 V. Reprinted with permission from Ref. [34]; copyright 2011 IEEE.

difference between them is simply the length of the movable portions of the rings. The measured frequency responses corresponding to the extreme switching states ("00" and "11") are also depicted in Figure 5.30. The tuning range is 12% for the filter of Figure 5.30a and 42% for the one depicted in Figure 5.30b. This difference is due to the larger capacitance variations experienced with the prototype that uses longer cantilevers. As compared to tunable stopband filters based on SRRs and varactor diodes, the filters of Figure 5.30 exhibit better insertion losses in the allowed bands. As compared to the filters based on tunable RF-MEMS based CSRRs of Figure 5.27, the approach presented in this subsection, based on cantilever-type SRRs, can provide better tuning ranges.

5.3.1.3 Metamaterial Transmission Lines Based on Ferroelectric Materials

Ferroelectric materials have been used for tuning purposes in several microwave applications [40], particularly using ferroelectric varactors [15, 41] and planar resonators [42]. Tunable SRRs based on ferroelectric materials were first reported in Ref.[43], but the resonance frequency of those SRRs was modified by varying the temperature, rather than by the application of an external voltage. In [44], the dependence of the dielectric permittivity of ferroelectric materials with an external electric field was exploited in order to modify the resonance frequency of SRRs by means of an applied voltage (the resonance frequency of the resonators is modified by tailoring the distributed capacitance between the two rings forming the resonator, similar to the capacitance variation in interdigital varactors). These tunable resonators were then used as loading elements in a host line for the implementation of tunable stopband filters [44, 45].

The tunable SRRs in Refs [44, 45] were implemented by using a BST ($Ba_{0.6}Sr_{0.4}TiO_3$) thick-film with a thickness of 3.5 μm, made by screen-printing a Fe-F co-doped BST paste on an alumina substrate and sintering at 1200°C [46]. Based on this ceramic, the structured metallization for the strip of the transmission line and the SRR was realized by a single lithography step and plating 2.5 μm thick Au electrode on a Cr/Au seed layer which was afterwards removed by wet etching. A cross-section scheme of the substrate is depicted in Figure 5.31a. The biasing requires the connection of DC feeding lines and pads to the rings forming the SRR as shown in Figure 5.31b, where the relevant dimensions of the structure are also included. The *Rogers RO3003* substrate was used as carrier substrate for the whole device including the biasing network. The carrier substrate was placed on a copper plate that acts as microstrip ground and ensures mechanical stability. Due to the topology of the SRRs, no DC-blocking elements are needed, whereas the RF signal is blocked by means of 100KΩ SMD-resistors. Using the cell depicted in Figure 5.31b, an order-3 tunable stopband filter was designed [45] (Fig. 5.32a). The filter responses for the untuned (0 V) and tuned (140 V) states are depicted in Figure 5.32b. In the simulations, the employed values for the relative permittivity of the BST layer are $\varepsilon_{UNTUNED} = 300$ and $\varepsilon_{TUNED} = 220$, which fit the measurement curves and are typical values of the permittivity for the considered dimensions, tuning voltages and used BST material.

The application of BST thick films to the implementation of tunable composite right-/left-handed (CRLH) transmission lines based on the combination of OSRRs and OCSRRs was reported in Ref. [47]. These artificial tunable lines were then used

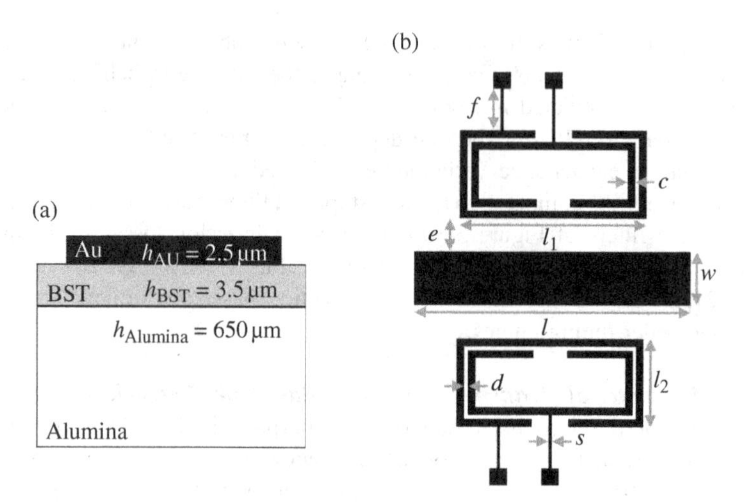

FIGURE 5.31 (a) Cross-section scheme of the substrate indicating layer thicknesses. (b) Layout of fabricated microstrip line and SRRs. Dimensions are $c = 100\,\mu m$, $d = 10\,\mu m$, $f = 590\,\mu m$, $s = 30\,\mu m$, $w = 520\,\mu m$, $l = 6.1$ mm, $l_1 = 3.1$ mm, $l_2 = 0.8$ mm, area $a = l_1 \times l_2 = 2.42\ mm^2 = 3.1\ mm \times 0.8\ mm = 0.07\lambda \times 0.02\lambda$ (λ is the guided wavelength at the resonance of the SRRs). Reprinted with permission from Ref. [44]; copyright 2009 IET.

for the implementation of tunable filters, phase shifters, and power dividers [48]. The structure reported in Ref. [47] (depicted in Fig. 5.33) consists of a CRLH unit cell implemented in CPW technology based on series connected OSRRs and a shunt OCSRR (with meandered lines to reduce dimensions). The fabrication process is similar to the one reported earlier, but the BST film has a thickness of 7.5 μm in this case. The relative permittivity of the BST in the untuned state is roughly $\varepsilon_r = 490$, and by applying a DC voltage, the permittivity can be reduced to approximately $\varepsilon_r = 220$, representing a material tunability of 55%. The equivalent circuit model of the structure is depicted in Figure 5.33c, where the variable capacitances are indicated. The series OSRRs are tuned by applying a voltage between the rings, through the feeding ports and the ground plane (the central strip and the ground plane of the CPW are electrically connected through the OCSRR). The tuning of the OCSRR requires two additional biasing lines inserted through the gaps of the rings since both rings are electrically connected. This biasing network is composed both of gold and chrome lines to avoid its RF influence on the biasing network, which could affect the band or the spurious response. The main advantage of this approach is that the biasing networks of the OCSRRs can be connected to the feeding ports, and hence the tuning of the whole artificial line can be achieved by applying only a single voltage between the feeding ports and the ground plane (which can be done directly from the probes). Further details are given in Refs [47, 48]. The measurement and EM simulations of the frequency response of the resulting topology are shown in Figure 5.33b. Notice that the CRLH line was not designed to satisfy given specifications. However the line is quasi-balanced, as revealed by the dispersion diagrams for the tuned and untuned states (Fig. 5.34). The reported structure exhibits good filtering behavior and tuning

(a)

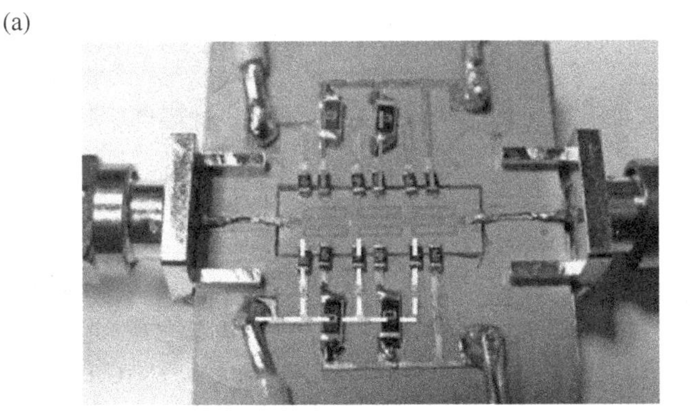

(b)

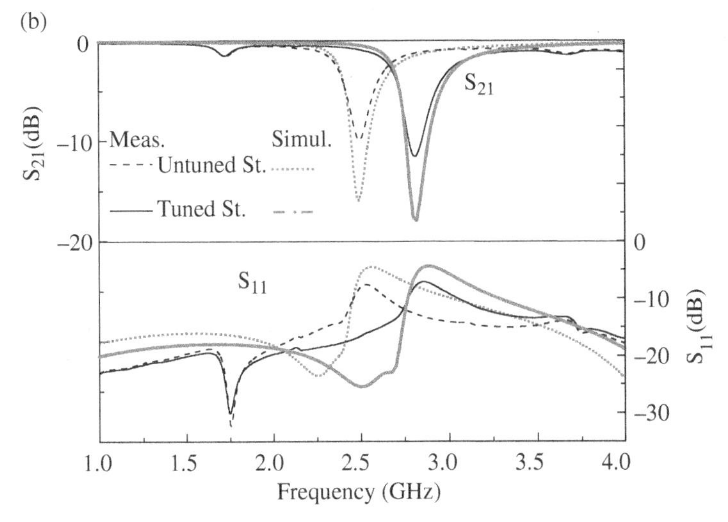

FIGURE 5.32 Photograph of the fabricated order-3 tunable BST-based filter (a), and measured and simulated frequency response of the device for the tuned (140 V) and the untuned (0 V) states (b). Two different parts of the biasing network are marked with dashed lines. For the tuned state, the one at the upper side corresponds to the 140 V, whereas the one at the lower side corresponds to 0 V. Reprinted with permission from Ref. [45]; copyright 2011 IET.

capability. However, to further control filtering characteristics, an independent control of the capacitance in the OSRRs and OCSRR is necessary. This aspect and the possibilities for the design of tunable filters and other microwave components based on CRLH lines are discussed in Ref. [48].

5.3.2 Tunable CL-Loaded Metamaterial Transmission Lines

Tunable CL-loaded metamaterial transmission lines can be designed by means of lumped varactors, ferroelectric varactors, RF-MEMS, and liquid crystals, among

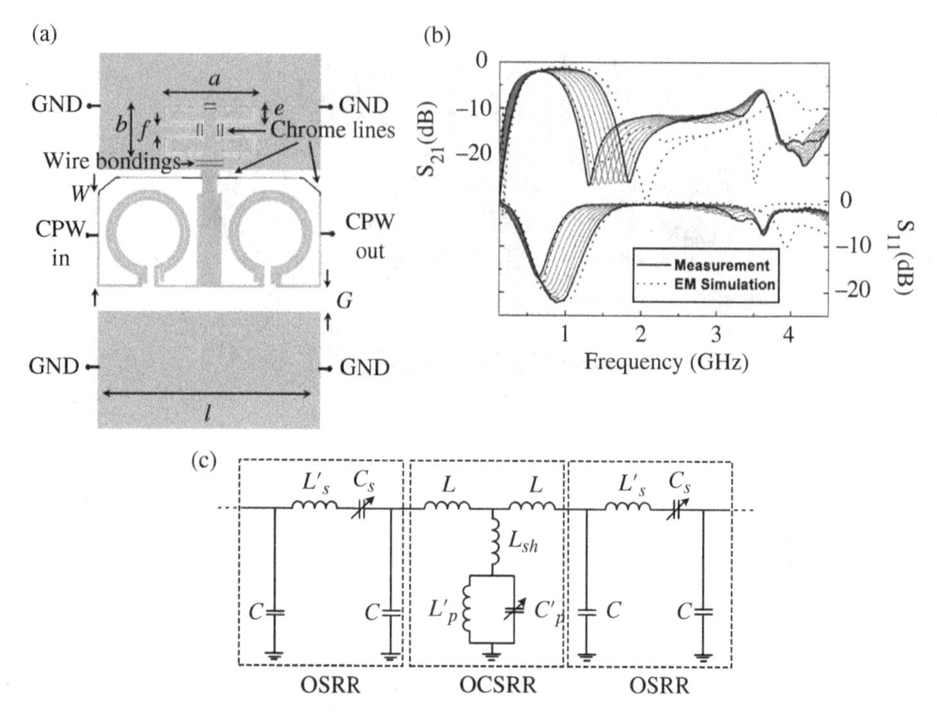

FIGURE 5.33 Layout (a) and measured response (b) of the tunable CRLH line, and equivalent circuit model (c). The tuning voltage is increased from 0 V up to 160 V in 8 steps. The indicated EM simulations, including losses, correspond to the maximum and minimum tuning states, considering the dielectric constants given in the text. The chrome lines are depicted in black. Wire bondings connecting the ground planes of the OCSRR are also needed to avoid slot modes. Dimensions are $l = 6.9$ mm, $W = 2.8$ mm, and $G = 0.72$ mm. For the OCSRR, $a = 3.05$ mm, $b = 1.66$ mm, $e = 0.7$ mm, $f = 0.26$ mm, $d = s = 60$ μm, and $c = 10$ μ m, s being the width of the biasing networks between the rings of the OCSRR. For the OSRR, $r_{ext} = 1.3$ mm, $d = 10$ μm, and $c = 0.16$ mm. Reprinted with permission from Ref. [47]; copyright 2011 IEEE.

others. In the previous subsection, devoted to tunable resonant-type metamaterial transmission lines, the considered structures were grouped by technologies (this was mainly motivated by the different tuning strategies/approaches for the resonant elements corresponding to each technology, although the tuning principle was indeed the same). In this subsection, we have opted for presenting two illustrative applications of CL-loaded CRLH tunable lines, as representative examples: this includes a tunable phase shifter, implemented by means of diode varactors, and a liquid crystal-based tunable CRLH leaky wave antenna.

5.3.2.1 Tunable Phase Shifters CRLH-based tunable phase shifters using lumped varactors [49, 50], ferroelectric varactors [51], and RF-MEMS [52, 53] have been reported. As an example, the design of a tunable phase shifter, implemented by

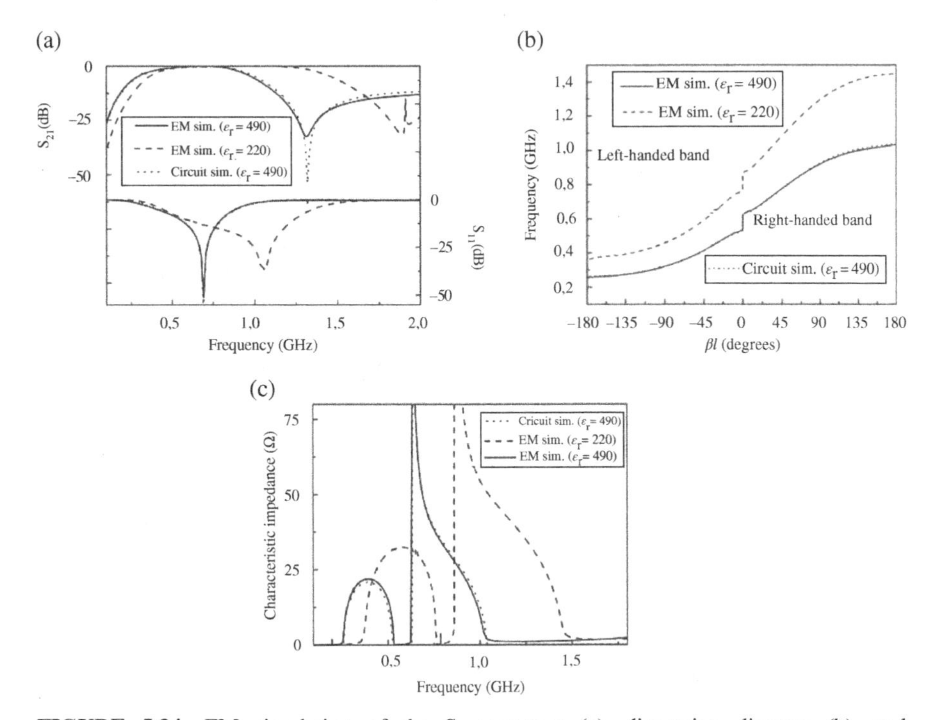

FIGURE 5.34 EM simulation of the S-parameters (a), dispersion diagram (b), and characteristic impedance (c) of the structure of Figure 5.33a for the maximum and minimum tuning states. The circuit simulation for the untuned state is also depicted. The equivalent circuit values are for the OSRR $C = 0.44$ pF, $L_s' = 5.25$ nH, $C_s = 16.87$ pF. For the OCSRR, $L = 0.27$ nH, $L_p' = 5.25$ nH, $C_p' = 10.27$ pF, and $L_{sh} = 1.93$ nH. Reprinted with permission from Ref. [47]; copyright 2011 IEEE.

means of varactor diodes, with separately tunable phase and line impedance [50], is reported. The independent tuning of phase and impedance reduces mismatch losses, allowing for a broadband behavior and a more constant insertion loss over all tuning states. The independent tuning principle is very simple. Assuming a lossless dual transmission line (Fig. 3.18a), in the homogeneous regime (long wavelength limit), the phase constant and the characteristic impedance are given by expressions (3.41) and (3.42). From these expressions, it follows that by tuning only one element (e.g., the capacitance), the phase shift can be tailored, but at the expense of a change in the characteristic impedance as well. However, since the line impedance is related to the quotient of the inductance and capacitance, it is possible to keep the line impedance constant by tuning both reactances by the same (dimensionless) factor, x, namely

$$\beta_L l(x) = -\frac{1}{\omega\sqrt{xL \cdot xC}} \tag{5.8}$$

$$Z_{\rm BL}(x) = \sqrt{\frac{xL}{xC}} = Z_{\rm BL} \tag{5.9}$$

Thus, an artificial line phase shifter that does not exhibit tuning induced mismatch is possible. However, a purely left-handed (PLH) line cannot be implemented in practice. Including the parasitics of the host line, a CRLH transmission line results, and the corresponding phase and characteristic impedance are given by expressions (3.51) and (3.52). In the long wavelength limit, these expressions can be written as follows:

$$\beta = \frac{s(\omega)}{l}\sqrt{\omega^2 L_{\rm R} C_{\rm R}\left(1-\frac{1}{\omega^2 L_{\rm R}\cdot xC_{\rm L}}\right)\left(1-\frac{1}{\omega^2 C_{\rm R}\cdot xL_{\rm L}}\right)} \tag{5.10}$$

$$Z_{\rm B} = \sqrt{\frac{L_{\rm R}}{C_{\rm R}}\frac{\left(1-\frac{1}{\omega^2 L_{\rm R}\cdot xC_{\rm L}}\right)}{\left(1-\frac{1}{\omega^2 C_{\rm R}\cdot xL_{\rm L}}\right)}} \tag{5.11}$$

where tuning, associated to the series capacitor and shunt inductor, has been included. Notice that (5.10), is identical to (3.59), but it has been expressed as shown, in order to appreciate the tuning elements. Inspection of (5.11) reveals that, in general, the terms in the parenthesis do not cancel. However, by balancing the line, that is, $L_{\rm R}C_{\rm L} = L_{\rm L}C_{\rm R}$, the characteristic impedance becomes constant over the tuning interval.

To avoid the use of tunable inductances, the strategy used in Ref. [50] consists of replacing such elements with tunable varactors in conjunction with transmission line impedance inverters. The circuit schematic of the tunable line is depicted in Figure 5.35, where the effective tuning voltages for the shunt and series varactors are U_1 and U_2, respectively. The inductors at the feeding points are RF chokes and are used to decouple the RF signals from DC polarization. The capacitors cascaded to the shunt varactors are virtual grounds for the RF signals. In Ref. [50], the 10nH inductor *LQW15AN10H00* from *Murata* for RF decoupling, and the 10 pF capacitor *500S100* from *American Technical Ceramics* for virtually grounding were used. The employed varactor diode was the *Infineon BB857* since it has a very low series inductance of 0.6 nH. Optimization of the transmission lines was done since it was found that the characteristic impedance of such lines controls the value of the realized inductance whereas the length controls the amount of variation of such inductances with frequency [50]. The details of prototype fabrication and calibration are out of the scope of this book and can be found in Ref. [50]. The photograph of the fabricated device is shown in Figure 5.36a. Figure 5.36b plots the transmission phase, considering $U_1 = U_2$, with tuning voltage steps of 4 V. The phase curves are equidistant for equidistant tuning voltages, so the phase is a linear function of the tuning voltage. The quality of the phase shifter is generally estimated from the following figure of merit (FoM), which relates the obtainable phase shift to the incurred transmission losses [50]:

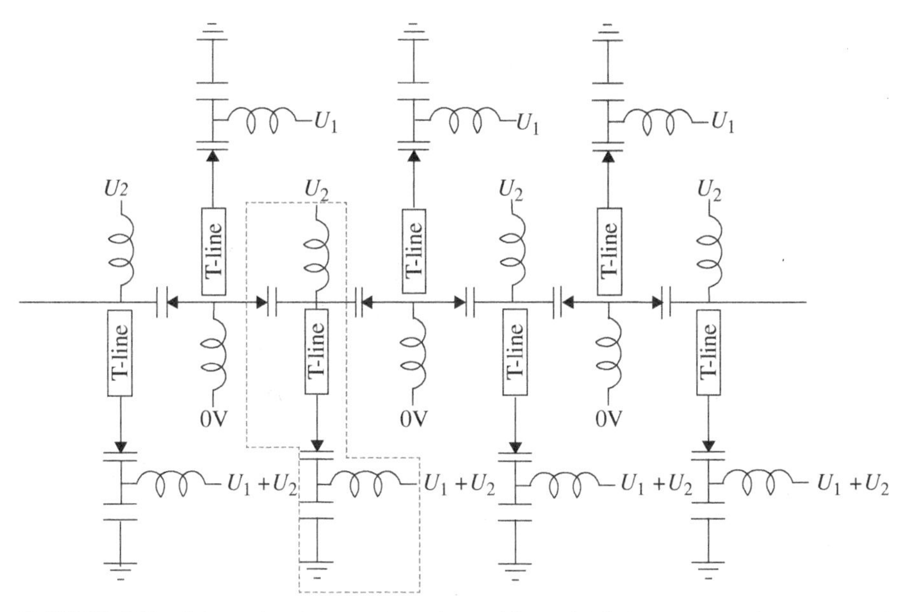

FIGURE 5.35 Schematic of the tunable phase shifter with independent tuning. Reprinted with permission from Ref. [50]; copyright 2006 EuMA.

$$\text{FoM} = \left| \frac{\varphi(C_2) - \varphi(C_1)}{\min\left(|S_{21}(x)|_{\text{dB}}\right)} \right| \quad x \in [C_1, C_2] \tag{5.12}$$

where C_1 and C_2 denote two tuning states and $\varphi(C_{1,2}) = \angle S_{21}(C_{1,2})$. Thus, the FoM is optimized if the phase shift variation experienced in the tuning range is large and the maximum attenuation ($\min(|S_{21}(x)|_{\text{dB}})$) occurring in the same tuning range is small. The measured FoM for tuning states with $U_1 = U_2$ is depicted in Figure 5.36c, where it can be appreciated that the FoM is larger than 60°/dB in a 30% bandwidth (around 6.5 GHz). In the same figure, the variation of the transmission magnitude over all tuning states from maximum to minimum differential phase shift is also included. In an ideal phase shifter the amplitude variation of the signal is zero, but the measured result (below 2.5 dB in the above cited bandwidth) is very reasonable.

5.3.2.2 Tunable Leaky Wave Antennas (LWA) Tunable LWAs have been implemented by means of CRLH lines using varactor diodes [54, 55], RF-MEMS [56], and liquid crystals [57, 58]. As an illustrative example, let us now focus on the LWA reported in Ref. [58], operative in the Ka-band, and able to provide beam steering for fixed operation frequency. Beam steering as a function of frequency (frequency tuning) in CRLH-based LWAs is consequence of the dispersive characteristic of the phase constant in CRLH lines (as it was discussed in Chapter 4). However, for many applications beam steering is desirable at a fixed frequency. Using varactor diodes, fixed frequency beam steering was achieved in Ref. [54]. However, semiconductor

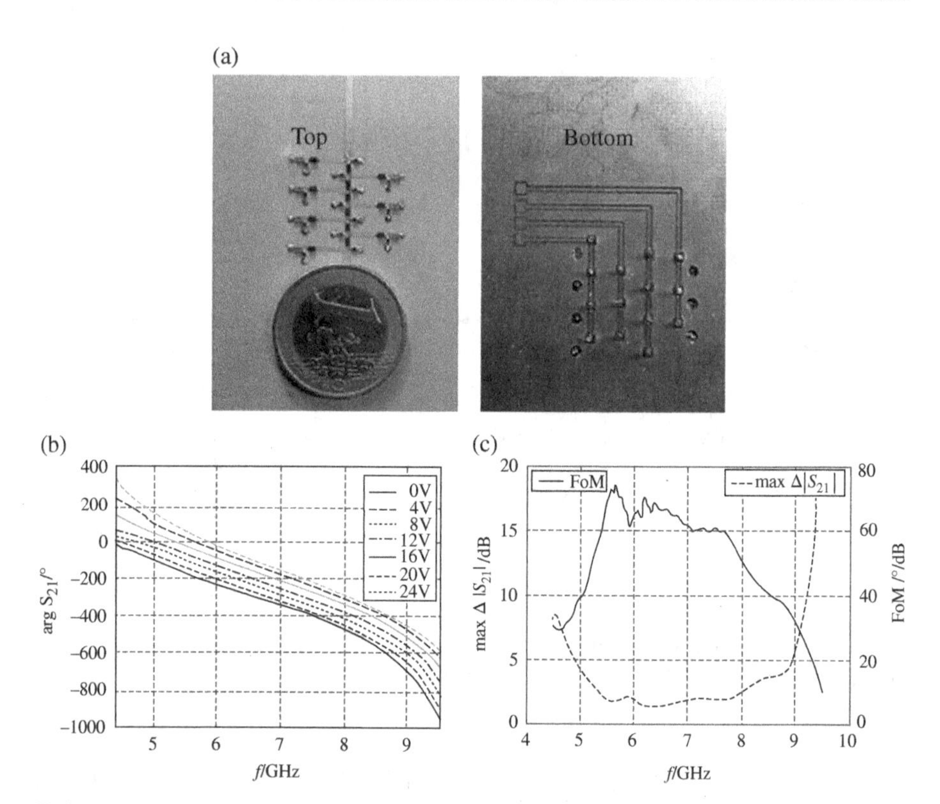

FIGURE 5.36 Photograph of the fabricated phase shifter (a), transmission phase (b), and FoM (c). The top layer is the RF part of the device with SMD components and signal lines, whereas the bottom layer is the DC feeding network. The ground plane for RF is sandwiched between these layers, so that the structure has three metal layers. The upper dielectric is the *Rogers RO3011* (0.51 mm thickness) and the bottom one is FR4 (1.4 mm thickness). Reprinted with permission from Ref. [50]; copyright 2006 EuMA.

varactors are frequency limited (i.e., they exhibit poor performance at high frequencies), and advanced technologies/materials, such as liquid crystals (acting as a continuously tunable anisotropic dielectric), are preferred for operation at high frequencies. The tuning of the phase constant, and hence the radiation angle, is achieved by tuning the liquid crystal (this can be done through magnetic or electric field actuation, as indicated in Section 5.2.4).

The unit cell layout of the antenna is depicted in Figure 5.37, and it consists of series-connected interdigital capacitors (C_L) alternating with meandered shunt inductors (L_L). At the extreme of the shunt inductors, interdigital (blocking) capacitors introduce an additional capacitance (C_{DC}) necessary for tuning by application of a bias voltage (in spite that tuning in Ref. [58] is carried out by means of magnetic field actuation). The parameters of the host line (L_R and C_R) complete the circuit model of the unit cell, also depicted in Figure 5.37. The capacitance C_{DC} introduces a transmission zero below the first LH band, but this transmission zero is not relevant for the

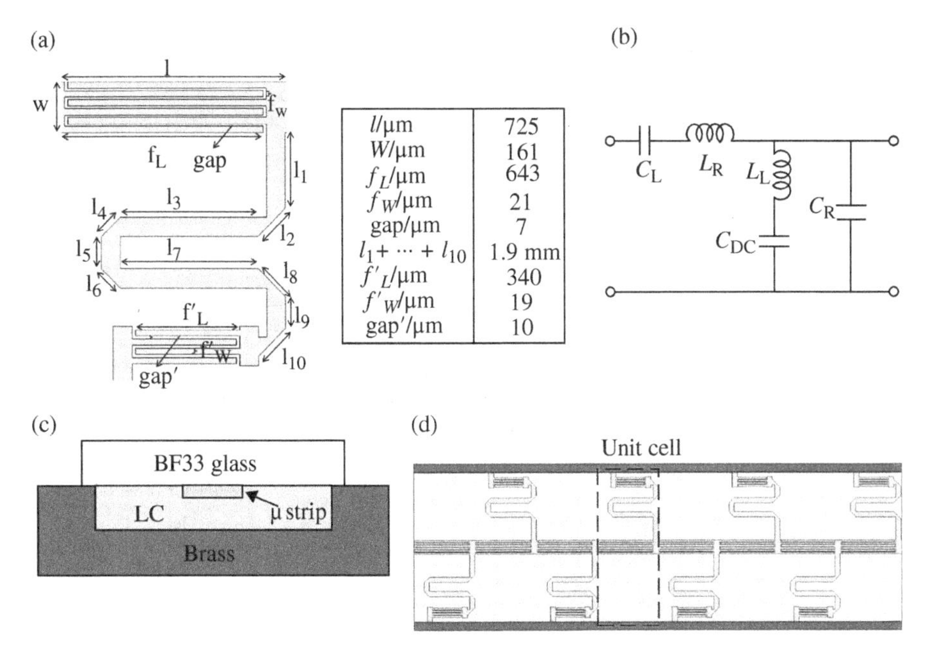

l/µm	725
W/µm	161
f_L/µm	643
f_W/µm	21
gap/µm	7
$l_1 + \cdots + l_{10}$	1.9 mm
f'_L/µm	340
f'_W/µm	19
gap'/µm	10

FIGURE 5.37 Unit cell layout, including dimensions (a), unit cell circuit model (b), cross section (c), and top view (d) of the liquid crystal-based tunable LWA. Reprinted with permission from Ref. [58]; copyright 2013 EuMA.

functionality of the structure as a tunable LWA.[8] Figure 5.37c and d show the cross sectional and top view of the LWA (actually the antenna is composed of 15 cells, although only eight cells are shown in the figure). As detailed in Ref. [58], the antenna was printed on the bottom side of a borofloat (BF33) glass substrate with dielectric constant $\varepsilon_r = 4.65$ and $\tan\delta = 0.008$. The metallic strips, made of gold, have a thickness of 2 µm. A polyimide layer was spin coated with a height of typically 10–100 nm and subsequently mechanically rubbed with a velvet cloth to anchor the liquid crystal molecules in the unbiased state parallel to the rubbing direction [59]. A metal block was used as ground plane and to provide mechanical stability. The glass substrate was glued with conducting glue on top of the metal block forming a 100 µm cavity. The cavity was filled with the liquid crystal *TUD-649* with $\varepsilon_{r\perp} = 2.43$, $\varepsilon_{r\parallel} = 3.22$, $\tan\delta < 0.0066$ and sealed with two-component epoxy glue. The orientation of the liquid crystal molecules and, therefore, its permittivity can be tuned either by applying static electric or magnetic field, although the latter was chosen in the proof-of-concept prototype presented in Ref. [58]. For that purpose, rare earth magnets with field strength of 0.3 T were used.

The description of the experimental setup and further details on the interface between the LWA and the connectors (necessary for far-field measurements) are

[8] Notice that with interdigital capacitors, it is very difficult to achieve the necessary high capacitance values to virtually ground the meandered inductors for RF signals.

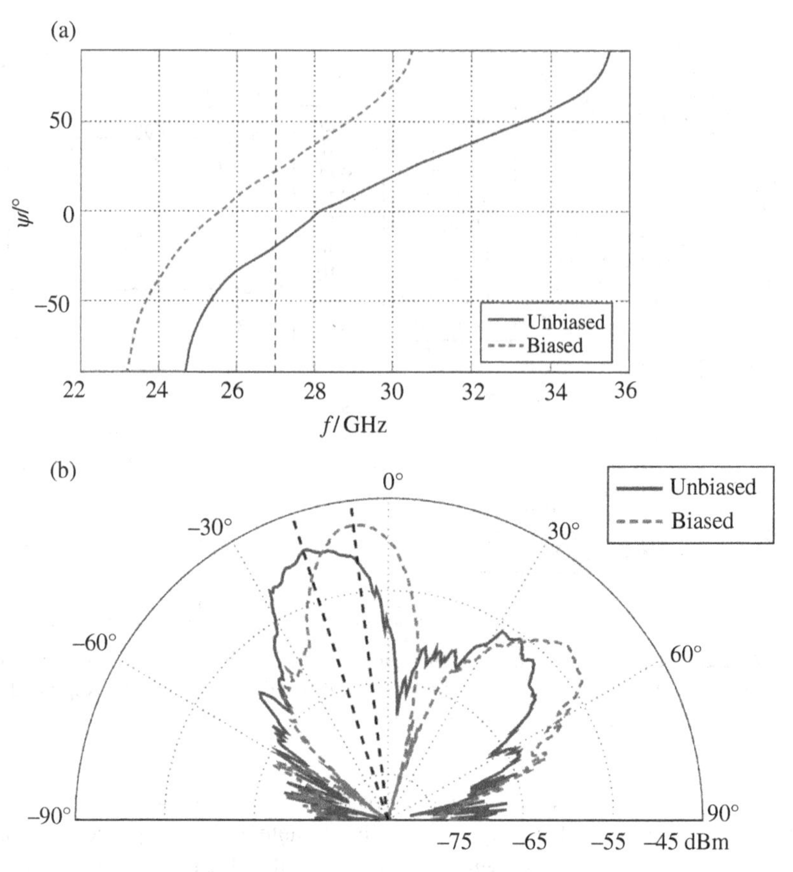

FIGURE 5.38 Radiation angle of the LWA inferred from the measured dispersion diagrams for the biased and unbiased states (a), and measured far-field radiation patterns at 27 GHz (b). Reprinted with permission from Ref. [58]; copyright 2013 EuMA.

given in Ref. [58]. From the measured dispersion for the biased and unbiased states, the radiation angle as a function of frequency was obtained (using expression 2.11). The results are depicted in Figure 5.38a, and indicate that for a fixed frequency of 27 GHz, it is possible to achieve a beam steering range from roughly −20° to +20°, going through the broadside direction if the liquid crystal orientation is continuously tuned. The measured far-field pattern at 27 GHz (depicted in Fig. 5.38b) reveals that for the unbiased state the radiation angle of the main beam is very close to the predicted value. However, for the biased state the main beam points to −7°. The good agreement of the unbiased state and the decreased scanning range indicate that the liquid crystal molecules are not completely oriented in the biased state in the far-field measurements. This was attributed in Ref. [58] to the larger distance of the rare earth magnets that had to be placed behind the antenna, in comparison to the on-wafer measurements (for measuring the dispersion diagrams), where the biasing magnets were placed directly under the glass substrate and thus, closer to the liquid crystal. Nevertheless, the tuning principle and the potentiality of liquid crystals for the implementation of high-frequency tunable LWAs based on CRLH lines are demonstrated.

To end this section, devoted to tunable and reconfigurable devices based on artificial transmission lines, we would like to mention that only a representative subset of the reported approaches, technologies, and applications has been presented. Many other tunable components (i.e., couplers, filters, impedance matching networks etc.) based on artificial transmission lines can be found in the literature (some additional references for those interested readers are Refs [60–66]).

5.4 NONLINEAR TRANSMISSION LINES (NLTLs)

NLTLs are periodic structures implemented by loading a transmission line with nonlinear elements, typically shunt-connected voltage-dependent capacitances. In the linear regime, these structures exhibit a slow-wave behavior (see Chapter 2), and the phase and group velocity can be tuned by virtue of the controllable effective capacitance of the line. In the nonlinear regime, NLTLs are very interesting structures that can be used to achieve pulse and impulse compression or harmonic generation (frequency multipliers) [67]. In NLTLs, there are two main effects that give rise to the rich phenomenology that these artificial lines exhibit: nonlinearity and dispersion. Dispersion is related to periodicity, whereas the voltage variable capacitances are responsible for nonlinearity. In the limit of weak dispersion,[9] wave propagation is dominated by the nonlinearity, with the effect of steepen the falling edge of a voltage waveform. This effect can be intuitively explained by considering the voltage-dependent capacitance of a varactor diode or a HBV (it decreases with the absolute value of the applied voltage). The instantaneous propagation velocity at any given point in space and time is inversely proportional to the square root of the effective capacitance. Since such capacitance is smaller for higher voltages, the result is that the points closer to the crest of the voltage waveform experience a faster propagation velocity and produce a shock-wave front. With this approach, fall times in the vicinity of 1 ps are possible [67].

However, the most intriguing aspect of NLTLs appears when both dispersion and nonlinearity are simultaneously present. Dispersion tends to spread out voltage pulses propagating in the line, whereas nonlinearity has the opposite effect. Under certain conditions both effects balance, giving rise to the propagation of electrical pulses with permanent profile, that is, solitons[10] [68]. Thus, solitons are electrical pulses that propagate in the line without distortion, and are the consequence of the combined effects of nonlinearity and dispersion. The concept of solitary wave was introduced by John Scott Russell in 1844 [69]. Solitons have been an object of research interest in several fields, including mathematics, applied physics (mainly optics [70, 71]), and electronics [67, 72, 73]. In the latter case, soliton propagation in NLTLs has been studied due to the ability of these lines to generate harmonics from the decomposition of the input

[9] Weak dispersion conditions are met if the frequency components of the waveform are small compared to the Bragg frequency. See Chapter 2 for more details.

[10] It is assumed that the NLTL is lossless. Otherwise, pulse amplitude progressively decreases.

signal into their constitutive solitons [67]. One fundamental property of NLTLs supporting soliton wave propagation is the fact that any input signal not being a soliton tends to progressively decompose in a set of solitons with different amplitudes and velocities. Therefore, frequency multiplication in NLTLs can be interpreted as due to the generation of various solitons per semicycle of the input waveform. Besides the above fundamental property, larger amplitude solitons are narrower and travel faster than lower amplitude solitons. Moreover, solitons of different amplitude (and hence pulse velocity) that collide preserve their identity after the collision.

In monolithic NLTLs, solitary pulses of picosecond duration can be supported. Hence, these lines are interesting for the generation of signals with high frequency harmonic content (i.e., up to hundreds of GHz and even THz) on the basis of frequency multipliers based on them. In the nineties, monolithic NLTLs were considered as firm candidates for the implementation of signal sources in the so-called THz gap [74–80] on the basis of harmonic generation from smaller frequency signals. One advantage of NLTLs as frequency multipliers is their potential for the suppression of high-order harmonics. This is due to the intrinsic low-pass filtering behavior of these lines, related to periodicity, which precludes wave propagation in a wide band above the Bragg frequency (see Chapter 2). For frequency triplers, for instance, it is possible to suppress the fourth- and higher-order harmonics by properly designing the lines (i.e., with the Bragg frequency set slightly above the third harmonic of the reference frequency).[11] On the other hand, the second and even-order harmonics can be suppressed by using nonlinear devices exhibiting a symmetric (even) capacitance-voltage characteristic, as occurs with HBVs (this was anticipated in Section 5.2.1). With such even nonlinear capacitance–voltage function, if the HBV is excited by a harmonic voltage, it produces a current with frequency components at the fundamental frequency and odd harmonics only. The reason is that the current can be expressed as an odd function of the applied voltage, and cancelation of the even harmonics results. A similar argument can be applied to explain even-order harmonic cancelation in NLTLs loaded with devices exhibiting a symmetric capacitance–voltage characteristic. This cancellation of undesired harmonics is important for conversion efficiency optimization (excellent HBV tripler results, demonstrating an efficiency of 12% at 247 GHz, were reported in Ref. [81]).

To gain more insight on soliton wave propagation in NLTLs, a simple model is presented in the next subsection, and then some illustrative results showing the decomposition of a feeding signal into solitary pulses are provided.

5.4.1 Model for Soliton Wave Propagation in NLTLs

Figure 5.39 shows the schematic of a NLTL together with the lumped-element equivalent circuit model, which has been used to study harmonic multiplication [79, 82] and pulse sharpening [83]. L and C_o are the per-section inductance and capacitance, respectively, of the line, while $C_D(V)$ represents the nonlinear device capacitance. The study of soliton propagation in nonlinear networks of the type of Figure 5.39b was already carried out by Hirota and Suzuki [84] in the seventies. They considered

[11] In the context of this section, the harmonic order n is related the fundamental frequency, f_o, by $f_n = nf_o$.

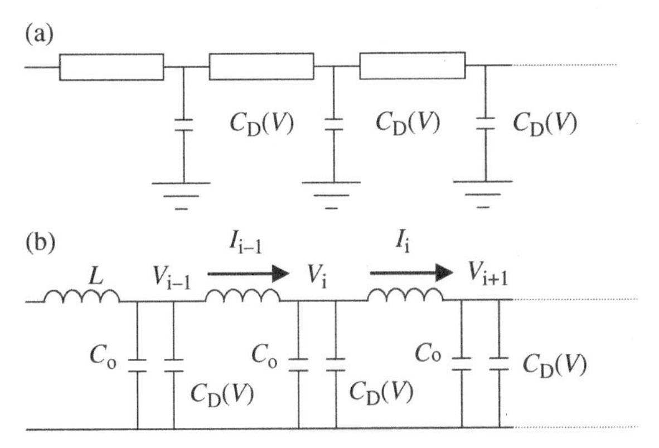

FIGURE 5.39 NLTL circuit schematic (a) and lumped-element equivalent circuit (b). Each transmission line section between nonlinear capacitances is modeled by a series inductor, L, and shunt capacitor, C_o.

a nonlinear shunt capacitance of the form $C(V) \propto 1/(V-V_o)$ (where V_o is a constant). In this particular case, the system is equivalent to the Toda lattice [85], for which soliton solutions are well known, and can be described by the Korteweg–de Vries (KdV) equation [86] in the long wave limit (i.e., nonlocalized solitons). Systems based on the KdV equation have been well studied, and the response of such systems to initial disturbances that break up into solitons has been described by the inverse scattering method [72]. However, for strong lumped solitons, or to study soliton propagation in nonlinear *LC* networks with arbitrary capacitance–voltage nonlinearity, a KdV approach can not generally be applied.

In this subsection, a simple procedure to obtain approximate soliton solutions in NLTLs described by its lumped-element equivalent circuit (Fig. 5.39b) is presented [87, 88], and applied to structures with HBV-like nonlinear devices (a similar model that considers MOS varactors to sharpen the rising and falling edge of a pulse was reported in Ref. [89]). The lumped model is accurate enough in the long wavelength limit, and it provides a reasonable description of NLTLs up to frequencies close to the cut-off (Bragg) frequency if $C_D(V) \gg C_o$, as it was discussed in Chapter 2 (and demonstrated in [87, 88]). A more sophisticated model [90] using the microwave circuit theory to explicitly simulate each transmission line section, has been developed to study high-frequency components of picosecond pulse generation with NLTLs. Nevertheless, the main interest of this Section is to analyze the effects of line parameters and nonlinearity on soliton characteristics (amplitude, width, and velocity). To this end, the proposed model suffices, and it is useful as guideline for the design of frequency multipliers based on NLTLs.

In view of Figure 5.39b, the voltage drop at the inductors with currents designated as I_{i-1} and I_i is

$$V_{i-1}(t) - V_i(t) = L \cdot \frac{dI_{i-1}(t)}{dt} \tag{5.13a}$$

$$V_i(t)-V_{i+1}(t)=L\cdot\frac{dI_i(t)}{dt} \tag{5.13b}$$

respectively. By subtracting the previous equations, it follows:

$$V_{i-1}(t)-2\cdot V_i(t)+V_{i+1}(t)=L\cdot\frac{d(I_{i-1}(t)-I_i(t))}{dt} \tag{5.14}$$

Taking into account that the I_{i-1}–I_i is the displacement current at the capacitance $C(V)=C_o+C_D(V)$, i.e.,

$$I_{i-1}(t)-I_i(t)=C(V_i(t))\frac{dV_i(t)}{dt} \tag{5.15}$$

expression (5.14) can be written as:

$$L\left(C(V_i(t))\frac{d^2V_i(t)}{dt^2}+\left(\frac{dV_i(t)}{dt}\right)^2\frac{dC(V)}{dV}\right)=V_{i-1}(t)-2\cdot V_i(t)+V_{i+1}(t) \tag{5.16}$$

In order to simplify this equation, we employ a standard continuum limit [72] for the voltage, that is, $V_i(t)\rightarrow V(x,t)$, where $x=i\cdot l$ and l is the distance between nonlinear capacitors. Within this limit, a Taylor expansion up to the fourth order can be carried out:

$$\begin{aligned}V_{i\pm1}(t)=V(x\pm l,t)\approx V(x,t)\pm\frac{\partial V(x,t)}{\partial x}\cdot l+\frac{1}{2}\frac{\partial^2V(x,t)}{\partial x^2}\cdot l^2\pm\frac{1}{3!}\frac{\partial^3V(x,t)}{\partial x^3}\cdot l^3\\+\frac{1}{4!}\frac{\partial^4V(x,t)}{\partial x^4}\cdot l^4\end{aligned} \tag{5.17}$$

Moreover, solitons moving at a propagation velocity v have a static profile in a new time coordinate system given by $s=t-x/v$. With this new variable, $\partial/\partial x=-1/v\cdot d/ds$ and $\partial/\partial t=d/dt$. In this way, Equation 5.16 can be rewritten as follows:

$$L\frac{d}{ds}\left(C(V)\frac{dV(s)}{ds}\right)=T^2\frac{d^2V(s)}{ds^2}+\frac{T^4}{12}\cdot\frac{d^4V(s)}{ds^4} \tag{5.18}$$

where T is the per-section propagation delay, that is, $T=l/v$. Integration of (5.18) gives:

$$\left(C(V)-\frac{T^2}{L}\right)\frac{dV(s)}{ds}=\frac{T^4}{12L}\frac{d^3V(s)}{ds^3} \tag{5.19}$$

where the independent term is null, provided odd order derivatives vanish at pulse maximum ($s = 0$), and we are seeking for localized solutions whose derivative tends to zero when $s \to \pm\infty$. Equation 5.19 can be integrated again to give:

$$F(V(s)) - \frac{T^2}{L}V(s) = \frac{T^4}{12L} \cdot \frac{d^2V(s)}{ds^2} \tag{5.20}$$

where $F(V)$ is a function responsible for non linearity defined as follows:

$$F(V) = \int_0^V C(V')dV' \tag{5.21}$$

In Figure 5.40, $F(V)$ obtained from the nonlinear capacitance given in the caption is depicted. The slope of this function asymptotically approaches C_o, the per-section capacitance of the line, while at zero bias the derivative is given by $C_o + C_D(0)$. An analysis of Equation 5.20 for $s \to \pm\infty$ indicates that the integration constant is zero since all terms vanish at this limit.

Inspection of Equation 5.20 reveals that solitary wave solutions are possible. To demonstrate this, the second term of the left-hand side of Equation 5.20 (for the conditions indicated in the caption) is also depicted in Figure 5.40. Since the pulse maximum occurs at $s = 0$, the left-hand side of (5.20) must be negative at this point (positive polarity pulses are considered). This is satisfied if the pulse amplitude is above the intersection voltage of Figure 5.40. However, this is not enough to obtain

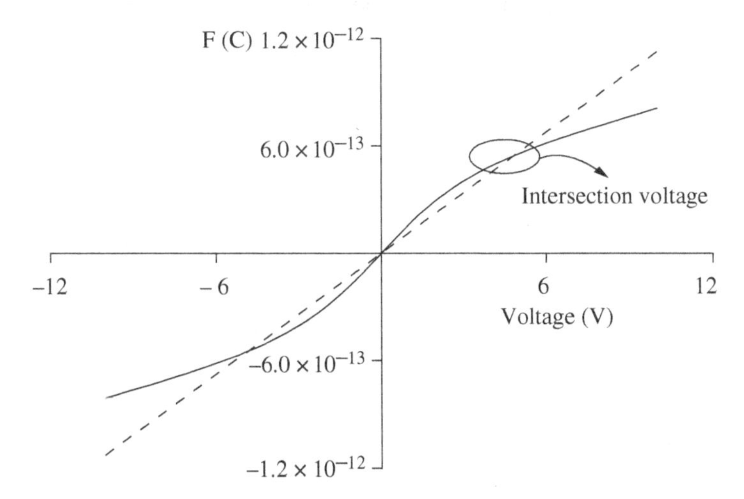

FIGURE 5.40 Representation of $F(V)$ obtained from (5.21) with $C_o = 43.5$ fF and $C_D(V) = 0.12\,\text{sech}(0.5\,V)$ pF (solid line). The second term of the left-hand side of equation (5.20) with $T = 3.5$ ps and $L = 108.7$ pH is also depicted (dashed line). Reprinted with permission from Ref. [87]; copyright 2001 AIP.

solutions of the form of solitons. Namely, as voltage decreases from maximum, the curvature decreases, changes polarity at the crossing point and, finally, an oscillatory behavior is found, unless the pulse amplitude takes the precise value that makes the solution vanish for $s \rightarrow \pm\infty$. In view of the previous argument and Figure 5.40, it is clear that solutions corresponding to solitary waves are only possible if there exits an intersection point between both terms of the left-hand side of Equation 5.20 above the origin. This limits the slope of the second term of the left-hand side to the range given by the asymptotic value of $F(V)$ and the derivative of $F(V)$ at the origin. This means that only solitons with per-section propagation delay T within the interval given by $[(LC_o)^{1/2}-(LC(0))^{1/2}]$ can propagate in the line. For a given T in this interval, a single solution exits, and soliton amplitude increases as T decreases. This behavior (in agreement with related literature [84]) can be deduced from Figure 5.40, where it is clearly seen that the intersection voltage increases as T approaches the lower limit of the earlier cited interval.

By numerically solving Equation 5.20, soliton solutions under arbitrary $C(V)$ characteristic can be obtained. Nevertheless, the dependence of soliton amplitude on T can be obtained without the need of solving (5.20). By multiplying Equation 5.20 by $dV(s)/ds$ and integrating in s, it follows:

$$\int_0^\infty \left[F(V(s)) - \frac{T^2}{L}V(s)\right]\frac{dV(s)}{ds}ds = \frac{T^4}{12L}\int_0^\infty \frac{d^2V(s)}{ds^2}\frac{dV(s)}{ds}ds \tag{5.22}$$

The right-hand side of (5.22) can be integrated by parts and found to be equal to zero for any function satisfying $dV(s)/ds = 0$ at $s = 0$ and $s = \infty$ (these are the boundary conditions for soliton solutions). By transforming the left-hand side of (5.22) to a voltage integral we finally obtain:

$$\int_0^{V_{max}} \left[F(V) - \frac{T^2}{L}V\right]dV = 0 \tag{5.23}$$

where it is assumed that $V(s \rightarrow \infty) = 0$, and $V(s = 0) = V_{max}$ is defined as the soliton amplitude. Integrating Equation 5.23, T can be isolated and compactly expressed as follows:

$$T = \frac{\sqrt{2 \cdot L \cdot G(V_{max})}}{V_{max}} \tag{5.24}$$

where $G(V)$ is related to $F(V)$ by

$$G(V) = \int_0^V F(V')dV' \tag{5.25}$$

By means of (5.24), the dependence of per-section propagation delay on soliton amplitude, or vice versa, can be easily obtained without the need to solve any differential equation. This relationship allows one to analyze the sensitivity of propagation velocity (or T) to soliton amplitude as a function of the NLTL parameters (including the nonlineariry). This is important for the optimization of NLTL-based frequency multipliers, since a strong sensitivity of T to soliton amplitude means that the decomposition of the feeding signal into solitons proceeds faster. This improves output power, since an important reduction of device (HBV) and transmission line losses is obtained by shortening NLTL length.

5.4.2 Numerical Solutions of the Model

The details of the numerical method to solve Equation 5.20 can be found in Refs [87, 88]. The advantage of such numerical method is that it is useful regardless of the specific nonlinearity of the variable capacitance device (including the possibility to consider experimental C(V) curves to analyze soliton propagation in actual structures). In Figure 5.41, soliton waveforms with different propagation delays that correspond to the network of Figure 5.39b demonstrate that higher amplitude solitons are faster. The nonlinear capacitance has been assumed to be given by $C_D(V) = C_M \cdot \text{sech}(kV)$ pF (with $C_M = 0.12$ pF and $k = 0.5\ V^{-1}$), which is a reasonable approximation to the capacitance dependence on bias for a typical HBV. Figure 5.42a depicts the interaction of two solitons obtained by numerical simulation of the network of Figure 5.39b. This figure demonstrates that solitons are stable and they preserve their identity after

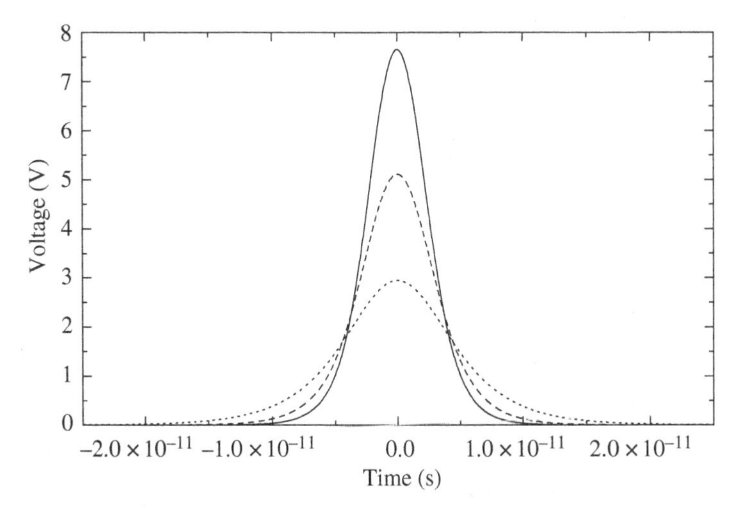

FIGURE 5.41 Soliton waveforms obtained using the approximate model for soliton propagation in NLTLs. Each curve corresponds to a different per-section propagation delay: $T = 3.5$ ps (solid line), $T = 3.75$ ps (dashed line), and $T = 4$ ps (dotted line). Model parameters indicated in Figure 5.40, which are reasonable for actual NLTLs, have been used. Reprinted with permission from Ref. [87]; copyright 2001 AIP.

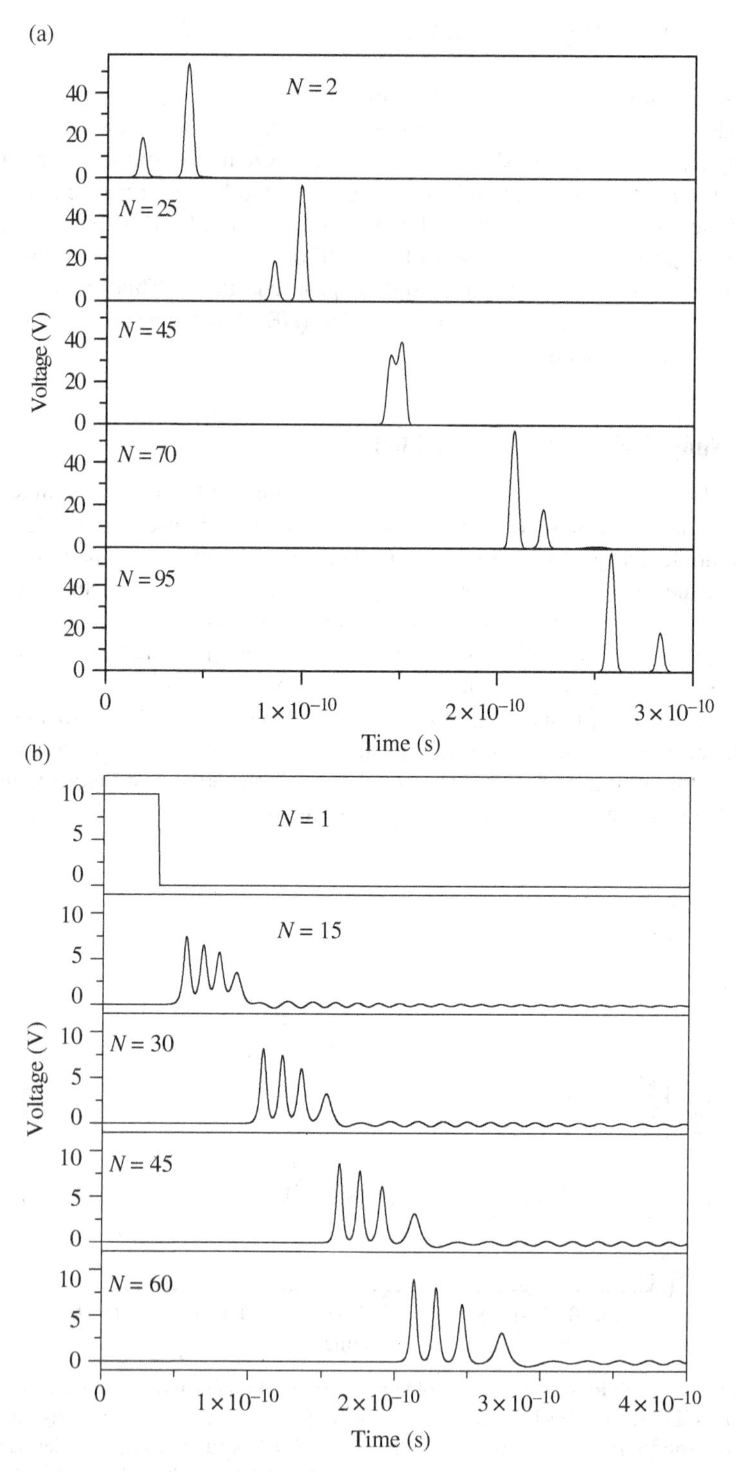

FIGURE 5.42 (a) Interaction of two solitons propagating in the same direction with different time delay ($T = 2.5$ ps for the high amplitude soliton and $T = 3$ ps for the small amplitude soliton). (b) Soliton decomposition of a 10 V square voltage pulse of 40 ps duration

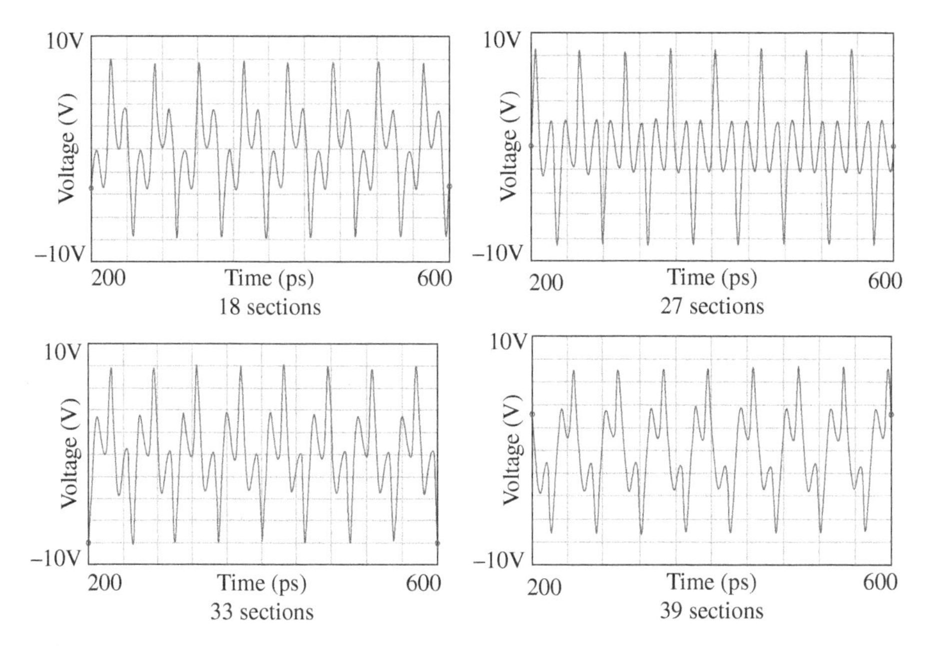

FIGURE 5.43 Voltage waveforms recorded at the output of NLTLs presenting different number of sections. Reprinted with permission from Ref. [93]; copyright 2002, Springer.

collision, as anticipated before. In Figure 5.42b, the response of the same network to a square feeding pulse indicates that the input signal progressively decomposes into solitons, and that larger amplitude solitons travel faster [91, 92].

The effects of device nonlinearity on soliton wave propagation were studied in Ref. [93], where it was pointed out that signal separation into solitons in NLTLs is enhanced by increasing device nonlinearity. Indeed, it was found in Ref. [93] that a figure of merit of the nonlinear device for harmonic generation is the area under the $C_D(V)$ curve. As this area increases, the number of sections required to optimize conversion efficiency to a given harmonic decreases, which means that soliton separation proceeds faster in structures with large $C_D(V)$ area devices. Figure 5.43 depicts the simulated waveforms obtained at the output of NLTLs with different number of sections, by considering an NLTL with $L = 108.7$ pH, $C_o = 43.5$ fF, $C_M = 0.12$ pF and $k = 0.5\ V^{-1}$, fed by a 20 GHz sinusoidal signal with 10 V amplitude. Under these conditions, the Bragg frequency is below 100 GHz. This means that significant power is only delivered to the third harmonic. The maximum dispersion

FIGURE 5.42 (*Continued*) expressed as temporal variation of voltage at several locations in the NLTL. The same NLTL parameters as in Figure 5.40 have been used. *N* indicates the section number of the line. (a) Reprinted with permission from Ref. [87]; copyright 2001 AIP; (b) Reprinted with permission from Ref. [91]; copyright 2001 Springer.

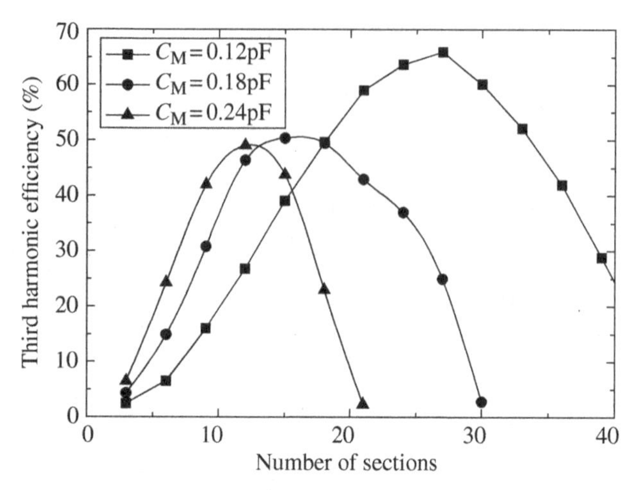

FIGURE 5.44 Effects of varying C_M on conversion efficiency, considering $k = 0.5\ \mathrm{V}^{-1}$. Reprinted with permission from Ref. [93]; copyright 2002, Springer.

of solitons is achieved after 27 sections, where it is expected that conversion efficiency is optimized. This is corroborated in Figure 5.44, where the effects of varying C_M on conversion efficiency are depicted. Losses have not been taken into account; and for this reason, conversion efficiency is much higher than the typical values of actual structures. As can be seen, by increasing C_M the number of sections giving maximum conversion efficiency decreases. This means that conversion efficiency can be optimized with a relatively small number of sections if the equilibrium capacitance is high enough. It can be observed that for $C_M = 0.12$ pF, the maximum conversion efficiency (obtained after 27 NLTL sections) is slightly above than for the other values of the equilibrium capacitance. However, in actual NLTLs it is expected to be an important degradation of output power after 27 sections due to losses. The effects of device losses have been studied in Ref. [93]. Figure 5.45 depicts the conversion efficiency with and without losses, for two different nonlinear capacitances having the same area. The results corroborate that the area under the $C_D(V)$ curve is the key factor for efficiency optimization, and that this effect is preserved even if losses are considered.

In this section, it has been shown that NLTLs, implemented by periodically loading a host (ordinary) line with shunt-connected voltage-dependent capacitors, can support solitons. These solitons are electrical (un-modulated) pulses that propagate along the line without distortion. The study of nonlinearity in LH (actually CRLH) transmission lines was carried out in the last decade. These lines are intrinsically dispersive and, combined with nonlinear capacitances, may also support soliton propagation. However, nonlinear LH lines support solitons similar to those in optical fibers, namely, envelope (modulated) solitons, which are solutions of the nonlinear Schrödinger equation [70, 71]. In particular, it was demonstrated in [94] that, depending on the capacitance-voltage characteristic of the capacitors, a nonlinear LH transmission line

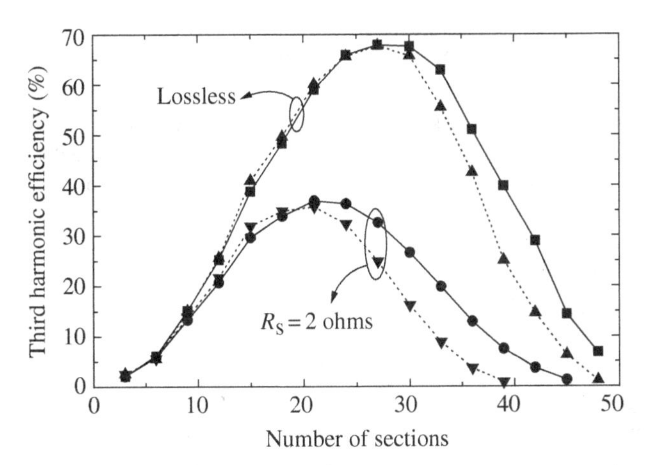

FIGURE 5.45 Effects of device losses on conversion efficiency. $C_M = 0.12$ pF, $k = 0.5$ V^{-1} (solid line); $C_M = 0.15$ pF, $k = 0.64$ V^{-1} (dotted line); the area effect is preserved when device losses are included. Reprinted with permission from Ref. [93]; copyright 2002, Springer.

can either support dark or bright[12] modulated solitons, both solutions of nonlinear Schrodinger-type equations. The analysis of soliton generation in nonlinear LH transmission lines is not straightforward, and is out of the scope of this book. Nevertheless, the author suggests references [94–103] to the interested reader on this topic.

REFERENCES

1. R. F. Pierret and G. Neudeck Editors, *The PN-Junction Diode*, Modular Series on Solid State Devices, vol. **2**, Addison Wesley, Reading, MA, 1982.
2. E. L. Kollberg and A. Rydberg, "Quantum barrier varactor for high efficiency millimeter-wave multipliers," *Electron. Lett.*, vol. **25**, pp. 1696–1698, 1989.
3. X. Melique, J. Carbonell, R. Havart, P. Mounaix, O. Vanbesien, and D. Lippens, "InGaAs/InAlAs/AlAs heterostructure barrier varactors for harmonic multiplication," *IEEE Microw, Guided Wave Lett.*, vol. **8**, pp. 254–256, 1998.
4. G. M. Rebeiz, *RF MEMS: Theory, Design, and Technology*, John Wiley, Hoboken, NJ, 2003.
5. B. Lacroix, A. Pothier, A. Crunteanu, C. Cibert, F. Dumas-Bouchiat, C. Champeaux, A. Cathérinot, and P. Blondy, "Sub-microsecond RF MEMS switched capacitors," *IEEE Trans. Microw. Theory Techn.*, vol. **55**, pp. 1314–1321, 2007.
6. B. Lakshminarayanan, D. Mercier, and G. Rebeiz, "High-reliability miniature RF-MEMS switched capacitors," *IEEE Trans. Microw. Theory Techn.*, vol. **56**, pp. 971–981, 2008.

[12] Bright solitons are pulse modulated solitons, whereas dark solitons exhibit a profile with a dip in a uniform background.

7. G. J. Papaioannou and J. Papapolymerou, "Dielectric charging mechanisms in RF-MEMS capacitive switches," *Proc. 37th European Microwave Conf.*, Munich, Germany, October 2007.
8. R. M. Young, J. D. Adam, C. R. Vale, T. T. Braggins, S. V. Krishnaswamy, C. E. Milton, D. W. Beve, L. G. Chorősinski, L.-S. Chen, D. E. Crocked, C. B. Freidhoff, S. H. Talisa, E. Capelle, R. Tranchini, J. R. Fended, J. M. Lorthioird, and A. R. Torres, "Low-loss bandpass RF filter using MEMS capacitance switches to achieve a one-octave tuning range and independently variable bandwidth," *IEEE MTT Int. Microwave Symp. Dig.*, Philadelphia, PA, June 2003, vol. **3**, pp. 1781–1784.
9. J. Papapolymerou, K. L. Lange, C. L. Goldsmith, A. Malczewski, and J. Kleber, "Reconfigurable double-stub tuners using MEMS switches for intelligent RF front-ends," *IEEE Trans. Microw. Theory Techn.*, vol. **51**, pp. 272–278, 2003.
10. S. Gevorgian, *Ferroelectrics in Microwave Devices, Circuits and Systems: Physics, Modeling, Fabrication and Measurements*, Springer-Verlag, London, 2009.
11. R. Jakoby, P. Scheele, S. Muller, and C. Weil, "Nonlinear dielectrics for tunable microwave components," *15th International Conference on Microwaves, Radar and Wireless Communications, MIKON*, 2004, vol. **2**, pp. 369–378.
12. P. Scheele, *Steuerbare passive Mikrowellenkomponenten auf Basis hochpermittiver ferroelektrischer Schichten*, PhD Thesis Dissertation, Technische Universität, Darmstadt, 2007.
13. J.-H. Jeon, "Effect of SrTiO3 concentration and sintering temperature on microstructure and dielectric constant of Ba(1-x)Sr(x)TiO3," *J. Eur. Ceram. Soc.*, vol. **24**, pp. 1045–1048, 2004.
14. A. Giere, P. Scheele, C. Damm, and R. Jakoby, "Optimization of uniplanar multilayer structures using nonlinear tunable dielectrics," *Proc. 35th Eur. Microw. Conf.*, Paris, France, Oct. 2005.
15. A. Giere, Y. Zheng, H. Gieser, K. Marquardt, H. Wolf, P. Scheele, and R. Jakoby, "Coating of planar Barium-Strontium-Titanate thick-film varactors to increase tunability," *Proc. 37th Eur. Microw. Conf.*, Munich, Germany, October 2007.
16. P. G. de Gennes and J. Prost, *The Physics of Liquid Crystals*, Oxford University Press, Oxford, 1995.
17. P. Collins, and M. Hird, *Introduction to Liquid Crystals: Chemistry and Physics*, Taylor & Francis, London, 1997.
18. I. W. Stewart, *The Static and Dynamic Continuum Theory of Liquid Crystals: a Mathematical Introduction*, Taylor & Francis, London, 2004.
19. I. Gil, J. García-García, J. Bonache, F. Martín, M. Sorolla, and R. Marqués, "Varactor-loaded split rings resonators for tunable notch filters at microwave frequencies," *Electron. Lett.*, vol. **40**, pp. 1347–1348, 2004.
20. I. Gil, J. Bonache, J. García-García, and F. Martín "Tunable metamaterial transmission lines based on varactor loaded split rings resonators," *IEEE Trans. Microw. Theory Techn.*, vol. **54**, pp. 2665–2674, 2006.
21. K. Aydin and E. Ozbay, "Capacitor-loaded split ring resonators as tunable metamaterial components," *J. Appl. Phys.*, vol. **101**, paper 024911, 2007.
22. J. B. Pendry, A. J. Holden, D. J. Robbins, and W. J. Stewart, "Magnetism from conductors and enhanced nonlinear phenomena," *IEEE Trans. Microw. Theory Techn.*, vol. **47**, pp. 2075–2084, 1999.

23. A. Vélez, J. Bonache, and F. Martín, "Varactor-loaded complementary split ring resonators (VLCSRR) and their application to tunable metamaterial transmission lines," *IEEE Microw. Wireless Compon. Lett.*, vol. **18**, pp. 28–30, 2008.

24. J. Bonache, M. Gil, I. Gil, J. García-García, and F. Martín, "On the electrical characteristics of complementary metamaterial resonators," *IEEE Microw. Wireless Compon. Lett.*, vol. **16**, pp. 543–545, 2006.

25. A. Vélez, F. Aznar, M. Durán-Sindreu, J. Bonache, and F. Martín, "Tunable coplanar waveguide (CPW) band-stop and band-pass filters based on open split ring resonators and open complementary split ring resonators," *IET Microw. Antennas Propag.*, vol. **5**, pp. 277–281, 2011.

26. J. S. Hong and M. J. Lancaster, *Microstrip Filters for RF/Microwave Applications*, John Wiley, New York, 2001.

27. A. Velez, F. Aznar, J. Bonache, M. C. Velázquez-Ahumada, J. Martel, and F. Martín, "Open complementary split ring resonators (OCSRRs) and their application to wideband CPW band pass filters," *IEEE Microw. Wireless Compon. Lett.*, vol. **19**, pp. 197–199, 2009.

28. A. Vélez, F. Aznar, M. Durán-Sindreu, J. Bonache, and F. Martín, "Stop-band and band-pass filters in coplanar waveguide technology implemented by means of electrically small metamaterial-inspired open resonators," *IET Microw. Antennas Propag.*, vol. **4**, pp. 712–716, 2010.

29. A. Vélez, J. Bonache, and F. Martín, "Metamaterial transmission lines with tunable phase and characteristic impedance based on complementary split ring resonators," *Microw. Opt. Technol. Lett.*, vol. **51**, pp. 1966–1970, 2009.

30. T. Hand and S. Cummer, "Characterization of tunable metamaterial elements using MEMS switches," *IEEE Antennas Wireless Propag. Lett.*, vol. **6**, pp. 401–404, 2007.

31. H. Tao, A. C. Strikwerda, K. Fan, W. J. Padilla, X. Zhang, and R. D. Averitt, "MEMS based structurally tunable metamaterials at terahertz frequencies," *J. Infrared Milli. Terahz. Waves*, vol. **32** pp. 580–595, 2011.

32. I. Gil, F. Martín, X. Rottenberg, and W. De Raedt, "Tunable stop-band filter at Q-band based on RF-MEMS metamaterials," *Electron. Lett.*, vol. **43**, p. 1153, 2007.

33. D. Bouyge, A. Crunteanu, M. Durán-Sindreu, A. Pothier, P. Blondy, J. Bonache, J. C. Orlianges, and F. Martín, "Reconfigurable split rings based on MEMS switches, and their application to tunable filters," *J. Opt.*, vol. **14**, paper 114001, 2012.

34. D Bouyge, D. Mardivirin, J. Bonache, A. Crunteanu, A. Pothier, M. Durán-Sindreu, P. Blondy, and F. Martín, "Split ring resonators (SRRs) based on micro-electro-mechanical deflectable cantilever-type rings: application to tunable stopband filters," *IEEE Microw. Wireless Compon. Lett.*, vol. **21**, pp. 243–245, 2011.

35. D. Mardivirin, A. Pothier, A. Crunteanu, B. Vialle, and P. Blondy, "Charging in dielectric less capacitive RF-MEMS switches," *IEEE Trans. Microw. Theory Techn.*, vol. **57**, pp. 231–236, 2009.

36. M. Fabert, A. Desfarges-Berthelemot, V. Kermène, A. Crunteanu, D. Bouyge, and P. Blondy, "Ytterbium-doped fibre laser Q-switched by a cantilever-type micro-mirror," *Opt. Express*, vol. **16**, pp. 22064–22071, 2008.

37. F. Martín, F. Falcone, J. Bonache, T. Lopetegi, R. Marqués, and M. Sorolla, "Miniaturized CPW stop band filters based on multiple tuned split ring resonators," *IEEE Microw. Wireless Compon. Lett.*, vol. **13**, pp. 511–513, 2003.

38. J.-S. Hong and M. J. Lancaster, "Couplings of microstrip square open-loop resonators for cross-coupled planar microwave filters," *IEEE Trans. Microw. Theory Techn.*, vol. **44**, pp. 2099–2109, 1996.
39. C.-M. Tsai, S.-Y. Lee, and C.-C. Tsai, "Performance of a planar filter using a 0° feed structure," *IEEE Trans. Microw. Theory Tech.*, vol. **50**, pp. 2362–2367, 2002.
40. A. K. Tagantsev, V. O. Sherman, K. F. Astafiev, J. Venkatesh, and N. Setter, "Ferroelectric materials for microwave tunable applications," *J. Electroceram.*, vol. **11**, pp. 5–66, 2003.
41. D. Kuylenstierna, A. Vorobiev, P. Linner, and S. Gevorgian, "Ultrawide-band tunable true-time delay lines using ferroelectric varactors," *IEEE Trans. Microw. Theory Techn.*, vol. **53**, pp. 2164–2170, 2007.
42. F. A. Miranda, F. W. Van Keuls, R. R. Romanofsky, and G. Subramanyam, "Tunable microwave components for Ku- and K-band satellite communications," *Integr. Ferroelectr.*, vol. **22** (1–4), pp. 789–798, 1998.
43. E. Ozbay, K. Aydin, S. Butun, K. Kolodziejak, and D. Pawlak, "*Ferroelectric based tuneable SRR based metamaterial for microwave applications*," *Proc. of the 37th Eur. Microw. Conf.*, Munich, Germany, October 2007, pp. 497–499.
44. M. Gil, C. Damm, A. Giere, M. Sazegar, J. Bonache, R. Jakoby, and F. Martín, "Electrically tunable split-ring resonators at microwave frequencies based on Barium-Strontium Titanate thick-film," *Electron. Lett.*, vol. **45**, pp. 417–419, 2009.
45. M. Gil, C. Damm, M. Maasch, M. Sazegar, A. Giere, F. Aznar, A. Vélez, J. Bonache, R. Jakoby, and F. Martín, "Tunable sub-wavelength resonators based on Barium-Strontium-Titanate thick-film technology," *IET Microw. Antennas Propag.*, vol. **5**, pp. 316–323, 2011.
46. F. Paul, A. Giere, W. Menesklou, J. R. Binder, P. Scheele, R. Jakoby, and J. Haußelt, "Influence of Fe-F-co-doping on the dielectric properties of Ba0.6Sr0.4TiO3 thick-films," *Int. J. Mater. Res.*, vol. **99**, pp. 1119–1128, 2008.
47. M. Durán-Sindreu, C. Damm, M. Sazegar, Y. Zheng, J. Bonache, R. Jakoby, and F. Martín, "Electrically tunable composite right/left handed transmission-line based on open resonators and barium-stronium-titanate thick films," *IEEE MTT-S Int. Microw. Symp. Dig.*, Baltimore, MD, June 2011.
48. M. Duran-Sindreu, C. Damm, M. Sazegar, Y. Zheng, J. Bonache, R. Jakoby, and F. Martín, "Applications of electrically tunable composite right/left handed transmission lines based on barium-stronium-titanate thick films and open resonators," *IET Microw. Antennas Propag.*, vol. **7**, pp. 476–484, 2013.
49. C. Damm, M. Schüßler, M. Oertel, and R. Jakoby, "Compact tunable periodically LC loaded microstrip line for phase shifting applications," *IEEE MTT-S Int. Microw. Symp. Dig.*, Long Beach, CA, June 2005, pp. 2003–2005.
50. C. Damm, M. Schüßler, J. Freese, and R. Jakoby, "Artificial line phase shifter with separately tunable phase and line impedance," *36th Eur. Microw. Conf.*, Manchester, UK, September 2006, pp. 423–426.
51. A. Giere, C. Damm, P. Scheele, and R. Jakoby, "LH phase shifter using ferroelectric varactors," *IEEE Radio Wireless Symp.*, San Diego, CA, October 2006, pp. 403–406.
52. S.-H. Hwang, T. Jang, Y.-S. Bang, J.-M. Kim, Y.-K. Kim, S. Lim, and C.-W. Baek, "Tunable composite right/left-handed transmission line with positive/negative phase tunability using integrated MEMS switches," *IEEE 23rd Int. Conf. MEMS*, Wanchai, Hong Kong, January 2010, pp. 763–766.

53. S.-H. Hwang, T. Jang, J.-M. Kim, Y.-K. Kim, S. Lim, and C.-W. Baek, "MEMS-tunable composite right/left-handed (CRLH) transmission line and its application to a phase shifter," *J. Micromech. Microeng.*, vol. **21**, paper 125022, 2011.
54. S. Lim, C. Caloz, and T. Itoh, "Electronically-controlled metamaterial based transmission line as a continuous-scanning leaky-wave antenna," *IEEE MTT-S Int. Microw. Symp. Dig.*, Fort-Worth, TX, June 2004, vol. **1**, pp. 313–316.
55. S. Lim, C. Caloz, and T. Itoh, "Metamaterial-based electronically controlled transmission-line structure as a novel leaky-wave antenna with tunable radiation angle and beamwidth," *IEEE Trans. Microw. Theory Techn.*, vol. **53**, pp. 161–173, 2005.
56. T. Kim, L. Vietzorreck, "Investigation of smart antennas using RF-MEMS based tunable CRLH-transmission lines," *Int. Conf. Electromag. Adv. Appl.*, Cape Town, South Africa, September 2012, pp. 266–267.
57. C. Damm, M. Maasch, R. Gonzalo, and R. Jakoby, "Tunable composite right/left-handed leaky wave antenna based on a rectangular waveguide using liquid crystals," *IEEE MTT-S Int. Microw. Symp. Dig.*, Anaheim, CA, May 2010, pp. 13–16.
58. M. Roig, M. Maasch, C. Damm, O. Hamza-Karabey, and R. Jakoby, "Liquid crystal based tunable composite right/left-handed leaky-wave antenna for ka-band applications," *43rd Eur. Microw. Conf.*, Nuremberg, Germany, October 2013, pp. 759–762.
59. F. Goelden, A. Gaebler, S. Mueller, A. Lapanik, W. Haase, and R. Jakoby, "Liquid-crystal varactors with fast switching times for microwave applications," *Electron. Lett.*, vol. **44**, pp. 480–481, 2008.
60. C. Damm, J. Freese, M. Schüßler, and R. Jakoby, "Electrically controllable artificial transmission line transformer for matching purposes," *IEEE Trans. Microw. Theory Techn.*, vol. **55**, pp. 1348–1354, 2007.
61. I. Vendik, D. Kholodnyak, P. Kapitanova, M.A. Hein, S. Humbla, R. Perrone, and J. Mueller, "Tunable dual-band microwave devices based on a combination of left/right-handed transmission lines," *38th Eur. Microw. Conf.*, Amsterdam, The Netherlands, October 2008, pp. 273–276.
62. A. S. Mohra and O. F. Siddiqui, "Tunable bandpass filter based on capacitor-loaded metamaterial lines," *Electron. Lett.*, vol. **45**, pp. 470–472, 2009.
63. A. L. Borja, J. Carbonell, J. D. Martinez, V. E. Boria, and D. Lippens, "A controllable bandwidth filter using varactor-loaded metamaterial-inspired transmission lines," *IEEE Antennas Wireless Propag. Lett.*, vol. **10**, pp. 1575–1578, 2011.
64. M. Morata, I. Gil, and R. Fernández-García, "Modeling tunable band-pass filters based on RF-MEMS metamaterials," *Int. J. Numer. Model. Electron. Networks, Devices Field*, vol. **24**, pp. 583–589, 2011.
65. I. Gil, M. Morata, R. Fernández-García, X. Rottenberg, and W. De Raedt, "Reconfigurable RF-MEMS metamaterials filters," *Prog. Electromag. Res. Symp. Proc.*, Marrakesh, Morocco, March 2011, pp. 1239–1242.
66. B. K. Kim and B. Lee, "Tunable bandpass filter with varactors based on the CRLH-TL metamaterial structure," *J. Electromag. Eng. Sci.*, vol. **13**, pp. 245–250, 2013.
67. M. G. Case, *Nonlinear Transmission Lines for Picosecond Pulse, Impulse and Millimeter-Wave Harmonic Generation*, PhD dissertation, University of California, Santa Barbara, July 1993.
68. A. C. Scott, F. Y. E. Chu, and D. W. McLaughlin, "The soliton: a new concept in applied science," *Proc. IEEE*, vol. **61**, pp. 1443–1483, 1973.

69. J. S. Russell, "*Report on Waves*," *Rep. 14th Meet. British Assoc. Adv. Sci.*, York, UK, September 1844, pp. 311–390.

70. P. J. Olver and D. H. Sattinger, *Solitons in Physics, Mathematics, and Nonlinear Optics*, Springer-Verlag, New York, 1990.

71. J. R. Tailor, *Optical Solitons—Theory and Experiment*, Cambridge University Press, Cambridge, 1992.

72. M. Remoissenet, *Waves Called Solitons: Concepts and Experiments*, Springer-Verlag, Berlin, Germany 1994.

73. M. J. W. Rodwell, M. Kamegawa, R. Yu, M. Case, E. Carman, and K. Giboney, "GaAs nonlinear transmission lines for picosecond pulse generation and millimeter-wave sampling," *IEEE Trans. Microw. Theory Techn.*, vol. **39**, pp. 1194–1204, 1991.

74. E. Carman, M. Case, M. Kamegawa, R. Yu, K. Giboney, and M. J. W. Rodwell, "V-band and W-band broad-band, monolithic distributed frequency multipliers," *IEEE Microw. Guided Wave Lett.*, vol. **2**, pp. 253–254, 1992.

75. H. Shi, W. M. Chang, C. W. Domier, N. C. Luhman, L. B. Sjogren, and H. L. Liu, "Novel concepts for improved transmission line performance," *IEEE Trans. Microw. Theory Techn.*, vol. **43**, pp. 780–789, 1995.

76. I. Ryjenkova, V. Mezentsev, S. Musher, S. Turitsyn, R. Hülsewede, and D. Jäger, "Millimeter wave generation on nonlinear transmission lines," *Ann. Telecommun.*, vol. **52**, pp. 134–139, 1997.

77. J. R. Thorpe, P. Steenson, and R. Miles, "Non-linear transmission lines for millimeter-wave frequency multiplier applications," *Proc. IEEE Sixth Int. Conf. on Terahertz Electron.*, Leeds, UK, 1998, p. 54.

78. M. Li and R.G. Harrison, "A fully distributed heterostructure barrier varactor nonlinear transmission line frequency tripler," *IEEE MTT-S Int. Microw. Symp. Dig.*, Baltimore, MD, June 1998, pp. 1639–1643.

79. E. Lheurette, M. Fernández, X. Melique, P. Mounaix, O. Vanbésien and D. Lippens, "Non linear transmission line quintupler loaded by heterostructure barrier varactors," *Proc. Eur. Microw. Conf.*, Munich, Germany. October 1999.

80. L. Dillner, M. Ingvarson, E. Kollberg, and J. Stake, "Heterostructure barrier varactor multipliers," *Proc. Eur. GaAs Conf.*, Paris, October 2000.

81. X. Melique, A. Maestrini, E. Lheurette, P. Mounaix, M. Favreau, O. Vanbesien, J. M. Goutoule, G. Beaudin, T. Nahri, and D. Lippens, "12% efficiency and 9.5 dBm output power from InP-based heterostructure barrier varactor triplers at 250 GHz," *IEEE MTT-S Int. Microwave Symp. Dig.*, Anaheim, CA, June 1999, vol. **1**, pp. 123–126.

82. E. Carman, K. Giboney, M. Case, M. Kamegawa, R. Yu, K. Abe, M. J. W. Rodwell, and J. Franklin, "28-39 GHz distributed harmonic generation on a soliton nonlinear transmission line," *IEEE Microw. Guided Wave Lett.*, vol. **1**, pp. 28–31, 1991.

83. M. M. Turner, G. Branch, and P. W. Smith, "Methods of theoretical analysis and computer modeling of the shaping of electrical pulses by nonlinear transmission lines and lumped-element delay lines," *IEEE Trans. Electron. Dev.*, vol. **38**, pp. 819–816, 1991.

84. R. Hirota and K. Suzuki, "Theoretical and experimental studies of lattice solitons in nonlinear lumped networks," *Proc. IEEE*, vol. **61**, pp. 1483–1491, 1973.

85. M. Toda, "Nonlinear lattice and soliton theory," *IEEE Trans. Circuits Syst.*, vol. **30**, pp. 542–554, 1983.

86. D. J. Korteweg and G. de Vries, "On the change of form of long waves advancing in a rectangular canal, and on a new type of long stationary waves," *Phil. Mag.*, vol. **39**, pp. 422–443, 1895.

87. F. Martín and X. Oriols "A simple model to study soliton wave propagation in periodically loaded nonlinear transmission lines," *Appl. Phys. Lett.*, vol. **78**, pp. 2802–2804, 2001.

88. X. Oriols and F. Martín "Analytical solitons in nonlinear transmission lines loaded with heterostructure barrier varactors," *J. Appl. Phys.*, vol. **90**, pp. 2595–2600, 2001.

89. E. Afshari and A. Hajimiri, "Nonlinear transmission lines for pulse shaping in silicon," *IEEE J. Solid-State Circuits*, vol. **40**, pp. 744–752, 2005.

90. X. Wang and R. J. Hwu, "Theoretical analysis and FDTD simulation of GaAs nonlinear transmission lines," *IEEE Trans. Microw. Theory Techn.*, vol. **47**, pp. 1083–1091, 1999.

91. F. Martín and X. Oriols "Understanding soliton wave propagation in nonlinear transmission lines for millimeter wave multiplication," *Int. J. Infrared. Milli. Waves*, vol. **22**, pp. 85–92, 2001.

92. F. Martín and X. Oriols "Effects of line parameters on soliton-like propagation in nonlinear transmission lines: application to the optimization of frequency triplers," *Int. J. Infrared Milli. Waves*, vol. **22**, pp. 225–235, 2001.

93. F. Martín, X. Oriols, J. A. Gil, and J. García-García, "Optimization of nonlinear transmission lines for harmonic generation: the role of the capacitance voltage characteristic and the area effect," *Int. J. Infrared Milli. Waves*, vol. **23**, pp. 95–103, 2002.

94. S. Gupta and C. Caloz, "Dark and bright solitons in left-handed nonlinear transmission line metamaterials," *IEEE MTT-S Int. Microw. Symp.*, Honolulu, HI, June 2007.

95. C. Caloz, I.-H. Lin, and T. Itoh, "Characteristics and potential applications of nonlinear left handed transmission lines," *Microw. Opt. Technol. Lett.*, vol. **40**, pp. 471–473, 2004.

96. A. B. Kozyrev and D. W. van der Weide, "Nonlinear transmission lines in left-handed media," *IEEE MTT-S Int. Microw. Symp.*, Fort Worth, TX, June 2004.

97. A. B. Kozyrev and D. W. van der Weide "Nonlinear wave propagation phenomena in left-handed transmission-line media," *IEEE Trans. Microw. Theory Techn.*, vol. **53**, pp. 238–245, 2005.

98. K. Narahara, T. Nakamichi, T. Suemitsu, T. Otsuji, and E. Sano, "Development of solitons in composite right- and left-handed transmission lines periodically loaded with Schottky varactors," *J. Appl. Phys.*, vol. **102**, pp. 024501–024504, 2007.

99. B. K. Alexander, and D. W. Van Der Weide, "Trains of envelope solitons in nonlinear left-handed transmission line media," *Appl. Phys. Lett.*, vol. **91**, pp. 254111–254113, 2007.

100. A. B. Kozyrev, and D. W. Van Der Weide, "Nonlinear left-handed transmission line metamaterials," *J. Phys. D: Appl. Phys.*, vol. **41**, pp. 173001–173010, 2008.

101. F. G. Gharakhili, M. Shahabadi, and M. Hakkak, "Bright and dark soliton generation in a left-handed nonlinear transmission line with series nonlinear capacitors," *Prog. Electromag. Res.*, vol. **96**, pp. 237–249, 2009.

102. Z. Wang, Y. Feng, B. Zhu, J. Zhao, and T. Jiang, "Schrödinger solitons and harmonic generation in short left-handed nonlinear transmission line metamaterial," *Proc. Asia Pacif. Microw. Conf.*, Singapore, December. 2009, pp. 1246–1249.

103. A. B. Kozyrev and D. W. van der Weide, "Nonlinear left-handed metamaterials," in *Metamaterials*, X.-Y. Jiang, Ed. InTech, May 2012.

6

OTHER ADVANCED TRANSMISSION LINES

6.1 INTRODUCTION

This chapter is focused on advanced topics not covered in the previous chapters, namely, magnetoinductive-wave (MIW) and electroinductive-wave (EIW) delay lines, differential transmission lines with common-mode suppression, wideband artificial transmission lines (including CRLH lines based on the lattice network unit cell and active—non-Foster—transmission lines), and substrate-integrated waveguides (SIWs). Rather than an in-depth treatment of each of these lines (all of them can be the subject of an intensive analysis), the aim of the chapter is to briefly analyze the main properties of such lines, their advantages, and some applications.

6.2 MAGNETOINDUCTIVE-WAVE AND ELECTROINDUCTIVE-WAVE DELAY LINES

MIW and EIW delay lines are waveguiding structures consisting of chains of coupled resonant elements that typically exhibit very small phase and group velocities. These lines can be an alternative to surface acoustic waves or ferrite-based lines as delay lines, as long as slow wave factors as high as $c/v_g \sim 100$ can be achieved. Although this value is not comparable to those achievable in surface acoustic or ferrite delay lines, the main advantage of MIW and EIW lines is their simple design/fabrication process, since these lines can be implemented in standard printed circuit boards (PCBs). Moreover, due to the scalability of planar circuits, MIW and EIW lines can be designed to be operative over a very wide frequency band of the microwave spectrum.

6.2.1 Dispersion Characteristics

The properties and dispersion characteristics of waves in magnetically coupled capacitively loaded loops were first studied by the group of Prof. Solymar in 2002 [1, 2]. Following these pioneering works, the same group published several papers on the topic of MIWs devoted to experimentally verify the dispersion characteristics [3], to highlight some potential applications (i.e., dividers and couplers) [4], and to study the interaction between MIWs and electromagnetic (EM) waves [5]. In 2004, it was demonstrated that MIWs can be supported by a chain of SRRs printed on a microwave substrate [6]. Indeed, MIWs are expected to be present in whatever periodic structure where the elements are magnetically coupled.

Without loss of generality, let us consider from now on MIW lines based on SRRs. Although, MIWs can be supported in one-dimensional chains of axially oriented or coplanar arranged SRRs (Fig. 6.1), the interest from the point of view of planar waveguiding structures is in the coplanar configuration. The dispersion relation in any of the structures of Figure 6.1 can be easily inferred from the circuit model of the unit cell, depicted in Figure 6.2a. L and C are the inductance and capacitance of the SRRs and M is the mutual inductance between adjacent SRRs (the first neighbor approximation is considered). The circuit model of Figure 6.2a can be transformed to

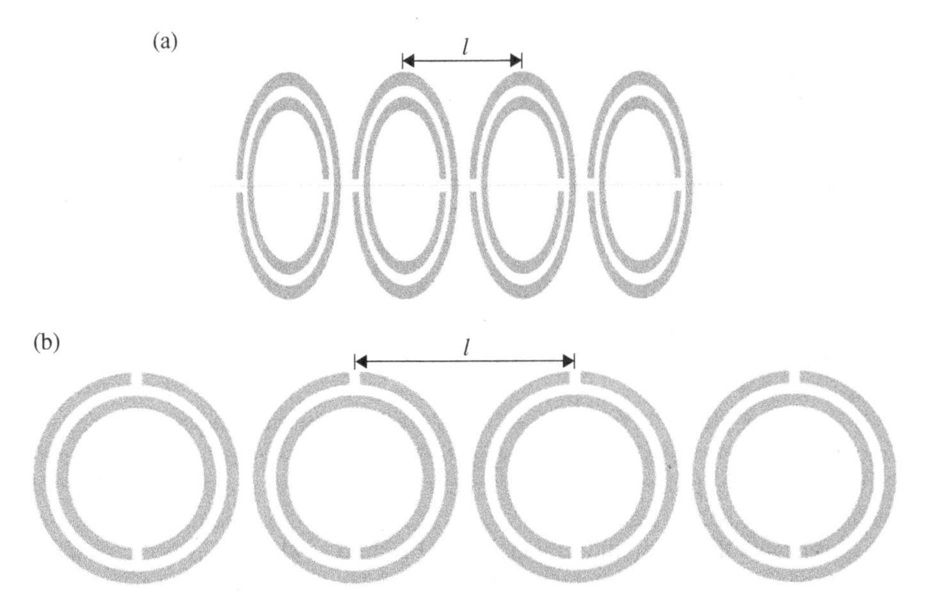

FIGURE 6.1 One-dimensional chain of inductively coupled SRRs. (a) Axial configuration and (b) coplanar configuration.

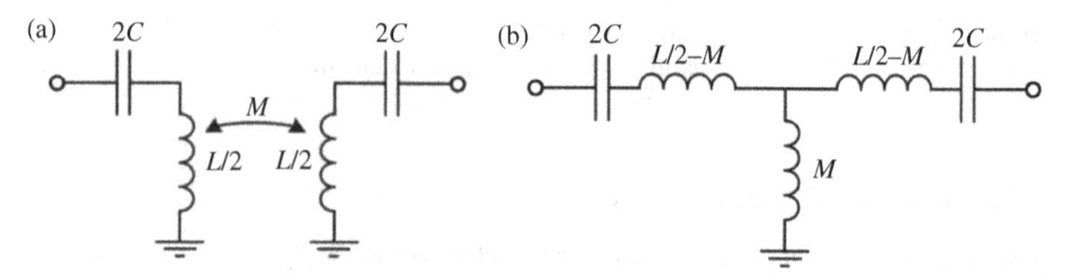

FIGURE 6.2 (a) Equivalent circuit model (unit cell) of the chain of SRRs of Figure 1, and (b) transformed model.

the T-circuit shown in Figure 6.2b. Application of expression (2.33) gives the dispersion relation, which is usually written as follows:

$$\frac{\omega_o^2}{\omega^2} = 1 \pm \frac{2|M|}{L}\cos(\beta l) \tag{6.1}$$

where

$$\omega_o = \frac{1}{\sqrt{LC}} \tag{6.2}$$

and the + and – sign corresponds to the axial and coplanar configuration, respectively. The difference arises from the different sign of the mutual inductance for the axial (positive) and coplanar (negative) configurations. According to (6.1), wave propagation is backward in the coplanar configuration, whereas it is forward for axially oriented SRRs. Notice that the limits of the transmission band are given by the frequencies

$$\omega_{\pm} = \frac{\omega_o}{\sqrt{1 \mp \frac{2|M|}{L}}} \tag{6.3}$$

Since the mutual inductance M is typically small as compared to the inductance L of the SRRs, the transmission band in MIW delay lines is very narrow, and the group velocity is very small.

By applying duality, it follows that a chain of closely located coupled complementary split-ring resonators or CSRRs (Fig. 6.3) can support electroinductive waves (EIWs) [7] (notice however that wave propagation in chains of electrically coupled resonators was already pointed out in the sixties [8]). In this case, the unit cell can be described by the circuit model depicted in Figure 6.4, which is indeed the circuit dual of the circuit model (unit cell) of the MIW delay line. In the model, C_M is the mutual capacitance between adjacent CSRRs, L_c is the inductance of the CSRR and the capacitance $C_G = C_c - 2C_M$, C_c being the capacitance of an isolated CSRR.

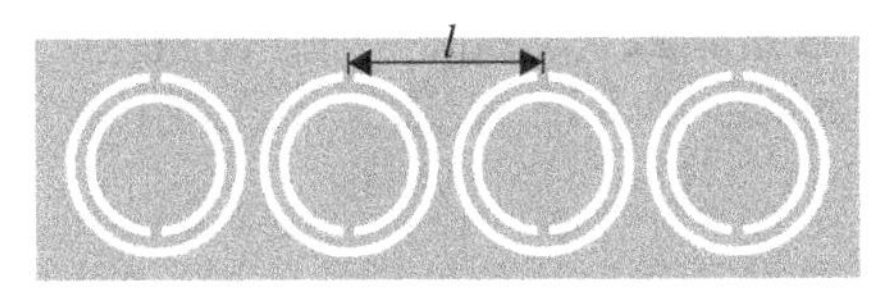

FIGURE 6.3 One-dimensional chain of electrically coupled CSRRs, acting as an EIW delay line.

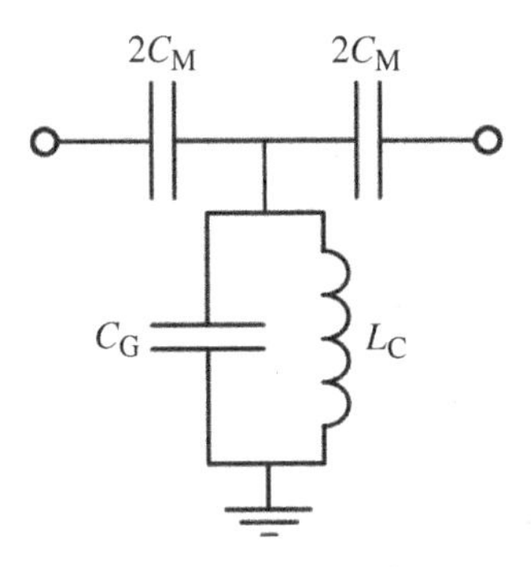

FIGURE 6.4 Equivalent circuit model (unit cell) of the chain of CSRRs of Figure 6.3.

In this case, the dispersion relation (under the first neighbor approximation) is found to be[1]:

$$\frac{\omega_o^2}{\omega^2} = 1 - \frac{2C_M}{C_c}\cos(\beta l) \tag{6.4}$$

where

$$\omega_o = \frac{1}{\sqrt{L_c(C_G + 2C_M)}} = \frac{1}{\sqrt{L_c C_c}} \tag{6.5}$$

The backward propagation band is delimited by the frequencies given by

$$\omega_{\pm} = \frac{\omega_o}{\sqrt{1 \mp \frac{2C_M}{C_c}}} \tag{6.6}$$

and bandwidth broadens as the electric coupling between adjacent CSRRs increases.

The dual behavior of SRR-based MIW (coplanar configuration) and CSRR-based EIW lines is apparent by comparing (6.4–6.6) with (6.1–6.3). Notice that the

[1] In Ref. [7], the dispersion relation of CSRR-based EIW lines is presented in a different form. However, both expressions are identical. As given earlier, (6.4) is formally identical to (6.1).

dispersion relations of both lines (considering the coplanar configuration of SRRs) and bandwidth are identical as long as $|M|/L = C_M/C_c$, as expected from duality considerations.

6.2.2 Applications: Delay Lines and Time-Domain Reflectometry-Based Chipless Tags for RFID

As long as MIW and EIW lines are waveguiding structures, microwave devices such as couplers, dividers, phase shifters, and so on, based on such lines can be implemented [3, 4, 9, 10]. However, probably the most relevant advantage of MIW and EIW lines is the possibility to achieve very high slow-wave factors in conventional fully planar technology (i.e., PCB technology). These lines are therefore interesting as delay lines [6], and for applications requiring delay lines with significant group delay. One of such applications is the implementation of passive chipless tags for radiofrequency identification (RFID) based on time-domain reflectrometry (TDR) [11].[2]

The tags are implemented with MIW delay lines comprising a periodic array of coupled square split-ring resonators (SSRRs) [11]. Tag encoding is achieved by introducing reflectors (which provide the identification signature) between the elements of the array. When the tags are interrogated with a pulse in time domain, they produce replicas at the positions where the reflectors are placed. Thanks to the slow group velocity of the MIW delay line, the replicas of the original pulse are not overlapped in time domain and can be demodulated, thus providing the identification code of the tag. First attempts to implement TDR-based chipless tags made use of left-handed (LH) lines [12, 13]. Such lines can be designed to exhibit small group velocities, but it is not possible to achieve slow-wave factors above 10. Thus, MIW-based lines appear to be a good alternative to LH lines for the implementation of chipless tags based on delay lines. The sketch of the proposed chipless RFID system is depicted in Figure 6.5. The printed antenna is used to communicate with the reader. The reflectors are simply strips located within the MIW delay line, as depicted in Figure 6.5.

In order to design the MIW delay lines for the chipless tags, it is necessary to compute the relation between the number of bits that can be stored and the physical parameters of the lines. The number of bits that can be stored in a delay line is

$$n_b = \frac{\Delta\tau}{\tau} \tag{6.7}$$

where $\Delta\tau$ is the total delay of the line working in reflection and τ is the temporal width of the pulse sent by the reader. The pulse width depends on the bandwidth of the system. In order to compute the total delay introduced by the MIW lines, the group velocity, given by

[2] TDR-based chipless RFID tags are an alternative to the RF bar codes considered in Chapter 4. The main difference is that RF bar codes operate in the frequency domain and require large bandwidths; whereas in TDR-based chipless tags, the ID code is inferred from the echoes of a pulse in time domain.

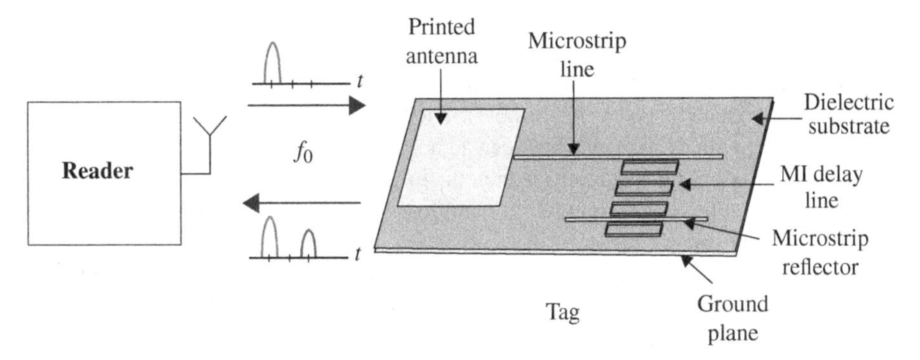

FIGURE 6.5 Sketch of the chipless RFID system proposed in [11]. Reprinted with permission from Ref. [11]; copyright 2012 IEEE.

$$v_g = \left(\frac{\partial\beta}{\partial\omega}\right)^{-1} = l\frac{|M|}{L}\frac{\omega^3}{\omega_o^2}\sqrt{1-\left[\frac{L}{2|M|}\left(\frac{\omega_o^2}{\omega^2}-1\right)\right]^2} \tag{6.8}$$

can be approximated by its value at the central operation frequency of the system, ω_o. Thus, the total delay is obtained as

$$\Delta\tau = 2\frac{n_{SSRR}l}{v_g|_{\omega_o}} = \frac{2L}{|M|}\frac{n_{SSRR}}{\omega_o} \tag{6.9}$$

where n_{SSRR} is the number of SSRRs of the MIW delay line. Introducing (6.9) in (6.7), the number of bits is found to be

$$n_b = 2\frac{n_{SSRR}l}{v_g|_{\omega_o}\tau} = \frac{2L}{|M|}\frac{n_{SSRR}}{\omega_o\tau} \tag{6.10}$$

and it increases with the number of SSRRs (as expected). However, it is desirable to use the smallest possible number of SSRRs, for a certain number of bits, in order to minimize tag dimensions. Since the central operation frequency and bandwidth are fixed by design or regulatory constraints, τ and L are not design parameters.[3] Thus, the only variable that can be modified in order to increment the number of bits is the mutual inductance, M, between the SRRs. This variable can be controlled by the separation between adjacent resonators. If this separation increases, the mutual inductance is reduced and, thus, the number of bits also increases. However, the distance

[3] The central frequency (6.2) is fixed by the inductance, L, and the edge capacitance of the SRRs, C, which is in turn given by the per-unit-length capacitance, C_{pul}. Usually, the per-unit-length capacitance is limited by fabrication restrictions, and the inductance is the only degree of freedom to adjust the central operation frequency, f_0. Thus, the inductance of the SRRs is fixed once these elements are designed to resonate at the central operation frequency.

between adjacent SSRRs is limited by three main reasons. First of all, the length of the delay line increases with the inter-SSRR distance, which is contrary to tag miniaturization. The second reason is that if the separation between SSRRs increases, the MIW bandwidth is reduced (see expression 6.3). The final one is due to propagation losses, which increase with the distance between adjacent resonators since they are intimately related to the coupling between resonators. In conclusion, a trade-off between the number of bits, line length, bandwidth, and losses, by choosing the appropriate line period, l, is necessary.

Figure 6.6 depicts the 2-bit chipless tags designed in Ref. [11] according to the aforementioned considerations. The distance between adjacent SRR is $l = 3.75$ mm since this represents a good trade-off between delay (16 ns), losses (5.6 dB), bandwidth (5.13%), and the total length of the line. The central frequency was chosen at $f_o = 2.45$ GHz (ISM band), and SSRRs dimensions are given in Figure 6.7. Notice that the number of bits ($n_b = 2$) is the value that results from (6.7) by considering a

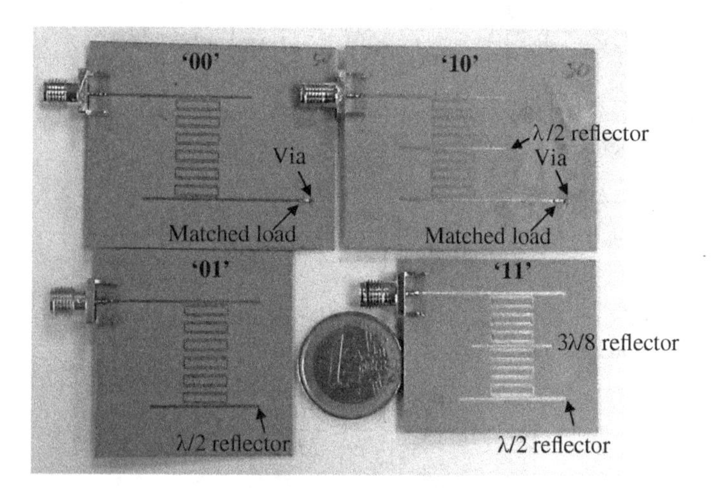

FIGURE 6.6 Photograph of the set of 2-bit chipless tags (the antenna is not included). Reprinted with permission from Ref. [11]; copyright 2012 IEEE.

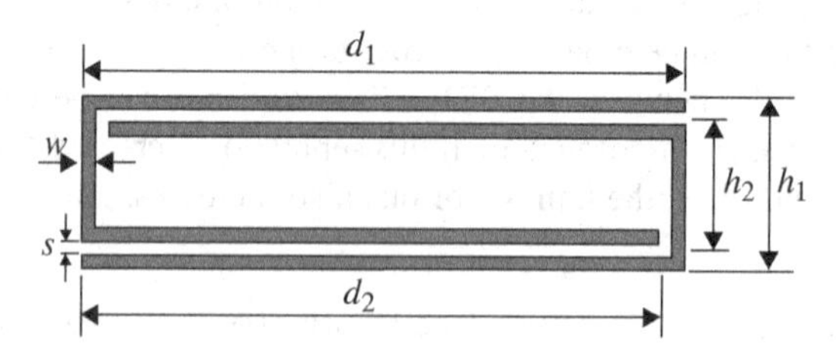

FIGURE 6.7 Geometry of the SSRR and relevant dimensions ($d_1 = 9.6$ mm, $h_1 = 2.6$ mm, and $s = w = 0.20$ mm). The substrate is the *Rogers RO3010* with dielectric constant $\varepsilon_r = 10.2$ and thickness $h = 635$ μm). Reprinted with permission from Ref. [11]; copyright 2012 IEEE.

pulse width of $\tau = 15$ ns (see Ref. [11] for more details). To avoid reflections at the end of the MIW line, which is necessary when the second bit is "0," a microstrip line ended with a matched load was coupled at the end of the MIW delay line. As mentioned earlier, the proposed approach to produce reflections consists of using open-ended 50-Ω microstrip lines placed between the SSRRs. For the case of the 2-bit tags, the first bit reflector is placed in the middle of the MIW delay line (between the third and fourth SSRRs in this implementation) and the second bit reflector is placed at the end of the line. When a total reflection is needed, the line acting as reflector must have a length of $\lambda/2$ at f_0. This is the situation when no more partial reflections are used (e.g., at the end of the line). In this case, the $\lambda/2$ lines provide a $\rho = -1$ reflection coefficient in the delay line, since the distance between the open ends and the center of the line is $\lambda/4$. If only a partial reflection is needed, the length of the reflector is made smaller, and only part of the signal is reflected back to the source. The transmitted signal may thus impinge on the additional reflectors if they are present.

Figure 6.8 shows the proposed setup to characterize the chipless tags and obtain their temporal responses. In this scheme, the generator sends a modulated Gaussian pulse at the central frequency of the system ($f_0 = 2.45$ GHz). After that, the modulated signal is transferred to the chipless tag and its response is obtained through the circulator. The output of the system is the envelope of the signal reflected back by the tag. The four tags were simulated by means of *Agilent ADS Momentum* and their temporal responses were obtained using the proposed setup in a transient simulation of *Agilent ADS*. Figure 6.9a shows the temporal response of the four tags. There are pulses at three different times. In all cases, an initial pulse is observed. This is due to certain mismatch at the tag input, so that part of the input signal is reflected back to the envelope detector without being transmitted to the tags. The second group of pulses is obtained after 16 ns approximately, which corresponds to the first identification bit. Finally, the third group of pulses is observed after 32 ns which is the total delay introduced by the MIW delay lines. These pulses correspond to the second bit. Considering −15 dB as the decision threshold, the four different possible codes ("00", "01", "10," and "11") are obtained.

The four fabricated tags were measured by means of a network analyzer. Then, their temporal responses were obtained through the *Agilent ADS* scheme proposed

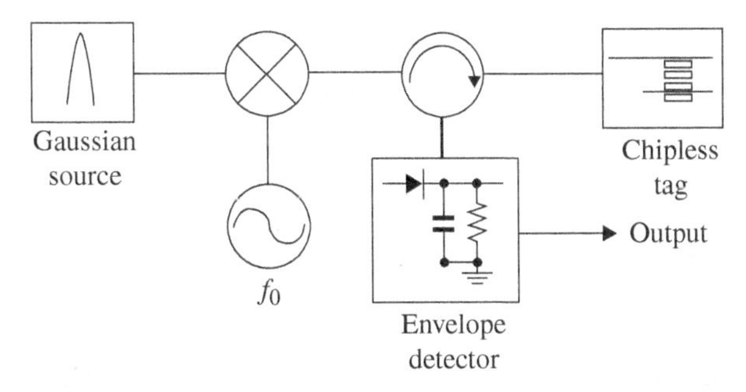

FIGURE 6.8 Setup to obtain the temporal response of the chipless tags. Reprinted with permission from Ref. [11]; copyright 2012 IEEE.

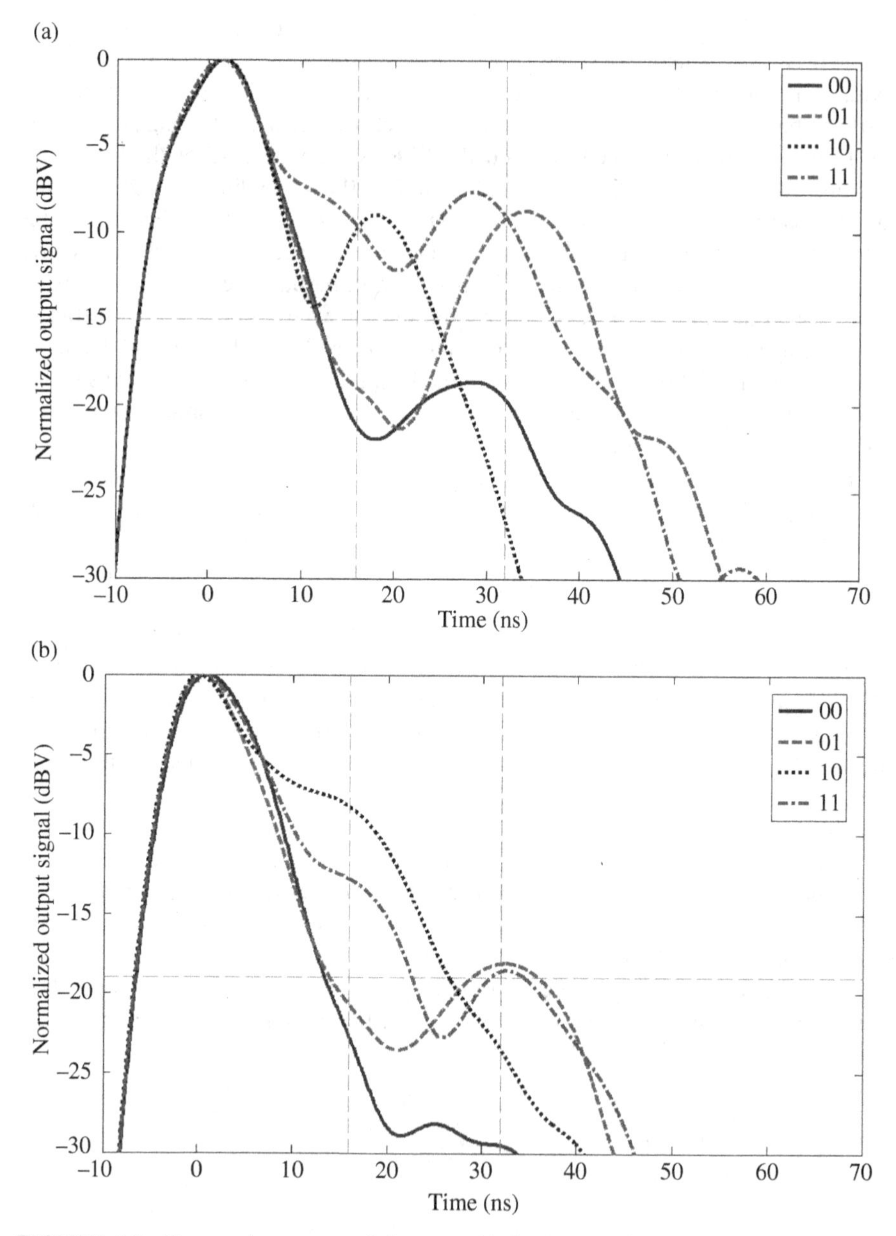

FIGURE 6.9 Temporal response of the tags. (a) Response inferred from simulation and (b) response inferred from the fabricated tags. Reprinted with permission from Ref. [11]; copyright 2012 IEEE.

in Figure 6.8 (see Fig. 6.9b). The output signals present a similar shape, although the insertion losses in the prototypes are higher than those predicted by the simulator. Hence, the detected power corresponding to the second bit is smaller because the propagation path is twice that of the first-bit path. Nevertheless, the groups of pulses corresponding to the first and second bits can be clearly identified at 16 ns and 32 ns, respectively. In this case, the decision threshold has been reduced to −19 dB. Considering this threshold, the four tags can be identified properly.

As compared to RF bar codes (see Subsection 4.3.2.3), TDR chipless tags based on MIW delay lines occupy much less bandwidth. However, very long delay lines are necessary to store a large amount of information, since the number of bits in TDR-based tags is proportional to the delay time.

6.3 BALANCED TRANSMISSION LINES WITH COMMON-MODE SUPPRESSION

Balanced (or differential) lines are of interest as interconnecting lines in high-speed digital and analog circuits due to their high immunity to environmental noise, electromagnetic interference (EMI), and crosstalk. However, in practical balanced systems, the presence of some level of common-mode noise is unavoidable. Therefore, the design of balanced lines able to suppress the common mode over a wide band, keeping the differential signals unaltered, has been a subject of interest in recent years. In this section, some strategies for common-mode noise rejection, with special emphasis on microstrip differential lines loaded with CSRRs are reviewed. A method for the design of common-mode filters based on CSRRs is proposed and used for the improvement of common-mode noise rejection in a differential bandpass filter. Finally, some examples of differential filters with inherent common-mode noise rejection are presented.

6.3.1 Strategies for Common-Mode Suppression

Several strategies to efficiently suppress the common mode while keeping the integrity of the differential signals in differential transmission lines have been reported [14–20]. The key idea to selectively suppress the common mode in the microwave domain is to load (or perturb) the differential line with elements acting as an efficient stopband filter for the common mode and, simultaneously, as an all-pass structure for the differential-mode in the frequency region of interest. According to Subsection 4.3.2, a transmission line loaded with a symmetric resonator with its symmetry plane aligned to that of the line is able to inhibit signal propagation (in the vicinity of resonance) if both symmetry planes are of the same EM nature. Conversely, if such planes are aligned and are of different EM nature, the line is transparent. According to the previous words, it follows that by symmetrically loading a differential microstrip line with symmetric resonators exhibiting a magnetic wall in its symmetry plane, selective mode suppression can be achieved. The reason is that the differential line exhibits a magnetic wall in its symmetry plane for the even mode, and an electric wall for the odd mode. Therefore, it is expected that the line loaded with such resonators is transparent for the differential mode and acts as a stopband filter for the common mode.

The common-mode filters based on low temperature co-fired ceramic (LTCC) technology reported in Ref. [14] or the negative permeability structures of Ref. [15] are compact and provide efficient common-mode rejection over wide frequency bands, but they are technologically complex since multilayer structures are considered in those designs. The common-mode suppression approaches reported here are limited to two types of defect ground structures in microstrip differential lines: dumbbell-shaped resonators [16] and CSRRs [18]. It is also remarkable the strategy reported in Ref. [20], where the ground plane is structured with capacitive patches and meandered inductors that effectively act as shunt-connected series resonators for the common mode, providing lowpass filtering functionality for that mode, and all-pass behavior for the differential mode.

6.3.1.1 Differential Lines Loaded with Dumbbell-Shaped Slotted Resonators Slotted dumbbell resonators symmetrically etched in the ground plane of a microstrip line can be modeled to a first-order approximation as series-connected parallel resonators [21]. These resonators are thus able to inhibit signal propagation in the vicinity of resonance. The symmetry plane of the line and resonator are magnetic walls, and this gives another interpretation to the stopband functionality of these lines, according to symmetry considerations (see earlier text). However, by symmetrically etching slotted dumbbell resonators in a differential microstrip line, it is expected that the common mode is efficiently suppressed, while signal propagation for the differential mode is preserved. In Ref. [16], the authors claim that the resonant elements open the return current path through the ground plane for the common mode, whereas the presence of resonators has small effect on the differential signals since relatively small current density returns through the ground plane for such signals. Regardless of the specific interpretation for selective common-mode suppression in differential lines loaded with slotted dumbbell resonators, such structures behave as efficient common-mode filters. Figure 6.10 depicts a specific common-mode filter topology and the frequency response. The structure is able to efficiently suppress the common-mode noise with broad stopband in the GHz range (Fig. 6.10b), yet keeping good signal integrity performance for the differential signals. In addition, mode conversion from differential-mode to common-mode, also shown in the figure, is very small.

6.3.1.2 Differential Lines Loaded with CSRRs CSRRs also exhibit a magnetic wall at their symmetry plane for the fundamental resonance. Therefore by aligning this symmetry plane with the symmetry plane of the differential microstrip line, common-mode suppression and preservation of signal integrity for the differential-mode are also expected. Notice that for the differential mode, there is not a net axial electric field in the inner metallic region of the CSRR able to excite the resonators, unless symmetry is truncated. Conversely, for the common mode the electric field below the transmission lines is co-directional, and the CSRRs are excited by this mode.

As discussed in Refs [18, 19], common-mode suppression bandwidth can be enhanced by tightly coupling the resonant elements.[4] Moreover, in order to enhance the rejection bandwidth (common mode) of an individual resonator, it is necessary to increase the coupling capacitance between the pair of lines and the CSRR, and to decrease the inductance and capacitance of the CSRR. This was achieved in [18] by increasing the rings width and

[4] Square-shaped or rectangular CSRR geometries favor interresonator coupling.

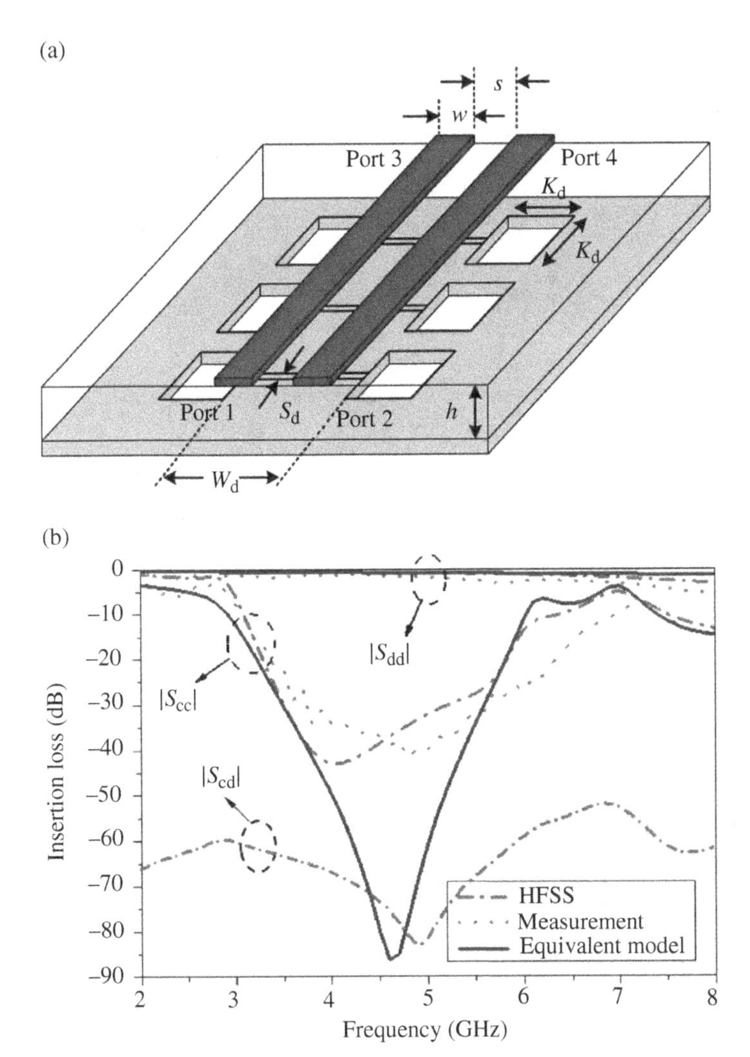

FIGURE 6.10 Differential microstrip line with common-mode suppression based on slotted dumbbell-shaped resonators (a) and frequency response (b). Insertion loss for the differential and common mode are denoted by S_{dd} and S_{cc}, respectively; mode conversion is determined by S_{cd}. Reprinted with permission from Ref. [16]; copyright 2008 IEEE.

separation. Figure 6.11a depicts a fabricated differential line loaded with symmetrically etched, square-shaped and tightly coupled CSRRs. The common-mode and differential-mode insertion loss are depicted in Figure 6.11b, and reveal that CSRRs are efficient elements to suppress the common-mode over a wide band, whilst the differential mode is transmitted in that band. A comparison between the CSRR-based approach and other approaches (excluding multilayered structures) indicates that the combination of size, design and fabrication simplicity, and common-mode rejection bandwidth in CSRR-loaded differential lines is very competitive [18].

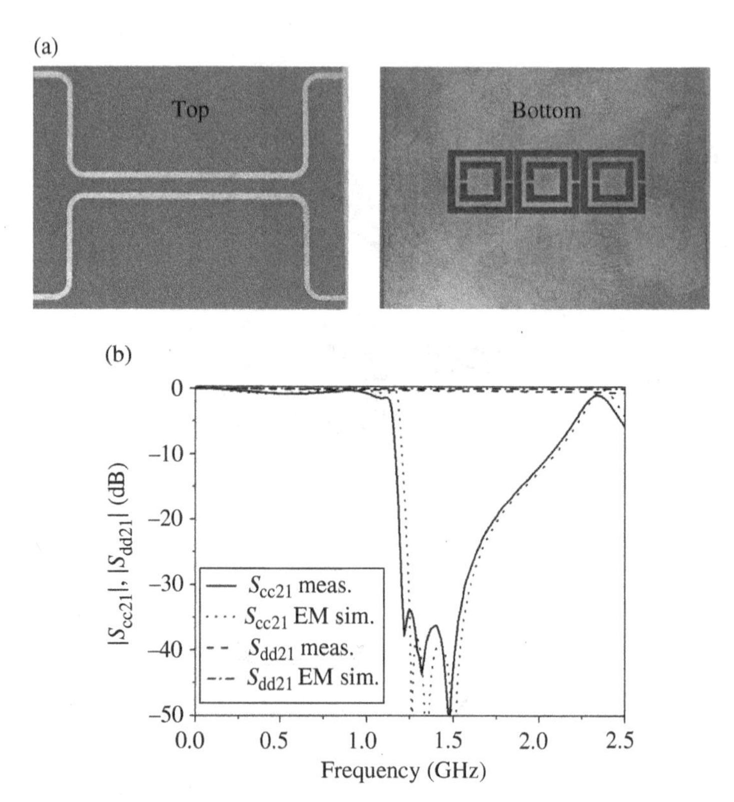

FIGURE 6.11 Photograph (a) and common-mode ($|S_{cc21}|$) and differential-mode ($|S_{dd21}|$) insertion loss (b) of a differential line with wideband common-mode rejection based on CSRRs. Reprinted with permission from Ref. [18]; copyright 2011 IEEE.

6.3.2 CSRR- and DS-CSRR-Based Differential Lines with Common-Mode Suppression: Filter Synthesis and Design

As discussed earlier, common-mode filters based on CSRRs provide significant rejection bandwidths for that mode. Actually, bandwidth requirements are dictated by the differential signals. Namely, in order to prevent unwanted common-mode noise in differential lines, it is necessary that the suppressed band for the common mode extends at least beyond the limits of the required band for differential signal transmission. Therefore, rather than pursuing a specific rejection bandwidth, common-mode filters must be typically designed to satisfy a predefined common-mode rejection level within a certain frequency band.

To design CSRR-based common-mode filters following a systematic design approach, a circuit model that describes the CSRR-loaded differential line, including inter-resonator coupling, is necessary. In this regard, two limitations arise. First of all, the convenient CSRRs to achieve a wide common-mode stopband (with wide slots and interslot distance) cannot be considered to be electrically small, and hence they cannot be described by an accurate circuit model over a wide band. Second, the CSRRs must be oriented with their symmetry plane aligned with the symmetry plane of the differential line. Under these conditions, mixed coupling (electric and magnetic) may arise, as discussed in Subsection 3.5.2.4, and the circuit model (provided the CSRRs are electrically small) is even more complex. This second aspect can be ignored to a first-order approximation, since a very accurate determination of the common-mode stop bandwidth is not required. Concerning the first aspect, if CSRRs are used as filtering elements, they must be electrically small in order to be accurately described by a lumped element equivalent circuit model. However, this is contrary to bandwidth enhancement. An intermediate solution is the use of DS-CSRRs (which is derived by application of duality to the DS-SRR introduced in Subsection 3.3.1.3—see Subsection 3.3.2.1), as pointed out in Ref. [19]. By driving the slot width and separation to the minimum value allowed by the available technology, DS-CSRRs can be considered to be electrically small, and the circuit model provides a reasonable description of the loaded differential lines. Such circuit model (unit cell), valid for electrically small CSRRs or DS-CSRRs, and neglecting magnetic coupling between the line and the resonators, is depicted in Figure 6.12a. In this model, the circuit parameters are identical to those of Figure 3.44b, except L_m and C_m that account for the magnetic and electric coupling, respectively, between the pair of lines.

The lumped-element equivalent circuit model for the even and odd modes are depicted in Figure 6.12b and c, respectively. For the differential mode, the resulting circuit model is simply the circuit model of an ordinary line with modified parameters. The circuit model for the even mode is formally identical to that of CSRR-loaded

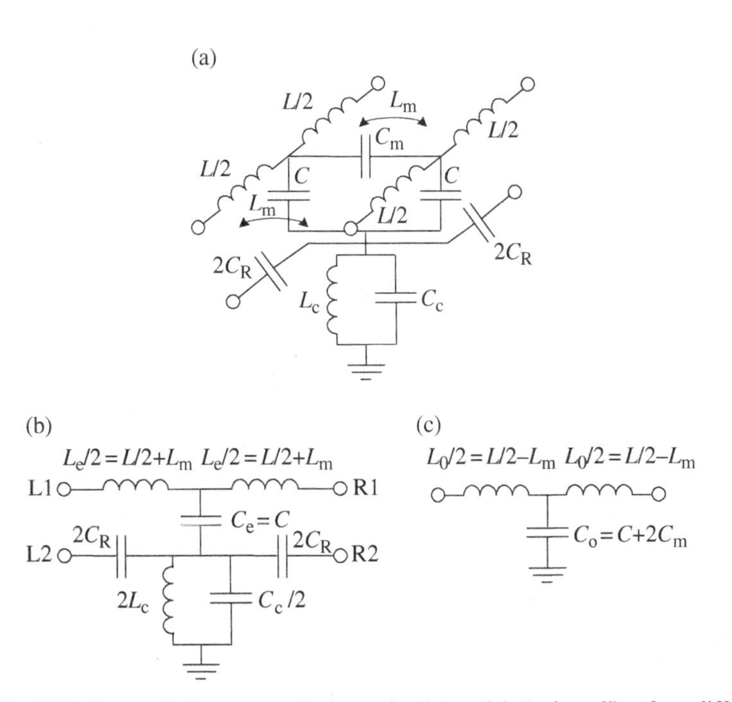

FIGURE 6.12 Lumped-element equivalent circuit model (unit cell) of a differential microstrip line loaded with CSRRs or DS-CSRRs (a), equivalent circuit model for the even-mode (b), and equivalent circuit model for the odd-mode (c).

single-ended microstrip lines with inter-resonator coupling (Fig. 3.44b). Therefore, the analysis carried out in Subsection 3.5.2 to obtain the dispersion relation in CSRR-loaded single-ended lines is also valid to obtain the dispersion relation for the common mode in CSRR- or DS-CSRR-loaded differential lines. In particular, the common-mode rejection bandwidth for an infinite structure (in practice for a differential line with a large number of cells) can be estimated if the elements of the circuit model for the common mode are known.

For common-mode filter design and estimation of the common-mode rejection bandwidth, the procedure is as follows. For the reasons explained before, relative to the validity of the models of Figure 6.12 and fabrication tolerances, the slot width and separation are set to a minimum implementable value with the technology in use (for instance $c = d = 200$ μm [19]). Notice that this reduces the degrees of freedom and eases the common-mode filter design. Square-shaped resonators (rather than circular) are considered, in order to enhance the electric coupling between the differential line and the resonators and between adjacent resonators as well. To further enhance inter-resonator coupling, the separation between adjacent resonators is also set to that limiting value (200 μm). In order to achieve a strong electric coupling between the pair of lines and the resonator, the lines must be fitted inside the CSRR (or DS-CSRR) region and must be as wide as possible, and hence as uncoupled as possible (line dimensions can easily be inferred from a transmission line calculator). The side length of the resonator is determined from the model of the CSRR reported in Ref. [22] (which gives an estimation of L_c and C_c) and the per-unit length capacitance of the coupled lines for the even mode (which gives C_e). The transmission zero frequency, given by

$$f_z = \frac{1}{2\pi\sqrt{2L_c(C_e + C_c/2)}} \tag{6.11}$$

is adjusted to the required filter central frequency, and this provides the CSRR side length (for a DS-CSRR the side length can also be determined by taking into account that the inductance is four times smaller than the inductance of the CSRR). Obviously, optimization of the resonator side length in order to fit the required central frequency is necessary.

With the previous procedure, the common-mode filter dimensions are perfectly determined. To predict the maximum achievable bandwidth, that is, the bandwidth obtained by considering an infinite number of cells, the dispersion relation given by (3.96) is used. However, since optimization at the layout level is required, it is necessary to extract the parameters of the circuit model following the procedure reported in Ref. [23] (and detailed in Appendix G). Thus, L_e, C_e, L_c, and C_c are first extracted by considering a single-cell structure, and then C_R is adjusted to fit the EM simulation of an order-2 common-mode filter. Once the circuit parameters are known, expression (3.96) can be evaluated, and the common-mode stopband can be estimated.

Following the previous approach, the maximum achievable rejection bandwidth for different CSRR and DS-CSRR common-mode filters was obtained. The rejection bandwidth was also obtained from full-wave EM simulations (the results are depicted in Fig. 6.13). The extracted circuit parameters and estimated fractional bandwidths are shown in Table 6.1.

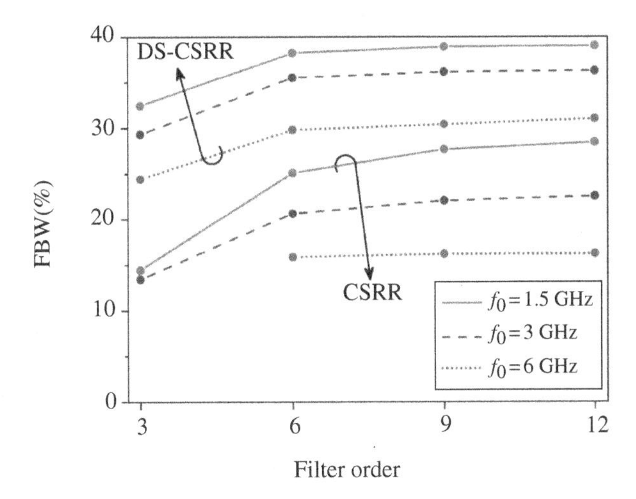

FIGURE 6.13 Fractional rejection bandwidth (*FBW*) at −20 dB for the common mode given by EM simulation for CSRR- and DS-CSRR-loaded differential lines. Dimensions are for the CSRRs and DS-CSRRs, $c = d = 0.2$ mm, and *inter-resonator distance* = 0.2 mm; for the CSRRs, *side length* = 7.3 mm (f_0 = 1.5 GHz), 4.3 mm (f_0 = 3 GHz), and 2.6 mm (f_0 = 6 GHz); for the DS-CSRRs, *side length* = 13.8 mm (f_0 = 1.5 GHz), 7.5 mm (f_0 = 3 GHz), and 4.3 mm (f_0 = 6 GHz); for the differential line, $2W + S = side\ length - 2(2c - d) + 0.4$ mm, exhibiting a 50 Ω characteristic impedance (odd mode). The considered substrate is the *Rogers RO3010* with thickness $h = 1.27$ mm, and dielectric constant $\varepsilon_r = 10.2$. Reprinted with permission from Ref. [19]; copyright 2012 IEEE.

TABLE 6.1 Extracted parameters and maximum fractional bandwidth inferred from the circuit model

f_0(GHz)	L_e(nH)	C_e(pF)	L_c(nH)	C_c(pF)	C_R(pF)	FBW(%)
CSRR-loaded differential lines						
1.5	6.3	1.1	2.1	3.2	0.05	30.7
3	4.2	0.5	1.0	2.0	0.03	26.4
6	3.0	0.2	0.4	1.4	0.04	20.7
DS-CSRR-loaded differential lines						
1.5	15.9	1.6	1.0	8.6	0.32	32.6
3	9.0	0.8	0.45	4.8	0.09	29.4
6	6.9	0.3	0.15	3.6	0.05	23.6

By comparing the fractional bandwidths predicted by the reported approach with the saturation values of Figure 6.13, it can be concluded that the reported approach is more accurate for CSRR-loaded lines, as expected. Figure 6.14 compares the circuit and EM simulation (common mode) of the order 1, 2, and 3 CSRR-based common-mode filters designed to exhibit a central frequency of 1.5 GHz (in the circuit simulation the inter-resonator capacitance at input and output ports has been left opened since the CSRRs of the input and output cells are not externally fed, resulting in a two-port circuit). There is good agreement between the circuit and EM simulation,

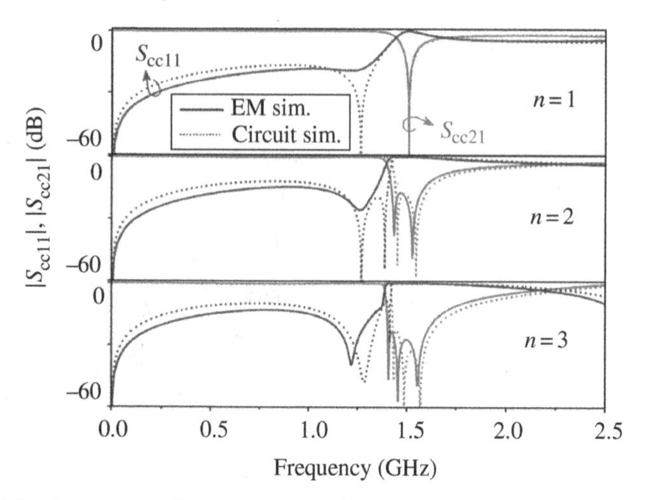

FIGURE 6.14 Common-mode return loss $|S_{cc11}|$ and insertion loss $|S_{cc21}|$ given by the EM and circuit simulation for the order 1, 2, and 3 common-mode filters (1.5 GHz central frequency) based on CSRRs. Dimensions and substrate are indicated in the caption of Figure 6.13. Circuit parameters are given in Table 6.1. Reprinted with permission from Ref. [19]; copyright 2012 IEEE.

pointing out the accuracy of the model for CSRR-loaded lines with narrow slots, c, and interslot distance, d.

As can be seen in Figure 6.13, six resonators are enough to nearly achieve the maximum rejection bandwidth. Obviously, common-mode filter size can be reduced by decreasing the number of resonators but at the expense of a reduced common-mode rejection bandwidth. Thus, following a systematic approach based on the circuit model of the common mode, it can be inferred whether a specified rejection bandwidth and central frequency can be roughly fulfilled or not. If the required bandwidth is wider, we are forced to consider resonators with wider slots (c) and interslot distance (d), or, alternatively, multiple-tuned resonators. In these cases, however, filter design and maximum bandwidth estimation are not so straightforward.

6.3.3 Applications of CSRR and DS-CSRR-Based Differential Lines

In this subsection, it is demonstrated that the suppression of the even mode in differential lines loaded with DS-CSRRs does not affect the integrity of the differential signals. Subsequently, it is shown that by cascading a DS-CSRR to the input and output port of a balanced bandpass filter, common-mode suppression within the differential filter pass band can be substantially improved.

6.3.3.1 Differential Line with Common-Mode Suppression In this subsection, the design procedure of a common-mode filter similar to that reported in Figure 6.11, but using DS-CSRRs, is reported. The target is to implement a common-mode filter roughly centered at 1.35 GHz and exhibiting at least 35% fractional bandwidth (at 20 dB rejection level). According to the previous methodology, these specifications cannot be fulfilled by using CSRRs with $c = d = 200$ μm. However, it is possible to achieve these filter requirements by means of DS-CSRRs. Indeed, the estimated maximum bandwidth for a common-mode filter centered at 1.35 GHz was found to be 37.3% (considering the substrate of Fig. 6.13), but we do expect a larger value since the model tends to slightly underestimate the maximum achievable bandwidth for DS-CSRR-loaded lines (this can be appreciated by comparing Table 6.1 and Fig. 6.13). Moreover, for comparison purposes, a rectangular-shaped DS-CSRR was considered with its transverse side length identical to that of the CSRR reported in Figure 6.11. This favors the electric coupling between the pair of lines and the DS-CSRRs, and hence the common-mode stopband expansion (the reason is that the DS-CSRR longitudinal side is longer than the transverse one, and this increases the coupling capacitance C_e as compared to that of a square-shaped DS-CSRR with identical transmission zero frequency). The longitudinal side length is thus the single-design parameter, and this has been determined following the same approach applied to square-shaped particles (the geometrical parameters of the structure are given in the caption of Fig. 6.15).

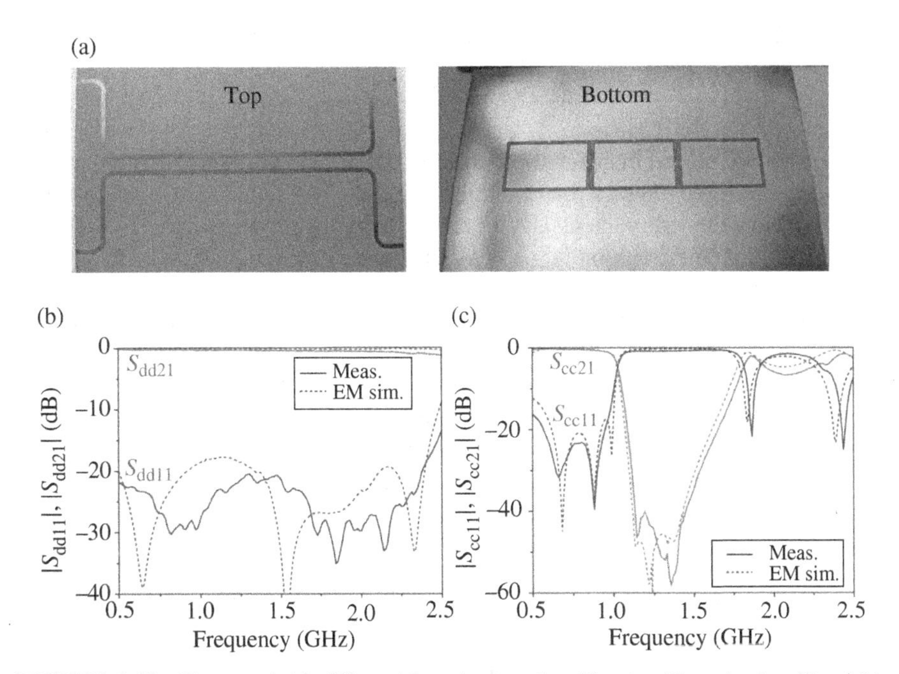

FIGURE 6.15 Photograph (a), differential-mode return loss $|S_{dd11}|$ and insertion loss $|S_{dd21}|$ (b), and common-mode return loss $|S_{cc11}|$ and insertion loss $|S_{cc21}|$ (c) of the designed differential line with common-mode suppression based on DS-CSRRs. Dimensions are for the DS-CSRRs, $c = d = 0.2$ mm, *longitudinal side length* = 17.6 mm, and *transverse side length* = 10.8 mm; *inter-resonator distance* = 0.2 mm; for the differential line, $W = 1$ mm, and $S = 2.5$ mm. The considered substrate is the *Rogers RO3010* with thickness $h = 1.27$ mm, dielectric constant $\varepsilon_r = 10.2$, and loss tangent $\tan\delta = 0.0023$. Reprinted with permission from Ref. [19]; copyright 2012 IEEE.

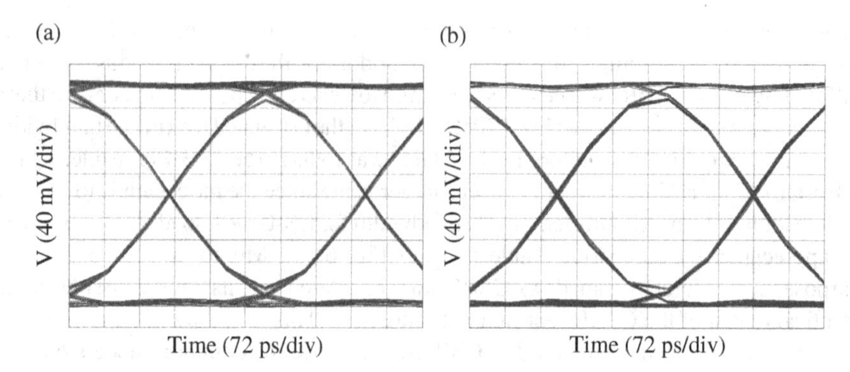

FIGURE 6.16 Measured differential eye diagrams for the differential line of Figure 6.15 with (a) and without (b) DS-CSRRs. Reprinted with permission from Ref. [19]; copyright 2012 IEEE.

TABLE 6.2 Measured eye parameters

	With DS-CSRRs	Without DS-CSRRs
Eye height	278 mV	281 mV
Eye width	371 ps	383 ps
Jitter (PP)	29.3 ps	16.9 ps
Eye-opening factor	0.76	0.76

The photograph and frequency responses of the device, an order-3 common-mode filter, are depicted in Figures 6.15a–c (this filter order was found to be enough to satisfy the bandwidth requirements). As it can be seen, the differential signal is almost unaltered, whilst the common mode is rejected within a fractional bandwidth (41%) comparable to that achieved in Figure 6.11 by using CSRRs with wide and widely spaced rings. The DS-CSRR-based structure is a bit larger than that reported in Figure 6.11, but the design was done following the systematic procedure explained in the previous subsection.

Figure 6.16 shows the measured differential eye diagrams[5] with the excitation of 0.2 V amplitude in 2.5 Gb/s for the differential line of Figure 6.15a with and without DS-CSRRs. The eye diagram quality in terms of eye height, eye width, jitter, and eye-opening factor is compared for these two structures (see Table 6.2). According to these

[5] An eye diagram is an indicator of the quality of signals in high-speed digital transmissions. The eye diagram is constructed from a digital waveform by folding the parts of the waveform corresponding to each individual bit into a single graph with signal amplitude on the vertical axis and time on horizontal axis. By repeating this construction over many samples of the waveform, the resultant graph will represent the average statistics of the signal and will resemble an eye. The parameters indicated in Table 6.2 are defined in many sources focused on signal integrity. In brief, eye height is a measure of the vertical opening of an eye diagram, and it is determined by noise, which "closes" the eye. Eye width is a measure of the horizontal opening of an eye diagram, and it is calculated by measuring the difference between the statistical mean of the crossing points of the eye. Jitter is the time deviation from the ideal timing of a data-bit event. To compute jitter, the time deviations of the transitions of the rising and falling edges of an eye diagram at the crossing point are measured.

results, the presence of the DS-CSRRs does not significantly degrade the differential mode. The peak-to-peak jitter varies notably, but it is still within very acceptable limits for the DS-CSRR-based structure. Moreover, the eye-opening factor, which measures the ratio of eye height and eye amplitude, is identical in both structures.

6.3.3.2 Differential Bandpass Filter with Enhanced Common-Mode Rejection Differential bandpass filters with common-mode rejection have been reported in the literature [24–36]. In this subsection, a balanced filter consisting of a pair of coupled folded stepped impedance resonators (SIRs) fed by a differential line is reported. The layout is depicted in Figure 6.17a (other balanced filters based on SIRs have been reported in Refs [31, 32, 35, 36]). The filter by itself rejects the common mode due to the symmetry of the structure, since the symmetry plane of the resonator exhibits an electric wall at the first SIR resonance. Therefore, such resonators cannot be excited by means of common-mode signals, and the even mode is reflected back to the source due to the presence of the slots between the pair of SIRs. However, the rejection level of the common mode in the region of interest is very limited since it depends on the distance between resonators and such inter-resonator distance is dictated by filter specifications. However, the common mode can be further rejected by introducing (cascading) DS-CSRRs.

The proposed differential filter, a second-order Chebyshev bandpass filter with a central frequency of 1.37 GHz, a fractional bandwidth of 10%, and 0.1 dB ripple was reported in Ref. [19]. The considered substrate is the *Rogers RO3010* with thickness $h = 1.27$ mm, and dielectric constant $\varepsilon_r = 10.2$. With these specifications and substrate, the layout of the filter is that depicted in Figure 6.17a (the design of the filter, out of the scope of this book, was done following the procedure described in Ref. [37]).

The frequency response of the filter (differential mode, S_{dd21} and S_{dd11}) is shown in Figure 6.17c. The common-mode insertion loss (S_{cc21}), also depicted in the figure, exhibits a rejection level of about 20 dB in the pass band region. In order to enhance the common-mode rejection, two identical DS-CSRRs, as shown in Figure 6.17b, were cascaded to the filter. Such DS-CSRRs were designed to generate a stopband for the common mode in the pass band region of the differential filter and are identical to those of Figure 6.15. The presence of the DS-CSRRs does not affect the filter response (odd mode). However, by merely introducing two DS-CSRRs, the common-mode rejection is roughly increased up to 50 dB in the region of interest (see Fig. 6.17c). These results point out that DS-CSRRs provide an efficient path to enhance the common-mode noise rejection in balanced filters.

6.3.4 Balanced Filters with Inherent Common-Mode Suppression

The common-mode suppressed balanced filter reported in the previous subsection exhibits inherent common-mode suppression, but the rejection level in the differential filter pass band is small; and for this reason, DS-CSRRs have been cascaded to the filter. In this subsection, balanced filters with a high common-mode rejection ratio (CMRR) without the need to cascade additional filter stages are reported. This high CMRR is achieved by introducing transmission zeros for the common mode in

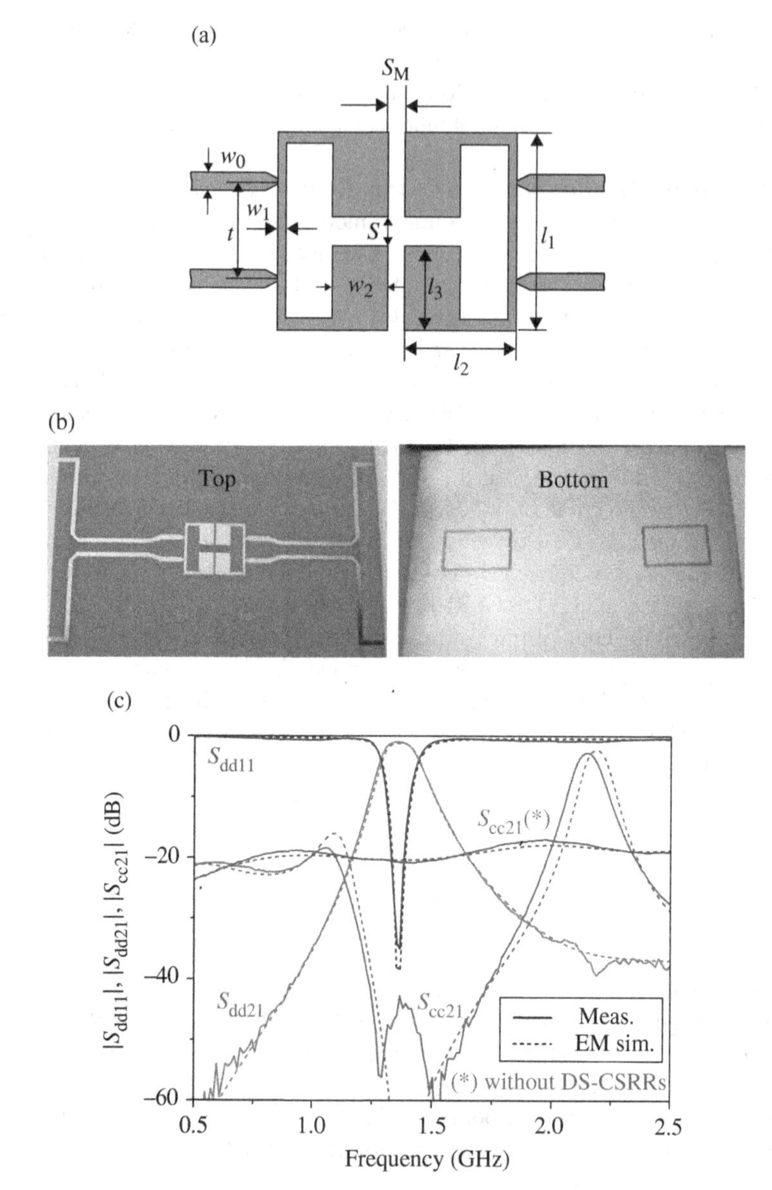

FIGURE 6.17 Layout (a), photograph (b), and frequency response (c) of the designed differential bandpass filter with improved common-mode rejection. Dimensions are for the DS-CSRRs, $c = d = 0.2$ mm, *longitudinal side length* = 17.6 mm, and *transverse side length* = 10.8 mm; for the differential line, $W = 1$ mm, and $S = 2.5$ mm; for the differential filter, $S_M = 0.5$ mm, $s = 2$ mm, $w_0 = 1.2$ mm, $w_1 = 0.7$ mm, $w_2 = 3.7$ mm, $l_1 = 12.8$ mm, $l_2 = 7.5$ mm, $l_3 = 5.4$ mm, and $t = 5.5$ mm. The considered substrate is the *Rogers RO3010* with thickness $h = 1.27$ mm, dielectric constant $\varepsilon_r = 10.2$, and loss tangent $\tan\delta = 0.0023$. Reprinted with permission from Ref. [19]; copyright 2012 IEEE.

the differential filter pass band. By this means, the filter inherently and efficiently suppresses the common mode as will be shown in the two reported examples. The first one [34] combines OSRRs and OCSRRs, and can be considered the balanced version of the filters reported in Subsection Bandpass Filters Based on CRLH Lines Implemented by Means of OSRRs and OCSRRs. The second one [35] is based on mirrored SIRs to generate transmission zeros for the common mode.

6.3.4.1 Balanced Bandpass Filters Based on OSRRs and OCSRRs Single-ended bandpass filters based on OSRRs and OCSRRs and consisting of a cascade of series-connected OSRRs alternating with shunt-connected OCSRRs were studied in Subsection 4.2.3.6. Since the OSRR and the OCSRR are described (to a first-order approximation) by means of a series and a parallel resonant tank, respectively, it follows that such structures synthesize the canonical circuit model of a bandpass filter. As it is discussed in [38], in microstrip technology, it is necessary to include metallic vias in the design in order to effectively ground the OCSRRs, and the ground plane is windowed in the regions beneath the OSRRs in order to obtain a more accurate description of the particle by means of a series resonator.

A typical layout (order 3) of the proposed OSRR/OCSRR-based differential bandpass filters is depicted in Figure 6.18. Notice that this structure is indeed the unit cell (π-model) of a balanced CRLH line. The structure is symmetric with respect to the indicated plane (dashed line). For the differential mode, where there is a virtual ground in that plane (electric wall), the OCSRRs are grounded (without the presence of vias), and the structure exhibits a pass band. If the distance between the host lines is large, no coupling effects take place between mirror elements and filter design is as simple as designing a single-ended bandpass filter.

The proposed strategy to suppress the common mode in the region of interest (i.e., the differential filter pass band) consists of tailoring the metallic region surrounding the OCSRRs. The lumped element equivalent circuit model of the four-port section corresponding to one pair of mirrored OCSRRs is depicted in Figure 6.19a [34].

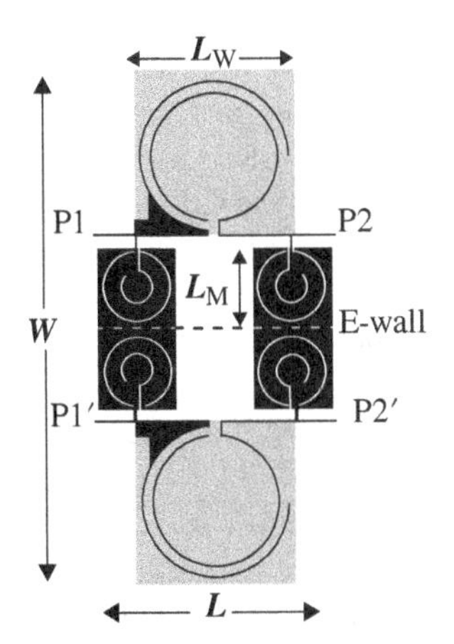

FIGURE 6.18 Typical layout of the proposed OSRR/OCSRR-based differential-mode bandpass filter (π-network, order-3) with common-mode suppression.

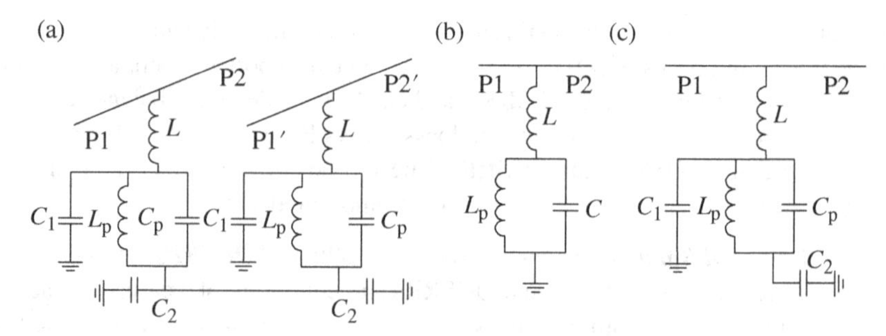

FIGURE 6.19 Equivalent circuit model of the mirrored OCSRR pair section. (a) complete model, (b) differential-mode model, and (c) common-mode model.

L_p and C_p model the OCSRR, L accounts for the inductive strip present between the microstrip lines and the center part of the OCSRRs, C_1 is the capacitance between the central metallic region of the OCSRR and the ground plane; finally, C_2 is the patch capacitance corresponding to the metallic region surrounding the OCSRRs. The models for the differential and common-modes are depicted in Figures 6.19b and c, respectively. Notice that the capacitance C_2 is grounded for the differential-mode and does not play any role for that mode. Thus, for this mode, the shunt OCSRR is described by a parallel resonator in series with an inductor (this inductor is useful to introduce a transmission zero above the differential filter pass band). However, for the common mode, the symmetry plane is an open circuit, and the effect of the capacitance C_2 is the presence of two transmission zeroes in the common-mode frequency response. Once the elements L, L_p, C_p, and C_1 are set to satisfy the differential-mode filter response (including the transmission zero above the pass band), C_2 must be adjusted to set the first transmission zero of the common-mode frequency response in the centre of the differential-mode pass band.

The reported example is a balanced filter with the following specifications [34]: order $n = 3$, Chebyshev response with fractional bandwidth FBW = 45%, central frequency $f_o = 1$ GHz, and 0.05 dB ripple. From these specifications, the elements of the canonical π-network order-3 bandpass filter can be obtained through well-known transformations from the lowpass filter prototype. Once these elements are known, the topology of the series-connected OSRRs is obtained by curve fitting the response of the series LC resonator giving the ideal Chebyshev response in the region of interest. For the OCSRRs, the elements of the model of Figure 6.19b (L, L_p, and $C = C_1 + C_p$) are derived from the susceptance slope at f_o (i.e., by forcing it to be equal to that of the LC tank giving the ideal Chebyshev response), and from the differential-mode transmission zero, given by

$$f_Z^{dd} = \frac{1}{2\pi}\sqrt{\frac{1}{C}\left(\frac{1}{L}+\frac{1}{L_p}\right)} \tag{6.12}$$

The parasitic capacitances of the π-model of the OSRR (which can be easily extracted from the EM simulation of the isolated particle) are small and have negligible effect on the differential filter response. Once L, L_p, and C are known, the OCSRRs are synthesized with the help of the model reported in Ref. [39] (analyzed in Subsections 3.3.2.2 and 3.5.2.5) and the parameter extraction procedure reported in Appendix G. From this model, the OCSRR capacitance, C_p, can be estimated and hence C_1 can be derived. Finally, C_2 is adjusted to the required value to force the common-mode transmission zero at f_o. The metallic region surrounding the OCSRRs is then expanded or contracted to adjust the common-mode transmission zero to that value (the initial size is inferred from the parallel plate capacitor formula). Following this procedure, the element values and the layout of the shunt branch were inferred.

The photograph of the designed and fabricated filter is shown in Figure 6.20a (dimensions are $0.15\lambda_g \times 0.30\lambda_g$, λ_g being the guided wavelength at the filter central frequency). Figure 6.20b shows the simulated and measured insertion and return loss for the differential mode, as well as the insertion loss for the common mode. The circuit simulations for the differential- and common-mode models are also included in the figure. Both the differential- and common-mode filter responses are in good agreement with the circuit simulations up to roughly 2 GHz. The agreement with the Chebyshev response is also good, although the selectivity and stop-band rejection above the pass band of the designed balanced filter are better due to the effects of the transmission zero (the first spurious appears at roughly $3f_o$). As expected, there is a transmission zero at f_o for the common mode (the CMRR at f_o being 53 dB), and the common-mode rejection within the differential filter pass band is better than 20 dB. The combination of filter size and common-mode rejection is very competitive, and this type of filters is especially suitable for wideband applications in differential systems.

6.3.4.2 Balanced Bandpass Filters Based on Mirrored SIRs An alternative approach to implement differential-mode bandpass filters with inherent common-mode suppression consists on using mirrored SIRs coupled through admittance inverters. The typical topology (order-3) for these filters is depicted in Figure 6.21, where the impedance inverters are implemented by means of 90° meandered lines for size reduction. The equivalent circuit model of the structure is depicted in Figure 6.22a, where $J_{i,i+1}$ denotes the admittance of the inverters, C_{pi} are the patch capacitances of the square transmission line sections cascaded to the inverters, L_{pi} are the inductances of the narrow strips of the mirrored SIRs, and finally, C_{zi} are the capacitances of the central patches. The circuit models for the differential and common modes are depicted in Figures 6.22b and c. For the differential mode, the symmetry plane is an electric wall, and the capacitances C_{zi} are grounded. The resulting structure is thus the canonical circuit of a bandpass filter, consisting of a cascade of parallel LC resonators coupled through admittance inverters. For the common mode, the symmetry plane is a magnetic wall, and the equivalent circuit exhibits a stopband behavior. Indeed, except for the presence of the capacitances C_{pi}, such circuit is the canonical circuit of a stopband filter. As long as the inverters exhibit their functionality in a narrow (or moderate) band, the synthesis of balanced filters with standard responses (Chebyshev or Butterworth) following this approach is bandwidth limited. Nevertheless, it is possible to generate balanced filters with significant bandwidths, as reported in the next example.

(a)

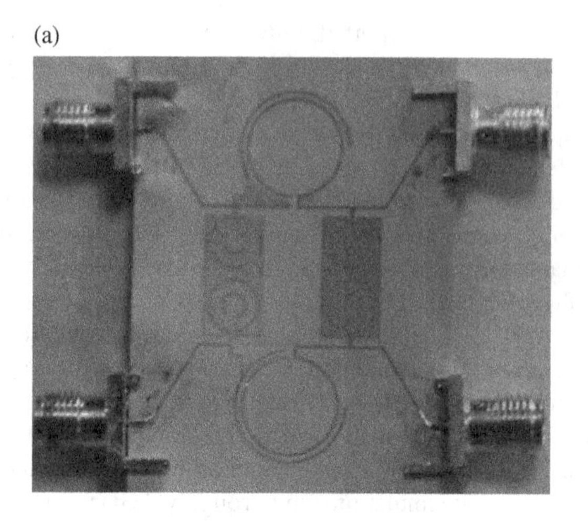

(b)

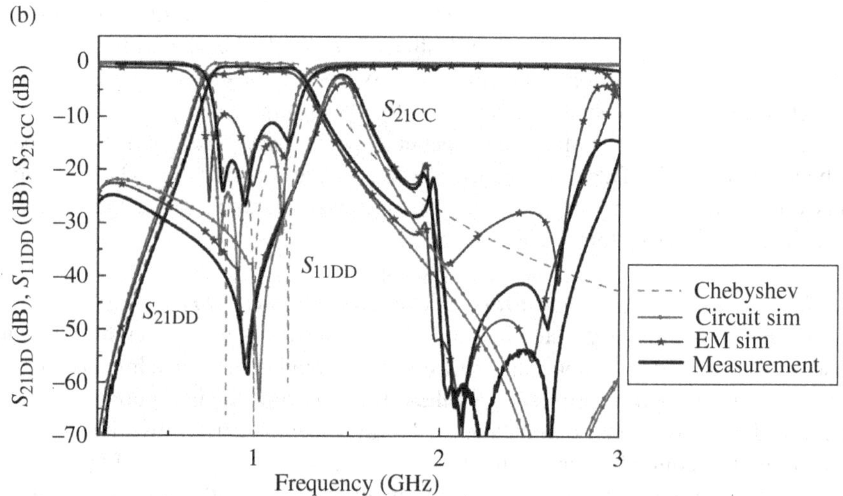

FIGURE 6.20 Photograph (a) and frequency response (b) of the fabricated balanced OSRR/OCSRR-based bandpass filter. The considered substrate is the *Rogers RO3010* with dielectric constant $\varepsilon_r = 10.2$ and thickness $h = 0.254$ mm. Dimensions (in reference to Fig. 6.18) are $L =$ 18.9 mm, $W = 37.8$ mm, $L_W = 12.6$ mm, and $L_M = 6$ mm. For the OCSRR, $r_{ext} = 2.7$ mm, $c =$ 0.2 mm, and $d = 1.2$ mm. For the OSRR, $r_{ext} = 5.8$ mm, $c = 0.2$ mm, and $d = 0.55$ mm. The 50 Ω microstrip lines have a width of 0.21 mm. For the π-model of the OSRR (Fig. 3.56e), the elements of the series LC tank are 22.28 nH and 1.22 pF, the values of the shunt capacitors are 1.35 pF (left) and 0.74 pF (right) and L_{m1} and L_{m2} have negligible values. The elements of the shunt branch are $L = 0.628$ nH, $L_p = 3.675$ nH, $C_p = 0.1$ pF, $C_1 = 6.85$ pF, and $C_2 = 5.65$ pF. Reprinted with permission from Ref. [34]; copyright 2013 IEEE.

Let us consider the implementation of an order-3 balanced common-mode suppressed Chebyshev bandpass filter with central frequency $f_o = 2.4$ GHz, 0.15 dB ripple, and fractional bandwidth FBW = 40% [35]. Identical LC parallel resonators (those of the differential-mode circuit, L_p and C_p) and admittance inverters with

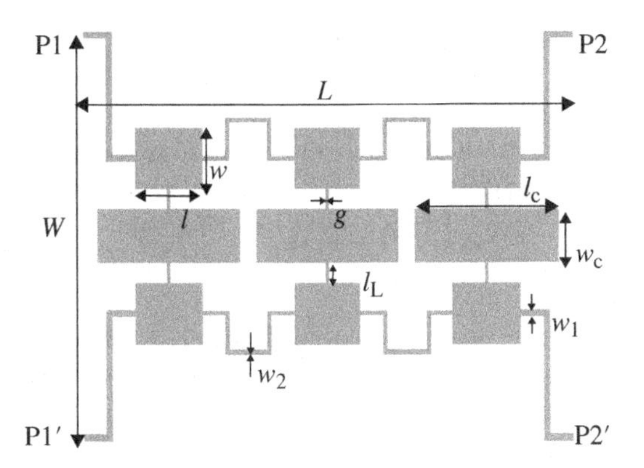

FIGURE 6.21 Typical layout (order-3) of a differential-mode bandpass filter with common-mode suppression based on mirrored SIRs and admittance inverters.

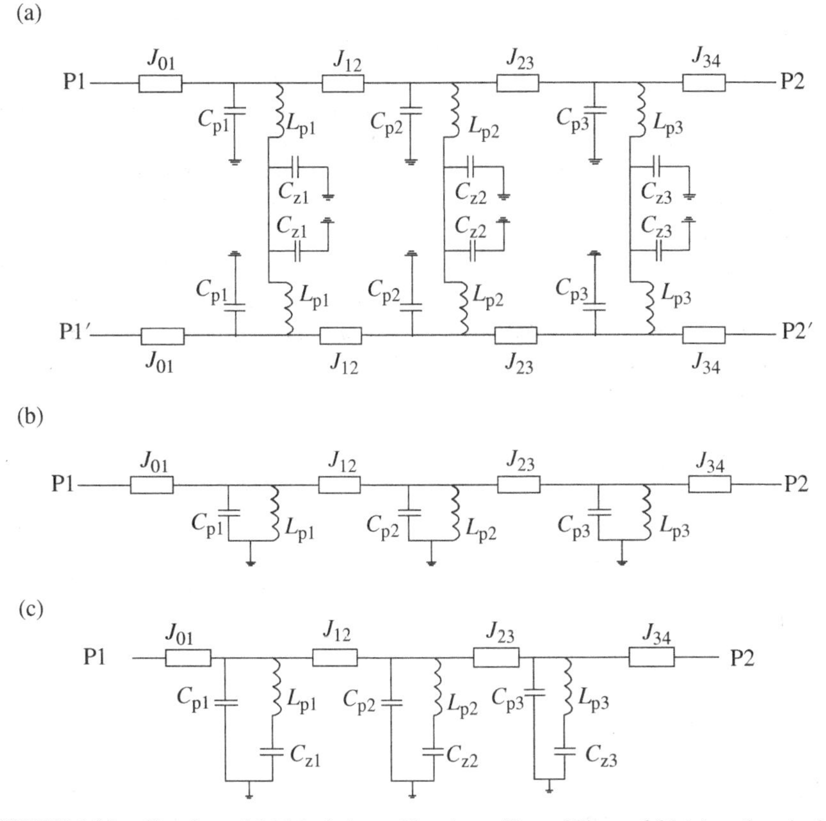

FIGURE 6.22 Circuit model of the balanced bandpass filter of Figure 6.21 (a), and equivalent circuits for the differential (b) and common (c) modes.

variable admittance have been considered. By applying the well-known transformations from the lowpass filter prototype [37], the element values are found to be $L_p = 1.465$ nH and $C_p = 3$ pF, and the admittances of the inverters are $J_{01} = J_{34} = 0.0178$S and $J_{12} = J_{23} = 0.0157$ S. For the suppression of the common mode, the transmission zeros, given by the resonances of the series resonators $L_{pi} - C_{zi}$, were forced to be identical and equal to the central frequency of the differential-mode response, f_o. Since the inductances L_p are determined by the differential-mode filter specifications, it follows that $C_z = 2.925$ pF.

To implement this filter, the widths of the transmission line sections corresponding to the calculated admittances of the inverters were obtained by means of a transmission line calculator (the considered substrate is the *Rogers RO3010*, with thickness $h = 0.635$ mm and dielectric constant $\varepsilon_r = 10.2$). Patch dimensions of the capacitances C_p were calculated from the well-known formula giving the capacitance of an electrically short low-impedance transmission line section (see Chapter 1). Finally, the dimensions of the pair of grounded resonators $L_p - C_z$ implemented through mirrored SIRs were calculated in order to accurately synthesize the required elements values up to at least $2f_o$, according to the procedure reported in Ref. [40]. The photograph of the fabricated device is shown in Figure 6.23a. Total filter size is 26.49 mm × 32.78 mm, that is, $0.51\lambda_g \times 0.63\lambda_g$, λ_g being the guided wavelength at f_o. The simulated (using the *Agilent Momentum* commercial software) and measured frequency responses of the filter for the differential- and common-mode are depicted in Figure 6.23b, together with the circuit simulations and the ideal Chebyshev response. The agreement between circuit simulation, EM simulation, and measurement is very good, and validates the circuit model and the proposed methodology for common-mode suppressed balanced filter design. The measured differential insertion loss is better than 0.5 dB, the maximum differential return losses are 28 dB in the pass band, and the differential stopband exhibits a rejection better than 12 dB up to $2f_o$. The measured common-mode rejection within the differential filter pass band is better than 45 dB and the CMRR at f_o is better than 60 dB. Although the filter of Figure 6.23 is electrically larger than the one of Figure 6.20, one key advantage of balanced filters based on the topology of Figure 6.21 is the absence of defects in the ground plane.

It is worth mentioning that by replacing the impedance inverters in the structure of Figure 6.21 with series resonant elements (e.g., implemented by inductive strips and interdigital capacitors), a compact balanced CRLH line with intrinsic common-mode suppression results.[6] Such balanced artificial lines have been applied to the design of dual-band balanced power splitters [41] and ultra wideband (UWB) bandpass filters [42].

[6] Indeed, the structure of Figure 6.21 cannot be considered to be an artificial transmission line. However, it is included here since by merely replacing the inverters with LC series resonators, a differential CRLH structure results.

(a)

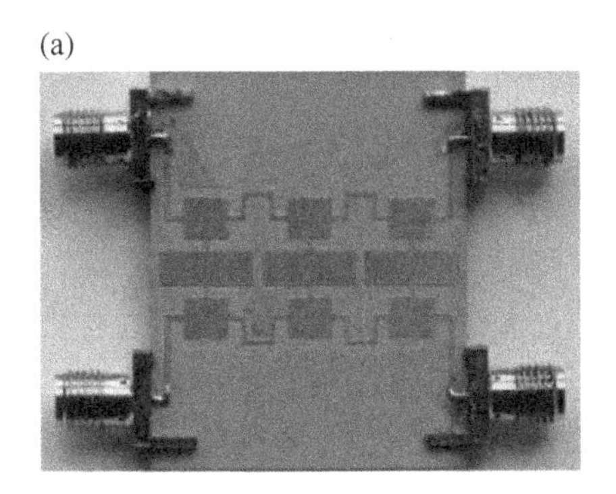

(b)

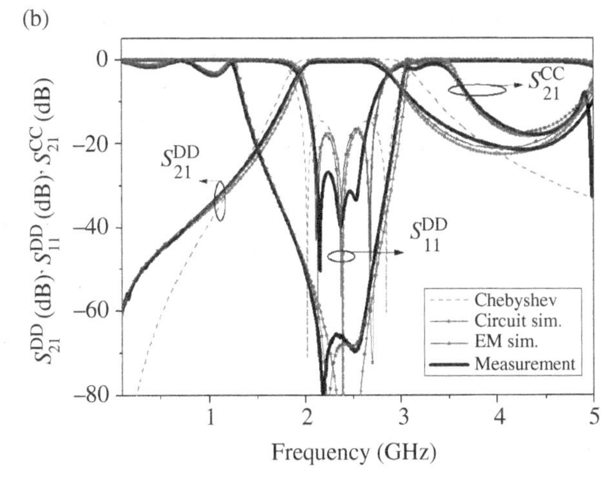

FIGURE 6.23 Photograph (a) and frequency response (b) of the designed order-3 balanced bandpass filter. Dimensions (in reference to Fig. 6.21) are $W = 26.49$ mm, $L = 32.78$ mm, $w = 4$ mm, $l = 4.475$ mm, $g = 0.2$ mm, $l_L = 1.32$ mm, $l_C = 9.6$ mm, $w_C = 3.4$ mm, $w_1 = 0.434$ mm, and $w_2 = 0.297$ mm. Reprinted with permission from Ref. [35]; copyright 2013 IEEE.

6.4 WIDEBAND ARTIFICIAL TRANSMISSION LINES

The CRLH and the generalized CRLH transmission lines studied in Chapter 3 support wave propagation in a limited frequency band. In this section, two strategies to widen the propagation band in artificial transmission lines are reviewed. The first one is based on the lattice network and other related unit cell topologies. The second one uses active elements in order to implement non-Foster reactances.

6.4.1 Lattice Network Transmission Lines

In this subsection, the dispersion and characteristic impedance of the generalized lattice network are derived. It is shown that all-pass CRLH transmission lines with arbitrary order can potentially be synthesized with these structures [43]. However,

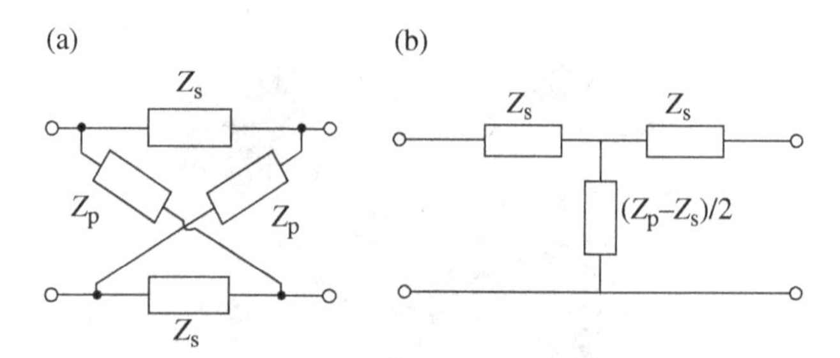

FIGURE 6.24 Lattice-network unit cell (a) and its equivalent T-circuit model (b).

the implementation of such artificial lines is not so straightforward. Practical implementation limitations are discussed, and some prototypes of lattice network CRLH lines are presented in this section. Finally, other networks related to the lattice network, such as the bridged-T network, are introduced as a more practical approach for the implementation of broadband CRLH transmission lines.

6.4.1.1 Lattice Network Analysis The artificial transmission lines considered in Chapter 3 are described either by a cascade of T or π unit cells. Let us now consider an X-type unit cell like the one shown in Figure 6.24a, that is, a lattice network [44],[7] with impedance Z_s in the series branches and Z_p in the cross diagonal arms. By calculating the elements of the impedance matrix, this two-port network can be transformed to its equivalent T-circuit model, depicted in Figure 6.24b. Using (2.33) and (2.30), the phase constant and characteristic (or iterative) impedance of the generalized lattice network of Figure 6.24a are given by[8]

$$\cos(\beta l) = \frac{Z_p + Z_s}{Z_p - Z_s} \tag{6.13}$$

$$Z_B = \sqrt{Z_s Z_p} \tag{6.14}$$

Let us now consider that the element of the series and cross branch is an inductor ($Z_s = j\omega L_R$) and a capacitor ($Z_p = -j/\omega C_R$), respectively (Fig. 6.25a). Evaluation of (6.13) and (6.14) gives

$$\cos(\beta l) = \frac{1 - L_R C_R \omega^2}{1 + L_R C_R \omega^2} \tag{6.15}$$

$$Z_B = \sqrt{\frac{L}{C}} \tag{6.16}$$

[7] Lattice networks have been used as phase equalizers, where the relative phase between the input and the output port varies with frequency, whereas the attenuation is constant at all frequencies. These structures have been also designated as all-pass lattice filters, and were already pointed out in 1923 by O. J. Zobel [44].
[8] Notice that (6.13) and (6.14) have been derived by identifying $(Z_p - Z_s)/2$, that is, the shunt impedance of the equivalent T-circuit model of the generalized lattice network, with Z_p in expressions (2.30) and (2.33).

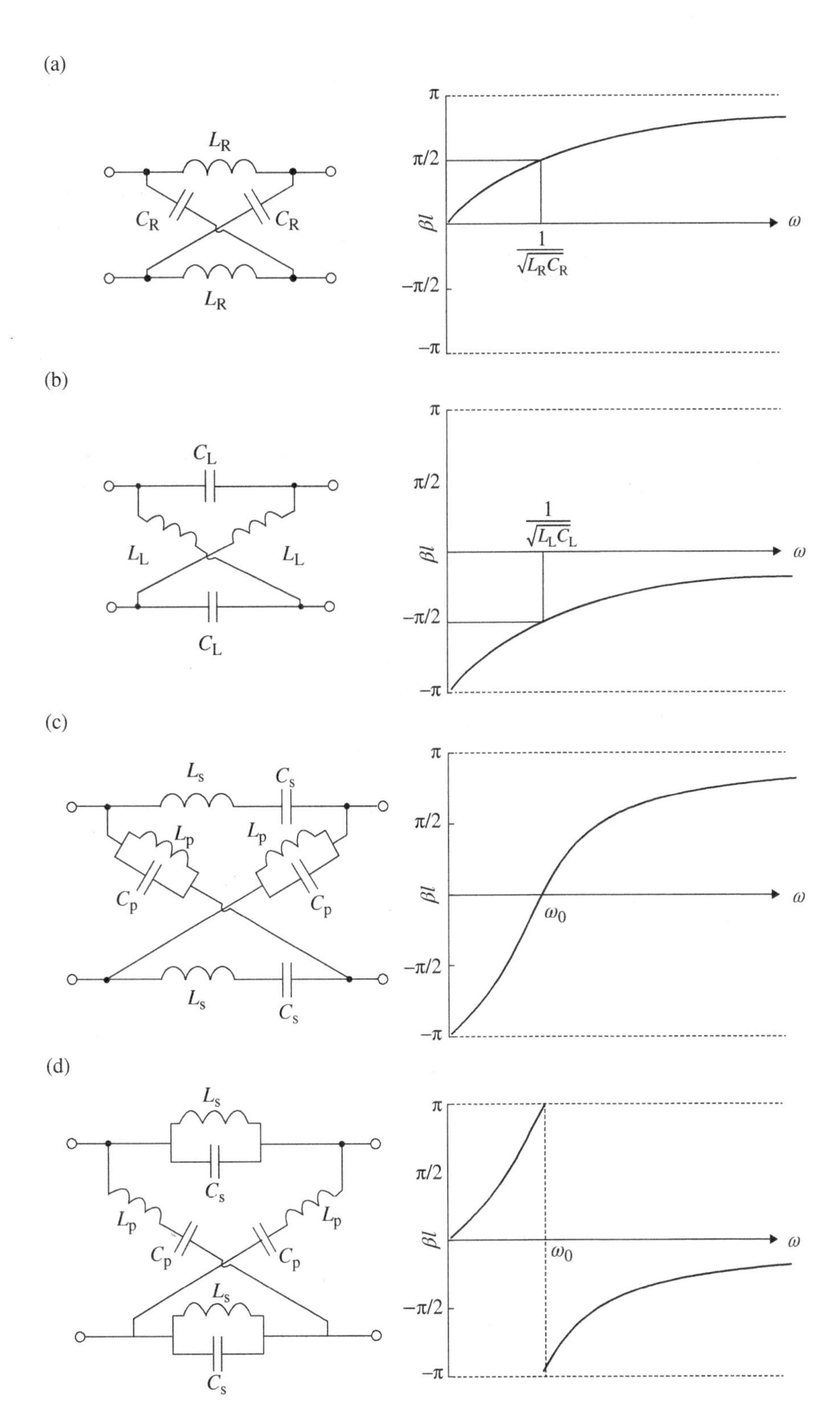

FIGURE 6.25 Examples of lattice networks and their corresponding dispersion curves. (a) order-1 RH, (b) order-1 LH, (c) order-2 CRLH, and (d) order-2 D-CRLH.

with $L = L_R$ and $C = C_R$ in (6.16). Inspection of (6.15) and (6.16) reveals that the network of Figure 6.25a is an all-pass structure (the characteristic impedance is real and frequency independent, like the one of an ordinary transmission line), and that wave propagation is forward from DC up to unlimited frequencies (the dispersion diagram is also depicted in Fig. 6.25a). If the inductor and capacitor are interchanged (Fig. 6.25b), the characteristic impedance is also constant and given by (6.16), with $L = L_L$ and $C = C_L$, whereas the phase constant is found to be

$$\cos(\beta l) = -\frac{1 - L_L C_L \omega^2}{1 + L_L C_L \omega^2} \tag{6.17}$$

and wave propagation is backward in the whole EM spectrum (see the dispersion diagram depicted in Fig. 6.25b). If the phase shift is small enough, the networks of Figure 6.25a and b mimic a conventional line and a purely left-handed (PLH) line, respectively.

By adequately choosing the reactive elements of the series and cross arms of the X-network, all-pass CRLH structures of arbitrary order can be obtained [43]. For instance, Figure 6.25c and d show the lattice networks corresponding to order-2 all-pass CRLH and dual-CRLH lines, respectively. To obtain all-pass structures, it is necessary that the zeros and poles of the reactance of the series branch coincide with the zeros and poles of the susceptance of the cross branch (see Appendix I). For the specific case of order-2 lattice networks, the condition is similar to the balance condition for T- or π-type CRLH transmission lines:

$$\omega_s \equiv \frac{1}{\sqrt{L_s C_s}} = \frac{1}{\sqrt{L_p C_p}} \equiv \omega_p = \omega_o \tag{6.18}$$

where L_s C_s and L_p C_p are the reactive elements of the series branch and cross branch, respectively, and ω_o is the transition frequency. For the network of Figure 6.25c, expression (6.13), subjected to (6.18), rewrites as follows:

$$\cos(\beta l) = \frac{1 - L_s C_p \omega^2 \left(1 - \frac{\omega_o^2}{\omega^2}\right)^2}{1 + L_s C_p \omega^2 \left(1 - \frac{\omega_o^2}{\omega^2}\right)^2} \tag{6.19}$$

whereas the dispersion relation for the network of Figure 6.25d is

$$\cos(\beta l) = -\frac{1 - L_p C_s \omega^2 \left(1 - \frac{\omega_o^2}{\omega^2}\right)^2}{1 + L_p C_s \omega^2 \left(1 - \frac{\omega_o^2}{\omega^2}\right)^2} \tag{6.20}$$

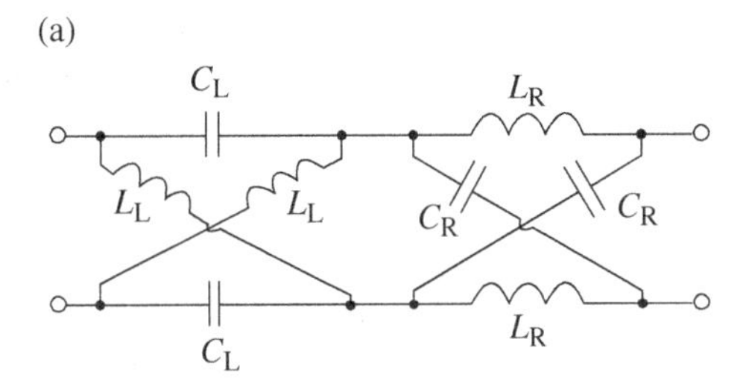

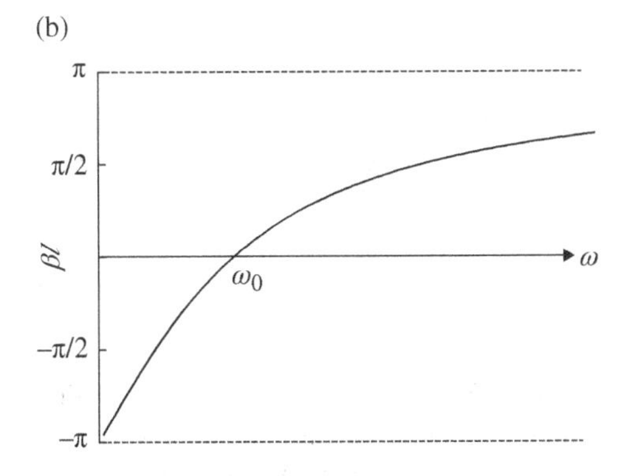

FIGURE 6.26 CRLH artificial transmission line unit cell consisting of two cascaded order-1 X-type RH and LH cells (a), and dispersion diagram (b). Elements are $C_L = C_R = 1.061$ pF and $L_L = L_R = 2.652$ nH (i.e., $f_o = \omega_0/2\pi = 3$ GHz).

The dispersion curves are also depicted in Figure 6.25c and d. In both cases, the characteristic impedance is given by (6.16), with $L = L_s$ and $C = C_p$.

From a practical viewpoint, the physical implementation of the order-2 lattice networks of Figure 6.25c and d is not straightforward. An alternative to implement all-pass order-2 CRLH lines is to combine PLH and PRH X-type cells, as depicted in Figure 6.26a. In this case, the following condition must be satisfied:

$$\sqrt{\frac{L_R}{C_R}} = \sqrt{\frac{L_L}{C_L}} \equiv Z_B \tag{6.21}$$

which gives the characteristic impedance of the whole all-pass network. The transition frequency is given by that frequency satisfying:

$$\left|\arccos\left(\frac{1-L_R C_R \omega_o^2}{1+L_R C_R \omega_o^2}\right)\right| - \left|\arccos\left(-\frac{1-L_L C_L \omega_o^2}{1+L_L C_L \omega_o^2}\right)\right| = 0 \tag{6.22}$$

namely:

$$\omega_o = \frac{1}{\sqrt{L_R C_L}} = \frac{1}{\sqrt{L_L C_R}} \tag{6.23}$$

In particular, if $L_R C_R = L_L C_L$ (which, according to 6.21, means that $L_R = L_L$ and $C_R = C_L$), the transition frequency can also be expressed as follows:

$$\omega_o = \frac{1}{\sqrt{L_R C_R}} = \frac{1}{\sqrt{L_L C_L}} \tag{6.24}$$

corresponding to the frequency where the PLH cell and the PRH cell experience a phase shift of $\beta l = -90°$ and $\beta l = +90°$, respectively (thus providing an overall phase shift of $\beta l = 0°$). Figure 6.26b shows the dispersion diagram corresponding to the structure of Figure 6.26a for the indicated element values.

It is worth mentioning that, despite that the networks of Figures 6.25c and 6.26a exhibit similar dispersion curves, there is not an element transformation that makes them equivalent. Notice that the dispersion relation (with the frequency variable in the x-axis) for the structure of Figure 6.26a is a concave function at all frequencies. However, for the order-2 CRLH X-type cell of Figure 6.25c, the dispersion relation is convex for frequencies below ω_o, and concave above that frequency (i.e., there is an inflexion point at ω_o). Although the synthesis of a cascaded order-1 LH and RH X-type cells is simpler than an order-2 CRLH lattice network, the former structure may present certain limitations due to the curvature of its dispersion relation. For instance, it is not possible to implement dual-band components based on dual-band (±90°) impedance inverters implemented with single-unit cells (like the one depicted in Fig. 6.26a) with a ratio of operating frequencies smaller than 3.

The synthesis of all-pass CRLH artificial lines can be further simplified by cascading an X-type LH unit cell with a transmission line section with identical characteristic impedance, as depicted in Figure 6.27a, or with a pair of transmission lines sections (at the input and output ports of the LH X-type cell). A typical dispersion curve for the structure of Figure 6.27a is depicted in Figure 6.27b. In this case, due to the presence of a distributed element, the phase of the structure grows indefinitely.[9]

6.4.1.2 *Synthesis of Lattice Network Artificial Transmission Lines*

X-type networks are balanced[10] structures with cross branches. Therefore, as mentioned earlier, their synthesis is not simple. CRLH artificial lines were implemented by Bongard *et al.* [45–47] by combining LH X-type cells and transmission line sections (the schematic is depicted in Fig. 6.28a). The structure reported in Ref. [47] utilizes paired strips technology (see Fig. 1.2) with two additional metal levels to implement the series capacitances of the LH X-type cells, and via holes to implement the cross inductances. The structure is shown in Figure 6.28b, and the S-parameters and phase of S_{21} are depicted in Figure 6.28c and d, respectively, in the range 2 GHz–9 GHz.

[9] Although the dispersion curves of the structures depicted in Figures 6.25 and 6.26 are limited to $\pm\pi$, in practical structures the presence of a parasitic transmission line (and other parasitic effects) cannot be avoided. Therefore, deviations of the dispersion diagram (with regard to the theoretical one) at high frequencies are expected.

[10] Notice that the term "balanced" is used here as synonymous of "differential." Do not confuse with the term "balanced" applied to CRLH lines with continuous transition between the LH and the RH band.

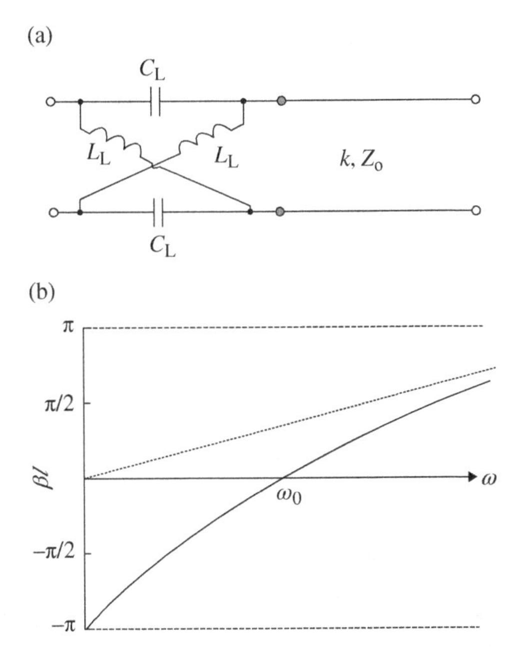

FIGURE 6.27 CRLH artificial transmission line unit cell consisting of an order-1 X-type LH cell cascaded to an ordinary transmission line section with phase constant k and characteristic impedance Z_o (a), and dispersion diagram (b).

The structure was designed to exhibit a transition frequency at $f_o = 6$ GHz, and it can be appreciated from Figure 6.28 that it can be considered all-pass, at least in a very wide region in the vicinity of f_o.

Using the unit cell structure of Figure 6.28a, a broad-band (matched from DC to millimeter waves) silicon-integrated CRLH transmission line using a monolayer CPW host line was proposed [48] (see Fig. 6.29). The measurements carried out in Ref. [48] demonstrate a balanced CRLH behavior from 5 GHz up to 35 GHz, with a transition frequency at $f_o = 20$ GHz. The interesting aspect of the structure of Figure 6.29 is that it is implemented using a single-metal layer. Since the series capacitances must be implemented in both conductors of the transmission line, the ground conductors of the host CPW must be of finite width. The required capacitances were realized by interdigital capacitors; whereas for the implementation of the crossed inductors, the solution proposed in Ref. [48] was to implement each of the inductances in a different slot of the CPW. Since the resulting structure is strongly asymmetrical, the excitation of the odd parasitic mode was prevented by the use of the bridges shown in dark gray in Figure 6.29. Thus, the structure indeed uses two metal levels. Other X-type LH, RH, and CRLH planar structures implemented by using two metal levels are reported in Ref. [49].

At this point, it is worth mentioning that from the equivalence between terminated coupled line sections and lattice networks [50], another approach for the physical

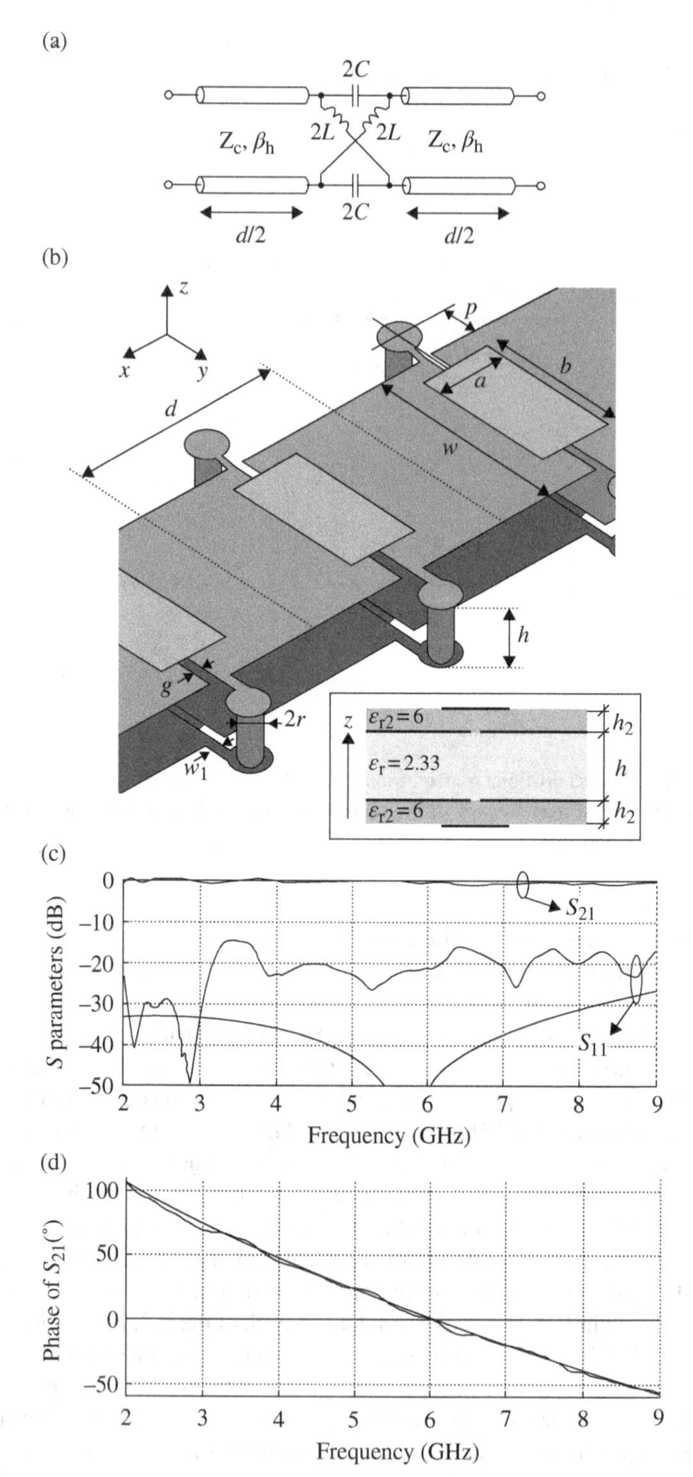

FIGURE 6.28 Schematic (unit cell) (a), structure (b), S-parameters (c) and phase of S_{21} (d) corresponding to a wideband CRLH transmission line implemented by means of X-type LH cells combined with transmission line sections. Reprinted with permission from Ref. [47]; copyright 2009 IEEE.

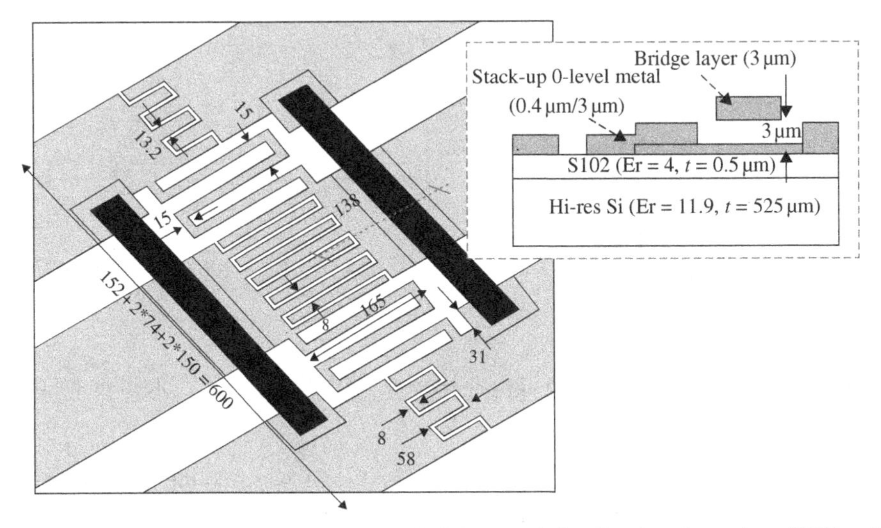

FIGURE 6.29 Structure of the CRLH artificial transmission line based on a host CPW and X-type LH cells reported in Ref. [48]. Reprinted with permission from Ref. [48]; copyright 2011 IEEE.

implementation of lattice network-based artificial CRLH lines was presented in Ref. [43]. The unit cell is based on the coupled-microstrip Schiffman section [51], which is an easily implementable structure that does not require the use of via-holes or air bridges, as opposed to the cells presented in Refs. [45–49]. Indeed, coupled-line sections were earlier used in Refs [52, 53] to obtain CRLH transmission lines, but with limited performance due to the different even- and odd-mode phase velocities of the coupled lines. This issue was satisfactorily solved in [43]. Thus, terminated coupled line sections offer a simple approach for the implementation of lattice network CRLH transmission lines. Obviously, this is a fully distributed approach that inherently increases line size as compared to lattice network CRLH lines based on lumped or semilumped (planar) elements. The author recommends Refs [43, 50, 52, 53] to those readers interested on multiband CRLH lines based on coupled line sections.

6.4.1.3 The Bridged-T Topology An interesting topology to achieve all-pass CRLH structures yet preserving the single-ended nature of the unit cell is the bridged-T (Fig. 6.30a). The equivalent T-circuit model of the bridged-T is depicted in Figure 6.30b. From it, the dispersion relation and characteristic impedance are found to be

$$\cos(\beta l) = \frac{2Z_1Z_2 + Z_2Z_3 + Z_1Z_3 + Z_1^2}{2Z_1Z_2 + Z_2Z_3 + Z_1^2} \tag{6.25}$$

$$Z_B = \frac{1}{2Z_1 + Z_3}\sqrt{Z_1^2Z_3^2 + 4Z_1^2Z_2Z_3 + 2Z_1Z_2Z_3^2 + 2Z_1^3Z_3} \tag{6.26}$$

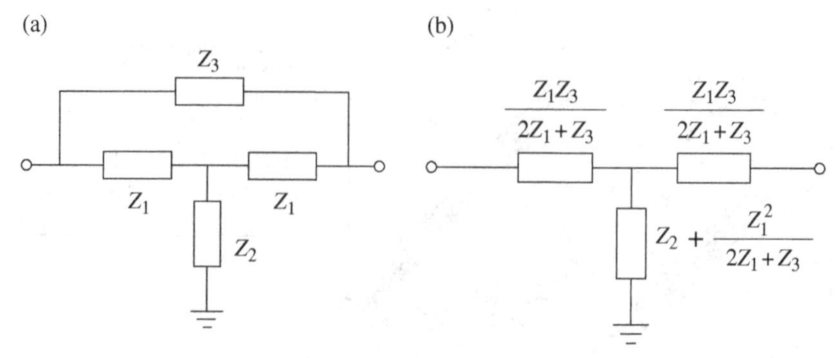

FIGURE 6.30 Unit cell of the bridged-T network (a) and equivalent T-circuit model (b).

Depending on the specific impedances of the bridged-T network, the previous equations may lead to cumbersome expressions. However, notice that if the impedances satisfy

$$Z_3 = 4Z_2 \tag{6.27}$$

The dispersion relation and the characteristic impedance simplify to

$$\cos(\beta l) = 1 + \frac{2}{1 + \frac{2Z_2}{Z_1} + \frac{Z_1}{2Z_2}} \tag{6.28}$$

$$Z_B = \sqrt{2Z_1Z_2} \tag{6.29}$$

and the structure is all-pass provided the zeros and poles of Z_1 and Y_2 coincide (see Appendix I).

As an example, let us consider $Z_1 = 1/C\omega j$, $Z_2 = L\omega j$ and $Z_3 = 4Z_2$. With these impedances, the dispersion relation is

$$\cos(\beta l) = 1 + \frac{2}{1 - 2LC\omega^2 - \frac{1}{2LC\omega^2}} \tag{6.30}$$

This bridged-T network and the corresponding dispersion are depicted in Figure 6.31. The structure is all-pass and it exhibits forward wave propagation for frequencies below $\omega_o = (2LC)^{-1/2}$ and backward wave propagation above that frequency. Notice that if the signs of the reactances of Z_1 and Z_2 are interchanged and $Z_1 = 2L\omega j$ and $Z_2 = 1/2C\omega j$, the dispersion relation is also given by (6.30).

Higher-order (quad-band) all-pass bridged-T networks have been demonstrated and implemented by Ryan *et al.* [54] in microstrip technology. The reported structure

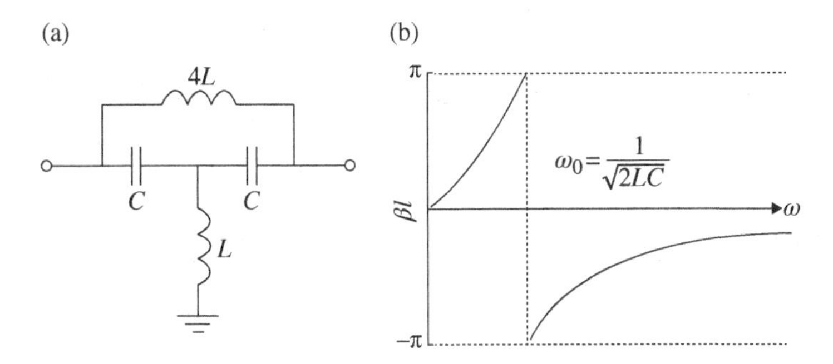

FIGURE 6.31 Bridged-T network with $Z_1 = 1/C\omega j$ and $Z_2 = Z_3/4 = L\omega j$ (a), and dispersion diagram (b).

exhibits a very wide bandwidth of 1–8 GHz, thus confirming the potential of these single-ended structures for the realization of all-pass networks with CRLH characteristics.

6.4.2 Transmission Lines Based on Non-Foster Elements

In Chapter 3, it was shown that passive media with negative permeability and/or permittivity are intrinsically dispersive.[11] This means that it is not possible to synthesize single negative (SNG) or LH passive media exhibiting a constant (i.e., frequency independent) permeability and/or permittivity (it is well known that SNG and LH media are either described by the Drude or Lorentz models). In metamaterial transmission lines that can be modeled by an equivalent T- or π-circuit wave propagation is analogous to plane wave propagation in isotropic and homogeneous dielectrics (see the mapping Eqs.3.45 and 3.46 leading to an effective permittivity and permeability). Therefore, metamaterial transmission lines exhibiting backward wave propagation (including CRLH lines) must be also dispersive and hence bandwidth limited.[12]

In the previous subsection, it was shown that bandwidth can be enhanced by using lattice networks, which do not have an analogous to passive isotropic and homogeneous dielectrics. Another potential approach for bandwidth enhancement by minimizing dispersion is to use active components, or, more precisely, non-Foster elements. Non-Foster elements are reactive elements that do not obey the Foster reactance theorem [55].[13] Thus, the conditions $\partial\chi/\partial\omega > 0$ and $\partial B/\partial\omega > 0$ (χ and B being reactance and susceptance, respectively) do not necessarily hold in such elements.

[11] Actually, passive plasma-like materials exhibiting positive permittivity and/or permeability satisfying $0 < \varepsilon < \varepsilon_o$ and/or $0 < \mu < \mu_o$ are also intrinsically dispersive (otherwise, expression (3.10) gives a density of energy smaller than the density of energy stored in vacuum). Such materials are usually designated as epsilon-near-zero (ENZ) and mu-near-zero (MNZ) metamaterials.

[12] Indeed bandwidth is also limited by the periodicity.

[13] The Foster theorem states that the reactance and susceptance of a passive, lossless two-terminal (one-port) network always monotonically increases with frequency.

This means that if a transmission line is implemented by means of non-Foster elements in the series and shunt branches, the effective permittivity and/or permeability of the line do not necessarily satisfy the conditions (3.12). Therefore, the design of dispersionless metamaterial transmission lines or artificial lines with engineered properties and small dispersion is potentially possible.

Negative inductances and capacitances are examples of non-Foster elements that can be implemented by means of negative impedance converters (NICs), that is, active electronic circuits [56].[14] In structures comprising such active components, the requirement of positive energy density (expression 3.11) no longer holds, and this leads also to violation of conditions (3.12). However, non-Foster elements may suffer from instability problems [57, 58]. For instance, let us consider a hypothetical transmission line consisting of a ladder network of series negative inductors and shunt negative capacitors. The effective permeability and permittivity of this line are simply given by (3.47) and (3.48). However, notice that the negative sign of the capacitance and inductance gives negative and constant values of the effective permittivity and permeability. In this structure, dispersion is only limited by periodicity (as occurs in the ladder network of a conventional transmission line). Indeed, the dispersion of this structure (derived from 2.33) is undistinguishable from the dispersion that results by considering positive element values. This result means that the phase and group velocities have the same sign in this line, which is in contradiction with LH wave propagation expected in structures with negative effective permittivity and permeability. This apparent contradiction is solved if one takes into account that such structure is unstable (under ordinary assumptions that both generator internal resistance and load resistance are positive), as it should occur in hypothetical materials with nondispersive negative parameters [57]. Notice also that, according to (3.6)–(3.9), stable backward wave propagation in LH media is guaranteed if the imaginary part of the effective permeability and permittivity are both negative (passive media), but these conditions are not necessarily satisfied in artificial materials with active inclusions.

According to the previous paragraph, instabilities might represent a limitation in non-Foster-based structures.[15] To the author knowledge, significant bandwidth enhancement in LH or CRLH lines by using non-Foster elements has not been experimentally demonstrated so far (nevertheless, theoretical and numerical investigations seem to confirm that broadband double-negative structures made of active inclusions are possible [59, 60]). However, very broadband ENZ transmission lines based on negative capacitances, exhibiting measured superluminal phase and group velocities over a broad band, have been reported [61, 62]. The principle for the implementation of such lines is simple: it consists of periodically loading a host transmission line with negative capacitors. The result is a fast-wave transmission line, as opposite to the

[14] The implementation of negative impedances through NICs is given in many textbooks and is out of the scope of this book.

[15] However, one should refrain from drawing a final conclusion about the feasibility of implementing dispersionless non-Foster-based structures, since the stability of transmission lines loaded with both ordinary "positive" elements and "negative" elements with gain is not well understood, and there are many effects that have not been studied so far.

slow-wave transmission lines studied in Chapter 2, that result by loading a host line with ordinary (positive) capacitors. As reported in Refs [61, 62], as long as the overall capacitance (i.e., the per-section capacitance of the line plus the negative capacitance) is positive, the structure is stable. Moreover, the designed and fabricated lines exhibit superluminal phase and group velocity in a very broad band since they were engineered to present a nondispersive (i.e., constant) effective permittivity smaller than the permittivity of vacuum (the effective permittivity can be made arbitrarily small by setting the absolute value of the negative capacitance close to the value of the line capacitance).

As pointed out in Refs. [61, 62], the superluminal group velocity is not related to the anomalous dispersion (a resonant narrowband phenomenon) that may arise in lossy media. On the contrary, the superluminal group (and phase) velocity in transmission lines loaded with negative capacitors is present in a very wide band, typically limited by the impossibility to generate the negative capacitance over a wide frequency band of the EM spectrum with practical NICs. However, as discussed in Ref. [62], and in many other sources, the fact that the group velocity is larger than c, the speed of light in vacuum, does not violate the theory of relativity and causality, since the group velocity is not identical to the energy velocity. The meaning of superluminal group velocity (and negative group velocity) is discussed in several papers (see, e.g., [63] and references therein, where this abnormal wave propagation phenomenon is reported). In brief, superluminal propagation means that the envelope of the wave-packet travels faster than c. However, this does not imply superluminal transmission of information, since under no circumstances the "front" velocity exceeds the speed of light in vacuum[16] (see Refs [61, 62] for further details).

To illustrate the potential of non-Foster elements for the implementation of broadband ENZ transmission lines, Figure 6.32 depicts the structure reported in Ref. [61], where a host line was loaded with three negative capacitors. The measured phase and group velocities, also depicted in Figure 6.32, indicate that superluminal propagation (with roughly frequency-independent phase and group velocity) is achieved in a very broad band (1:20). This means that the effective permittivity (real part) is nondispersive and smaller than the permittivity of vacuum in that band. Despite that the ENZ transmission line depicted in Figure 6.32 exhibits superluminal propagation at relatively small frequencies, the latest non-Foster ENZ transmission lines with superluminal phase and group velocities developed in the Group of Prof. Hrabar to date span the bandwidth 1MHz-400MHz (i.e., more than seven octaves) [64].

6.5 SUBSTRATE-INTEGRATED WAVEGUIDES AND THEIR APPLICATION TO METAMATERIAL TRANSMISSION LINES

Substrate-integrated waveguides (SIWs) are planar structures able to guide EM waves in a similar fashion to rectangular waveguides. In SIWs, the lateral metallic walls of the rectangular waveguides are replaced with a periodic array of metallic vias

[16] The carrying information "front" arises as long as any physically realizable signal is time limited; namely, any generated EM signal must have a beginning in time (i.e., a "front").

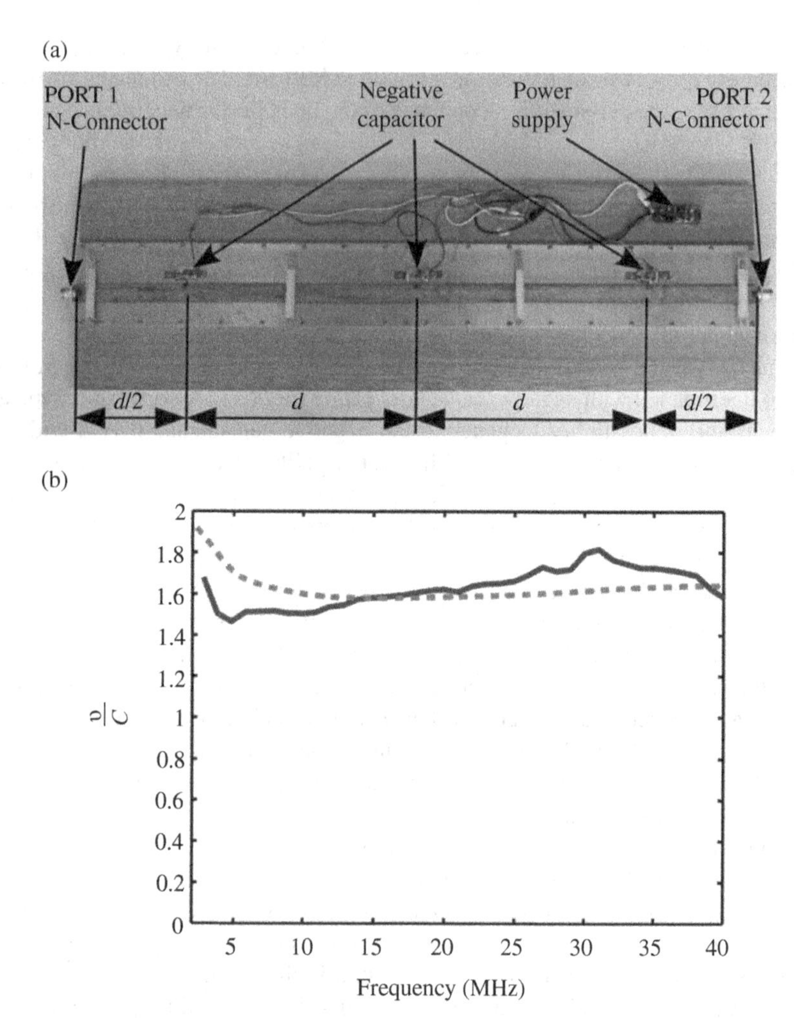

FIGURE 6.32 Experimental realization of an ENZ transmission line operative in the 2–40 GHz frequency range (a) and measured phase (dashed line) and group (solid line) velocities (b). The negative capacitances (−60 pF) were implemented through NICs based on the ultra-high-speed operational amplifier *AD8099*. The host transmission line was designed to exhibit a characteristic impedance of 31.5 Ω, corresponding to a distributed capacitance of 106 pF/m. These values were chosen to achieve a distributed capacitance of 45 pF/m in the loaded line, corresponding to an equivalent dielectric constant of 0.3 and to a 50 Ω characteristic impedance. Reprinted with permission from Ref. [61]; copyright 2011 AIP.

(see Fig. 6.33). Hence, the fabrication of SIWs is fully compatible with planar technology, and the advantages of such technology (mainly low cost and easy integration) are combined with the well-known advantages of conventional rectangular waveguides, namely, high-Q factor and high power capacity [65–67]. Thus, SIWs are promising candidates for a new generation of low-cost PCB interconnects for ultra

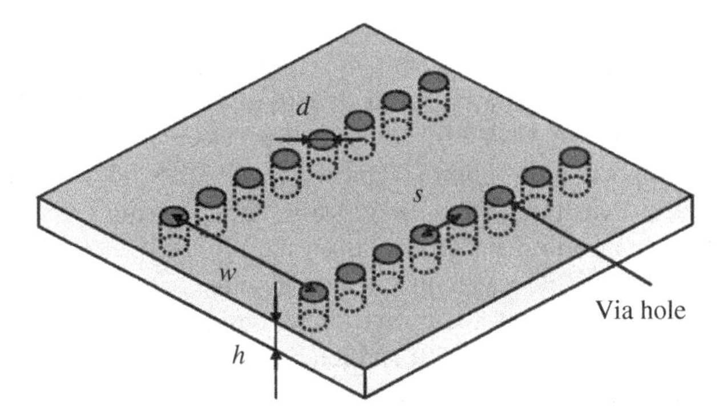

FIGURE 6.33 Sketch of a SIW and its fundamental physical parameters. Reprinted with permission from Ref. [66]; copyright 2005 IEEE.

high-speed digital applications, and can also be used as building blocks for the implementation of microwave circuits (among them, those based on metamaterial concepts will be the ones considered in this section).

Essentially, SIWs emulate dielectric-filled rectangular waveguides whose lateral metallic walls are formed through rows of vias sufficiently close to each other. However, due to the absence of metal between adjacent vias, only TE_{n0} modes are supported by SIWs[17] [66]. Indeed, the dispersion characteristics for these modes are the same as those of its equivalent rectangular waveguide [66]. Design formulas for the SIW, including the width, W (essential to control the cutoff frequency of the dominant TE_{10} mode), as well as limiting values of via diameter, d, and separation, s (important to avoid radiation losses), can be found in Refs [66–68] and are summarized in the following:

$$W = 0.5\left[a + \sqrt{(a+0.54d)^2 - 0.4d^2}\right] + 0.27d \tag{6.31}$$

$$d \le \frac{2a}{5\sqrt{n^2-1}} \tag{6.32}$$

$$s \le 2d \tag{6.33}$$

where a is the effective width of the SIW, related to the cutoff frequency of the TE_{10} mode by:

$$a = \frac{c}{2f_{c10}\sqrt{\varepsilon_r}} \tag{6.34}$$

[17] Since TM modes impose longitudinal surface currents on the lateral walls of a rectangular waveguide, and considering that only vertical surface currents can flow in the SIW through the vertical metallic vias, such TM modes are not supported by SIWs. By the same reason, TE_{nm} with m not being equal to zero modes cannot be supported. Thus, only the TE_{n0} modes can be preserved in SIWs.

c and ε_r being the speed of light in free space and the substrate dielectric constant, respectively. In (6.32), n is the order of the first higher-order mode propagating along the SIW, which in turn determines the bandwidth of the fundamental TE_{10} mode.[18] The previous formulas (6.31–6.34) are useful for design purposes, although they are empirical, and post optimization is typically required for both the SIW structure and the microstrip transitions, in order to satisfy design requirements (cutoff frequency, bandwidth, insertion and return losses, etc.) [68, 69].

There are many available papers that demonstrate the potential of SIWs for high-speed interconnects and planar microwave circuit design (some former examples of applications can be found in Ref. [70] and references therein). Indeed, SIWs can be considered to be a type of artificial transmission lines in the sense that their structure and functionality/performance are different than those of conventional lines. However, microwave circuit design using SIW technology is very similar to the design of waveguide-based circuits, and this is out of the scope of this book. Nonetheless, it has been demonstrated that by introducing further elements to SIWs, it is possible to implement artificial lines with functionalities similar to those of metamaterial transmission lines [71]. In particular, it is possible to implement CRLH lines and many circuits based on them, as well as planar filters/diplexers based on SIWs loaded with electrically small resonant elements (i.e., CSRRs). In the next subsections, these applications are briefly reviewed.

6.5.1 SIWs with Metamaterial Loading and Applications to Filters and Diplexers

Rectangular waveguides loaded with SRRs tuned below the cutoff frequency of the dominant TE mode and oriented perpendicularly to the magnetic field exhibit an LH pass band below that frequency [72]. Unfortunately, this configuration cannot be implemented in planar form by means of SIW technology due to the vertical orientation of the SRRs. Rectangular waveguides with CSRRs (and other complementary resonators) etched in the horizontal walls were used in Ref. [22] in order to experimentally obtain the resonance frequencies of such particles from the resulting notches in the transmission coefficient. In these experiments, the resonant elements were designed to exhibit the fundamental resonance frequency above the cutoff frequency of the waveguide. If an array of CSRRs is etched in the waveguide, it is expected that the structure exhibits a stopband behavior. Such stopband can be interpreted as due to the negative effective permittivity of the CSRRs (oriented perpendicular to the electric field). One advantage of CSRR-loaded rectangular waveguides (over SRR-loaded waveguides) is that the single vertically oriented metallic surfaces are the lateral walls of the waveguide. Therefore, it is apparent that by replacing such walls with metallic vias, CSRR-loaded SIWs exhibiting a stopband above cutoff will result. In CSRR-loaded microstrip lines, a similar stopband behavior is obtained.

[18] The microstrip transition plays an important role in determining the first higher-order mode propagating along the SIW (see Refs [68, 69] for further details), but this aspect is out of the scope of this book.

In order to switch the response to a pass band, series gaps are introduced in the strip of the microstrip line, and wave propagation in the first resulting pass band is backward. However, there is an alternative to obtain a pass band in CSRR-loaded microstrip lines, that is, to load the line with shunt inductive elements [73]. In this case, wave propagation is forward and occurs in the frequency region where the reactance of the shunt branch of the unit cell is capacitive (in such band, the CSRR array exhibits a positive permittivity that compensates the negative permittivity of the inductive elements).

In SIW technology, we can take advantage of the inherent negative permittivity below cutoff for the TE_{10} mode (like in conventional rectangular waveguides). Thus, by tuning the resonance frequency of the CSRR below cutoff, a band pass behavior (with forward wave propagation) is expected. This has been corroborated and applied to the design of filters and diplexers in SIW technology [71, 74–76]. These filters are small as long as CSRR are electrically small resonators. As compared to CSRR-based filters in microstrip technology, the back substrate side is kept unaltered in these CSRR-loaded SIW filters. As an illustrative example, Figure 6.34 reports an order-3 bandpass filter first published in Ref. [74] (where the design methodology is explained in detail) and then included in the review paper [71]. An alternative approach considers the implementation of SIW filters by etching pairs of single loop CSRRs at both sides of the SIW (one on top of the other and rotated 180°) in a configuration that was designated as broadside coupled CSRR (BC-CSRR) by the authors [77].

6.5.2 CRLH Lines Implemented in SIW Technology and Applications

Let us now consider that the SIW is loaded with a transverse meandered slot on the top metallic surface, as depicted in Figure 6.35a [78, 79]. Such slot acts like a series capacitor. Since the SIW below cutoff behaves as a negative effective permittivity medium, and this behavior can be modeled by a shunt inductance, it follows that the circuit model of the unit cell of a slot-loaded SIW is the canonical CRLH transmission line model of Figure 3.20. Thus, backward and forward transmission bands are expected in slot-loaded SIWs, and wave propagation below cutoff is possible by properly tailoring the geometry. Obviously, by optimizing the structure, it is also possible to implement balanced CRLH lines.

Size reduction in SIWs can be achieved by simply cutting the structure by the longitudinal symmetry plane. The resulting structure is the so-called half-mode SIW (HM-SIW) [80, 81]. The field distribution in a HM-SIW for the dominant mode is very similar to that present in one of the halves of a conventional SIW. The reason is that the symmetric plane along the transmission direction is equivalent to a magnetic wall. Hence, half of the SIW will keep exactly the half field distribution unchanged if the cutting plane is a perfect magnetic wall. Actually, the open side aperture of the HM-SIW is nearly equivalent to a perfect magnetic wall due to the high ratio of width to height in the SIW. Thus, we can take benefit of this fact and implement SIWs with significantly reduced transverse dimensions, yet keeping the main properties of the original structure. The HM-SIW concept can also be applied to the implementation of CRLH structures based on transverse meandered slots. The typical unit cell of

(a)

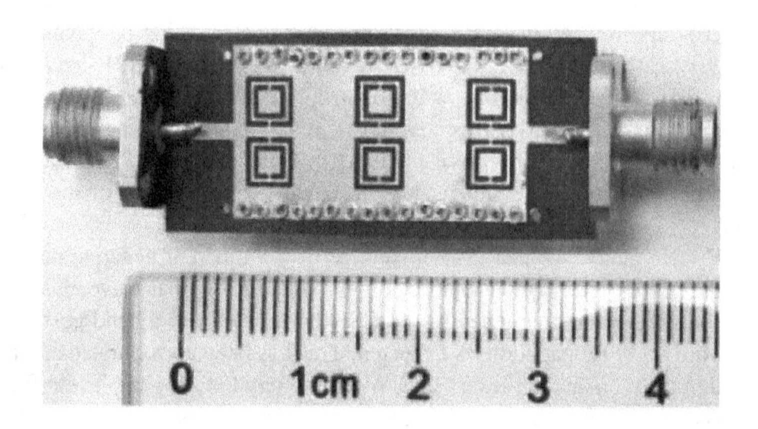

(b)

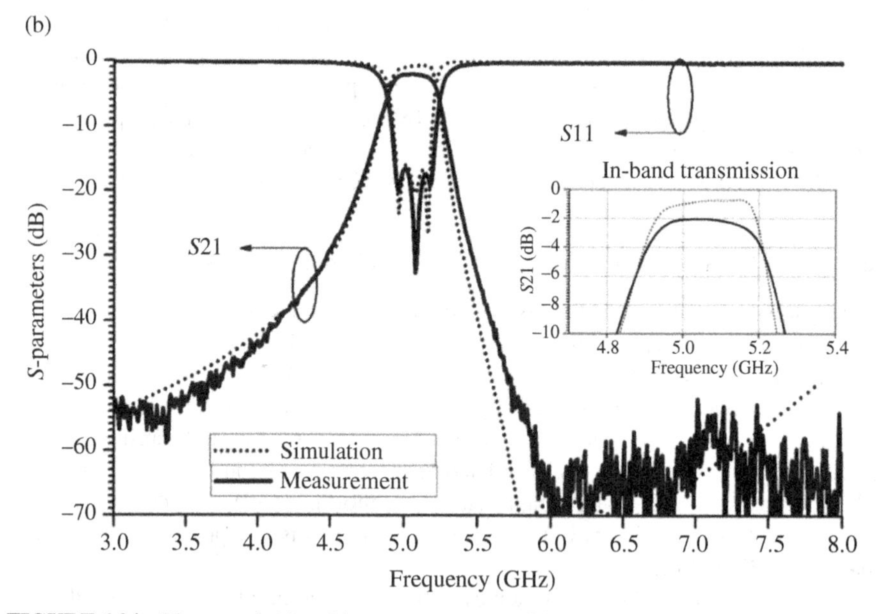

FIGURE 6.34 Photograph (a) and frequency response (b) of a CSRR-loaded SIW designed to exhibit bandpass filter functionality. Reprinted with permission from Ref. [74]; copyright 2009 IEEE.

a CRLH HM-SIW transmission line is depicted in Figure 6.35b and c depicts a three-stage structure, including microstrip transitions [78].

In Ref. [78], it was demonstrated through simulation and experiment that CRLH SIW and their equivalent CRLH HM-SIW transmission lines roughly exhibit the same characteristics. Figure 6.36 depicts the dispersion diagrams and the frequency responses for the structure of Figure 6.35c by considering an unbalanced and a balanced case. As anticipated, the structures exhibit a CRLH behavior with a continuous

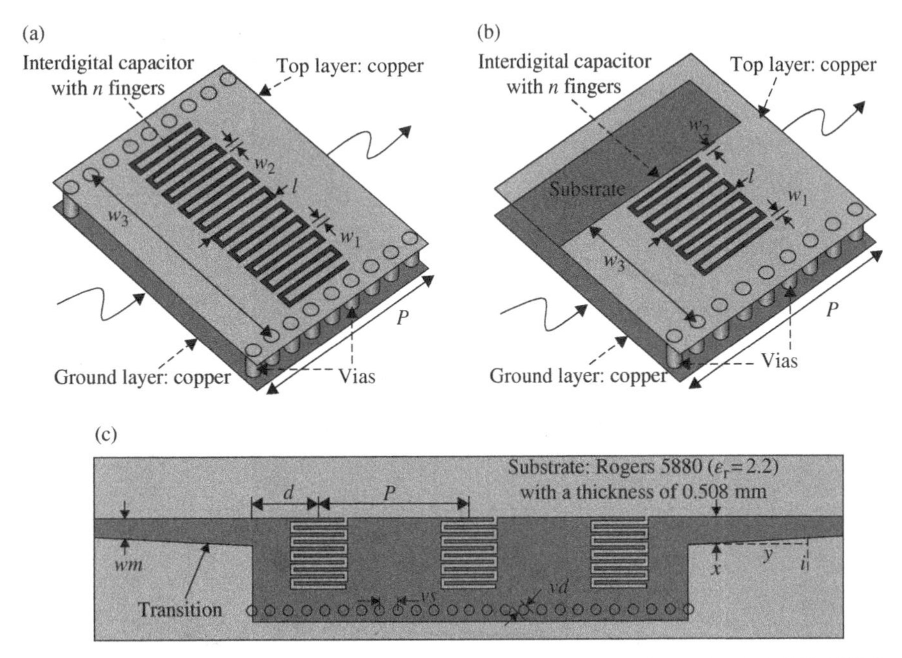

FIGURE 6.35 Structure (unit cell) of a slot-loaded CRLH SIW (a), slot-loaded CRLH HM-SIW (b), and three-stage CRLH HM-SIW transmission line (c). Reprinted with permission from [78]; copyright 2009 IEEE.

transition between the LH and the RH band for the balanced case. The transmission coefficient of the SIW structure (without slot), also depicted in the figure, indicates that transmission below cutoff arises by introducing the slots.

CRLH SIW and CRLH HM-SIW transmission lines have been applied to the design of several microwave components, including couplers [78, 82], leaky wave antennas (LWAs) [79, 83], slot antennas [84], and so on. As an illustrative example, Figure 6.37 depicts a dual-band rat-race coupler based on CRLH HM-SIW lines [82], where the dual-band functionality is achieved thanks to the controllability of the dispersion diagram, as discussed in Subsection 4.2.2. The complete characterization of the device requires too many curves so that we include in Figure 6.37 the return loss, coupling, isolation, and phase balance by considering only the Δ port as the input port (the complete characterization and design process can be found in Ref. [82]). This approach is fully planar and does not require etching the back metallic side of the substrate.

Higher-order CRLH transmission lines implemented in SIW technology can also be implemented. In particular, order-4 CRLH SIW lines have been reported and applied to the design of quad-band splitters [85], dual-band filters [86], and dual-band LWAs [87]. Such lines were designated as extended CRLH (E-CRLH) SIW transmission lines in Refs [85–87]. A typical topology and equivalent circuit model of the proposed E-CRLH SIW transmission line unit cell is shown in Figure 6.38. The equivalent circuit is the canonical T-circuit of the E-CRLH transmission line basic cell with

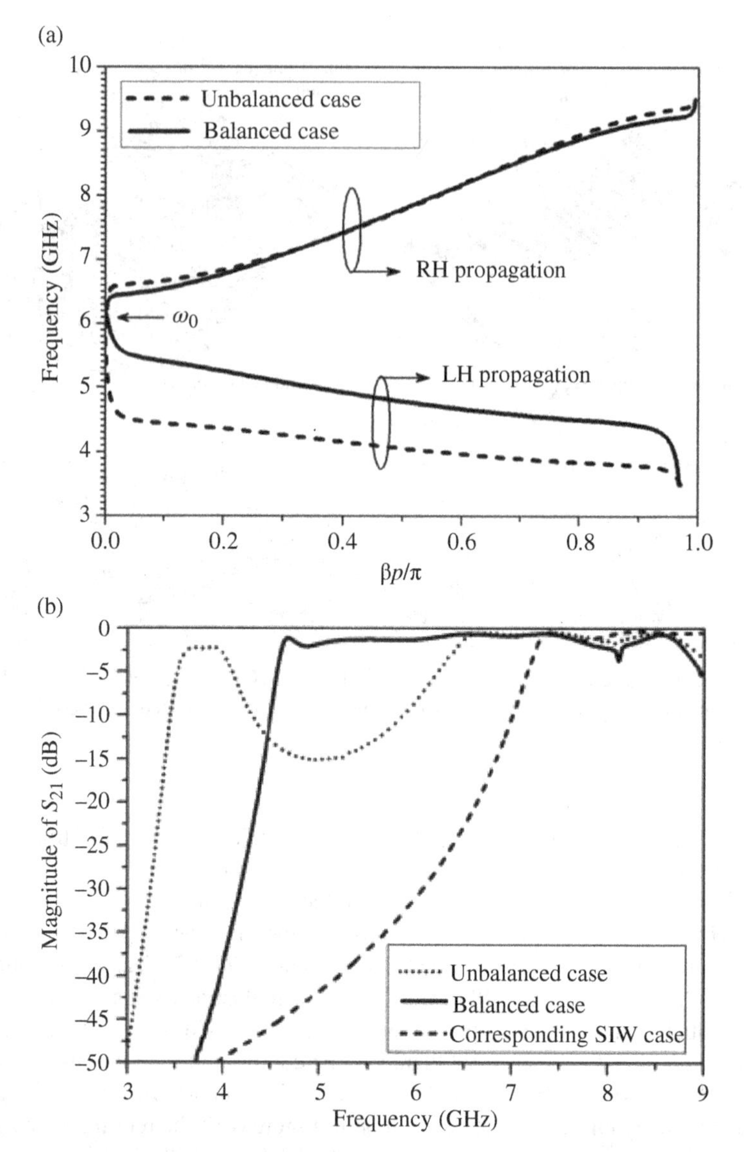

FIGURE 6.36 Dispersion diagram (a) and response (b) of the CRLH SIW unit cells. Balanced case: $w_1 = 0.35$ mm, $w_2 = 0.33$ mm, $w_3 = 15$ mm, $n = 19$, $p = 12.5$ mm, and $l = 4.6$ mm. Unbalanced case: $w_1 = 0.35$ mm, $w_2 = 0.2$ mm, $w_3 = 15$ mm, $n = 19$, $p = 12.5$ mm, and $l = 4.8$ mm. The considered substrate is the *Rogers 5880* with thickness $h = 0.508$ mm and a relative permittivity $\varepsilon_r = 2.2$. All the metallic via holes have a diameter of 0.8 mm and a center-to-center spacing of 1.45 mm. Reprinted with permission from Ref. [78]; copyright 2009 IEEE.

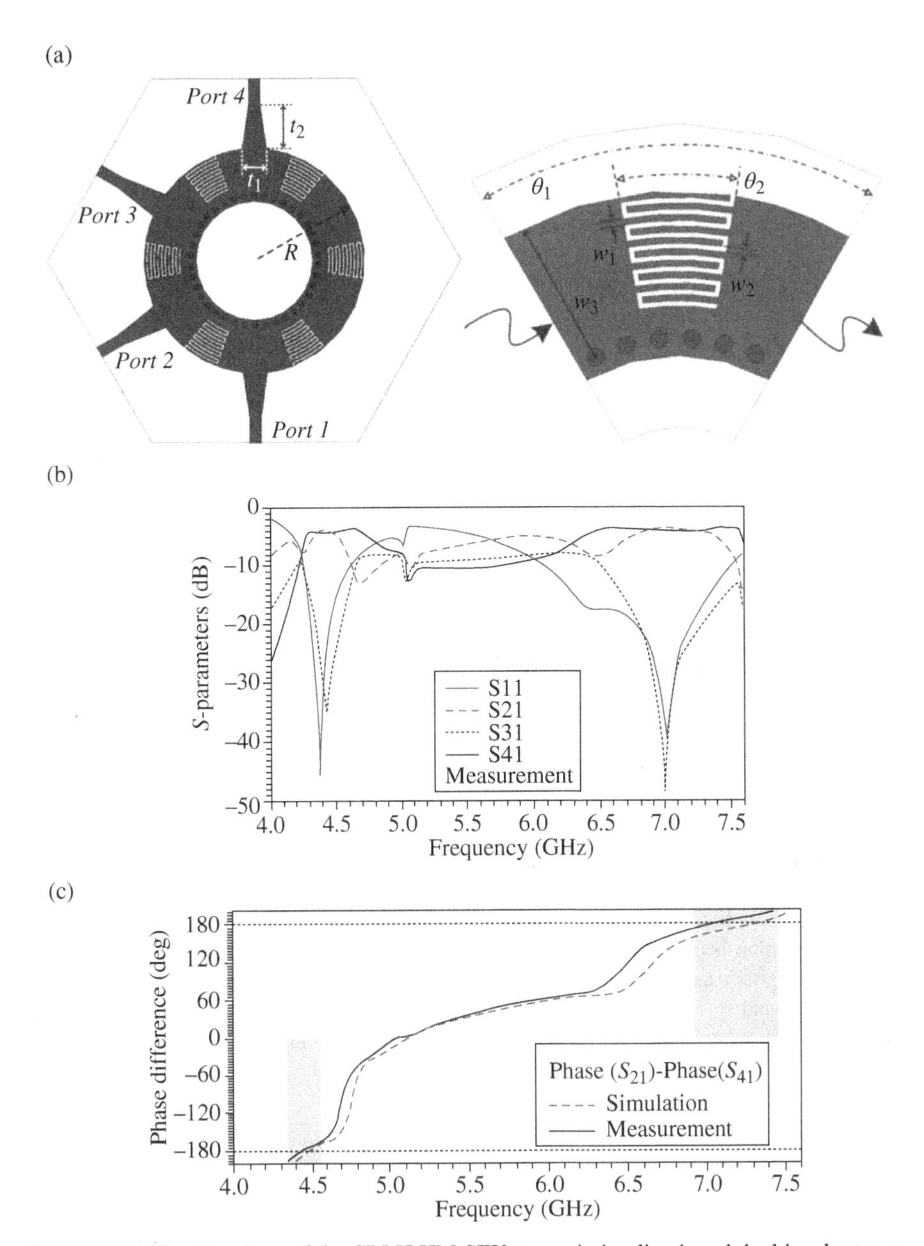

FIGURE 6.37 Topology of the CRLH HM-SIW transmission line-based dual-band rat race coupler and detail of the unit cell (a), frequency response for the Δ port (b), and phase balance for the Δ port (c). The considered substrate is the *Rogers 5880* with a thickness $h = 0.508$ mm and a relative permittivity $\varepsilon_r = 2.2$. All the metallic via holes have a diameter of 0.8 mm and a center-to-center spacing of 1.45 mm. $w_1 = 0.3$ mm, $w_2 = 0.21$ mm, $w_3 = 6.55$ mm, $t_1 = 3.783$ mm, $t_2 = 7.04$ mm, $\theta_1 = 60°$, and $\theta_2 = 19.6°$. Reprinted with permission from Ref. [82]; copyright 2010 IEEE.

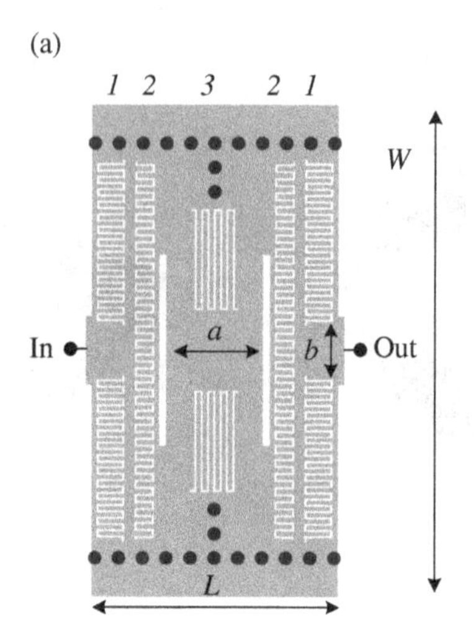

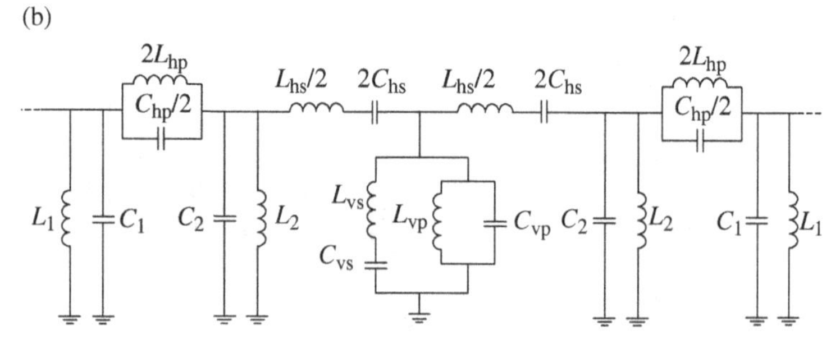

FIGURE 6.38 Topology (a) and circuit model (b) of the SIW based E-CRLH transmission line. The top metal is indicated in gray, whereas the vias are indicated in black. Dimensions are $a = 5.66$ mm, $b = 3.34$ mm, $W = 28.34$ mm, and $L = 15.1$ mm. All the interdigital capacitors have a separation between fingers (gap) of 0.12 mm. The fingers width (w_i) and length (l_i) of the interdigital capacitors are for the series capacitors (from left to right): $w_1 = 0.24$ mm, $l_1 = 1.52$ mm and $w_2 = 0.285$ mm, $l_2 = 1.02$ mm; for the shunt branch, $w_3 = 0.25$ mm, and $l_3 = 5.72$ mm. The vertical slot has a width of 0.41 mm and length of 11.04 mm. The vias have a radius of 0.4 mm and a center-to-center distance between vias of 1.4 mm. Reprinted with permission from Ref. [85]; copyright 2012 EuMA.

additional shunt elements (L_1, C_1, L_2, C_2) to account for the distributed behavior of the SIW host line. The shunt capacitance and inductance $L_{vp} - C_{vp}$ are mainly determined by the SIW host line width W and substrate thickness h. The series resonator $L_{hs} - C_{hs}$ models the etched series interdigital capacitors (labeled 2 in Fig. 6.38a). These element values can be controlled through the width and length of the fingers. By facing the same resonator to the grounded vias, the shunt $L_{vs} - C_{vs}$ series resonator is obtained (labeled 3). Moreover, considering this latter resonator the values $L_{vp} - C_{vp}$ can be further controlled by the vertical vias and length a. If the same series resonator with a central inductive path is added (controlled through the width b), a shunt

(a)

(b)

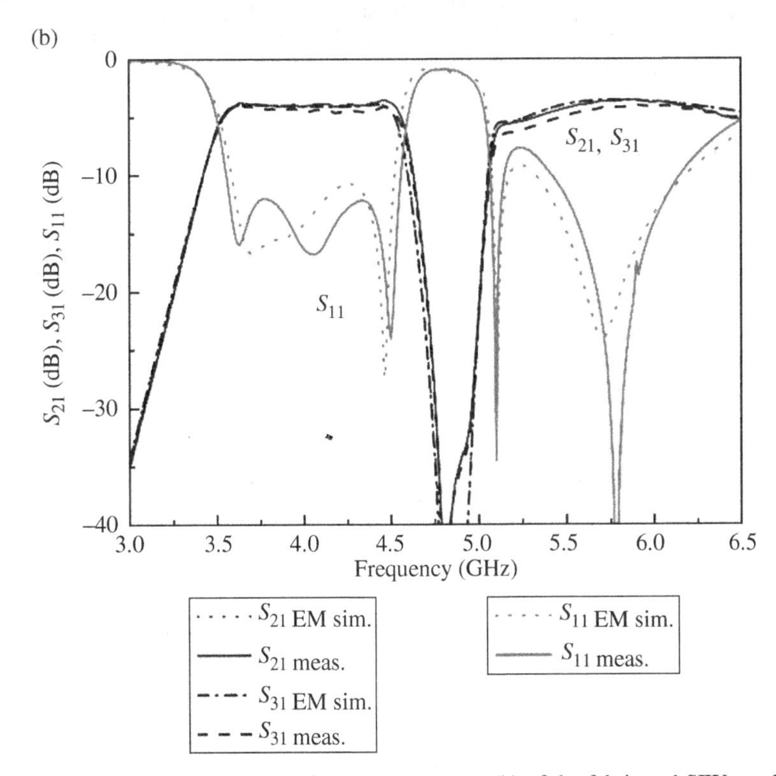

FIGURE 6.39 Photograph (a) and frequency response (b) of the fabricated SIW quad-band Y-junction power divider based on the quad-band impedance inverter of Figure 6.38. Reprinted with permission from Ref. [85]; copyright 2012 EuMA.

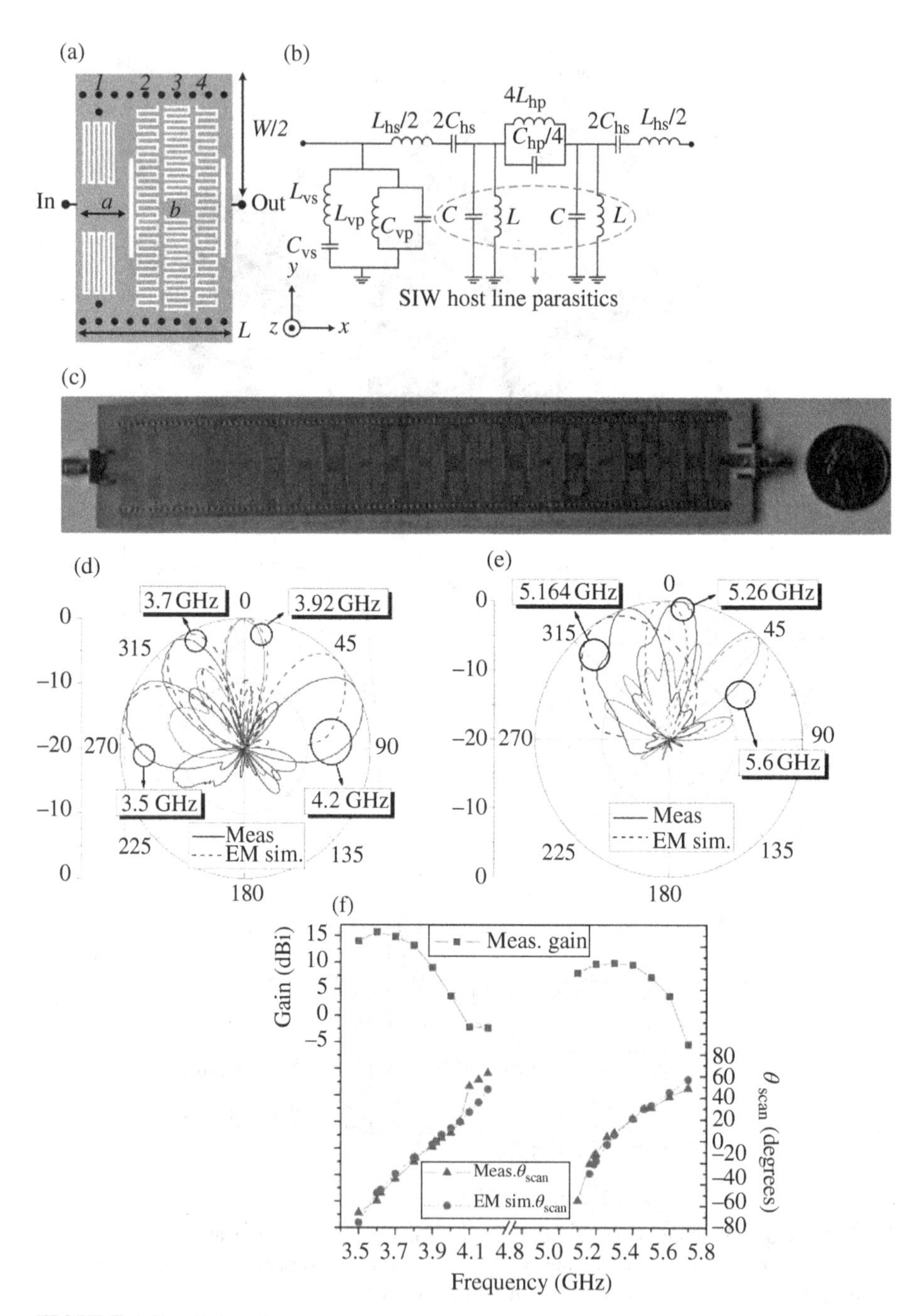

FIGURE 6.40 Unit cell topology (a), circuit model (b), photograph (c), normalized radiation patterns (x–z plane) for the first (d) and second (e) bands, and gain/scan angle (f) of the dual-band E-CRLH SIW LWA reported in Ref. [87]. Top numbers indicate the subindex

resonator in the series branch is obtained, corresponding to the $L_{hp} - C_{hp}$ resonator (labeled 1 in Fig. 6.38a). Finally, if increasing the series inductance L_{hs} is required, additional slots adjacent to the series resonators can be etched (see Fig. 6.38a). This topology has a similar phenomenology as the canonical E-CRLH unit cell implemented in microstrip or CPW technology.

The topology of Figure 6.38 indeed corresponds to a designed quad-band 35.35 Ω impedance inverter, which was in turn used to implement a quad-band power splitter operative at 3.75 GHz, 4.46 GHz, 5.15 GHz and 5.9 GHz [85]. The fabricated device (implemented in the *Rogers 5880* substrate with dielectric constant $\varepsilon_r = 2.2$, thickness $h = 1.27$ mm, and loss tangent $\tan\delta = 0.001$) and its measured characteristics are depicted in Figure 6.39 [85] (device dimensions are given in the caption of Fig. 6.38). The design process and performance of this device are similar to those corresponding to quad-band inverters and dividers implemented by means of E-CRLH microstrip or CPW transmission lines. However, let us insist on the fact that the backside metal is not used for design purposes in the device of Figure 6.39.

The last example to illustrate the potential of E-CRLH SIW lines is a dual-band LWA, where continuous beam scanning with frequency from backward to forward leaky wave radiation is achieved in two predefined bands [87] (such CRLH dual-band LWAs were formerly proposed in microstrip technology [88]).[19] Figure 6.40 shows the photograph of the fabricated antenna, where the unit cell (also included) is similar to that depicted in Figure 6.38a, but described by the canonical L-type E-CRLH circuit model (except by the presence of the parasitics). The normalized radiation patterns, gains, and scan angles for different frequencies at both bands are also included in the figure (see Ref. [87] for further details on the design of this dual-band antenna).

In summary, the combination of SIW technology and metamaterial concepts is a powerful approach for the implementation of planar microwave components with novel functionalities, small size, low cost, low loss, and backside isolation.

FIGURE 6.40 (*Continued*) for the dimensions of each interdigital capacitor. Dimensions are $a = 5.83$ mm, $b = 2.14$ mm, $W = 28.34$ mm, and $L = 17.2$ mm. The fingers width (w_i), length (l_i) and separation between fingers (gap g_i) of the interdigital capacitors are: for the shunt branch: $w_1 = 0.25$ mm, $g_1 = 0.28$ mm, $l_1 = 6$ mm; for the series capacitors (from left to right): $w_2 = w_4 = 0.45$ mm, $l_2 = l_4 = 1.94$ mm, $g_2 = g_4 = 0.26$ mm, $w_3 = 0.35$ mm, $l_3 = 2.3$ mm and $g_3 = 0.28$ mm; The vertical slot has a width of 0.5 mm and length of 10.49 mm. The vias have a radius of 0.4 mm and a horizontal and vertical center-to-center distance between vias of 1.72 mm and 1.84 mm, respectively. The considered substrate is the *Rogers RT/duroid 5880LZ* with permittivity $\varepsilon_r = 1.96$ and thickness $h = 1.27$ mm. Reprinted with permission from Ref. [87]; copyright 2013 IEEE.

[19] To achieve a continuous scan from backward to forward radiation at both bands, the E-CRLH SIW line must be balanced, that is, it must exhibit a continuous transition between the first LH band and the first RH band, and also a continuous transition between the second LH band and the second RH band.

REFERENCES

1. E. Shamonina, V. A. Kalinin, K. H. Ringhofer, and L. Solymar, "Magneto-inductive waveguide," *Electron. Lett.*, vol. **38**, pp. 371–373, 2002.
2. E. Shamonina, V. A. Kalinin, K. H. Ringhofer, and L. Solymar, "Magneto-inductive waves in one, two and three dimensions," *J. Appl. Phys.*, vol. **92**, pp. 6252–6261, 2002.
3. M. C. K. Wiltshire, E. Shamonina, I. R. Young, and L. Solymar, "Dispersion characteristics of magneto-inductive waves: comparison between theory and experiment," *Electron. Lett.*, vol. **39**, pp. 215–217, 2003.
4. E. Shamonina and L. Solymar, "Properties of magnetically coupled metamaterial elements," *J. Magn. Magn. Mater.*, vol. **300**, pp. 38–43, 2006.
5. R. R. A. Syms, E. Shamonina, V. Kalinin, and L. Solymar, "A theory of metamaterials based on periodically loaded transmission lines: interaction between magnetoinductive and electromagnetic waves," *J. Appl. Phys.*, vol. **97**, paper 064909, 2005.
6. M. J. Freire, R. Marqués, F. Medina, M. A. G. Laso, and F. Martin, "Planar magnetoinductive wave transducers: theory and applications," *Appl. Phys. Lett.*, vol. **85**, pp. 4439–4441, 2004.
7. M. Beruete, F. Falcone, M. J. Freire, R. Marqués, and J. D. Baena, "Electroinductive waves in chains of complementary metamaterial elements," *Appl. Phys. Lett.*, vol. **88**, paper 083503, 2006.
8. J. Shefer, "Periodic cylinder arrays as transmission lines," *IEEE Trans. Microw. Theory Techn.*, vol. **11**, pp. 55–61, 1963.
9. E. Shamonina and L. Solymar, "Magneto-inductive waves supported by metamaterials elements: components for a one-dimensional waveguide," *J. Phys. D: Appl. Phys.*, vol. **37**, pp. 362–367, 2004.
10. I. S. Nefedov, and S. A. Tretyakov, "On potential applications of metamaterials for the design of broadband phase shifters," *Microw. Opt. Technol. Lett.*, vol. **45**, pp. 98–102, 2005.
11. F. J. Herraiz-Martínez, F. Paredes, G. Zamora, F. Martín, and J. Bonache, "Printed magnetoinductive-wave (MIW) delay lines for chipless RFID applications," *IEEE Trans. Antenna Propag.*, vol. **60**, pp. 5075–5082, 2012.
12. M. Schüßler, C. Damm, and R. Jakoby, "Periodically LC loaded lines for RFID backscatter applications," *1st Int. Congress. Adv. Electromagn. Mater. Microw. Opt. (Metamater.)*, Rome, Italy, October 2007, pp. 103–106.
13. M. Schüßler, C. Damm, M. Maasch, and R. Jakoby, "Performance evaluation of left-handed delay lines for RFID backscatter applications," *IEEE MTT-S Int. Microw. Symp. Dig.*, Atlanta, GA, June 2008, pp. 177–180.
14. B. C. Tseng and L. K. Wu, "Design of miniaturized common-mode filter by multilayer low-temperature co-fired ceramic," *IEEE Trans. Electromagn. Compat.*, vol. **46**, pp. 571–579, 2004.
15. C.-H. Tsai and T.-L. Wu, "A broadband and miniaturized common-mode filter for gigahertz differential signals based on negative-permittivity metamaterials," *IEEE Trans. Microw. Theory Techn.*, vol. **58**, pp. 195–202, 2010.
16. W. T. Liu, C.-H. Tsai, T.-W. Han, and T.-L. Wu, "An embedded common-mode suppression filter for GHz differential signals using periodic defected ground plane," *IEEE Microw. Wireless Compon. Lett.*, vol. **18**, pp. 248–250, 2008.
17. S.-J. Wu, C.-H. Tsai, T.-L. Wu, and T. Itoh, "A novel wideband common-mode suppression filter for gigahertz differential signals using coupled patterned ground structure," *IEEE Trans. Microw. Theory Techn.*, vol. **57**, pp. 848–855, 2009.

18. J. Naqui, A. Fernández-Prieto, M. Durán-Sindreu, J. Selga, F. Medina, F. Mesa, and F. Martín, "Split rings-based differential transmission lines with common-mode suppression," *IEEE MTT-S Int. Microw. Symp. Dig*, Baltimore, MD, June 2011.
19. J. Naqui, A. Fernández-Prieto, M. Durán-Sindreu, F. Mesa, J. Martel, F. Medina, and F. Martín, "Common mode suppression in microstrip differential lines by means of complementary split ring resonators: theory and applications," *IEEE Trans. Microw. Theory Techn.*, vol. **60**, pp. 3023–3034, 2012.
20. A. Fernández-Prieto, J. Martel, J. S. Hong, F. Medina, S. Qian, and F. Mesa, "Differential transmission line for common-mode suppression using double side MIC technology," *Proc. 41st Eur. Microw. Conf. (EuMC)*, Manchester, UK, October 2011, pp. 631–634.
21. D. Ahn, J.-S. Park, C.-S. Kim, J. Kim, Y. Qian, and T. Itoh, "A design of the low-pass filter using the novel microstrip defected ground structure," *IEEE Trans. Microw. Theory Techn.*, vol. **49**, pp. 86–93, 2001.
22. J. D. Baena, J. Bonache, F. Martín, R. Marqués, F. Falcone, T. Lopetegi, M. A. G. Laso, J. García, I. Gil, and M. Sorolla, "Equivalent circuit models for split ring resonators and complementary split rings resonators coupled to planar transmission lines," *IEEE Trans. Microw. Theory Techn.*, vol. **53**, pp. 1451–1461, 2005.
23. J. Bonache, M. Gil, I. Gil, J. Garcia-García, and F. Martín, "On the electrical characteristics of complementary metamaterial resonators," *IEEE Microw. Wireless Compon. Lett.*, vol. **16**, pp. 543–545, 2006.
24. C. H. Wu, C. H. Wang, and C. H. Chen, "Novel balanced coupled-line bandpass filters with common-mode noise suppression," *IEEE Trans. Microw. Theory Techn.*, vol. **55**, pp. 287–295, 2007.
25. A. Saitou, K. P. Ahn, H. Aoki, K. Honjo, and K. Watanabe, "Differential-mode bandpass filters with four coupled lines embedded in self complementary antennas," *IEICE Trans. Electron.*, vol. **E90-C**, pp. 1524–1532, 2007.
26. C.-H. Wu, C.-H. Wang, and C. H. Chen, "Balanced coupled-resonator bandpass filters using multisection resonators for common-mode suppression and stopband extension," *IEEE Trans. Microw. Theory Techn.*, vol. **55**, pp. 1756–1763, 2007.
27. T. B. Lim and L. Zhu, "A differential-mode wideband bandpass filter on microstrip line for UWB applications," *IEEE Microw. Wireless Compon. Lett.*, vol. **19**, pp. 632–634, 2009.
28. J. Shi and Q. Xue, "Novel balanced dual-band bandpass filter using coupled stepped-impedance resonators," *IEEE Microw. Wireless Compon. Lett.*, vol. **20**, pp. 19–21, 2010.
29. J. Shi and Q. Xue, "Dual-band and wide-stopband single-band balanced bandpass filters with high selectivity and common-mode suppression," *IEEE Trans. Microw. Theory Techn.*, vol. **58**, pp. 2204–2212, 2010.
30. J. Shin and Q. Xue, "Balanced bandpass filters using center-loaded half-wavelength resonators," *IEEE Trans. Microw. Theory Techn.*, vol. **58**, pp. 970–977, 2010.
31. C.-H. Wu, C.-H. Wang, and C. H. Chen, "Stopband-extended balanced bandpass filter using coupled stepped-impedance resonators," *IEEE Microw. Wireless Compon. Lett.*, vol. **17**, pp. 507–509, 2007.
32. C.-H. Lee, C.-I. G. Hsu, and C.-C. Hsu, "Balanced dual-band BPF with stub-loaded SIRs for common-mode suppression," *IEEE Microw. Wireless Compon. Lett.*, vol. **20**, pp. 70–73, 2010.
33. X.-H. Wu and Q.-X. Chu, "Compact differential ultra-wideband bandpass filter with common-mode suppression," *IEEE Microw. Wireless Compon. Lett.*, vol. **22**, pp. 456–458, 2012.

34. P. Vélez, J. Naqui, A. Fernández-Prieto, M. Durán-Sindreu, J. Bonache, J. Martel, F. Medina, and F. Martín, "Differential bandpass filter with common mode suppression based on open split ring resonators and open complementary split ring resonators," *IEEE Microw. Wireless Compon. Lett.*, vol. **23**, pp. 22–24, 2013.
35. P. Vélez, J. Naqui, A. Fernánde Prieto, M. Durán-Sindreu, J. Bonache, J. Martel, F. Medina, and F. Martín, "Differential bandpass filters with common-mode suppression based on stepped impedance resonators (SIRs)," *IEEE MTT-S Int. Microw. Symp. Dig.*, Seattle, WA, June 2013.
36. A. Fernandez-Prieto, J. Martel-Villagran, F. Medina, F. Mesa, S. Qian, J.-.S Hong, J. Naqui, and F. Martin, "Dual-band differential filter using broadband common-mode rejection artificial transmission line," *Prog. Electromagn. Res.*, vol. **139**, pp. 779–797, 2013.
37. J.-S. Hong and M. J. Lancaster, *Microstrip Filters for RF/Microwave Applications*, Wiley, Hoboken, NJ, 2001.
38. M. Durán-Sindreu, A. Vélez, F. Aznar, G. Sisó, J. Bonache, and F. Martín, "Application of open split ring resonators and open complementary split ring resonators to the synthesis of artificial transmission lines and microwave passive components," *IEEE Trans. Microw. Theory Techn.*, vol. **57**, pp. 3395–3403, 2009.
39. A. Velez, F. Aznar, J. Bonache, M.C. Velázquez-Ahumada, J. Martel, and F. Martín, "Open complementary split ring resonators (OCSRRs) and their application to wideband CPW band pass filters," *IEEE Microw. Wireless Compon. Lett.*, vol. **19**, pp. 197–199, 2009.
40. J. Naqui, M. Durán-Sindreu, J. Bonache, and F. Martín, "Implementation of shunt connected series resonators through stepped-impedance shunt stubs: analysis and limitations," *IET Microw. Antenna Propag.*, vol. **5**, pp. 1336–1342, 2011.
41. P. Vélez, M. Durán-Sindreu, A. Fernández-Prieto. J. Bonache, F. Medina, and F. Martín, "Compact dual-band differential power splitter with common-mode suppression and filtering capability based on differential-mode composite right/left handed transmission line metamaterials," *IEEE Ant. Wireless Propag. Lett.*, vol. **13**, pp. 536–539, 2014.
42. P. Vélez, J. Naqui, A. Fernández-Prieto, J. Bonache, J. Mata-Contreras, J. Martel, F. Medina, and F. Martín, "Ultra-compact (80 mm^2) differential-mode ultra-wideband (UWB) bandpass filters with common-mode noise suppression," *IEEE Trans. Microw. Theory Techn.*, vol. **63**, pp. 1272–1280, 2015.
43. J. Esteban, C. Camacho-Peñalosa, J. E. Page, and T. M. Martín-Guerrero, "Generalized lattice network-based balanced composite right-/left-handed transmission lines," *IEEE Trans. Microw. Theory Techn.*, vol. **60**, pp. 2385–2393, 2012.
44. O. J. Zobel, "Theory and design of uniform and composite electric wave filters," *Bell Syst. Technol. J.*, vol. **2**, pp. 1–46, 1923.
45. F. Bongard and J. R. Mosig, "A novel composite right/left-handed unit cell and potential antenna applications," *Proc. IEEE Antennas Propag. Soc. Int. Symp.*, July 2008, pp. 1–4.
46. F. Bongard, J. Perruisseau-Carrier, and J. R. Mosig, "A novel composite right/left-handed unit cell based on a lattice topology: theory and applications," *2nd Int. Congr. Adv. Electromagn. Mater. Microw.Opt.* (*Metamater.*), Pamplona, Spain, September 2008, pp. 338–340.
47. F. Bongard, J. Perruisseau-Carrier, and J. R. Mosig, "Enhanced CRLH transmission line performances using a lattice network unit cell," *IEEE Microw. Wireless Compon. Lett.*, vol. **19**, pp. 431–433, 2009.

48. J. Perruisseau-Carrier, F. Bongard, M. Fernandez-Bolaños, and A. M. Ionescu, "A microfabricated 1-D metamaterial unit cell matched from DC to millimeter-waves," *IEEE Microw. Wireless Compon. Lett.*, vol. **21**, pp. 456–458, 2011.

49. P. Vélez, M. Durán-Sindreu, J. Bonache, and F. Martín, "Compact right-handed (RH) and left-handed (LH) lattice-network unit cells implemented in monolayer printed circuits," *Asia Pacific Microw. Conf.*, Melbourne, VIC, December 2011, pp. 534–537.

50. J. E. Page, J. Esteban, and C. Camacho-Peñalosa, "Lattice equivalent circuits of transmission-line and coupled-line sections," *IEEE Trans. Microw. Theory Techn.*, vol. **59**, pp. 2422–2430, 2011.

51. B. M. Schiffman, "A new class of broadband microwave 90-degree phase shifters," *IRE Trans. Microw. Theory Techn.*, vol. **MTT-6**, pp. 232–237, 1958.

52. A. M. E. Safwat, "Microstrip coupled line composite right/left-handed unit cell," *IEEE Microw. Wireless Compon. Lett.*, vol. **19**, pp. 434–436, 2009.

53. A. E. Fouda, A. M. E. Safwat, and H. El-Hennawy, "On the applications of the coupled-line composite right/left-handed unit cell," *IEEE Trans. Microw. Theory Techn.*, vol. **58**, pp. 1584–1591, 2010.

54. C. G. M. Ryan, and G. V. Eleftheriades, "A single-ended all-pass generalized negative-refractive-index transmission line using a bridged-T circuit," *IEEE-MTT-S Int. Microw. Symp. Dig.*, Montreal, QC, June 2012.

55. R. M. Foster, "A reactance theorem," *Bell Syst. Technol. J.*, vol. **3**, no. 2, pp. 259–267, 1924.

56. J. G. Linvill, "Transistor negative-impedance converters," *Proc. IRE*, vol. **41**, pp. 725–729, 1953.

57. S. A. Tretyakov and S. I. Maslovski, "Veselago materials: what is possible and impossible about the dispersion of the constitutive parameters," *IEEE Antenna Propag. Mag.*, vol. **49**, pp. 37–43, 2007.

58. E. Ugarte-Muñoz, S. Hrabar, D. Segovia-Vargas, and A. Kiricenko, "Stability of non-Foster reactive elements for use in active metamaterials and antennas," *IEEE Trans. Antenna Propag.*, vol. **60**, pp. 3490–3494, 2012.

59. S. A. Tretyakov, "Meta-materials with wideband negative permittivity and permeability," *Microw. Opt. Technol. Lett.*, vol. **31**, pp. 163–165, 2001.

60. T. P. Weldon, K. Miehle, R. S. Adams, and K. Daneshva, "A wideband microwave double-negative metamaterial with non-Foster loading," *Proc. 2012 IEEE SoutheastCon*, Orlando, FL, March 2012.

61. S. Hrabar, I. Krois, I. Bonic, and A. Kiricenko, "Negative capacitor paves the way to ultra-broadband metamaterials," *Appl. Phys. Lett.*, vol. **99**, paper 254103, 2011.

62. S. Hrabar, I. Krois, I. Bonic, and A. Kiricenko, "Ultra-broadband simultaneous superluminal phase and group velocities in non-Foster epsilon-near-zero metamaterial," *Appl. Phys. Lett.*, vol. **102**, paper 054108, 2013.

63. M. Mojahedi, K. J. Malloy, G. V. Eleftheriades, J. Woodley, and R. Y. Chiao, "Abnormal wave propagation in passive media," *IEEE J. Sel. Top. Quant. Electron.*, vol. **9**, pp. 30–39, 2003.

64. S. Hrabar, I. Krois, I. Bonic, and A. Kiricenko, "Superluminal propagation in metamaterials: anomalous dispersion versus non-Foster approach," *7th Int. Congr. Adv. Electromagn. Mater. Microw. Opt. – Metamater. 2013*, Bordeaux, France, September 2013.

65. D. Deslandes and K. Wu, "Integrated microstrip and rectangular waveguide in planar form," *IEEE Microw. Wireless Compon. Lett.*, vol. **11**, pp. 68–70, 2001.
66. F. Xu and K. Wu, "Guided-wave and leakage characteristics of substrate integrated waveguide," *IEEE Trans. Microw. Theory Techn.*, vol. **53**, pp. 66–73, 2005.
67. D. Deslandes and K. Wu, "Design consideration and performance analysis of substrate integrated waveguide components," *Proc. Eur. Microw. Conf.*, Milan, Italy, September 2002, pp. 881–884.
68. J. E. Rayas-Sanchez and V. Gutierrez-Ayala, "A general EM-based design procedure for single-layer substrate integrated waveguide interconnects with microstrip transitions," *IEEE MTT-S Int. Microw. Symp. Dig.*, Atlanta, GA, June 2008.
69. J. E. Rayas-Sanchez, "An improved EM-based design procedure for single-layer substrate integrated waveguide interconnects with microstrip transitions," *IEEE MTT-S Int. Microw. Workshop Ser. Signal Integr. High-Speed Interconnects*, Guadalajara, Mexico, February 2009.
70. K. Wu, D. Deslandes, and Y. Cassivi, "The substrate integrated circuits—a new concept for high-frequency electronics and optoelectronics," *6th Int. Telecommun. Modern Satellite, Cable, Broadcast. Service. Conf.*, Nis, Serbia and Montenegro, October 2003, pp. P-III–P-X.
71. Y. Dong and T. Itoh, "Promising future of metamaterials," *IEEE Microw. Mag.*, vol. **13**, pp. 39–56, 2012.
72. R. Marques, J. Martel, F. Mesa, and F. Medina, "Left-handed-media simulation and transmission of EM waves in subwavelength split ring resonator-loaded metallic waveguides," *Phys. Rev. Lett.*, vol. **89**, pp. 183901–183904, 2002.
73. G. Sisó, M. Gil, J. Bonache, F. Martín, "Applications of resonant-type metamaterial transmission lines to the design of enhanced bandwidth components with compact dimensions," *Microw. Opt. Technol. Lett.*, vol. **50**, pp 127–134, 2008.
74. Y. Dong, T. Yang, and T. Itoh, "Substrate integrated waveguide loaded by complementary split-ring resonators and its applications to miniaturized waveguide filters," *IEEE Trans. Microw. Theory Techn.*, vol. **57**, pp. 2211–2223, 2009.
75. Y. Dong and T. Itoh, "Miniaturized dual-band substrate integrated waveguide filters using complementary split-ring resonators," *IEEE MTT-S Int. Microw. Symp. Dig.*, Baltimore, MD, June 2011.
76. Y. Dong and T. Itoh, "Substrate integrated waveguide loaded by complementary split-ring resonators for miniaturized diplexer design," *IEEE Microw. Wireless Compon. Lett.*, vol. **21**, pp. 10–12, 2011.
77. L. Huang, I. D. Robertson, N. Yuan, and J. Huang, "Novel substrate integrated waveguide bandpass filter with broadside-coupled complementary split ring resonators," *IEEE MTT-S Int. Microw. Symp. Dig.*, Montreal, QC, June 2012.
78. Y. Dong and T. Itoh, "Composite right/left-handed substrate integrated waveguide and half-mode substrate integrated waveguide," *IEEE MTT-S Int. Microw. Symp. Dig.*, Boston, MA, June 2009, pp. 49–52.
79. Y. Dong and T. Itoh, "Composite right/left-handed substrate integrated waveguide and half mode substrate integrated waveguide leaky-wave structures," *IEEE Trans. Antenna Propag.*, vol. **59**, pp. 767–775, 2011.
80. W. Hong, B. Liu, Y. Q. Wang, Q. H. Lai, and K. Wu, "Half mode substrate integrated waveguide: a new guided wave structure for microwave and millimeter wave

application," *Proc. Joint 31st Int. Infrared Millimeter Waves Conf./14th Int. Terahertz Electron. Conf.*, Shanghai, China, September 2006, p. 219.

81. B. Liu, W. Hong, Y. Zhang, H. J. Tang, X. Yin, and K. Wu, "Half mode substrate integrated waveguide 180° 3-dB directional couplers," *IEEE Trans. Microw. Theory Techn.*, vol. **55**, pp. 2586–2592, 2007.
82. Y. Dong and T. Itoh, "Application of composite right/left-handed half-mode substrate integrated waveguide to the design of a dual band rat-race coupler," *IEEE MTT-S Int. Microw. Symp. Dig.*, Anaheim, CA, May 2010, pp. 712–715.
83. Y. Dong and T. Itoh, "Substrate integrated composite right-/left-handed leaky-wave structure for polarization-flexible antenna application," *IEEE Trans. Microw. Theory Techn.*, vol. **60**, pp. 760–771, 2012.
84. Y. Dong and T. Itoh, "Miniaturized substrate integrated waveguide slot antennas based on negative order resonance," *IEEE Trans. Antenna Propag.*, vol. **58**, pp. 3856–3864, 2010.
85. M. Durán-Sindreu, J. Bonache, F. Martín, and T. Itoh, "Novel fully-planar extended-composite right/left handed transmission line based on substrate integrated waveguide for multi-band applications," *Proc. Eur. Microw. Conf.*, Amsterdam, Netherlands, October 2012.
86. M. Durán-Sindreu, J. Bonache, F. Martín, and T. Itoh, "Single-layer fully-planar extended-composite right/left handed transmission lines based on substrate integrated waveguides for dual-band and quad-band applications," *Int. J. Microw. Wireless Technol.*, vol. **5**, pp. 213–229, 2013.
87. M. Durán-Sindreu, J. Choi, J. Bonache, F. Martín, and T. Itoh, "Dual-band leaky wave antenna with filtering capability based on extended-composite right/left-handed transmission lines," *IEEE MTT-S Int. Microw. Symp. Dig.*, Seattle, WA, June 2013.
88. C. G. M. Ryan and G. V. Eleftheriades, "A dual-band leaky-wave antenna based on generalized negative-refractive-index transmission-lines," *IEEE Antenna. Propag. Soc. Int. Symp. (APS-URSI)*, July 2010, pp. 1–4.

Appendix A

EQUIVALENCE BETWEEN PLANE WAVE PROPAGATION IN SOURCE-FREE, LINEAR, ISOTROPIC, AND HOMOGENEOUS MEDIA; TEM WAVE PROPAGATION IN TRANSMISSION LINES; AND WAVE PROPAGATION IN TRANSMISSION LINES DESCRIBED BY ITS DISTRIBUTED CIRCUIT MODEL

Let us consider a source-free, linear, isotropic, and homogeneous dielectric medium. Assuming a time dependence of the form $e^{j\omega t}$, the Maxwell's curl equations can be expressed as follows:

$$\nabla \times \vec{E} = -j\omega\mu \vec{H} \tag{A.1a}$$

$$\nabla \times \vec{H} = j\omega\varepsilon \vec{E} \tag{A.1b}$$

where $\vec{E}$ and $\vec{H}$ are the electric and magnetic field intensities, respectively, and μ and ε are the permeability and permittivity, respectively, of the considered medium. Taking the curl of (A.1a), and using (A.1b), gives

Artificial Transmission Lines for RF and Microwave Applications, First Edition. Ferran Martín.

$$\nabla \times \nabla \times \vec{E} = -j\omega\mu \nabla \times \vec{H} = \omega^2 \mu\varepsilon \vec{E} \tag{A.2}$$

Equation (A.2) can be simplified by using the following vector identity (where $\vec{A}$ is an arbitrary vector):

$$\nabla \times \nabla \times \vec{A} = \nabla(\nabla \cdot \vec{A}) - \nabla^2 \vec{A} \tag{A.3}$$

Using (A.3), and taking into account that in a source-free region $\nabla \cdot \vec{E} = 0$, the equation for the electric field (known as wave equation or Helmholtz equation) is found to be

$$\nabla^2 \vec{E} + \omega^2 \mu\varepsilon \vec{E} = 0 \tag{A.4}$$

Similarly, the wave equation for the magnetic field is

$$\nabla^2 \vec{H} + \omega^2 \mu\varepsilon \vec{H} = 0 \tag{A.5}$$

In the previous expressions, we can define the wavenumber, or propagation constant, as follows:

$$k = \omega\sqrt{\varepsilon\mu} \tag{A.6}$$

If the medium is lossless, μ and ε are real and k is also real. Note the analogy between the wavenumber and the phase constant of a transmission line (given by expression 1.6) for the lossless case. If we now consider plane wave propagation in the z-direction, with the electric field polarized in the x-direction, (A.4) reduces to

$$\frac{\partial^2 E_x}{\partial z^2} + k^2 E_x = 0 \tag{A.7}$$

and the general solution of this equation is formally identical to (1.7), that is,

$$E_x(z) = E_o^+ e^{-jkz} + E_o^- e^{jkz} \tag{A.8}$$

Introducing (A.8) in (A.1a), the magnetic field intensity is found to be

$$H_x = H_z = 0 \tag{A.9a}$$

$$H_y = -\frac{1}{j\omega\mu}\frac{\partial E_x}{\partial z} = \frac{k}{\omega\mu}\left(E_o^+ e^{-jkz} - E_o^- e^{jkz}\right) \tag{A.9b}$$

Defining the ratio of the $\vec{E}$ and $\vec{H}$ fields as the wave impedance of the medium, η, it is apparent from (A.8) and (A.9) that

$$\eta = \frac{\omega\mu}{k} = \sqrt{\frac{\mu}{\varepsilon}} \tag{A.10}$$

On the other hand, the phase velocity is given by

$$v_p = \frac{\omega}{k} = \frac{1}{\sqrt{\varepsilon\mu}} \tag{A.11}$$

In a lossy medium, with real and imaginary part of the permittivity and small (but finite) conductivity, the wave equation that results from the Ampere–Maxwell law (1.62) is as follows:

$$\nabla^2 \vec{E} + \omega^2 \mu \left(\varepsilon' - j \frac{\sigma + \omega\varepsilon''}{\omega} \right) \vec{E} = 0 \tag{A.12}$$

and the complex propagation constant is defined in this case as

$$\gamma = \alpha + j\beta = j\omega \sqrt{\mu \left(\varepsilon' - j \frac{\sigma + \omega\varepsilon''}{\omega} \right)} \tag{A.13}$$

The solutions of (A.12) for the electric field are similar to those given by (A.8), that is,

$$E_x(z) = E_o^+ e^{-\gamma z} + E_o^- e^{\gamma z} \tag{A.14}$$

corresponding to exponentially decaying travelling waves with phase velocity $v_p = \omega/\beta$. The associated magnetic field is

$$H_y = -\frac{1}{j\omega\mu} \frac{\partial E_x}{\partial z} = \frac{\gamma}{j\omega\mu} \left(E_o^+ e^{-\gamma z} - E_o^- e^{\gamma z} \right) \tag{A.15}$$

and the wave impedance is

$$\eta = \frac{j\omega\mu}{\gamma} \tag{A.16}$$

Plane wave propagation in a source-free, linear, isotropic, and homogeneous medium is a particular case of TEM wave propagation. The field solutions given earlier (A.8 and A.9 for the lossless case, or A.14 and A.15 for the lossy case) are identical to those that result by considering a hypothetical parallel plate transmission line with infinitely wide plates (or at least very wide as compared to their separation) oriented with the line axis in the z-direction and the plates parallel to the z-y plane, and filled with a source-free, linear, isotropic, and homogeneous dielectric.

For a general transmission line supporting TEM wave propagation[1] along the z-axis (positive direction), the fields can be expressed as follows:

$$\vec{E}\,(x,y,z)=\vec{E}_t(x,y)\cdot e^{-j\beta z} \tag{A.17a}$$

$$\vec{H}\,(x,y,z)=\vec{H}_t(x,y)\cdot e^{-j\beta z} \tag{A.17b}$$

where $\vec{E}_t(x,y)$ and $\vec{H}_t(x,y)$ are the transverse electric and magnetic field components, respectively, and β is the propagation constant[2]. Introducing (A.17) in (A.1), the following equations result:

$$j\beta E_y=-j\omega\mu H_x \tag{A.18a}$$

$$-j\beta E_x=-j\omega\mu H_y \tag{A.18b}$$

$$\frac{\partial E_y}{\partial x}-\frac{\partial E_x}{\partial y}=0 \tag{A.18c}$$

$$j\beta H_y=j\omega\varepsilon E_x \tag{A.18d}$$

$$-j\beta H_x=j\omega\varepsilon E_y \tag{A.18e}$$

$$\frac{\partial H_y}{\partial x}-\frac{\partial H_x}{\partial y}=0 \tag{A.18f}$$

where it has been taken into account that $E_z=H_z=0$ and the $e^{-j\beta z}$ z-dependence of the field components. Combining (A.18a) and (A.18e), or (A.18b) and (A.18d), one obtains the following:

$$\beta=\omega\sqrt{\varepsilon\mu}=k \tag{A.19}$$

and the wave impedance for the TEM mode is found to be

$$Z_{\mathrm{TEM}}=\frac{E_x}{H_y}=-\frac{E_y}{H_x}=\frac{\omega\mu}{\beta}=\sqrt{\frac{\mu}{\varepsilon}}=\eta \tag{A.20}$$

as results from (A.18a) and (A.18b), or (A.18d) and (A.18e). Notice that from (A.20), it follows that the transverse fields are related by

$$\vec{H}_t(x,y)=\frac{1}{Z_{\mathrm{TEM}}}\vec{z}\times\vec{E}_t(x,y) \tag{A.21}$$

[1] For TEM wave propagation, the line must have at least two conductors.
[2] If losses are present, then $j\beta$ must be replaced with γ.

where $\vec{z}$ is the unit vector in the z-direction. Thus, the propagation constant and the wave impedance in TEM transmission lines are given by identical expressions as those for plane waves in a source-free, linear, isotropic, and homogeneous media. By mapping the material parameters with the line parameters according to

$$\varepsilon = C' \tag{A.22a}$$

$$\mu = L' \tag{A.22b}$$

the wave equation (A.7) for the electric field is equivalent to Equation 1.4a for the line voltage, considering a lossless line (i.e., $\gamma^2 = -\beta^2$). Analogously, the wave equation for the magnetic field (formally identical to (A.7)) is equivalent to Equation 1.4b for the line current. Hence, there is a clear link between plane wave propagation in source-free, linear, isotropic, and homogeneous media, TEM wave propagation in transmission lines and wave propagation in transmission lines as described by the distributed approach (this link is pointed out in Section 1.1).

The solutions for the transverse field components depend on the specific geometry of the line and are given by the Helmholtz equations (A.4) and (A.5). Both equations can be expressed as a pair of order-2 differential equations, that is,

$$\left(\frac{\partial^2}{\partial x^2} + \frac{\partial^2}{\partial y^2} + \frac{\partial^2}{\partial z^2} + k^2\right) E_x = 0 \tag{A.23a}$$

$$\left(\frac{\partial^2}{\partial x^2} + \frac{\partial^2}{\partial y^2} + \frac{\partial^2}{\partial z^2} + k^2\right) E_y = 0 \tag{A.23b}$$

$$\left(\frac{\partial^2}{\partial x^2} + \frac{\partial^2}{\partial y^2} + \frac{\partial^2}{\partial z^2} + k^2\right) H_x = 0 \tag{A.23c}$$

$$\left(\frac{\partial^2}{\partial x^2} + \frac{\partial^2}{\partial y^2} + \frac{\partial^2}{\partial z^2} + k^2\right) H_y = 0 \tag{A.23d}$$

Taking into account the $e^{-j\beta z}$ z-dependence of the field components, the previous equations can be simplified and expressed in a compact form as follows:[3]

$$\nabla_t^2 \vec{E}_t(x, y) = 0 \tag{A.24a}$$

$$\nabla_t^2 \vec{H}_t(x, y) = 0 \tag{A.24b}$$

[3] Notice that $\partial^2 E_x/\partial z^2 = -\beta^2 E_x = -k^2 E_x$ and $\partial^2 E_y/\partial z^2 = -\beta^2 E_y = -k^2 E_y$. The same applies to the transverse components of the magnetic field.

where $\nabla_t^2 = \partial^2/\partial x^2 + \partial^2/\partial y^2$ is the transverse Laplacian operator. The solutions of (A.24) are determined by the transverse geometry of the line and boundary conditions. Nevertheless, since the transverse fields satisfy the Laplace equation, these fields are the same as the static fields between the two conductors of the TEM line. In particular, the electric field can be expressed as the gradient of a scalar potential, that is,[4]

$$\vec{E}_t(x,y) = -\nabla_t V(x,y) \tag{A.25}$$

Since the divergence of the electric field is zero ($\nabla_t \cdot \vec{E}_t(x,y) = 0$), the scalar potential also satisfies the Laplace equation:

$$\nabla_t^2 V(x,y) = 0 \tag{A.26}$$

and the voltage between the two conductors is found as the path integral of the field

$$V_{12} = -\int_1^2 \vec{E}_t(x,y) \cdot d\vec{l} \tag{A.27}$$

Finally, the current in the conductor can be found from the Ampere law as follows:

$$I = \oint_C \vec{H}_t(x,y) \cdot d\vec{l} \tag{A.28}$$

where C is the cross-sectional contour of the conductor.

For the particular case of a parallel plate transmission line (Fig. A.1) with width W and height h (and $W >> h$ to neglect the fringing fields at the lateral sides), the solution

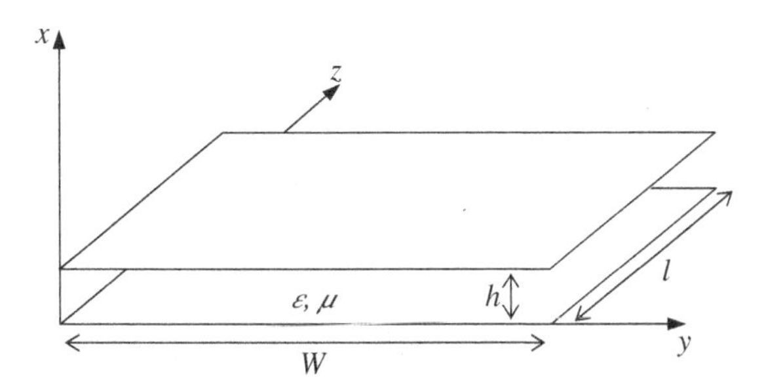

FIGURE A.1 Section of a parallel-plate transmission line of length l (the other relevant dimensions are also indicated). TEM waves propagate in the z-direction.

[4] This is valid if the two-dimensional curl of $\vec{E}_t(x,y)$, defined as $\partial E_y/\partial x - \partial E_x/\partial y$, is null. Equation A.18c reveals that this is the case.

of (A.26) subjected to the boundary conditions $V(x=0,\ y)=0$ and $V(x=h,\ y)=V_o$ gives

$$V(x,y)=\frac{V_o x}{h} \tag{A.29}$$

The total voltage and electric field are thus

$$V(x,y,z)=\frac{V_o x}{h}e^{-jkz} \tag{A.30}$$

$$\vec{E}(x,y,z)=-\vec{x}\frac{V_o}{h}e^{-jkz} \tag{A.31}$$

and the voltage between the two conductors is simply

$$V_{12}\equiv V=V_o e^{-jkz} \tag{A.32}$$

Using (A.21), the magnetic field is found to be

$$\vec{H}(x,y,z)=-\vec{y}\frac{V_o}{\eta h}e^{-jkz} \tag{A.33}$$

and the current is obtained from (A.28), where the z-dependence is introduced

$$I=\int_0^W \vec{y}\frac{V_o}{\eta h}e^{-jkz}\cdot\vec{y}\,dy=\frac{V_o W}{\eta h}e^{-jkz} \tag{A.34}$$

Finally the characteristic impedance is given by

$$Z_o=\frac{V}{I}=\eta\frac{h}{W} \tag{A.35}$$

It is worth mentioning that (1.9) can be obtained by calculating the per unit length capacitance C' and inductance L' of the parallel plate transmission line. Using the general expression of the time-averaged density of energy[5]

$$U_{\text{nd}}=\frac{1}{4}\left(\varepsilon\left|\vec{E}\right|^2+\mu\left|\vec{H}\right|^2\right) \tag{A.36}$$

it follows that the electric and magnetic energy stored in a volume V are given by

[5] Notice that this expression in only valid for nondispersive media. The general expression for both dispersive and nondispersive media is given in Chapter 3.

$$E_{\mathrm{e}} = \frac{1}{4}\int \varepsilon \left|\vec{E}\right|^2 dV \tag{A.37a}$$

and

$$E_{\mathrm{m}} = \frac{1}{4}\int \mu \left|\vec{H}\right|^2 dV \tag{A.37b}$$

respectively. Introducing (A.31) in (A.37a) and (A.33) in (A.37b), and integrating over a volume of the parallel plate transmission line delimited by its width W, its height h and an arbitrary length l, the stored electric and magnetic energy are found to be:

$$E_{\mathrm{e}} = \frac{\varepsilon}{4} V_{\mathrm{o}}^2 \frac{Wl}{h} \tag{A.38a}$$

$$E_{\mathrm{m}} = \frac{\mu}{4} \frac{V_{\mathrm{o}}^2}{\eta^2} \frac{Wl}{h} \tag{A.38b}$$

On the other hand, the energy stored by the capacitance C and inductance L of the line corresponding to a section of length l are, according to circuit theory, $CV_{\mathrm{o}}^2/4$ and $LI_{\mathrm{o}}^2/4$, respectively. Hence, the per-unit-length inductance and capacitance of the line are found to be

$$C' = \varepsilon \frac{W}{h} \tag{A.39a}$$

$$L' = \mu \frac{1}{\eta^2} \frac{V_{\mathrm{o}}^2}{I_{\mathrm{o}}^2} \frac{W}{h} = \varepsilon \frac{V_{\mathrm{o}}^2}{I_{\mathrm{o}}^2} \frac{W}{h} \tag{A.39b}$$

Combining (A.39a) and (A.39b), the characteristic impedance $Z_{\mathrm{o}} = V_{\mathrm{o}}/I_{\mathrm{o}}$ as expressed by (1.9) is obtained.

The calculation of the fields, characteristic impedance, and propagation constant for other TEM transmission lines (e.g., stripline and coaxial line) and quasi-TEM lines (e.g., microstrip and CPW) is out of the scope of this appendix (see Ref. [1–8] of Chapter 1 for an in-depth study of these lines), which is only devoted to demonstrating the equivalence between the field theory and the distributed approach for the analysis of TEM transmission lines. For quasi-TEM lines, wave propagation can be approximated by TEM wave propagation (as described in this appendix) by simply replacing the permeability μ and permittivity ε (the parameters of the homogeneous dielectric material) with effective parameters (μ_{eff} and $\varepsilon_{\mathrm{eff}}$) that take into account the transverse nonuniformity (nonhomogeneity) of the dielectric or the presence of air (in open lines).[6]

[6] The effective permeability or permittivity of an ordinary quasi-TEM transmission line depends on the material composition of the substrate and transverse geometry and should not be confused with the effective permeability and permittivity of metamaterials and metamaterial transmission lines (introduced in Chapter 3). It is also worth mentioning that in Chapter 2 nonuniform transmission lines are defined as transmission lines where the transverse geometry and/or material composition of the substrate change along the propagation direction z. In these lines, the effective permittivity and permeability are a function of z.

Appendix B

THE SMITH CHART

The Smith chart is a graphical means to simultaneously visualize the reflection coefficient and impedance of a certain load (i.e., a lumped element, a terminated transmission line, or even a complex one-port system). It is very useful for the analysis and synthesis of circuits based on transmission lines and stubs, and most microwave commercial CAD tools and test equipment allow for the visualization of simulation and measurement results on a Smith chart. The Smith chart (see Fig. B.1) is essentially the representation of the reflection coefficient in polar coordinates, $\rho = |\rho|e^{j\theta}$, where $-\pi \leq \theta \leq \pi$, and the origin of θ is the right-hand side of the horizontal axis. For any passive load ($|\rho| \leq 1$), the reflection coefficient is given by a single point within the circle of unit radius, which is the considered region for the representation of the reflection coefficient in the chart.

However, the main relevant use of the Smith chart is the direct conversion from reflection coefficients to normalized impedances, or admittances, and vice versa. Let us consider for the moment the conversion to normalized impedances, defined by the quotient between a given impedance Z and a reference impedance Z_o, which is typically the characteristic impedance of a transmission line:

$$\bar{Z} = \frac{Z}{Z_o} \tag{B.1}$$

Considering Z the load impedance of a transmission line, the reflection coefficient can be expressed as follows:

$$\rho = \frac{\bar{Z} - 1}{\bar{Z} + 1} \tag{B.2}$$

Artificial Transmission Lines for RF and Microwave Applications, First Edition. Ferran Martín.

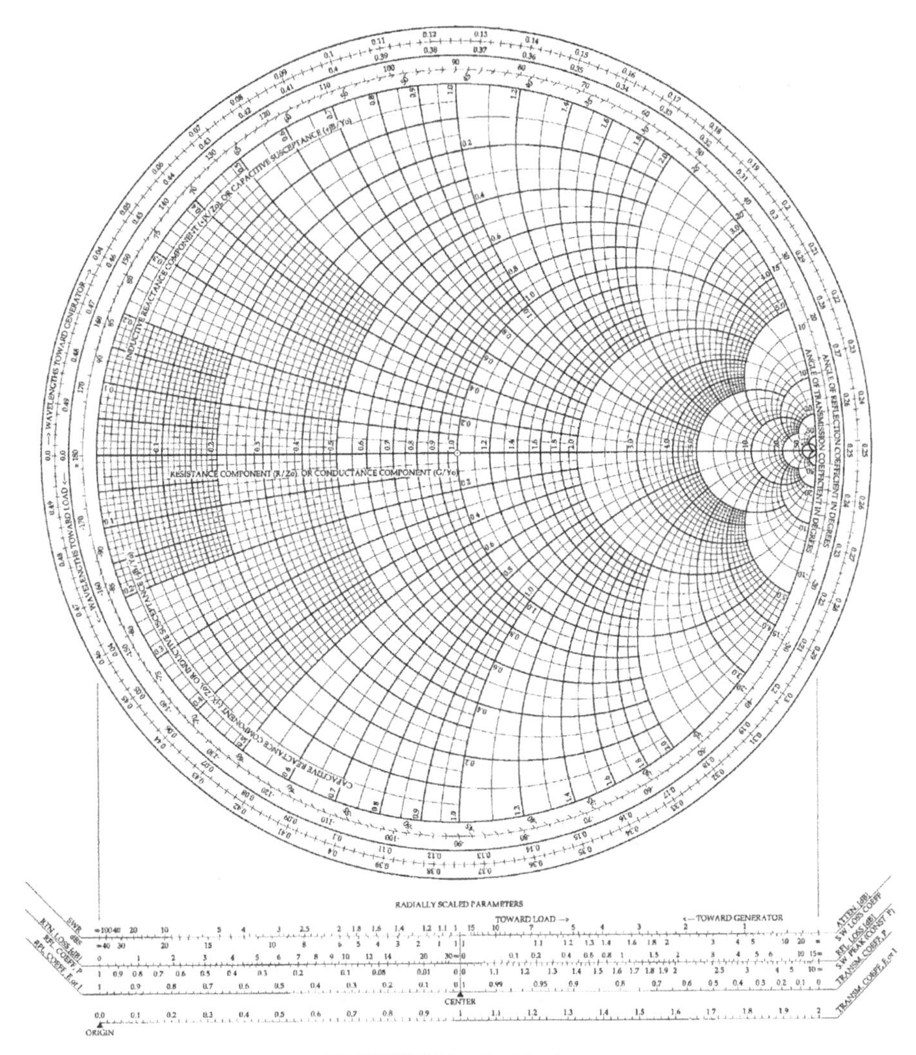

FIGURE B.1 Smith chart.

and there is a univocal correspondence between ρ and $\bar{Z}$. From B.2, the normalized impedance can be isolated and expressed as follows:

$$\bar{Z} = \frac{1+\rho}{1-\rho} \tag{B.3}$$

or

$$\bar{R} + j\bar{\chi} = \frac{1+\rho_{\mathrm{r}}+j\rho_{\mathrm{i}}}{1-\rho_{\mathrm{r}}-j\rho_{\mathrm{i}}} \tag{B.4}$$

where $\bar{Z}$ and ρ have been decomposed into the real and imaginary parts. In the previous complex equation, both the real and the imaginary parts must be equal. This gives

$$\left(\rho_{\rm r}-\frac{\bar{R}}{1+\bar{R}}\right)^2+\rho_{\rm i}^2=\left(\frac{1}{1+\bar{R}}\right)^2 \tag{B.5a}$$

$$(\rho_{\rm r}-1)^2+\left(\rho_{\rm i}-\frac{1}{\bar{\chi}}\right)^2=\left(\frac{1}{\bar{\chi}}\right)^2 \tag{B.5b}$$

Considering $\bar{R}$ constant, (B.5a) is the equation of a circumference in the $\rho_{\rm r}$–$\rho_{\rm i}$ plane. Thus, (B.5a) is a family of circumferences parameterized by $\bar{R}$, and each circumference has its center in the $\rho_{\rm i}=0$ axis. These circumferences, all contained within the Smith chart, are called constant resistance circumferences, and the value of $\bar{R}$ is indicated in the horizontal axis of the Smith chart. Notice that the $\bar{R}=0$ circumference coincides with the unit radius circumference, as one expects for a purely reactive load, where $|\rho|=1$. As $\bar{R}$ increases, the radius of the constant resistance circumferences decrease, and the circumference degenerates in a single point $\rho_{\rm r}=1$, $\rho_{\rm i}=0$ (or $|\rho|=1$, $\theta=0$) for $\bar{R}\to\infty$.

By contrast, (B.5b) is the equation of a family of circumferences parameterized by the normalized reactance (constant reactance circumferences). However, for any constant reactance circumference, only a portion of it lies within the Smith chart (the centers of these circumferences lie in the vertical line $\rho_{\rm r}=1$ of the chart). For increasing values of $\bar{\chi}$, the radius of the reactance circumferences decreases and $\rho_{\rm r}=1$, $\rho_{\rm i}=0$ for $\bar{\chi}\to\infty$. For $\bar{\chi}=0$, the constant reactance curve degenerates in the straight line $\rho_{\rm i}=0$, as expected on account of the real reflection coefficient for a purely resistive load. The normalized reactance values that label the different reactance circumferences are indicated in the Smith chart (along the whole external circumference).

Notice that the constant resistance and reactance circumferences are orthogonal. Some relevant curves and normalized impedances are indicated in the Smith chart of Figure B.2.

Given a reflection coefficient, represented by a unique point in the Smith chart, the normalized impedance can be immediately visualized by directly reading in the chart. The Smith chart has many applications, but one relevant use of the Smith chart concerns the graphical solution of the input impedance, $Z_{\rm in}$, of a terminated transmission line (expression 1.31). Since the reflection coefficient at the input of the terminated line is given by (1.27), it is clear that increasing the line length is equivalent to a clockwise rotation of an angle $\theta=2\beta l$ (l being the line length) from the point of the load ($\rho_{\rm L}$) with center on the center of the Smith chart. Notice that for a $\lambda/2$ line, $\theta=2\pi$, corresponding to a complete rotation and $Z_{\rm in}=Z_{\rm L}$ (the input impedance does not experience any change for transmission line lengths that are multiple of half wavelength). To facilitate the solution of this type of transmission line problems, the Smith chart has scales around its periphery calibrated in terms of the wavelength toward (clockwise) or away from (counterclockwise) the generator. Figure B.3 plots the location of

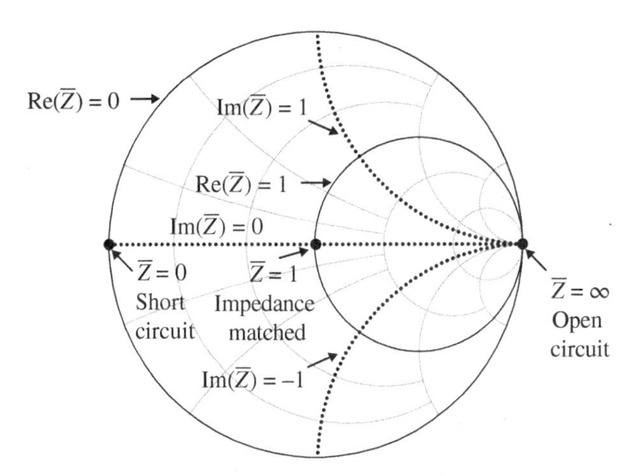

FIGURE B.2 Some relevant constant resistance/reactance curves and normalized impedances plotted in the Smith chart. The constant resistance circumference that crosses the center of the Smith chart (Re($\bar{Z}$) = 1) is called unit resistance circumference.

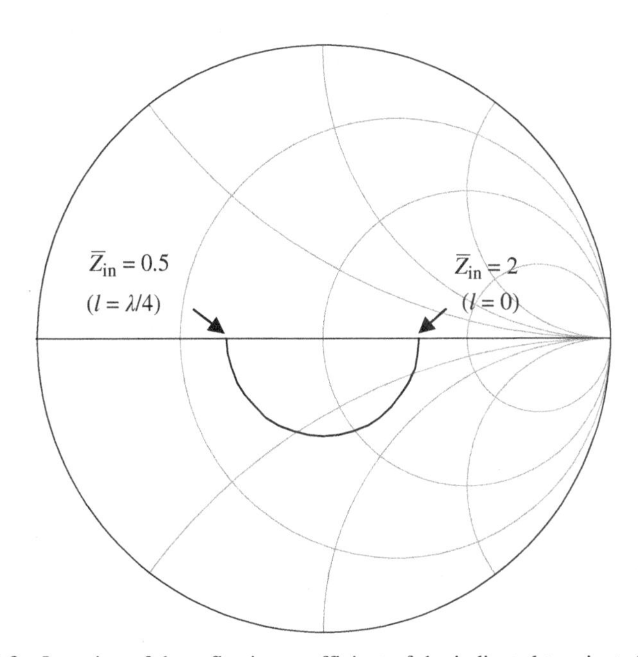

FIGURE B.3 Location of the reflection coefficient of the indicated terminated line for line length increasing from 0 to $\lambda/4$. Notice that the left extreme of the plot gives the normalized admittance of the load ($\bar{Y} = 0.5$), which coincides with the normalized input impedance of the line.

the reflection coefficient at the input port of a transmission line terminated with a purely resistive load ($\bar{Z}=2$) when the line length is increased from $l=0$ to $l=\lambda/4$.

Following a similar procedure, a Smith chart for the normalized admittance can be constructed. However, the normalized admittance of a given load can be directly visualized in the impedance Smith chart. The reason is that the normalized input impedance of a $\lambda/4$ terminated line (impedance inverter) is

$$\overline{Z_{\text{in}}} = \frac{1}{\overline{Z_{\text{L}}}} \tag{B.6}$$

which is the normalized admittance of the load. Since a $\lambda/4$ transformation is equivalent to a 180° rotation in the Smith chart, the normalized admittance of the load can be simply inferred by imaging the impedance point across the center of the Smith chart. Thus, the impedance Smith chart can be used to deal with normalized impedances or admittances, indistinctly. Figure B.3, indicates that the normalized admittance of the load considered in the example is $\bar{Y}=0.5$.

Since the reflection coefficient of a matched load is null, that is, it is located in the centre of the Smith chart, the Smith chart is a useful tool to evaluate the level of matching of a given load and its dependence with frequency (strong deviations from the center of the chart indicate significant mismatch). In reactive loads, departure from the zero-resistance circumference must be attributed to the presence of a non-negligible-resistive component in the load impedance, and gives an indication of the deviation from the purely reactive nature of the load.

To finalize this appendix, let us consider an illustrative example to show how the Smith chart can be used to obtain information on reactive two-port circuits loaded with resistive impedances. Let us consider the lumped-element circuit of Figure B.4a (this circuit describes the unit cell of a particular type of artificial transmission line, as discussed in Chapter 3), and let us assume that the series inductance L and the elements of the shunt branch (C, C_c, and L_c) are unknown. Let us represent the dependence of the input impedance (or reflection coefficient) with frequency in the Smith chart, considering that the output port is loaded with the reference impedance Z_o (Fig. B.4b). There are two singular frequencies in the circuit: (1) the resonance frequency of the L_c–C_c tank, f_o and (2) the frequency that nulls the shunt reactance, f_z. At f_z, given by

$$f_z = \frac{1}{2\pi\sqrt{L_c(C+C_c)}} \tag{B.7}$$

the impedance seen from the input port is simply $j\omega L$ since the shunt branch is short-circuited to ground. This means that the input impedance must be tangent to the zero-resistance circle of the Smith chart at f_z. Thus, from the value of the reactance at this point, the inductance L can be inferred, and from the value of f_z, a condition for the parameters of the shunt branch is inferred. A second condition comes from f_o,

$$f_o = \frac{1}{2\pi\sqrt{L_c C_c}} \tag{B.8}$$

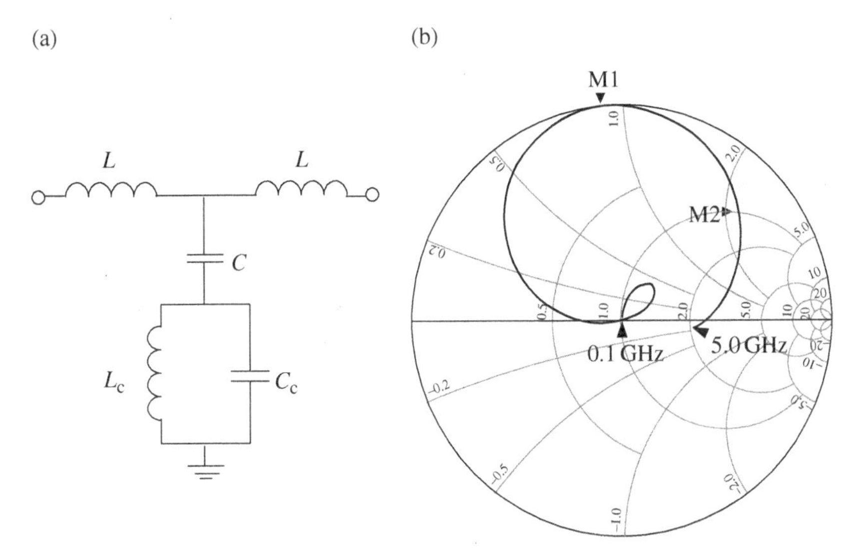

FIGURE B.4 Lumped-element-reactive two-port network (a), and its reflection coefficient considering that the output port is terminated with the reference impedance Z_o (b). The relevant frequencies are $f_o = 3.56$ GHz (M2) and $f_z = 3$ GHz (M1). The normalized impedance at f_z is $\bar{Z} = j0.946$, giving a value for the series inductance of $L = 2.5$ nH. The other element values are $C = 0.8$ pF, $C_c = 2$ pF, and $L_c = 1$ nH. The frequency swept covers the range 0.1–5.0 GHz.

This frequency can be easily identified because at this frequency the shunt branch opens and the normalized input impedance can be expressed as $\overline{Z_{in}} = 1 + j\bar{\chi}$. In other words, f_o is that frequency where the input impedance crosses the so-called unit resistance circumference, and the second condition is given by expression (B.8). To univocally determine the parameters of the shunt branch an additional condition is necessary, for instance, the reactance value at a given frequency. However, a more elegant procedure to extract the parameters of structures described by this and other similar circuits is detailed in Appendix G.

Appendix C

THE SCATTERING MATRIX

Let us consider an arbitrary network with N ports and the corresponding reference planes (Fig. C.1). This network can be characterized by means of the impedance (**Z**) or the admittance (**Y**) matrix, where $\mathbf{V} = \mathbf{Z} \cdot \mathbf{I}$ (**V** and **I** being column vectors composed of the voltages and currents, respectively, at the ports of the network) and $\mathbf{Y} = \mathbf{Z}^{-1}$. At microwave frequencies, a variation (displacement) of the reference plane of the ports modifies the elements of the Z- and Y-matrix in a so complex form, that it might be difficult (not to say impossible) to identify two identical networks with the ports located at different planes.

Microwave networks are usually (although not exclusively) described by means of the scattering matrix (S-matrix). Let us consider that the incident and reflected waves[1] at the reference plane of port i are characterized by the voltages and currents (V_i^+, I_i^+) and (V_i^-, I_i^-), respectively, and that the characteristic impedance of port i is Z_{oi}. We can define the normalized voltages as follows:

$$a_i = \frac{V_i^+}{\sqrt{Z_{oi}}}; \; b_i = \frac{V_i^-}{\sqrt{Z_{oi}}} \tag{C.1}$$

[1] Notice that the meaning of incident and reflected waves simply refers to waves impinging externally to the port (i.e., entering the network) and waves coming from the network (i.e., traveling to the external region of the network). Thus, the incident waves are not necessarily generated by an external source (they can be generated by internal port reflection caused by a mismatched load). Similarly, the reflected waves are not necessarily caused by external reflection of a source connected to the port (they can be generated by external sources connected to other ports of the network).

Artificial Transmission Lines for RF and Microwave Applications, First Edition. Ferran Martín.

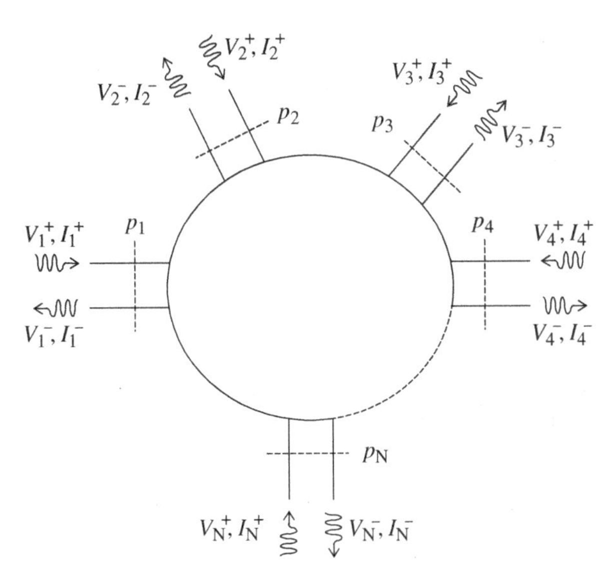

FIGURE C.1 Arbitrary microwave N-port network. The reference planes of the ports are labeled p_i.

so that the total voltage and current at port i can be expressed as follows:

$$V_i = V_i^+ + V_i^- = \sqrt{Z_{oi}}(a_i + b_i) \tag{C.2a}$$

$$I_i = I_i^+ - I_i^- = \frac{1}{\sqrt{Z_{oi}}}(a_i - b_i) \tag{C.2b}$$

and, from (1.17), the power delivered to the i-th port is

$$P_i = \frac{1}{2}\left(|a_i|^2 - |b_i|^2\right) \tag{C.3}$$

The S-matrix relates the normalized voltages of the incident and reflected waves, that is, $\mathbf{b} = \mathbf{S} \cdot \mathbf{a}$, where $\mathbf{a}$ and $\mathbf{b}$ are column vectors composed of the normalized voltages of the incident and reflected waves, respectively, and the elements of the S-matrix are given by

$$S_{ij} = \left.\frac{b_i}{a_j}\right|_{a_k = 0,\ \text{for } k \neq j} \tag{C.4}$$

Notice that if all the port impedances are identical, the S-matrix elements can be written in terms of the incident and reflected port voltages as follows:

$$S_{ij} = \left.\frac{V_i^-}{V_j^+}\right|_{V_k^+ = 0,\ \text{for } k \neq j} \tag{C.5}$$

The diagonal elements of the S-matrix are the reflection coefficients at the different ports looking into the network, when no incident waves are present on the other ports. To guarantee that no incident waves impinge on a port, that port must be terminated with a matched load. For networks with identical port impedances, the elements S_{ij} with $i \neq j$ can be interpreted as the transmission coefficients between port j and port i when all the ports except the port j are terminated in matched loads. In two-port networks, such as microwave filters, S_{11} and S_{21} are usually expressed in decibel and referred to as return and insertion loss, respectively, rather than reflection and transmission coefficients.[2]

For a given network, the calculation of the S-matrix may be tedious, as compared to the Z- or Y-matrix. However, since these matrices completely characterize the network, they must be related. In other words, the S-matrix can be expressed in terms of the Z- or Y-matrix, and vice versa. Let us consider that the impedances of the ports are all identical (Z_o), which is the most usual situation. Introducing the incident and reflected wave variables in the matrix equation $\mathbf{V} = \mathbf{Z} \cdot \mathbf{I}$, the following expression results:

$$\mathbf{V}^{+} + \mathbf{V}^{-} = \mathbf{Z}(\mathbf{I}^{+} - \mathbf{I}^{-}) \tag{C.6}$$

or

$$\mathbf{V}^{+} + \mathbf{S}\mathbf{V}^{+} = \mathbf{Z}\left(\frac{\mathbf{V}^{+}}{Z_o} - \frac{\mathbf{V}^{-}}{Z_o}\right) \tag{C.7}$$

$$Z_o(\mathbf{1} + \mathbf{S})\mathbf{V}^{+} = \mathbf{Z}(\mathbf{1} - \mathbf{S})\mathbf{V}^{+} \tag{C.8}$$

where $\mathbf{1}$ is the identity matrix. Therefore, the normalized impedance matrix ($\bar{\mathbf{Z}} = \mathbf{Z}/Z_o$) can be expressed in terms of the S-matrix as follows:

$$\bar{\mathbf{Z}} = (\mathbf{1} + \mathbf{S})(\mathbf{1} - \mathbf{S})^{-1} \tag{C.9}$$

From (C.8), we can isolate the S-matrix:

$$\mathbf{S} = (\bar{\mathbf{Z}} + \mathbf{1})^{-1}(\bar{\mathbf{Z}} - \mathbf{1}) \tag{C.10}$$

Since $\mathbf{Y} = \mathbf{Z}^{-1}$, the relations between the Y- and the S-matrix can be easily inferred.

For two-port networks, the transmission $ABCD$ matrix is a very useful matrix to describe the structure. In reference to the two-port network of Figure C.2a, and the total voltages and currents in the ports, the $ABCD$ matrix is given by

$$\begin{pmatrix} V_1 \\ I_1 \end{pmatrix} = \begin{pmatrix} A & B \\ C & D \end{pmatrix} \begin{pmatrix} V_2 \\ -I_2 \end{pmatrix} \tag{C.11}$$

[2] Normally, port 1 and 2 are the input and output ports, respectively.

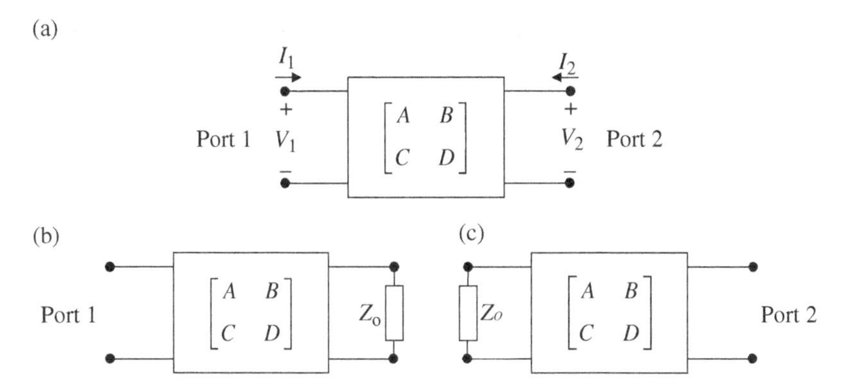

FIGURE C.2 Arbitrary two-port network (a). The loaded networks used to express S_{11}, S_{21}, and S_{22}, S_{12} as a function of the *ABCD* parameters are shown in (b) and (c), respectively.

The main advantage of the transmission *ABCD* matrix over the *S*-, *Z*-, or *Y*-matrix is that if several two-port networks are cascaded, the *ABCD* matrix of the resulting two-port network is given by the product of the *ABCD* matrices of the individual networks. Thus, if several two-port networks are cascaded, finding first the *ABCD* matrix and then transforming to the *S*-, *Z*-, or *Y*-matrices is the convenient way to obtain these matrices for the complete network.

Let us show how to express the *S*-matrix in terms of the *ABCD* matrix. $S_{11}$is the reflection coefficient from port 1 when port 2 is terminated with the impedance Z_o (which is assumed to be the impedance of both ports). Such reflection coefficient can be calculated from the impedance, Z_{in}, seen from port 1, given by

$$Z_{in} = \frac{V_1}{I_1} = \frac{A + B/Z_o}{C + D/Z_o} \tag{C.12}$$

as results from the following expressions, given by the *ABCD* matrix for the structure of Figure C.2b:

$$V_1 = AV_2 + B\frac{V_2}{Z_o} \tag{C.13a}$$

$$I_1 = CV_2 + D\frac{V_2}{Z_o} \tag{C.13b}$$

From (C.12) and (1.20) (with $Z_L = Z_{in}$), S_{11} is found to be

$$S_{11} = \frac{A + B/Z_o - CZ_o - D}{A + B/Z_o + CZ_o + D} \tag{C.14}$$

The transmission coefficient S_{21} can be obtained from the circuit of Figure C.2b, taking into account that $V_2^- = V_2$ and $V_1 = (1+S_{11})V_1^+$:

$$S_{21} = \left.\frac{V_2^-}{V_1^+}\right|_{V_2^+=0} = \frac{V_2}{V_1}(1+S_{11}) \tag{C.15}$$

From (C.13a) and (C.14),

$$S_{21} = \frac{2}{A+B/Z_o+CZ_o+D} \tag{C.16}$$

To determine S_{22}, port 1 is loaded with Z_o (Fig. C.2c). The impedance seen from port 2 can be expressed in terms of the *ABCD* matrix as follows:

$$Z_{out} = \frac{V_2}{I_2} = \frac{D+B/Z_o}{C+A/Z_o} \tag{C.17}$$

and the reflection coefficient is

$$S_{22} = \frac{-A+B/Z_o-CZ_o+D}{A+B/Z_o+CZ_o+D} \tag{C.18}$$

Finally, the transmission coefficient from port 2 to port 1 is given by

$$S_{12} = \left.\frac{V_1^-}{V_2^+}\right|_{V_1^+=0} = \frac{V_1}{V_2}(1+S_{22}) \tag{C.19}$$

From the first algebraic equation inferred from the *ABCD* matrix applied to the circuit of Figure C.2c, that is

$$V_1 = AV_2 - B\frac{V_2}{Z_{out}} \tag{C.20}$$

and (C.17) and (C.18), we obtain

$$S_{12} = \frac{2(AD-BC)}{A+B/Z_o+CZ_o+D} \tag{C.21}$$

Since for reciprocal networks[3] $S_{21} = S_{12}$, comparison of (C.16) and (C.21) gives $AD - BC = 1$.

[3] Reciprocal networks, defined in Chapter 1, exhibit a symmetric *S*-, *Y*-, and *Z*-matrix.

The dependence of the *ABCD* matrix elements with the *S*-parameters can be inferred from (C.14), (C.16), (C.18), and (C.21):

$$A = \frac{(1+S_{11})(1-S_{22})+S_{12}S_{21}}{2S_{21}} \tag{C.22a}$$

$$B = Z_o \frac{(1+S_{11})(1+S_{22})-S_{12}S_{21}}{2S_{21}} \tag{C.22b}$$

$$C = \frac{1}{Z_o}\frac{(1-S_{11})(1-S_{22})-S_{12}S_{21}}{2S_{21}} \tag{C.22c}$$

$$D = \frac{(1-S_{11})(1+S_{22})+S_{12}S_{21}}{2S_{21}} \tag{C.22d}$$

The conversion between those pairs of matrices not given above can be indirectly inferred. Nevertheless, a complete conversion table for two-port network parameters can be found in Ref. [1].

REFERENCE

1. D. M. Pozar, *Microwave Engineering*, Addison Wesley, New York, 1990.

Appendix D

CURRENT DENSITY DISTRIBUTION IN A CONDUCTOR

Let us consider a sinusoidal current flowing in a homogeneous conducting half space with conductivity σ and permeability μ (see Fig. D.1), and let us assume that the current density is parallel to the surface, oriented in the x-direction, and dependent only on the coordinate orthogonal to the surface (z), that is,

$$\vec{J}(x,y,z) = J_x(z)\,\vec{x} \tag{D.1}$$

In order to obtain the distribution of current in the conductor, we use the Maxwell's curl equations, taking into account that the displacement current can be neglected, that is,

$$\nabla \times \vec{E} = -j\omega\mu\,\vec{H} \tag{D.2a}$$

$$\nabla \times \vec{H} = \vec{J} \tag{D.2b}$$

Since $\vec{J} = \sigma\,\vec{E}$, expression (D.2a) can be written as follows:

$$\nabla \times \vec{J} = -j\omega\mu\sigma\,\vec{H} \tag{D.3}$$

From the Biot–Savart law and symmetry considerations, it follows that there is only a y component of the magnetic field, and Equations (D.2) and (D.3) give

Artificial Transmission Lines for RF and Microwave Applications, First Edition. Ferran Martín.

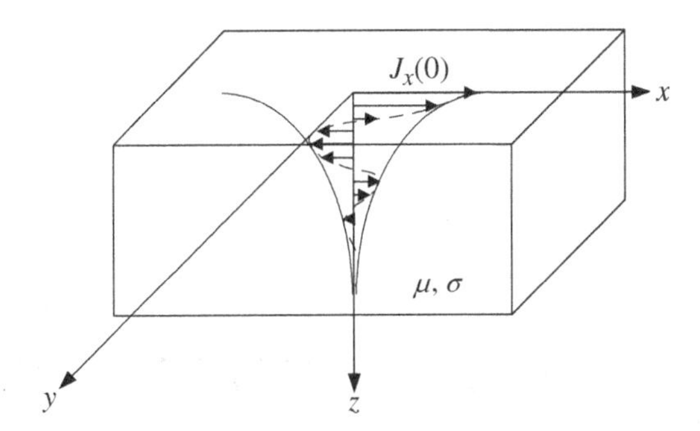

FIGURE D.1 Portion of the homogeneous conducting half space (extending everywhere for $z > 0$), and instantaneous current density distribution.

$$\frac{\partial J_x}{\partial z} = -j\omega\mu\sigma H_y \tag{D.4a}$$

$$-\frac{\partial H_y}{\partial z} = J_x \tag{D.4b}$$

Combining the previous equations, the following second-order differential equation for J_x is obtained:

$$\frac{\partial^2 J_x}{\partial z^2} = j\omega\mu\sigma J_x \tag{D.5}$$

The general solution of this equation is of the following form:

$$J_x(z) = J_1 e^{\gamma z} + J_2 e^{-\gamma z} \tag{D.6}$$

with

$$\gamma = \sqrt{j\omega\mu\sigma} = \frac{1+j}{\delta_p} \tag{D.7}$$

and δ_p defined in (1.79). Notice that γ is identical to the propagation constant given in (1.78). Obviously, the particular solution must have $J_1 = 0$ since the current cannot increase indefinitely from the surface. Considering that the current in the surface is given by the surface field as $J_x(z=0) = \sigma E_o$, the current can be expressed as follows:

$$J_x(z) = \sigma E_o e^{-z/\delta_p} \cdot e^{-jz/\delta_p} \tag{D.8}$$

which is identical to expression (1.84).

Appendix E

DERIVATION OF THE SIMPLIFIED COUPLED MODE EQUATIONS AND COUPLING COEFFICIENT FROM THE DISTRIBUTED CIRCUIT MODEL OF A TRANSMISSION LINE[1]

According to expressions (1.3), the voltage and current in a transmission line obey the following expressions:

$$\frac{dV}{dz} = -Z' \cdot I \tag{E.1a}$$

$$\frac{dI}{dz} = -Y' \cdot V \tag{E.1b}$$

where Z' and Y' are the per-unit length series impedance and shunt admittance, respectively, of the transmission line. These parameters are related to the phase constant and characteristic impedance of the line according to (see expressions 1.5 and 1.8):[2]

$$j\beta = \sqrt{Z'Y'} \tag{E.2}$$

[1] This appendix has been co-authored with Txema Lopetegi (Public University of Navarre, Spain).
[2] Actually expression (E.2) is valid for a lossless line, where $\alpha = 0$.

Artificial Transmission Lines for RF and Microwave Applications, First Edition. Ferran Martín.

$$Z_o = \sqrt{\frac{Z'}{Y'}} \tag{E.3}$$

Notice that all these variables Z', Y', β, and Z_o vary with the axial distance in a periodic (nonuniform) transmission line. Let us now express the total voltage and current in a transmission line in terms of the complex amplitudes of the modes:[3]

$$V = V^+ + V^- = (a^+ + a^-)\sqrt{Z_o} \tag{E.4a}$$

$$I = I^+ + I^- = (a^+ - a^-)\frac{1}{\sqrt{Z_o}} \tag{E.4b}$$

Introducing (E.4a) and (E.4b) in (E.1a) and (E.1b), respectively, and using (E.2) and (E.3), one obtains

$$\frac{da^+}{dz} + \frac{da^-}{dz} + \frac{1}{2Z_o}(a^+ + a^-)\frac{dZ_o}{dz} = -j\beta(a^+ - a^-) \tag{E.5a}$$

$$\frac{da^+}{dz} - \frac{da^-}{dz} - \frac{1}{2Z_o}(a^+ - a^-)\frac{dZ_o}{dz} = -j\beta(a^+ + a^-) \tag{E.5b}$$

and adding and subtracting the previous equations, the simplified form of the coupling mode equations is obtained, that is

$$\frac{da^+}{dz} + \frac{1}{2Z_o}\frac{dZ_o}{dz}a^- = -j\beta a^+ \tag{E.6a}$$

$$\frac{da^-}{dz} + \frac{1}{2Z_o}\frac{dZ_o}{dz}a^+ = +j\beta a^- \tag{E.6b}$$

Notice that these expressions are formally identical to (2.47) if

$$K = -\frac{1}{2Z_o}\frac{dZ_o}{dz} \tag{E.7}$$

Therefore, it is demonstrated that the coupling coefficient is given by (2.50).

[3] According to (2.55), $V^+ = Ca^+$ and $V^- = Ca^-$. Similarly, the currents verify $I^+ = C'a^+$ and $I^- = -C'a^-$, where the signs are consistent with the signs of the proportionality factors in the voltages. Notice that for a forward travelling wave, $V^+/I^+ = Z_o$. On the other hand, the complex power flow carried out by the mode is $P^+ = |a^+|^2/2$, which is also expressed in terms of the voltages and currents as $P^+ = V^+I^{+*}/2$. Therefore, the proportionality constants are $C = \sqrt{Z_o}$ and $C' = 1/\sqrt{Z_o}$.

Appendix F

AVERAGING THE EFFECTIVE DIELECTRIC CONSTANT IN EBG-BASED TRANSMISSION LINES[1]

The approximate analytical solutions of the coupled mode equations given in Section 2.4.3 have been obtained by neglecting the dependence on z of the phase constant, β, or the effective dielectric constant, ε_{re} (these parameters are related by 2.49). However, it is possible to improve the accuracy of the solutions by considering the z-dependence of β. To this end, let us introduce a new spatial variable, ζ, defined as follows:

$$\zeta(z) = \int_0^z \sqrt{\varepsilon_{re}(z')} \cdot dz' \tag{F.1}$$

in the first of the coupled-mode equations (2.47a). With this new variable, (2.47a) can be written as follows:

$$\frac{da^+}{d\zeta}\frac{d\zeta}{dz} + j\beta_o \frac{d\zeta}{dz} \cdot a^+ = K(\zeta)\frac{d\zeta}{dz} \cdot a^- \tag{F.2}$$

where $\beta_o = \omega/c$ and

$$\frac{d\zeta(z)}{dz} = \sqrt{\varepsilon_{re}} \tag{F.3}$$

[1] This appendix has been co-authored with Txema Lopetegi (Public University of Navarre, Spain).

Artificial Transmission Lines for RF and Microwave Applications, First Edition. Ferran Martín.

as results by differentiating (F.1). The coupling coefficient in terms of this new variable is simply

$$K(z) = -\frac{1}{2Z_o}\frac{dZ_o}{d\zeta}\frac{d\zeta}{dz} = K(\zeta)\frac{d\zeta}{dz} \quad \text{(F.4)}$$

The derivative of ζ with respect to z can be eliminated in (F.2), and the resulting equation is

$$\frac{da^+}{d\zeta} + j\beta_o a^+ = K(\zeta)a^- \quad \text{(F.5a)}$$

Similarly, the second coupled mode equation (2.47b) can be expressed in terms of ζ as follows:

$$\frac{da^-}{d\zeta} - j\beta_o a^- = K(\zeta)a^+ \quad \text{(F.5b)}$$

Notice that Equations F.5 are formally identical to Equations 2.47 if β is considered to be constant. Since this approximation (i.e., $\beta \neq \beta(z)$) has been used in order to obtain analytical solutions of the coupled-mode equations, it follows that the solutions of (F.5) must be the same (provided β_o is constant), simply replacing z with ζ. In particular, the frequency of maximum reflectivity, corresponding to $\Delta\beta = 0$, can be inferred from

$$\beta_{o,\max} = \frac{n\pi}{\zeta(z=l)} \quad \text{(F.6)}$$

or

$$\frac{2\pi f_{\max}}{c} = \frac{n\pi}{\int_0^l \sqrt{\varepsilon_{re}(z')}.dz'} \quad \text{(F.7)}$$

Comparing (F.7) and (2.84), the averaging value of (2.57) is justified.

Appendix G

PARAMETER EXTRACTION

Parameter extraction is a technique for the determination of the element values of the circuit model describing certain electromagnetic (EM) structure. The number of required conditions to find these values is equal to the number of elements of the circuit model, and these conditions are inferred either from the EM simulation or from the measured response of the structure. In this appendix, the specific methods for parameter extraction in SRR-, CSRR-, OSRR-, and OCSRR-loaded lines are reported. The lossless EM simulation is the considered response in all the cases; and coherently, the circuit models do not include the effect of losses.

G.1 PARAMETER EXTRACTION IN CSRR-LOADED LINES

The parameters of the circuit of Figure 3.38a can be extracted from the EM simulation of the CSRR-loaded (unit cell) line according to the following procedure. First, the reflection coefficient, S_{11}, is represented in the Smith chart. At the intercept of S_{11} with the unit resistance circle, the shunt branch opens and hence the resonance frequency of the CSRR can be determined:

$$f_o = \frac{1}{2\pi\sqrt{L_c C_c}} \tag{G.1}$$

The series reactance at this frequency, χ, directly provides the value of the inductance L, namely

Artificial Transmission Lines for RF and Microwave Applications, First Edition. Ferran Martín.

$$L = \frac{\chi}{2\pi f_o} \tag{G.2}$$

To univocally determine the three-circuit elements of the shunt branch, two additional conditions, apart from expression (G.1), are needed. One of them is the transmission zero frequency, f_z,

$$f_z = \frac{1}{2\pi\sqrt{L_c(C + C_c)}} \tag{G.3}$$

that can be easily determined from the representation of the magnitude of S_{21} with frequency. The third required condition can be derived from (2.33), which can be rewritten as follows:

$$\cos(\phi) = 1 + \frac{Z_s(\omega)}{Z_p(\omega)} \tag{G.4}$$

where $Z_s(\omega)$ and $Z_p(\omega)$ are the series and shunt impedance of the T-circuit model of the unit cell. Forcing $\phi = \beta l = \pi/2$, it follows that

$$Z_s(\omega_{\pi/2}) = -Z_p(\omega_{\pi/2}) \tag{G.5}$$

where $\omega_{\pi/2}$ is the angular frequency where $\phi = \pi/2$, which means that the phase of the transmission coefficient is $\phi_{21} = -\pi/2$ and can be easily computed. Thus, from (G.1), (G.3), and (G.5) the three reactive element values that contribute to the shunt impedance can be determined.

The circuit of Figure 3.38b has an additional parameter (C_g). Therefore, an additional condition to fully determine the circuit parameters of CSRR/gap-loaded lines is needed. This condition can be the resonance frequency of the series branch

$$f_s = \frac{1}{2\pi\sqrt{LC_g}} \tag{G.6}$$

which is given by the intercept of S_{11} with the unit conductance circle, where the impedance of the series branch nulls. Notice that for CSRR/gap-loaded lines, (G.5) rewrites as follows:

$$Z_s(\omega_{-\pi/2}) = -Z_p(\omega_{-\pi/2}) \tag{G.7}$$

$\omega_{-\pi/2}$ being the angular frequency of the left-handed (LH) band, where $\phi = -\pi/2$, that is, $\phi_{21} = \pi/2$.[1] Moreover, notice that for CSRR/gap-loaded lines, the series reactance at

[1] Although condition (G.4), as it is, can in principle be used to determine the element values of CSRR/gap-loaded lines, its use implies that a condition in the second (right-handed) transmission band is imposed. Since the model of CSRR-based lines is not very accurate in the second (right-handed) band, it is more convenient to use (G.6), which is a condition for an angular frequency ($\omega_{-\pi/2}$) within the first (LH) band.

f_o, inferred from the Smith chart, gives a condition involving L and C_g (not only involving L as given by expression G.2). Combining this condition with (G.6) allows us to determine L and C_g.

The parameter extraction method for CSRR- and CSRR/gap-loaded lines was first reported in Ref. [1], where the effects of losses were also considered (by adding a parallel resistance to the L_c-C_c tank), and parameter extraction was inferred from the measured responses of the considered CSRR-based lines.

G.2 PARAMETER EXTRACTION IN SRR-LOADED LINES

Parameter extraction in SRR-loaded lines, described by the circuit of Figure 3.34d (if the magnetic wall concept is applied) or by the circuit of Figure 3.35b by excluding L'_p (in this case, the magnetic wall concept is not applied), follows a similar procedure (the method was first reported in Ref. [2]). Since parameter extraction is obtained from the EM simulation of real structures (i.e., including the two halves), the considered model of SRR-loaded lines for parameter extraction is the one depicted in Figure 3.35b with the exclusion of L'_p. From the representation of the reflection coefficient, S_{11}, in the Smith chart, two conditions are obtained. On the one hand, the frequency that nulls the series reactance, f_s, is inferred from the intercept of S_{11} with the unit conductance circle. This frequency is given by the following expression:

$$f_s = \frac{1}{2\pi}\sqrt{\frac{1}{L'_s C'_s} + \frac{1}{L' C'_s}} \tag{G.8}$$

On the other hand, the susceptance, B, of the unit cell seen from the ports at f_s, which can be inferred from the Smith chart, directly gives C:

$$C = \frac{B}{2\pi f_s} \tag{G.9}$$

Another condition concerns the parallel resonator of the series branch. At the resonance frequency of this resonator, given by

$$f_z = \frac{1}{2\pi}\sqrt{\frac{1}{L'_s C'_s}} \tag{G.10}$$

the series branch opens, and the transmission coefficient exhibits a transmission zero at this frequency. The fourth condition is simply expression (G.5), applied to the circuit model of Figure 3.35b without L'_p.

For SRR/strip-loaded lines, an additional (fifth) condition is required since there are five elements in the circuit model of Figure 3.35b, namely, the shunt inductance L'_p is now included. The fifth condition is given by the frequency that opens the shunt branch, which can be inferred from the intersection of the S_{11} trace in the Smith chart with the unit resistance circle. This frequency is given by

$$f_p = \frac{1}{2\pi}\sqrt{\frac{2}{L'_p C}} \tag{G.11}$$

Moreover, for SRR/strip-loaded lines, condition (G.7), rather than (G.5), must be used (similar to CSRR/gap-loaded lines).

G.3 PARAMETER EXTRACTION IN OSRR-LOADED LINES

The parameters of the circuit model of a CPW loaded with an OSRR (Fig. 3.53e) can be extracted from the EM simulation of the structure following a straightforward procedure (notice that only three parameters are involved in the circuit model) [3]. From the intercept of S_{11} with the unit conductance circle in the Smith chart, we can directly infer the value of the shunt capacitance according to

$$C = \frac{B}{4\pi f_s} \tag{G.12}$$

where B is the susceptance at the intercept point. The frequency at this intercept point is the resonance frequency of the series branch:

$$f_s = \frac{1}{2\pi}\sqrt{\frac{1}{C_s L'_s}} \tag{G.13}$$

To determine the two element values of this branch, another condition is needed. This condition comes from the fact that at the reflection zero frequency, f_z, (maximum transmission) the characteristic impedance of the structure is $Z_o = 50\,\Omega$ (the typical value of the reference impedance of the ports). In this π-circuit, the characteristic impedance is given by (3.84). Thus, by forcing this impedance to $50\,\Omega$, the second condition results. By inverting equations (G.13) and (3.84), the element values of the series branch can be determined. The following results are obtained:

$$C_s = \left[\frac{f_z^2}{f_s^2} - 1\right] \cdot \left\{\frac{1}{8\pi^2 Z_o^2 f_z^2 C} + \frac{C}{2}\right\} \tag{G.14}$$

$$L'_s = \frac{1}{4\pi^2 f_s^2 C_s} \tag{G.15}$$

G.4 PARAMETER EXTRACTION IN OCSRR-LOADED LINES

The parameters of the circuit model of a CPW loaded with an OCSRR (Fig. 3.54e) can be extracted following a similar procedure. In this case, the intercept of S_{11} with the unit resistance circle in the Smith chart gives the value of the series inductance:

$$L = \frac{\chi}{4\pi f_{\mathrm{p}}} \tag{G.16}$$

where χ is the reactance at the intercept point. The shunt branch resonates at this frequency; that is,

$$f_{\mathrm{p}} = \frac{1}{2\pi}\sqrt{\frac{1}{L'_{\mathrm{p}}C'_{\mathrm{p}}}} \tag{G.17}$$

Finally, at the reflection zero frequency (f_{z}), the characteristic impedance, given by (2.30) must be forced to be 50 Ω. From these two latter conditions, we finally obtain the following:

$$L'_p = \left[\frac{f_{\mathrm{z}}^2}{f_{\mathrm{p}}^2} - 1\right] \cdot \left\{\frac{Z_{\mathrm{o}}^2}{8\pi^2 f_{\mathrm{z}}^2 L} + \frac{L}{2}\right\} \tag{G.18}$$

$$C'_{\mathrm{p}} = \frac{1}{4\pi^2 f_{\mathrm{p}}^2 L'_p} \tag{G.19}$$

and the element values are determined.

If the wideband model of Figure 3.55a, including the inductance L_{sh}, is considered, an additional condition for the determination of the elements of the shunt branch, such as (G.5), is required. Alternatively, the transmission zero frequency, given by the frequency that nulls the shunt impedance, can also be used.

REFERENCES

1. J. Bonache, M. Gil, I. Gil, J. Garcia-García, and F. Martín, "On the electrical characteristics of complementary metamaterial resonators," *IEEE Microw. Wireless Compon. Lett.*, vol. **16**, pp. 543–545, 2006.
2. F. Aznar, M. Gil, J. Bonache, J. D. Baena, L. Jelinek, R. Marqués, and F. Martín, "Characterization of miniaturized metamaterial resonators coupled to planar transmission lines," *J. Appl. Phys.*, vol. **104**, paper 114501, 2008.
3. M. Durán-Sindreu, A. Vélez, F. Aznar, G. Sisó, J. Bonache, and F. Martín, "Application of open split ring resonators and open complementary split ring resonators to the synthesis of artificial transmission lines and microwave passive components," *IEEE Trans. Microw. Theory Tech.*, vol. **57**, pp. 3395–3403, 2009.

Appendix H

SYNTHESIS OF RESONANT-TYPE METAMATERIAL TRANSMISSION LINES BY MEANS OF AGGRESSIVE SPACE MAPPING

Resonant-type metamaterial transmission lines are modeled by lumped-element equivalent circuit models that describe their behavior to a good approximation, as has been reviewed in Section 3.5.2. These circuit models are useful for design purposes, that is, the element values can be adjusted in order to satisfy certain requirements or specifications. However, the synthesis of these structures, namely, the determination of the layout from the elements of the equivalent circuit models, is not simple. The synthesis is the opposite process to parameter extraction. Although there are models that link the geometry of SRRs, CSRRs and their open counterparts to the element values of the equivalent circuits, such models are valid under conditions that do not hold when these resonators are coupled to transmission lines. Thus, the synthesis of resonant-type metamaterial transmission lines has been a process mainly based on the experience of microwave engineers or researchers involved in this type of structures. Obviously, parameter extraction has represented a useful tool to aid the synthesis of resonant-type metamaterial transmission lines. However, synthesis tools able to automatically provide the layout from the element values of the circuit models are highly desirable. In this appendix, it is shown how space mapping (SM) optimization can be applied to the synthesis of resonant-type metamaterial transmission lines. Specifically, the guidelines for the implementation of an aggressive space mapping (ASM) algorithm applied to the automated synthesis of CSRR-loaded microstrip

Artificial Transmission Lines for RF and Microwave Applications, First Edition. Ferran Martín.

lines are provided (further details of the technique and the application of it to the synthesis of CSRR/gap-loaded microstrip lines can be found in Ref. [1]).

H.1 GENERAL FORMULATION OF ASM

SM is a technique extensively used in the optimized design process of microwave devices, which makes proper use of two simulation spaces [2–4]. In the optimization space, $\mathbf{X_c}$, the variables are linked to a coarse model, which is simple and computationally efficient, although not accurate. On the other hand, the variables corresponding to the validation space, $\mathbf{X_f}$, are linked to a fine model, typically more complex and CPU intensive, but significantly more precise. In each space, a vector containing the different model parameters can be defined. Let us denote such vectors as $\mathbf{x_f}$ and $\mathbf{x_c}$ for the fine model and coarse model parameters, respectively. Using the same nomenclature, $\mathbf{R_f}(\mathbf{x_f})$ and $\mathbf{R_c}(\mathbf{x_c})$ will denote the responses of the fine and coarse models, respectively. For microwave applications, the model response is related to the evaluation of the device behaviour, for instance, a scattering parameter, such as $|S_{11}|$ or $|S_{21}|$, computed in a certain frequency range.

The key idea behind the space mapping algorithm is to generate an appropriate parameter transformation

$$\mathbf{x_c} = \mathbf{P}(\mathbf{x_f}) \tag{H.1}$$

mapping the fine model parameter space to the coarse model parameter space such that

$$\|\mathbf{R_f}(\mathbf{x_f}) - \mathbf{R_c}(\mathbf{x_c})\| \leq \eta \tag{H.2}$$

in some predefined region, $\|\cdot\|$ being a certain suitable norm and η a small positive number close to zero. If $\mathbf{P}$ is invertible, then the inverse transformation:

$$\mathbf{x_f} = \mathbf{P}^{-1}(\mathbf{x_c}^*) \tag{H.3}$$

is used to find the fine model solution, which is the image of the coarse model solution, $\mathbf{x_c}^*$, that gives the target response, $\mathbf{R_c}(\mathbf{x_c}^*)$.

The determination of $\mathbf{P}$ according to the procedure reported in Ref. [2] follows an iterative process that is rather inefficient. However, the efficiency of the method can be improved by introducing quasi-Newton type iteration [3]. This method aggressively exploits each fine model EM analysis with the result of a faster convergence. Hence, the new approach was called ASM [3], and it is the optimization procedure considered for the synthesis of CSRR-based lines. Essentially, the goal in ASM is to solve the following set of nonlinear equations:

$$\mathbf{f}(\mathbf{x_f}) = \mathbf{P}(\mathbf{x_f}) - \mathbf{x_c}^* = 0 \tag{H.4}$$

For a better understanding of the iterative optimization process, a superscript is added to the notation that actually indicates the iteration number. Hence, let us assume that $\mathbf{x_f}^{(j)}$ is the jth approximation to the solution of (H.4) and $\mathbf{f}^{(j)}$ the error function corresponding to $\mathbf{f}(\mathbf{x_f}^{(j)})$. The next vector of the iterative process $\mathbf{x_f}^{(j+1)}$ is obtained by a quasi-Newton iteration according to

$$\mathbf{x}_{\mathbf{f}}^{(j+1)} = \mathbf{x}_{\mathbf{f}}^{(j)} + \mathbf{h}^{(j)} \tag{H.5}$$

where $\mathbf{h}^{(j)}$ is given by

$$\mathbf{h}^{(j)} = -(\mathbf{B}^{(j)})^{-1}\mathbf{f}^{(j)} \tag{H.6}$$

and $\mathbf{B}^{(j)}$ is an approach to the Broyden matrix [3]:

$$\mathbf{B}^{(j+1)} = \mathbf{B}^{(j)} + \frac{\mathbf{f}^{(j+1)}\mathbf{h}^{(j)T}}{\mathbf{h}^{(j)T}\mathbf{h}^{(j)}} \tag{H.7}$$

which is also updated at each iterative step. In (H.7), $\mathbf{f}^{(j+1)}$ is obtained by evaluating (H.4) using a certain parameter extraction method providing the coarse model parameters from the fine model parameters, and the superindex T stands for transpose.

The implementation of the ASM algorithm is well reported in Ref. [3]. If the fine and coarse models involve the same space parameters, then the first vector in the fine space is typically set equal to the target vector in the coarse space, and the Broyden matrix is initialized by forcing it to be the identity. However, this is not the case of the resonant-type metamaterial transmission lines considered in this appendix (CSRR-based lines), with different variables in both spaces. Hence, a different approach is considered, as will be seen in Section H.4.

H.2 DETERMINATION OF THE CONVERGENCE REGION IN THE COARSE MODEL SPACE

Let us first describe the two spaces involved in the ASM algorithm for CSRR-loaded microstrip lines. The optimization space, $\mathbf{X_c}$, is the circuit model, and the space variables are the circuit elements appearing in the equivalent circuit of Figure 3.38a. The validation space, $\mathbf{X_f}$, is the electromagnetic (EM) model of the physical structure that provides the frequency response from the geometry and substrate parameters. The substrate parameters are set to certain values, and therefore these parameters are not considered as variables of the fine model. Losses are neither considered in the circuit model nor in the EM model. Some constraints are applied to the geometry parameters in order to reduce the degrees of freedom and thus work with the same number of variables in both spaces (the matrix $\mathbf{B}$ is thus square and invertible, and the computational effort is minimized).

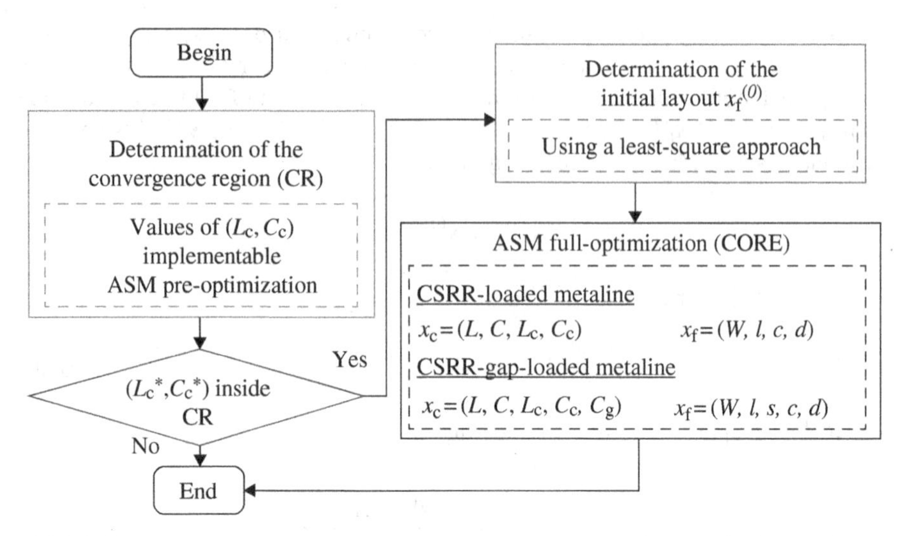

FIGURE H.1 Schematic of the two-step ASM algorithm, including the ASM preoptimizer and the ASM full-optimizer (core). The algorithm can be used for the synthesis of CSRR- and CSRR/gap-loaded lines, although the appendix is focused only on CSRR-loaded lines. Reprinted with permission from Ref. [1]; copyright 2013 IEEE.

From a practical point of view, a synthesis tool should be able to guarantee if there is a mapping between the optimal coarse model solution (i.e., a set of circuit parameters giving the target response) and a fine model point providing a physically implementable layout. Thus, the first step is the determination of a convergence region in the coarse model space. This involves a preoptimization ASM scheme; and therefore, the proposed synthesis tool for CSRR-loaded lines is based on a two-step ASM approach (see Fig. H.1).

The geometrical variables are the microstrip line length, l, and width, W, and the width, c, and separation, d, of the slot rings (see Fig. H.2). In order to deal with the same number of variables in the coarse and fine model spaces, we consider $s_{split} = c$ and $l = 2r_{ext}$, where r_{ext} is the external radius of the CSRR. This latter choice is justified to avoid distributed effects (not accounted for by the circuit model) that may appear if l is substantially larger than $2r_{ext}$, and to ensure that the ports of the structure are accessible. Nevertheless, the circuit model of Figure 3.38a is also valid if l is slightly larger than $2r_{ext}$.

Given a set of circuit (target) variables, L^*, C^*, L_c^*, and C_c^*, a procedure to determine if this set of variables has a physically implementable layout is needed. According to the circuit model, L^* and C^* are physically realizable if these values are not too extreme and a microstrip line with reasonable width and length results. The important aspect is thus, given a pair of implementable values of L^* and C^*, to determine the convergence region for the circuit values modeling the CSRR (L_c and C_c). The strategy consists in calculating the line geometry that provides the element values L^* and C^* under four different scenarios corresponding to the extreme values of c and d,

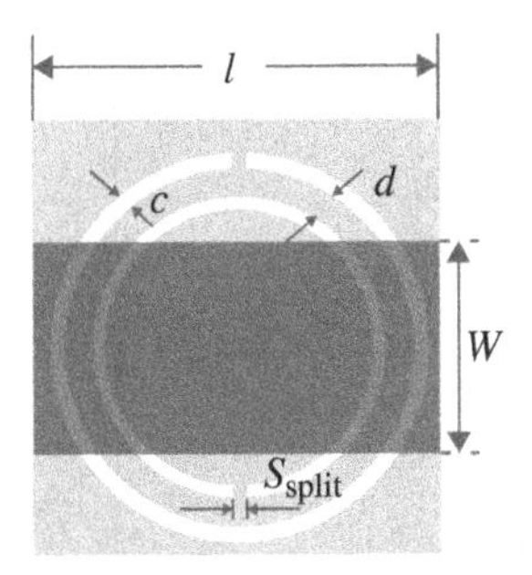

FIGURE H.2 Typical topology and geometrical variables of a CSRR-loaded microstrip line. The ground plane, where the CSRR is etched, is depicted in light gray.

namely, c_{min}–d_{min}, c_{max}–d_{max}, c_{min}–d_{max}, and c_{max}–d_{min}. The parameters c_{min} and d_{min} are the minimum achievable slot and strip widths, respectively, with the technology in use (typically, the minimum values are set to $c_{min} = d_{min} = 0.15$ mm). On the other hand, c_{max} and d_{max} are set to a reasonable (maximum) value that guarantees the validity of the model in the frequency region of interest. For values exceeding 0.4–0.5 mm, the coupling between the slot rings is very limited, and hence the maximum values are set to $c_{max} = d_{max} = 0.4$–0.5 mm (larger values expand the convergence region, but at the expense of less accuracy in the final solutions).

For each case, the single geometrical variables are l and W (c and d are fixed and $r_{ext} = l/2$). These variables must be optimized with the goal of recovering the L and C values corresponding to the target (L^* and C^*). The extraction of the elements of the circuit model from the EM simulation of the CSRR-loaded line layout has been reported in Appendix G.

For the determination of l and W, an ASM optimization scheme is considered. This is very simple since only two variables in each space are involved. The initial values of l and W are obtained from the characteristic impedance and phase constant resulting from L^* and C^*, with the help of a transmission line calculator. With the resulting line geometry, the parameters of the circuit model are extracted, and the Broyden matrix is initialized (the initiation of the Broyden matrix follows a similar approach to that corresponding to the core ASM algorithm, which is reported later). Then the process iterates until convergence is achieved. In this preoptimization ASM process, the stopping criterion is usually tighter than the one considered for the full-ASM optimization, that is, smaller values of the error functions are forced. This more restrictive criterion is chosen not only to accurately determine the vertices of the convergence region but also to obtain a better estimate of the initial layout of the core ASM algorithm (however, this does not require much computational effort due to the very small number of terms—coarse model parameters—involved in the error calculation).

Once the l and W values corresponding to the target L^* and C^* for a given c and d combination (e.g., c_{min}–d_{min}) are found, the whole geometry is known, and the element values L_c and C_c can be obtained by means of the parameter extractor. These element values (L_c and C_c) correspond to the considered CSRR geometry (c_{min}–d_{min}),

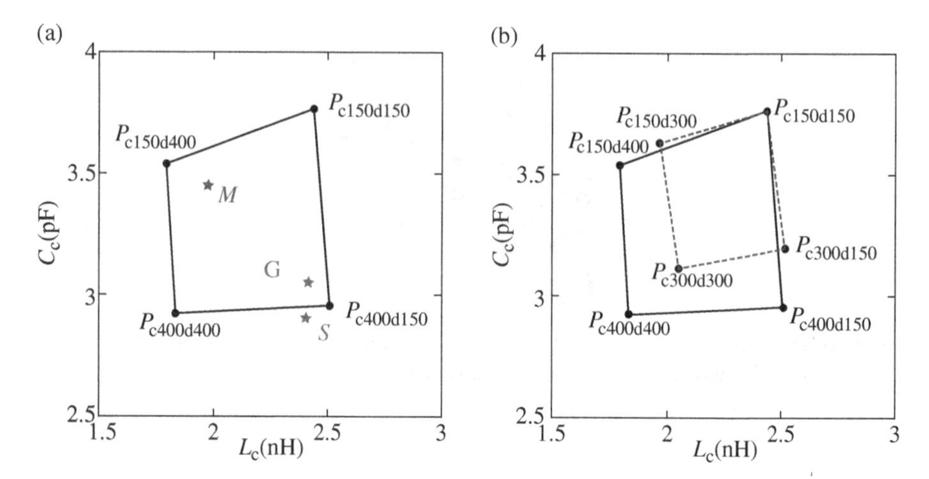

FIGURE H.3 (a) Convergence region for a CSRR-loaded line model with $L^* = 4.86$ nH and $C^* = 1.88$ pF, defined by four points in the L_c–C_c subspace, and considering $c_{min} = d_{min} = 0.15$ mm and $c_{max} = d_{max} = 0.4$ mm. (b) Various regions defined with different constraints for the same values of L^* and C^*. The thickness of the substrate is $h = 1.27$ mm and the dielectric constant is $\varepsilon_r = 10.2$. Reprinted with permission from Ref. [1]; copyright 2013 IEEE.

and actually define the first vertex, $P_{c150d150}$, of the polygon which defines the convergence region (see Fig. H.3). Notice that the nomenclature used for identifying the vertices indicates (subscript) the values of c and d in microns.

Then, the same process is repeated for obtaining the next points, that is, the L_c and C_c pairs corresponding to c_{max}–d_{max}, c_{min}–d_{max}, and c_{max}–d_{min} with the target L^* and C^* (for these cases, the initial values of l and W are set to the solutions of the previous vertex, since it provides a much better approach to the optimal final solution). As a result, a set of points in the coarse model subspace of L_c and C_c is obtained. These points define a polygon (see Fig. H.3) which is a rough estimate for the region of convergence in the L_c–C_c subspace, for the target values L^* and C^*. The criterion to decide if the target element values can be physically implemented is the pertinence or not of the point (i.e., L_c^* and C_c^*) to the region enclosed by the polygon. It is possible to refine the convergence region by calculating more points, meaning further values/combinations of c and d (e.g., those corresponding to a combination of an extreme and an intermediate value).

As a representative example of a CSRR-loaded line model, the target values $L^* = 4.86$ nH and $C^* = 1.88$ pF are considered. The convergence region for the subspace L_c–C_c calculated following the procedure explained earlier is depicted in Figure H.3a. Obviously, decreasing the maximum allowable value of c_{max} and d_{max} has the effect of reducing the area of the polygon, as Figure H.3b illustrates. However, $c_{max} = d_{max} = 0.4$ mm is a reasonable value that represents a tradeoff between accuracy (related to the frequency responses of the fine and coarse models) and size of the convergence region.

H.3 DETERMINATION OF THE INITIAL LAYOUT

Let us consider the synthesis of a given set of circuit parameters of the circuit model (target coarse model solution) for a CSRR-loaded metamaterial transmission line, and let us assume that the previous analysis reveals that such target parameters in the coarse model space have an implementable fine model solution. The next step is the determination of the initial layout, unless L_c^* and C_c^* coincide with any of the vertices of the converging polygon in the L_c–C_c subspace (in this case, the layout is already known and hence no further optimization is necessary). From the previous analysis, it is expected that the dimensions of the CSRR after ASM optimization depend on the position of the L_c^*–C_c^* point in the convergence region. Namely, if the L_c^*–C_c^* point is close to a vertex, it is expected that c and d are similar to the values corresponding to that vertex.

The procedure to determine the initial layout, necessary to start-up the core ASM algorithm, was first proposed in Ref. [1], and it is based on a least squares approach (as compared to previous procedures based on the use of analytic expressions [5, 6], it provides an initial layout which is very close to the final solution). The aim is to express any of the geometrical variables (c, d, l, or W) as a function of L_c and C_c. To obtain the initial value of each geometrical dimension involved in the optimization process, a linear dependence with L_c and C_c is assumed. For instance, the initial value of c (for the other variables, identical expressions are used) will be estimated according to

$$c = f(L_c, C_c) = (A + BL_c)(C + DC_c) \tag{H.8}$$

The previous expression can be alternatively written as follows:

$$c = f(L_c, C_c) = a_0 + a_1 L_c + a_2 C_c + a_3 L_c C_c \tag{H.9}$$

where the constants a_i determine the functional dependence of the initial value of c with L_c and C_c. To determine the constants a_i, four conditions are needed. Let us consider the following error function:

$$f_{\text{error}} = \sum_{j=1}^{N_v} \left(c_j - f\left(L_{c_j}, C_{c_j}\right) \right)^2 \tag{H.10}$$

where the subscript j is used to differentiate between the different vertices, and hence c_j is the value of c in the vertex j, and L_{cj}, C_{cj} the corresponding values of L_c and C_c for that vertex. Four vertices are considered (corresponding to CSRR-loaded lines), but the formulation can be generalized to a higher number of vertices (e.g., it has been found that eight vertices are necessary for the determination of the convergence region in the L_c–C_c subspace for CSRR/gap-loaded lines [1]). Expression (H.10) can then be written as follows:

$$f_{\text{error}} = \sum_{j=1}^{4} c_j^2 - 2\sum_{j=1}^{4} c_j \cdot \left(a_0 + a_1 L_{c_j} + a_2 C_{c_j} + a_3 L_{c_j} C_{c_j}\right) + \sum_{j=1}^{4}\left(a_0 + a_1 L_{c_j} + a_2 C_{c_j} + a_3 L_{c_j} C_{c_j}\right)^2 \quad \text{(H.11)}$$

To find the values of the constants a_i, the partial derivatives of the previous error function with regard to a_i are obtained, and they are forced to be equal to zero [7]:

$$\frac{\partial f_{\text{error}}}{\partial a_i} = -2\sum_{j=1}^{4} c_j \frac{\partial f\left(L_{c_j}, C_{c_j}\right)}{\partial a_i} + 2\sum_{j=1}^{4}\left[f\left(L_{c_j}, C_{c_j}\right)\frac{\partial f\left(L_{c_j}, C_{c_j}\right)}{\partial a_i}\right] = 0 \quad \text{(H.12)}$$

for $i = 1, 2, 3, 4$. Following this least-squares approach, four independent equations for the constants a_i are obtained. Such equations can be written in matrix form as follows:

$$\begin{pmatrix} 4 & \sum_{j=1}^{4} L_{c_j} & \sum_{j=1}^{4} C_{c_j} & \sum_{j=1}^{4} L_{c_j} C_{c_j} \\ \sum_{j=1}^{4} L_{c_j} & \sum_{j=1}^{4} L_{c_j}^2 & \sum_{j=1}^{4} L_{c_j} C_{c_j} & \sum_{j=1}^{4} L_{c_j}^2 C_{c_j} \\ \sum_{j=1}^{4} C_{c_j} & \sum_{j=1}^{4} L_{c_j} C_{c_j} & \sum_{j=1}^{4} C_{c_j}^2 & \sum_{j=1}^{4} C_{c_j}^2 L_{c_j} \\ \sum_{j=1}^{4} L_{c_j} C_{c_j} & \sum_{j=1}^{4} L_{c_j}^2 C_{c_j} & \sum_{j=1}^{4} C_{c_j}^2 L_{c_j} & \sum_{j=1}^{4} C_{c_j}^2 L_{c_j}^2 \end{pmatrix} \begin{pmatrix} a_o \\ a_1 \\ a_2 \\ a_3 \end{pmatrix} = \begin{pmatrix} \sum_{j=1}^{4} c_j \\ \sum_{j=1}^{4} c_j L_{c_j} \\ \sum_{j=1}^{4} c_j C_{c_j} \\ \sum_{j=1}^{4} c_j L_{c_j} C_{c_j} \end{pmatrix} \quad \text{(H.13)}$$

Once the constants a_i are obtained (solving the previous equations), the initial value of c is inferred from (H.9). The process is repeated for d, W, and l, and the initial geometry necessary for the initiation of the ASM algorithm is thus obtained.

H.4 THE CORE ASM ALGORITHM

Once the initial geometry is calculated, the response of the fine model is obtained through EM analysis, the circuit parameters are extracted, and the error function (H.4) can be obtained. To initiate the Broyden matrix, each geometrical variable is slightly perturbed from the value corresponding to the initial layout, and the circuit parameters resulting from each geometry variation are obtained. The relative changes can be expressed in a matrix form as follows:

$$\mathbf{B}^{(1)} = \begin{pmatrix} \frac{\delta L}{\delta W} & \frac{\delta L}{\delta l} & \frac{\delta L}{\delta c} & \frac{\delta L}{\delta d} \\ \frac{\delta C}{\delta W} & \frac{\delta C}{\delta l} & \frac{\delta C}{\delta c} & \frac{\delta C}{\delta d} \\ \frac{\delta L_c}{\delta W} & \frac{\delta L_c}{\delta l} & \frac{\delta L_c}{\delta c} & \frac{\delta L_c}{\delta d} \\ \frac{\delta C_c}{\delta W} & \frac{\delta C_c}{\delta l} & \frac{\delta C_c}{\delta c} & \frac{\delta C_c}{\delta d} \end{pmatrix} \tag{H.14}$$

which corresponds to the initial Broyden matrix. Once the Broyden matrix is known, the geometry of the following iteration can be derived from (H.5), and the process is iterated until convergence is obtained. To avoid that the variables in the fine model space exceed the limits of the implementable range of values with the available technology, geometrical constraints, and a shrinking factor δ are introduced. As can be seen in Figure H.4, the vector $\boldsymbol{x}_\mathbf{f}$ is tested; and if some of the geometrical variables exceed the geometrical constraints, the algorithm uses the shrinking factor in order to obtain a new vector $\boldsymbol{x}_\mathbf{f}$ inside the limits.

The iterative algorithm can be controlled, for instance, by the MATLAB commercial software as the core tool of the implemented ASM algorithm. The three main building blocks are the initial geometry calculator (that previously determines the convergence regions in the L_c–C_c subspace as explained before), the EM solver, and the parameter extractor. For the EM simulation, commercial EM simulators such as the *Agilent ADS Momentum*, or the *Ansoft Designer*, among others, can be used. The EM solver creates the geometry of the structure, which is then exported to the

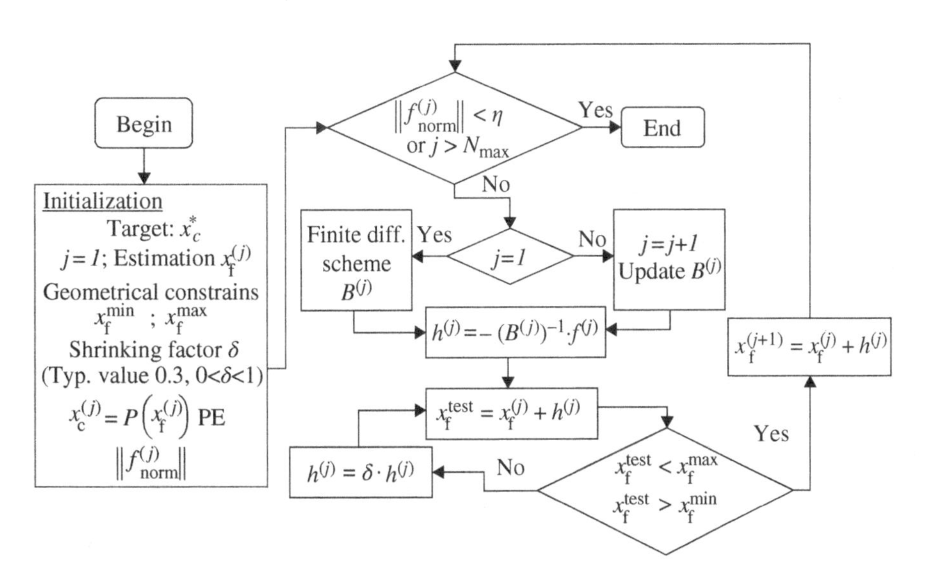

FIGURE H.4 Flow diagram of the ASM algorithm. Reprinted with permission from Ref. [1]; copyright 2013 IEEE.

commercial EM tool to carry out the EM simulation. The results of this simulation (the S-parameters) are then imported back to MATLAB and processed by the parameter extraction module, which allows us to extract the circuit parameters. These values are compared with the target parameters, and the error function is obtained according to

$$||f_{\text{norm}}|| = \sqrt{\left(1-\frac{L}{L^*}\right)^2 + \left(1-\frac{C}{C^*}\right)^2 + \left(1-\frac{L_c}{L_c^*}\right)^2 + \left(1-\frac{C_c}{C_c^*}\right)^2} \quad \text{(H.15)}$$

From this, the next geometry of the iterative process is inferred following the update procedure described before. The flow diagram of the complete ASM algorithm (including constraints to avoid unwanted solutions) is depicted in Figure H.4.

H.5 ILLUSTRATIVE EXAMPLES AND CONVERGENCE SPEED

To illustrate the potential of the proposed ASM algorithm to synthesise CSRR-loaded lines in few steps, let us consider the optimal coarse model parameters detailed in Table H.1. The location of these three different cases (they have the same L^* and C^* values) is shown in Figure H.3a. Notice that in the case that the point is out of the convergence region but close to it (S point), the algorithm is left to continue, whilst the limiting constrains of c and d are modified.

The number of iterations that the ASM algorithm needs to converge to the final layout dimensions and its corresponding evaluated error are summarized in Table H.2, for the three different examples considered (η has been set to 0.01).

TABLE H.1 Optimal coarse solution

	x_c^*			
Case	L (nH)	C (pF)	L_c (nH)	C_c (pF)
S	4.860	1.880	2.407	2.902
G	4.860	1.880	2.417	3.053
M	4.860	1.880	1.980	3.450

TABLE H.2 Dimensions of final layouts

	x_{em}^*					Error
Case	l^a	W^a	c^a	d^a	Iteration no.	$\|\|f_{\text{norm}}\|\|$
S	8.41	2.28	0.41	0.16	1	0.0066
G	8.47	2.22	0.36	0.17	1	0.0098
M	8.59	2.20	0.18	0.33	2	0.0084

[a] All dimensions in millimeter.

Convergence is very fast since the initial layout, inferred from the least-squares approach explained before, is very close to the final solution.

The final layout obtained for the first example (S) is slightly out of the upper boundary ($c = 0.41$ mm), as it is expected since the target S is placed out of the convergence region (see Fig. H.3a). However, the algorithm has been left to continue toward convergence by relaxing the limiting values of c and d. Thus, the fact that the target is out of the convergence region does not necessarily mean that convergence is not possible, but that the resulting geometric values might be beyond or below the considering limits of c and d. For the S point, the close proximity to the line of the

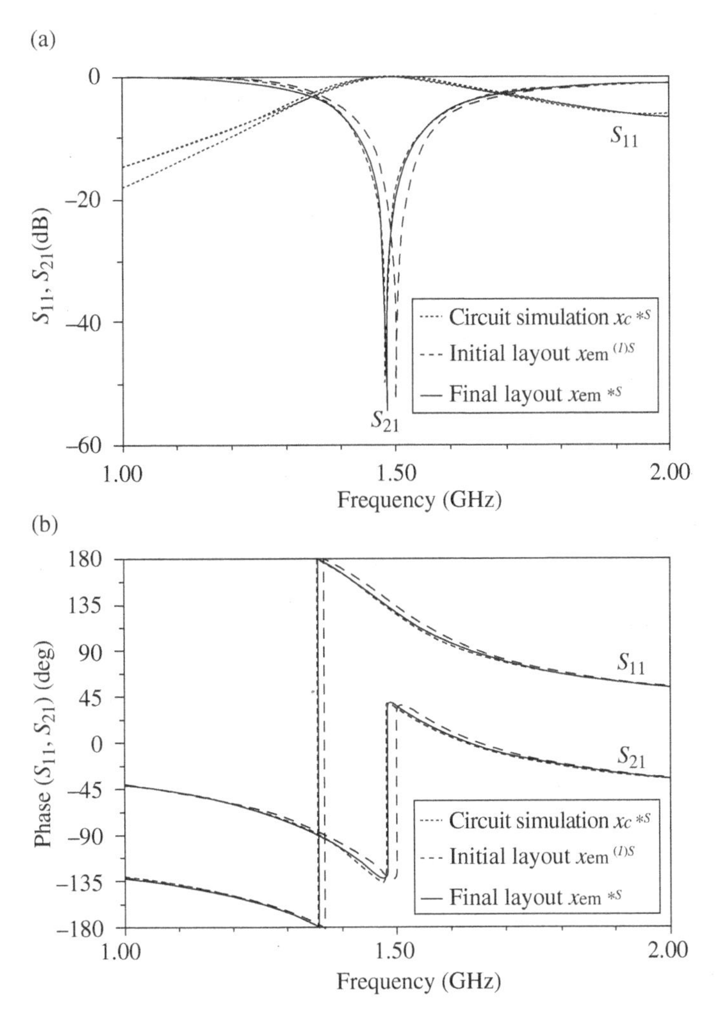

FIGURE H.5 Magnitude (a) and phase (b) of the scattering parameters S_{21} and S_{11} at initial solution $x_{em}^{(1)}$, x_{em}^{*}, and circuit simulation of the x_c^{*} for point S. Reprinted with permission from Ref. [1]; copyright 2013 IEEE.

polygon corresponding to $c = 0.4$ mm (Fig. H.3) explains that the final value of c is 0.41 mm. The agreement between the frequency response obtained by EM simulation of the final layout (obtained after a single iteration) and the circuit simulation of the target parameters is excellent, as it can be appreciated in Figure H.5. This is reasonable since the value of c is not far from the considering limiting value that guarantees that the CSRRs are accurately described by the model depicted in Figure 3.38. For cases G and M, the agreement between circuit simulation (target parameters) and EM simulation of the final layout is also very good [1].

REFERENCES

1. J. Selga, A. Rodríguez, V.E. Boria, and F. Martín, "Synthesis of split rings based artificial transmission lines through a new two-step, fast converging, and robust aggressive space mapping (ASM) algorithm," *IEEE Trans. Microw. Theory Tech.*, vol. **61**, pp. 2295–2308, 2013.
2. J. W. Bandler, R. M. Biernacki, S. H. Chen, P. A. Grobelny, and R. H. Hemmers, "Space mapping technique for electromagnetic optimization," *IEEE Trans. Microw. Theory Tech.*, vol. **42**, pp. 2536–2544, 1994.
3. J. W. Bandler, R. M. Biernacki, S. H. Chen, R. H. Hemmers, and K. Madsen, "Electromagnetic optimization exploiting aggressive space mapping," *IEEE Trans. Microw. Theory Tech.*, vol. **43**, pp. 2874–2882, 1995.
4. J. W. Bandler, Q. S. Cheng, S. A. Dakroury, A. S. Mohamed, M. H. Bakr, K. Madsen, and J. Søndergaard, "Space mapping: the state of the art," *IEEE Trans. Microw. Theory Tech.*, vol. **52**, pp. 337–361, 2004.
5. J. Selga, A. Rodríguez, M. Gil, J. Carbonell, V. E. Boria, and F. Martín, "Synthesis of Planar Microwave Circuits through Aggressive Space Mapping using commercially available software packages," *Int. J. RF Microw. Comput.-Aided Eng.*, vol. **20**, pp. 527–534, 2010.
6. J. Selga, A. Rodríguez, M. Gil, J. Carbonell, J. Bonache V. E. Boria, and F. Martín, "Towards the automatic layout synthesis in resonant-type metamaterial transmission lines," *IET Microw. Ant. Propag.*, vol. **4**, pp. 1007–1015, 2010.
7. D. M. Bates and D. G. Watts, *Nonlinear Regression Analysis and its Applications*, Wiley, New York, 1998.

Appendix I

CONDITIONS TO OBTAIN ALL-PASS X-TYPE AND BRIDGED-T NETWORKS

The purpose of this appendix is to demonstrate that in order to obtain all-pass X-type or bridged-T networks, it suffices to satisfy a simple condition regarding the reactances or susceptances of the branch impedances.

Let us first consider the lattice network of Figure 6.24a, with dispersion relation given by (6.13). This expression can be rewritten as follows:

$$\cos(\beta l) = 1 + \frac{2}{Z_p Y_s - 1} \tag{I.1}$$

where Y_s is the admittance of the series branch. In terms of the reactance, χ_p, and susceptance, B_s, the previous expression adopts the form

$$\cos(\beta l) = 1 - \frac{2}{\chi_p B_s + 1} \tag{I.2}$$

where $Z_p = j\chi_p$ and $Y_s = jB_s$. If the zeros and poles of χ_p coincide with the zeros and poles of B_s, the product $\chi_p B_s$ is a real and positive number, and the right-hand side of (I.2) is comprised in the range [−1, +1]. Therefore, the phase constant has a real solution at all frequencies and the network is an all-pass structure.

Artificial Transmission Lines for RF and Microwave Applications, First Edition. Ferran Martín.

Let us now consider the bridged-T topology of Figure 6.30a and that the impedances satisfy (6.27), so that the dispersion relation is given by (6.28). In terms of reactances and susceptances, the dispersion relation can be expressed as follows:

$$\cos(\beta l) = 1 + \frac{2}{1 - 2\chi_2 B_1 - (1/2\chi_2 B_1)} \tag{I.3}$$

where $Z_2 = j\chi_2$ and $Y_1 = jB_1$. Again, if the zeros and poles of χ_2 coincide with the zeros and poles of B_1, the product $\chi_2 B_1$ is a real and positive number. A simple mathematical analysis reveals that under these conditions, the right-hand side of (I.3) is comprised in the range [−1, +1], and the network is an all-pass structure.

ACRONYMS

AC	alternating current
ARLH	alternate right-/left-handed
ASM	aggressive space mapping
BC-CSRR	broadside-coupled complementary split-ring resonator
BC-SRR	broadside-coupled split-ring resonator
BJT	bipolar junction transistor
BST	barium strontium titanate
BW	bandwidth
CAD	computer-aided design
CDL	chirped delay line
CMOS	complementary metal–oxide–semiconductor
CMRR	common-mode rejection ratio
CPS	coplanar strips
CPU	Central Processing Unit
CPW	coplanar waveguide
CRLH	composite right-/left-handed
CSIR	complementary stepped impedance resonator
CSR	complementary spiral resonator
CSRR	complementary split-ring resonator
DC	direct current
D-CRLH	dual composite right-/left-handed
DCS	Digital Cellular Service

Artificial Transmission Lines for RF and Microwave Applications, First Edition. Ferran Martín.

DNG	double negative
DPS	double positive
DS-CSRR	double-slit complementary split-ring resonator
DS-SRR	double-slit split-ring resonator
EBG	electromagnetic bandgap
E-CRLH	extended composite right-/left-handed
EIRP	equivalent isotropic radiated power
EIW	electroinductive wave
ELC	electric LC
EM	electromagnetic
EMI	electromagnetic interference
ENG	epsilon negative
ENZ	epsilon-near-zero
FBW	fractional bandwidth
FCC	Federal Communications Commission
FET	field-effect transistor
FMCW-GPR	frequency-modulated continuous-wave ground-penetrating radar
FoM	figure of merit
GPS	Global Positioning System
GSM	Global System for Mobile Communications
HBV	heterostructure barrier varactor
HM-SIW	half-mode substrate-integrated waveguide
HRL	high resistive line
IL	insertion loss
ISM	industrial, scientific, and medical
LH	left-handed
LTCC	low-temperature co-fired ceramic
LWA	leaky wave antenna
MIWs	magnetoinductive waves
MLA	meander line antenna
MLC	magnetic LC
MMIC	monolithic microwave-integrated circuit
MNG	mu negative
MNZ	mu-near-zero
NB-CSRR	nonbianisotropic complementary split-ring resonator
NB-SRR	nonbianisotropic split-ring resonator
NIC	negative impedance converters
NLTL	nonlinear transmission line
NRI	negative refractive index
OCSRR	open complementary split-ring resonator
OSRR	open split-ring resonator
PBG	photonic bandgap
PC	photonic crystal

PCB	printed circuit board
PEC	perfect electric conductor
PECVD	plasma-enhanced chemical vapor deposition
PIN	P-type/intrinsic/N-type (diode)
PLH	purely left-handed
PMC	perfect magnetic conductor
PMGI	polymethylglutarimide
PN	P-type/N-type (diode)
PRH	purely right-handed
RF	radiofrequency
RFID	RF identification
RF-MEMS	radiofrequency-microelectromechanical system
RH	right-handed
RL	return loss
SIR	stepped impedance resonator
SISS	stepped impedance shunt stub
SIW	substrate-integrated waveguide
SM	space mapping
SMD	surface mount device
SMT	surface mount technology
SNG	single negative
SR	spiral resonator
SRR	split-ring resonator
SSRR	square split-ring resonator
SWR	standing wave ratio
TDR	time-domain reflectometry
TE	transverse electric
TEM	transverse electromagnetic
TM	transverse magnetic
UHF	ultra-high frequency
UWB	ultra-wideband
VCO	voltage-controlled oscillator
VLCSRR	varactor-loaded complementary split-ring resonator
VLOCSRR	varactor-loaded open complementary split-ring resonator
VLOSRR	varactor-loaded open split-ring resonator
VLSRR	varactor-loaded split-ring resonator

INDEX

Artificial Transmission Lines for RF and Microwave Applications, First Edition. Ferran Martín.

WILEY SERIES IN MICROWAVE AND OPTICAL ENGINEERING

Kai Chang, Series Editor

Texas A&M University

FIBER-OPTIC COMMUNICATION SYSTEMS, Fourth Edition • *Govind P. Agrawal*

ASYMMETRIC PASSIVE COMPONENTS IN MICROWAVE INTEGRATED CIRCUITS • *Hee-Ran Ahn*

COHERENT OPTICAL COMMUNICATIONS SYSTEMS • *Silvello Betti, Giancarlo De Marchis, and Eugenio Iannone*

PHASED ARRAY ANTENNAS: FLOQUET ANALYSIS, SYNTHESIS, BFNs, AND ACTIVE ARRAY SYSTEMS • *Arun K. Bhattacharyya*

HIGH-FREQUENCY ELECTROMAGNETIC TECHNIQUES: RECENT ADVANCES AND APPLICATIONS • *Asoke K. Bhattacharyya*

RADIO PROPAGATION AND ADAPTIVE ANTENNAS FOR WIRELESS COMMUNICATION LINK NETWORKS: TERRESTRIAL, ATMOSPHERIC, AND IONOSPHERIC, Second Edition • *Nathan Blaunstein and Christos G. Christodoulou*

COMPUTATIONAL METHODS FOR ELECTROMAGNETICS AND MICROWAVES • *Richard C. Booton, Jr.*

ELECTROMAGNETIC SHIELDING • *Salvatore Celozzi, Rodolfo Araneo, and Giampiero Lovat*

MICROWAVE RING CIRCUITS AND ANTENNAS • *Kai Chang*

MICROWAVE SOLID-STATE CIRCUITS AND APPLICATIONS • *Kai Chang*

RF AND MICROWAVE WIRELESS SYSTEMS • *Kai Chang*

RF AND MICROWAVE CIRCUIT AND COMPONENT DESIGN FOR WIRELESS SYSTEMS • *Kai Chang, Inder Bahl, and Vijay Nair*

MICROWAVE RING CIRCUITS AND RELATED STRUCTURES, Second Edition • *Kai Chang and Lung-Hwa Hsieh*

MULTIRESOLUTION TIME DOMAIN SCHEME FOR ELECTROMAGNETIC ENGINEERING • *Yinchao Chen, Qunsheng Cao, and Raj Mittra*

DIODE LASERS AND PHOTONIC INTEGRATED CIRCUITS, Second Edition • *Larry Coldren, Scott Corzine, and Milan Masanovic*

EM DETECTION OF CONCEALED TARGETS • *David J. Daniels*

RADIO FREQUENCY CIRCUIT DESIGN • *W. Alan Davis and Krishna Agarwal*

RADIO FREQUENCY CIRCUIT DESIGN, Second Edition • *W. Alan Davis*

FUNDAMENTALS OF OPTICAL FIBER SENSORS • *Zujie Fang, Ken K. Chin, Ronghui Qu, and Haiwen Cai*

MULTICONDUCTOR TRANSMISSION-LINE STRUCTURES: MODAL ANALYSIS TECHNIQUES • *J. A. Brandão Faria*

PHASED ARRAY-BASED SYSTEMS AND APPLICATIONS • *Nick Fourikis*

SOLAR CELLS AND THEIR APPLICATIONS, Second Edition • *Lewis M. Fraas and Larry D. Partain*

FUNDAMENTALS OF MICROWAVE TRANSMISSION LINES • *Jon C. Freeman*

OPTICAL SEMICONDUCTOR DEVICES • *Mitsuo Fukuda*

MICROSTRIP CIRCUITS • *Fred Gardiol*

HIGH-SPEED VLSI INTERCONNECTIONS, Second Edition • *Ashok K. Goel*

FUNDAMENTALS OF WAVELETS: THEORY, ALGORITHMS, AND APPLICATIONS, Second Edition • *Jaideva C. Goswami and Andrew K. Chan*

HIGH-FREQUENCY ANALOG INTEGRATED CIRCUIT DESIGN • *Ravender Goyal (ed.)*

RF AND MICROWAVE TRANSMITTER DESIGN • *Andrei Grebennikov*

ANALYSIS AND DESIGN OF INTEGRATED CIRCUIT ANTENNA MODULES • *K. C. Gupta and Peter S. Hall*

PHASED ARRAY ANTENNAS, Second Edition • *R. C. Hansen*

STRIPLINE CIRCULATORS • *Joseph Helszajn*

THE STRIPLINE CIRCULATOR: THEORY AND PRACTICE • *Joseph Helszajn*

LOCALIZED WAVES • *Hugo E. Hernández-Figueroa, Michel Zamboni-Rached, and Erasmo Recami (eds.)*

MICROSTRIP FILTERS FOR RF/MICROWAVE APPLICATIONS, Second Edition • *Jia-Sheng Hong*

MICROWAVE APPROACH TO HIGHLY IRREGULAR FIBER OPTICS • *Huang Hung-Chia*

NONLINEAR OPTICAL COMMUNICATION NETWORKS • *Eugenio Iannone, Francesco Matera, Antonio Mecozzi, and Marina Settembre*

FINITE ELEMENT SOFTWARE FOR MICROWAVE ENGINEERING • *Tatsuo Itoh, Giuseppe Pelosi, and Peter P. Silvester (eds.)*

INFRARED TECHNOLOGY: APPLICATIONS TO ELECTROOPTICS, PHOTONIC DEVICES, AND SENSORS • *A. R. Jha*

SUPERCONDUCTOR TECHNOLOGY: APPLICATIONS TO MICROWAVE, ELECTRO-OPTICS, ELECTRICAL MACHINES, AND PROPULSION SYSTEMS • *A. R. Jha*

TIME AND FREQUENCY DOMAIN SOLUTIONS OF EM PROBLEMS USING INTEGTRAL EQUATIONS AND A HYBRID METHODOLOGY • *B. H. Jung, T. K. Sarkar, S. W. Ting, Y. Zhang, Z. Mei, Z. Ji, M. Yuan, A. De, M. Salazar-Palma, and S. M. Rao*

OPTICAL COMPUTING: AN INTRODUCTION • *M. A. Karim and A. S. S. Awwal*

INTRODUCTION TO ELECTROMAGNETIC AND MICROWAVE ENGINEERING • *Paul R. Karmel, Gabriel D. Colef, and Raymond L. Camisa*

MILLIMETER WAVE OPTICAL DIELECTRIC INTEGRATED GUIDES AND CIRCUITS • *Shiban K. Koul*

ADVANCED INTEGRATED COMMUNICATION MICROSYSTEMS • *Joy Laskar, Sudipto Chakraborty, Manos Tentzeris, Franklin Bien, and Anh-Vu Pham*

MICROWAVE DEVICES, CIRCUITS AND THEIR INTERACTION • *Charles A. Lee and G. Conrad Dalman*

ADVANCES IN MICROSTRIP AND PRINTED ANTENNAS • *Kai-Fong Lee and Wei Chen (eds.)*

SPHEROIDAL WAVE FUNCTIONS IN ELECTROMAGNETIC THEORY • *Le-Wei Li, Xiao-Kang Kang, and Mook-Seng Leong*

MICROWAVE NONCONTACT MOTION SENSING AND ANALYSIS • *Changzhi Li and Jenshan Lin*

COMPACT MULTIFUNCTIONAL ANTENNAS FOR WIRELESS SYSTEMS • *Eng Hock Lim and Kwok Wa Leung*

ARITHMETIC AND LOGIC IN COMPUTER SYSTEMS • *Mi Lu*

OPTICAL FILTER DESIGN AND ANALYSIS: A SIGNAL PROCESSING APPROACH • *Christi K. Madsen and Jian H. Zhao*

THEORY AND PRACTICE OF INFRARED TECHNOLOGY FOR NONDESTRUCTIVE TESTING • *Xavier P. V. Maldague*

METAMATERIALS WITH NEGATIVE PARAMETERS: THEORY, DESIGN, AND MICROWAVE APPLICATIONS • *Ricardo Marqués, Ferran Martín, and Mario Sorolla*

OPTOELECTRONIC PACKAGING • *A. R. Mickelson, N. R. Basavanhally, and Y. C. Lee (eds.)*

OPTICAL CHARACTER RECOGNITION • *Shunji Mori, Hirobumi Nishida, and Hiromitsu Yamada*

ANTENNAS FOR RADAR AND COMMUNICATIONS: A POLARIMETRIC APPROACH • *Harold Mott*

INTEGRATED ACTIVE ANTENNAS AND SPATIAL POWER COMBINING • *Julio A. Navarro and Kai Chang*

ANALYSIS METHODS FOR RF, MICROWAVE, AND MILLIMETER-WAVE PLANAR TRANSMISSION LINE STRUCTURES • *Cam Nguyen*

LASER DIODES AND THEIR APPLICATIONS TO COMMUNICATIONS AND INFORMATION PROCESSING • *Takahiro Numai*

FREQUENCY CONTROL OF SEMICONDUCTOR LASERS • *Motoichi Ohtsu (ed.)*

INVERSE SYNTHETIC APERTURE RADAR IMAGING WITH MATLAB ALGORITHMS • *Caner Özdemir*

SILICA OPTICAL FIBER TECHNOLOGY FOR DEVICE AND COMPONENTS: DESIGN, FABRICATION, AND INTERNATIONAL STANDARDS • *Un-Chul Paek and Kyunghwan Oh*

WAVELETS IN ELECTROMAGNETICS AND DEVICE MODELING • *George W. Pan*

OPTICAL SWITCHING • *Georgios Papadimitriou, Chrisoula Papazoglou, and Andreas S. Pomportsis*

MICROWAVE IMAGING • *Matteo Pastorino*

ANALYSIS OF MULTICONDUCTOR TRANSMISSION LINES • *Clayton R. Paul*

INTRODUCTION TO ELECTROMAGNETIC COMPATIBILITY, Second Edition • *Clayton R. Paul*

ADAPTIVE OPTICS FOR VISION SCIENCE: PRINCIPLES, PRACTICES, DESIGN AND APPLICATIONS • *Jason Porter, Hope Queener, Julianna Lin, Karen Thorn, and Abdul Awwal (eds.)*

ELECTROMAGNETIC OPTIMIZATION BY GENETIC ALGORITHMS • *Yahya Rahmat-Samii and Eric Michielssen (eds.)*

INTRODUCTION TO HIGH-SPEED ELECTRONICS AND OPTOELECTRONICS • *Leonard M. Riaziat*

NEW FRONTIERS IN MEDICAL DEVICE TECHNOLOGY • *Arye Rosen and Harel Rosen (eds.)*

ELECTROMAGNETIC PROPAGATION IN MULTI-MODE RANDOM MEDIA • *Harrison E. Rowe*

ELECTROMAGNETIC PROPAGATION IN ONE-DIMENSIONAL RANDOM MEDIA • *Harrison E. Rowe*

HISTORY OF WIRELESS • *Tapan K. Sarkar, Robert J. Mailloux, Arthur A. Oliner, Magdalena Salazar-Palma, and Dipak L. Sengupta*

PHYSICS OF MULTIANTENNA SYSTEMS AND BROADBAND PROCESSING • *Tapan K. Sarkar, Magdalena Salazar-Palma, and Eric L. Mokole*

SMART ANTENNAS • *Tapan K. Sarkar, Michael C. Wicks, Magdalena Salazar-Palma, and Robert J. Bonneau*

NONLINEAR OPTICS • *E. G. Sauter*

APPLIED ELECTROMAGNETICS AND ELECTROMAGNETIC COMPATIBILITY • *Dipak L. Sengupta and Valdis V. Liepa*

COPLANAR WAVEGUIDE CIRCUITS, COMPONENTS, AND SYSTEMS • *Rainee N. Simons*

ELECTROMAGNETIC FIELDS IN UNCONVENTIONAL MATERIALS AND STRUCTURES • *Onkar N. Singh and Akhlesh Lakhtakia (eds.)*

ANALYSIS AND DESIGN OF AUTONOMOUS MICROWAVE CIRCUITS • *Almudena Suárez*

ELECTRON BEAMS AND MICROWAVE VACUUM ELECTRONICS • *Shulim E. Tsimring*

FUNDAMENTALS OF GLOBAL POSITIONING SYSTEM RECEIVERS: A SOFTWARE APPROACH, Second Edition • *James Bao-yen Tsui*

SUBSURFACE SENSING • *Ahmet S. Turk, A. Koksal Hocaoglu, and Alexey A. Vertiy (eds.)*

RF/MICROWAVE INTERACTION WITH BIOLOGICAL TISSUES • *André Vander Vorst, Arye Rosen, and Youji Kotsuka*

InP-BASED MATERIALS AND DEVICES: PHYSICS AND TECHNOLOGY • *Osamu Wada and Hideki Hasegawa (eds.)*

COMPACT AND BROADBAND MICROSTRIP ANTENNAS • *Kin-Lu Wong*

DESIGN OF NONPLANAR MICROSTRIP ANTENNAS AND TRANSMISSION LINES • *Kin-Lu Wong*

PLANAR ANTENNAS FOR WIRELESS COMMUNICATIONS • *Kin-Lu Wong*

FREQUENCY SELECTIVE SURFACE AND GRID ARRAY • *T. K. Wu (ed.)*

PHOTONIC SENSING: PRINCIPLES AND APPLICATIONS FOR SAFETY AND SECURITY MONITORING • *Gaozhi Xiao and Wojtek J. Bock*

ACTIVE AND QUASI-OPTICAL ARRAYS FOR SOLID-STATE POWER COMBINING • *Robert A. York and Zoya B. Popoviać (eds.)*

OPTICAL SIGNAL PROCESSING, COMPUTING AND NEURAL NETWORKS • *Francis T. S. Yu and Suganda Jutamulia*

ELECTROMAGNETIC SIMULATION TECHNIQUES BASED ON THE FDTD METHOD • *Wenhua Yu, Xiaoling Yang, Yongjun Liu, and Raj Mittra*

SiGe, GaAs, AND InP HETEROJUNCTION BIPOLAR TRANSISTORS • *Jiann Yuan*

PARALLEL SOLUTION OF INTEGRAL EQUATION-BASED EM PROBLEMS • *Yu Zhang and Tapan K. Sarkar*

ELECTRODYNAMICS OF SOLIDS AND MICROWAVE SUPERCONDUCTIVITY • *Shu-Ang Zhou*

MICROWAVE BANDPASS FILTERS FOR WIDEBAND COMMUNICATIONS • *Lei Zhu, Sheng Sun, and Rui Li*

FUNDAMENTALS OF MICROWAVE PHOTONICS • *Vincent Jude Urick Jr., Jason Dwight McKinney, and Keith Jake Williams*

RADIO-FREQUENCY INTEGRATED-CIRCUIT ENGINEERING • *Cam Nguyen*

ARTIFICIAL TRANSMISSION LINES FOR RF AND MICROWAVE APPLICATIONS • *Ferran Martín*